4th International Conference on Advanced Materials, Mechanics and Structural Engineering (AMMSE 2017)

IOP Conference Series: Materials Science and Engineering
Volume 269

Tianjin, China
22 – 24 September 2017

ISBN: 978-1-5108-5268-6
ISSN: 1757-8981

Printed from e-media with permission by:

Curran Associates, Inc.
57 Morehouse Lane
Red Hook, NY 12571

Some format issues inherent in the e-media version may also appear in this print version.

Printed by Curran Associates, Inc. (2017)

For permission requests, please contact the Institute of Physics
at the address below.

Institute of Physics
Dirac House, Temple Back
Bristol BS1 6BE UK

Phone: 44 1 17 929 7481
Fax: 44 1 17 920 0979

techtracking@iop.org

Additional copies of this publication are available from:

Curran Associates, Inc.
57 Morehouse Lane
Red Hook, NY 12571 USA
Phone: 845-758-0400
Fax: 845-758-2633
Email: curran@proceedings.com
Web: www.proceedings.com

Table of contents

Volume 269

The 2017 4th International Conference on Advanced Materials, Mechanics and Structural Engineering (4th AMMSE 2017)

22–24 September 2017, Tianjin, China

Accepted papers received: 14 November 2017
Published online: 29 November 2017

Preface

Papers

Preface

The 2017 4th International Conference on Advanced Materials, Mechanics and Structural Engineering (4th AMMSE 2017) was held in Tianjin, China, during September 22-24, 2017.

4th AMMSE 2017 Conference is an annual conference aimed at presenting contemporary study research related to engineering technology, materials and mechanical science. The idea of the conference is for the scientists, scholars, engineers and students from universities, research institutes and industries all around the world to present ongoing research activities.

On behalf of the conference organizing committee, we would like to express our sincere appreciation to the keynote speakers, for their excellent and inspirational speeches, the reviewers, for their valuable inputs for improving the scientific content of these papers. We also would like to thank every technical committee member and author who has contributed to the success of the conference.

The Organizing Committee of AMMSE 2017

Committee Chair

Prof. Mosbeh Kaloop

Incheon Disaster Prevention Research Center

IOP Conf. Series: Materials Science and Engineering **269** (2017) 011001 doi:10.1088/1757-899X/269/1/011001

Conference Photograph

IOP Conf. Series: Materials Science and Engineering **269** (2017) 011001 doi:10.1088/1757-899X/269/1/011001

Sponsors

Incheon Disaster Prevention Research Center (IDPRC)	IDPRC 인천대학교 방재연구센터 Incheon Disaster Prevention Center

Recently, various efforts to prevent and prepare are vitally needed for prevention of Disasters and calamities. So we understand the necessity for technology of disaster and we built up the Incheon Disaster Prevention Research Center (IDPRC) in Incheon National University (INU) in 1997.

Accordingly Incheon Disaster Prevention Research Center (IDPRC) in Incheon National University has progressed of research on the prevention of disasters and calamities through the various seminars, conference and lectures. These research could be conducted cooperation with Incheon National University (INU) in various fields structure, soil, hydraulic and environment.

Incheon Disaster Prevention Research Center (IDPRC) will try to be a leader in the disaster of industry through the various research activities and global conference.

Incheon National University	INU 인천대학교 INCHEON NATIONAL UNIVERSITY

Incheon National University established a global campus by integrating two city council funded colleges into one in March 2010 and became a national university operated by the legal entity. With the start of attracting the branch of Lawrence Berkeley Research Center that is a world famous US national policy research center for education and research, Incheon National University will attract St. Petersburg University in Russia, Polymers University in Britain, and Kent University in Belgium and other foreign universities and research centers. These efforts will make INU spring to a world-wide competitive university.

As a local development leading university, Incheon National University established INU VISON 2020 and improved its competition actively through investment in selection and concentration. In compliance with the geographic feature and areal specialty, it will concentrate on the international trade, goods distribution, applied technology convergence, BNT-based life science, urban science and the sinology- based local humanity fields.

IOP Conf. Series: Materials Science and Engineering **269** (2017) 011001 doi:10.1088/1757-899X/269/1/011001

Lists of Committees

Committee name	Organizer name	Organization
Honorary Chair	Mosbeh Kaloop	Incheon National University, South Korea
Conference General Chairs	Jongwon Jung	Chungbuk University, South Korea
Honorary Chair	Seungho Hong	West Virginia University, L.A.
Technical Committees	Iman Mansouri	Birjand University of Technology
	Mohamed Eldessouki	Technical University of Liberec
	Trongphuoc Huynh	Can Tho University, Vietnam

IOP Conf. Series: Materials Science and Engineering **269** (2017) 011002 doi:10.1088/1757-899X/269/1/011002

Peer review statement

All papers published in this volume of *IOP Conference Series: Materials Science and Engineering* have been peer reviewed through processes administered by the proceedings Editors. Reviews were conducted by expert referees to the professional and scientific standards expected of a proceedings journal published by IOP Publishing.

IOP Conf. Series: Materials Science and Engineering **269** (2017) 012001 doi:10.1088/1757-899X/269/1/012001

Study on the Vehicle Dynamic Load Considering the Vehicle-Pavement Coupled Effect

H L Xu[1], L He[1], D An[1]

[1]School of Civil Engineering, North China University of Technology, Beijing 100144, China

Abstract. The vibration of vehicle-pavement interaction system is sophisticated random vibration process and the vehicle-pavement coupled effect was not considered in the previous study. A new linear elastic model of the vehicle-pavement coupled system was established in the paper. The new model was verified with field measurement which could reflect the real vibration between vehicle and pavement. Using the new model, the study on the vehicle dynamic load considering the vehicle-pavement coupled effect showed that random forces (centralization) between vehicle and pavement were influenced largely by vehicle-pavement coupled effect. Numerical calculation indicated that the maximum of random forces in coupled model was 2.4 times than that in uncoupled model. Inquiring the reason, it was found that the main vibration frequency of the vehicle non-suspension system was similar with that of the vehicle suspension system in the coupled model and the resonance vibration lead to vehicle dynamic load increase significantly.

1. Introduction

The interaction between vehicle and pavement included two research systems: the vehicle system and the pavement system. Existing pavement design methods simplified vehicle load as a constant static load. In fact, the vehicle load was random dynamic load which was excited by the pavement roughness. The vehicle dynamic load caused the pavement deformation caused by dynamic load is larger than the static load, the extra deformation increase the vibration of the vehicle. The study on the vehicle dynamic load should carry out the vehicle and the pavement together as an integrated system. Pavement roughness as a random excitation, the whole system should be studied in the theory of random vibration.

At present, the research on random vibration of the vehicle and the pavement was carried out usually by two ways: (1) Using the simplified vehicle model, the dynamic load of the vehicle was calculated. Then using the dynamic load, the response of the pavement structure was studied [1-3]; (2) Assumed the dynamic load of the vehicle as a fluctuating load, and then the response of the pavement structure was studied [4, 5]. The above methods did not completely reflect the actual situation of the vehicle-pavement coupled vibration.

2. The establishment of vehicle-pavement coupled linear elastic model

2.1. Traditional vehicle-pavement model

Previous research considered that the pavement deformation under the vehicle load was small and the pavement deformation relative to pavement roughness could be neglected in the research. So the 1/4 vehicle model was usually used in the research [3, 6, 7] which was shown in figure 1.

The main ideas of the traditional model research: using the 1/4 vehicle model, the random vibration load of the vehicle caused by the pavement roughness could be obtained; Then using this load as the input for the pavement vibration response research.

In the traditional model, the vehicle and the pavement were considered separately, and the pavement was assumed to be rigid in the random vibration analysis, which was not in conformity with the actual situation. For the random vibration problem, the coupled effect of the vehicle and the pavement would affect the frequency distribution of the freedom degree. The influence of the change of frequency distribution on vibration amplitude can not be ignored. It is necessary to carry out comparative study on the influence of the coupling effect of the vehicle-pavement system, especially on the influence of the vehicle road interaction.

2.2. Vehicle-pavement coupled linear elastic model

In view of the shortcomings of the previous research model, this paper introduced the pavement structure system into the traditional vibration model, and then established a new vehicle-pavement coupled model. In this model, the pavement structure was stimulated by the mass-spring-damping elements, and the influence of the adjacent masses block of the pavement structure was stimulated by the shear spring and the shear damping elements. The vehicle-pavement coupled model was shown in figure 2.

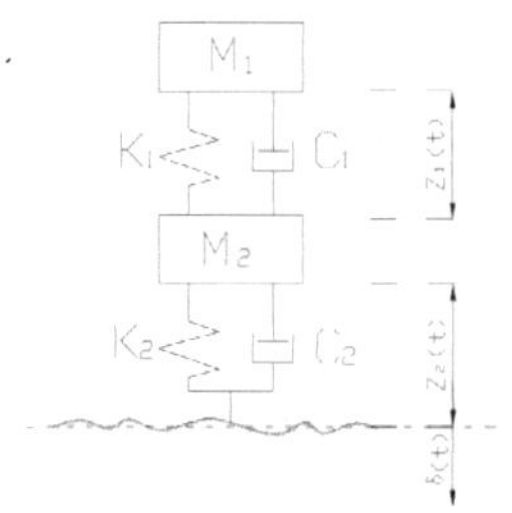

Figure 1. 1/4 Vehicle Model.

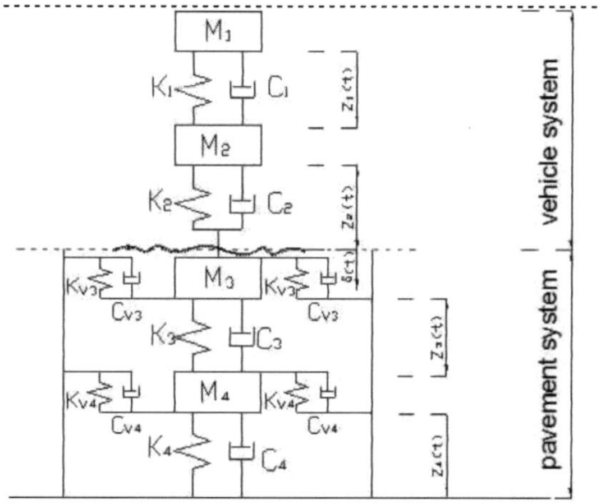

Figure 2. The Linear Model of Vehicle-Pavement coupled System.

2.3. Vehicle-pavement coupled model solution

2.3.1. Random vibration of vehicle road coupled model solution. According to D'A lembert theory, the vibration equation of the vehicle-pavement coupled linear model could be expressed in matrix form:

$$[M]\{Z''(t)\}+[C]\{Z'(t)\}+[K]\{Z(t)\}=[K_f]\{\delta''(t)\} \tag{1}$$

According to the random vibration theory, the relationship between the excitation and response was the following:

$$Z_i(\omega)=H_{ij}(\omega)\delta_j(\omega) \tag{2}$$

In the formula, $\delta_j(\omega)$-frequency function for excitation, $Z_i(\omega)$-frequency function for response, $H_{ij}(\omega)$-frequency response function, which could be referred to $H_i(\omega)$ for a single incentive input. The formula (1) was transformed with the Fourier transform, using the formula (2), $H_i(\omega)$ ($i=1,2,3,4$) could be obtained.

Based on random vibration theory, the relationship between the excitation spectral density function $S_\delta(\omega)$ and the response spectral density function was obtained as follows:

$$S_i(\omega) = |H_i(\omega)|^2 S_\delta(\omega) \tag{3}$$

$$S_{\dot{i}}(\omega) = \omega^2 S_i(\omega) \tag{4}$$

$$S_{i''}(\omega) = \omega^4 S_i(\omega) \tag{5}$$

In the formula: $S_{\dot{i}}(\omega)$ was velocity spectral density function of response function and $S_{i''}(\omega)$ was acceleration spectral density function of response function.

2.3.2. Vehicle dynamic load solution. From figure 2, the vehicle-pavement interaction force could be expressed as

$$F(t) = K_2 Z_2(t) + C_2 Z_2'(t) \tag{6}$$

In the formula: $F(t)$—dynamic load of vehicle, which was a centralized random process.

The load power density function $S_F(\omega)$:

$$S_F(\omega) = (K_2^2 + C_2^2\omega^2) S_{Z_2}(\omega) \tag{7}$$

3. Pavement roughness model

At present, the spectral density of typical pavement had various expressions, this paper adopted the following form [8, 9]:

$$G_q(\Omega) = G_q(\Omega_0)(\frac{\Omega}{\Omega_0})^{-w} \tag{8}$$

In the formula, Ω_0 -reference space frequency, Ω_0 =0.1c/m; $G_q(\Omega_0)$ -power spectrum density of pavement under the reference space frequency, called the pavement roughness coefficient; W-frequency index, general W=2.

v -the vehicle speed, Ω -the spatial frequency of pavement roughness, ω -the circular frequency, f -the natural frequency.

From the formula (8) available showed as follows:

$$S_\delta(\omega) = \frac{1}{2\pi v} G_q(\Omega = \frac{\omega}{2\pi v}) \tag{9}$$

$$S_\delta(f) = 2\pi S_\delta(\omega = 2\pi f) \tag{10}$$

4. Validation of vehicle-pavement coupled model

In the reference [10], the field study on the interaction between vehicle and pavement obtained the acceleration value at wheel axle of the vehicle. The first peak appeared near 2.3Hz, and the amplitude was 0.317; the second peak appeared near 17.5Hz, and the amplitude was 0.385.

The vehicle-pavement coupled model in this paper was used to simulate the field test. The simulation used the test vehicle data and the pavement structure data valued using the method [11].

The main parameters were as follows: M_1 =12.98×10^3 kg, M_2 =5.0×10^2 kg, M_3 =3.5×10^2 kg, M_4 =1.5×10^3 kg, K_1 =5.0×10^5 N/m, K_2 =1.5×10^6 N/m, K_3 =1.0×10^9 N/m, K_4 =2.5×10^7 N/m, K_{V3} = 1.0×10^8 N/m, K_{V4} =4.0×10^5 N/m, C_1 =1.5×10^4 $N \cdot s/m$, C_2 =5.0×10^3 $N \cdot s/m$, C_3 =3.0×10^4 $N \cdot s/m$,

$C_4 = 5.0\times10^3\ N\cdot s/m$, $C_{V3} = 2.0\times10^4\ N\cdot s/m$, $C_{V4} = 3.0\times10^3\ N\cdot s/m$, $v = 70\ km/s$. The field test is on the Qing-Huang Freeway, the simulation used the A grade pavement spectrum.

Using the vehicle-pavement coupled model, the acceleration spectrum at wheel axle of the vehicle was shown in figure 3.

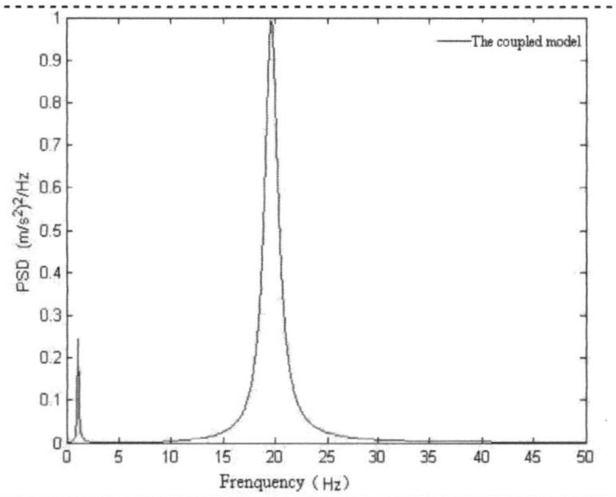

Figure 3. Simulation Result Using the coupled Model.

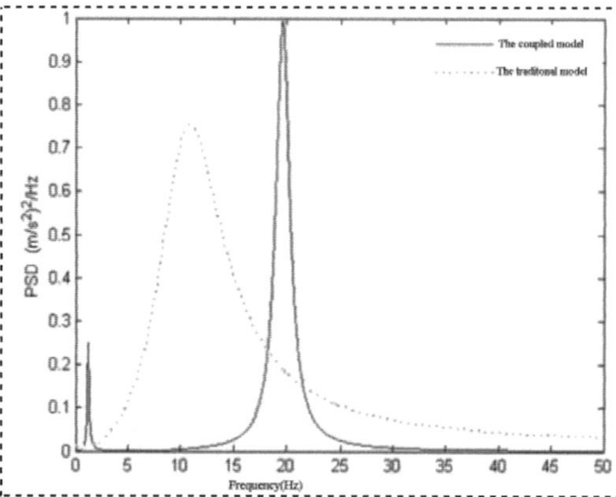

Figure 4. Vertical Acceleration Power Spectral Density Comparison Diagram of Vehicle Non-suspension System.

It could be seen from the simulation results that there were two peaks in the acceleration spectrum, the first peak appeared near 2Hz, and the amplitude was 0.26; the second peak appeared near 20Hz, and the amplitude was 0.98. It could be found that the results of the coupled model could reflect the vibration of the vehicle and match with the field data. The difference in amplitudes was mainly determined by the input data error in measurement process.

Using the same parameters, the traditional model was also used to simulate the field test. And the simulation results were compared with the coupled model, as shown in figure 4. Using the traditional model, there was only one peak in the acceleration spectrum, which appeared near 10Hz.In contrast, the result of the vehicle-pavement coupled model was better match with the field data than that of the traditional model.

5. Analysis on the influence of vehicle-pavement coupled effect on the dynamic load of vehicle

In the research on the pavement structure vibration response, the dynamic load of the vehicle is a key factor. In order to analyse the influence of vehicle-pavement coupled effect on the dynamic load, a comparative study was carried out on the dynamic load of the vehicle, using the vehicle-pavement coupled model and the traditional model.

The same parameters were used in the vehicle-pavement coupled model and the traditional model. The simulation used the C grade pavement spectrum. The parameters were as follows: $M_1 = 4.5\times10^3\ kg$, $M_2 = 5.0\times10^2\ kg$, $M_3 = 3.5\times10^2\ kg$, $M_4 = 1.5\times10^3\ kg$, $K_1 = 5.0\times10^5\ N/m$, $K_2 = 1.5\times10^6\ N/m$, $K_3 = 1.0\times10^9\ N/m$, $K_4 = 2.5\times10^7\ N/m$, $K_{V3} = 1.0\times10^8\ N/m$, $K_{V4} = 4.0\times10^5\ N/m$, $C_1 = 1.5\times10^4\ N\cdot s/m$, $C_2 = 5.0\times10^3\ N\cdot s/m$, $C_3 = 3.0\times10^4\ N\cdot s/m$, $C_4 = 5.0\times10^3\ N\cdot s/m$, $C_{V3} = 2.0\times10^4\ N\cdot s/m$, $C_{V4} = 3.0\times10^3\ N\cdot s/m$, v (speed)=$30\ m/s$.

The results of the power spectral density of the dynamic load (the forces between vehicle and pavement) were as shown in figure 5. In figure 5, it could be found that the peaks all appeared near 1.5Hz, but the peak amplitudes of the two models differed greatly. The peak amplitude of the coupled model was 8 times more than that of the traditional model. To evaluate the maximum force between vehicle and pavement as (was mean square deviation), the maximum force of the coupled model was 2.4 times than that of the traditional model (The maximum force was central, not including the static load of the vehicle). So it was obvious that the vehicle-pavement coupled effect had a great influence on the interaction between the vehicle and the pavement.

The research results under different speed, different grade pavement spectrum and different vehicle load all showed that the maximum force between vehicle and pavement of the coupled model was several times higher than that of the traditional model. So the vehicle-pavement coupled effect was an important factor that could not be ignored in the research on the force between vehicle and pavement.

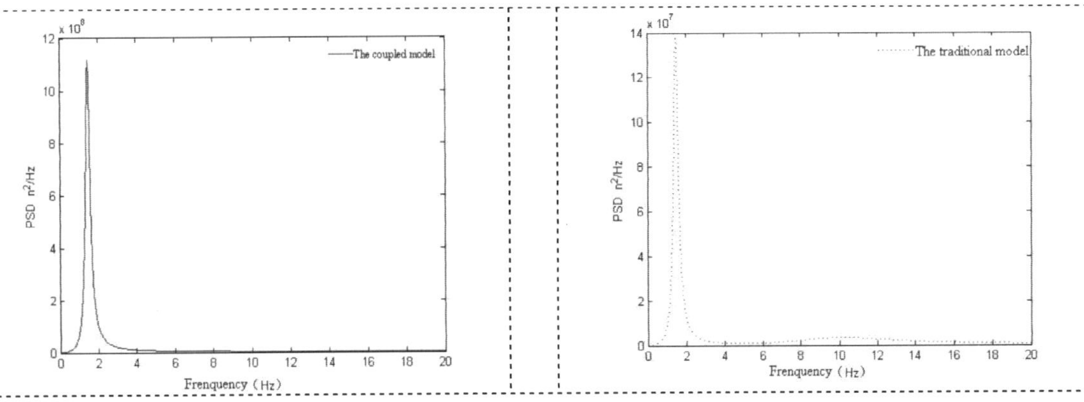

Figure 5. Power Spectral Density Comparison Diagram of the Forces between Vehicle and Pavement.

6. Analysis on the influence of vehicle-pavement coupled effect on the dynamic load of vehicle

In order to analyse the reason for the influence of vehicle-pavement coupled effect, the acceleration spectral density of the vehicle suspension system and the vehicle non-suspension system were obtained using the coupled model and the traditional model separately.

The curve of the coupled model almost completely coincided with that of the traditional model. The peaks all appeared near 1.5Hz. It could be seen that the vehicle-pavement coupled effect on the vibration of the vehicle suspension is not very seriously. So the research on the vehicle vibration was feasible to use the traditional model. In figure 7, it also could be seen that the peak frequency of the vehicle suspension appeared near 1.5Hz as the same as the peak frequency of the force between vehicle and pavement. It means that the force between vehicle and pavement mainly controlled by the suspension system of the vehicle, this phenomenon was in accordance with the actual observation.

The vibration of vehicle non-suspension of the coupled model varied significantly with that of the traditional model. Using the coupled model, there were two peaks which appeared near 1.5Hz and 20Hz; using the traditional model, there was only one peak which appeared near 10Hz. So the vehicle-pavement coupled effect had very important influence on the frequency distribution of the vehicle non-suspension. Using the coupled model, one of vibration peaks frequency of the vehicle non-suspension was 1.5Hz, which was close to the vibration peak frequency of the vehicle suspension. So the vehicle suspension system and the non-suspension system had occurred resonance at the frequency 1.5Hz, thus the maximum amplitude of the force between vehicle and pavement increased many times than that using the traditional model.

7. Conclusion

A new linear elastic model of the vehicle-pavement coupled system was established. The new model was verified with the data measured in field and the new model could reflect the real vibration between vehicle and pavement. Using the new model, the study on the vehicle dynamic load considering the vehicle-pavement coupled effect showed that the vehicle dynamic load (centralization) was influenced largely by vehicle-pavement coupled effect. Numerical calculation indicated that the maximum of random forces using coupled model was 2.4 times than that using uncoupled model. Inquiring the reason, it was found that the vibration peak frequency of the vehicle non-suspension system was similar with the vibration frequency of the vehicle suspension system using the coupled model and resonance vibration lead to vehicle dynamic load increase significantly.

References

[1] Cebon D 1985 *The Dynamic Interaction between Wheeled Vehicles and Road Surface*. Cambridge University Engineering Department, Ph. D Dissertation.
[2] Magdy Y M 1996 *Three Dimensional Mechanistic Analysis of Vehicle-Pavement Interaction*. Arizona State University, Ph. D Dissertation.
[3] Sun L, Deng X J 1997 *China Civil Eng. J.* **30** 34-40.
[4] Zhang Y M 2001 *Study on dynamic characteristics of Expressway Subgrade*. Lanzhou: Lanzhou Railway Institute, Ph. D Dissertation.
[5] Chen H 2004 *Dynamic finite element analysis of highway subgrade under traffic load*. Lanzhou University of Technology, Ph. D Dissertation.
[6] Deng X J, Sun L. 1998 *Vehicle-ground structure system dynamics*. Beijing: People's Communications Press.
[7] Wang Z M, Zhang T Q, Wu X G 2007 *J. Zhejiang Univ.* **41(6)** 1007-11.
[8] Zhao J H, Wang Z R, Guan Ch L 2000 *Measurement analysis and application of pavement roughness*. Beijing: Beijing Institute of Technology press.
[9] State Standard of the People's Republic of China. *Vehicle vibration input method for road surface roughness* (GB7031-86), 1986.
[10] Chen J 2002 *A Basic Study on Interaction between Vehicle and Roadway*, Jilin University, Ph.D. Dissertation.
[11] Chen G 2000 *The analysis on random vibration of vehicle/track coupling system*. SouthWest JiaoTong University, Ph.D. Dissertation.
[12] Wu S Q, Law S S 2011 *Probabil. Eng. Mech.* **26(2)** 281-93.
[13] Zhang Y W, Lin J H, Zhao Y 2010, *J. Sound Vibr.* **329(21)** 4440-54.

IOP Conf. Series: Materials Science and Engineering **269** (2017) 012002 doi:10.1088/1757-899X/269/1/012002

Research on the Total Leakage Rate Test of the Spacecraft and its Uncertainty Evaluation

Y Wang[1], R Yan[2], L Sun[2]

[1]Senior Engineer, Beijing Institute of Spacecraft Environment Engineering, Beijing, China
[2]Professor, Beijing Institute of Spacecraft Environment Engineering, Beijing, China

Abstract. Leak testing is the major type of non destructive testing for spacecrafts and the sealing performance of the spacecraft is an important parameter to evaluate the quality of the spacecraft. So it is necessary to study the total leakage rate test of the spacecraft and its uncertainty evaluation. In this paper, the basic principle of the total leakage rate test of the spacecraft was analyzed firstly from the theoretical aspect. Secondly, the various influencing factors on the results of total leakage rate test had been systemically analyzed. Finally, through the uncertainty evaluation theory, the uncertainty components induced by these influencing factors were respectively quantified and combined. And the expanded uncertainty of the result of total leakage rate test was given.

1. Introduction

As we all know, leak testing is the major type of non destructive testing for spacecrafts and the sealing performance of the spacecraft is an important parameter to evaluate the quality of the spacecraft. A tiny leak may cause huge loss. So, in the AIT process of the spacecraft, it needs the strict leak testing for spacecrafts [1]. Nowadays, the non-vacuum accumulation leak testing method and vacuum leak testing method are widely used for testing the total leakage rate of spacecrafts, such as the hermetic cabin, propulsion pipeline system. The vacuum leak testing method is often used for Russian spacecrafts. But the above two kinds of method are both used for American spacecrafts [2]. Generally speaking, the non-vacuum accumulation leak testing method is used for spacecrafts which have low leakage rate demand, such as satellites and Node 1 of International Space Station [2-4]. While the vacuum leak testing method is used for these spacecrafts which have high leakage rate demand, such as Airlock Module and Laboratory Module of ISS [2]. In China, the total leakage rate test mainly adopts the non-vacuum accumulation leak testing method. Its testing system often includes accumulation chamber, leak detector, atmospheric reference gas, leaking gas sampling system and leakage rate calibration system. The basic principle is that: Helium gas is filled into the spacecraft using the Helium filling device. And the spacecraft is placed into the accumulation chamber. The initial value u_1 of the spacecraft is firstly measured with the leak detector, and after the accumulation time t, the final value u_2 is also measured. In order to quantify the total leakage rate of the spacecraft, a quantitative helium gas w is filled into the accumulation chamber, and the sampling value u_3 is measured after the gas in the accumulation chamber is mixed homogeneously. So the total leakage rate of the spacecraft Q is

$$Q = \frac{w(u_2 - u_1)}{t(u_3 - u_2)} \tag{1}$$

The sketch of the non-vacuum accumulation leak testing method is shown in figure 1.

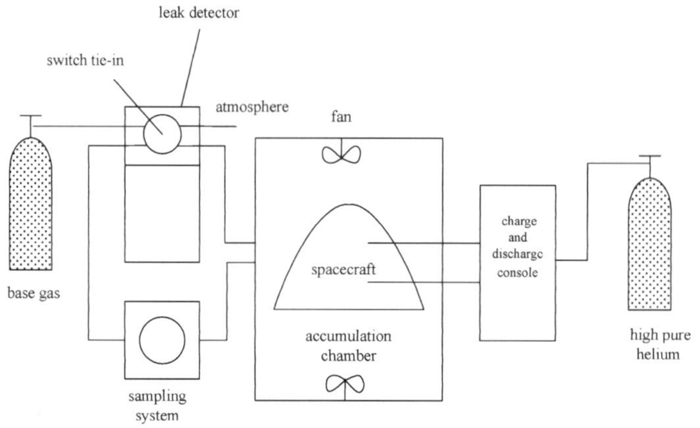

Figure 1. The sketch of the non-vacuum accumulation leak testing method.

All of the previous papers mainly focused on the application of kinds of leak testing methods but rarely referred to the uncertainty evaluation [2-4]. This paper tries to give the uncertainty evaluation process of the leak testing result.

2. Theoretical research on the non-vacuum accumulation leak testing method

The helium gas flow rate entering into the leak detector is

$$Q_{He} = \gamma Q_z = \gamma p_e S_z \tag{2}$$

where Q_z is the inlet mixed gas flow of the leak detector, S_z is the pumping speed of the leak detector, γ is the concentration of the mixed gas and p_e is the inlet pressure of the leak detector.

By literature [5], the measured value u of leak detector can be described by

$$u = kp_e S_z \gamma \tag{3}$$

where u is the measured value of the leak detector, k is the amplificatory factor of the leak detector. p_e can be adjusted by a fine adjustment valve. When p_e is fixed, the pumping speed S_z is also fixed.

Usually, k, p_e, S_z can be fixed, so the measured value of the leak detector is proportional to the measured helium concentration of the mixed gas.

Equation 3 can also be expressed by

$$u_i = kp_e S_z \gamma_i \quad (i = 1, 2, 3) \tag{4}$$

where i is free index, and denotes the initial value, final value and sampling value respectively.

The total leakage rate of the spacecraft Q can be described by

$$Q = \frac{p_{atm} V(\gamma_2 - \gamma_1)}{t} \tag{5}$$

where p_{atm} is atmospheric pressure, V is the effective volume which equals to the difference volume between the accumulation chamber and the spacecraft, γ_2, γ_1 denote the final and initial helium concentration in the accumulation chamber respectively and t is the accumulation time.

Because the total leakage rate of the spacecraft is very small and the effective volume is big, the pressure in the accumulation chamber changes little and almost equals to p_{atm}.

After accumulation time t, the accumulation chamber is filled by quantitative helium gas $w = p_0V_0$, then the helium concentration changes from γ_2 to γ_3. So we can obtain.

$$\frac{p_0V_0}{p_{atm}V} = \gamma_3 - \gamma_2 \tag{6}$$

From equations 5, 6, the total leakage rate of the spacecraft can be obtained by

$$Q = \frac{p_0V_0(\gamma_2 - \gamma_1)}{t(\gamma_3 - \gamma_2)} \tag{7}$$

With equation 4, the concentration can be instead by the measured value of the leak detector and we can obtain

$$Q = \frac{p_0V_0(u_2 - u_1)}{t(u_3 - u_2)} = \frac{w(u_2 - u_1)}{t(u_3 - u_2)} \tag{8}$$

The above derivation process is the quantitative principle of the non-vacuum accumulation leak testing method.

3. The analysis on uncertainty resources of the test results

Through the above analysis, the influence factors of the non-vacuum accumulation leak testing method are shown as figure 2. Among them, the white parts are the systemic factors and the yellow part is the random factor.

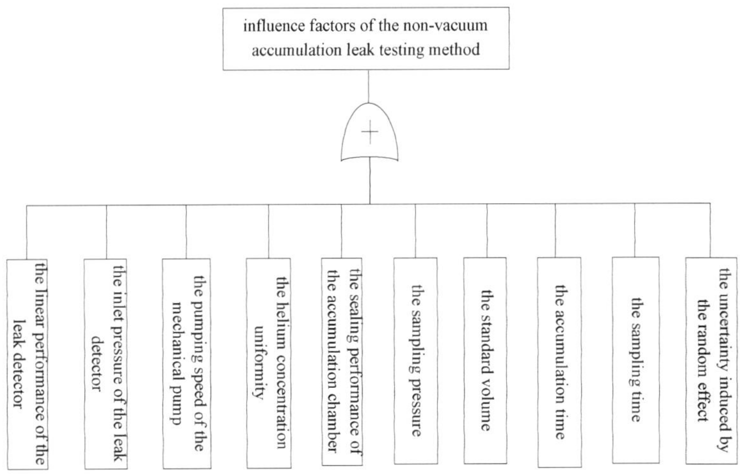

Figure 2. The influence factors of the non-vacuum accumulation leak testing method.

4. Uncertainty evaluation of the test result

4.1 Evaluation of the uncertainty induced by the systemic effect

4.1.1 Uncertainty component induced by the linear performance of the leak detector. With the type B evaluation method, the uncertainty component induced by the linear performance of the leak detector is about 5% [6].

4.1.2 Uncertainty component induced by the inlet pressure of the leak detector. In order to study the inlet pressure change of the leak detector, the following experiments are made in 30 minutes and the test data are shown in table 1.

Table 1. The test data of the inlet pressure change of the leak detector.

Number	The test value (Pa)	Maximum value (Pa)	Minimum value (Pa)	Average value (Pa)	Change value (Pa)	Relative change value
1	29.10, 28.54, 29.66, 27.98	29.66	27.98	28.82	1.68	5.83%
2	29.10, 28.54, 29.66, 27.98	29.66	27.98	28.82	1.68	5.83%

From table 1, the uncertainty component induced by the inlet pressure of the leak detector is $5.83\% / \sqrt{3} = 3.37\%$ if the hypothesis of uniform distribution is adopted.

4.1.3 Uncertainty component induced by the pumping speed of the mechanical pump. From the linear relationship between the inlet pressure and pumping speed, the relative change rate between them can be seen as the same. So the uncertainty component induced by the pumping speed of the mechanical pump is also 3.37%.

4.1.4 Uncertainty component induced by the helium concentration uniformity in the accumulation chamber. From the literature [7], the maximum deviation caused by the helium concentration uniformity is about 4.45%. So the uncertainty component induced by the helium concentration uniformity is $4.45\% / \sqrt{3} = 2.57\%$ if the hypothesis of uniform distribution is adopted.

4.1.5 Uncertainty component induced by the sealing performance of the accumulation chamber. In the design process of the accumulation chamber, the sealing performance usually demands that the pressure drop is less than 300Pa when the gas of 2KPa pressure is filled into the accumulation chamber. In practice, its sealing performance is often better than this demand. Conservatively, the design value is adopted to evaluate the uncertainty component induced by the sealing performance of the accumulation chamber. That is $300 / (100000 \times \sqrt{3}) = 0.17\%$. So the uncertainty component induced by the sealing performance of the accumulation chamber is 0.17%.

4.1.6 Uncertainty component induced by the sampling pressure. The level of the pressure sensor is 0.1, namely the corresponding relative error is 0.1%. So the uncertainty component induced by the sampling pressure is $0.1\% / \sqrt{3} = 0.06\%$ if the hypothesis of uniform distribution is also adopted.

4.1.7 Uncertainty component induced by the standard volume. From the calibration document, the uncertainty component induced by the standard volume is 0.001%.

4.1.8 Uncertainty component induced by the accumulation time. Usually the accumulation time of non-vacuum accumulation leak testing method is about 24 hours, while its error can be controlled in minute range, so the uncertainty component induced by the accumulation time is less than 0.1%. Conservatively, the uncertainty component induced by the accumulation time can be adopted by 0.1%.

4.1.9 Uncertainty component induced by the sampling time. From the literature [1], the uncertainty component induced by the sampling time is 2%.

From the above analysis, it can be found that the uncertainty component induced by the linear performance of the leak detector was biggest. So if we want to improve the correctness of the testing result, the most effective way is to improve the leak detector. The uncertainty induced by the systemic effect is

$$u_{Br} = \sqrt{5\%^2 + 3.37\%^2 + 3.37\%^2 + 2.57\%^2 + 0.17\%^2 + 0.06\%^2 + 0.001\%^2 + 0.1\%^2 + 2\%^2} = 7.64\% \quad (9)$$

4.2 Evaluation of the uncertainty induced by the random effect.
A tested object was tested by the non-vacuum accumulation leak testing method and the six results were given in table 2.

Table 2. The test results.

Number	1	2	3	4	5	6
Leakage rate ($Pa\cdot m^3/s$)	1.86×10^{-5}	1.65×10^{-5}	1.98×10^{-5}	1.82×10^{-5}	1.92×10^{-5}	2.00×10^{-5}

So, the arithmetic average value is $\bar{Q}=\frac{1}{n}\sum_{i=1}^{n}Q_i=1.87\times10^{-5}$.

The standard deviation is $s(Q)=\sqrt{\frac{1}{n-1}\sum_{i=1}^{n}(Q_i-\bar{Q})^2}=0.1284\times10^{-5}$.

The standard uncertainty is $u_A(\bar{Q})=\frac{s(Q)}{\sqrt{n}}=0.05244\times10^{-5}$.

The relative standard uncertainty is $u_{Ar}(\bar{Q})=\frac{u_A(\bar{Q})}{\bar{Q}}\times100\%=2.8\%$.

4.3 The combined standard uncertainty

$$u_{cr}=\sqrt{{u_{Br}}^2+{u_{Ar}}^2}=\sqrt{7.64\%^2+2.8\%^2}=8.14\% \tag{10}$$

4.4 The expanded uncertainty

$U=ku_c=2\times8.14\%=16.28\%$ $k=2$, where k is the coverage factor.

5. Conclusion

In this paper, the basic principle of the total leakage rate test of the spacecraft was analyzed firstly from the theoretical aspect. Secondly, the various influencing factors on the results of total leakage rate test had been systemically analyzed. Finally, through the uncertainty evaluation theory, the uncertainty components induced by these influencing factors were respectively quantified and combined. And the expanded uncertainty of the result of total leakage rate test was given. The results indicated that the non-vacuum accumulation leak testing method was accurate enough to test the total leakage rate of the spacecraft and the expanded uncertainty of the testing result was only about 16.28%(k=2).

References
[1] Wang Y and Yan R 2010 *Chinese Space Sci. Tech.* **4** 71.
[2] Underwood S and Lvovsky O 2007 NASA-20070022572.
[3] Underwood S and Smith L 2003 *40th Space Congress(Huntsille)*, 312.
[4] Dario B and Michele C 2007 58th International Astronautical Congress(Hyderabad), 115.
[5] Wang Y et al. 2012 *Chinese J. Vac. Sci. Tech.* **2** 118.
[6] Information on http://www.chinesevacuum.com/ ShowArticle.aspx?id=26850
[7] Wang Y et al 2013 *Chinese J. Vac. Sci. Tech.* **2** 149.

Acknowledgements
This work was supported by the National Natural Science Foundation of China through Grant Number U1537109 and National Defence Technique Foundation of China through Grant Number JSJL2015203B019.

IOP Conf. Series: Materials Science and Engineering **269** (2017) 012003 doi:10.1088/1757-899X/269/1/012003

Deformation Control by VSR Technique on Al alloy Thin-Walled Components

G H Xiong[2], K Liao[1], X Y Chang[2], J A Liu[2]

[1] Senior Production Editor, School of Mechanical and Electronic Engineering, Central South University of Forestry and Technology, Changsha 410004, China
[2] Production Assistant, School of Mechanical and Electronic Engineering, Central South University of Forestry and Technology, Changsha 410004, China

Abstract. Initial residual stress and machining stress of Al-alloy materials are the main factors that cause the deformation of Al-alloy thin-walled components. According to the theory of plastic yield and analysis of the stress-deformation characteristics in materials, the component is treated by using vibration stress relief (VSR) technology under three different aging of 30 min, 60 min and 90 min, after that the deformation are measured by coordinate machine respectively. The results of experiment present that, based on the balance of structural stress, after VSR aging 30 min, the component comes to deformation and becomes flatter than one before VSR. Similarly, as for complicated thin-walled component, the VSR aging time to achieve this effect mentioned above is about 90 min. The research manifests that VSR has a positive impact on deformation control of component even if structural stress level is extremely low in material.

1. Introduction

The large-scale aluminum structure components are widely used in aircraft manufacturing, such as frame, column, rib and wing. In the machining of slab, caused by the combined action of nonuniform field like milling force, fixture force, milling heat, macroscopical deformation happens in the structure components. Vibration stress relief (VSR) can change the residual stress by generating alternating stress in materials. At present, VSR is mainly used for ferrous metal. Stress concentration of the casting and welding parts can be significantly decreased and homogenized to satisfy performance requirements. Some meaningful results have been found by the researchers, such as improving the dimensional stability and reducing residual stress of bellows by aging [1], the stress release of welding parts by high frequency vibration [2, 3]. However, the research of the application of VSR in nonferrous metals is limited, because the residual stress is much small in nonferrous metals, it is very difficult for the material to yield by stress superposition. Shen et al. [4] studied the effects of process parameters of VSR for aluminum thick plate. Hu [5] studied stress release by VSR in different frequency and found the positive effects for stress homogenization of Al alloy thick plate. Cui, Teng and Lin [6] applied the ultrasonic method in aluminum welding components. The experiment results indicated the effects on the residual stress release. The VSR method can be applied to the surface characteristics of components to study its positive role in the material surface. In this paper, the micro mechanism is explained by dislocation theory.

IOP Conf. Series: Materials Science and Engineering **269** (2017) 012003 doi:10.1088/1757-899X/269/1/012003

2. Mechanism analysis

2.1. Stress distribution of the structure

The stress distribution of thin wall frame structures is very different with thick plates. Because of the milling process, the stress around thin walls is released. At the same time, high machining stress is generated on the machining surfaces. Generally, the stress on the bottom part will be strengthened due to the redistribution caused by the deformation. A stress area with high stress gradient generates on the machining surface. The initial stress is about ±20 MPa. However, the internal stress on the machining surface has reached 140 MPa just in the depth of 80 μm. This high stress gradient leads to the unstable internal energy in this area and needs to be released and homogenized. The section of the thin wall structure after machining is shown in figure 1, where the dark area is the machining stress area. Although VSR cannot significantly release the stress in the whole structure, in some stress concentration area, some positive effects can be observed by using VSR. Therefore, VSR can be applied to release the unstable internal energy caused by stress concentration in the machining stress area.

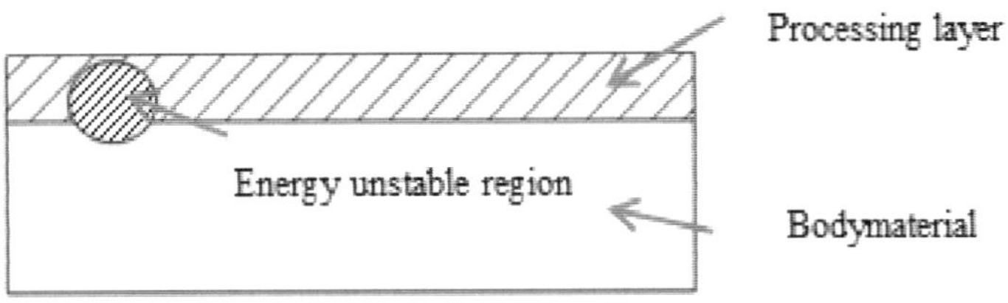

Figure 1. Profile of component.

2.2. Micro-mechanism analysis

In the view of dislocation, the micro yield condition, mechanism of the material and the relationship of aging, micro yield and stress homogenization can be analysed by combining dislocation motion and the change of microstructure. Dislocation can be seen as a line defect in materials. Amounts of dislocation pinning aggregation can be observed around grain boundary area in deformation. Therefore, in the high elastic energy area in the machining layer, the dislocation can be inspired and propagation under alternating stress caused by vibration. If amount of dislocation aggregation in machining layer, materials will be strengthen and the state will change from body material, which will lead to elastic deformation to reach the balance of stresses. In the other aspect, nonuniform distribution of organization always exists in materials. Stress concentration may also exist in some part of the material or micro areas. Under the effect of alternating stress, the structure will deform due to micro yield.

3. Experiment

3.1. Experimental materials

The material is Al alloy thick plate (7075-T651). The dimension is 500 mm×230 mm×60 mm. The thickness of machining wall is 2 mm. VSR is applied to the structures in different aging time. The micro structures of deformed parts are tested, observed and compared by electron microscope. Two types of structures are obtained by milling, as shown in figure 2(a) and figure 2 (b). In order to reduce the addition deformation caused by the fixture, the part is fixed on the bottom by vacuum adsorption.

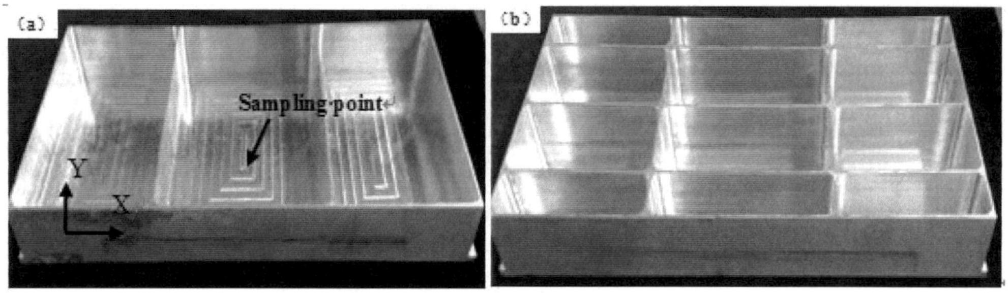

Figure 2. (a) Simple component; (b) complicated component.

A test sample is cut from the bottom of the specimen. The cross section of the specimen is tested by using TEM scanning electron microscope to obverse the generation of dislocation and grain deformation.

3.2. Vibration equipment and specimen.
The specimen is fixed on the vibration table. Rubber blankets are used as the elastic support of the vibration table. The rubber blankets are placed on the nodal points of vibration of the table, as shown in figure 3. The first order natural frequency acceleration of the vibration system is 67.0 m/s^2. Therefore, the exciting force is 2.7 KN (54.0 m/s^2), which makes the vibration system work in sub-frequency resonance area. Three aging times (30 min, 60 min, 90 min) are applied to compare the results.

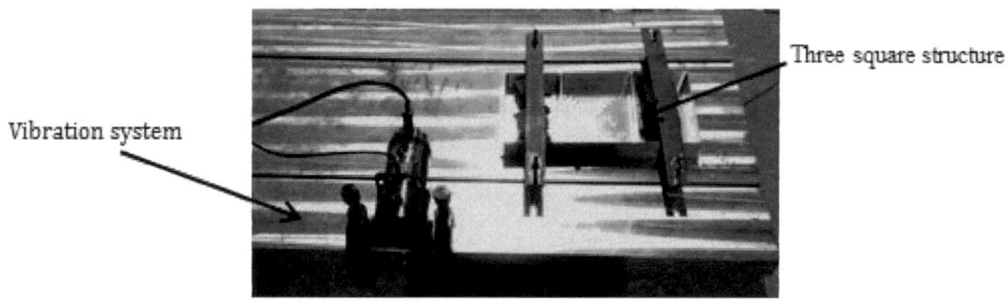

Figure 3. Vibration equipment and fixture.

4. Experiment analysis

4.1. Distribution of the deformation
The deformation of the specimen before and after VSR is measured to compare the influences along the length direction. Figure 4(a) shows the deformation of X direction of three square structures with different aging time: 30 min, 60 min, 90 min. It can be seen that after 30 min VSR aging, the deformation increased 40 μm averagely, which indicates that the deformation is intensified in this period. After 60 min VSR aging, the deformation decreased 20 μm, which indicates that continuous aging decreased the deformation. The deformation curve becomes flat, this indicates that the bending and twisting of the structure are released. When the aging time extended to 90 min, the deformation of the structure increased obviously and became a bend curve. It can be concluded preliminarily that for this three square structure, 30 min is the best aging time to increase the dimensional stability.

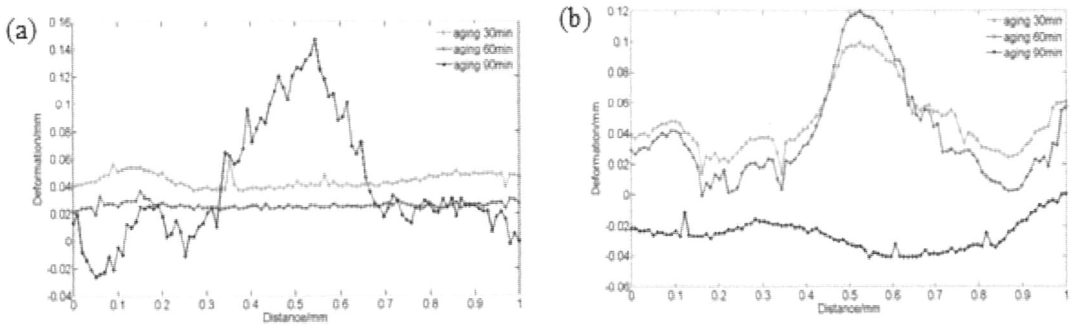

Figure 4. Relative deformation along with X direction. (a) three square and (b) multi-square structure.

The comparison of the same experiment of multi-square structure is shown in figure 4(b). After 30 min and 60 min aging, the deformation increased significantly, which indicates the internal stress redistributed in this period. When the aging time extended to 90 min, spring back deformation can be observed, which indicates that the deformation became stable with the redistribution of internal stress. Therefore, 90min is the best aging time for multi-square structure to improve the stability. From the results, it can be seen that the deformation of the structure varies with different VSR aging time. Theoretically, the total intensity of initial stress and vibration stress cannot reach to plastic yield, so the deformation should be caused by the changes in micro areas.

4.2. metallographic analysis

Figure 5(a)-(d) shows the dislocation distribution of machining area by TEM.

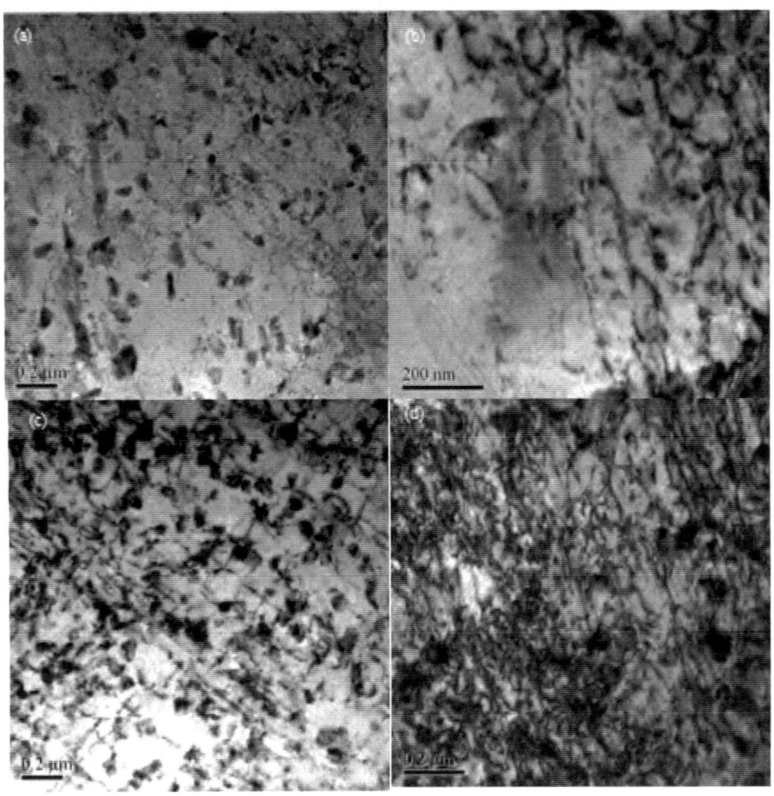

Figure 5. Dislocation along with X direction profile
(a) initial state, (b) 30min aging, (c) 60min aging and (d) 90min aging.

Figure 4 shows that dislocation density around grain boundary increases with the aging time. The vibration energy excited and increased the micro dislocation. The dislocation can be easily observed in the second phase. Microscopic hardening caused by the dislocation pinning in this precipitation strengthening phase increase the difficulty of plastic deformation. At the same time, this material hardening also leads to deterioration in machining layer and material behaviour change in some local place. This behaviour change is the main reason of elastic deformation of the structure.

Firstly, the vibration energy excited the internal dislocation which exists in most of manufacture process of plates with micro defect and inhomogeneity of structure. The observed dislocation pinning around machining layer corresponds to this process. Secondly, the nonuniform distribution of dislocation density leads to the nonuniform of material behaviour and the nonuniform of the micro intensity distribution in cross section. Thirdly, the dimensional stability is different for different shapes of structures in the same aging time by VSR. The aging time and process of VSR should be adjusted for different shapes to obtain the optimal dimensional stability.

5. Summary
The experiment results demonstrated the effectiveness of VSR to improve the dimensional stability of aluminum thin wall structures. The optimal aging time for different structures depends on both the structure shape and initial internal stress state.

6. Reference
[1] Liu X D, Tao S H and Han Z Q 2015 *J. Vibr. Shock* **34(4)** 171-4.
[2] Rao D L and Zhu Z Q 2005 *J. Vibr. Shock* **24(2)** 140-2.
[3] Jiang G. 2007 *Simulation and Experimental Research on Welding Residual Stress Relieving by High Frequency Vibratory Stress Relief Technology*. Hangzhou, Zhejiang University.
[4] Shen H L and Wu Y X 2009 *J. Vibr. Shock* **28(8)** 191-4.
[5] Hu Y H 2011 *Experimental Research on Stress Relaxation Behavior of Aviation Aluminum-alloy Plate*. Changsha: Central South University.
[6] Cui G J and Teng J Z 2009 *Machinery* **1(36)** 70-3.

Acknowledgements
This paper is funded by National Natural Science Foundation of China (Grant No. 51475483), Aid Program for Science and Technology Innovative Research Team in Higher Educational Institutions of Hunan Province (Grant No. 2014207), and Scientific Research Fund of CSUFT (Grant No. Q2012015A).

Functional-to-form mapping for assembly design automation

Z G Xu[1], W M Liu[1], W D Shen[2], D Y Yang[2], T T Liu[2]

[1]Shenzhen Research Institue of Shandong University, 518057, High-tech Zone A301, Virtual University Part, Shenzhen, China.
[2]School of Mechanical Eng, Shandong University, 250061, No 17923, Jingshilu, Lixiaqu, Jinan, Shandong, China

Abstract. Assembly-level function-to-form mapping is the most effective procedure towards design automation. The research work mainly includes: the assembly-level function definitions, product network model and the two-step mapping mechanisms. The function-to-form mapping is divided into two steps, i.e. mapping of function-to-behavior, called the first-step mapping, and the second-step mapping, i.e. mapping of behavior-to-structure. After the first step mapping, the three dimensional transmission chain (or 3D sketch) is studied, and the feasible design computing tools are developed. The mapping procedure is relatively easy to be implemented interactively, but, it is quite difficult to finish it automatically. So manual, semi-automatic, automatic and interactive modification of the mapping model are studied. A mechanical hand F-F mapping process is illustrated to verify the design methodologies.

1. Introduction

So far, many design automation tools are embedded in present CAD systems, the model creation, modification, maintenance and propagation become more convenient with the introduction of the latest 2D and 3D parametric design technologies [1] (Hoffmann, 2001). ICAD systems are becoming even more powerful, when PDM (product data model), or ERP (enterprise resource planning) and MRP II (materials resource process planning) database as well as advanced HCI (human computer interface), VR (virtual reality), and web-based technologies etc are rapidly incorporated. However, it may be pointed out that it only represents a vision which is almost identical to the earliest visions of CAD systems in view of assembly-level design automation. The major difference now is that we have the opportunity through our improved knowledge of AI and cognitive science to take important strides towards delivering CAD systems closer to these visions.

Some warned that the opportunity may be missed if we concentrate on the wrong issues [2]. Many others may suggest the appropriate approaches [3]. In this paper, several assembly-level "enhancement procedures" are testified.

A specific function-to-form mapping model (F-F mapping model for short) is proposed to cope with divergent exploration, automatic transformation and convergent exploitation. Decomposition of design domains in computational design context helps to set up the domain specific knowledge base (KB).

Automatic mapping among design domains could dramatically increase exploration proficiency. And reconstitution algorithms evolving default logic is to cope with ambiguous information occurred in conceptual design stage, which is also effective exploration tool for concept creation and visualization, called exploitation. The F-F mapping model cooperating axiomatic design theory [4] is effective to integrate requirement, function and form domains for a creative design tool.

2. Mathematical model for assembly stage mapping

Let $\boldsymbol{U}$ be a functional domain (Function Requirements, referred to as **FRs**), $\boldsymbol{V}$ is the domain (Design Properties, referred to as **DPs**), $\boldsymbol{P(U)}$, $\boldsymbol{P(V)}$ is the power set of $\boldsymbol{U}$, $\boldsymbol{V}$, $\boldsymbol{FRs} \subset \boldsymbol{P(U)}$, $\boldsymbol{DPs} \subset \boldsymbol{P(V)}$. $f_{FRs \to DPs}$, where, ***FRs*** is called the definition domain,

Definition 1, functional carrier, if there exist transitional domains, between functional domain and form domain, called ***CRs***, make $\boldsymbol{x} \subset \boldsymbol{CRs}$, if and only when $\boldsymbol{x}$ satisfies the followed two terms:

- A simple, intuitive, deterministic correspondence that satisfies ***FRs*** to ***CRs***
- At the same time, mapping from ***CRs*** to ***DPs*** satisfies the 'deterministic' criterion, where 'deterministic' means configurability.

Non-manifold entities are the "carrier" from the function-to-form mapping. For example, in a gearbox, the shifting fork is to "shift the direction of the two gears", the Non-manifold entities of the shifting fork is composed of three functional faces, as in figure 1.

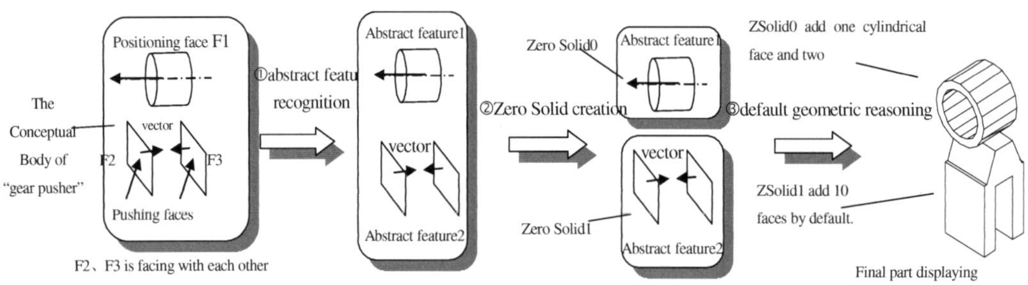

Figure 1. Function-to-form mapping of the shifting fork.

3. Definition of assembly level function

Functional domain is classified into three groups, quantitative, abstract, and assembly. We believe that functional definition could be correlated to the problems concerned; product-level, assembly-level or component-level functions are to be taken into considerations.

As long as the assembly structural design is concerned, the assembly level function should be summarized, classified and analyzed. As in figure 2, five levels of functional definitions are arranged in parallel rows, where the assembly-level function is summarized as a connection link model, each node of which is an information unit to express the prerequisite assembly requirement, like positioning, transmission, supporting and lubrication etc in a network structure.

Table 1. Hierarchy of FRs.

Hierarchical function definitions	**Function contents**	**Source of function requirements**
Product level functions	Product design requirements	Market analysis, user requirements, etc.
Mechanism level sub function	The performance requirements of the product, mainly refers to the relative translation or rotation between parts	Behavioral performance requirements of products produced by demand analysis
Component assembly stage sub function	Functions such as movement, transmission, blocking, positioning, clamping, sealing, lubrication, connecting and fastening between parts	Obtained by the construction principle of the mechanism

The information unit could be correlated with many alternatives of functional carriers, what-ever it might be, a conceptual face, functional feature, component or mechanism etc. The hierarchical definition of the overall functional requirements is in table 1.

Prior to the first phase of design, there exists a requirement management and mapping process. Requirement domain is dealing with product specifications, user's intent, market information, product life cycle analysis/assessment (LCA) inventory, as well as information about the Production Company etc.

It would provide an information bulk necessary to start the development of a new product. Requirement domain is associated six research areas.

1) capturing of customer's need effectively [5]; 2) creation of product specifications elicited from customers need [6]; 3) computer-based representation of product specifications; 4) testing of product performance against the product specification. 5) design intent management with LCA inventories, where design intent management and the interface for acquisition, representation, analyzing, recording and retrieving of the user's intent information from the product life cycle design perspectives are outlined.

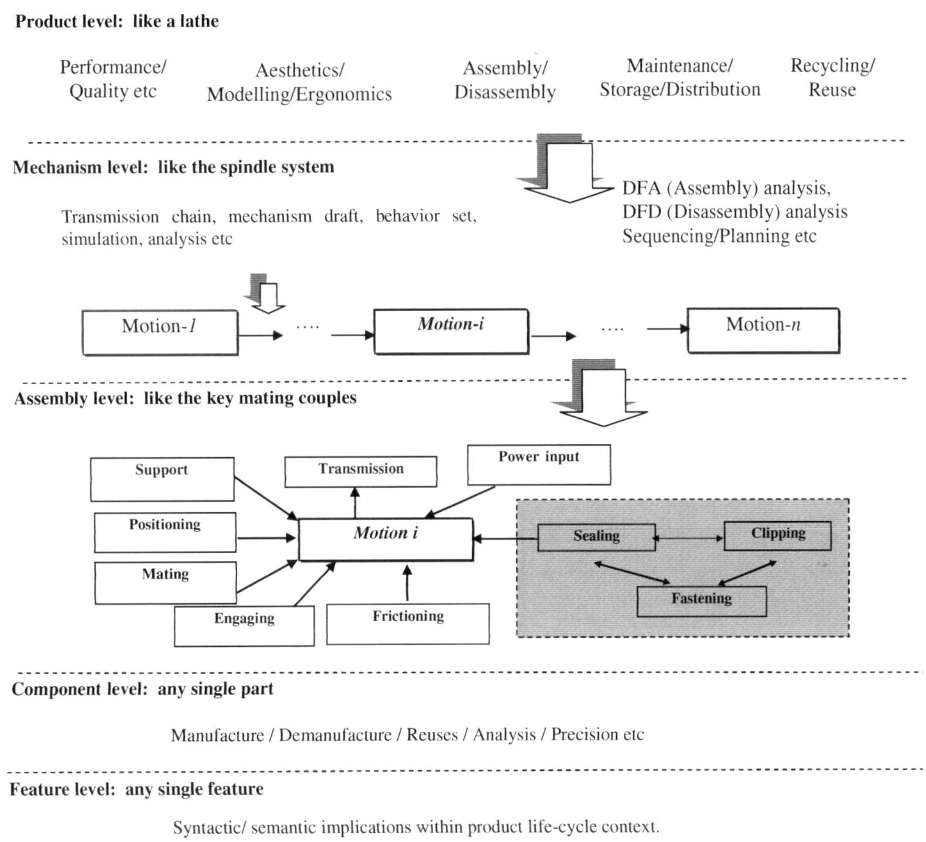

Figure 2. Function definitions.

After the Design requirement is captured, recorded and analyzed. The next step is to propagate the factors to the downstream design activities.

4. Generalized positioning for mechanism

Generalized positioning: Traditionally, a part is thought to be totally positioned by six-point, i.e. translations in X, Y, Z, and rotations in A (around X), B (around Y), C (around Z). However the concept of twelve-point positioning is introduced for better understanding of part dynamics as well as static positioning. One part is thought to be totally positioned by 12 degrees of freedom, i.e. translations in X, -X, Y, -Y, Z, -Z, and rotations in A, -A, B, -B, C, -C, is called generalized positioning, and the degrees of freedom is called generalized DOF (GDOF for short).

The reason of the proposal of 12 GDOF is that it is easier to set up the mapping correlation of the positioning requirement i.e., the positioning requirement of the GDOF number of the component with respect to a certain feature.

For example, the following slot feature could be thought of being restricted in the user coordinate system (UCS) of the –Y, +X and –X , +A and –A, +B and –B directions, and the other GDOFs are free, which will be encoded in a string of hex code like in table 2.

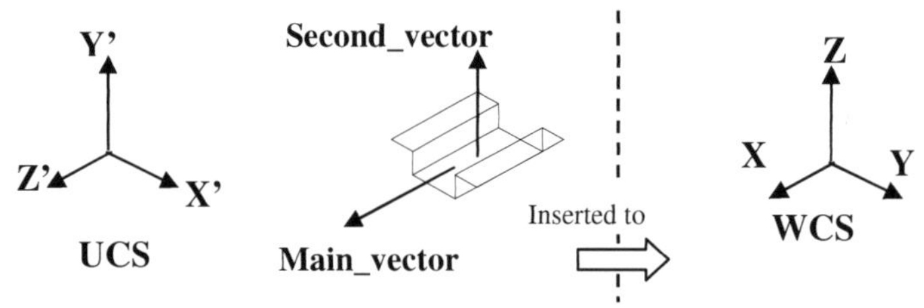

Figure 3. Slot feature in UCS with respect to WCS.

Table 2 gives the GDOF representation of the slot feature (figure 3). The GDOF code in UCS is F73, while the GDOF code in world coordinate system (WCS) should be modified according to the inserting orientation of the UCS with respect to WCS, at this insertion point, the GDOF code in WCS should be 3F7. Table 2 is the GDOF representation of slot feature with respect to the UCS and WCS.

Table 2. The GDOF representation of slot feature with respect to the UCS and WCS.

WCS	Y		B		Z		C		X		A	
UCS	X'		A'		Y'		B'		Z'		C'	
	+	-	+	-	+	-	+	-	+	-	+	-
Binary	1	1	1	1	0	1	1	1	0	0	1	1
Hex	F				7				3			

✱ Hex coding is accepted by GPAL_KN due to its simple and explicit format.

5. Product assembly model based on functional features

After function-to-form mapping (**F-F mapping** for short), the product conceptual model has been established, the conceptual product model is composed of functional faces and their correlations. There are three levels in a assembly model of one product.

According to the functional requirement of support, stop, fixing or positioning etc, different features are mapped to the corresponding parts, so one specific feature is mapped to two or more corresponding parts, and the relationship between parts in a product are naturally established, as in figure 4.

The Sub-functions in figure 4 mainly refers to the functions of drive, location, clamping, sealing, blocking, etc. for motion function motioni, after **F-F** mapping, a meta_element is formed, defined as meta_elementArray:

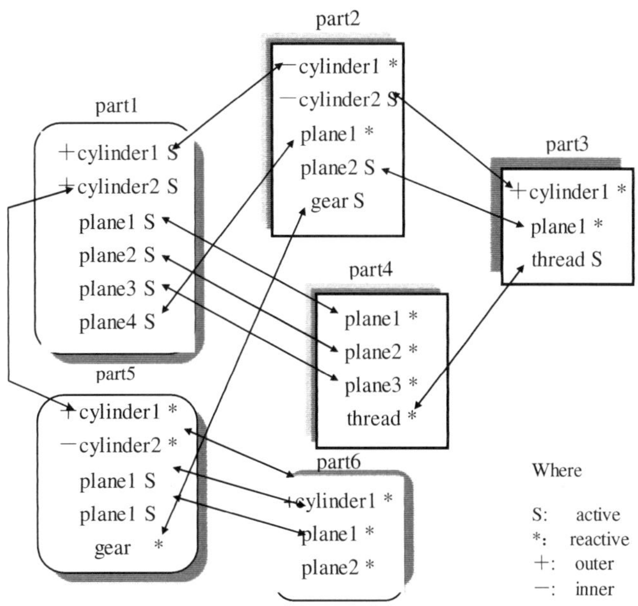

Figure 4. Product assembly model.

```
class meta_element
{ ……
  enum Function_Type              // assembly level functional definition
  enum element_type;              // include: FuncFace, FuncFaceArray
                                  // and standard_part .
  MfuncFace * facept;             // functional faces
  AcDbVoidPtrArray face_array;    // functional faces sets
  Mpart * standard_part           // pointer to a standard part
  ………
}
AcDbVoidPtrArray meta_elementArray // meta_element.
```

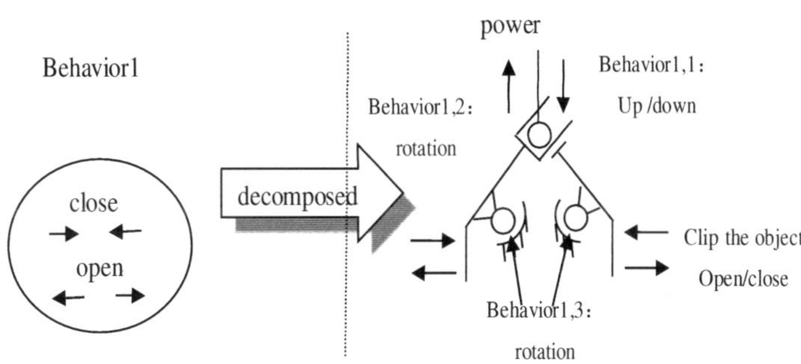

Figure 5. Functional requirement of the mechanical hand.

IOP Conf. Series: Materials Science and Engineering **269** (2017) 012004 doi:10.1088/1757-899X/269/1/012004

Based on the product assembly model, the features are expressed in the assembly, and the relationship between features and features, parts and features, and parts and parts etc, with the information of location, size, constraints etc, are defined in a data structure to describe the three types of relations, as in figure 4.

6. Case study

To clamp one work-piece, the mechanical hand have the opening and closing movement, the Behavior1 can be furtherly decomposed into a transmission principle diagram as shown in figure 5.

Stepping motors is firstly selected interactively. The stepper motor is related to several parts as, drive parts, nuts, screw drive components. In combination with figure 5, the 3-D transmission chain of the mechanical hand is deduced, as in figure 6, and the corresponding parts of the product are created at the same time.

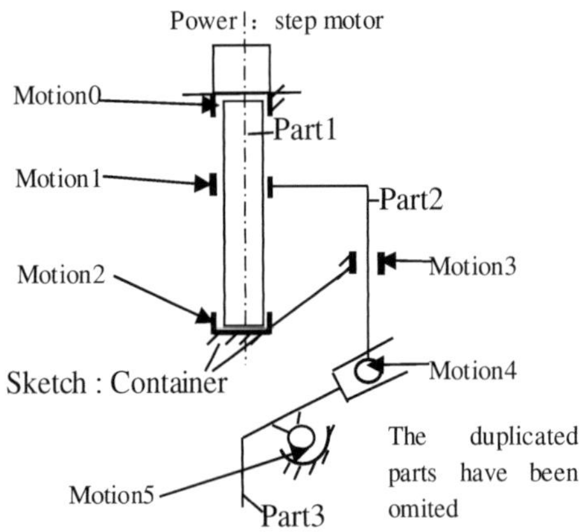

Figure 6. 2-D Functional mode of the mechanical hand.

As shown in figure 6, from the driving principle of the mechanical hand, four correlated parts created, namely, Part1, Part2, Part3 and Container.

Table 3. The relationship between the created parts and the resulting meta bodies.

function	meta_element	Parts involved
Motion0	F01(Cylinder),F02 (plane)	Part1, Container
Motion1	F11(thread)	Part1, part2
Motion2	F21(Cylinder),F22(plane)	Part1, container
Motion3	F31,F32 F33,F34(plane)	Part2, container
Motion4	F41(double),F42、F43(Plane)	Part2(F411,F412(plane)) part3(F413(cylinder))
Motion5	F51(cylinder),F52、F53(plane)	Part3, container

Table 3 illustrates the five motion in the mechanical hand, meta_elements are firstly deducted. For example, the meta_element F41 is deduced by the motion4, F41 is mapped on Part2, and the corresponding feature, F411 and F412 are created, in the mean while F412 is created on Part3.

And the relationship between the two parts, i.e. Part2, Part3, and the corresponding features F411, F412 and F413 are established. The functional surface sets of Part1, part2, and part3 formed by the function structure mapping are shown in figure6. And the mapping process and the mapping results are shown in figure 7.

IOP Conf. Series: Materials Science and Engineering **269** (2017) 012004 doi:10.1088/1757-899X/269/1/012004

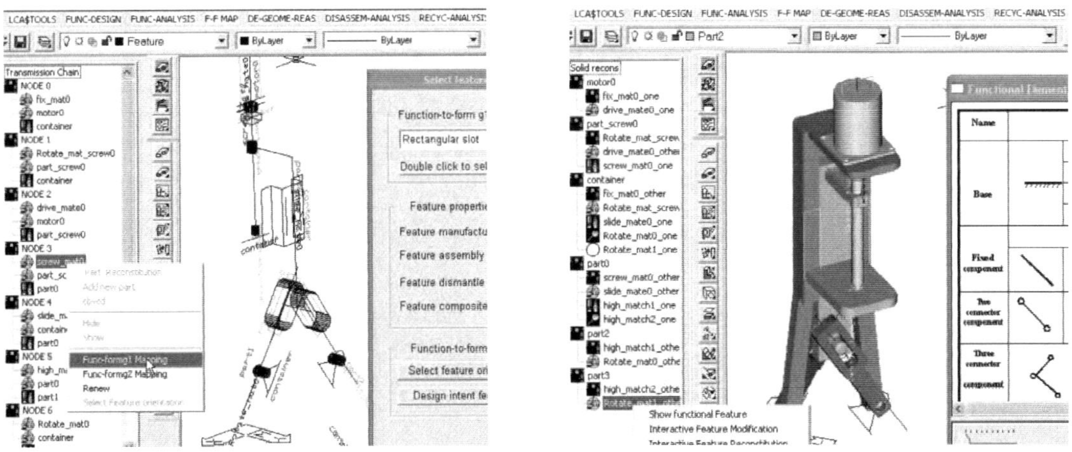

(a) 3-D transmission chain of the mechanical hand (b) 3-D Model of the mechanical hand

Figure 7. F-F mapping of the mechanical hand and the result.

7. Conclusion

An effective assembly level function-to-form mapping model is proposed in this paper. It could be concluded that, the **decomposition**, F-F **mapping** and **reconstitution** model could be a common exploration, transformation and exploitation procedure for management of computational design tools, therefore helpful for creative work, specifically the synthesizing of form domain is not simply re-arrangement of physical elements, rather it needs geometrical as well as algebraic reasoning on un-manifold polyhedral. Novel mathematical as well as AI technologies were introduced. Several packages of design tools have been developed to testify the effectiveness of the design methodology.

8. References

[1] Hoffmann C M and Kim K J 2001 *Comput. Aid. Des.* **33(1)** 81-90.
[2] MacCallum K J 1990 *Artific. Intell. Eng.* **5(2)** 55-64.
[3] Faltings B and Sun K 1996 *Comput. Aid. Des.* **28(3)** 207-16.
[4] Suh N P 1990 *The principles of Design*. Oxford University Express, New York.
[5] Stauffer L and Morris L 1992 *A new program to enhance the development of product requirements*. Presented at NSF Design and Manufacturing Systems Conference, Atlanta, Georgia,
[6] Grothe M T 1998 *Visual modeling of products and life-cycle activities*. Presented at DETC'98 ASME Design Engineering Technical Conference, Atlanta, Georgia,

Acknowledgment

The above-mentioned research work is supported by the Science, Technology and Innovation Commission of Shenzhen Municipality, JCYJ20160510165328965 and the Chinese NSFC, 61272017, P.R. China.

Influence of Inclusions on Low Cycle Fatigue Life of PM Superalloy

Zhao Kai[1, 2, 3], Liu Xin-ling[1, 2, 3], He Yu-huai[1, 2, 3]

[1]AVIC Failure Analysis Center, Beijing Institute of Aeronautical Materials, Beijing 100095, China
[2]Beijing Key Laboratory of Aeronautical Materials Testing and Evaluation, Beijing 100095, China
[3]Aviation Key Laboratory of Science and Technology on Materials Testing and Evaluation, Beijing 100095, China

Abstract. In the range of operating the resistance to low cycle fatigue (LCF) of P/M nickel-based superalloy is one of the most important performances. In this paper the fracture features of LCF on P/M nickel-based superalloy were investigated. The study focused on macroscopic and microscopic features of LCF fractures under 650°C, $\triangle\ \varepsilon$=0.82%, R=0.05, 0.33Hz test condition by optical microscope and SEM. The types and morphologies of failure origins were observed. The effect of position and size of inclusions on LCF were researched. A new life prediction model appropriated for this test condition was provided. The result shows that the LCF life and the size of inclusions in this test condition follow the Weibull distribution. The LCF life exponentially decreases with the inclusion size increases.

1. Introduction

As a key material of high thrust-weight ratio engine, P/M nickel-based superalloy has been mainly applied in the fields of aeronautics and astronautics [1-5]. In industry, many treasures have been taken to eliminate the inclusions from powder metallurgical (P/M) superalloy; however, there is a long way to go to get rid of them completely. So the research on the influence of inclusions on LCF is necessary [6].

The study focused on macroscopic and microscopic features of LCF fractures by optical microscope and SEM. The types and morphologies of failure origins were observed. The influence of size of inclusions on LCF life was researched. A new life prediction model appropriated for this test condition was provided.

2. Experimental

P/M nickel-based superalloy was used to machining samples (figure 1). The alloy in this investigation is an advanced nickel-base PM superalloy. The chemical composition of P/M nickel-based superalloy is listed in table 1. Before machining samples, the location of inclusions was tested by the ultrasonic equipment. When samples were machining, inclusions can be stayed in the working part of samples.

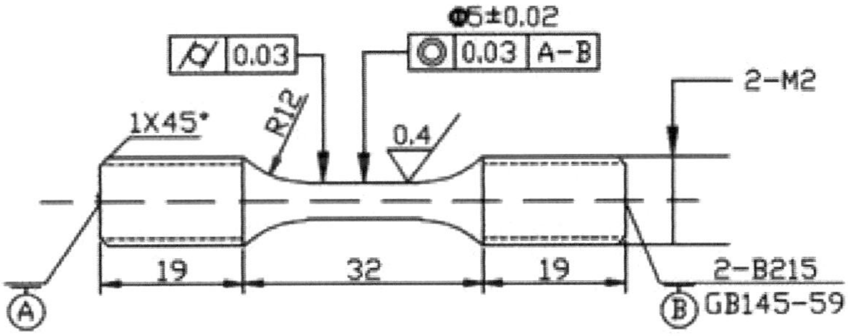

Figure 1. LCF sample.

Table 1. Main chemical composition of P/M nickel-based superalloy in wt/%.

C	Al	Co	Cr	Mo	Nb	Ta	Ti	W	Zr	B
0.05	3.34	20.86	13.02	3.78	0.98	2.48	3.67	2.25	0.038	0.036

The experiment was processed in the MTS fatigue-test machine under 650°C and the $\Delta\ \varepsilon$=0.82%, R=0.05, f=0.33Hz.

Conventional metallographic and analytical electron microscopy procedures were adopted for micro-structural characterization and analysis. Microstructural analyses of the sample were performed by optical microscopy, scanning electron microscopy (SEM) and field-emission scanning electron microscopy.

3. Results and discussions

3.1. Macrostructure

The origin which is a point source locates in the interior of the sample. The fatigue extension area is blue because of the high temperature and it is vertical to the axis of the sample. This area accounts for 30% of the total. The final rupture region which is about 45° to the axis appears gray (figure 2).

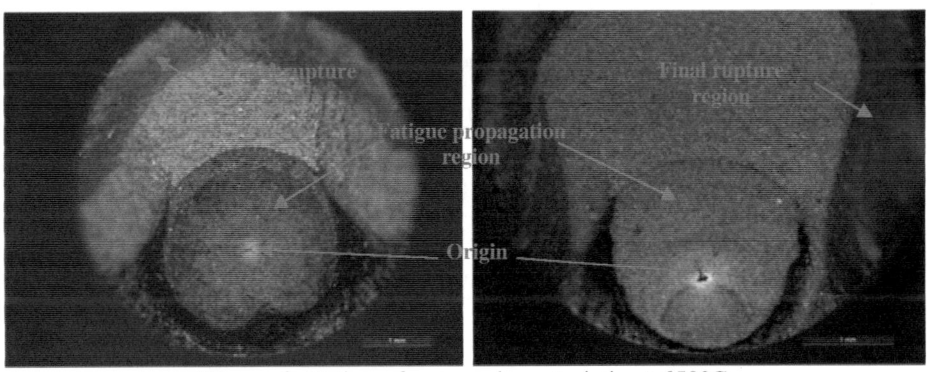

Figure 2. Fatigue fracture characteristic at 650°C.

3.2. Microstructure

Inclusions present different shapes and there are about five shapes, ellipse, polygon, half-moon, strip, triangle (figure 3). The ellipse and polygon are in the majority (figure 4). The constituent of every inclusion is analysed by the EDS. There are three kinds, Al_2O_3, MgO, SiO_2 and the Al_2O_3 is in the majority (figure 5).

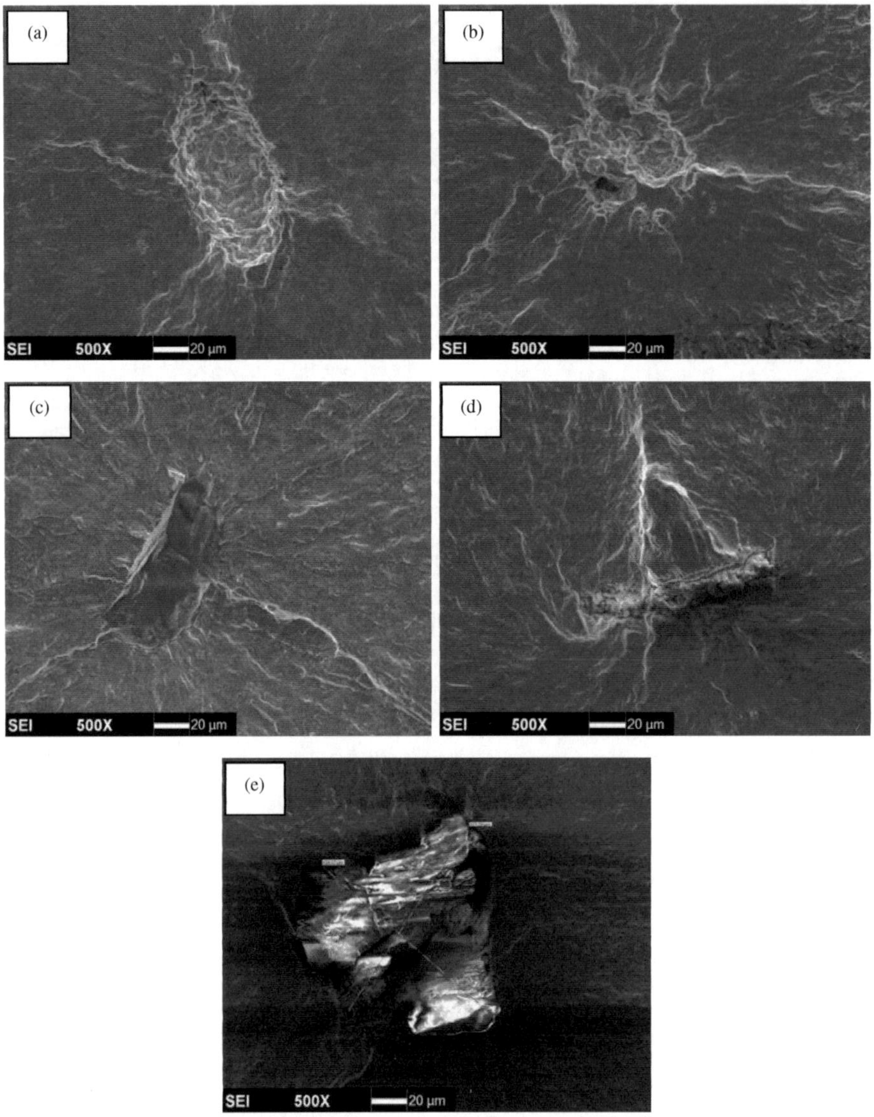

Figure 3. Crack originated from an inclusion
(a) Ellipse (b) Polygon (c) Half-moon (d) Strip (e) Triangle.

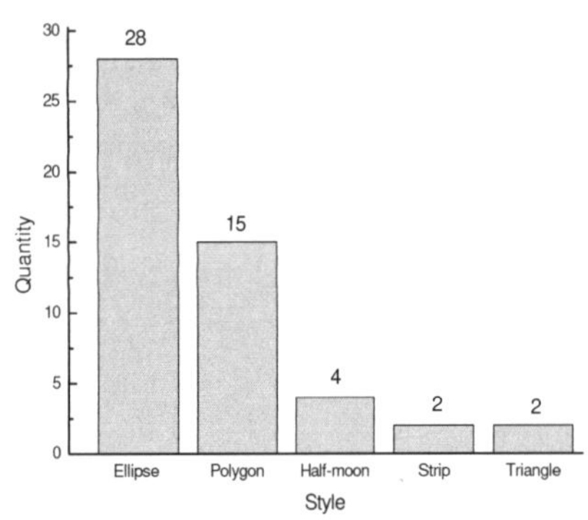

Figure 4. Quantity of different styles inclusions.

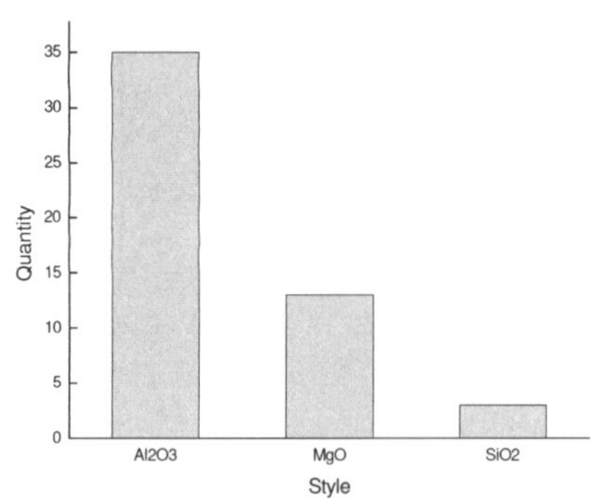

Figure 5. Quantity of different components inclusions.

The LCF life of all the samples and the area of inclusions are statisticed and statistic results form a Weibull distribution (figure 6 and figure 7).

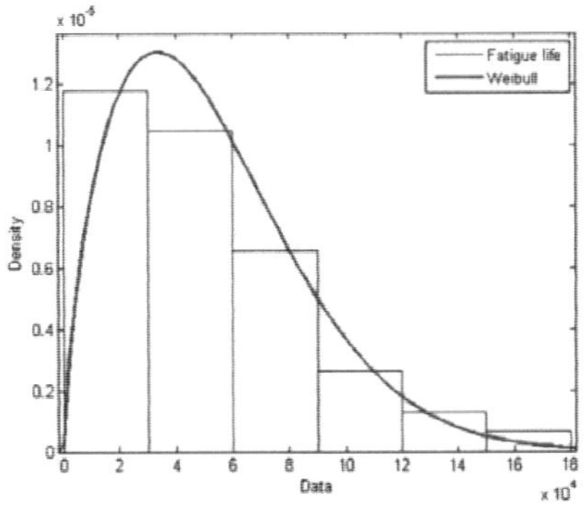

Figure 6. Weibull distribution of fatigue life.

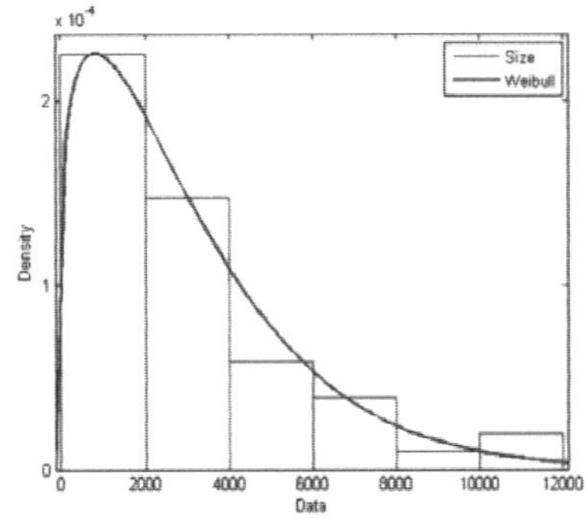

Figure 7. Weibull distribution of inclusions' size.

3.3. Life prediction model

The distance between the inclusion and the surface of sample is measured and results range from 0.1 to 2.33mm. The functional relationship graphics are plotted between inclusions' size and the LCF life (figure 8). The LCF life exponentially decreases with the inclusion's size increases. The formula is showed as blow.

$$y = 1.284\times10^{4} + 1.024\times10^{5} e^{-4.187\times10^{-4}x}$$

x is the area of the inclusion/μm^2, y is the LCF life.

The lives of the other samples are predicted by using of the new formula. The results show that most points of the predicted lives are in the double error band. So this LCF model can provide satisfying accuracy (figure 9).

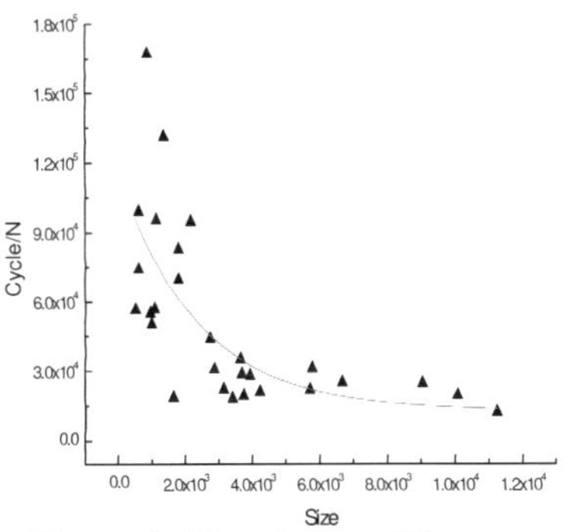

Figure 8. Size-fatigue life curves.

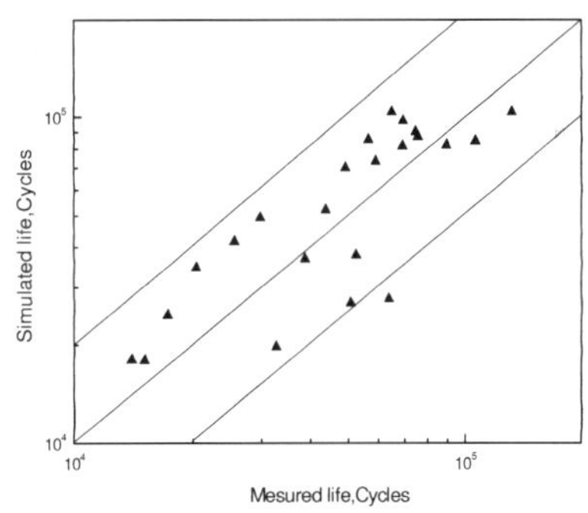

Figure 9. The relationship between simulated and measured life.

4. Conclusions

1) The LCF life of all the samples and the area of inclusions are statisticed. The LCF life and the size of inclusions in this test condition follow the Weibull distribution.

2) The constituent of every inclusion is analyzed by the EDS. The constituent of inclusions are three kinds, Al_2O_3, MgO, SiO_2 and the Al_2O_3 is in the majority.

3) The distance between the inclusion and the surface of sample is measured and results range from 0.1 to 2.33mm. The LCF life exponentially decreases with the inclusion's size increases.

4) The lives of the other samples are predicted by using of a new formula. Most points of the predicted lives are in the double error band.

References

[1] Zhou L, Wang H, Zhang G D, et al 2009 *J. Plast. Eng.* **16(5)** 149-54.

[2] Krueger D D, Kissinger R D and Menzies R G 1992 *Development and Introduction of a Damage Tolerant High Temperature Nickle-base Disk Alloy Rene'88 DT Superalloys,* Proceedings of the Seventh International Symposium on Superalloys, 1992 277
[3] Raujol S, Pettinari F, Locq D, et al 2004 *Mater. Sci. Eng.* **A(387-389)** 678-82.
[4] Sophile dubiez-Le G, Raphael C, Lure G, et a1. 2008 *Mater. Sci. Eng.* **A(483-484)** 598-601.
[5] Hu C Y, Liu X L and Chen X, 2014 *Acta Aeronaut. Astronaut. Sin.* **35(2)** 461-8.
[6] Li X, Zhang L and Dong J X, 2001 *J. Univ. Sci. Tech. Beijing*, **8(4)** 286-9.

Acknowledgements
The authors would like to acknowledge the financial support of the National High-tech R&D Program (863 Program) (2015AA034401).

Application and Mechanics Analysis of Multi-Function Construction Platforms in Prefabricated-Concrete Construction

Meihua Wang[1*], Rongshuai Li[1], Wenze Zhang[2]

[1]Shanghai Construction Engineering Group Co., Ltd, Shanghai 200080, China
[2]Structural Engineering, Tongji University, Shanghai 200092, China

Abstract. Multi-function construction platforms (MCPs) as an "old construction technology, new application" of the building facade construction equipment, its efforts to reduce labour intensity, improve labour productivity, ensure construction safety, shorten the duration of construction and other aspects of the effect are significant. In this study, the functional analysis of the multi-function construction platforms is carried out in the construction of the assembly building. Based on the general finite element software ANSYS, the static calculation and dynamic characteristics analysis of the MCPs structure are analysed, the simplified finite element model is constructed, and the selection of the unit, the processing and solution of boundary are under discussion and research. The maximum deformation value, the maximum stress value and the structural dynamic characteristic model are obtained. The dangerous parts of the platform structure are analysed, too. Multiple types of MCPs under engineering construction conditions are calculated, so as to put forward the rationalization suggestions for engineering application of the MCPs.

1. Introduction

Since the 21st century, especially in the last five years, with the sustained and rapid development of Chinese economy, China's urbanization and urban modernization process accelerated, energy-saving and environmental protection requirements increased, labour costs continued to grow. In China, the research and application of precast concrete assembly construction technology and Building gradually warming up, and promoting the building assembly is the general trend [1]. The State Council issued the "Guidance on the development of fabricated buildings" and put forward eight tasks to implement the five development concept of fabricated buildings by the Chinese government, not only the strongest voice to vigorously develop the assembly of the building , but also the solemn commitment to the world [2]. From a global perspective, the three major trends for the development of the construction industry include greenization, informatization and industrialization. The assembly structure is in line with this trend, and it is also an important part of the Paris Declaration on Climate Change and the China-US Joint Statement [3].

Zhou, J. H. [3], who is Academician of Chinese Academy of Engineering (CAE), pointed out that to develop assembly-style buildings, building standardization is the premise. The structural components industrialization and new building materials is the material basis, on-site assembly and construction mechanization is a necessary mean. Shen, Z.Y., (Academician of CAE) [3] proposed to achieve a new type of construction industry needs research and development, including structural system, enclosure system, building parts system and other integrated decorative technology. Building assembly is now widely carried out in China, but still faces many problems, mainly including as the following: (1) The design and construction are still separated because the investors, designers, manufacturers of the components and the construction side are independent to each other, it is not conducive to the promotion of assembly building. (2) Assembly technology research is immature. The structural design of prefabricated wall has been preliminary solved, in addition, the thermal performance of exterior shear wall joints, waterproof performance, the joints shrinkage cracks for the new and old concrete and other issues are still not solved well. (3) The existing standards are imperfect. Design and construction concepts are based on the concept of cast-in-place, and the concept of assembly is not formed.

At present, in the construction of the facade of the assembled building, the traditional steel scaffolding or electric basket as the main construction equipment are basically used. During the erection of steel scaffolding, the working procedures are various and the operations are complex, which leads to the waste of labours, serious resource consumption and low production efficiency. Electric basket is high altitude equipment, the use of its process also has many issues, such as single function, unsafe construction, poor stability, etc. Therefore, it is necessary to develop a kind of fast, safe and efficient construction equipment, which is suitable for prefabricated construction in order to meet the rapid development of assembly buildings. Multi-function construction platforms (MCPs), as a kind of portable and efficient facade construction equipment, can be quickly assembled to form a construction platform for the construction of the building facade. It can be used as a structural external protection system, became a strong guarantee for construction ([4-9]). In order to ensure the safe operation of MCPs, it is very significant to study the mechanical characteristics of MCPs before application.

In this paper, taking the construction loads, operators and operating load, wind loads and other environmental loads into consideration, combined with the finite element theory, the three-dimensional finite element model of MCPs is established by using ANSYS finite element software. On this basis, the static and dynamic characteristics of MCPs model are analyzed respectively, which are used to evaluate the safety and reliability of MCPs. By static calculation of various types of MCPs under differences conditions of engineering application, the maximum deformation and the maximum stress value, and analysis of the dangerous part of MCPs structure are obtained. Furthermore, the natural frequency and mode of MCPs structure are calculated by calculating the dynamic characteristics of the model. And finally put forward reasonable suggestions for the application of MCPs in engineering.

2. Function analysis of multi-function construction platform

Multi-functional construction platforms (MCPs), according to the characteristics of the construction platforms and the functional requirements of the assembly building construction, can be divided into several types: such as a tool-based protection system, as for hoisting of the prefabricated concrete (PC) components, as auxiliary external facade construction, etc.

(1) As a tool-based protection system: MCPs climb to the operating layer of the structure. Its work platform and railings are used as a part of the protection system of the facade construction, forming a movable protection system for building external facade construction. Application of MCPs in prefabricated construction can remove the movable enclosures with the fixed bolt, the ordinary steel pipe scaffold and the tools combination of falsework system and, etc. Taking the assembly structure of PC external wall as example, conventional construction needs to set the external safety protection that required steel scaffolding, climbing scaffolding or special enclosure system, the use of conventional enclosure system will cause a lot of waste of materials and labors with bad economy.

(2) As for hoisting of the prefabricated concrete (PC) components: PC components' installation is often accompanied by its positioning, measuring, fixing and correcting, by steels' banding, templates' lapping, prefabricated wall joints processing and other details operations. According to China's domestic projects research, the author found that the above-mentioned multi-channel detail process was a special technology for the construction of the assembly building. Because the current facade erection did not develop a unique construction equipment, the construction process suffering a lot of high-altitude operations and non-standard construction phenomenon.

(3) As auxiliary external facade construction: the external facade construction is one of the core functions of MCPs. Prefabricated building facade construction includes the following contents: concrete surface wall leveling, Playing the priming-coat surface plaster, External wall insulation, facade multi-channel paint whitewash, external-wall facing brick, facade by the dry hanging stone, installation of the outside window, facade rainwater pipe layout and other construction technologies. The process has the characteristics of wide range, involving the numerous processes. Compared to the erection of scaffolding or hanging-basket equipment, MCPs has obvious advantages. The platform can be docked at any desired position, which is comfortable for operator's operation, reducing labor intensity, improving work efficiency. Different from the hanging-basket equipment, MCPs can be installed from the bottom of the structure in the construction. With the progress of the construction schedule, it can be lifted or lowered to do the decoration operations and does not interfere with other processes. Therefore, the construction period will be guaranteed

3. MCPS structure and numerical model

3.1 Introduction of multifunctional construction platforms (MCPs)

Multi-function construction platforms (MCPs) are divided into single-mast form and double-mast form, and the whole equipment is mainly composed of mast frame device, working platform device and power system. The mast frame device includes vertical standard section, climbing guide rail, base frame, operating platform fixed frame, wall-attaching equipment; The working platform device was comprised by horizontal standard section, railing and connecting pieces; Power system is to control the lifting or lowering of the working platform, the base according to the requirements of the building can be set stationary and mobile. Between the mast device and the building, the upper and lower groups of attachment devices are arranged at a certain distance. According to functional requirements, MCPs can be set as single-stage or multi-level. The Multi-level of MCPs structure diagram is showed in figure 1.

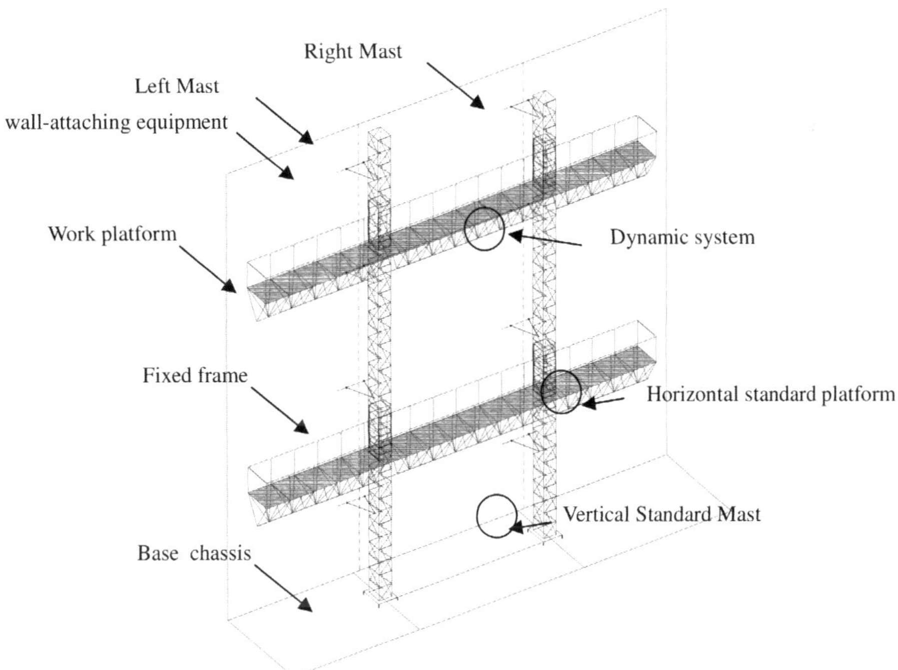

Figure 1. A design diagram of multi - functional construction platforms (MCPs).

3.2 Numerical model of Multi-functional Construction Platforms (MCPs)

3.2.1. Model simplification. In order to ensure the accuracy of calculation, reduce the scale of calculation and make the model to reflect the real mechanical properties of MCPs, in the modeling process, the actual structure of MCPs needs to be simplified. This article mainly focuses on the mechanical properties of MCPs structure under different loads conditions. Therefore, the influence of local stress concentration on the overall structure is neglected. According to their structural characteristics and working conditions, we can make the following simplifications.

1) Vertical standard section, between the wall device and the prefabricated wall and so on are bolted. This paper assumed that the bolts strength is sufficient, no damage, so simplified can be considered as the rigid connection;

2) Ignore the small structural defects that have little impact on structural analysis, such as all bolt holes, pin holes, etc.

3) Regardless of the change in the properties of the weld material, it is considered that the material properties of the weld are the same as those of the adjacent structure.

3.2.2. Options of the finite element type. MCPs belong to three-dimensional (3D) space truss structure. In ANSYS, the 3D beam elements include BEAM4, BEAM188, BEAM189. Among them, BEAM4 is a single-axis force unit that can be used to withstand tension, pressure, bending moment and torque, used to simulate the 3D general beam. The units of BEAM188 and BEAM189 not only maintain the main features of the beam element with the characteristics of a one-dimensional component that approximates to the three-dimensional solid structure, but also enhance its cross-section definition function and improve the visualization of the other two dimensions of the beam member. In addition, the element takes the effects of shear deformation into account. In the element interpolation function, the deflection and cross section rotation are individually interpolated. Furthermore, BEAM188 is a linear beam element with a stronger nonlinear analysis function [10]. So the beam components of MCPs all select element of BEAM188 unit to simulate. In the mast device and work platform device, the web member of the truss structure are consisted of reinforced, and welded with the main frame, can transmit axial force and bending moment, so we can choose PIPE16. Since the pedal of horizontal operating platform device is a steel wire mesh, which has no influence on the overall force of MCPs. Therefore, the steel wire mesh can be converted into loads acting on the construction platform. The section type of element and the real constant are shown in table 1.

Table 1. Section type of element and real constant.

Component	Unit type	real constant and ID
Mast device	BEAM188	□40×40×3; ID=1
	BEAM188	∟40×40×3; ID=2
	PIPE16	O Ø16×8; ID=3
Work platform	BEAM188	□50×50×4; ID=4
	BEAM188	□30×80×2; ID=5
	PIPE16	O Ø16×8; ID=6
Wall-attaching equipment	BEAM188	□40×40×3; ID=7
Fixed frame	PIPE16	O Ø40×10; ID=8

3.2.3. Loads and boundary conditions. According to the maximum allowable loads of MCPs to calculate, the construction platform factory takes the construction load based on the construction platform design requirements: Q_l=15kN,With uniform loads to consider. Criterions [11] provided that the ground above the six winds do not carry out the platform operations, installation and demolition, so $W_{k=}0.11KN/m^2$. The quality of each person is 80kg and the quality of each carrying device is 40kg, assuming that the manual operation force is 20kg. Take the operator loads of Q_2= 1.4kN, to consider as the concentration of load. The specification [11] sets that Single guide platform are no more than 2 peoples, multi-guide platform is for 4 peoples.

According to the provisions of the codes [11-13], MCPs loads can be divided into: ① rated uniform loads;② 125% rated uniform loads (to withstand dynamic test loads);③ The telescopic beam of the working platform extends 1.5m; ④ Lateral loads (including wind loads and man-made lateral loads);⑤ Concentrated forces bias.

This paper mainly analyses the following four kinds of load conditions:

Case1:① ④ ⑤ ;

Case2:① ③ ④ ⑤ ;

Case 3:② ④ ⑤ ;

Case 4:② ③ ④ ⑤ .

The connection between the wall-attaching equipment and the wall generally use embedded welding or reserved hole through the wall bolts and so on, so the connection-nodes can pass the bending moment and shear force, therefore the point can be treated as a fixed point.

3.2.4. Establishment of Finite Element Model. Taking the MC-36/15 double-mast / single-mast MCPs as an example, the finite element analysis are established. The endpoints of the components are defined as the key point, and the finite element model of MCPs are established by dividing their units. There are a total of 1838 units and 1014 nodes, as shown in figure 2. In the modeling, loading and solving process, only the elastic situation is considered, regardless of the impact of materials plasticity. Q235 steel is used in the steel frame and pedal of MCPs.

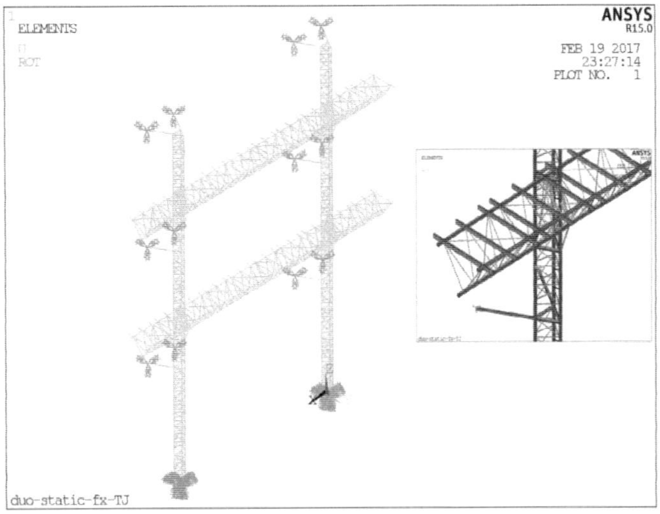

Figure 2. The finite element model of multi-functional construction platforms (MCPs).

4. Structural solution and analysis

4.1 Analysis of static characteristics

In this paper, the static analysis of MCPs is carried out. The stress, strain and displacement of the structure of MCPs are mainly calculated under four load cases. In other words, the topic is to discuss the stress, strain and deformation of MCPs structure after being subjected to external forces in order to check its strength. In this paper, the stress unit in the stress nephogram is MPa, and the displacement unit of the deformation nephogram is mm. In order to make the model truly reflect the actual structure of MCPs, the self-weight of MCPs are applied to each unit in the form of inertial forces, and the gravitational acceleration is in the opposite direction of the Z-axis of MCPs.

According to the functional requirements of the application of the assembled building construction, MCPs will be used in four types: double mast, single mast, single layer and multiple layers. In this paper, the four combinations of application types are respectively calculated under the four kinds of working conditions in the static stage, in order to check the strength and stiffness, and provide suggestions for construction application. Table 2 shows the results of stress and deformation in four conditions.

Table 2. The results of stress and deformation in four conditions.

	Double-mast & Single-layer			Double-mast & Multi-layers			Single-mast & Single-layer		Single-mast & Multi-layers	
	1.2,2.4,1.2	2.4,0,2.4	4.8,0,0	1.2,2.4,1.2	2.4,0,2.4	4.8,0,0	1.2,1.2	2.4,0	1.2,1.2	2.4,0
Case 1	54.4	57.8	67.1	59.2	60.5	69.1	70.8	101	76.3	128
	4.89	4.425	5.219	5.399	4.397	5.559	4.948	29.669	7.86	28.759
Case 2	162	162	162	162	162	171	162	158	167	161
	25.586	24.874	24.068	27.825	26.34	31.579	28.216	59.478	32.35	58.772
Case 3	62.8	63.3	75.2	69.2	69.5	80.5	81.8	160	80.4	159
	4.858	4.039	5.333	6.032	5.001	6.782	5.865	32.137	6.605	32.137
Case 4	200	202	202	202.6	202.9	213.9*	202	201.8	202.3	201.2
	33.334	28.878	35.935	33.408	32.689	39.237	34.892	58.029	35.52	58.947

Note: the calculation results under different cases: the first line is stress, the second line is deformation, where 1.2 indicates that the operator applies the concentrated load, and the unit is KN.

The Reference code [13] specifies the mechanical properties of the structural steel, the proportional limit is 200MPa and the yield point is 235MPa. When the working platform is under loads, the maximum deflection value of the main beam in the bottom of platform shall be less than 1/500 of the length of platform, and the secondary beam shall be no more than 1/250 of the length of the platform. So the maximum deflection allowed by the platform is 30mm or 60mm.

As shown in table 2, under the four working conditions of MCPs in addition to the individual (Double-mast & multi-layers in case 4), the four types of MCPs all meet the required strength and deflection requirements. According to the assumptions of load cases, the working cases from one to four are all fully loaded. The above data shows that the strength and stiffness have obvious differences in the different types of MCPs under different bias loads.

As shown in table 2, in the same type of MCPs under the same load bias effect, the difference of maximum stress value has very small in the case 1 and case 3, and those values are also very close in the case 2 and case 4. The case1 and the case 3 under the same load bias, the maximum displacement of MCPs is 4.039mm to 10mm, which is generally small. Yet, The maximum displacement of the case 2 and the case 4 reaches 58.947mm, which is generally large and occurs at the end of the telescopic beam cantilever of working platform.

Combined with figure 3, under the case 1 and case 2 within the same load bias effect, the maximum stress value of MCPs is from 59.2MPa to 162MPa, and the change is significant. Figure 4 shows that under four concentrated loads, the maximum stress occurs on the cantilever solid end of the telescopic beam. Therefore, the reason for stress value and displacement value are larger in the case 2 and case 4, due to the platform telescopic beam extension 1.5m, and the concentrated load produces larger bias. As shown in figure 4, the webs of the triangular trusses shows that the webs are presented as different colors monochromatic tie rods, which indicates that the webs of the space truss units have the axial force distribution.

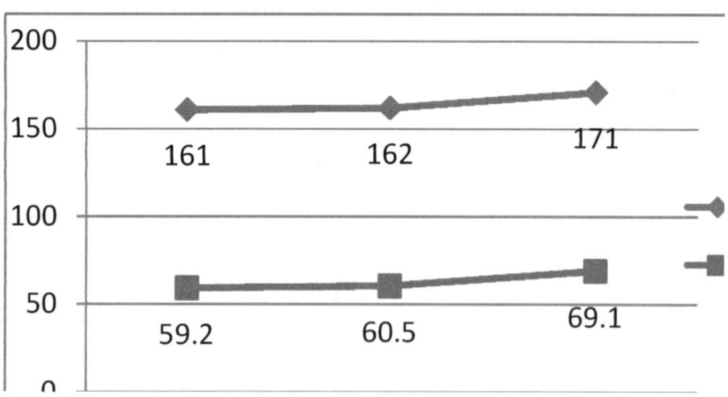

Figure 3. Contrast curve of the maximum stress for Double-mast & Multi-layers under case 2 and case 4.

IOP Conf. Series: Materials Science and Engineering **269** (2017) 012006 doi:10.1088/1757-899X/269/1/012006

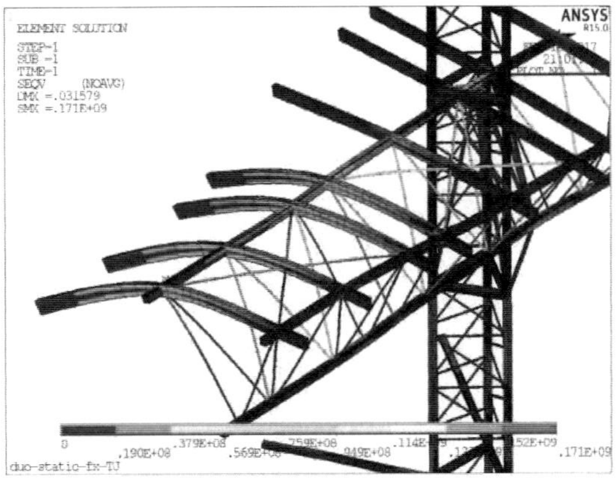

Figure 4. Local stress nephogram of Double-mast & multi-layers under case 2.

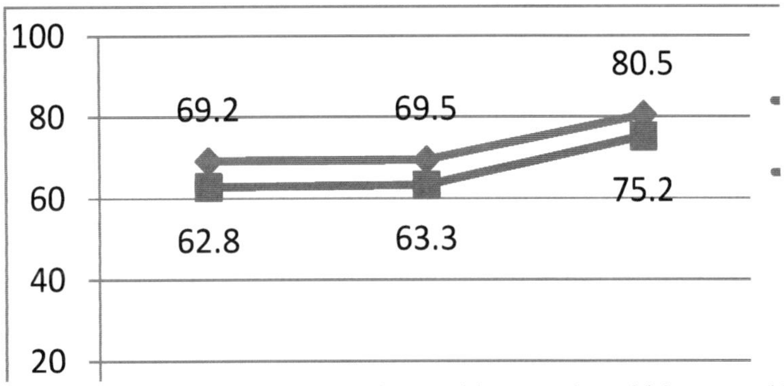

Figure 5. Contrast curve of the maximum stress for Double-mast & multi-layers and Double-mast & Single-layer

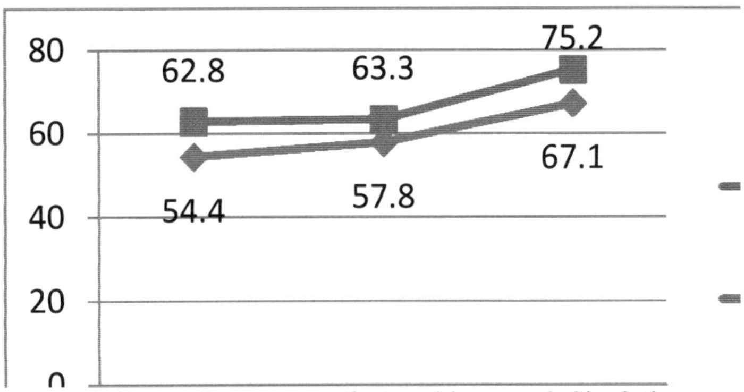

Figure 6. Contrast curve of the maximum stress for Double-mast & Single-layer under the rated load.

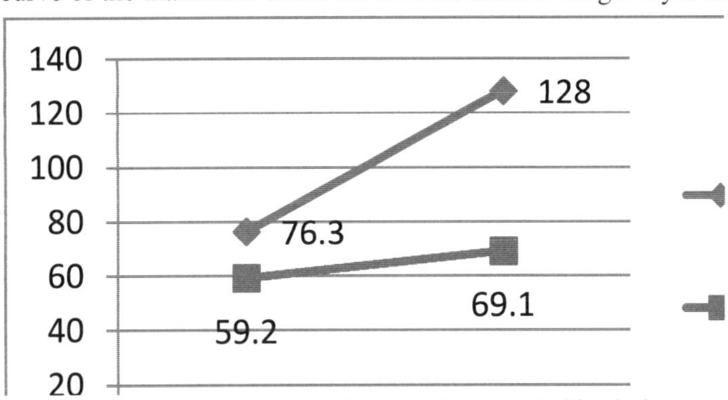

Figure 7. Contrast curve of the maximum stress for Double-mast & Single-layer under the rated load.

IOP Conf. Series: Materials Science and Engineering **269** (2017) 012006 doi:10.1088/1757-899X/269/1/012006

It can be concluded from figure 5 to figure 7, under the four working conditions, the maximum stress value and the deformation value of MPCs are changed to some extent with the change of the concentrated load. Figure 4 compares the maximum stress curves under different uniform load ratings, which indicates that the change of the maximum stress is obvious with the change of the concentrated load and the rated load. Generally speaking, MCPs are more sensitive to concentrated loads, uniform loads and load bias. Figure 5 compared the maximum stress of the single mast and double mast under the same load bias. Under the same conditions, the maximum stress between the single mast and double mast is very in large difference. The single mast is more sensitive than double mast.

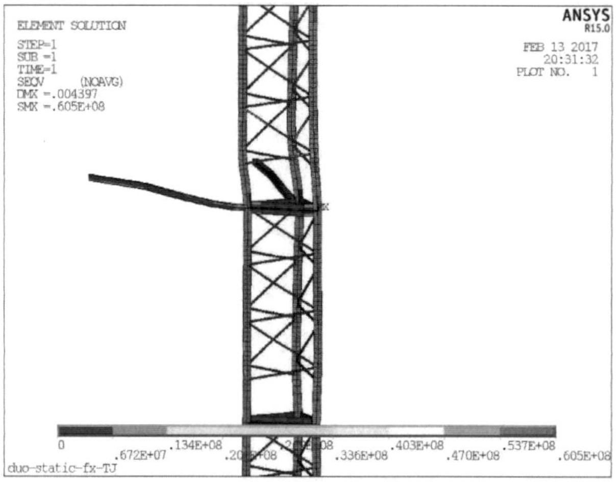

Figure 8. Stress nephogram of the fixed frame for Double-mast & multi-layers under the case 1.

As shown in figure 8, the maximum stress of the Double-mast & multi-layers of MCPs occurs at the junction between the guide mast device and the wall-attaching device. According to the stress nephogram of the guide mast device, the stress at the upper end of the wall-attaching equipment is 33.6MPa to 60.5MPa, the stress of the lower end is 20MPa to 33.6MPa. It can be concluded that the wall-attaching equipment bears the majority of the bending moment and the partial axial force, which is transmitted from the guide mast column, and the attachment device has the function of optimizing the internal force of the guide mast.

4.2. Analysis of structural dynamic characteristics

In the process of operation, MCPs are often accompanied by starting-braking, moving load, sudden loading or unloading and multi-level platforms relative movement and other dynamic conditions. Therefore, it is necessary to calculate and analyze the dynamic characteristics of each type of MCPs before use, in order to check its stability. The analysis of dynamic characteristics is modal analysis (MA), and the essence of MA is to calculate the eigenvalues and eigenvectors from the structural vibration equation. The former frequency of the structure is lower, which has a great influence on the dynamic characteristics of the structure, and the larger the order, the greater the error. So the MA only extracts the results of the former 10 natural frequency and its corresponding vibration mode, which can meet the engineering requirements well. Taking guide rails and the wall-attaching devices with 6m pitch as an example, four types of MCPs were analyzed respectively, to obtain the natural frequencies and mode shapes of the 10 orders, as shown in table 3.

Table 3. The former 10-order natural frequency (Hz) for the MCPs.

number	Order	1	2	3	4	5	6	7	8	9	10
1	Double-mast & Single-layer	8.1512	9.175	9.9545	15.307	17.786	19.029	20.947	21.019	21.1	21.54
2	Double-mast & Multi-layers	7.5398	8.446	8.837	9.2367	10.264	10.686	14.379	15.544	16.5	17.73
3	Single-mast & Single-layer	0.95605	1.093	3.3976	3.5474	7.1455	7.9594	8.1867	9.491	13.8	16.36
4	Single-mast & Multi-layers	1.0234	3.463	7.6667	8.9478	15.469	18.346	18.504	20.986	27.4	29.07

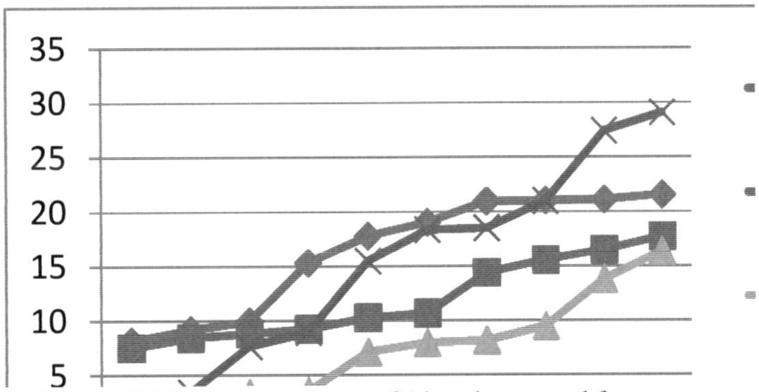

Figure 9. The parison charts of 10-order natural frequency.

According to table 2 and figure 7, the dynamic characteristics of the structure are different with the different types of the MCPs, and the natural frequency of each order varies greatly.

Among them, in the double-mast & multi-layers MCPs, the guide mast vibrates front and rear along the Y direction in the first three orders, the guide mast vibrates with the platform from the forth order to the seventh order, the platform vibrates up and down along the Z direction in the seventh order to the ninth order, and the device twists around the X axis in the tenth order; In the Single-mast & Single-layer MCPs, the platform swing left and right in the first two orders, the guide mast vibrates front and rear along the Y direction from the fifth order to the eighth order, the platform twists along X axis; Due to the working of the Single-mast & Single-layer MCPs vibrates left and right first at low frequencies of 0.9-3.5Hz which is vulnerable to wind load in the construction process and easy to lead to the Pendulum resonance. So it is necessary to set connection measures between the platform device and guide mast device or shorten the length of work platform. Low-order natural frequency and vibration characteristics are obtained through the dynamic characteristics of modal analysis, which indicates that the structural stability and safety of the Double-mast MCPs are high, the natural vibration frequency of the single-mast MCPs is low, and the platform vibrates after swinging, which leads to poor operation stability and poor operation comfort according to the data in table 2, for the Double-mast & multi-layers, Double-mast & Single-layer and the Single-mast & Single-layer, the difference between the frequency of each order is small, and the frequency of the partial order appears the phenomenon of aggregation. So the frequency density needs further analysis. In paper [14], it is pointed out that there is no uniform standard in the discrimination about the structural frequency density. Gregory et al [15] revealed that the frequency density can be measured by the density index δ1.

$$\delta_1 = (f_{i+1} - f_i)/(f_{i+1} + f_i)$$

As shown in table 4, the frequency density of the four types of MCPs.

As shown in table 4, in addition to individual conditions (the Double-mast & Single-layer MCPs under the third order, the Single-mast & Single-layer MCPs under the first, second and forth order, and the Single-mast & multi-layers under the second, forth and eighth order), The frequency-intensive indexes of MCPs are all small. Liu and Fan et al [16], concluded that the structures with symmetric, quasi-symmetrical or locally symmetric tend to appear in varying degrees of intensity. The data in table 4 are consistent with the descriptions in the literature.

Table 4. The frequency density of the former for the four types of MCPs.

Frequency density δ_1	1	2	3	4	5	6	7	8	9
Double-mast & Single-layer	0.0591	0.0408	0.2119	0.0749	0.0338	0.0480	0.0017	0.0016	0.0107
Double-mast & Multi-layers	0.0567	0.0226	0.0221	0.0527	0.0201	0.1473	0.0389	0.0296	0.0362
Single-mast & Single-layer	0.5438	0.3777	0.0771	0.2671	0.0851	0.0043	0.0629	0.1331	0.0290
Single-mast & Multi-layers	0.0667	0.5133	0.0216	0.3365	0.0539	0.0141	0.0738	0.1855	0.0842

5. Conclusions

(1) Through the calculation of the different MCPs, the maximum stress and the maximum deformation under different working conditions are obtained, and the structural strength and stiffness meet the requirements. The most dangerous parts are often caused by the local stress for the concentrated load offset in the cantilever of platform. The larger the displacement of the cantilever, the greater the inclination of the platform is. Therefore it is recommended that the construction platform should not be allowed to be overloaded or bear too much concentrated load, and the materials should be placed on both sides of the guide mast.

(2) The platform telescopic beam is extended by 1.5m in case 2 and case 4. The maximum stress and the maximum deformation under different working conditions of MCPs are very close. Therefore, it is significant to strictly control the load of the telescopic beam section. The telescopic part does not allow stacking of materials, with only personnel to perform on-site operations.

(3) MCPs are generally load-sensitive aerial work construction equipment. The load sign need to be clearly specified before application. The material should be stacked strictly according to the platform load distribution chart.

(4) MCPs are space truss system. The rods are with stress distribution. Therefore, when the construction platform is used, no one can remove the platform components and rods arbitrarily.

(5) According to the dynamic analysis of structure, the stability of double mast is better than that of single mast. Suggested that the double mast of MCPs should be preferred in construction.

(6) MCPs are a multi-symmetrical structure with the characteristics of large flexibility, low frequency mode dense, high degree of modal coupling and so on. The vibration needs to be controlled before the construction. That is to say, the necessary anti-vibration measures need to be set to reduce the damage to the construction platform.

Reference

[1] Jiang Q J 2010 *Arch. Tech.* **41(12)** 1074-7.
[2] Liu K 2015 *Build. Tech. Dev.* **42(1)** 7-15.
[3] 2016 the 2nd session of the national construction and engineering structure of industrial construction technology exchange.
[4] Wang M H, Zhang W Z and Li R S 2016 *Build. Constr.* **38(9)** 1263-19.
[5] Zhang H and Li S L 2011 *Constr. Mech.* **3** 19-24.
[6] Chen J P, Li H M and Zhao Z Z et al 2010 *Constr. Mech.* **9** 13-5.
[7] Zhou F X 2011 *Arch. Tech.* **43(8)** 692-4.
[8] Lu Z 2011 *Shenyang Architecture University.*
[9] Chen J P, Li H M and Zhao Z Z et al 2015 *Constr. Mech.* **3** 23-8.
[10] Shang X J, Qiu F et al 2006 *Advanced finite element analysis method of ANSYS structure and its application,* Beijing: China Water Power Press.
[11] GB/T 27547-2011 2011 *Elevating work platforms-mast-climbing work platforms* Beijing: Standards Press of China.
[12] JGJ202-2010 2010 *Technical code for safety of implementation scaffold practice in construction* Beijing: Standards Press of China.
[13] GB50017-2014 2014 *Code for design of steel structure* Beijing: Standards Press of China.
[14] Xie F X and Sun L M 2009 *J. Shandong Univ. (Eng. Sci.)* **39(5)** 101-5.
[15] Corporation Acncal-Schwendler MSC/NASTRAN user's manual[M/OL] 1983[2009-04-05] http://www mscsoftwarecom/ products/new cfm ? PI =7
[16] Liu L J, Fan J L, Zhang Z Y et al 2007 *J. Vibr. Shock* **26(4)** 109-5.

Acknowledgements

The authors would like to gratefully acknowledge the research grants from the Science and Technology Commission of Shanghai Municipality (15DZ1203505) and China Green Building Council (China GBC).

IOP Conf. Series: Materials Science and Engineering **269** (2017) 012007 doi:10.1088/1757-899X/269/1/012007

Longitudinal Fracture Analysis of a Two-Dimensional Functionally Graded Beam

V Rizov[1]

[1]Professor, Department of Technical Mechanics, University of Architecture, Civil Engineering and Geodesy, 1046 - Sofia, Bulgaria

Abstract. Longitudinal fracture in a two-dimensional functionally graded beam is analyzed. The modulus of elasticity varies continuously in the beam cross-section. The beam is clamped in its right-hand end. The external loading consists of one longitudinal force applied at the free end of the lower crack arm. The longitudinal crack is located in the beam mid-plane. The fracture is studied in terms of the strain energy release rate. The solution derived is used to elucidate the effects of material gradients along the height as well as along the width of the beam cross-section on the fracture behaviour. The results obtained indicate that the fracture in two-dimensional functionally graded beams can be regulated efficiently by employing appropriate material gradients.

1. Introduction

The application of the functionally graded materials has been constantly increasing for the last thirty years due mainly to the fact that by gradually changing the composition of the constituent materials during manufacturing, one can optimize the performance of structural members to the external loading [1-4]. Since the material properties of the functionally graded materials are distributed in one or more spatial directions, the study of fracture behaviour is especially significant [5-11].

The present study is motivated by the fact that the functionally graded materials can be built up layer by layer [12], which is a premise for appearance of longitudinal cracks between layers. The basic purpose of the present paper is to obtain a solution to the strain energy release rate for a longitudinal crack in a two-dimensional functionally graded beam. The solution is applied to elucidate the effects of material gradients on the fracture behaviour.

2. Analysis of the strain energy release rate

The functionally graded beam configuration analyzed in the present paper is shown in figure 1. A longitudinal crack of length, a, is located in the beam mid-plane. The beam length is l. The beam is loaded by a longitudinal force, F, applied at the free end of the lower crack arm. It is obvious that the upper crack arm is free of stresses. The right-hand end of the beam is clamped. The beam cross-section is a rectangle of width, b, and height, $2h$. The material is two-dimensional functionally graded in the beam cross-section. The modulus of elasticity, E, varies continuously in the beam cross-section according to the following law:

$$E(y_3, z_3) = E_D + \frac{E_H - E_D}{b}\left(\frac{b}{2} + y_3\right) + \frac{E_K - E_D}{4h^2}(z_3 + h)^2 \qquad (1)$$

where E_D, E_H and E_K are material properties (E_H and E_K govern the material gradient along the width and the height of the beam cross-section, respectively). The centroidal axes, y_3 and z_3, are shown in figure 1.

The strain energy release rate, G, is obtained by differentiating the beam strain energy, U, with respect to the crack area, A [13].

$$G = \frac{dU}{dA} \tag{2}$$

where

$$dA = bda \tag{3}$$

In (3), da is an elementary increase of the crack length.

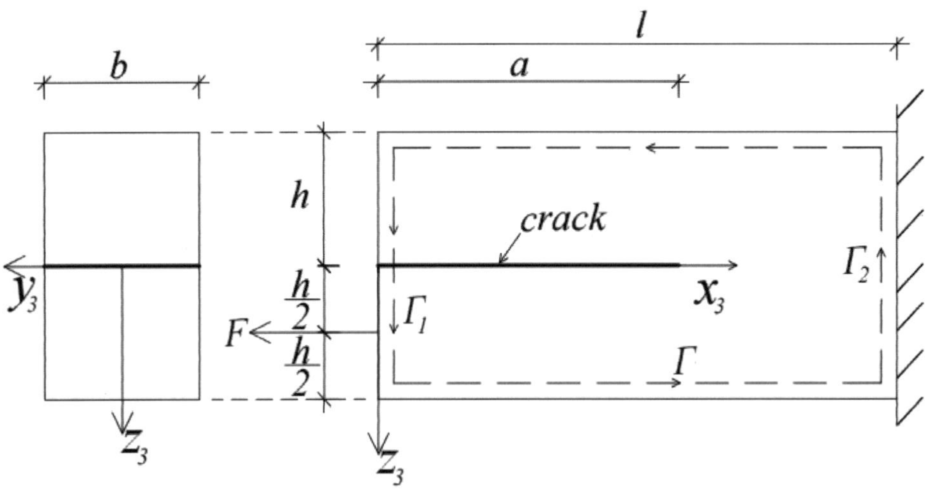

Figure 1. A two-dimensional functionally graded beam with a longitudinal crack.

The beam strain energy is derived by integrating the strain energy density in the lower crack arm (the upper crack arm is free of stresses) and in the un-cracked beam portion $(a < x_3 < l)$

$$U = a\int_{-\frac{h}{2}}^{\frac{h}{2}}\left(\int_{-\frac{b}{2}}^{\frac{b}{2}} u_L dy_1\right)dz_1 + (l-a)\int_{-h}^{h}\left(\int_{-\frac{b}{2}}^{\frac{b}{2}} u_U dy_2\right)dz_2 \tag{4}$$

where u_L and u_U are the strain energy densities in the lower crack arm and the un-cracked beam portion, respectively. The centroidal axes of the lower crack arm cross-section are marked by y_1 and z_1. In (4), y_2 and z_2 are the centroidal axes of the cross-section of the un-cracked beam portion.

By substituting of (3) and (4) in (2), one arrives at

$$G = \frac{1}{b}\int_{-\frac{h}{2}}^{\frac{h}{2}}\left(\int_{-\frac{b}{2}}^{\frac{b}{2}} u_L dy_1\right)dz_1 - \frac{1}{b}\int_{-h}^{h}\left(\int_{-\frac{b}{2}}^{\frac{b}{2}} u_U dy_2\right)dz_2 \tag{5}$$

The strain energy density in the lower crack arm is written as

$$u_L = \frac{1}{2} E \varepsilon^2 \tag{6}$$

where ε is the distribution of the longitudinal strains.

The strains are analyzed by applying the Bernoulli's hypothesis for plane sections, since the span to height ratio of the beam under consideration is large. Besides, since the beam (Fig. 1) is loaded in eccentric tension, the only non-zero strains are the longitudinal strains. Therefore, according to the small strains compatibility equations, ε is distributed linearly in the beam cross-section. Thus, ε in the lower crack arm is written as

$$\varepsilon = \varepsilon_{C_1} + \kappa_{y_1} y_1 + \kappa_{z_1} z_1 \tag{7}$$

where ε_{C_1} is the longitudinal strain in the lower crack arm cross-section centre, κ_{y_1} and κ_{z_1} are the curvatures of lower crack arm in the $x_1 y_1$ and $x_1 z_1$ planes, respectively.

By using (1), the distribution of E in the lower crack arm cross-section is written as

$$E(y_1, z_1) = E_D + \frac{E_H - E_D}{b}\left(\frac{b}{2} + y_1\right) + \frac{E_K - E_B}{4h^2}(z_1 + r)^2 \tag{8}$$

where $r = 3h/2$.

By substituting of (7) and (8) in (6), one obtains

$$u_L = \frac{1}{2}\left[E_D + \frac{E_H - E_D}{b}\left(\frac{b}{2} + y_1\right) + \frac{E_K - E_B}{4h^2}(z_1 + r)^2 \right]\left(\varepsilon_{C_1} + \kappa_1 y_1 + \kappa_2 z_1\right)^2 \tag{9}$$

The following equations for equilibrium of the cross-section of the lower crack arm are used to determine ε_{C_1}, κ_{y_1} and κ_{z_1}:

$$N_1 = \int_{-\frac{h}{2}}^{\frac{h}{2}} \left(\int_{-\frac{b}{2}}^{\frac{b}{2}} E\varepsilon dy_1 \right) dz_1 \tag{10}$$

$$M_{y_1} = \int_{-\frac{h}{2}}^{\frac{h}{2}} \left(\int_{-\frac{b}{2}}^{\frac{b}{2}} E\varepsilon z_1 dy_1 \right) dz_1 \tag{11}$$

$$M_{z_1} = \int_{-\frac{h}{2}}^{\frac{h}{2}} \left(\int_{-\frac{b}{2}}^{\frac{b}{2}} E\varepsilon y_1 dy_1 \right) dz_1 \tag{12}$$

where N_1 is the axial force, M_{y_1} and M_{z_1} are the bending moments about y_1 and z_1, respectively. It is obvious that $N_1 = F$, $M_{y_1} = 0$ and $M_{z_1} = 0$ (figure 1). After substituting of (7) and (8) in (10), (11) and (12) the equations obtained should be solved with respect to ε_{C_1}, κ_{y_1} and κ_{z_1} by using the MatLab computer program.

The strain energy density in the un-cracked beam portion, u_U, can be obtained by formula (9). For this purpose, y_1, z_1, r, ε_{C_1}, κ_{y_1} and κ_{z_1} should be replaced with y_2, z_2, h, ε_{C_2}, κ_{y_2} and κ_{z_2}, respectively (ε_{C_2} is the longitudinal strain in the centre of the cross-section of the un-cracked beam portion, κ_{y_2} and κ_{z_2} are the curvatures of the un-cracked beam portion in the x_2y_2 and x_2z_2 planes, respectively). The same replacements should be done in equations (10), (11) and (12) in order to determine ε_{C_2}, κ_{y_2} and κ_{z_2}. Besides, M_{y_2} should be substituted equal to $Fh/2$ in (11).

After substituting of u_L and u_U in (5), the integration should be performed by using the MatLab computer program.

The solution to the strain energy release rate is verified by analyzing the longitudinal crack with the help of the *J*-integral [14]. The solution of the *J*-integral is obtained by using an integration contour, Γ, which coincides with the beam contour (figure 1). Thus, the *J*-integral has non-zero values only in the segments, Γ_1 and Γ_2, of the integration contour (Γ_1 coincides with the free end of the lower crack arm, Γ_2 coincides with the clamping). Therefore, the *J*-integral solution is written as $J = J_{\Gamma_1} + J_{\Gamma_2}$, where J_{Γ_1} and J_{Γ_2} are the *J*-integral values in segments Γ_1 and Γ_2, respectively. The integration of the *J*-integral is carried-out by using the MatLab computer program. The result obtained matches exactly the solution to the strain energy release rate. This fact is a verification of the fracture analysis performed in the present paper.

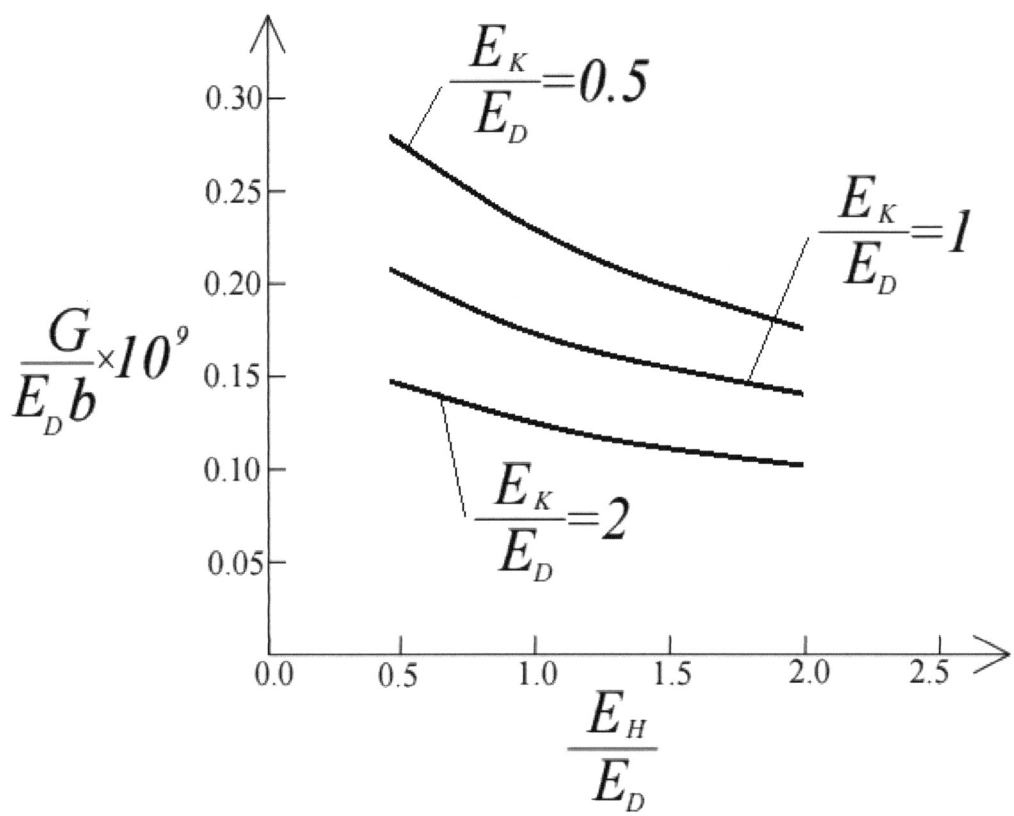

Figure 2. The strain energy release rate in non-dimensional form plotted against E_H / E_D ratio at three E_K / E_D ratios.

The solution to strain energy release rate obtained in the present paper is applied to elucidate the effects of material gradients on the fracture behaviour of the two-dimensional functionally graded beam configuration shown in figure 1. It is assumed that $b = 0.02$ m, $h = 0.006$ m and $F = 40$ N. The material gradients along the width and height of the beam are characterized by E_H / E_D and

E_K / E_D ratios, respectively. It should be noted that E_D is kept constant in the calculations. Thus, E_H and E_K are varied in order to generate various E_H / E_D and E_K / E_D ratios. The strain energy release rate is presented in non-dimensional form by using the formula $G_N = G / (E_D b)$. The strain energy release rate in non-dimensional form is plotted against E_H / E_D ratio in figure 2 at three E_K / E_D ratios. The curves in figure 2 indicate that the strain energy release rate decreases with increasing of E_H / E_D and E_K / E_D ratios. This finding is attributed to the increase of the beam stiffness.

3. Conclusions

The longitudinal fracture behaviour of a two-dimensional functionally graded cantilever beam configuration is studied in terms of the strain energy release rate. The beam is loaded by one longitudinal force applied at the free end of the lower crack arm. The modulus of elasticity varies non-linearly in the beam cross-section. In order to verify the solution obtained, the fracture is analyzed also by the *J*-integral approach. The solution obtained is applied to elucidate the effects of material gradients along the width as well as along the height of the beam on the fracture. The analysis reveals that the strain energy release rate decreases with increasing of E_H / E_D and E_K / E_D ratios. It can be summarized that the fracture behaviour of two-dimensional functionally graded beams can be efficiently regulated by employing appropriate material gradients in the design stage of the beams.

References

[1] Neubrand A and Rödel J 1997 *Zeit. f Met.* **88** 358-71.
[2] Suresh S and Mortensen A 1998 *Fundamentals of functionally graded materials* (IOM Communications Ltd, London).
[3] Hirai T and Chen L 1999 *Mater. Sci. Forum* **308-311** 509-14.
[4] Nemat-Allal M M, Ata M H, Bayoumi M R, Khair-Eldeen W 2011 *Mater. Sci. Appl.* **2** 1708-18.
[5] Erdogan F 1995 *Comput. Eng.* **5** 753-70.
[6] Paulino G C 2002 *Engng. Fract. Mech.* **69** 1519-30.
[7] Tilbrook M T, Moon R J and Hoffman M 2005 *Comp. Sci. Tech.* **65** 201-20.
[8] Carpinteri A and Pugno N 2006 *Eng. Fract. Mech.* **73** 1279-91.
[9] Upadhyay A K and Simha K R Y 2007 *Int. J. Fract.* **144** 209-13.
[10] Monfared M M, Ayatollahi M, Mousavi S M 2016 *Arch. Appl. Mech.* **86** 713-28.
[11] Brajesh P, Goutam P 2016 *J. Mech. Eng. Sci.* **230** 1468-97.
[12] Bohidar S K, Sharma R and Mishra P R 2014 *Int. J. Res.* **1** 289-301.
[13] Alexandrov A 1990 *Fundamentals of theory of elasticity and plasticity* (M: Vishaya shkola).
[14] Broek D 1986 *Elementary engineering fracture mechanics* (Springer).

Acknowledgments

The present study was supported financially by the Research and Design Centre (CNIP) of the University of Architecture, Civil Engineering and Geodesy (UACEG), Sofia (Contract BN – 198/2017).

IOP Conf. Series: Materials Science and Engineering **269** (2017) 012008 doi:10.1088/1757-899X/269/1/012008

Delamination Analysis of a Multilayered Two-Dimensional Functionally Graded Cantilever Beam

V Rizov[1]

[1]Professor, Department of Technical Mechanics, University of Architecture, Civil Engineering and Geodesy, 1046 - Sofia, Bulgaria

Abstract. Delamination fracture behaviour of a multilayered two-dimensional functionally graded cantilever beam is analyzed in terms of the strain energy release rate. The beam is made of an arbitrary number of layers. Perfect adhesion is assumed between layers. Each layer has individual thickness and material properties. Besides, the material is two-dimensional functionally graded in the cross-section of each layer. There is a delamination crack located arbitrary between layers. The beam is loaded by a bending moment applied at the free end of the lower crack arm. The upper crack arm is free of stresses. The solution to strain energy release rate derived is applied to investigate the influence of the crack location and the material gradient on the delamination fracture. The results obtained can be used to optimize the multilayered two-dimensional functionally graded beam structure with respect to the delamination fracture behaviour.

1. Introduction

The functionally graded materials are manufactured by mixing of two material constituents. These new un-homogeneous materials permit tailoring of their properties in one or more spatial directions during manufacturing. This fact is one of the basic advantages of the functionally graded materials over the traditional structural materials. Multilayered functionally graded structural members and components are made by bonding of layers of different functionally graded materials. Usually, the delamination fracture, i.e. the separation of layers, is the earliest failure mode of the multilayered structures [1, 2]. The delamination reduces the strength and stiffness, complicates the post-buckling behaviour of structural members loaded in compression and may lead even to catastrophic failure of the entire structure.

The main objective of the present paper is to develop an analysis of the delamination fracture behaviour of a multilayered two-dimensional functionally graded beam in terms of the strain energy release rate. The solution obtained is used to investigate the influence of crack location and material gradient on the delamination fracture.

2. Solution to the strain energy release rate

The present paper reports a delamination fracture analysis of the multilayered two-dimensional functionally graded cantilever beam shown in figure 1. The beam is made of an arbitrary number of horizontal layers. Perfect adhesion is assumed between layers. The material is two-dimensional functionally graded in the cross-section of each layer. Also, the layers may have different thickness and material properties. A delamination crack of length, a, is located arbitrary between layers. The external loading consists of a bending moment, M, applied at the free end of the lower crack arm. Obviously, the upper crack arm is free of stresses. The thicknesses of the lower and the upper crack arms are h_1 and h_2, respectively. The beam length is l. The beam has a rectangular cross-section of width, b, and height, $2h$. The right-hand end of the beam is clamped.

The strain energy release rate, G, is expressed as [3].

$$G = \frac{dU}{dA} \tag{1}$$

where

$$dA = bda \tag{2}$$

is an elementary increase of the crack area, A. In (2) and (3), U is the beam strain energy, da is an elementary increase of the crack length.

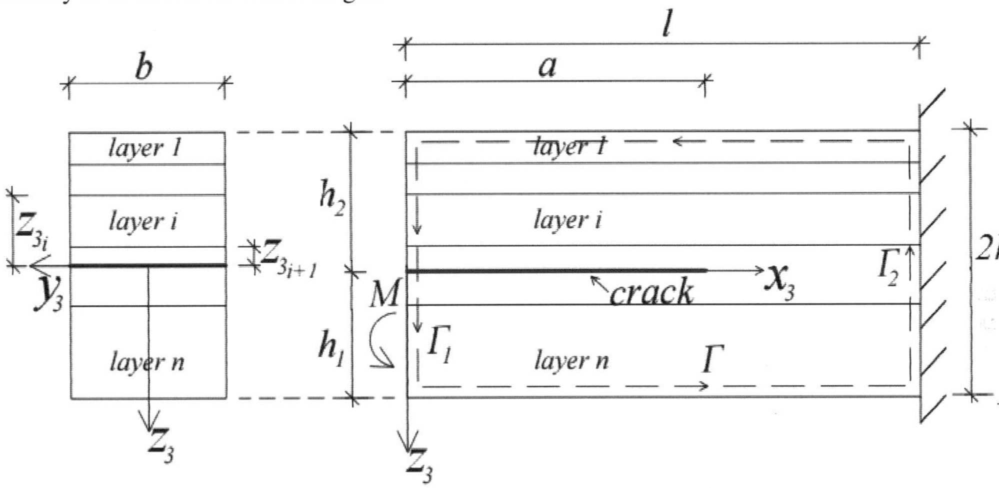

Figure 1. Geometry and loading of a multilayered two-dimensional functionally graded cantilever beam.

By integrating the strain energy density in the lower crack arm (the upper crack arm is free of stresses) and the un-cracked beam portion $(a < x_3 < l)$, the beam strain energy is written as

$$U = a\sum_{i=1}^{i=n_L} \int_{z_{1i}}^{z_{1i+1}} \left(\int_{-\frac{b}{2}}^{\frac{b}{2}} u_{L_i} dy_1 \right) dz_1 + (l-a) \sum_{i=1}^{i=n} \int_{z_{2i}}^{z_{2i+1}} \left(\int_{-\frac{b}{2}}^{\frac{b}{2}} u_{U_i} dy_2 \right) dz_2 \tag{3}$$

where n_L and n are numbers of layers in the lower crack arm and the un-cracked beam portion, respectively. The strain energy densities in the i-th layer of the lower crack arm and the un-cracked beam portion are denoted by u_{L_i} and u_{U_i}, respectively. In (3), z_{1i} and z_{1i+1} are the coordinates, respectively, of the upper and lower surfaces of the i-th layer of the lower crack arm, y_1 and z_1, are the centroidal axes of the cross-section of the lower crack arm (z_1 is directed downwards). The centroidal axes of the cross-section of the un-cracked beam portion are denoted by y_2 and z_2.

By substituting of (2) and (3) in (1), one obtains

$$G = \frac{1}{b}\sum_{i=1}^{i=n_L} \int_{z_{1i}}^{z_{1i+1}} \left(\int_{-\frac{b}{2}}^{\frac{b}{2}} u_{L_i} dy_1 \right) dz_1 - \frac{1}{b} \sum_{i=1}^{i=n} \int_{z_{2i}}^{z_{2i+1}} \left(\int_{-\frac{b}{2}}^{\frac{b}{2}} u_{U_i} dy_2 \right) dz_2 \tag{4}$$

The strain energy density in the i-th layer of the lower crack arm is expressed as

$$u_{L_i} = \frac{1}{2} E_i \varepsilon^2 \tag{5}$$

where E_i is the distribution of the modulus of elasticity in the cross-section of the same layer, ε is the distribution of the longitudinal strain.

The modulus of elasticity is distributed in the cross-section of the i-th layer according to the following law:

$$E_i(y_1, z_1) = E_{B_i} + \frac{E_{T_i} - E_{B_i}}{b^2}\left(\frac{b}{2} + y_1\right)^2 + \frac{E_{S_i} - E_{B_i}}{z_{1i+1} - z_{1i}}(z_1 - z_{1i}) \tag{6}$$

where E_{B_i}, E_{T_i} and E_{S_i} are material properties (E_{T_i} and E_{S_i} govern the material gradient along the width and thickness of the layer, respectively).

The distribution of the longitudinal strain is analyzed by applying the Bernoulli's hypothesis for plane sections since the span to height ratio of the beam under consideration is large. It should also be mentioned that since the beam is loaded in pure bending (figure 1), the only non-zero strain is ε. Therefore, according to the small strain compatibility equations, ε is distributed linearly in the cross-section. Thus, the distribution of the longitudinal strain in the cross-section of the lower crack arm is written as

$$\varepsilon = \varepsilon_{C_1} + \kappa_{y_1} y_1 + \kappa_{z_1} z_1 \tag{7}$$

where ε_{C_1} is the strain in the lower crack arm cross-section centre, κ_{y_1} and κ_{z_1} are the curvatures of lower crack arm in the $x_1 y_1$ and $x_1 z_1$ planes, respectively.

By combining of (5), (6) and (7), one arrives at

$$u_{L_i} = \frac{1}{2}\left[E_{B_i} + \frac{E_{T_i} - E_{B_i}}{b^2}\left(\frac{b}{2} + y_1\right)^2 + \frac{E_{S_i} - E_{B_i}}{z_{1i+1} - z_{1i}}(z_1 - z_{1i}) \right](\varepsilon_{C_1} + \kappa_{y_1} y_1 + \kappa_{z_1} z_1)^2 \tag{8}$$

The quantities, ε_{C_1}, κ_{y_1} and κ_{z_1}, in (8) are determined from the following equilibrium equations of the cross-section of the lower crack arm:

$$N_1 = \sum_{i=1}^{i=n_L} \int_{z_{1i}}^{z_{1i+1}} \left(\int_{-\frac{b}{2}}^{\frac{b}{2}} E_i \varepsilon dy_1 \right) dz_1,\ M_{y_1} = \sum_{i=1}^{i=n_L} \int_{z_{1i}}^{z_{1i+1}} \left(\int_{-\frac{b}{2}}^{\frac{b}{2}} E_i \varepsilon z_1 dy_1 \right) dz_1,\ M_{z_1} = \sum_{i=1}^{i=n_L} \int_{z_{1i}}^{z_{1i+1}} \left(\int_{-\frac{b}{2}}^{\frac{b}{2}} E_i \varepsilon y_1 dy_1 \right) dz_1, \tag{9}$$

where the axial force and the bending moments for y_1 and z_1 are denoted by N_1, M_{y_1} and M_{z_1}, respectively. Figure 1 indicates that $N_1 = 0$, $M_{y_1} = M$ and $M_{z_1} = 0$. After substituting of (6) and (7) in (9) the equations obtained should be solved with respect to ε_{C_1}, κ_{y_1} and κ_{z_1} by using the MatLab computer program.

Formula (8) can be applied also to obtain the strain energy density distribution, u_{U_i}, in the cross-section of the i-th layer in the un-cracked beam portion. For this purpose, y_1, z_1, z_{1i}, z_{1i+1}, ε_{C_1}, κ_{y_1} and κ_{z_1} should be replaced, respectively, with y_2, z_2, z_{2i}, z_{2i+1}, ε_{C_2}, κ_{y_2} and κ_{z_2}, where ε_{C_2}, κ_{y_2} and κ_{z_2} are the longitudinal strain in the centre of the cross-section of the un-cracked beam portion and the curvatures of the un-cracked beam portion in the x_2y_2 and x_2z_2 planes, respectively. Analogous replacements should be done in (9) to determine ε_{C_2}, κ_{y_2} and κ_{z_2}.

After substituting of u_{L_i} and u_{U_i} in (4), the integration should be performed by using the MatLab computer program.

The *J*-integral approach [4] is applied in order to verify the solution to the strain energy release rate derived in the present paper. The integration contour, Γ, which coincides with the beam contour, is used (figure 1). It can be observed in figure 1 that the *J*-integral is non-zero only in the segments, Γ_1 and Γ_2, of the integration contour (Γ_1 coincides with the free end of the lower crack arm, Γ_2 coincides with the clamping). Therefore, the *J*-integral value is obtained by summation, i.e. $J = J_{\Gamma_1} + J_{\Gamma_2}$, where the *J*-integral values in segments Γ_1 and Γ_2 are denoted by J_{Γ_1} and J_{Γ_2}, respectively. The MatLab computer program is used to perform the integration of the *J*-integral. The result obtained matches exactly the solution to the strain energy release rate, which is a verification of the fracture analysis of the two-dimensional functionally graded multilayered beam developed in the present paper.

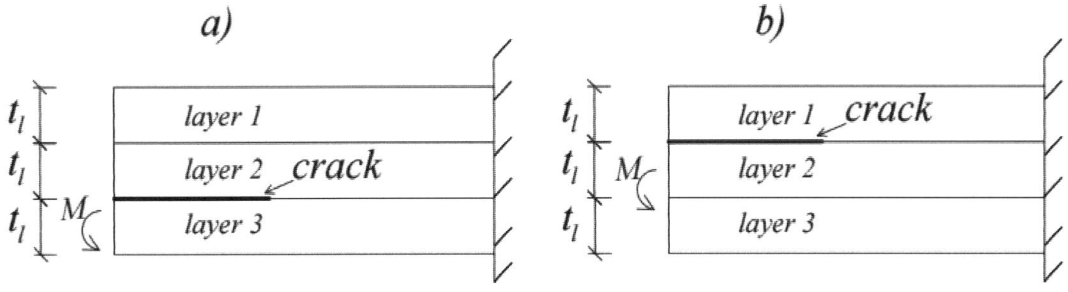

Figure 2. Two three-layered beam configurations.

The influence of material gradient and crack location on the delamination fracture behaviour is investigated.

For this purpose, calculations are carried-out by using the solution to the strain energy release rate (4). Two three-layered beam configurations are analyzed in order to evaluate the influence of the crack location (figure 2). Each layer has thickness of t_l (figure 2). The crack is located between layers 2 and 3 in the beam configuration shown in figure 2a. A beam with a crack between layers 1 and 2 is also analyzed (figure 2b). It is assumed that $b = 0.025$ m, $h = 0.0045$, $t_l = 0.003$ m and $M = 70$ Nm. The strain energy release rate is presented in non-dimensional form by using the formula $G_N = G/(E_{B_3}b)$. The material gradient in layer 3 is characterized by E_{S_3}/E_{B_3} ratio. The strain energy release rate is plotted against E_{S_3}/E_{B_3} ratio in figure 3 for the two three-layered beam configurations. The curves in figure 3 indicate that the strain energy release rate decreases with increasing of E_{S_3}/E_{B_3} ratio (this is due to the increase of the beam stiffness). It can also be observed in figure 3 that the strain energy release rate decreases when the crack location is changed from this shown in figure 2a to that in figure 2b. This finding is attributed to the increase of the lower crack arm thickness.

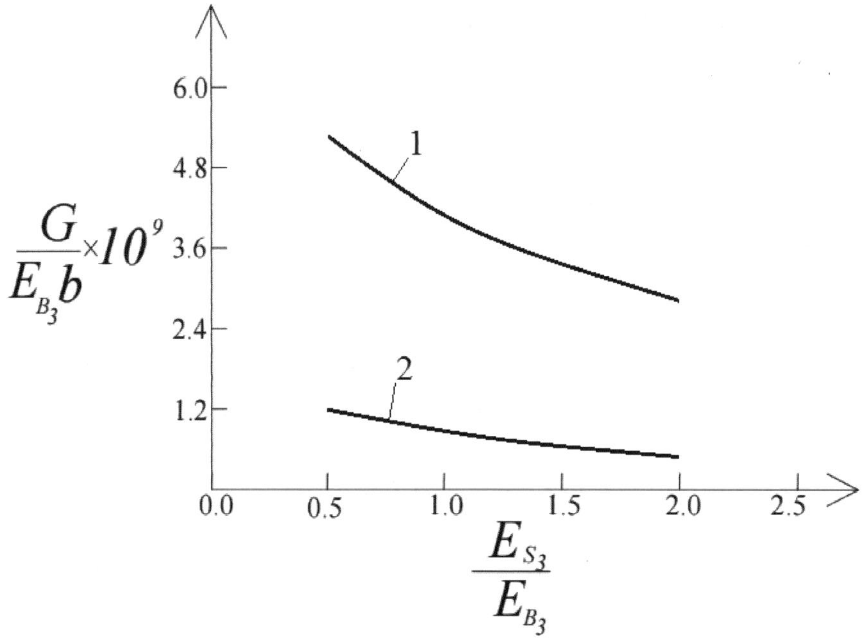

Figure 3. The strain energy release rate in non-dimensional form presented as a function of E_{S_3} / E_{B_3} ratio (curve 1 – for the beam configuration shown in figure 2a, curve 2 – for the beam configuration shown in figure 2b).

3. Conclusions

An analysis of the delamination in a multilayered two-dimensional functionally graded beam is performed. A solution to the strain energy release rate is derived. The solution is applicable for a beam made of an arbitrary number of horizontal layers. Perfect adhesion is assumed between layers. Each layer may have individual thickness and material properties. The material is two-dimensional functionally graded in the cross-section of each layer (the modulus of elasticity varies continuously along the thickness as well as along the width of each layer). The delamination is located arbitrary between layers. The solution obtained is verified by analyzing the delamination fracture with the help of the *J*-integral. The influence of the material gradient and the crack location on the strain energy release rate is investigated. It is found that the strain energy release rate decreases with increasing the thickness of the lower crack arm (the upper crack arm is free of stresses). The increase of E_{S_3} / E_{B_3} ratio leads also to decrease of the strain energy release rate. The results obtained can be applied for optimization of multilayered two-dimensional functionally graded beams with respect to the delamination fracture behaviour.

References

[1] Dahan I, Admon U, Sarei J, Yahav B, Amar M, Frage N, Dariel MP 1999 *Mater. Sci. Forum* **308-311** 923-9.
[2] Hsueh C H, Tuan W H, Wei W C J 2009 *Script. Mater.* **60** 721-4.
[3] Alexandrov A 1990 *Fundamentals of theory of elasticity and plasticity* (M: Vishaya shkola).
[4] Broek D 1986 *Elementary engineering fracture mechanics* (Springer).

Acknowledgments

The present study was supported financially by the Research and Design Centre (CNIP) of the University of Architecture, Civil Engineering and Geodesy (UACEG), Sofia (Contract BN – 198/2017).

Experimental Study on Vacuum Dynamic Consolidation Treatment of Soft Soil Foundation

Ni Fu-lai[1], WEN Xin[2, 3], Zhang Xiao-bin[3], Li Wei[1]

[1]China Railway twenty-one Bureau six company, Beijing, China;
[2]Institute of Civil Engineering, Hebei University of Technology, Tianjin 300130, China
[3]Tangshan Industrial Vocational Technical College, Tangshan, Hebei, China

Abstract. In view of the deficiency of the saturated silt clay foundation reinforced by the dynamic consolidation method, combination the project of soft foundation treatment test area in Tangshan, the reaserch analysed indexes, included groundwater level, pore water pressure, settlement about soil layer and so on, by use of field tests and indoor geotechnical tests, The results showed that the whole reinforcement effect with vacuum dynamic compaction method to blow fill foundation is obvious, due to the result of vacuum precipitation, generally, the excess pore water pressure can be dissipated by 90% above in 2 days around and the effective compaction coefficient can reached more than 0.9,the research work in soft foundation treatment engineering provide a new method and thought to similar engineering.

1. Introduction

The dynamic compaction method is a kind of conventional foundation treatment method, which is widely used in gravel soil, sandy soil, cohesive soil and low saturation silt soil. However, it is not ideal for the treatment of soft clay foundation with high water table, especially the soft soil foundation with high water content, large void ratio, low permeability and low intensity, It is considered that the excess pore water pressure, which is formed in soft soil foundation under the condition of compaction, is difficult to dissipate in time, the result is the loss of shear strength of soil [1-3]. As the active composite drainage consolidation measures, the vacuum dynamic consolidation takes into account the advantages of the dynamic compaction method and the vacuum preloading method, avoiding the damage to the drainage system, the application range of the soft soil foundation treatment method in the coastal area is widened by the accelerated dissipation of excess pore water pressure and the effective consolidation of soft soil.

In view of the existing dynamic foundation reinforcement in the process of excess pore pressure dissipation and hard rubber soil problems of the phenomenon, and based series of achievements has been made in the research on the reinforcement mechanism [4], permeability coefficient variation law of [5, 6], Zhou Jian [7] puts forward the method of dynamic drainage consolidation by combining dynamic compaction and well point dewatering, based on the field test, the mechanism of soil reinforcement is analyzed and discussed, Wang Liming and [8] have also proved that the vacuum dynamic consolidation can achieve good reinforcement effect.

2. Vacuum dynamic consolidation treatment of saturated soft soil foundation

Vacuum dynamic consolidation is suitable for the reinforcement of coastal muddy clay, and the vacuum drainage effect is directly related to the reinforcement effect, due to the impact of compaction and vacuum negative pressure, the pressure difference will be more obvious, then, through multiple times of high vacuum drainage and dynamic consolidation, the water content of saturated soft soil is reduced. The consolidation mechanism of vacuum dynamic consolidation method is mainly composed of three parts:

(1) Vacuum drainage. The saturated soft clay foundation is treated by using vacuum pumping equipment, the saturation of the soil and the excess pore water pressure are reduced by the high vacuum, so that the pore water and pore gas can be discharged quickly, and then the dynamic consolidation efficiency can be enhanced, and the soil consolidation can be accelerated.

(2) Dynamic consolidation. Dynamic consolidation is mainly realized by dynamic compaction, which is based on the construction of high vacuum drainage, and pore water and excess pore water pressure can be discharged and dissipated through the permeable channel, therefore, the effect of dynamic compaction is obvious, and the compactness of soil can be increased rapidly.

(3) Formation of "over consolidated hard shell layer". The saturated soft clay foundation will accelerate consolidation on the basis of the "negative pressure" produced by high vacuum and the excess pore water pressure generated by dynamic compaction, and low water content, high density features of the over consolidated crust layer is vacuum drainage and dynamic consolidation multiple interactions in saturated soft clay ground surface results can meet the bearing capacity of building foundation soil and settlement.

3. Engineering geological condition

The foundation treatment is located in Caofeidian area of Tangshan. The site engineering geological condition is complex, the upper part of dredger fill in unconsolidation state, and the uneven distribution of site poor engineering properties, engineering geological survey measured soil properties are shown in table 1.

Table 1. Soil properties index.

Soil number	Soil name	Bearing capacity characteristic value (kPa)	Compression modulus E_s (MPa)
②	Plain sand	74	1.8~4.4
	Silty clay	52~89	2.1~3.9
② -3	Silt	71~89	4.2~5.0
③	sand	102~128	5.1~10.0
③ -3	Silt	92~108	4.3~5.0
④	sand	151~179	14.1~15.9
⑤	Silty clay	120~150	4.0~7.0
⑤ -1	Clay	120~150	4.0~5.0

4. Process research and test scheme design

4.1 Main process flow

Vacuum drainage and dynamic compaction are the basic requirements of the vacuum dynamic consolidation method, the consolidation of saturated soft soil foundation requires two processes of cycle (figure 1).

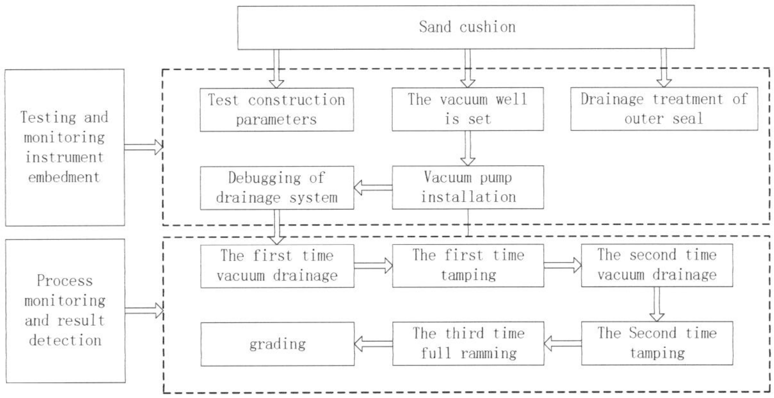

Figure 1. Vacuum dynamic consolidation construction flow.

4.2 Test scheme

Test area of 2500m^2, the foundation treatment using three down three ram, that is, the vacuum drainage and the three times the compaction of alternating, dynamic compaction construction should follow the principle of "from light to heavy, less hit multiple" principle, the depth of reinforcement is controlled by 5~7 m, and the designed load characteristic value is between 160~200 kPa, the distance between tamping points is 4 m, and the interval time is 2~3 days. The other process parameters of vacuum dewatering and dynamic compaction in the test area are shown in table 2.

Table 2. Test scheme design of vacuum dynamic consolidation.

Vacuum dewatering required depth (below ground m)			The number of tamping per tamping and tamping energy (kN - M)		
First time	Second times	Third times	First 4~5 hit	Second 6~8 hits	Third 8~10 hits
2.5	3.0	3.0	1000	1500	2000

Figure 2 shows the vacuum drainage by deep well (length 6 m) and Asakai (length 3 m) spaced construction scheme, the vertical well spacing of 4 m, the transverse horizontal pipe spacing of 4 m, Considering that the groundwater level of the test site is obviously affected by the flood tide, the pulling out of the vacuum well point pipe should be kept synchronized with the tamping. In order to ensure the drainage of the pore water and the effect of excess pore water pressure dissipation, the well point pipe shall be re inserted and continuously operated for 72 h at the end of each tamping end. The construction time of dynamic compaction shall be controlled, and 3 water level observation holes shall (6 meters deep) be installed on the test site, the observation hole has a filter length of 2 m and it is guaranteed to be observed at least 2 times a day.

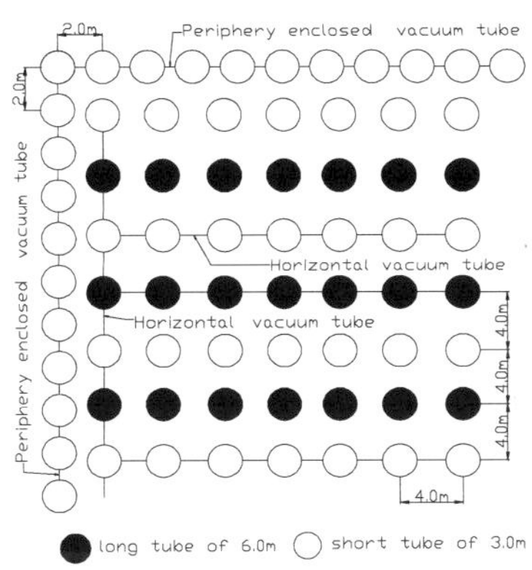

Figure 2. Schematic diagram of vacuum drainage system (1/4 area).

Figure 3. Vacuum precipitation process (first time).

The compaction process adopts medium tamping, and the energy can be considered for the convenience of comparing the foundation reinforcement effect, the selection of round hole hammer can avoid the energy dissipation caused by the overlap of the hammer landing. The size of the hammer is 2.5 m, the weight of hammer is 11.5 T, and the distance of the hammer is controlled at 9 m~20 m. The observation of the water table by the pore water pressure meter determines the time interval after tamping, the next dynamic compaction shall be carried out after the excess pore water pressure of the foundation soil is dissipated by 85%~90%.

5. Monitoring and results analysis

The test of vacuum dynamic consolidation process in the test area is divided into three stages: pre strengthening, strengthening and consolidation, the standard penetration test (B4, B5, B6) and static cone penetration test (J4, J5, J6) were carried out before strengthening; pore pressure observation, deep deformation observation and ground acceleration test were carried out in the experiment. The pore water pressure gauge selects the steel string type, altogether 8 (K31~K34, K41~K44), to be buried at different depths (3 m, 5 m, 7 m, 9 m) depending on the need. In order to ensure the validity and integrity of the observation data, and better reflect the characteristics of the deep deformation of the foundation treatment, three groups of deep targets (S1, S2, S3) were embedded in the test area. The ground acceleration test is selected for the test area; After the completion of the vacuum dynamic consolidation test, the test site was carried out through the standard penetration test (hole B4 ', B5', B6') and static cone penetration test (hole J3', J4 ', J5'), the specific hole layout shown in figure 4.

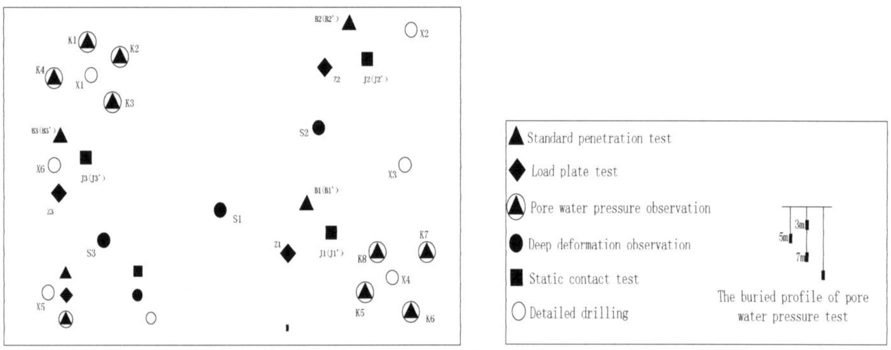

Figure 4. Layout of inspection points in test area.

5.1 Comparative analysis of standard penetration test

In the test area, the "Three times precipitation, three times ramming" was used to control the impact energy, which was mainly controlled by M 2000 kN, the observation results of standard penetration test before and after vacuum dynamic consolidation (figure 5) show that the number of shallow soil SPT in the foundation soil is obviously improved, and the increase range is mainly between the range of 100%~200%, however, with the deepening of soil layer, the extent of the increase gradually decreased until the deep layer of foundation soil was lost. The strength of foundation soil is increased in the range of 4~9 m after vacuum dynamic consolidation, the depth and bearing capacity of foundation soil can be improved by increasing the depth of precipitation and increasing the energy of compaction, therefore, the depth of precipitation and the corresponding energy level should be arranged reasonably according to the thickness of the soil layer, so as to meet the requirements of the foundation bearing capacity and deformation.

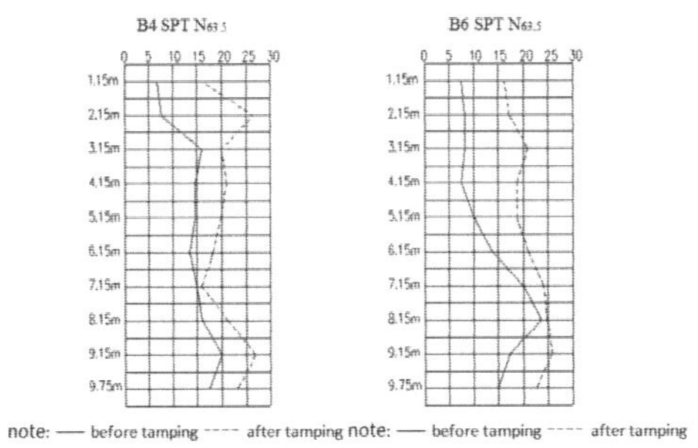

a. Before tamping b. After compaction

Figure 5. Comparison Chart of standard penetration curve.

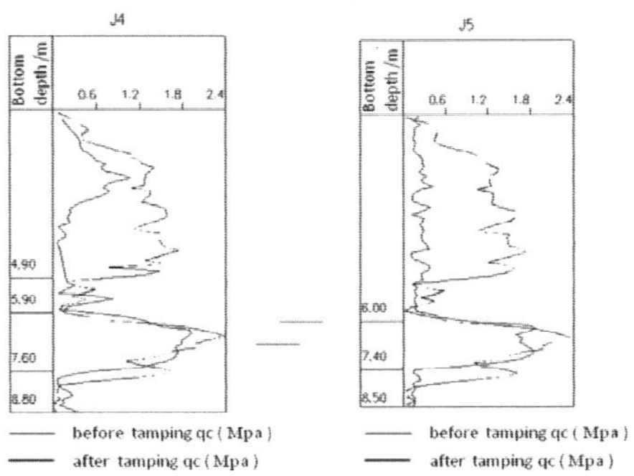

a. Before tamping b. After compaction

Figure 6. Comparison curve of cone head resistance.

5.2 Analysis of static sounding test results

The resistance curve of the cone head before and after the vacuum dynamic consolidation in the test area (figure 6) shows that the resistance value of the saturated soft clay ground cone in a certain depth range is improved obviously, the strength and compactness of soft soil foundation increased significantly. Before and after the vacuum dynamic consolidation, the resistance curves of the cone head approximately intersect at the depth of 6~7 m, which indicates that the resistance of the cone head before and after reinforcement is not significant, as a result, the effective depth of the foundation reinforcement of the vacuum dynamic consolidation soft soil foundation with "Three times precipitation, three times ramming" is considered as 6~7 m. Because the surface soil is easy to be disturbed in the construction process, which leads to the best reinforcement effect of the soil under the scope of 2~3 m.

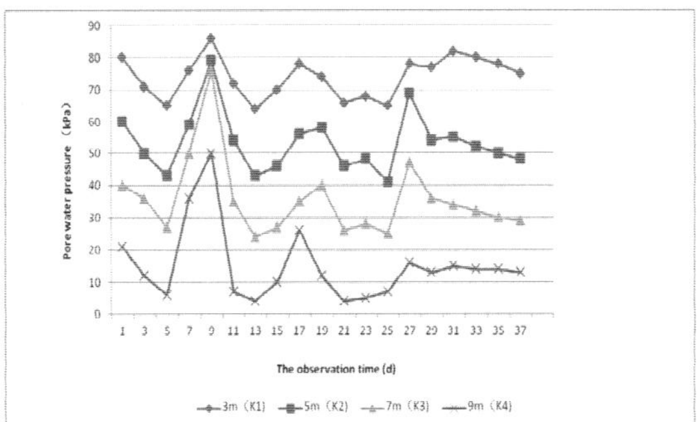

Figure 7. The time history curve of peak pore water pressure (K31~K34).

5.3 Analysis of pore water pressure variation

The dynamic time history curves of the peak pore water pressure of the pre buried pore water pressure gauges (K31, K32, K33, K34) are shown in figure 7, It is found that during the vacuum dynamic consolidation process, the foundation soil is partly removed from the pore water in the vacuum dewatering process (January 23 ~ 25), the initial pore water pressure decreases and the saturation of the soil decreases, the efficiency of tamping is improved, and there is no rubber soil. The peak value of pore water pressure at different depths increased obviously during the dynamic compaction (January 25~28), the increase of pore water pressure increases gradually with the increase of depth, which can be reflected in the curve; The excess pore water pressure was dissipated in the two vacuum active precipitation process (January 28~31), the excess pore water pressure in the shallow layer of the foundation soil dissipates rapidly, and the excess pore water pressure generated by the dynamic compaction can be dissipated more than 90% in 2 days, the reason is that the "positive pressure" produced by the excess pore water and the "negative pressure" produced by vacuum dewatering and drainage in the foundation soil have a gradually increasing "pressure difference". The drainage effect of the project is improved, the gap between the dynamic compaction and the construction period is shortened; Pore water pressure variation peak in the subsequent two times of compaction (February 1~2, February 6~14) and again vacuum precipitation (February 4~6) in the construction of the same similar, but then two times of compaction caused by excess pore water pressure decreases.

5.4 Analysis of site deformation characteristics

5.4.1 Test result analysis of tamping pit. The observation and analysis of 4 groups of different tamping energy in the test area were carried out (figure 8): ground settlement and hammer number showed the proportion of positive change, but the change rate decreases gradually, the overall consolidation settlement occurs in the main components of the tamping number is less than 6 when it is struck, therefore, to determine the best overall implementation of foundation treatment of tamping hammer blow number per times contribute to better. The crater near without obvious uplift indicated that soil compaction is approximately equivalent to the crater volume, thus reflecting the soil compaction effect can be directly used to measure the crater volume can reach more than 0.9, calculation of the effective coefficient of compaction. Local failure deformation law of foundation soil showed extensive damage, after the final failure characteristics of elastic deformation and that in times of compaction conditions, with the increase of soil compaction number was significantly increased.

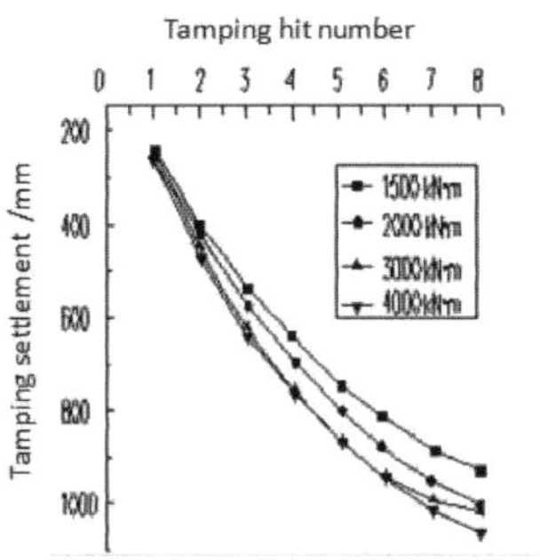

Figure 8. Relation curve of tamping number and tamping quantity.

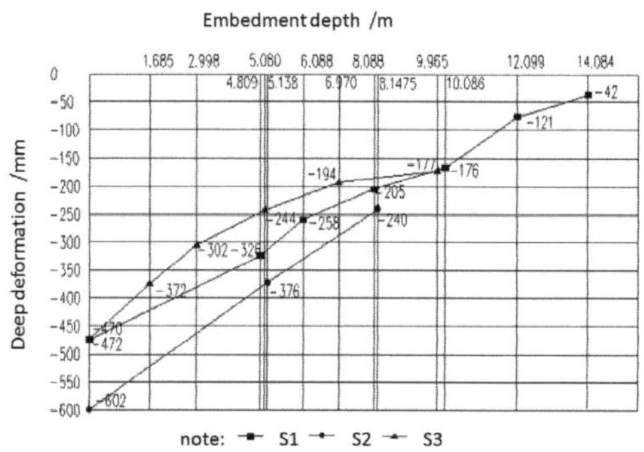

Figure 9. Variation curves of deep foundation displacement (S1, S2, S3).

5.4.2 Deep deformation analysis. Figure 9 shows the deep foundation soil standard Determination of S1, S2 and S3 three groups of deep displacement, through the complete process of vacuum dynamic consolidation settlement observation, the overall soil depth attenuation characteristics of deformation with convergence. The settlement of S1 deep scale soil under 14.084 m is less than 8.9% of the total settlement, but the change of settlement is obvious at the above position (from 472 mm to 14.084 m of the surface soil mass, 42 mm at the depth); S2 deep scale settlement is attenuated from 602 mm (ground surface) to 240 mm (8.1475 m depth), The calculation shows that the settlement below the soil layer still reaches 39.9% of the total settlement; The settlement of S3 deep scale soil decreased from 470 mm (surface) settlement to 177mm (9.965 m depth), and the soil settlement still accounted for 37.7% of the total settlement under 9.965 m. The analysis of the observed curve can be that the depth of soil's S1 standard position has basically reached the vacuum dynamic consolidation effect, while the S2 and S3 standard has not yet been laid in place, because the bottom still has a large settlement, and the soil below this layer is affected by the compaction of vacuum dynamic consolidation

The observation results show that, in the process of vacuum dynamic consolidation in shock wave and dynamic stress of foundation soil under the action of force more dense, the strength of the foundation soil, stiffness and compression modulus increased significantly, while the soil compression volume growth rate showed reverse decreasing trend, the soil tends to be more dense

6. Conclusion

With the development of the integration of Beijing, Tianjin and Hebei, and as well as the continuous implementation of Caofeidian area. The strengthening of large saturated soft clay foundation is also highlighted. In the paper, the influence of excess pore water pressure, ground deformation and tamping energy on the soil mass change before and after foundation treatment in the experimental area are systematically monitored and analyzed. The following conclusions have been obtained:

(1) The excess pore water pressure and water content in foundation soil can be dissipated and reduced by the combination of vacuum drainage and dynamic compaction. The compaction efficiency is improved obviously and the construction period is shortened.

(2) The soil strength and modulus of deformation in the experimental area are obviously improved after vacuum dynamic consolidation, and the risk of soil liquefaction is completely eliminated.

(3) The number of vacuum precipitation and tamping energy are closely related to the strengthening effect of foundation soil, and the effect of reinforcement increases obviously with the increase of precipitation number and tamping energy. In the range of influence, the settlement of the tamping pit and the total settlement of the ground increase with the increase of tamping energy and tamping number.

Reference

[1] Pritchard M A, Savigny K W 1990 *Canadian Geotech. J. Pritch,* **27(6)** 823-34.
[2] Menard L and Broise Y 1975 *Geotech.* **23(1)** 3-18.
[3] Chow Y K et al., 1991 *Comput. Meth. Adv. Geomech.* **1** 232-42.
[4] Meng Q S, Wang R and Lei X W 2004 *Geotech. Mech.* **22(4)** 427-31.
[5] Gong X N and Cen Y R 2002 *Found. Treat.* **13(4)** 7-11.
[6] Yang B, Li Y and Zhou W 2015 *J. Guangxi Univ.* **40(2)** 325-30.
[7] Zhou J. Cao Y and Jia M C 2003 Rock. Soil. Mech. Dewater. Comb. Dyn. Compact. **24(3)** 376-80.
[8] Wang L M, Han X J and Zhu Y W 2006 *J. Nanjing Univ. Tech.* **28(6)** 57-61.

Acknowledgement

Research project of China Railway twenty-one Bureau (17B-4); Science and technology research foundation of Hebei province (QN2014335); Hebei Natural Science Foundation of China (E2014209221).

IOP Conf. Series: Materials Science and Engineering **269** (2017) 012010 doi:10.1088/1757-899X/269/1/012010

Finite Element Analysis and Vibration Control of Lorry's Shift Mechanism

Li Qiangwei[1]

[1]Engineering training center, Soochow University, Suzhou, China

Abstract. The transmission is one of the important parts of the automobile's transmission system, Shift mechanism's main function of transmission is to adjust the position of the shift fork, toggle the synchronizer's tooth ring, so that the gears are separated and combined to achieve the shift. Therefore, in order to ensure the reliability and stability of the shift process, the vibration characteristics of the shift mechanism cannot be ignored. The static analysis of the shift fork is carried out, and the stress distribution of the shift fork is obtained according to the operating characteristics of the shift mechanism of the lorry transmission in this paper. The modal analysis of the shift mechanism shows the low-order vibration frequencies and the corresponding modal vibration shapes, and the vibration control analysis is carried out according to the simulation results. The simulation results provide the theoretical basis for the reasonable optimization design of the shift mechanism of the lorry transmission.

1. Introduction

The transmission is a device that provides a gearshift operation for the vehicle, and the traditional mechanical manual transmission is still the most widely used in transmissions for light trucks. As an important mechanism in the transmission mechanism, Transmission shift mechanism's strength and dynamic characteristics influence the security and stability in the processing of gearshift. In recent years, the developed countries have adopted advanced virtual simulation system to simulate the reliability and rationality of mechanical structure, and provide a reference for the optimal design [1]. The Ford Motor Company carries out the static analysis by the method of finite element to the automobile transmission's structure in the study of its force situation, and then the stress distribution of the transmission is obtained. The Nissan Motor Company analyzes the structure and dynamic characteristics of transmission box, input shaft, output shaft and intermediate shaft with the computer simulation technology, and then optimizes the design of the transmission box, input shaft, output shaft and intermediate shaft [2]. The fatigue life analysis with the method of finite element of the transmission shift fork is carried out in reference [3], and the shift fork assemblies' fatigue lives are obtained. Based on the theory of finite element static and modal analysis, the finite element analysis and vibration control of lorry's shift mechanism are carried out in this paper, the stress distribution of the shift fork, the low-order vibration frequencies and modes of the shift mechanism are obtained, and then the results can provide a guidance for the later design and optimization. Figure 1 shows the structure of lorry's shift mechanism.

IOP Conf. Series: Materials Science and Engineering **269** (2017) 012010 doi:10.1088/1757-899X/269/1/012010

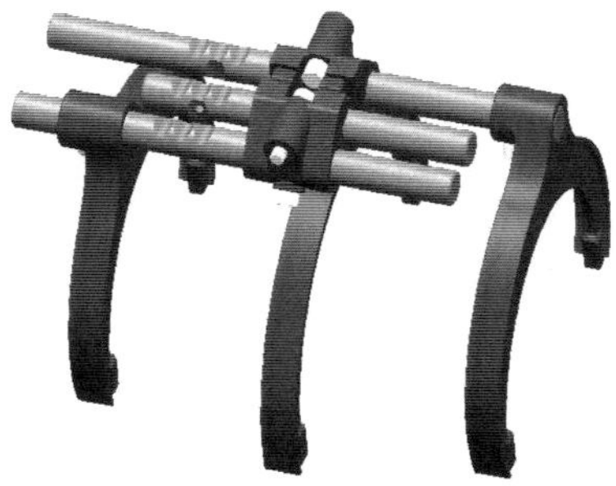

Figure 1. The structure of lorry's shift mechanism.

2. Method and Theory

The vibration characteristics of the mechanical structure mainly include the natural frequency and vibration mode. Through the static analysis of the lorry's forklift and the modal analysis of the gearshift mechanism, the stress distribution of the lorry's forklift and the shift mechanism's low-order vibration frequencies and vibration modes in the preload state can be obtained [4].

The vibration equation of the multi degree of freedom system is

$$[M]\{\ddot{u}\}+[C]\{\dot{u}\}+[K]\{u\}=\{F(t)\} \tag{1}$$

[M], [F] and [C] are the mass matrix, damping matrix and stiffness matrix, {F (t)} are the excitation vectors, $\{\ddot{u}\},\{\dot{u}\},\{u\}$ are the acceleration vector, the velocity vector and the displacement vector in (1).

The natural frequency of the machine is the inherent characteristic of the mechanical system, which is not related to the application of the external load, and the influence of the small damping on the natural frequency and the mode of the structure is not significant [4]. Therefore, when the influence of the damping is neglected the free vibration equation with no damping multiple degrees of freedom is:

$$[M]\{\ddot{u}\}+[K]\{u\}=\{0\} \tag{2}$$

[M] is the total mass matrix of the structure, and [K] is the total stiffness matrix of the structure in (2). The corresponding eigenvalue equation is:

$$([K]-\omega^2[M])\{u\}=\{0\} \tag{3}$$

ω is the natural frequency, $\{u\}$ is the corresponding vibration mode in (3).

3. Modeling and Pre-processing

Model of the lorry's shift mechanism is established in the professional three-dimensional modeling software, and then the three-dimensional model is imported into the finite element software by using the intermediate format between the 3D software and the finite element software for pre-processing. In the pre-processing process, the small fillets and thread feature are optimization by filling, parts which are easily lead to the calculation of divergent should be amended. The shift mechanism is assembled in the finite element software. The nonlinear part is linearized to reduce the computation time on the basis of ensuring the accuracy of the calculation. The finite element model of the shift fork is shown in figure 2a and the finite element model of the shift mechanism is shown in figure 2b.

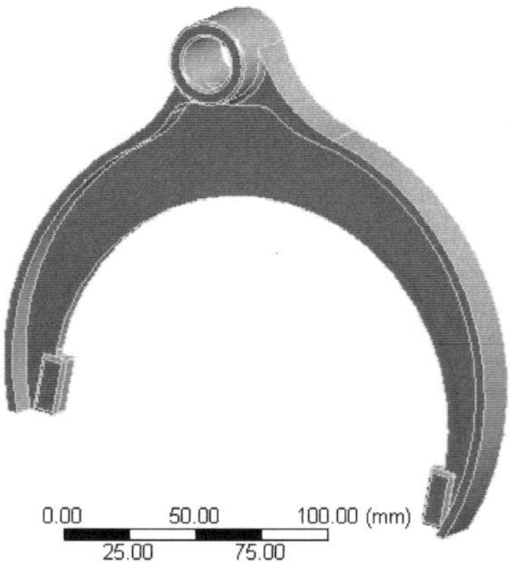

Figure 2a. The finite element model of the shift fork.

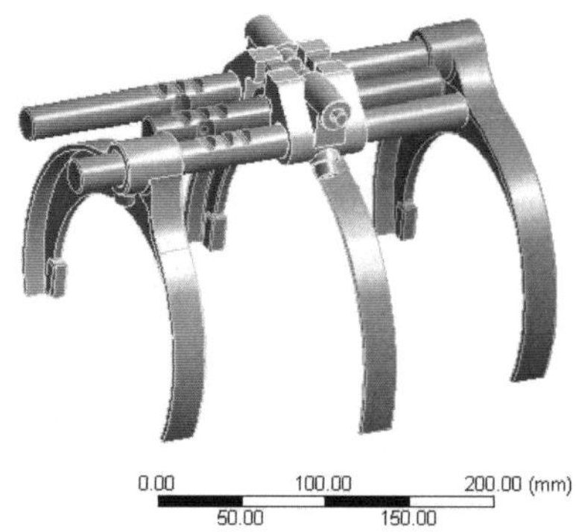

Figure 2b. The finite element model of the shift mechanism.

Lorry's shift mechanism is mainly composed of 45 steel, so material properties for the simulation are steel, the elastic modulus of 200GPa, poisson's ratio of 0.3, and the density of 7850kg/m^3. Meshing is an important part of finite element calculation. The grid segmentation affects the calculation accuracy and calculation time [5]. Therefore, reasonable and effective meshing is very necessary. Static analysis and modal analysis of the grid division requirements are different in the finite element analysis, and static analysis requires high grid partitioning [6]. Therefore, the meshing of the shift fork is mainly based on the tetrahedron-based with free division, the precision level is Fine, and the grid correlation is 0. Holes and other parts are refined [7]. The number of nodes is 14816, and the number of elements is 7854. Meshing of shift fork as shown in figure 3a. As a result of that the static analysis requires a higher meshing precision, so the grid quality inspection of the shift fork is checked after dividing the grid, and the main part of the grid distortion below 0.88, therefore, the meshing of the shift fork is in accordance with the finite element calculation requirements. The shift mechanism is meshed in the condition of satisfying the calculation precision. The division method is the same as that of the shift fork. The number of nodes is 91150, and the number of elements is 48333. Meshing of shift mechanism as shown in figure 3b.

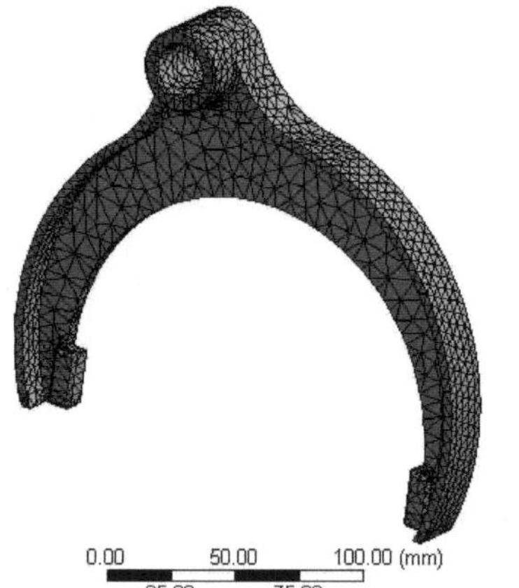

Figure 3a. Meshing of shift fork.

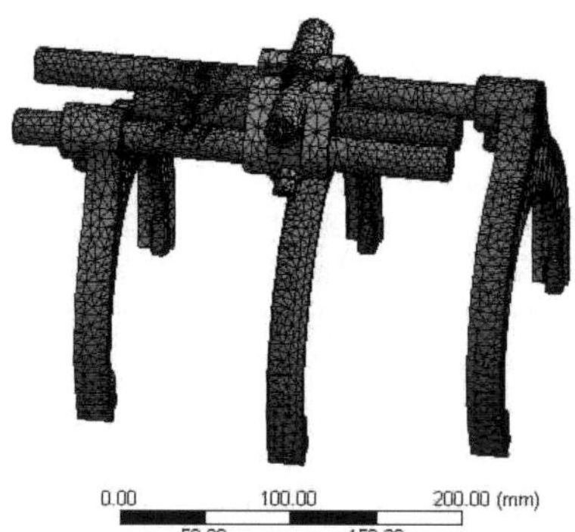

Figure 3b. Meshing of shift mechanism.

IOP Conf. Series: Materials Science and Engineering **269** (2017) 012010 doi:10.1088/1757-899X/269/1/012010

Under normal circumstances, the force of people to change shift under the sitting position is between 0N and 300N. The handle force takes 300N in the case of clutch failure shift. The force of shift fork is checked five times by the handle force, so the force is 1500N, acting on the contaction between shift fork and synchronizer. The displacement constraint of the shift fork uses the constraint type of displacement to constrain the movement of the input shaft port in the Y and Z directions. The boundary constraint of the shift fork is shown in figure 4a. Since the shift mechanism is an assembly, so it is necessary to carry out the assembly restriction of the parts in the finite element software. Due to the large number of parts of the assembly, the contact type of the shift mechanism is assembled by bonded to improve the calculation efficiency. The fixed constraint type is used at the contact between the shift mechanism and the handle mechanism, and the displacement in the three directions of X, Y and Z is constrained. The boundary constraint of the shift mechanism is shown in figure 4b.

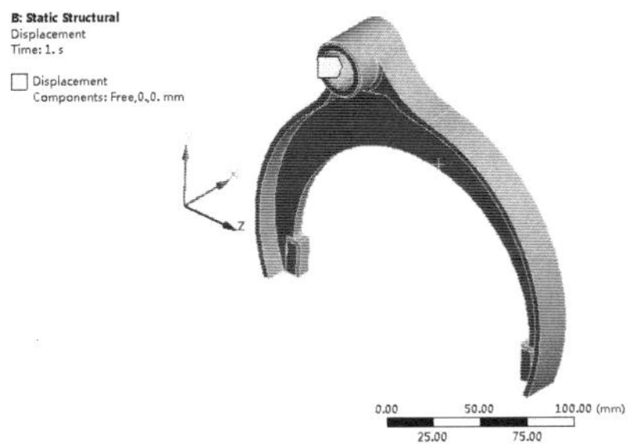

Figure 4a. The boundary constraint of the shift fork.

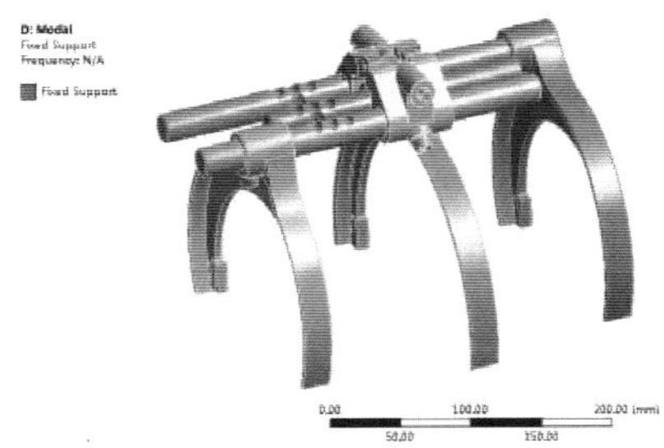

Figure 4b. The boundary constraint of the shift fork.

After the static analysis of the shift fork, the stress distribution of the shift fork is shown in figure 5.

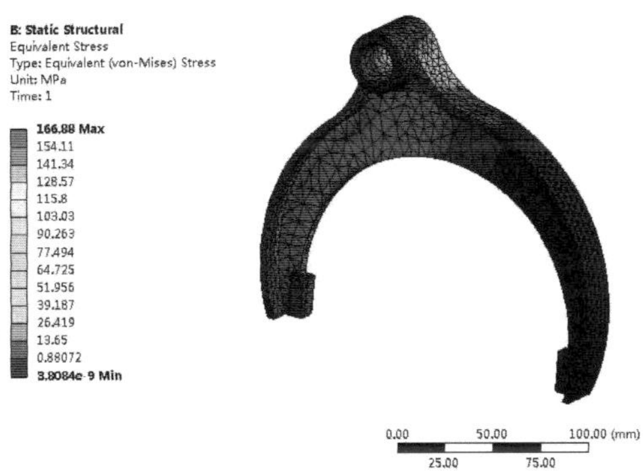

Figure 5. The stress distribution of operating platform.

It can be seen from figure 5 that the load of the shift fork is up to 166.9MPa under the effect of clutch failure, the yield strength of 45 steel with quenched is 840Mpa, the maximum stress of shift fork stress is far less than the yield strength of the material, so the shift fork meets the job requirements.

The modal analysis of the shift mechanism is conducted after the pre-treatment and the low-order vibration frequencies and modes of the shift mechanism can be obtained. The low-order vibration frequencies of the shift mechanism is shown in table 1, and the corresponding modal vibration shapes of the shift mechanism are shown in figure 6.

Table 1. The low-order vibration frequencies of the shift mechanism.

Mode	Frequency (Hz)
1	172.7
2	178.6
3	185.1
4	191.2
5	263.3
6	268.3

Figure 6. The corresponding modal vibration shapes of the shift mechanism.

From the results of modal analysis, it can be seen that the vibration frequency of the shift mechanism is between 172.7Hz and 268.3Hz, the overall performance is the vibration of the shift fork:

(1) The first-order vibration frequency is 172.7Hz. The vibration performance is the shift forks of the first two gears rotate around left and right along the input shaft's axis. The maximum vibration appears in the contact front end between the shifts forks of the first two gears and the synchronizer, and the maximum displacement is 73.9mm;

(2) The second-order vibration frequency is 178.6Hz. The vibration performance is the shift fork of the fifth gear rotates around left and right along the input shaft's axis. The maximum vibration appears in the contact back end between the shift fork of the fifth gear and the synchronizer, and the maximum displacement is 84.5mm;

(3) The third-order vibration frequency is 185.1Hz. The vibration performance is the shift forks of the first two gear rotate around left and right along the input shaft's axis. The maximum vibration appears in the contact back end between the shift forks of the first two gear and the synchronizer, and the maximum displacement is 87.1mm;

(4) The fourth-order vibration frequency is 191.2Hz. The vibration performance is the shift fork of the fifth gear rotates around left and right along the input shaft's axis. The maximum vibration appears in the

contact front end between the shift fork of the fifth gear and the synchronizer, and the maximum displacement is 84.7mm;

(5) The fifth-order vibration frequency is 263.3Hz. The vibration performance is the shift forks of the third and fourth gear rotate around left and right along the input shaft's axis. The maximum vibration appears in the contact back end between the shifts forks of the third and fourth gear and the synchronizer, and the maximum displacement is 100.3mm;

(6) The sixth-order vibration frequency is 268.3Hz. The vibration performance is the shift forks of the third and fourth gear rotate around left and right along the input shaft's axis. The maximum vibration appears in the contact both ends between the shift forks of the third and fourth gear and the synchronizer, and the maximum displacement is 103.2mm.

4. Vibration Control

Through the finite element modal analysis of the shift mechanism, it can be concluded that the low-order vibration frequencies of the shift mechanism is within 172.7Hz ~ 268.3Hz, and mainly vibration performance is the torsional vibration of the shift fork. Therefore, the vibration characteristics and strengthen of shift fork should be focused in the latter part of the optimization process to prevent the vibration caused by the suspension in the course of driving.

5. Conclusions

The stress distribution of the shift fork and the low-order vibration frequencies and the corresponding modal vibration shapes of the shift mechanism are obtained through the finite element analysis of lorry's shift mechanism. The vibration control scheme is proposed according to the simulation results, and the results could provide the theoretical basis for the optimal design of lorry's shift mechanism.

References

[1] Qian Y D 2013 *The Study on performance and test analysis of automobile mechanical transmission*, Wuhan University of Technology.
[2] Xiao H Y 2015 *Static and dynamic analysis and kinematics simulation of truck transmission*, Shandong Jianzhu University.
[3] Jia Y, Luo D G, et al., 2014 *Tianjin Auto* **(8)** 48-50, 59.
[4] Dong X, Zheng Z C, et al., 2014 *Appl. Mech. Mater.* **685** 265-70.
[5] Cheng L, Liu Y W, et al., 2012 *Mach. Des. Manuf.* **1** 147-9.
[6] Wang J H, Wang Z C, Gao Y, et al., 2013 *Int. Comb. Eng. Power Plant.* **30(1)** 29-34.
[7] Shu B, Yu D Y, Wang D and Zhang S Q, 2012 *J. Mod. Manuf. Eng.* **2** 71-73, 121.

IOP Conf. Series: Materials Science and Engineering **269** (2017) 012011 doi:10.1088/1757-899X/269/1/012011

Effect law of Damage Characteristics of Rock Similar Material with Pre-Existing Cracks

S G Li[1, 2], X Y Cheng[1, 2] and C Liu[1, 2]

[1]College of Safety Science and Engineering, Xi'an University of Science and Technology, Xi'an 710054, China
[2]Key Laboratory of Western Mine Exploitation and Hazard Prevention Ministry of Education, Xi'an University of Science and Technology, Xi'an 710054, China

Abstract. In order to further study the failure mechanism for rock similar materials, this study established the damage model based on accumulative AE events, investigated the damage characteristics for rock similar material samples with pre-existing cracks of varying width under uniaxial compression load. The equipment used in this study is the self-developed YYW-II strain controlled unconfined compression apparatus and the PCIE-8 acoustic emission (AE) monitoring system. The influences of the width of the pre-existing cracks to the damage characteristics of rock similar materials are analyzed. Results show that, (1) the damage model can better describe the damage characteristics of rock similar materials; (2) the tested samples have three stages during failure: initial damage stage, stable development of damage stage, and accelerated development of damage stage; (3) with the width of pre-existing cracks vary from 3mm to 5mm, the damage of rock similar materials increases gradually. The outcomes of this study provided additional values to the research of the failure mechanism for geotechnical similar material models.

1. Introduction

Rock is a common natural complicated geological body. Rock has defects such as cracks, beddings, and joins, which heavily influence its stability. The cracks' geometrical parameters have crucial effects to damage characteristics of rock [1-3]. Therefore, the study of damage characteristics of rock with cracks has great importance and significance to the geotechnical research of the rock mass failure mechanism.

Many researchers have studied the damage characteristics of rock mass. Some of the typical researches are summarized here. Heiple and Carpenter [4] conducted a number of AE experiments and concluded that the AE ringing counts can reflect the material internal damage process. Jansen et al. [5] used the AE technology to investigate the damage accumulation and development during rock failure. Based on electromagnetic radiation characteristics, Jin et al. [6] studied the damage evolution of rock under uniaxial compression. Zhao et al. [7] established the damage-based constitutive model for rock under impacting load. Zong et al. [8] studied the mechanical and damage evolution properties of sandstone under triaxial compression. From a micro perspective and based on the AE technology, Liu et al. [9] obtained the failure development characteristics for coal and rock under uniaxial compression through the study of the acoustic emission. By analyzing the AE characteristics and failure development for salt rock under uniaxial compression, Zhou et al. [10] compared the damage model based on ringing counts and that based on energy parameters.

From the above literature review, one can see that the researches on the damage characteristics of rock have achieved abundant outcomes. However, there are still rooms for improvement. The most objects of study are native rock, thus the research on rock similar materials is limited. However, similar material simulation experimental study is one of the important means in the research of geotechnical engineering, which have been widely used. Therefore, there is a need for the systematical study of the rock similar material. Most studies for rock damage characteristics are based on the AE ringing counts and established the damage model, and few researches have taken crack parameters into account. The AE ringing count is not a sufficient representation of rock damage, and crack parameters is one of the important factors that impact rock damage characteristics. For this reason, it is necessary to explore more appropriate factors that represent rock damage, and based on which establish new damage model for continued in-depth study.

In order to establish a more suitable damage model for rock similar material, and further study the damage characteristics of rock similar material with cracks. This study selected accumulative AE events as representation of damage, established the damage model, and selected crack width as one geometrical variable parameter, tested rock similar material samples with different pre-existing crack width. By using the self-developed compression apparatus and high precision acoustic emission monitoring system, the influence of crack width to the damage characteristics of rock similar materials are investigated. The outcomes of this research can be used for supplementation and improvement of the current deformation damage mechanism research for rock similar materials.

2. The damage model based on accumulative AE events

Among other AE characteristic parameters, the accumulative AE events represent local change to the material identified by the AE signal collision. This is directly related to material failure and is a good indicator for the damage to the material. Therefore, the damage model is established based on the accumulative AE events for rock similar material under uniaxial compression.

Kachanov [11] proposed the concept of damage variable (*D*), and defined it as:

$$D = \frac{A_d}{A} \tag{1}$$

where A_d is damage plane surface area, and A is the initial cross sectional area of the material.

Due to the fat that it is challenging to determine the effective bearing area when the material is damaged, Lemaitre [12] proposed the equivalent strain hypothesis, which indirectly measures material damage based on the effective stress shown in the following equation.

$$\sigma' = \frac{\sigma}{1 - D} \tag{2}$$

where σ' is equivalent uniaxial stress, and σ is the actual uniaxial stress.

According to elastic-plastic mechanics, the strain for an undamaged material can be expressed as:

$$\varepsilon = \frac{\sigma'}{E} \tag{3}$$

where E is modulus of elasticity, and ε is the strain.

Combining equiation (2) with (3), the uniaxial stress can be expressed as:

$$\sigma = \sigma'(1 - D) = E\varepsilon(1 - D) \tag{4}$$

Assume the number of accumulative AE events is N_m for complete damage of the surface (A) for the nondestructive material, the generated AE events per unit area N_w can be calculated as:

$$N_w = \frac{N_m}{A} \tag{5}$$

The accumulative AE events generated when the damage area is A_d can be calculated as:

$$N_d = N_w A_d = \frac{N_m}{A} A_d \tag{6}$$

Combing equation (1) with equation (6), the damage variable can be expressed as a function of the accumulative AE events:

$$D = \frac{A_d}{A} = \frac{N_d}{N_m} \tag{7}$$

Rock material cannot be completely damaged when compressed, which means a residual strength exist and the damage variable cannot reach to 1. Thus the damage variable can be modified to:

$$D = D_u \frac{N_d}{N_m} \tag{8}$$

where D_u is damage critical value, N_m is the number of accumulative AE events when the damage variable is D_u. Using the linear function transformation method, the damage critical value can be normalized as:

$$D_u = 1 - \frac{\sigma_c}{\sigma_p} \tag{9}$$

where σ_p is the peak stress, and σ_c is the residual strength.

When $\sigma_c=\sigma_p$, D_u=0, and thus D=0 in equation (8). Such rock similar material can be treated as ideal elastoplastic material and compression cannot cause damage. When σ_c=0, D_u=1 in equation (8), which means the rock similar material can be completely damaged under compression. However, a certain level of damage will occur when compression is applied, and rock similar material has a residual strength. Therefore, in this study, $D_u \in (0,1)$.

Combing Equation (4), (8), with (9), we achieve the following damage model for rock similar material under uniaxial compression, which is based on the accumulative AE events.

$$\sigma = E\varepsilon(1 - D) = E\varepsilon\left(1 - \left(1 - \frac{\sigma_c}{\sigma_p}\right)\frac{N_d}{N_m}\right) \tag{10}$$

3. Experimental designs

3.1. Specimen preparation

To prepare the rock similar material, sand is used as the aggregate, cement and starch are used as cementing agent, and water is the dissolvent. The sand to cement weight ratio is 10:1, and water amount is 1/9 of the total weight of the similar material. After uniformly mixing of the materials, it is put in a standard mould (Φ50mm×100mm). The specimens are taken out of the mould after 1-2 minutes compaction, and cured in room temperature and condition for 5-7 days. The prepared specimens are shown in figure 1.

Several groups of tests were conducted. Due to limited space, only analyses of one group of specimens are presented in this paper. The results can represent the overall results for other test groups. The three specimens in this group are numbered W-1, W-2, and W-3. A hacksaw blade was used to make the cracks on each specimen. The length of the cracks is 40mm, and the angle is 45°. The crack width for specimen W-1, W-2, and W-3 are 3mm, 4mm, and 5mm, respectively. The completed specimens with different width of pre-existing cracks are shown in figure 2.

Figure 1. The prepared rock similar material specimens.

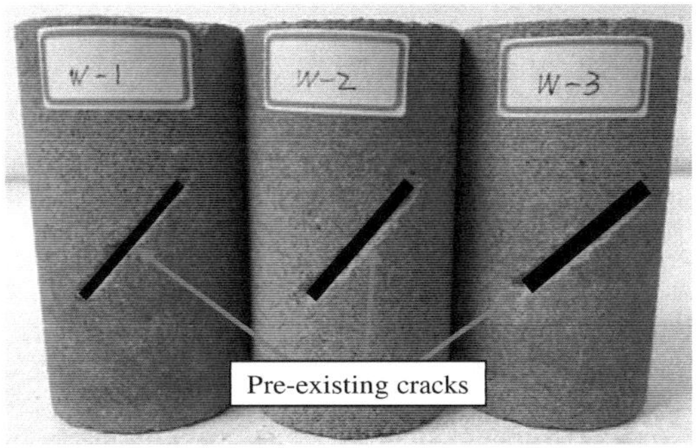

Figure 2. Prepared specimens with different width of pre-existing cracks.

3.2. Loading equipment and conditions

The loading equipment used in this study is the self-developed YYW-II strain controlled unconfined compression apparatus, as shown in figure 3. Its major components include a loading ring, an electrical motor, a lifting board, an axial force meter, and an axial displacement meter. This apparatus is suitable for conducting the uniaxial compression test for similar materials, monitoring the stress-strain in the testing process. The deformation rate controlled method is used in this research, and the loading rate is 0.87mm/min.

3.3. AE monitoring system and monitoring design

The AE system used is the PCIE-8 system from the Physical Acoustics. The system achieves high maximum sustained processing speeds, better de-noising ability, reliability and stability. The threshold value is set to be 40dB, the transmission gain for the preamplifier is 40dB, the sampling frequency is 10MHz, and the resonant frequency for the sensor is 1-100kHz.

As shown in figure 4, six sensor arrays were used to collect the AE signals. To improve contact between the sensors and the specimen and reduce signal loss, Vaseline is applied in between the sensors and the specimen, and the sensors are fixed to the specimen by sticky tapes. Lead-break tests were conducted close to each sensor to calibrate the sensors before the actual experiments.

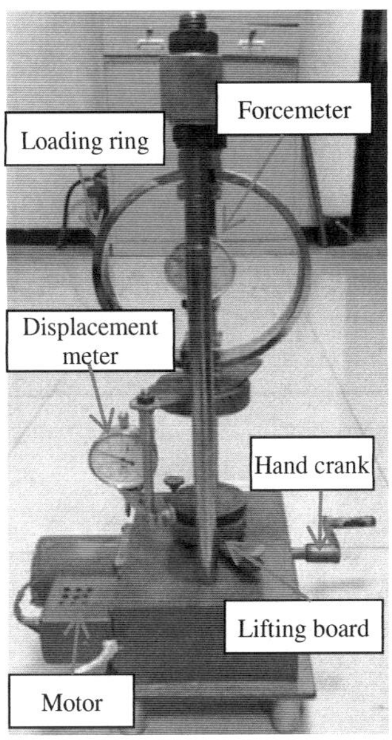

Figure 3. YYW-II strain controlled unconfined compression apparatus.

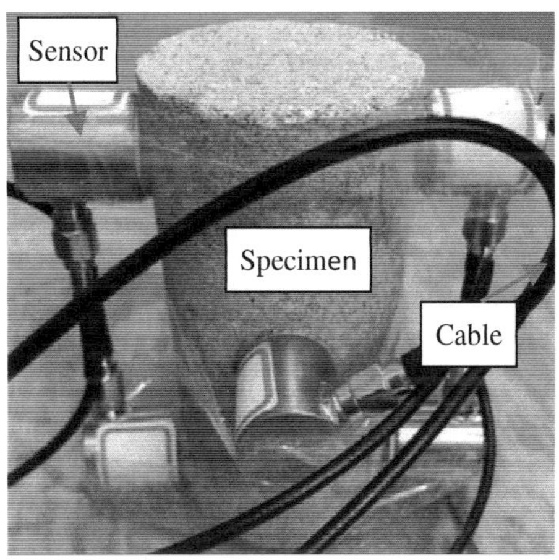

Figure 4. Sensor arrangement.

IOP Conf. Series: Materials Science and Engineering **269** (2017) 012011 doi:10.1088/1757-899X/269/1/012011

4. Damage characteristics analysis

Based on the experimental results and the damage model derived in the above section, the parameters for specimen W-1 are σ_p=518KPa, σ_c=129KPa, E=51944KPa, N_m=323, for W-2 are σ_p=479KPa, σ_c=119KPa, E=30357KPa, N_m=252, and for W-3 are σ_p=411KPa, σ_c=102KPa, E=46846KPa, N_m=134. The damage-strain curve can be plotted according to equation (8) and equation (9), as shown in figure 5. The theoretical stress-strain curves are derived according to equation (10), and plotted together with the actual experimental stress-strain curve in figure 6. As can be seen, the curves have very similar trends, and the theoretical curves are generally below the actual measured curves. This is due to the stress loss in the actual experiments. The close match of the two curves has proved the validity of the normalized damage variable based on the accumulative AE events in equation (8).

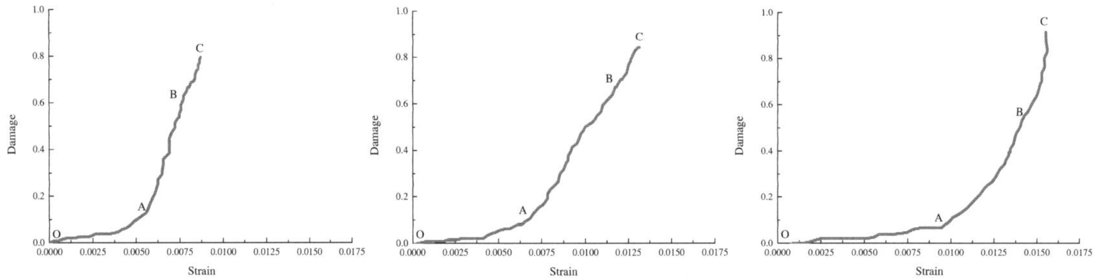

(a) Specimen W-1 (b) Specimen W-2 (c) Specimen W-3

Figure 5. The damage-strain curves.

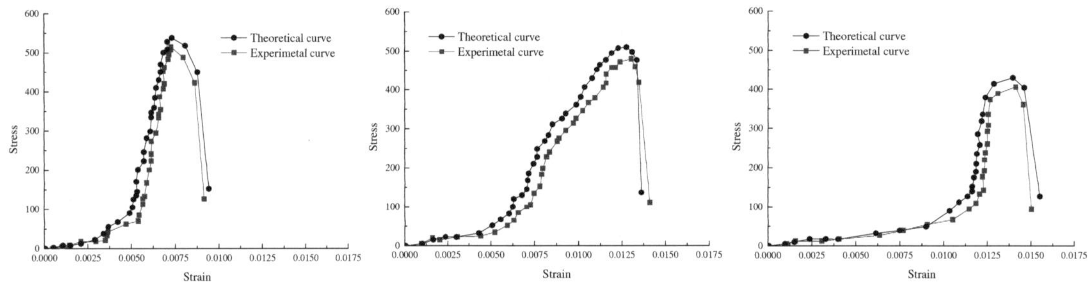

(a) Specimen W-1 (b) Specimen W-2 (c) Specimen W-3

Figure 6. The comparison of the theoretical and experimental stress-strain curves.

As can be seen in figure 5, the damage development processes for the three tested specimens are similar, and they are closely related to their load and deformation, and AE development. The damage development process can be divided into three stages.

The first stage is the initial damage stage (OA). In this stage, only minimum damage has occurred and the level is relatively low. This phase is in correspondence to the initial compression stage, and the slow accumulation of AE events stage. The original micro cracks and micro pores are compacted to close, and only few AE events are generated. The damages are due to the internal particle friction and grain sliding.

The second stage is the damage stable development stage (AB), in which the damage variables increase steadily. It is in correspondence to the elastic deformation stage, and the steady growth of AE events stage. New micro cracks are generated, and they are gradually extended and connected. The number of AE events increases steadily, which indicates stable damage.

The last phase is the damage rapid development stage (BC) and the damage variable rapidly increases to the maximum value. It is in correspondence to the plastic deformation stage, and the rapid increase of AE events stage. In this stage, micro cracks are rapidly generated, extended, and connected. The number of AE events accelerates at a growing rate until eventually causes specimen failure. The damages development is also non stable. The above analysis supports the conclusion that the width of the pre-existing cracks does not have influence to the characteristics of specimen damage in different phases.

Further look at figure 5, we can see that the maximum damage variables for W-1, W-2, and W-3 are 0.791, 0.856, and 0.913, respectively. This means a larger pre-existing crack width increases the damage of specimen.

As can be seen from the three damage development stages, the damage evolution is a nonlinear process and it is not possible to use a continuous function to represent the damage mechanism. Based on polynomial curves fitting of the curves in figure 5, the damage evolution equations are obtained as equations (11) to (13). The degree of fitting is between 0.98263-0.9967, indicating a preferable fitting result.

$$\text{W-1}: D = \begin{cases} 2.45182 \times 10^{-4}\varepsilon^2 - 0.00542\varepsilon + 0.00645 \; (0 \le \varepsilon \le 0.0037) \\ -7.17421 \times 10^{-6}\varepsilon^2 + 0.00541\varepsilon - 0.08407 \; (0.0037 \le \varepsilon \le 0.0075) \\ -6.54744 \times 10^{-4}\varepsilon^2 + 0.15742\varepsilon - 18.15782 \; (0.0075 \le \varepsilon \le 0.0088) \end{cases} \tag{11}$$

$$\text{W-2}: D = \begin{cases} 5.56461 \times 10^{-5}\varepsilon^2 - 3.21685 \times 10^{-4}\varepsilon + 0.00675 \; (0 \le \varepsilon \le 0.0063) \\ -8.24851 \times 10^{-6}\varepsilon^2 + 0.00356\varepsilon - 0.45212 \; (0.0063 \le \varepsilon \le 0.0113) \\ 5.17245 \times 10^{-4}\varepsilon^2 - 0.45218\varepsilon + 18.54782 \; (0.0113 \le \varepsilon \le 0.0138) \end{cases} \tag{12}$$

$$\text{W-3}: D = \begin{cases} 4.42056 \times 10^{-5}\varepsilon^2 - 2.14824\varepsilon - 0.00517 \; (0 \le \varepsilon \le 0.0101) \\ 9.01485 \times 10^{-6}\varepsilon^2 + 0.00145\varepsilon - 0.25634 \; (0.0101 \le \varepsilon \le 0.0137) \\ -3.14245 \times 10^{-4}\varepsilon^2 + 0.05485\varepsilon - 12.14521 \; (0.0137 \le \varepsilon \le 0.0163) \end{cases} \tag{13}$$

Further derive the above equations according to equation (10), we can get the damage constitutive equations for the three tested specimens.

$$\text{W-1}: \sigma = 51944\varepsilon \begin{bmatrix} 2.45182 \times 10^{-4}\varepsilon^2 - 0.00542\varepsilon + 0.00645 \; (0 \le \varepsilon \le 0.0037) \\ -7.17421 \times 10^{-6}\varepsilon^2 + 0.00541\varepsilon - 0.08407 \; (0.0037 \le \varepsilon \le 0.0075) \\ -6.54744 \times 10^{-4}\varepsilon^2 + 0.15742\varepsilon - 18.15782 \; (0.0075 \le \varepsilon \le 0.0088) \end{bmatrix} \tag{14}$$

$$\text{W-2}: \sigma = 30357\varepsilon \begin{bmatrix} 5.56461 \times 10^{-5}\varepsilon^2 - 3.21685 \times 10^{-4}\varepsilon + 0.00675 \; (0 \le \varepsilon \le 0.0063) \\ -8.24851 \times 10^{-6}\varepsilon^2 + 0.00356\varepsilon - 0.45212 \; (0.0063 \le \varepsilon \le 0.0113) \\ 5.17245 \times 10^{-4}\varepsilon^2 - 0.45218\varepsilon + 18.54782 \; (0.0113 \le \varepsilon \le 0.0138) \end{bmatrix} \tag{15}$$

$$\text{W-3}: \sigma = 46846\varepsilon \begin{bmatrix} 4.42056 \times 10^{-5}\varepsilon^2 - 2.14824\varepsilon - 0.00517 \; (0 \le \varepsilon \le 0.0101) \\ 9.01485 \times 10^{-6}\varepsilon^2 + 0.00145\varepsilon - 0.25634 \; (0.0101 \le \varepsilon \le 0.0137) \\ -3.14245 \times 10^{-4}\varepsilon^2 + 0.05485\varepsilon - 12.14521 \; (0.0137 \le \varepsilon \le 0.0163) \end{bmatrix} \tag{16}$$

5. Conclusions

(1) The damage model based on accumulative AE events for rock similar material under uniaxial compression has been established, and the damage model can better describe the damage characteristics of rock similar material.

(2) The pre-existing cracks do not have influence to the characteristics of specimen damage in different phases. All specimens have three stages during failure: initial damage stage, stable development of damage stage, and accelerated development of damage stage.

(3) With the width of pre-existing cracks vary from 3mm to 5mm, the influence of pre-existing cracks to damage variables increases, the damage of rock similar material increases gradually. The damage evolution equations and the damage constitutive equations for rock similar material have been established.

References

[1] Bahaaddini M, Sharrock G and Hebblewhite B K 2013 *Comput. Geotech.* **49** 206–25.
[2] Camones L A M, do Amaral Vargas Jr E and de Figueiredo R P 2013 *Eng. Geol.* **153** 80–94.
[3] Manouchehrian A, Sharifzadeh M, Marji M F and Gholamnejad J 2014 *Arch. Civ. Mech. Eng.* **14** 40–52.
[4] Heiple C R and Carpenter S H 1983 *Acoustic emission from dislocation motion* (New York: Gordon and Breach Science Publishers).
[5] Jansen D P, Carlson S R and Young R P 1993 *J. Geophy. Res.* **B12** 22231–43.
[6] Jin P J, Wang E Y, Liu X F, Huang N and Wang S H 2013 *Int. J. Min. Sci. Tech.* **23** 213–9.
[7] Zhao G M, Xie L X and Meng X R 2014 *Int. J. Min. Sci. Tech.* **24** 505–11.
[8] Zong Y J, Han L J, Wei J J and Wen S Y 2016 *Int. J. Min. Sci. Tech.* **26** 601–7.
[9] Liu B X, Huang J L and Wang Z Y 2009 *Chin. J. Rock Mech. Eng.* **28** 3234–8.
[10] Zhou Z W, Liu J F and Zou H 2016 *J. Yangtze Riv. Sci. Res. Inst.* **33** 63-8.
[11] Kachanov L M 1958 *Izv. AN SSSR Otd. Tekhn. Nauk.* **8** 26–31.
[12] Lemaiture J 1972 *Proc. ICM-1* (Kyoto) pp 540–9.

Research on the Helium Permeability of Graphene Oxide Membranes

G.H Ren[1], D.H Meng[2], R.X Yan[3], C.W Guo[4]

[1]Engineer, Beijing Institute of Spacecraft Environment Engineering, Beijing, China
[2]Senior Engineer, Beijing Institute of Spacecraft Environment Engineering, Beijing, China
[3]Professor, Beijing Institute of Spacecraft Environment Engineering, Beijing, China
[4]Engineer, Beijing Institute of Spacecraft Environment Engineering, Beijing, China

Abstract. In order to meet the sealing performance requirements, extra-high sensitive mass spectrometer leak detection method is developed. So the leak rate of 10^{-15}Pa•m^3 / s on the order of the standard leakage is demanded. Increasing the number of holes in the graphene on the artificial control, the leak rate of grapheme would be improved. Based on this idea, a certain defective graphene as a penetrating element, using in the ultra-sensitive leak detection can be made. In this paper the relationship among the permeability of helium and the pressure difference and the thickness of the GO membrane were studied. The permeation mechanism of GO membrane of the minimal leak rate was discussed, which provides a reference for the study of ultra-sensitive leak detection technology.

1. Introduction

The sealing performance is an important parameter to evaluate the quality of vacuum Components. With the continuous development of electronic technology, the size of vacuum electronic devices is getting smaller, life expectancy is longer, and reliability requirement is getting higher. Especially for long-life optical device, it is also crucial to maintain the internal gas composition constant for long periods of time. In order to meet the sealing performance requirements, extra-high sensitive mass spectrometer leak detection method is developed. The sensitivity of the leak detection method reached the order of 10^{-15}Pa·m^3 / s, so the leak rate of 10^{-15}Pa·m^3 / s on the order of the standard leakage is demanded [1]. In vacuum technology, the leak rate of quartz infiltration helium standard leakage hole commonly used generally can only reach the order of 10^{-10}Pa·m^3 / s. Lots of research work to carry out for a new standard leak, try to get the lower limit of 10^{-15}Pa·m^3 / s vacuum standard leaks in recent years. Hajime Yoshida using a stainless steel sintered filter with a pore size of less than 1 μm as the standard leak, get the standard leak whose rate is 10^{-8}Pa·M^3 / s ~ 10^{-6}Pa·m^3 / s [2]. Yongheng Zhao used standard porosity of 70 nm porous alumina production to make the standard leak. The leakage rate of standard leak is 8.55×10^{-9}Pa · m^3 / s [2].

Pure graphene, due to its perfect hexagonal structure, is not permeable to any gas molecule. Increasing the number of holes in the pure graphene on the artificial control, the leak rate of grapheme would be improved. Based on this idea, a certain defective graphene as a penetrating element, the ultra-sensitive leak detection standard leak can be made. The graphene oxide (GO) is a kind of separation membrane material with good engineering application prospect. Nair has studied the permeability of GO membrane with thickness of 1μm. Hang Li has studied the gas permeability of GO membranes with thickness of 1.8nm ~ 180nm. The results show that the permeability and the thickness of the membranes are exponentially related to the thickness and have good gas selectivity

[3-5]. Zhiping Xu uses molecular dynamics to simulate the process of passing gases through GO membranes [6].

Based on the previous research, this paper studies the relationship between the permeability and the pressure difference and the thickness of the GO membrane. The permeation mechanism of the minimal leakage rate GO membrane is discussed, that provides a reference for the study of ultra-sensitive leak detection technology.

2. Experiments

The modified Hummer method was used to prepare the graphene nanosilver dispersion, and the GO membrane was prepared by vacuum filtration. In this paper, a GO membrane with a thickness of 1.8nm ~ 2μm was obtained with a volume of 2.7 μl to 3000 μl and a 0.2% of the mass percent graphene solution. The based material of the GO membrane was made of an anodic alumina membrane (AAO) having a pore size of 0.2 μm produced by Whatman Corporation, having an outer diameter of 2.5 cm and an effective alumina diameter of 2.0 cm. In order to prevent damage to GO on the seal in the test experiment, a layer of aluminum foil is covered on the GO and the final effective permeate area is about 1 cm in diameter. The actual situation is shown in figure 1.

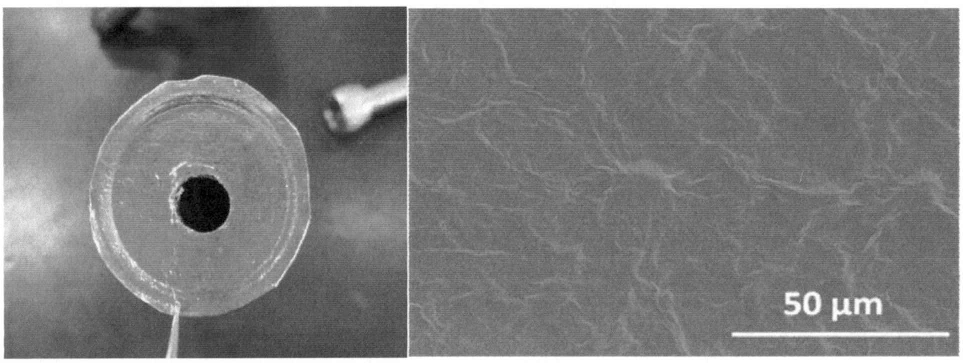

a) Picture of GO b) SEM view of GO membrane

Figure 1. Picture of GO membrane.

The principle of helium permeability measurement of GO membrane by helium mass spectrometer leak detector is shown in figure 2. The inflatable console provides high-purity helium gas to the upstream of the GO membrane, and the helium flow through the membrane downstream of the GO membrane is leaked by the helium mass spectrometer. The calibration of the helium mass spectrometer leak detection system is carried out using standard leaks to avoid errors in the signal drift of the helium mass spectrometer leak detector. Based on this system, the effects of pressure difference on both sides of the membrane and the effect of membrane thickness on the infiltration rate of GO were carried out.

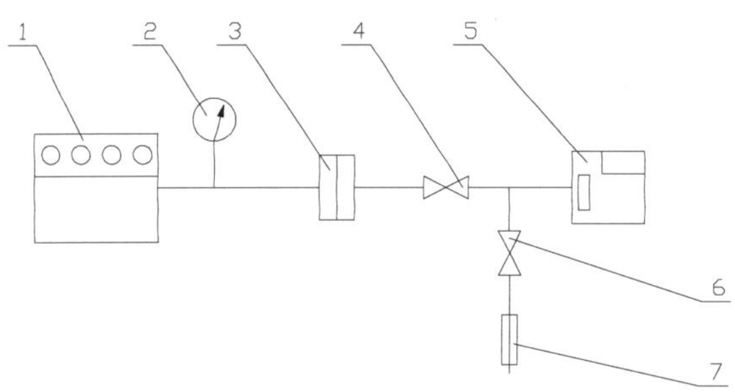

1-Inflatable console, 2-pressure gauge, 3-membrane permeation module, 4-leak detecting valve, 5-helium mass spectrometer leak detector, 6-calibration valve, 7-standard leak

Figure 2. Schematic diagram of helium permeability test system.

3. Experimental results and analysis

3.1 Relationship between the infiltration helium rate and the pressure difference on both sides of the membrane

Using the experimental methods shown in figure 2, the relationship between the infiltration rate of the GO membrane and the pressure difference on both sides of the film was tested. The top view of the GO membrane to be measured is shown in figure 1(b). During the experiment, the pressure downstream of the membrane has been kept below 10Pa, and the upstream pressure of the membrane has changed in the range of 100kPa (absolute pressure) and 500kPa (absolute pressure). The helium permeation rate of 11 kinds of thickness of the membranes in the range of 200nm ~ 2μm was measured, that is shown in figure 3. The experimental results show that the flow conductance of GO membranes with thickness of 2μm reached $10^{-13}m^3/s$. For a variety of GO membranes, the helium permeability is proportional to the pressure difference between the two sides of the film. That is to say, as the pressure difference on both sides of the film changed, the flow conductance of the film remains stable. The experimental results show that the flow of helium in the membrane is molecular flowing in the pressure range of 100 kPa ~ 500 kPa for the GO membrane with thickness of 200 nm ~ 2μm.

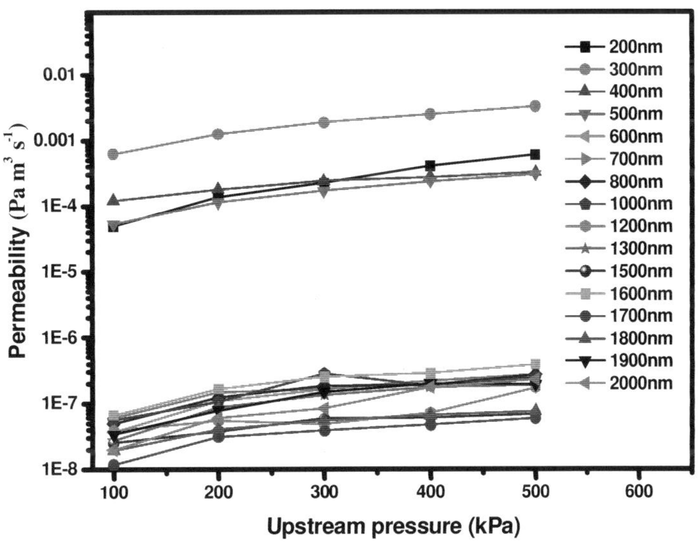

Figure 3. Relationship between flow conductance of GO membrane and pressure.

All experiments in this paper were carried out at room temperature of 20 °C, and the free path of any gas molecule at 20°C was calculated by the following equation (1) [7].

$$\bar{\lambda}=\frac{9.108\times10^{-22}}{p\sigma^2} \quad (1)$$

where, $\bar{\lambda}$ - mean free-path of gaseous molecules, m; p -gas pressure, Pa; σ^2 -gas molecular diameter, m.

From the above formula can be obtained, helium molecular at a pressure of 1atm molecular average free path is 192nm at 20 ℃. The GO membrane is made of monolithic layer GO sheets. The gas molecules pass through the gap among monomolecular layers. The GO membrane samples used in this paper are measured by XRD, and the interlayer spacing can be calculated by the equation 2 [7].

$$2d\sin\theta=n\lambda \quad (2)$$

where, λ-wavelength, nm; d- the interlayer spacing, nm; θ-angle. And the interlayer spacing is 0.88nm, as shown in figure 4. The average molecular free range of helium is much larger than that of the GO, and the helium flow through the GO membrane always keeps the molecular flow state. As a result, the flow conductance of a thin membrane remains constant for helium under different pressure differences. In the case of constant temperature, the mean free path of a gas is inversely proportional to the pressure. It is calculated that the mean free path of the molecule is 1.92nm when the helium pressure reaches 10MPa at 20 ℃. It can be deduced that for the GO membrane, the permeation of helium in the film can be maintained in the molecular flow state when the upstream pressure of the film is in the range of 0-10MPa.

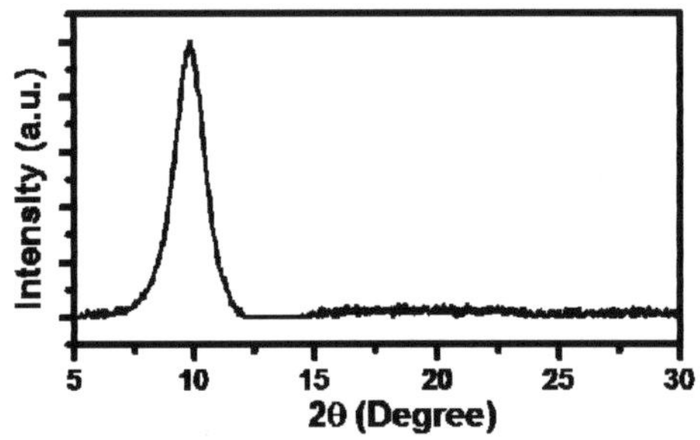

Figure 4. Interlayer spacing of the GO samples was tested by XRD.

3.2 Relationship between helium permeability and membrane thickness

As for the above experimental data, the relationship between the helium permeability and the thickness of the GO membrane can be obtained, and the thicknesses of the GO membrane were measured by SEM method. The cutaway view is shown in figure 5. Under the pressure difference of 100kPa ~ 500kPa, the relationship between the helium permeability and the thickness of the membrane is shown in figure 6. The results of more than 100 times experiments shown that, when the pressure on both sides of GO membrane remains unchanged, the permeability of the helium is exponentially related to the thickness of the membranes, which is consistent with the results reported in Reference [7]. In the vacuum leak detection technology, based on the characteristics of the relationship between the permeability and the thickness of the GO membrane, the change of the leak rate can be achieved by the small change of the thickness, so that the wide range permeability of standard leakage could be obtained.

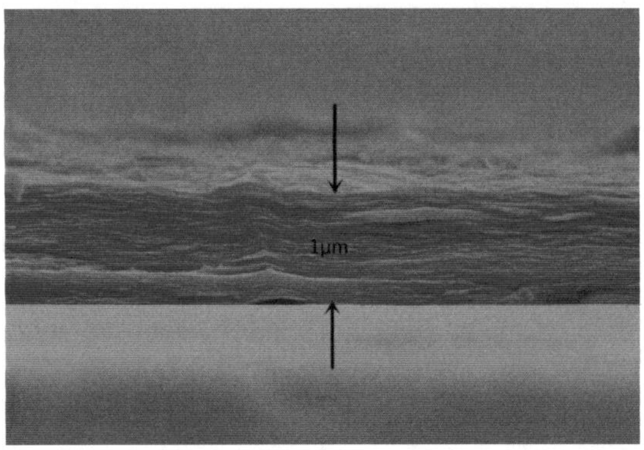

Figure 5. The sectional view of 1μm thickness GO membrane.

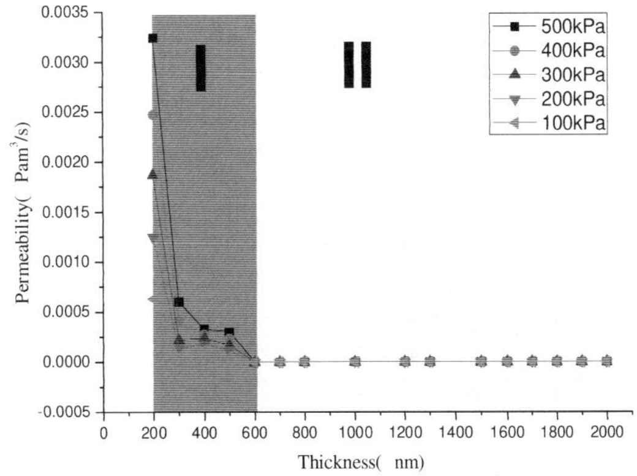

(a) Relationship between helium permeability and thickness of GO membranes

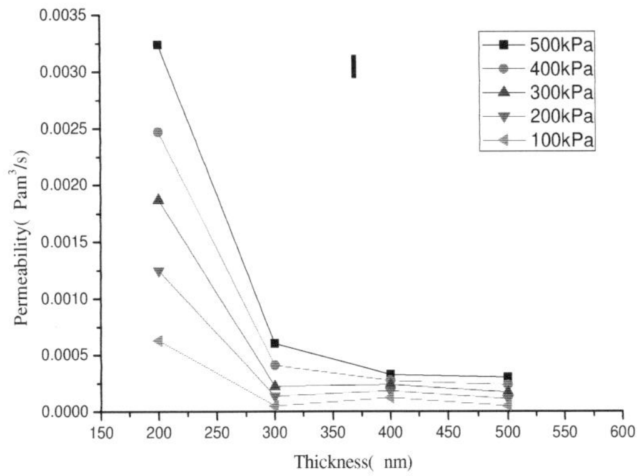

(b) The helium permeability of 200nm-500nm of GO membranes

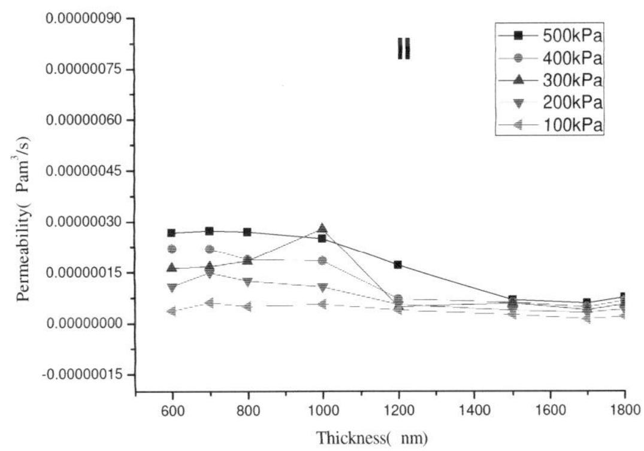

(c) The helium permeability of 200nm-500nm of GO membranes

Figure 6. Relationship between helium permeability and thickness of GO membranes.

4. Conclusion

In this paper, an improved Hummer method was used to prepare GO membranes with various thicknesses, which range from 200nm to 2μm. The relationship between the permeation rate and the pressure difference and thickness were studied. When the upstream pressure of GO membranes is lower than 500kPa, the permeation process of helium in the membranes belongs to the molecular flow state, and the permeability of the membranes to helium varies exponentially with the thickness. The permeability of the helium is exponentially related to the thickness of the membranes. These characteristics provide a suggestion that GO membranes could be used to prepare Permeable helium-type standard leak with a small leak rate or a large leak rate.

Reference

[1] Chen P, Chen X, Jin Q J and Cha L Z, 2002 *Presented at AVS 49th International Symposium,* Nov. 4-8, 2002, Denver, Colorado, USA.
[2] Hajime Y, Kenta A, Masahiro H and Hitoshi A 2012 *Vacuum* **86** 838-42.
[3] Nair R R, Wu H A, Jayaram P N, Grigorieva I V and Geim A K, 2012 *Science* **335** 442–4.
[4] Song Z N, Zhang X J, et al. 2013 *Science* **342** 95-98.
[5] Kim H W, Yoon h W, Yoon S M, Yoo B M, et al., 2013 *Science* **342** 91-5.
[6] Jiao S P and Xu Z P, 2015 *Appl. Mater. Interf.* **7** 9052-9.
[7] Lafferty J M 1998 *Foundations of vacuum science and technology*. A wiley-inter science publication, 10-110.

Acknowledgements

This work was supported by the National Natural Science Foundation of China through Grant Number U1537109 and National Defence Technique Foundation of China through Grant Number JSJL2015203B019.

IOP Conf. Series: Materials Science and Engineering **269** (2017) 012013 doi:10.1088/1757-899X/269/1/012013

Design of a Soft Robot with Multiple Motion Patterns Using Soft Pneumatic Actuators

Yu Miao[1], Wei Dong[1] and Zhijiang Du[1]

[1]State Key Laboratory of Robotics and System, Harbin Institute of Technology, China
E-mail: dlmiaoyu@outlook.com

Abstract. Soft robots are made of soft materials and have good flexibility and infinite degrees of freedom in theory. These properties enable soft robots to work in narrow space and adapt to external environment. In this paper, a 2-DOF soft pneumatic actuator is introduced, with two chambers symmetrically distributed on both sides and a jamming cylinder along the axis. Fibers are used to constrain the expansion of the soft actuator. Experiments are carried out to test the performance of the soft actuator, including bending and elongation characteristics. A soft robot is designed and fabricated by connecting four soft pneumatic actuators to a 3D-printed board. The soft robotic system is then established. The pneumatic circuit is built by pumps and solenoid valves. The control system is based on the control board Arduino Mega 2560. Relay modules are used to control valves and pressure sensors are used to measure pressure in the pneumatic circuit. Experiments are conducted to test the performance of the proposed soft robot.

1. Introduction

Robots have been widely used in the fields of industrial manufacturing, agricultural production and medical service. Traditional robots normally consist of rigid modules and can assure the high precision of mechanical motion. However, these robots have limited movement ability in narrow space, which has limited their application in some situations. By comparison, soft robots have low stiffness and high flexibility. They can change their shapes to adapt to the external environment. Due to these advantages, soft robots have wide prospect in application.

Many researchers have paid attention to soft robotics and kinds of soft robots have been developed. Based on the bionics principle, Michael Wehner et al. designed a soft robot, Octobot [1]. It is the first robot with a fully soft body and can move its arms like an octopus. Hernando Leon-Rodriguez et al. proposed a soft micro robot based on the bio-inspired motion of the Amoeba [2]. Ferrofluid material is used to fabricate the robot and external electromagnetic coils are used to actuate and control the soft robot. Hongliang Ren et al. presented a human-compliant body-attached soft robot using four parallel pneumatic soft actuators [3]. The purpose of the soft robotic system is to steer the ultrasound probe cooperatively and replace the manual operation. Nicholas W. Bartlett et al. reported a combustion powered soft robot using 3D-printed techniques [4]. Actuated by the combustion of oxygen and butane, the soft robot is able to jump.

Currently, most soft robots are proposed from the aspect of bionics, e.g. starfish-like soft robots, worm-like soft robots and soft robotic octopus arms [5-7]. New materials, e.g. shape memory alloy (SMA), shape memory polymer (SMP), ionic polymer metal composite (IPMC) and dielectric elastomer (DE), have been used to develop soft robots. Although there are still many challenges to tackle with before wide application of soft robots, new manufacturing technologies and materials have provided fresh ideas for the development of soft robotics.

In this paper, a 2-DOF soft pneumatic actuator is investigated. Based on the proposed soft actuator, a soft robotic system is designed, including actuation system and control system. Experiments are carried out to test the performance of the soft actuator and the soft robot.

2. Design of soft robot

2.1. Soft actuator

The soft actuator includes four parts, a flexible body, a jamming cylinder, two chambers and fibers. The flexible body is made of Ecoflex 0030, a kind of soft materials. The structure of the soft actuator is illustrated in figure 1. The flexible body has two chambers on opposite sides. When one chamber is pressured, the soft actuator will bend to the opposite side. When both two chambers are actuated, the soft actuator will extend along the axis. Along the axis is a jamming cylinder, a membrane with coarse coffee inside. When it is vacuumed, the stiffness of the cylinder will increase. Kevlar fibers on the outer surface of the flexible body can limit inflation in case that balloon effect occurs.

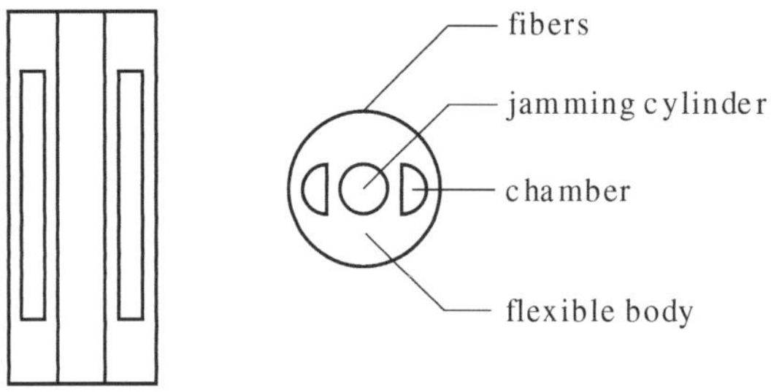

Figure 1. Structure of soft actuator.

2.2. Soft robot

The soft robot is shown in figure 2. The proposed soft robot has four legs. Four soft actuators are connected to a 3D-printed board. The distance between two adjacent soft actuators is 70mm. By controlling the bending state of four actuators, the soft robot can complete different tasks, e.g. grasping and walking.

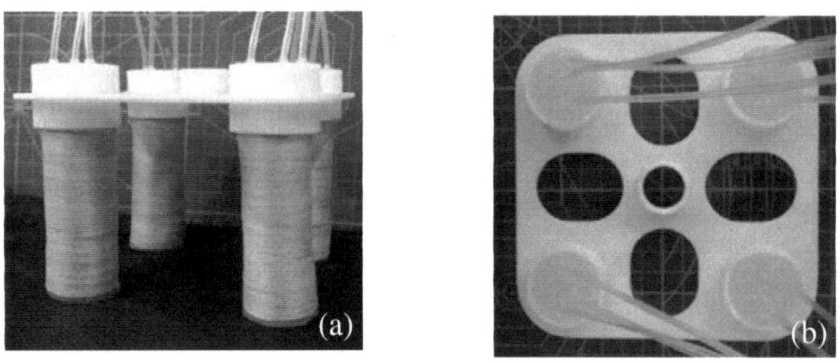

Figure 2. Soft robot.

3. Design of soft robotic system

As shown in figure 3, the soft robot system includes two parts, actuation section and control section. The actuation section contains the soft robot, solenoid valves, pumps, regulation boards and a DC power. The control section is designed based on Arduino Mega 2560, including host PC, Arduino board, relay modules and pressure sensors.

Each air channel is connected to the air pump or the vacuum pump via a 3/2 solenoid valve and a 2/2 solenoid valve. Regulation boards are used to control gas flow. By controlling solenoid valves, each channel can be actuated, released or be kept in the current state. Therefore, the motion of the soft robot can be controlled.

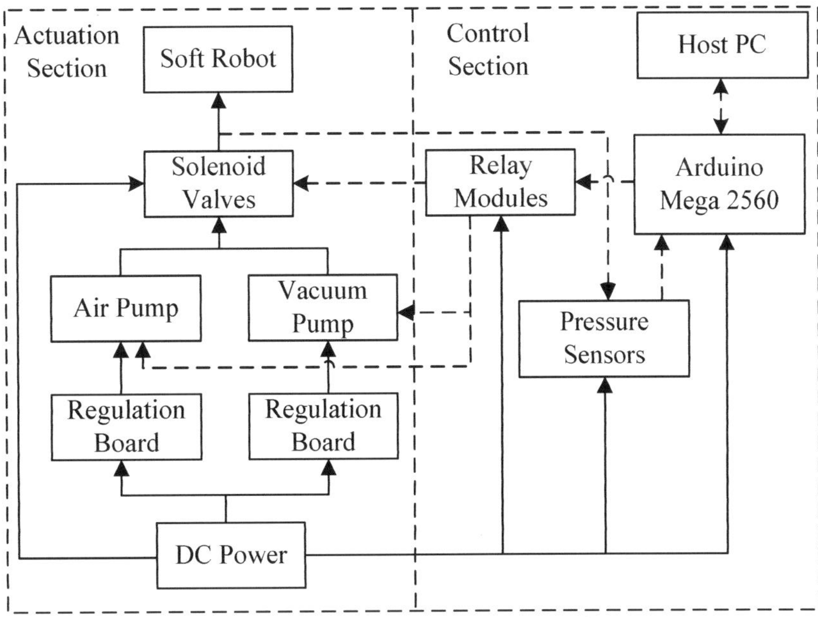

Figure 3. Schematic diagram of soft robotic system.

The user interface on host PC is developed by Matlab GUI and Matlab Arduino Support Package is used to establish communication between host PC and Arduino board. Through user interface, users can send commands to Arduino board from host PC. Thus, relay modules can be controlled to control the motion of the soft robot. Arduino board also collects data from pressure sensors. After processing, pressure values of air channels are displayed on the user interface. Figure 4 shows the soft robot experiment platform.

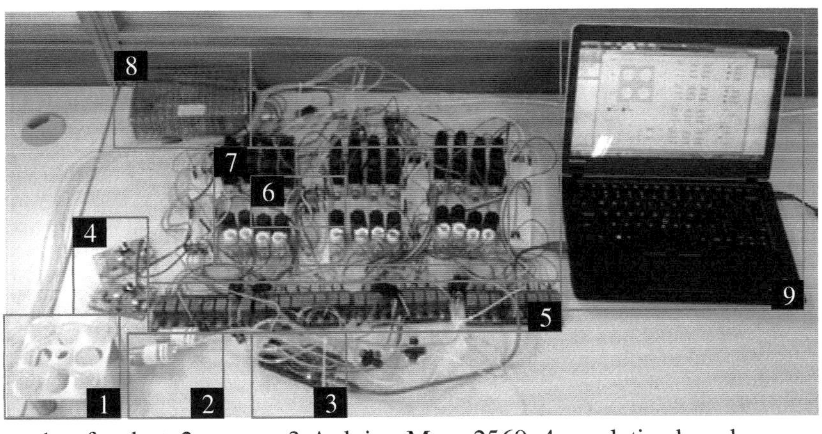

1-soft robot; 2-pumps; 3-Arduino Mega 2560; 4-regulation boards; 5-relay modules; 6-pressure sensors; 7-solenoid valves; 8-DC power; 9-host PC

Figure 4. Soft robot experiment platform.

IOP Conf. Series: Materials Science and Engineering **269** (2017) 012013 doi:10.1088/1757-899X/269/1/012013

4. Experiments

4.1. Characteristics of soft actuator

The characteristics of the soft actuator have a direct impact on the performance of the soft robot. The motion of a soft actuator consists of bending and elongation, so it is important to understand these characteristics.

The soft actuator will bend or extend along the axis when it is actuated. As figure 5 shows, the bending angle of the soft actuator is defined as the angle between the central axis of the end surface and the central axis of the fixed end. When one chamber is actuated, the soft actuator will bend. The relationship between bending angle and air pressure is shown in figure 6. The bending angle will increase with the increase of air pressure. When the jamming cylinder is jammed, the stiffness of the soft actuator will increase. And the bending angle is smaller than that of unjammed status. The bending angle becomes more sensitive to pressure in high pressure region since the wall thickness of the flexible body decreases as the pressure increases.

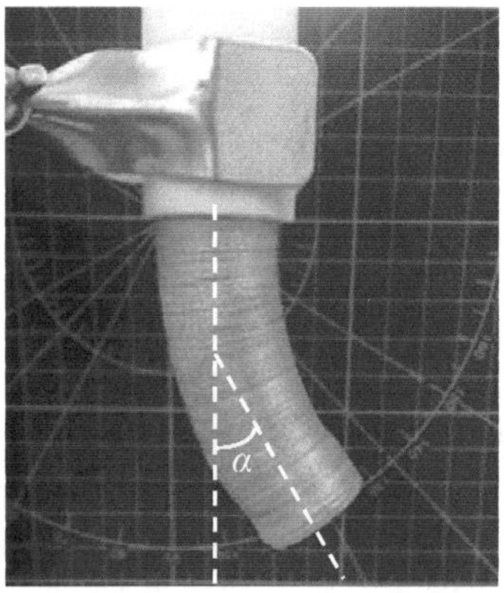

Figure 5. Definition of bending degree.

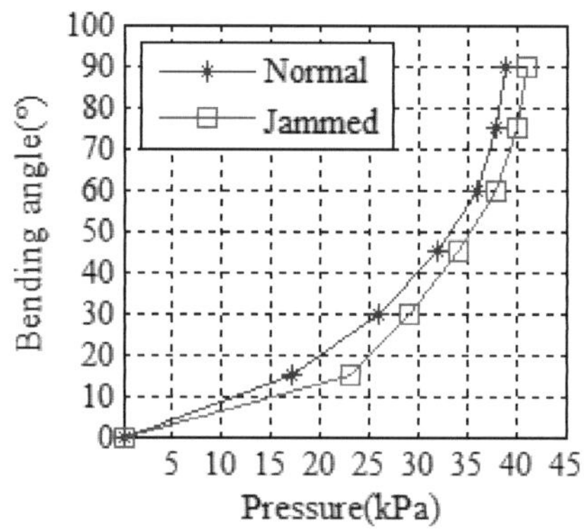

Figure 6. Relationship between bending degree and pressure.

When two chambers are actuated simultaneously, the soft actuator will be elongated in the central axis direction. Figure 7 shows the states of the actuator before and after it is actuated. The relationship between elongation and pressure in two chambers is shown in figure 8. The extension length increases with the increase of pressure. The elongation becomes more sensitive to pressure in high pressure region. This is because the cross-sectional area becomes smaller. When the jamming cylinder is vacuumed, the extension length is smaller than that of normal status under the same pressure. This indicates that jammed status can increase the stiffness of the soft actuator.

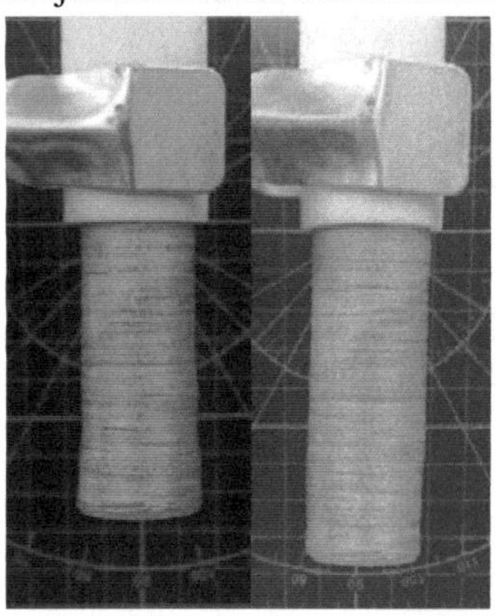

Figure 7. Elongation status.

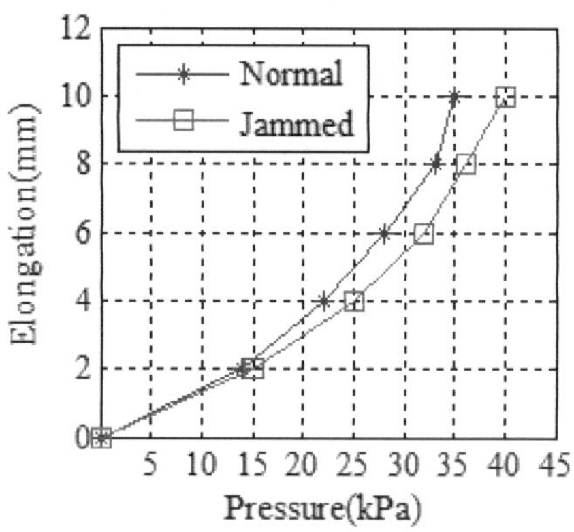

Figure 8. Relationship between elongation and pressure.

IOP Conf. Series: Materials Science and Engineering **269** (2017) 012013 doi:10.1088/1757-899X/269/1/012013

4.2. Grasping experiment

By controlling the states of chambers in each actuator, the proposed soft robot can work as a gripper. As show in figure 9(a), when four chambers along the outside are actuated, four actuators will bend to the inside and an object can be grasped from its external surface. Instead, an object can be grabbed from the inner surface when four chambers on the inside are actuated in the way shown in figure 9(b).

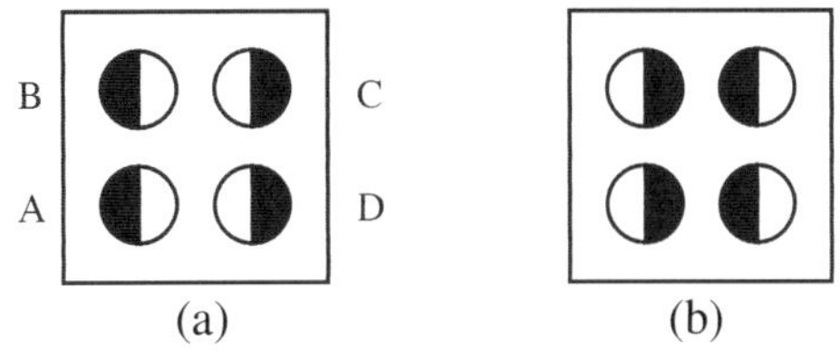

Figure 9. Actuation method.

Objects of different shapes and sizes are used to test the performance of the gripper. Experimental results are shown in figure 10. Grasping modes can be classified in to three types. As shown in figure 10(a) and (b), the tip of each actuator will be used to fetch small objects. For medium size objects, both the side surface and the tip of the actuator will contact the object, as shown in figure 10(c) and (d). Figure 10(e) and (f) illustrate the way to grasp objects with large external dimensions. The gripper can pick up these objects by contacting with their inner surfaces.

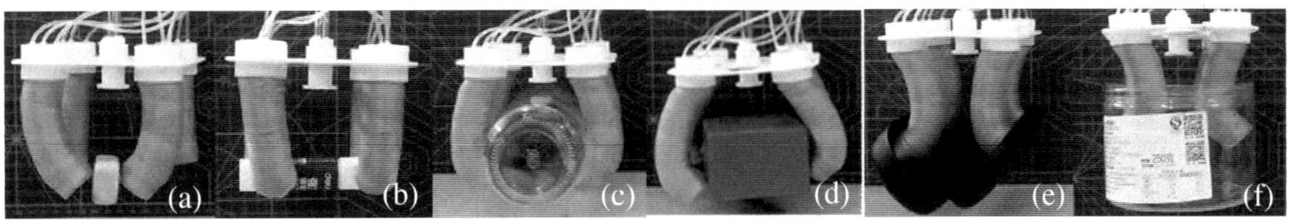

Figure 10. Grasping experiment.

4.3. Walking experiment

Soft actuators are the legs of the soft robot. By controlling the action order and the status of actuators, the four-legged the soft robot can walk on a horizontal surface. As shown in figure 11, a walking circle contains following steps:

- One chamber in leg A and one chamber in leg C are actuated. Leg A and leg C will bend, and the soft robot will move forward.
- All the chambers in leg B and leg D are actuated and chambers in leg A and leg C are released. Leg B and leg D will extend, and leg A and leg C will recover to normal state.
- Leg B and Leg D are released and will recover to normal state.
- One chamber in leg B and one chamber in leg D are actuated. Leg B and leg D will bend, and the soft robot will move forward.
- All the chambers in leg A and leg C are actuated and chambers in leg B and leg D are released. Leg A and leg C will extend, and leg B and leg D will recover to normal state.
- Leg A and Leg C are released and will recover to normal state.

Thus, the soft robot completes a walking circle.

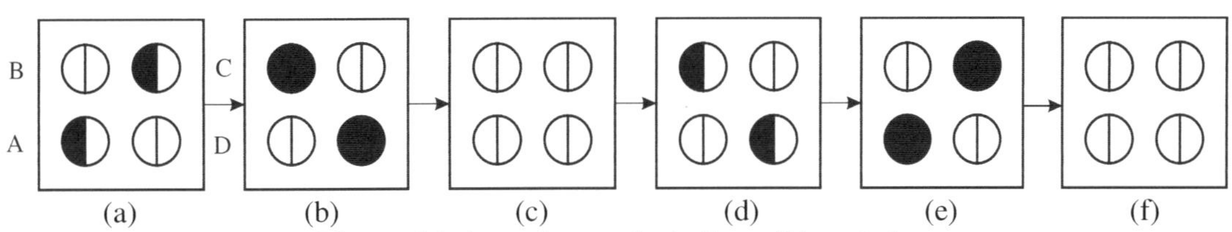

Figure 11. Actuation method of a walking circle.

When four soft pneumatic actuators are actuated in the way shown in figure 11, the soft robot can walk on a horizontal surface. Figure 12 shows a waking circle of the soft robot. The experimental results show that soft actuators have good controllability. The robot can complete preconcerted tasks and walk as expected. The step length of the soft robot is about 35mm.

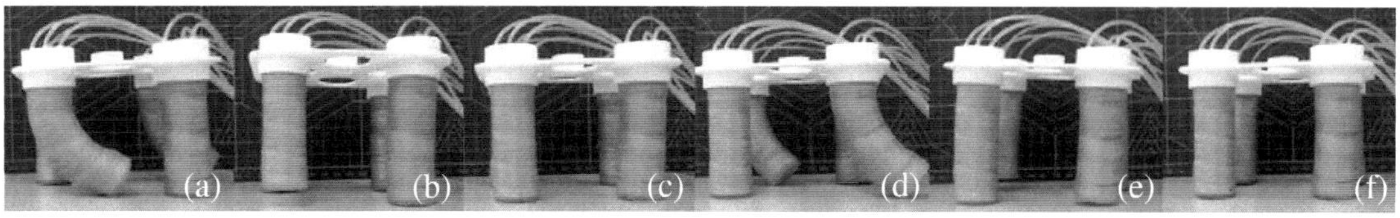

Figure 12. Walking experiment.

5. Conclusion

In this paper, a soft pneumatic actuator with two chambers and a jamming cylinder is introduced. A soft robotic system is developed based on this soft actuator and Arduino board. The soft robot is made of four soft actuators. Four soft actuators are distributed at four corners of a 3D-printed board. Experiments show that the soft pneumatic actuator has good flexibility and controllability. Elongation and bending degree of the soft actuator increase with the pressure and are more sensitive to the change of pressure in high pressure region. When the jamming cylinder along the axis is vacuumed, the stiffness of the soft actuator will increase. By controlling the pressure in each chamber, the proposed soft robot can grasp objects and walk. It can grasp objects of different sizes and its walking step length is about 35mm.

References

[1] Wehner M, Truby R L, Fitzgerald D J, Mosadegh B, Whitesides G M, Lewis J A and Wood R J 2016 *Nature* **536(7617)** 451.

[2] Leon-Rodriguez H, Le V H, Ko S Y and Park J O 2015 *Int. Conf. Control, Autom. Systems (Busan)* (New York: IEEE) pp 1833–8.

[3] Ren H, Gu X and Tan K L 2016 *Int. Conf. Comput. Sup. Cooperat. Work in Design (Nanchang).*

[4] Bartlett N W, Tolley M T, Overvelde J T, Weaver J C, Mosadegh B, Bertoldi K, Whitesides G M, Wood R J 2015 *Science* **349(6244)** 161.

[5] Mao S, Dong E, Zhang S, Yang J, Jin H, Xu M, Zhang S, Yang J and Low K H 2014 *J. Bionic Eng.* **11(3)** 400.

[6] Horchler A D, Kandhari A, Daltorio K A, Moses K C, Andersen K B, Bunnelle H, Kershaw J, Tavel W H, Bachmann R J, Chiel H J and Quinn R D 2015 *Lect. Notes. Comput. Sci.* **9222** 26.

[7] Guglielmino E, Zullo L, Cianchetti M, Follador M, Branson D and Caldwell D G 2012 *Int. Conf. Robot. Autom. (Saint Paul)* (New York: IEEE) **20** 5277-82.

Acknowledgments

This research was financially supported by the State Key Laboratory of Robotics and System, Harbin Institute of Technology under Grant No. SKLRS201501A03.

Comparative Experimental Study on Ionic Polymer Mental Composite based on Nafion and Aquivion Membrane as Actuators

B Luo[1], Z Chen[2]

[1]State key laboratory of mechanical structural strength and vibration, Xi'an Jiaotong University, Xi'an, 710049, China;
[2]School of Mechanical Engineering, Shaoyang University, Shaoyang, 422200, China

Abstract. Most ionic polymer mental composites employ Nafion as the polymer matrix, Aquivion can also manufactured as ionic polymer mental composite while research was little. This paper researched on two kinds of ionic polymer mental composite based on Aquivion and Nafion matrix with palladium electrode called Aquivion-IPMC and Nafion-IPMC. The samples were fabricated by the same preparation process. The current and deformation responses of the samples were measured at voltage to characterize the mechano-electrical properties. The experimental observations revealed that shorter flexible side chains in Aquivion-IPMC provide a larger force than Nafion-IPMC, while the displacement properties were similar in two different samples. The results also showed that Aquivion membrane can also replace Nafion to reproduce IPMC application in soft robots, MEMS, and so on.

1. Introduction

An ionic polymer-metal composite (IPMCs) consists of a thin perfluorinated ionomer membrane with electrodes plated on both faces. When the membrane was hydrated, a device of this type will undergo a large bending motion as a low level electric field was applied across the two electrodes [1-3]. IPMC actuators can be readily miniaturized and typically have a low density and high mechanical flexibility, such that they showed promise for a wide variety of applications, ranging from MEMS actuators to soft robots [4-6]. One of the most common ionic polymers in use today is DuPont's Nafion developed in the 1960s and commercialized in the 1970s, since then Nafion is a Teflon TM -based polymer with pendant sulfonic acid side groups. Aquivion was known in literature as short side chain ionomer in comparison to Nafion that was indicated as long side chain ionomer developed by Solvay Solexis [7]. These perfluorosulfonate ionomers consist of a polytetrafluoroethylene (PTFE) backbone and double ether perfluoro side chains terminating in a sulfonic acid group as illustrated in figure 1. The short flexible side chain in Aquivion provides a lower equivalent (980g/mol) mass than Nafion (1100g/mol). Although in the literature, most of ionic polymer-metal composite employed Nafion as the polymer matrix, Aquivion were employed to manufacture IPMC can also be feasible, but the research was little. In this study, ionic polymer mental composite based on Nafion and Aquivion membrane with palladium-electroded were manufactured to conduct experiment called as Nafion-IPMC and Aquivion-IPMC. We focus on the different mechanical and electrical performance between Nafion-IPMC and Aquivion-IPMC under same preparation process. A series of SEM pictures on the electrodes surface and cross-section of theses samples have been captured, and the mechano-electrical responses of IPMC under DC voltage have been measured. Through experimental comparison, it is investigated the ways that how the side chain-ionomer influenced the mechanical and electrical parameter and corresponding mechano-electrical properties. The results of this research help to explore whether the Aquivion can replace Nafion to reproduce IPMC because of Aquivion' price lower than Nafion [7].

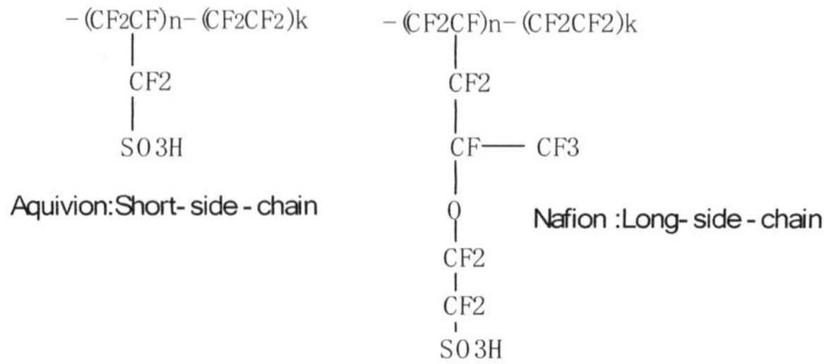

Figure 1. The molecular structure of short side chain Aquivion and long side chain Nafion.

Table 1. Comparison of Aquivion and Nafion membrane parameters [8-10].

Membrane	Equivalent mass/g/eq	Density /g/cm^3	Thickness /um	Elastic modulus/MPa	Ionic conductivity/ ms/.cm	Water content (%)
Nafion117	1100	1.97	183	249	60-140	38%
Aquivion9815S	980	1.93	150	290	>160	20%

2. Experimental

2.1. Materials preparation

Nafion-IPMC and Aquivion-IPMC was fabricated through an effective process to manufacture palladium-electroded IPMC by our lab, mainly by impregnation-reduction plating (IRP) and autocatalytic plating (ACP) [3]. In our experiment, Nafion 117 with a thickness of 180 μm, purchased from Dupont in USA, and Aquivion (EW98015S) with a thickness of 150 μm purchased from Solvay Solexis in Japan. The noble metal Pd, which has similar physical and chemical properties to Pt and is less expensive (approximately one-third cheaper), was selected as the electrode material. Na ion was exchanged into the IPMC as actuating ions, while water acted as solvent. Strip-shaped IPMCs with the dimensions of 35mm×5mm were tested and analyzed. The thickness of Aquivion-IPMC was about 0.20mm, wihle the Nafion-IPMC was about 0.22mm due to different thickness of membrane.

2.2. Characterization.

2.2.1 Morphology observation. The surface and cross-section morphology of samples were examined by means of SEM (TESCAN-VEGA \XMU VG3210677). Scanning electron microscopy (SEM) was performed at an accelerating voltage of 20.0kV. All sample cross-sections were obtained by low-temperature cracking, being placed in liquid nitrogen for 5min and then broken into pieces.

2.2.2. Surface resistance. The sheet resistances of the samples were measured by a low-resistivity meter with a measurement range of 10^{-2}–10^{6} Ω (the Loresta-EP handheld unit with a four-pin probe). Membranes were cut into discs with a 10 mm radius size and immersed in de-ionized (DI) water for 2h prior to testing. The same samples from the same batch were tested five times and the average value was recorded. All operations were carried out at room temperature and room humidity.

2.2.3. Bending stiffness measures. The bending stiffness of fully hydrated IPMC samples with the dimensions of 5mm (width)×35 mm (free length) were estimated using the free oscillation attenuation method which were described in [3].

2.2.4. Water content measures. The two IPMC samples were soaked in DI water for 12 hours to absorb water completely. After being picked up from the DI water and cleaning the surface water, the sample were put on an electric balance. The transient mass was m_{wet} recorded until it reached the equilibrium state. Then IPMC samples were dried at 80 °C in an oven for 12 hours to sufficiently remove the inside water. After cooling its were put on the electric balance to measure the transient mass m_{dry}, the water content can be calculated by the following formula:

$$w = \frac{m_{wet} - m_{dry}}{m_{dry}} \tag{1}$$

2.2.5. Electrochemical properties of IPMCs. Cyclic voltammetry (CV) and electrochemical impedance spectroscopy (EIS) were employed to evaluate the interfacial surface of the IPMC electrodes in terms of capacitance. In the (CV) and EIS measurements for the IPMC specimens, the electrochemical VersaStudio (VersaSTAT 3, Princeton, USA) was used to obtain the CV curves and Nyquist plots. The CV measurements were carried out with potential range between −1.0 and 1.0 V at a scan rate of 50 mV s $^{-1}$, and the EIS measurements were carried out over the frequency range from 100 kHz to 1Hz with applied current of 100 mA. All the measurements were performed as [11]. The charging-discharging capacitance was determined from the measured cyclic voltammograms according to the relation [12]:

$$C = \frac{|I^{+}| + |I^{-}|}{2dV/dt} \tag{2}$$

where I + and I - are the polarization currents at 0 V, and dV/dt is the potential scan rate.

2.2.6. Mechano-electrical properties of test. The measurement setup for mechano-electrical properties testing as shown in figure 2. IPMC samples were clamped between gold electrodes to form a cantilever. DC analog signal generated by PC was amplified by a power amplifier HM8143 to stimulate the IPMC sample. The deformation at 25 mm from the clamp was measured with a laser displacement sensor, the current was measured by a self-made circuit board. The displacement and current signals were recorded in a PC through an ADLINK DAQ2214 data acquisition card.

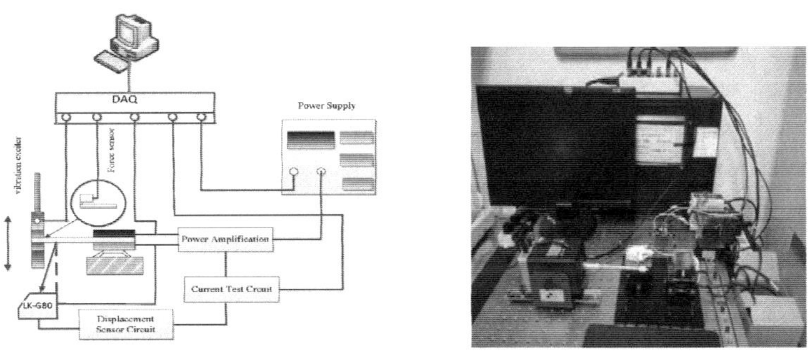

Figure 2. Measurement setup for mechano-electrical and sensing properties testing.

3. Results and discussion

3.1. Morphology of surface electrode and cross-section

Form figure 3 (a), we can see that Nafion-IPMC appears as a flat, smooth and compact surface electrode with few cracks, the Pd particle distribute on the surface uniformly to form good electrical conductivity. Most of Pd ions were reduced on the surface, cross-section morphologies of sample shows that the thickness of upper and lower electrode layers is about 10um respectively. It's worth noting that a small number of particles penetrate into the substrate to increase the pseudo-capacitor of

Nafion-IPMC. Form figure 4(a), we also know that Aquivion-IPMC with a compact surface electrode due to autocatalytic plating. The cross-section morphologies of Aquivion-IPMC in figure 3(b) gives an evidence of fewer particles penetrate into the substrate than Nafion-IPMC.

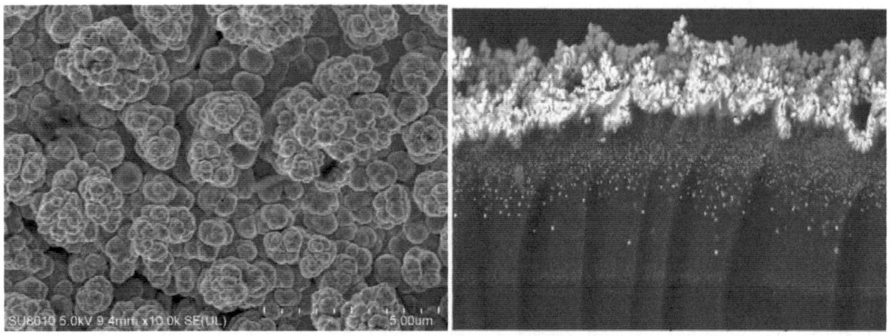

Figure 3. SEM image of Nafion-IPMC:
(left) the electrode surface morphology and cross-section morphology (right).

Figure 4. SEM image of Aquivion-IPMC:
(left) the electrode surface morphology and cross-section morphology (right).

3.2. Electro-mechanical parameters
Previously, we have known that some key factors such as surface resistivity, elastic modulus, water content, and specific capacitance affect the mechano-electrical properties [13]. These parameters were measured shown in table 2. In this work, we focused on the effect of different matrix membranes on the physical parameters and investigated the reasons for these differences. Nafion-IPMC and Aquivion-IPMC both have low surface resistance, Nafion-IPMC has lower elastic modulus than Aquivion-IPMC because of the different form membrane and number of particles penetrates into membrane. The water content of Nafion-IPMC higher than the Aquivion-IPMC due to equivalent mass of the two was different. Specific capacitance measuring charge discharge capacity. Obviously, the specific capacitance of Nafion-IPMC higher than the Aquivion-IPMC. These parameters affect the driving performance will be discussed in the following.

Table 2. Electro-mechanical parameters of Nafion-IPMC and Aquivion-IPMC (in wet state).

Sample type	Surface resistivity (Ω/□)	Elastic modulus/MPa	Water content	Specific capacitance/mF/cm^2
Nafion-IPMC	0.268	275	23%	0.1684
Aquivion-IPMC	0.190	365	21%	0.0518

3.3. Electrochemical properties of IPMCs
Electricalchemical performances of the IPMC samples have bee investigated employing CV and EIS measurements. Figure (5) shows the CV curves of the IPMC samples measured with potential range between −1 and +1V and scan rate of 50mVs^{-1}. The results indicate that the charge/discharge proceeds capacitively for samples in this potential range. As can be seen, the current density of all the Nafion-IPMC was significantly higher than Aquivion-IPMC. The higher in current density also indicated

greater capacitive behavior and more efficient electrolyte ion transport of Nafion-IPMC, which would result in a better the electro-mechanical performances.

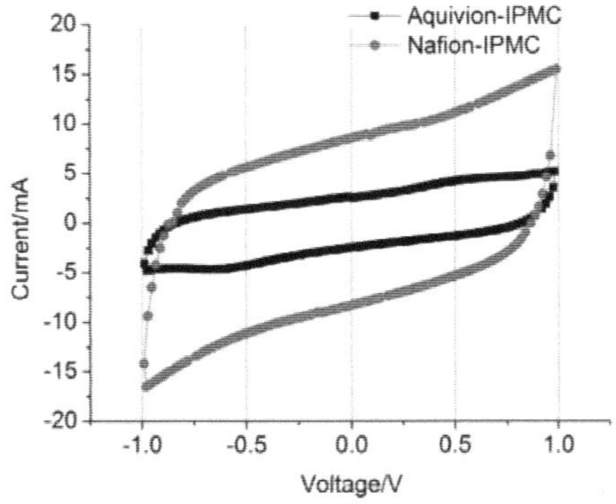

Figure 5. CV curves of Nafion-IPMC and Aquivion-IPMC.

The IPMC's equivalent circuit can be viewed as RC loop. The interlayer of IPMCs can be viewed as an ionconductive material and modeled by a capacitor (C) and a membrane resistor (Rm) in parallel [14]. The electrode can be modeled by two resistors (Rpd/2), The value of Rm, Rpd, C can be fitting out by (EC-Lab V9.24) calculated the Nyquist plots of the impedance spectra of the IPMC directly as shown in table 3. Form table 3, we known that the membrane resistance of Aquivion is larger than Nafion. Also the Surface resistance including interface resistance of Nafion-IPMC is lower than Aquivion-IPMC. Capacitance of the two type IPMCs show great difference consistent with the result was estimated by formula 2.

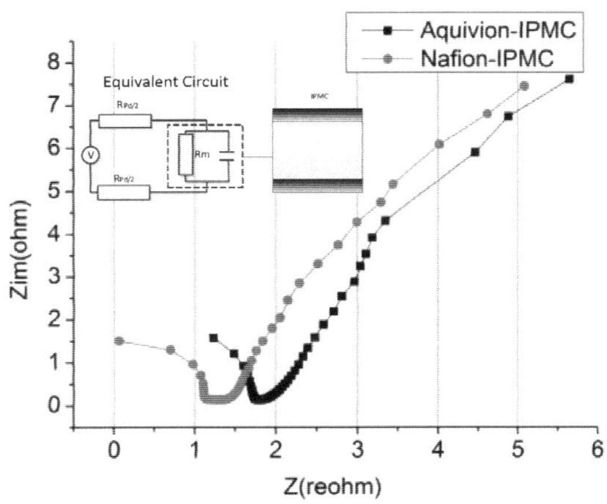

Figure 6. Nyquist plots of the impedance spectra of the IPMC samples.

Table 3. Equivalent electrical parameters corresponding to IPMCs.

Sample type	Membrane resistance/Rm	Surface resistance /Rpd	Capacitance/mF/cm^2
Nafion-IPMC	0.25	1.337	0.1835
Aquivion-IPMC	0.51	1.915	0.062

3.4. Mechano-electrical properties of IPMCs

IPMC's tip displacement with respect to time when working under 2V DC was shown in figure 7. Nafion-IPMC and Aquivion-IPMC's anode deformations were basically the same. While relaxation amplitude of two sample were different. Nafion-IPMC's relaxation amplitude larger than quivion-IPMC's, even more than the initial equilibrium. The anode deformations is usually ascribe to fast cations distribution under the electric field, relaxation deformation was caused by the reverse diffusion of free water [16]. The different relaxation amplitude correlates with the water content which has been proven in ref [17].

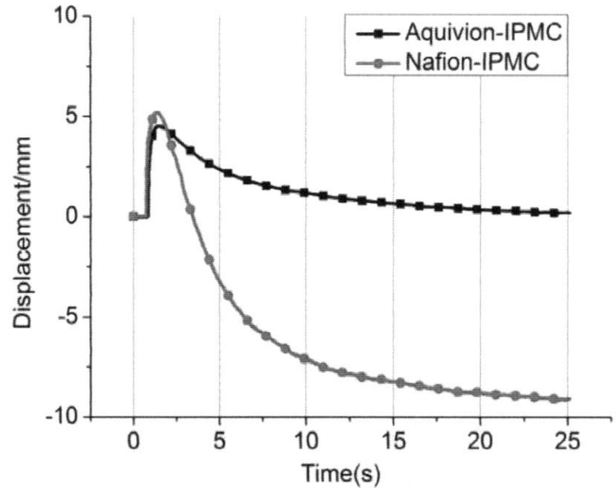

Figure 7. Displacement of IPMCs under DC 2 V.

IPMC's force responses of the two IPMC samples were plotted under 2V DC in figure 8. Obviously, Aquivion-IPMC generated a higher blocking force than Nafion-IPMC due to a higher elastic modulus.

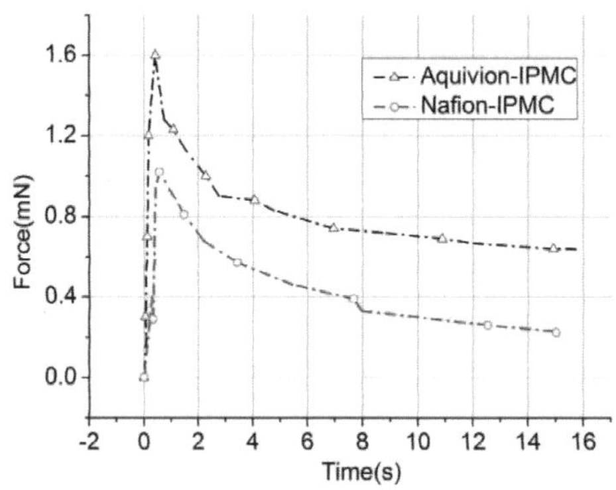

Figure 8. Blocking force of IPMCs under DC 2 V.

4. Conclusions and future work

In this paper, a series of experiments were performed with the aim of giving the comparison between the Nafion-IPMC and Aquivion-IPMC in mechano-electrical properties under the same condition. Firstly, the Nafion/Aquivion membranes and corresponding IPMCs were prepared and characterized. Then, electrochemical properties were measured. In view of the above results and discussions, the following conclusions can be drawn.

(1) Aquivion-IPMC's displacement deformation lower than Nafion-IPMC under 2V voltage because of Aquivion-IPMC own higher membrane resistance than Nafion and a lower capacitance compared to Nafion-IPMC

(2) Output force of Aquivion-IPMC larger than Nafion-IPMC's because of the Aquivion-IPMC's elastic modulus higher than Nafion-IPMC's.

As a result, The IPMC based on Aquivion can used as actuator identical with IPMC based on Nafion in aerospace, medical, robots,and so on.

5. References

[1] Shahinpoor M and Kim K J 2005 *Smart Mater. Struct.* **14** 197.
[2] Chen Z and Tan X B, 2010 *Sens. Actuators,* A **157** 246.
[3] Chang L, Chen H, Zhu Z and Li B 2012 *Smart Mater. Struct.* **21** 065018.
[4] Shahinpoor M 2003 *Electrochim. Acta* **48** 2343.
[5] Dogruer D, Tiwari R and Kim K J 2007 *Electroact. Polym. Actuat. Dev. (EAPAD)* 65241C.
[6] Punning A, Anton M, Kruusmaa M and Aabloo A 2004 *A biologically inspired raylike underwater robot with electroactive polymer pectoral fins.* International IEEE Conference on Mechatronics and Robotics 2004 (MechRob'04), Aachen, Germany, pp. 241-5.
[7] Lin J H, Liu Y, Zhang Q M 2011 *Polymer* **52** 540.
[8] Fischer R. 2010 *Properties of stretched 830EW Aquivion.* Vanderbilt University, 10.
[9] http://www.solvay.cn/zh/markets-and-products/featured-products/Aquivion.html
[10] Shahram Z 2005 *Microtech. Mems.* **74** 673.
[11] Palmre V, Pugal D and Leang K K 2013 *The effects of electrode surface morphology on the actuation performance of IPMC.* SPIE Smart Structures and Materials+Nondestructive Evaluation and Health Monitoring. International Society for Optics and Photonics, 86870W-86870W-10.
[12] Ru J, Wang Y J and Chen H L, et al. 2016 *Smart Mater. Struct* **25** 095006.
[13] Wang Y J, Zhu Z C and Chen H L, et al. 2014 *Smart Mater. Struct.* **23** 125015.
[14] Kanno R, Tadokoro S, Takamori T, et al. 1996 *Linear approximate dynamic model of ICPF (ionic conducting polymer gel film) actuator Proceedings of the 1996 IEEE International Conference on Robotics & Automation, Minneapolis, Minnesota,* 219-25.
[15] Nemat-Nasser S, and Wu Y 2003 *J. Appl. Phys.* **93** 5255.
[16] Zhu Z C, Chen H and Wang Y Q 2011 *EPL(Europhysics Letters)* **96** 27005.
[17] Shahinpoor M and Kim K J 2000 *Smart Mater. Struct.* **9** 543.

IOP Conf. Series: Materials Science and Engineering **269** (2017) 012015 doi:10.1088/1757-899X/269/1/012015

Effect of Sr Additive Amount and Holding Time on Microstructure of A390 Aluminum Alloy

J H Zhang[1], S M Xing[1], Q Y Han[2], Q Guo[3], R F Wang[3]

[1]School of Mechanical and Electronic Control Engineering, Beijing Jiaotong University, China
[2]Department of Materials Engineering, Purdue University, USA
[3]Tianjin Lizhong alloy Group Company Limited, China

Abstract. The microstructure of A390 alloy under different Sr additive amounts and holding times was studied by means of direct reading spectrum analysis, energy spectrum analysis, optical microscope and electron microscope. The results show that Sr has a good modification to eutectic Si, while it has a negative effect on primary silicon. The Sr addition will increase the size of primary silicon. When the addition amount of Al-10Sr alloy is 0.6%, the modification of eutectic silicon is the optimum. The Sr has a short incubation period and a fine modification at 10min, but it is more serious burning rate in small furnace smelting, and the modification effect disappears basically after 100min.

1. Introduction

A390 hypereutectic Al-Si alloy has many advantages, such as small density, high temperature strength, good wear resistance, low thermal expansion coefficient and good thermal conductivity [1, 2]. It is an ideal wear-resistant and heat-resistant material which is widely used in engine piston and swash plate of automobile air conditioning compressor. In the conventional casting process, the microstructure of A390 is mainly composed of coarse primary silicon bulk phase, long needle eutectic silicon phase and dendritic primary α phase composition, in which the size and distribution of hard silicon phase points directly affect the mechanical properties, wear resistance and machinability of the alloy. Adding modifier can effectively refine the microstructure of alloy, which has been widely used in industrial production because of its simple process and easy operation [3].

Sr is usually used as metamorphic eutectic silicon, which can transform long acicular eutectic silicon into granular or fibrous form. Because of its stable modification effect and long effective time, it has become an effective eutectic silicon modifier [4, 5]. In the existing studies, Sr is usually added to hypoeutectic Al Si alloy, but its effect on hypereutectic Al Si alloy is seldom studied. In this paper, Sr was added to A390 hypereutectic Al alloy in the form of Al-10Sr master alloy, and the effects of Sr addition and holding time on the residual content and microstructure of Sr in A390 aluminum alloy were investigated.

2. Experimental materials and methods

The experimental material is A390 aluminum alloy, and its chemical composition is shown in table 1. Al-10Sr master alloy is used as a modifier.

Table 1. The chemical composition of A390 aluminum alloy (wt.%, the same below).

Element	Si	Cu	Mg	Zn	Fe	P	Sr	Ca	Al
Content	16.5	4.8	0.52	0.035	0.016	0.0035	0.0025	0.0010	Bal.

The test equipment is a well type graphite crucible resistance furnace which is equipped with KSW-5-12A regulating temperature measuring controller. The A390 aluminium alloy crucible furnace was heated and melted at 760℃ for refining and slag skimming. The experiment was divided into two groups. The first set of experiments as follows: adding 0.2%, 0.4%, 0.6%, 0.8% and 1.0% Al-10Sr master alloy respectively, mixing and holding 10min, then pouring into cylindrical steel mould in casting preheating 100°C, which size is φ50×40mm. The second set of experiments as follows: adding 0.6% Al-10Sr master alloy, stiring fully and holding 10min, 40min, 70min, 100min and 130min respectively, then pouring into cylindrical steel mould in casting preheating 100°C, which size is φ50×40mm. The 10mm location from the bottom of the casting was taken for component inspection and organization analysis. The samples were groud, polished and corroded in the mixed acid solution (2.5ml HNO3+1.5ml HCl+1.0ml HF+95ml H2O). The tissue characteristics were observed by DM2000 type metallographic microscope and S-4800 scanning electron microscope.

3. Results and discussion

3.1. Cast microstructure of A390 aluminum alloy

In this experiment, the composition of A390 alloy is Al-16.5Si-4.8Cu-0.5Mg, and the two element phase diagram of Al-Si alloy under this component is shown in figure 1. The dashed line in figure 1 is A390 alloy. From the dotted line, it is found that in the equilibrium solidification of A390 aluminum alloy, the precipitated phases are mainly composed of primary silicon phase, α-Al matrix phase, (α-Al+Si) eutectic phase and the second phase precipitated at the grain boundary [6].

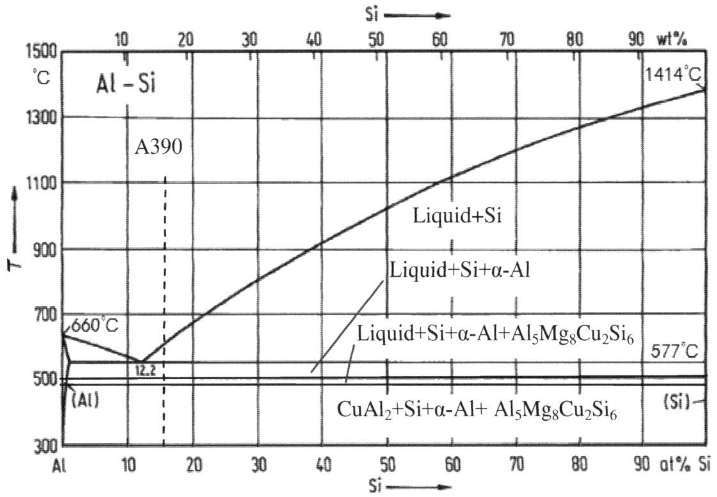

Figure 1. Al-Si binary alloy phase diagram.

The microstructure of the untreated A390 aluminum alloy is shown in figure 2. The microstructure of A390 is composed of dark gray block facies A, pale grey needle-lamellar phase B, white bright near globular phase C, white bright irregular phase D and matrix E. In order to further confirm the phases in A390 aluminum alloy, EDS analysis is carried out. A point components 98.7%Si and 1.3%Al, which can be identified as the primary silicon phase. B point components 53.1%Al, 45.8%Si, 0.9%Cu and 0.2%Mg, which can determine the eutectic silicon phase. C point components 64.8%Al, 31.3%Cu, 2.3%Si, 1.4%Mg and 0.2%Fe, which can be identified as $CuAl_2$ intermetallic compounds, also known as θ phase. D point components 76.7%Al, 15.4%Cu, 4.1%Si and 3.8%Mg, which can be identified as

$Al_5Mg_8Cu_2Si_6$ intermetallic compounds, also known as the Q phase. E point components 97.7%Al, 1.4%Si, 0.7%Cu and 0.2%Mg, which can be identified as α-Al phase.

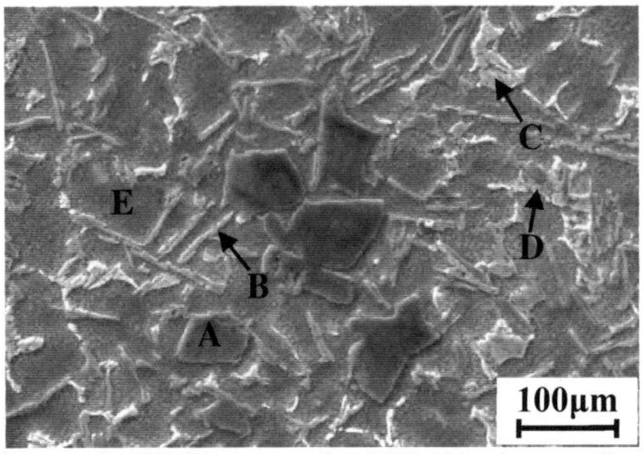

Figure 2. SEM image of A390 aluminum alloy.

The above analysis results show that the microstructure of A390 aluminum alloy consists of α-Al phase, primary silicon phase, eutectic silicon phase, intergranular binary phase θ and quaternary element phase Q. As Al-Si alloy belongs to a small plane-non small plane alloy, high melting point of silicon phase is easy to generate a small plane, while the low aluminum low melting point phase is easy to generate non small plane, so the primary silicon phase appears polygonal massive, eutectic silicon phase appears flaky, silicon phase distribution inside α-Al matrix phase, also θ phase and Q phase present in the interdendritic of α-Al phase.

3.2. Effect of Sr additive amount on A390 alloy

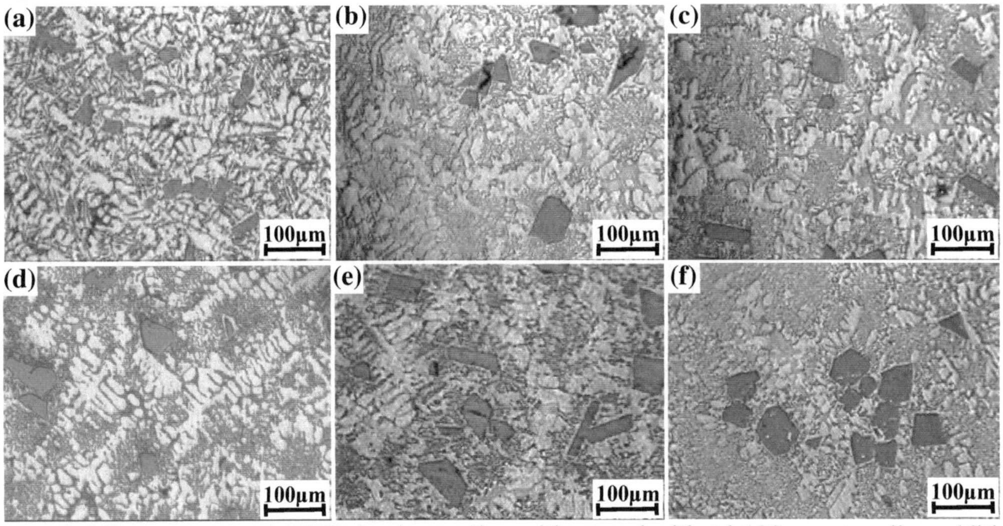

Figure 3. Microstructure of A390 aluminum alloy without and with Al-10Sr master alloy addition
(a) no addition; (b) 0.2%Al-10Sr; (c) 0.4%Al-10Sr;
(d) 0.6%Al-10Sr; (e) 0.8%Al-10Sr; (f) 1.0%Al-10Sr.

Figure 3 shows the microstructural photographs of the Al-10Sr intermediate alloy A390 with different additions of Al-10Sr master alloy. It can be seen from figure 3(a) that the primary silicon is polygonal and the eutectic silicon is long acicular, and the α-Al phase exhibits dendritic morphology in the unmodified A390 alloy. Because the magnification is not high enough, the phase of θ and Q is not observed in the metallographic picture. After adding 0.2% and 0.4% Al-10Sr master alloy to A390 alloy (figure 3b, 3c), primary silicon is still polygonal massive, which size increases obviously. While eutectic silicon starts to change into fiber shape, and α-Al phase has no obvious change. When the Al-10Sr intermediate alloy content reaches 0.6% (figure 3d), eutectic silicon presents fibrous, small size and uniform distribution. The morphology of eutectic silicon is optimum. At the same time, the size of the primary silicon no further increases, α-Al phase is still no significant change. When the amount of Al-10Sr master alloy is more than 0.8%, as shown in figure 3(e, f), the eutectic silicon is large in size, its boundary is not clear and the distribution is uneven. The eutectic silicon begins to appear modification. The above analysis shows that the addition of Al-10Sr master alloy can significantly change the morphology of eutectic silicon, but will cause coarsening of primary silicon, which may be due to the melt trace P and Al reaction of AlP as heterogeneous nucleation of primary silicon, and the addition of Sr generates Al_2Si_2Sr which has the poison effect on AlP particles, resulting in nucleation rate decreases and the size of primary silicon phase increases [7].

3.3. Effect of holding time on A390 alloy

As shown in figure 4, the Sr residual content varies with the holding time after the A390 alloy adding with the 0.6%Al-10Sr master alloy. As can be seen from figure 4, in the case of 720°C preserve heat, with the increasing of holding time, the Sr residual content in A390 decreases gradually and declines obviously. When the holding time is 10min, the residual Sr content is 0.0235% and the Sr loss rate is 60.8% which is much higher than the 30% Sr the burning rate in industrial applications, this may be due to stirring the melt in order to ensure the sufficient dissolution of Sr causes to the acceleration of Sr oxidation. When the melting time is 100min, the Sr residual content is 0.0025%, which has dropped to the original Sr content of A390 alloy, and the external Sr has been completely burnt out. Under the experimental conditions, the effective metamorphic time of Sr is less than 100min.

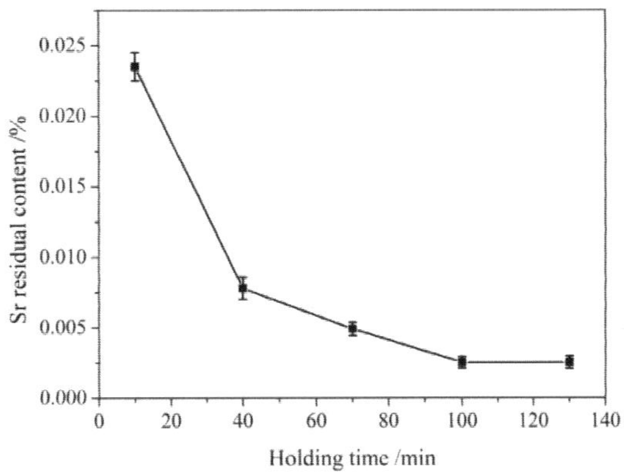

Figure 4. Effect of holding time on Sr residual content of A390 alloy with 0.6%Al-10Sr addition.

Figure 5 shows the micrographs of A390 alloy with 0.6% Al-10Sr master alloy under different holding times. As can be seen from figure 5, the holding time has a significant influence on the modification effect. When the holding time is 10min (figure 5b) compared with unmodified A390 alloy (figure 5a), the eutectic silicon appeared typical metamorphic organization, which morphological changes obviously. Most of the coarse acicular eutectic silicon phases translate into fine micro morphology, the particle size is relatively uniform, and the phenomenon of agglomeration occurs. But the primary silicon is still polyhedron. The size of primary silicon increases obviously, which average grain size is 50μm, meanwhile, the phase of α-Al is dendritic. When the holding time increases, as shown in figure 5(c, d), the primary silicon size decreases, and the eutectic silicon gradually changes from fiber to needle like, the eutectic silicon cluster phenomenon disappears gradually, and the phase of α-Al has no obvious change. When the holding time increases to more than 130min (figure 5e, 5f), the eutectic silicon appears long acicular, and the modification effect of Sr is almost disappeared. The change rule of microstructure is corresponding to the change rule of Sr residual content in the melt. When the holding time is 10~100min, the residual content of Sr decreases gradually and the modification effect decreases in response. When the holding time is more than 100min, the deterioration effect of Sr is almost disappeared because the content of Sr is completely burnt out.

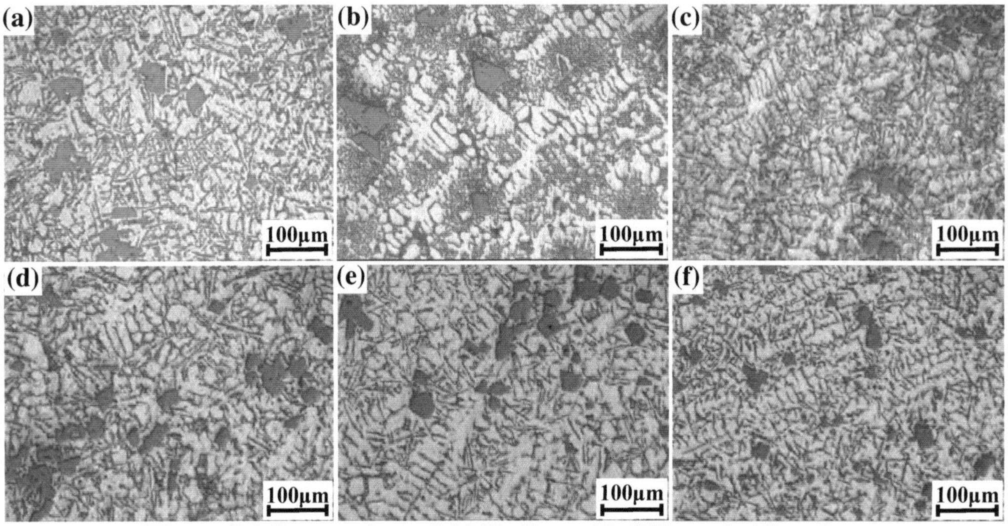

Figure 5. Microstructure of A390 alloy with different holding time
(a) no addition, 10min; (b) 0.6%Al-10Sr addition, 40min;
(c) 0.6%Al-10Sr addition, 90min; (d) 0.6%Al-10Sr addition, 130min;
(e) 0.6%Al-10Sr addition, 160min; (f) 0.6%Al-10Sr addition, 190min.

4. Conclusion

(1) The addition of Sr has a good modification effect on eutectic Si, but it is detrimental to primary silicon and will increase the size of primary silicon. When the addition amount of Al-10Sr alloy is 0.6%, the modification of eutectic silicon is the optimum.

(2) With the increase of holding time, the residual content of Sr in melt decreases gradually, and the modification effect gradually decreases. The metamorphic incubation period of Sr is relatively short, and it has well modification at 10min. However, the burning is more serious in small furnace smelting, and the modification effect disappears basically after 100min.

References

[1] Cui X L, Wu Y Y and Liu X F 2015 *J. Mater. Sci. Technol.* **31** 1027.
[2] Rahvard M M, Tamizifar M, Boutorabi S M A and Shiri S G 2014 *Mater. Des.* **56** 105.
[3] Rathi S K, Sharma A and Sabatino M D 2017 *Eng. Fail. Anal.* **79** 592.
[4] Marzouk M, Jain M and Shankar S 2014 *Mater. Sci. Technol.* **598** 577.
[5] Yu J M, Wanderka N, Miehe G and Banhart J 2016 *Intermetallics* **72** 53.
[6] Tebib M, Morin J B, Ajersch F and Chen X G 2010 *T. Nonferr. Metal. Soc.* **20** 1743.
[7] Eiken J, Apel M, Liang S M and Schmid-Fetzer R 2015 *Acta Mater.* **98** 152.

Acknowledgement

Project support: Program of International S&T Cooperation (2014DFA53050).

Improvement and Application of the Softened Strut-and-Tie Model

Guoxi Fan[1], Debin Wang[2], Yuhong Diao[3], Huaishuai Shang[4], Xiaocheng Tang[5], Hai Sun[1]

[1]School of Engineering, Ocean University of China, Qingdao, Shandong, 266100, China
[2]School of Civil and Safety Engineering, Dalian Jiaotong University, Dalian, Liaoning, 116028, China;
[3]Shandong Zhuoyuan Architectural Design Co.Ltd, Qingdao, Shandong, 266011, China;
[4]School of Civil Engineering, Qingdao University of Technology, Qingdao, Shandong, 266033, China;
[5]School of Civil Engineering and Architecture, Northeast Electric Power University, Jilin, Jilin, 132012, China

Abstract. Previous experimental researches indicate that reinforced concrete beam-column joints play an important role in the mechanical properties of moment resisting frame structures, so as to require proper design. The aims of this paper are to predict the joint carrying capacity and cracks development theoretically. Thus, a rational model needs to be developed. Based on the former considerations, the softened strut-and-tie model is selected to be introduced and analyzed. Four adjustments including modifications of the depth of the diagonal strut, the inclination angle of diagonal compression strut, the smeared stress of mild steel bars embedded in concrete, as well as the softening coefficient are made. After that, the carrying capacity of beam-column joint and cracks development are predicted using the improved softened strut-and-tie model. Based on the test results, it is not difficult to find that the improved softened strut-and-tie model can be used to predict the joint carrying capacity and cracks development with sufficient accuracy.

1. Introduction

At present, numerous theoretical models have been proposed to investigate the mechanical properties of beam-column joints. Paulay [1] pointed out that 2 kinds of shear resistant mechanisms including diagonal strut mechanism and truss mechanism existed in the joint core area. However, this model does not consider the constraint effect of stirrups on the concrete in the joint core area. Subsequently, the softened truss model was proposed which was one of five kinds of models belonging to the reinforced concrete unified theory, and satisfied equilibrium, compatibility and constitutive laws. Hsu [2] proposed the rotating angle softened truss model, while Pang and Hsu [3] proposed the fixed angle softened truss model. Ji [4] pointed out that the rotating angle softened truss model could not explain the existence of shear stress due to aggregate interlock and dowel action of reinforcement bars, and was no longer suitable for higher accuracy structure calculation using the constitutive relation of bare reinforcement bars. In regard to the fixed angle softened truss model, a problem was ignored that the direction of cracks was not only related with the external stress, but also related with the ratio of two perpendicular directions of reinforcement bars. In contrast, the softened strut-and-tie model is often termed as a rational model for determining the shear carrying capacity of beam-column joints [5, 6]. The softened strut-and-tie model originates from the strut-and-tie concept and satisfies equilibrium, compatibility, and constitutive laws of cracked reinforced concrete, which has been proposed for determining the shear strength of beam-column joints.

Among those models, the softened strut-and-tie model can be used as a tool to clarity the roles of different parameters, besides the particular use in the strength prediction of discontinuity regions. The aims of this paper are to predict the joint carrying capacity and cracks development theoretically. Based on the former considerations, four adjustments for the softened strut-and-tie model are made to get better predicting of the test results. And the application of improved beam-column joint model is deeply discussed.

2. Modification of the softened strut-and-tie model

The softened strut-and-tie model consists of the diagonal, horizontal, and vertical mechanisms. On the basis of the softened strut-and-tie model derived previously, Hwang and Lee [7] proposed a simple predicting procedure, which dealt with the carrying capacity prediction of reinforced concrete discontinuity regions failing in diagonal compressions without getting lost in trivialities. This simplified approach is a useful and practical tool for determining the shear carrying capacity of the discontinuity regions failing in diagonal compressions, especially for that of beam-column joints.

By comparing the experimental results and the calculation results of beam-column joints at home and abroad, it is found that the following problems exist in the softened strut-and-tie model. (1) Test results show that, because the neglect of a_b (the height of concrete compression zone of beam section) in computing a_s (the height of effective cross section of diagonal compression strut) is assumed, the proposed model underestimates the joint shear carrying capacity for the cases with lower axial loads [6]. (2) The average value of the reduction factor of concrete strength estimated by the simplified method is lower than that computed by the general method [7]. (3) The column axial load plays a beneficial role in the shear carrying capacity of beam-column joints in a certain range, because it increases the height of effective cross section of diagonal compression strut. But on the contrary, the column axial load provides an adverse effect on the joint shear carrying capacity due to the early crushed of the concrete in the core area of the joint when the column axial load is too large. The softened strut-and-tie model does not consider the adverse effect of the larger column axial load. (4) The smeared stress of mild steel bars embedded in concrete f_{sy} is lower than the yield stress of the bare steel bars f_y [8]. However, the stress-strain relationship of the bare steel bars is adopted in the softened strut-and-tie model. In view of the above problems, four adjustments for the softened strut-and-tie model are made.

(1) The relative height of concrete compression zone of beam section can be obtained by the ductile design control criterion of reinforced concrete structures, which is approximate 0.2 [9].

(2) The inclination angle θ of diagonal compression strut can be calculated by equation (1) for the range of axial compression ratio less than 0.1. Otherwise, it is defined as equation (2).

$$\theta = \tan^{-1}\left(\frac{h_b^{''}}{h_c^{''}}\right) \tag{1}$$

$$\theta = \tan^{-1}\left(\frac{h_b^{''}}{h_c - \frac{2a_c}{3}}\right) \tag{2}$$

(3) The smeared stress of mild steel bars embedded in concrete f_{sy} is lower than the yield stress of the bare steel bars f_y [8]. It is expressed as:

$$f_{sy} = f_y(0.93 - 2B) \tag{3}$$

$$B = \frac{(f_{cr} / f_y)^{1.5}}{\rho} = \frac{(0.31\sqrt{f_c^{'}} / f_y)^{1.5}}{\rho} \tag{4}$$

where ρ is the reinforcement steel ratio (limited to a minimum of 0.25%).

(4) The average value of ζ estimated by the simplified method is lower than that computed by the general method [7]. So the softening coefficient can be computed by equation (5).

$$\zeta = \frac{3.65}{\sqrt{f_c^{'}}} \leq 0.52 \tag{5}$$

3. Application of improved beam-column joint model

3.1 Prediction of the carrying capacity of beam-column joints

Based on the test results of references [10-16], 19 specimens are selected to verify the adaptability of the improved beam-column joint model. The specimens selected encompass a wide range of material properties, reinforcement detailing and cross section geometry. Figure 1 shows that satisfactory results are obtained for the comparison of measured and computed joint shear carrying capacity. After previously mentioned adjustments made, the average ratio of the carrying capacity of beam-column joints computed by the softened strut-and-tie model and experimental results is 0.8847, which is very close to the reduction factor. On the other hand, it also proves the correctness of the adjustments mentioned above. Thus, it is not difficult to find that the improved softened strut-and-tie model can be used to predict the joint carrying capacity with sufficient accuracy.

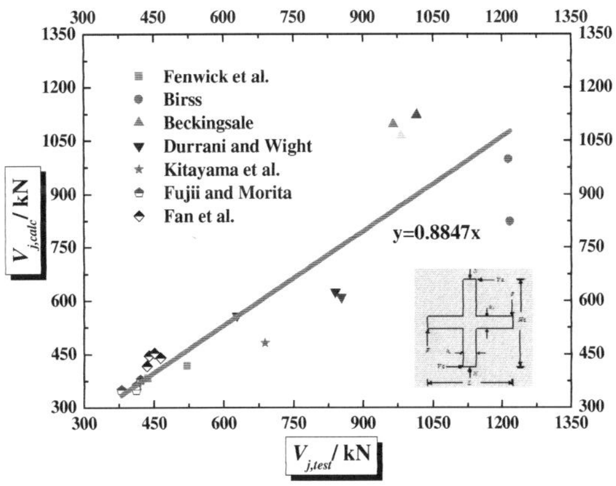

Figure 1. Correlation of experimental and predicted joint shear carrying capacity.

3.2 Prediction of cracks development of beam-column joints

The prediction process of cracks development can be determined under the premise of known material properties, reinforcement detailing, axial compression ratio and cross section geometry. Based on the value of θ (equation (1)) and the geometric relation, the inclination angle α of flat strut and the ratio of tensile force provided by the horizontal tie to the joint horizontal shear force γ_h can be obtained according to equation (6) and (7), respectively.

$$\tan\alpha = \frac{1}{2}\tan\theta \tag{6}$$

$$\gamma_h = \frac{2\tan\theta - 1}{3} \tag{7}$$

Under the action of horizontal shear V_{jh}, the compression force of the flat strut D_1, the tensile force of the horizontal tie F_h and the compression force of the diagonal strut D can be calculated according to equations (8), (9) and (10), respectively.

$$D_1 = \frac{F_h}{\cos\alpha} \tag{8}$$

IOP Conf. Series: Materials Science and Engineering **269** (2017) 012016 doi:10.1088/1757-899X/269/1/012016

$$F_h = \gamma_h \times V_{jh} \tag{9}$$

$$D = \frac{(1-\gamma_h) \times V_{jh}}{\cos\theta} \tag{10}$$

The effective area of the diagonal strut A_{str} is defined as:

$$A_{str} = a_s \times b_s \tag{11}$$

where b_s is the width of the effective cross section of diagonal strut, which is equal to the effective width of the beam-column joint. a_s is the depth of the effective cross section of diagonal strut, which can be determined as:

$$a_s = \sqrt{a_b^2 + a_c^2} \tag{12}$$

The ultimate carrying capacity of the diagonal strut D_c and the ultimate carrying capacity of the horizontal tie F_{hs} can be obtained by material strength, geometry dimension of the cross section and equation (13).

$$a_c = (0.25 + 0.85\frac{N}{A_g f_c^{'}})h_c \tag{13}$$

If the contrast result is $D_c \geq F_{hs}$, the cracking carrying capacity of the diagonal compression strut D_{cr} and the yielding carrying capacity of the horizontal tie F_{hy} should be determined by equations (14)-(16).

$$f_{cr} = 0.31\sqrt{f_c^{'}} \tag{14}$$

$$D_{cr} = f_{cr} \times A_{str} \tag{15}$$

$$F_{hy} = 2 \times (1 + 0.5 \times n) \times f_{yv} \times A_s \tag{16}$$

where f_{cr} is the cracking strength of concrete. $f_c^{`}$ is the compressive strength of concrete cylinder. n represents the number of transverse reinforcements in the joint core area, in addition to the transverse reinforcements within the center half of the joint core. f_{yv} is the yield stress of transverse reinforcement. A_s is the sectional area of transverse reinforcement. Then, the sequence order of the generation of cracks in the diagonal compression strut and the yield of transverse reinforcements can be determined. If the contrast result is $D_c < F_{hs}$, the following conclusion can be draw that cracks occur in the diagonal compression strut before the yielding of transverse reinforcements. Under the action of horizontal shear V_{jh}, the position of the first crack can be estimated by the carrying capacity of the diagonal compression strut and the flat strut.

Take the specimen of reference [17] for example, the calculation results are shown in table 1. By comparison, it can be found that the contrast result is $D_c \geq F_{hs}$. In addition, the horizontal tie bears the greater horizontal shear force before yielding, so that the horizontal tie yielding before the diagonal strut failure. According to the calculation results, it is not difficult to find that cracks occur in the area of the strut before transverse reinforcement yielding. Under the action of horizontal shear V_{jh}, the pressure beared by the flat strut is greater than that beared by the diagonal strut. However, the softened strut-and-tie model provides that the cracking carrying capacity of the flat strut is smaller than that of the diagonal strut due to the smaller effective cross sectional area. Therefore, the first crack occurs in the flat strut region. Test results show when the first crack occurs, the strain of transverse reinforcement within the center half of the joint core is 660.5×10^{-6}, which is less than the yielding strain of transverse reinforcement (1433.3×10^{-6}). And the first crack occurs in the flat strut region, as shown in figure 2. It also proves the adaptability of the improved strut-and-tie model in predicting cracks development, on the other hand.

Table 1. Calculation results of improved softened strut-and-tie model.

$\theta(°)$	$\alpha(°)$	γ_h	D_1	F_h	D	D_c(kN)	F_{hs}(kN)	D_{cr}(kN)	F_{hy}(kN)
55.9	36.4	0.65	0.81 V_{jh}	0.65V_{jh}	0.62V_{jh}	985.99	196.12	69.92	169.36

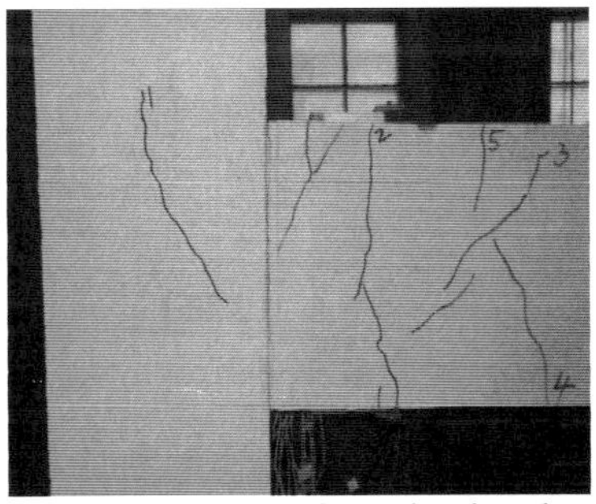

Figure 2. Test results of cracks development.

4. Conclusions

Reinforced concrete beam-column joints play an important role in the mechanical properties of moment resisting frame structures, so as to require proper design. The main objective of this paper is to predict the joint carrying capacity and cracks development theoretically. For this purpose, the softened strut-and-tie model is selected to be introduced and analyzed. Based on the analytical results present in this paper, the following conclusions can be drawn:

(1) Four adjustments for the softened strut-and-tie model including modifications of the depth of the diagonal strut, the inclination angle θ of diagonal compression strut, the smeared stress of mild steel bars embedded in concrete, as well as the softening coefficient are made. It has been proved by test data that predicted results by the improved softened strut-and-tie model are consistent with the test data and conservative. The average carrying capacity ratio of predicted results by the improved model and experimental results is 0.8847.

(2) Based on the test results, it is not difficult to find that the improved softened strut-and-tie model can be used to predict the joint carrying capacity with sufficient accuracy. Test results also show that the improved strut-and-tie model can be used to predict cracks development.

5. References

[1] Paulay T, Park R and Priestley M J N. 1978 *ACI J. Proc.* **75(11)** 585–93.
[2] Hsu T T C 1988 *ACI Struct. J.* **85(6)** 624-35.
[3] Pang X B D and Hsu T T C 1996 *ACI Struct. J.* **93(2)** 197-207.
[4] Ji T, Zheng Z Q and Zheng J L 2001 *J. Build. Struct.* **22(1)** 69-75.
[5] Hwang S J and Lee H J 1999 *ACI Struct. J.* **96(5)** 846-57.
[6] Hwang S J and Lee H J 2000 *ACI Struct. J.* **97(1)** 35-44.
[7] Hwang S J and Lee H J 2002 *J. Struct. Eng.* **128(12)** 1519-26.
[8] Belarbi A and Hsu T T C 1994 *ACI Struct. J.* **91(4)** 465–74.
[9] Fan G X and Song Y P 2014 *Magaz. Concr. Res.* **66(23)** 1219-36.
[10] Fan G X, Song Y P and Wang L C 2014 J. *Reinf. Plast. Comp.* **33(7)** 601-18.
[11] Fenwick R C and Irvine H M 1977 Bull. *New Zealand Nat. Soc. Earthq. Eng.* **10(4)** 174-85.
[12] Birss G R 1978 *The elastic behaviour of earthquake resistant reinforced concrete beam-column joints.* Report, University of Canterbury, Christchurch, New Zealand.
[13] Beckingsale C W 1980 *Post elastic behaviour of reinforced concrete beam-column joints.* Report, University of Canterbury, Christchurch, New Zealand.
[14] Durrani A J and Wight J K. 1985 *ACI J. Proc.* **82(3)** 343–9.
[15] Kitayama K, Otani S and Aoyama H 1991 *ACI Spec. Publ.* **123** 97–123.
[16] Fujii S and Morita S 1991 *ACI Spec. Publ.* **123** 145-66.
[17] Fan G X, Wang D B and Yang S T, et al. 2015 *J. Harbin Eng. Univ.* **36(10)** 1326-30, 34.

IOP Conf. Series: Materials Science and Engineering **269** (2017) 012016 doi:10.1088/1757-899X/269/1/012016

Acknowledgements
This work was supported by the Fundamental Research Funds for the Central Universities (Grant No.201713032), the Postdoctoral Application Research Program of Qingdao City (Grant No.861605040024), the Outstanding Young Teacher Cultivation Program, and the National Natural Science Foundation for Young Scientists of China (Grant No.51408093). The authors would like to express their gratitude for the support.

IOP Conf. Series: Materials Science and Engineering **269** (2017) 012017 doi:10.1088/1757-899X/269/1/012017

Study on Parameter Identification of Assembly Robot based on Screw Theory

Shi Yun[1], **Zhang Xiaodong**[1]*
[1]School of Mechanical and Automotive Engineering, West Anhui University, Lu'an, Anhui, 237012, China
e-mail:710667612@qq.com

Abstract. The kinematic model of assembly robot is one of the most important factors affecting repetitive precision. In order to improve the accuracy of model positioning, this paper first establishes the exponential product model of ER16-1600 assembly robot on the basis of screw theory, and then based on iterative least squares method, using ER16-1600 model robot parameter identification. By comparing the experiment before and after the calibration, it is proved that the method has obvious improvement on the positioning accuracy of the assembly robot.

1. Introduction

The accuracy of the assembly robot is one of the important indicators of its performance. The precision of the assembly robot refers to the accuracy of the final movement of the end actuator to the specified desired coordinates, which is an important criterion for measuring the robot [1]. Under normal circumstances the assembly of the robot on the movement accuracy requirements higher than the handling of robots and arc welding robot [2].

At present, the commonly used kinematic model has D-H model [3], which establishes a coordinate system for each link, the base system is connected with 0 link, and the tool coordinate system is built on the end of the assembly robot, and the kinematic model. But there are three problems in the calibration process: the calculation process is more cumbersome; cannot arbitrarily set the base coordinate system and tool coordinate system; the model will cause a certain singularity, does not have continuity and completeness. Compared with the D-H model, the use of spin theory to study the kinematics of the robot has three advantages [4, 5]: the overall description of the robot movement, to avoid the description of the singularity caused by DH; movement described clearly, simplifies the institutional Analysis; problem description and resolution simple and unified, easy and other methods of conversion. Therefore, the study of robot parameter identification is becoming more and more concerned [6-8].

In [9], the kinematic parameter calibration method of the product of exponential (POE) is proposed first, but it is inconvenient to use. In [10] improved the model of [9], given the explicit calibration model. [11] Using the spin theory to differentiate the kinematic parameters, the approximate model was established, but the model was not accurate. In [12], the joint rotation error is regarded as the generalized kinematic motion, and the error model is established, but the consideration of the error source is not comprehensive. In this paper, the calibration parameters of assembly robot parameters are completed based on the explicit calibration model of Screw theory. The experimental results show that the assembly robot positioning accuracy after calibration is improved obviously.

2. Screw theory

Screw theory was first introduced by Brockett [13] into the kinematics modeling of robots. Assuming that a rigid body rotates about the angle θ about the rotation axis s, the corresponding rigid body transformation can be expressed as a matrix exponent:

$$\mathrm{R} = \exp(\hat{s}\theta) \quad (1)$$

where $\hat{s}$ is an antisymmetric matrix, R is a rotation matrix. Formula (1) can be expanded to:

$$\mathrm{R} = \exp(\hat{s}\theta) = I + \hat{s}t + (\hat{s}t)^2/(2\,!) + (\hat{s}t)^3/(3\,!) + \cdots = I + \hat{s}\sin\theta + \hat{s}^2(1-\cos\theta) \quad (2)$$

When the rigid body to do the spiral movement, the rigid body on a point P rotation angle θ, and then move in the direction of the axis d = hθ after the final coordinates can be expressed as:

$$\mathrm{p}(\theta,\mathrm{h}) = \mathrm{r} + \exp(\hat{s}\theta)\,(\mathrm{p}(0)-\mathrm{r}) + \mathrm{hs}\theta, \mathrm{s} \neq 0 \quad (3)$$

This equation holds for any $\mathrm{p}(0) \in \mathbb{R}^3$, so the form of homogeneous coordinates is:

$$\mathrm{T} = \begin{pmatrix} \exp(\hat{s}\theta) & r(1-\exp(\hat{s}\theta)) + hs\theta \\ 0 & 1 \end{pmatrix}, s \neq 0 \quad (4)$$

According to the relative theorem of screw theory, any rigid body transform T can be written as a matrix of exponential form of a motion helix:

$$\mathrm{T} = \exp(\hat{\xi}\theta) = \begin{pmatrix} \exp(\hat{s}\theta) & r(1-\exp(\hat{s}\theta)) + hs\theta \\ 0 & 1 \end{pmatrix}, s \neq 0 \quad (5)$$

where $\hat{\xi} = \begin{pmatrix} \hat{s} & -s \times r \\ 0 & 0 \end{pmatrix}$

When the rigid body is pure rotation h = 0, this time:

$$\mathrm{T} = \exp(\hat{\xi}\theta) = \begin{pmatrix} \exp(\hat{s}\theta) & r(1-\exp(\hat{s}\theta)) \\ 0 & 1 \end{pmatrix} = \begin{pmatrix} R & t \\ 0 & 1 \end{pmatrix}, s \neq 0 \quad (6)$$

3. Robot kinematics model

The forward kinematics of the assembly robot can be synthesized by the rigid body motion caused by each joint. According to the screw theory, the motion of each joint is generated by the kinematic rotation of the associated joint axis, which can give its kinematic geometric description. If ξ is used to represent the unit motion axis coordinate of a joint axis, the rigid body motion along the joint axis can be expressed as:

$$\mathrm{T}(\theta) = G_1 G_2 \cdots G_n T_{st}(0) = \exp(\hat{\xi}_1\theta_1)\exp(\hat{\xi}_2\theta_2)\cdots\exp(\hat{\xi}_n\theta_n)\,T_{st}(0) \quad (7)$$

$$G_i = \exp(\hat{\xi}\theta) = \begin{pmatrix} \exp(\hat{\xi}_i\theta_i) & r_i(1-\exp(\hat{\xi}_i\theta_i)) \\ 0 & 1 \end{pmatrix} = \begin{pmatrix} R_i & t_i \\ 0 & 1 \end{pmatrix}, i = 1,2\cdots 6 \quad (8)$$

$$\hat{\xi}_i = \begin{pmatrix} \hat{s}_i & -s_i \times r_i \\ 0 & 0 \end{pmatrix} \; i = 1,2\cdots 6 \quad (9)$$

where T(θ) represents the position of the axis after the rotation θ angle, $T_{st}(0)$ indicates that the robot is in the initial position. $T_{st}(0)$ can be written as a constant rotation of the exponential form, the formula (7) rewritten as:

$$\mathrm{T}(\theta) = \exp(\hat{\xi}_1\theta_1)\exp(\hat{\xi}_2\theta_2)\cdots\exp(\hat{\xi}_n\theta_n)\exp(\hat{\Gamma}) \quad (10)$$

The equation (10) is obtained by:

$$\delta\mathrm{TT}^{-1} = \left(\frac{\partial T}{\partial \xi}\delta\xi + \frac{\partial T}{\partial \Gamma}\delta\Gamma\right)T^{-1} \quad (11)$$

The relationship between the theoretical value ξ^n of the rotation joint coordinate and the actual value ξ^a can be expressed by a fixed rigid body transformation $\exp(\hat{\eta}) = \begin{bmatrix} R & q \\ 0 & 1 \end{bmatrix}$, and the relationship between the kinematic rotations of the rotational force is expressed as:

$$\xi^a = Ad(\exp\hat{\eta})\xi^n = \exp(\hat{\eta})\,\xi^n\exp(-\hat{\eta}) = \begin{bmatrix} R & 0 \\ \hat{q}R & R \end{bmatrix}\begin{bmatrix} \omega^n \\ v^n \end{bmatrix} = \begin{bmatrix} R\omega^n \\ \hat{q}R\omega^n + Rv^n \end{bmatrix} \tag{12}$$

where $Ad(\exp\hat{\eta})\xi^n$ is the adjoint matrix of spin factor ξ^n, the rotation joint satisfies the constraint condition:

$$||\omega|| = ||\omega^a|| = ||R\omega^n|| = ||\omega^n|| = 1, \omega^T v = 0 \tag{13}$$

$$(\omega^a)^T v^a = (R(\omega^n)^T)(\hat{p}R\omega^n + Rv^n) = 0 \tag{14}$$

According to the formula (12), the kinematic equation (10) can be written as the concomitant transformation of the theoretical joint rotation:

$$T^a(\theta) = e^{(\hat{\xi}_1^n\theta_1 e^{\hat{\eta}_1}e^{-\hat{\eta}_1})}e^{\hat{\xi}_2^n\theta_2 e^{\hat{\eta}_2}e^{-\hat{\eta}_2}}\cdots e^{\hat{\xi}_n^n\theta_n e^{\hat{\eta}_n}e^{-\hat{\eta}_n}}e^{\hat{\Gamma}^a} \tag{15}$$

The kinematics equation (15) is substituted into equation (11), and the simplification is obtained:

$$(\delta TT^{-1})^{\vee} = \left(\left(\frac{\partial T}{\partial \xi}\delta\xi\right)T^{-1}\right)^{\vee} = (\delta(\mathrm{A}_1\cdots\mathrm{A}_n)(\mathrm{B}_1\cdots\mathrm{B}_n))^{\vee} =$$

$$(\delta(\mathrm{A}_1\mathrm{B}_1))^{\vee} + \cdots + \quad \mathrm{Ad}\left(\textstyle\prod_{i=1}^{n-1}\mathrm{A}_i\right)\cdot(\delta(\mathrm{A}_n\mathrm{B}_n))^{\vee} \tag{16}$$

where $\mathrm{A}_i = e^{\hat{\eta}_i}e^{\hat{\xi}_i^n\theta_i}e^{-\hat{\eta}_i}, \mathrm{B}_i = e^{\hat{\eta}_i}e^{-\hat{\xi}_i^n\theta_i}e^{-\hat{\eta}_i}, \mathrm{i} = 1,2\cdots\mathrm{n}$

Since the identity $\delta(e^{\hat{\eta}})e^{-\hat{\eta}} + e^{\hat{\eta}}\delta(e^{-\hat{\eta}}) = 0$, we obtain:

$$\left(\delta\mathrm{A}_i\mathrm{B}_i^{-1}\right)^{\vee} = (I_6 - Ad(\mathrm{A}_i))\left(\delta e^{\hat{\eta}_i}e^{-\hat{\eta}_i}\right)^{\vee} \tag{17}$$

According to the explicit expression of the exponential mapping in the literature [14] on the screw differential:

$$(\delta(e^{\hat{\eta}})e^{-\hat{\eta}})^{\vee} =$$

$$\left(\mathrm{I} + \frac{4-||\beta_i||\sin||\beta_i||-4\cos||\beta_i||}{2||\beta_i||^2}\Psi_i + \frac{4||\beta_i||-5\sin||\beta_i||-||\beta_i||\cos||\beta_i||}{2||\beta_i||^3}\Psi_i^2 + \frac{2-||\beta_i||\sin||\beta_i||-2\cos||\beta_i||}{2||\beta_i||^4}\Psi_i^3 + \frac{2||\beta_i||-3\sin||\beta_i||-||\beta_i||\cos||\beta_i||}{2||\beta_i||^5}\Psi_i^4\right)\delta\eta_i = \mathrm{K}_i\delta\eta_i \tag{18}$$

where $||\beta_i|| = \sqrt{\beta_{1i}^2 + \beta_{2i}^2 + \beta_{3i}^2}, \eta_i = (\beta_i \quad v_i)^T, \Psi_i = \begin{pmatrix} \hat{\beta}_i & 0 \\ \hat{v}_i & \hat{\beta}_i \end{pmatrix}$

The actual position of the robot end effector is T^a, and its theoretical position is T^n, then δTT^{-1} is the deviation from T^a to T^n, and the deviation is small enough to be obtained by the first order approximation:

$$\delta TT^{-1} = \log(T^a(T^n)^{-1}) \tag{19}$$

For a set of pose error data, the formula (18) is written as follows:

$$\mathrm{y} = \mathrm{Jx} \tag{20}$$

where $J_i = \begin{cases} (-\hat{q} \quad I_3)(I_n - \mathrm{Ad}A_1)T_1 & i = 1 \\ (-\hat{q} \quad I_3)Ad\left(\prod_{k=1}^{n-1}\mathrm{A}_k\right)\cdot(I_n - \mathrm{Ad}A_i)T_i & 1 < i \le 6 \\ R_g & i = 7 \end{cases}$

Assuming that the robot moves k coordinate points in the workspace, the corresponding k actual pose is obtained by Kinect color camera, and the theoretical pose of each coordinate point is calculated by the robot controller. Then by the formula (19) can be obtained k group of geometric error equation, the simultaneous equations can be obtained for the equations:

$$\mathrm{Y} = \mathrm{Lx} \tag{21}$$

where $\mathrm{Y} = [y_1 \quad \cdots \quad y_k]^T, \mathrm{L} = [J_1 \quad \cdots \quad J_k]^T$.

The corresponding geometric parameter error x can be obtained by solving the iterative least squares method:

$$x = (L^T L)^{-1} L^T Y \tag{22}$$

4. Parameter identification experiment

Parameter identification experimental equipment is the research group ER16-1600 type assembly robot (figure 1), figure 2 is its structural diagram, the experiment used in the measurement equipment Kinect color camera. Parameter identification results shown in table 1, the experimental compensation before and after the robot positioning accuracy shown in figure 3. The average positioning accuracy of the assembled robot is 3.8558 mm, and the average positioning accuracy of the assembled robot is reduced to 0.70035mm. After the parameter identification, the positioning accuracy of the assembly robot is obviously improved, and the validity of the compensation method is proved.

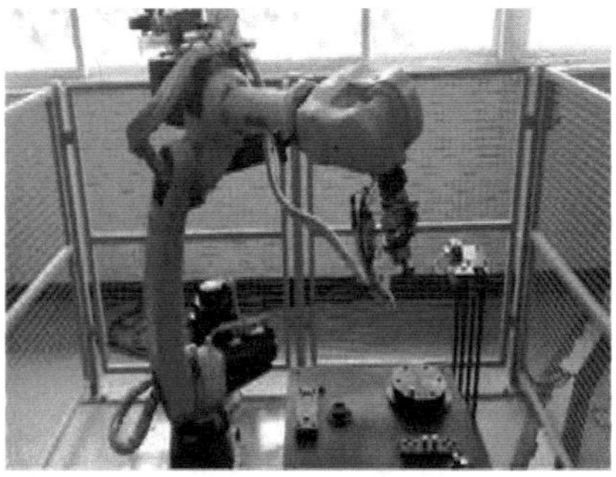

Figure 1. ER16-1600 assembly robot.

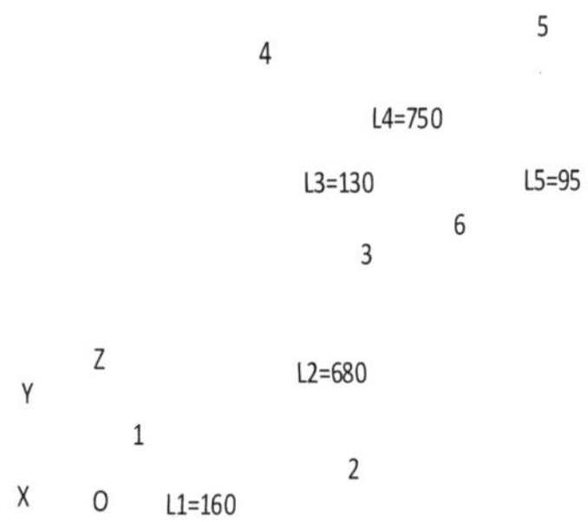

Figure 2. Motion diagram.

Table 1. Kinematic parameters.

Parameter	Nominal value	Identification results
ξ_1	$(0,0,0,1,0,0)^T$	$(-0.0028,-0.0009,0.0024,0.9902,,0.1551,0.0115)^T$
ξ_2	$(0,0,-160,0,1,0)^T$	$(0.0016,0.0101,-159.9909,0.0010,1.0002,0.0083)^T$
ξ_3	$(-680,0,-160,0,1,0)^T$	$(-680.1043,0.0015,-160.0002,0.0073,1.0005,0.0008)^T$
ξ_4	$(0,680,0,1,0,0)^T$	$(0.0005,680.0035,0.0073,1.0004,0.1018,0.0074)^T$
ξ_5	$(-680,0,-910,0,1,0)^T$	$(-679.9508,0.0019,-910.5511,0.2131,1.0108,0.0158)^T$
ξ_6	$(0,680,0,1,0,0)^T$	$(0.2041,680.0426,0.0112,1.0017,1.0254,2.1086)^T$
T_{st}	$(-1050,0,810,0,0,0)^T$	$(-1051.0152,0.8304,809.7386,0.0001,0.0009,1.0005)^T$

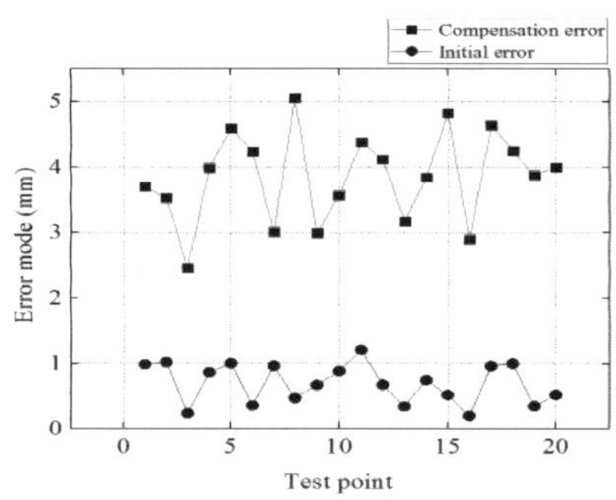

Figure 3. Errors' module.

5. Conclusion

The kinematic model of the 6-DOF assembly robot is established by using the spin matrix and the exponential product. The modeling method is simple and has obvious geometric intuition. On this basis, the iterative least squares algorithm is combined with spin theory. Compared with the traditional D-H parameter method, the kinematic derivation process of this method is simpler and the geometric significance is clearer. The algorithm is parallel to the common axis of the 6-DOF series robot, and the algorithm has a certain degree of versatility.

References

[1] Chen Z L 2016 *Four Degrees of Freedom of Assembly Robot Trajectory Planning and the Posture Error Analysis*. Chang'an University, 1-6.
[2] Peng S X, Zhang P and Tang R C 1988 *Robot*. **2(1)** 14-8.
[3] Cai Z X 2009 *Robotics*. Beijing: Tsinghua University press, 35-58.
[4] Wang D S 2006 *Calibration Techniques Research of industrial robot*. Northeastern University, 1-14.
[5] Rong P X, Yang J and Hu L G 2007 *Electr. Mach. Contr.* **11(3)** 303-6.
[6] Hou S J 2012 *Research on the Structural Parameter Identification and Positioning Calibration of the Industrial Robot*. Nanjing University of Aeronautics and Astronautics, 3-8.
[7] Li B F 2011 *Workspace Analysis and Parameter Identification of Space Robot*. Beijing University of Posts and Telecommunications, 5-16.
[8] He R B, Zhao Y J, Yang S N, et al. 2010 *IEEE Trans. Robot.* **26(3)** 411-23.
[9] Okamura K and Park E 1996 *Robot* **14(04)** 415-21.
[10] He R B, Zhao Y J et.a1. 2011 *Robot*, **33(001)** 35-9.
[11] Tan Y S, Sun H X, Jia Q X, Shao Z Y et a1. 2006 J. Beijing Univ. Aeronaut. Astronaut. **32(9)** 1104-8.
[12] Huang Y G and Du L 2010 J. Harbin Inst. Tech. **42(3)** 484-9.
[13] Brockett R 1984 *Robotic manipulators and the product of exponentials formula*. Mathematical theory of networks and systems: Springer, 120-9.
[14] Selig J 2004 *Geometic fundamentals of robotics*. Springer.

Acknowledgment

This work was supported by the natural science key project of Anhui Province (No. KJ2010A329), and by the National Natural Science Fund of China (No. 61302179).

A Study of Fatigue Property for FV520B-I at Different Loading Frequencies

Q C Zhao[1], Y L Zhang[1], J L Wang[1], Y H Shen[1] and S J Liu[1]

[1]School of Mechanical Engineering, Dalian University of Technology, Dalian, China

Abstract. Fatigue property of FV520B-I can be affected by the changing of the loading frequency. However, few theories about fatigue property for metal at different loading frequencies has ever been proposed. Experimental data are obtained from the conducted traditional experiment (140 Hz) and ultrasonic experiment (20 kHz). Fatigue property of high strength steel FV520B-I obtained by ultrasonic high-frequency fatigue testing method is higher than that by traditional low frequency testing. FV520B-I fatigue strength conversion model are established with comprehensive use of a fitting algorithm based on the combination of experimental data and classic three-parameter-model. The fatigue strength conversion coefficient C_0 for FV520B-I is proposed. A clear understand of the effect of loading frequency on the fatigue property of FV520B-I is novel and has an important significance in guaranteeing the accuracy of the actual fatigue analysis of FV520B-I.

1. Introduction

High strength metal FV520B-I has numerous good mechanical properties including high strength, high corrosion resistance, high abrasive resistance and good welding characteristics [1]. These properties make FV520B-I widely adopted in the manufacturing of centrifugal compressor vanes. The manufactured vanes are generally used under cyclic loading condition. So, the fatigue life of the used FV520B-I should be over 10^7 cycles and even reaches 10^9 or 10^{10} cycles, a “giga-cycle” fatigue level [2-4]. Fatigue failure can cause serious accidents easily [5-7], which will lead to tremendous economic loss, significant impact on the mechanical system and even can threaten human life.

Traditional fatigue test system frequency is usually tens of hertz to several hundred hertz, which makes it hard to study fatigue properties in ultra-high cycle range [8, 9]. The emergence of ultrasonic fatigue test system makes the fatigue test loading frequency directly increased to 20KHz, the efficiency increased by 200 times, so that for various types of metal materials, ultra-high cycle fatigue research can be carried out. However, it was found that [10, 11] the fatigue properties of the materials obtained by the ultrasonic fatigue test method were higher than those under the traditional low-frequency fatigue conditions, and the fatigue failure mechanism under different loading frequencies was different. So the fatigue properties of the materials obtained under ultra-high frequency loading cannot be directly applied to the actual fatigue analysis. In this paper, ultrasonic fatigue test (20kHz) and traditional fatigue test (140Hz) were carried out for high strength steel FV520B-I respectively. A fatigue strength conversional model for FV520B-I at different loading frequencies was established. Using this model, the fatigue strength of the material obtained under the ultra-high frequency loading condition can be converted into the fatigue strength of the material under the conversional condition. So it is an important work which has important engineering significance in practical fatigue analysis.

2. Experiment

The ultrasonic fatigue experiment is carried out in USF-2000 ultrasonic fatigue test system with an operating frequency of 20 kHz, and the traditional fatigue experiment is carried out in PLG-100 with an operating frequency of 140Hz. The experiments are generally carried out under room temperature (20°C) and the mean-stress was zero which presented that the stress ratio $r = -1$, the stress amplitude interval is 25MPa. Standard fatigue test specimen must be used in two experiments, the geometrical shape of the specimen used in the test are shown in figure 1 (a) and figure 1 (b).

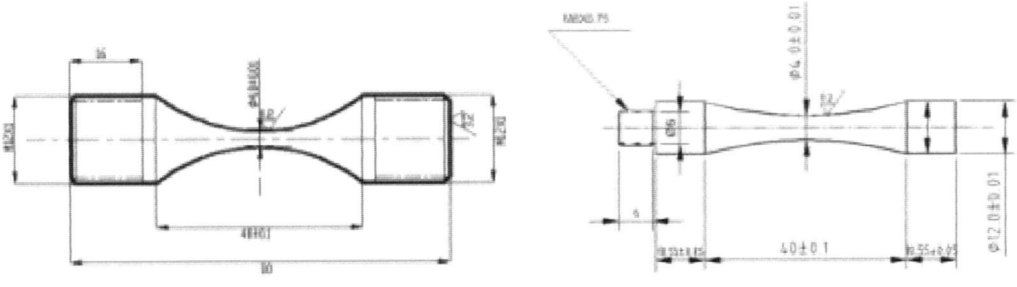

(a) the traditional frequency; (b) the ultrasonic frequency

Figure 1. Specimen dimensions.

The chemical compositions and mechanical property of the FV520B-I specimens are displayed in table 1 and table 2 as below:

Table 1. Chemical compositions of the specimen.

Chemical Elements	Si	C	Ni	Cu	Mn	S	Cr	Mo	Nb
Weight Percent/%	0.15-0.7	0.02-0.07	5-6	1.3-1.8	0.3-1	<0.025	13-14.5	1.3-1.8	0.25-0.45

Table 2. Mechanical property of FV520B-I.

Mechanical property	E/GP	Rm/MPa	Rp0.2/MPa	HV/kgf.mm-2	A/%	ρ/kg.m-3
Value	194	1180	1029	380	16.07	7.82x103

Note: E—elastic modulus, Rm—tensile strength, Rp0.2—yield strength, HV—Vickers hardness, A—shrinkage ratio, ρ—density.

3. Result and observation

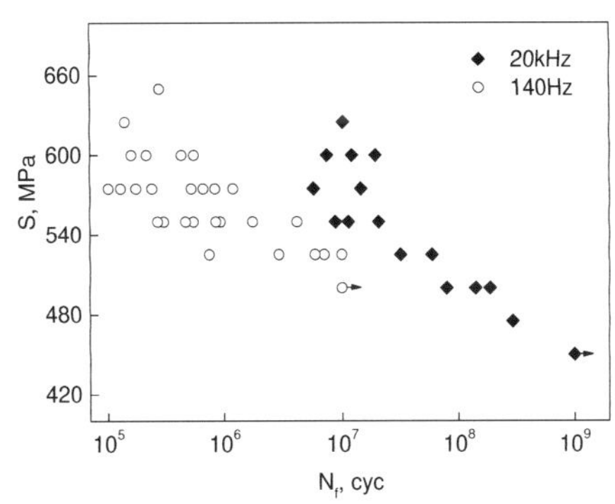

Figure 2. Experimental data.

The experimental data obtained from the traditional fatigue test (140 Hz) and the ultrasonic fatigue test (20 kHz) are shown in figure 2. It shows that when the fatigue life is at the same level, the ultrasonic fatigue experiment will obtain the higher fatigue strength than the traditional one. The observations can be captured as shown in figure 3(a) and figure 3(b).

Figure3 shows that the fatigue life is about 7×10^6 cycles but the corresponding stress amplitudes are 525MPa (140 Hz) and 600MPa (20 KHz). GBF region and "fish-eye" can be observed clearly in the figure 3(b), which means that the fatigue failure was caused by internal inclusion. However, no such observations can be obtained from the traditional experimental fracture section, and the primary factor result in the surface fatigue failure is the surface defect, especially the surface roughness [12].

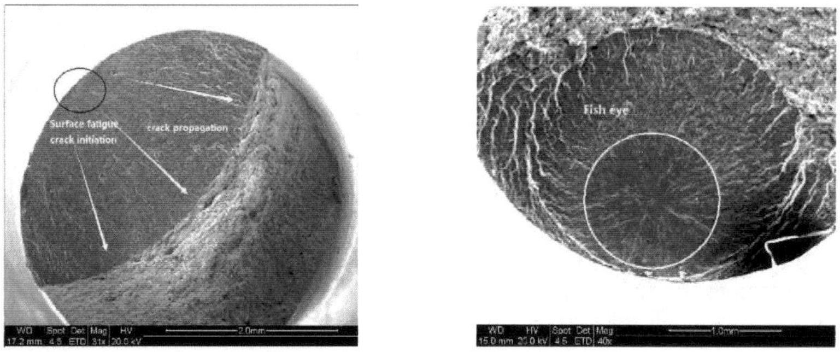

(a) The traditional observation (525MPa, 7.12e6); (b) the ultrasonic observation (600MPa, 7.44e6)

Figure 3. Specimen observation.

The observations and the experimental data show that the primary factors leading to the fatigue failure under 140 Hz and 20 KHz loading frequencies are different that imply that the FV520B-I fatigue properties under different loading frequencies are also different [13]. Therefore, the obtained fatigue properties data must be corrected under the impact of loading frequency before it can be applied to the fatigue analysis under the traditional condition.

4. The influence of loading frequency on FV520B-I fatigue property

In the study of high-cycle and even very-high cycle fatigue problems, the three-parameter model [14] is widely used in the fitting of fatigue life S-N curve by virtue of the form flexibility and the ability of fitting data. The three-parameter model at the loading frequency of 20KHz and 140Hz can be expressed as follow:

$$\begin{cases} (S_H - S_{0H})^{\alpha_H} N_{fH} = C_H \text{——} 20KHz \\ (S_L - S_{0L})^{\alpha_L} N_{fL} = C_L \text{——} 140Hz \end{cases} \tag{1}$$

where S_{0H} and S_{0L} are the theoretical fatigue strength for 20KHz and 140Hz; α_H, C_H, α_L, C_L are the undetermined coefficients for 20KHz and 140Hz; S_H and S_L are loading stress amplitudes (or fatigue strengths) for 20KHz and 140Hz; N_{fH} and N_{fL} are fatigue life for 20KHz and 140Hz. Six model parameters (S_{0H}, α_H, C_H, S_{0L}, α_L, C_L) for the material FV520B-I at 20KHz and 140Hz are obtained according to the experimental data in figure 2 and the equation (1) with the use of MATLAB curve fitting tool. The fitting results are shown in table 3.

Table 3. Model parameters for FV520B-I.

Coefficient	S_{0H}	α_H	C_H	S_{0L}	α_L	C_L
Fitted Value	432.1	3.182	8.3×10^{13}	476.7	4.187	7.65×10^{13}

By substituting the parameters into the equation (1), the empirical three-parameter model formulas for FV520B-I with 20 KHz and 140 Hz can be written as:

$$\begin{cases} (S_H - 432.1)^{3.182} N_{fH} = 8.3\times10^{13} \text{——} 20KHz \\ (S_L - 476.7)^{4.187} N_{fL} = 7.65\times10^{13} \text{——} 140Hz \end{cases} \tag{2}$$

As shown in figure 4, the S-N curve of the FV520B-I ultrasonic fatigue test and the traditional fatigue test are obtained by the equation (2) after the results of the parameter fitting in table 3.

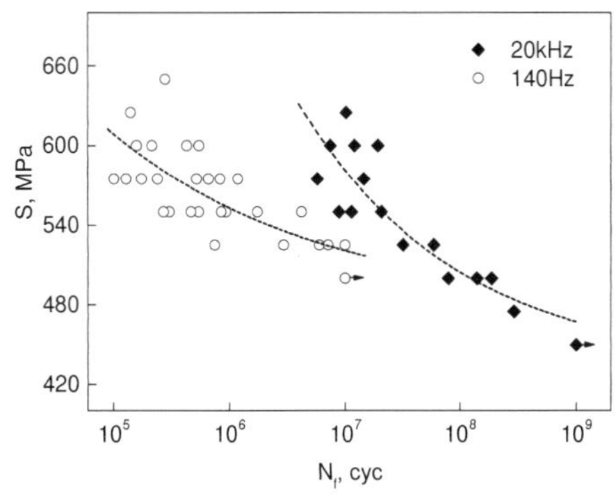

Figure 4. S-N curve from three-parameter model.

As the fatigue life for 20kHz is similar to the fatigue life for 140Hz, that is $N_{fL} \approx N_{fH}$, according to equation (1), the relationship between the corresponding fatigue strength (S_H and S_L) can be expressed as:

$$(S_L - S_{0L})^{\alpha_L} = \frac{C_L}{C_H}(S_H - S_{0H})^{\alpha_H} \tag{3}$$

The ratio between the determined coefficients (C_L and C_H) is the fatigue strength conversion factor, which can be expressed as:

$$C_0 = \frac{C_L}{C_H} = 0.922 \tag{4}$$

The final fatigue strength conversion model can be expressed as:

$$(S_L - 476.7)^{4.187} = 0.922(S_H - 432.1)^{3.182} \tag{5}$$

So the converted S_L can be calculated by substituting the ultrahigh loading frequency experimental data (S_H) into equation (5), the results and the errors between their related fatigue life (N_{fH} and N'_{fL}) are shown in the table 4.

Table 4. Fatigue strength conversion and the errors.

N_{fH}/cyc	3.21e7	2.07e7	1.45e7	7.44e6
S_H	525	550	575	600
S_L	507.4	513.5	519.3	524.8
N'_{fL}	4.54e7	2.13e7	1.15e7	6.90e6
Errors	41.43%	2.90%	20.69%	7.26%

Note: N_{fH}—Fatigue life experimental data for 20 kHz, S_H—Fatigue strength for 20 kHz, S_L —converted fatigue strength from 20 kHz to 140 Hz, —Fatigue life for 140 Hz related to $S_L$according to equation (2).

It can be seen from the data in table 4 that the errors between the fatigue life (N_{fH} and N_{fL}') obtained from three-parameter model in the low-frequency and the experimental fatigue life in the ultrahigh loading frequency are within the acceptable range, which satisfied the assumption that the fatigue life for 20kHz is similar to the fatigue life for 140Hz. Thus the fatigue strength conversion model (equation (3)) is established which can be used to convert the fatigue strength with different loading frequencies.

5. Conclusion

(1) The fatigue performance of FV520B-I obtained under ultrasonic fatigue test is better than that under the low frequency fatigue condition when the fatigue life is at the same range.

(2) Changes in loading frequency causes changes in cause of fatigue failure. Fatigue failure in ultrahigh loading frequency (20KHz) is mainly originated in the internal non-metallic inclusions, and in traditional loading frequency (140Hz) is mainly caused by surface roughness defects.

(3) Six model parameters (S_{0H}, α_H, C_H, S_{0L}, α_L, C_L)for FV520B-I are obtained which can be used in the fatigue analysis for the material FV520B-I.

(4) The fatigue strength conversion model of metal material at different loading frequencies is established (equation (3)), and the fatigue strength conversion coefficient C_0 is proposed.

References

[1] Zhang M, Wang W Q, Wang P F, Liu Y N and Li J F 2014 *Fatigue behavior and mechanism of FV520B-I in ultrahigh cycle regime* C 20th European Conference on Fracture. 2035-41.
[2] Suh C M, Suh M S and Kim S G 2012 *Fatig. Fract. Eng. Mater. Struct.* **35(1)** 30-6.
[3] Jones R 2014 *Fatig. Fract. Eng. Mater. Struct.* **37(5)** 463-83.
[4] Chen F, Wang F and Cui W 2012 *Fatig. Fract. Eng. Mater. Struct.* **35(3)** 278-90.
[5] Tsutsumi N, Murakami Y and Doquet V 2009 *Fatig. Fract. Eng. Mater. Struct.* **32(6)** 473-83.
[6] Furuya Y, Torizuka S, Takeuchi E, Bacher-Hochst M and Kuntz M 2012 *Mater. Des.* **37** 515-20.
[7] Shalvandi M, Hojjat Y, Abdullah A and Asadi H 2013 *Mater. Des.* **46** 713-23.
[8] Zhu M L, Liu L L and Xuan F Z 2015 *Int. J. Fatig.* **77** 166-73.
[9] Zhao A, Xie J J, Sun C Q, Lei Z G and Hong Y S 2012 *Int. J. Fatig.* **38** 46-56.
[10] Nonaka I, Setowaki S and Ichikawa Y 2014 *Int. J. Fatig.* **60** 43-7.
[11] Zhang Y Y, Duan Z and Shi H J 2013 *Sci. China Phys. Mech. Astron.* **56(3)** 617-23.
[12] Wang J L, Zhang Y L, Sun Q C, Liu S J, Shi B W and Lu H T 2016 *Mater. Des.* **89** 1024-38.
[13] Zhang J W, Song Q P, Zhang N, Lu L T, Zhang M T, Cui G D 2014 *Int. J. Fatig.* **70** 235-40.
[14] Fu H M 1993 *Acta Aeronaut. Astronaut. Sin.* **14(3)** 173-6.

Acknowledgement

The authors gratefully acknowledge the research funding support from the National Natural Science Foundation of China (No.51375074) and the Fundamental Research Funds for Central Universities (No. DUT17ZD230 & No. DUT17GF214).

Investigation on the Activity Activation and Cementitious Property of Coal Gangue with High Iron and Silica Contents

Hong Wu[1,*], Yu Li[1], Min Teng[1], Yu Yang[1]

[1]School of Chemical and Materials Engineering, Liupanshui Normal University, Liupanshui, Guizhou 553004, PR China

Abstract. The activity of coal gangue by thermal activation and composite activation technologies was investigated. The crystal composition, framework structure and morphology change were analyzed by XRD, FT-IR and SEM, respectively. The cementitious property of coal gangue was measured by strength test. The results showed that thermal activation decomposed kaolinite in coal gangue, and formed the metastable structure with a porous state, multiple internal broken bonds and large specific surface areas. Based on thermal activation, the added lime provided the alkaline environment, then this reduced the bond energy of reactant particles and the degree of crystallinity of quartz in coal gangue. The two activation methods could effectively improve the cementitious property of coal gangue based unburned bricks, and that the composite activation technology was superior performance.

1. Introduction

Coal gangue is a complex industrial solid waste excavated in the mining production and washed in the preparation process [1]. The stockpile of coal gangue can cause serious pollution which included local environment and the health of residents in the mining area. However, coal gangue contains clay minerals, quartz, carbonate minerals and pyrite, which can be utilized as resources [2, 3]. The application of coal gangue in building materials has attracted widespread attentions because of its high consumption. But its ratio of use as a cement additive is less than 15%, this is because the stable crystal structure of coal gangue has low pozzolanic activity and weak cementitious property [4]. How to effectively improve the activity, changing the structural state and phase composition, and enhancing the cementitious property of coal gangue are the key of the efficient utilization of coal gangue in building materials.

Generally, coal gangue was mainly activated by mechanical activation, thermal activation, chemical activation and a combination of two and more than two kinds of activation method, namely, composite activation. Li et al. [5] studied the conversion mechanism of thermal activation process, the results showed that calcined component $Al_2O_3•2SiO_2•2H_2O$ was converted into $Al_2O_3•2SiO_2$, and amorphous SiO_2 and Al_2O_3, which were the main sources of the activity of coal gangue. Gu and Wang et al. [6] implemented the comparative study on the activation of coal gangue and coal gangue cement by chemical activators. The results showed that the strength of coal gangue cement increased significantly by the activators with 2 %Na_2SO_4 and 4 %$Ca(OH)_2$.

Liu et al. [7] studied the mechanical activation of calcined coal gangue and found that mechanical activation gradually made the kaolin crystal into amorphous state, and crystalline degree of α-quartz was decreased. Coal gangue after the different activation process, and the cementitious property was markedly enhanced.

These activation methods of mechanical activation, thermal activation, chemical activation and composite activation can effectively enhance the activity of coal gangue. Whether these methods can effectively enhance the activity of coal gangue with high iron and silica contents remains unknown, which is activation method better to prepare cementitious materials. In this paper, the coal gangue of Liupanshui areas was used as the raw material. We adopted the thermal activation and composite activation to determinate the coal gangue influence of mineral composition, crystal structure, morphology change. The cementitious property of coal gangue based unburned bricks were further measured. This way provided scientific and technical references for the preparation of cementitious materials by coal gangue with high iron and silica contents.

2. Experimental procedures

2.1 Raw materials

Coal gangue used in this experiment was obtained from Liupanshui city. Slag was collected from Shuicheng Iron and Steel Group Co., Ltd. Lime was industrial grade, and the effective calcium oxide content was 80%. Cement (P.O.32.5) was supplied from Liupanshui cement plant. Sodium sulfate as analytical reagent was produced by Xiya Reagent Factory.

2.2 Coal gangue activation procedure and preparation of coal gangue based unburned bricks

Coal gangue was grinded by GJ- III pulverizer and screened by 120 mesh sieve. The coal gangue was named as R-gangue. Coal gangue powder grinded to 120 meshes was placed in XL-III muffle furnace, heated to 800°C, preserved for four hours, and cooled to room temperature. The activated coal gangue was named as T-gangue. Lime and coal gangue powder grinded to 120 meshes (mass ratio 1:9) was heated from room temperature to 800°C, preserved for four hours, and cooled to room temperature. The mixture was grinded to 180 meshes by XMB-70 rod mill. The active coal gangue was named as CTM-gangue.

The activated coal gangue, slag, cement, sodium sulfate and lime were mixed at the mass ratio of 37:6:5:1:1. The moisture content of mixture was 22%. The 40 mm × 40 mm × 160 mm test bricks were prepared in the ZT-96 cement mortar swing table, and conducted with steam curing for 12 hours at 90°C, then taken out after natural cooling.

2.3 Testing conditions

The crystal phase composition of samples was analysed using X'Pert PRO X-ray diffractometer (PANalytical, Netherlands). The framework of samples was analyzed by the Nicolet FTIR-IS50 spectrometer (Nicolet, USA). Morphology of samples was analyzed using Zeiss Sigma X-Max 20 field-emission electron microscope (ZEISS, Germany). The strength tests were analyzed using YAW-2000 compression- testing machine (HST/ST, China).

3. Results and discussion

3.1 XRD analysis

XRD patterns of coal gangue by different activation methods were presented in figure 1. We could find that the mineral compositions in coal gangue mainly included quartz, kaolinite, mica, siderite and anorthite, etc. The characteristic peaks of kaolinite disappeared in thermal activation coal gangue, and the peak intensity of quartz and anorthite was weakened obviously. Kaolinite belonged to a layered structure, and the dehydration and decomposition reaction occurred at 700-1000°C. Quartz belonged to the trigonal system, and anorthite belonged to triclinic system. They had stable chemical nature, and were not likely to be activated. In addition to the above characteristics of thermal activation, composite activation coal gangue also showed the characteristic peaks of lime. Compared with raw coal gangue, the activated coal gangue had sharp peaks at 20.6°, 29.2°, and 32.9°, which was caused by crystal transformation of the mineral components.

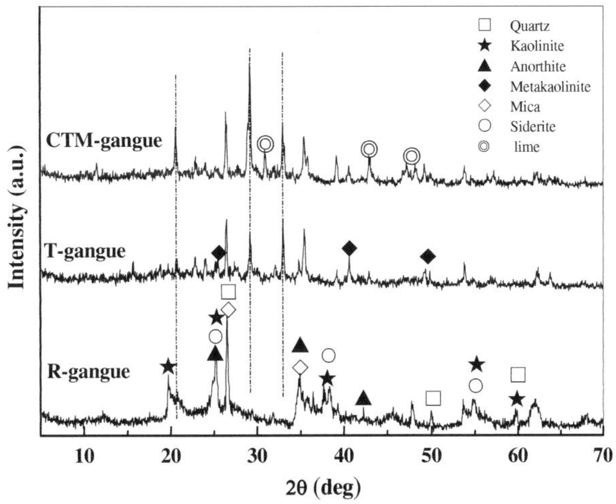

Figure 1. XRD patterns of coal gangue with different activation methods.

3.2 FT-IR analysis

Figure 2 displayed the FT-IR of coal gangue by different activation methods. The absorption peak at 1122cm^{-1} of R-gangue corresponded to the Si-O-Si stretching vibration absorption peak. The absorption peak at 1627cm^{-1} corresponded to Si-O-Si asymmetric stretching vibration and O-Si-O stretching vibration absorption peak. 876cm^{-1} was attributed to Si-O-Si bending vibration absorption peak, 3439cm^{-1} was related to O-H stretching vibration absorption peak. These absorption peaks indicated that coal gangue contained kaolinite mineral [8]. 1431cm^{-1} corresponded to CO_3^{2-} stretching vibration absorption peak, indicating that the coal gangue contained carbonate. 469cm^{-1} was the characteristic bands of quartz. The sharp degree of the stretching vibration absorption peak at 469cm^{-1} of the activated coal gangue was decreased. That suggested that the degree of crystallinity of quartz in the activated coal gangue was reduced. An absorption peak appeared at 558cm^{-1} after the activated coal gangue, and this was a characteristic spectra of metakaolinite [8, 9]. It resulted from that kaolinite was decomposed in the process of activation, and metakaolinite was formed. This was consistent with XRD analysis results. In addition, O-H stretching vibration absorption peak of $Ca(OH)_2$ occurred at 3623cm^{-1} and 3698cm^{-1}, this showed that the added CaO was carbonized and slaked in the composite activation process.

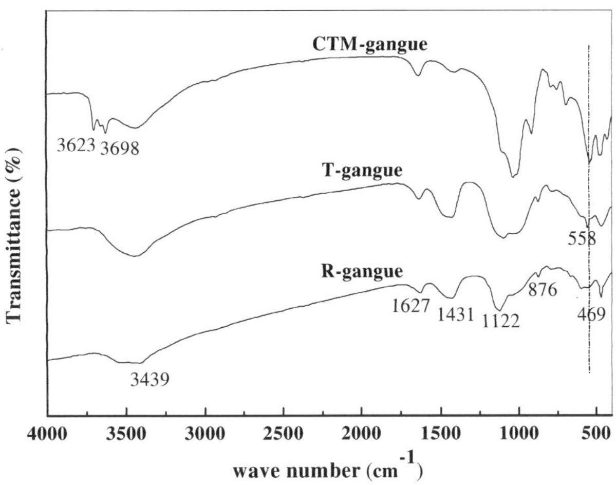

Figure 2. FT-IR patterns of coal gangue with different activation methods.

3.3 SEM analysis

SEM patterns of coal gangue by different activation methods were showed in figure 3. Coal gangue in figure 3(a) had compact structure. Combined with XRD, it could be seen that it was composed of massive quartz and granular kaolinite. In figure 3(b), the structure of coal gangue presented a loose state, and a large number of microspores occurred. This was due to the decomposition of kaolinite and organic carbon during the thermal activation process. The structures were connected with the porous state, multiple broken bonds, soluble matters, high internal energy [10]. Composite activation coal gangue in figure 3(c) also showed some molten states. The added lime as the mineralizer reduced the reaction bonding between particles, increased the chemical activity, and promoted the solid phase reaction. Meanwhile, the liquid phase property was changed, which was conductive to enhance the cementitious property of building materials prepared by coal gangue.

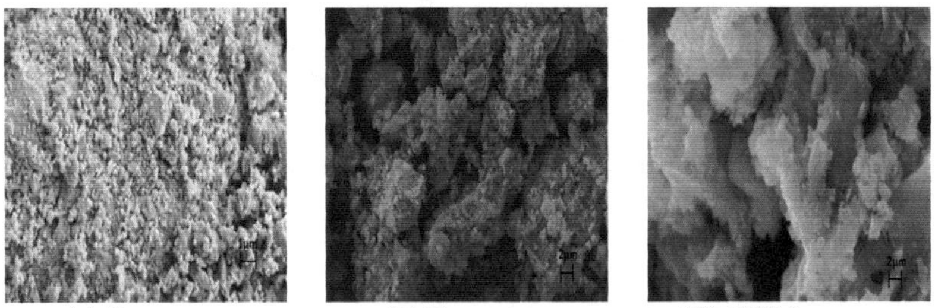

(a) R-gangue (b) T-gangue (C) CTM-gangue

Figure 3. SEM patterns of coal gangue with different activation methods.

Based on the above analysis, decomposition of kaolinite was the main source of the activity of coal gangue. After heat activation, the kaolinite removed the hydroxyl and the original layer structure was kept. However, the atoms had undergone large dislocation, forming metakaolinite with poor crystallinity. Metakaolinite could be further transformed into amorphous SiO_2 and Al_2O_3. The mineral structure was in metastable state with porous state, multiple internal broken bonds and large specific areas. On the one hand, coal gangue decreased the particle size by mechanical activation and increased the specific areas, the particle surface free energy increased. On the other hand, the lime which was added to coal gangue provided an alkaline environment promoting the decomposition of mineral and reducing the bond energy between the reactant particles, then the reaction activity was improved [11].

3.4 Cementitious property

In order to compare the activity of the activated coal gangue, the coal gangue based unburned bricks were prepared according to the method in 1.2. The compressive and flexural strengths were tested at 3d, 7d, and 28d as shown in figure 4 and 5. The strength standard of test bricks was accordance with JC/T422-2007 MU15 standard.

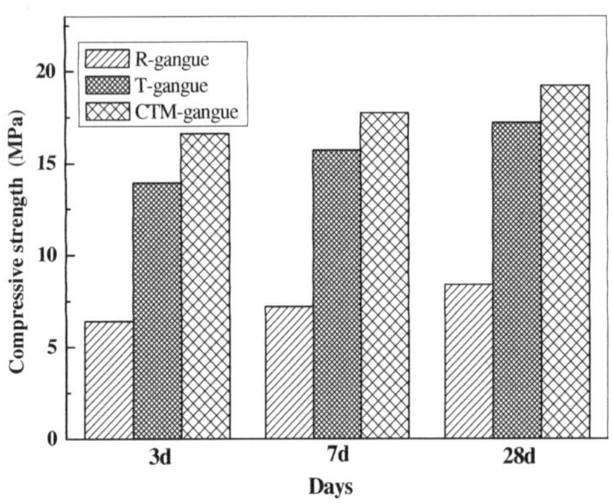

Figure 4. Compressive strength of coal gangue based unburned bricks.

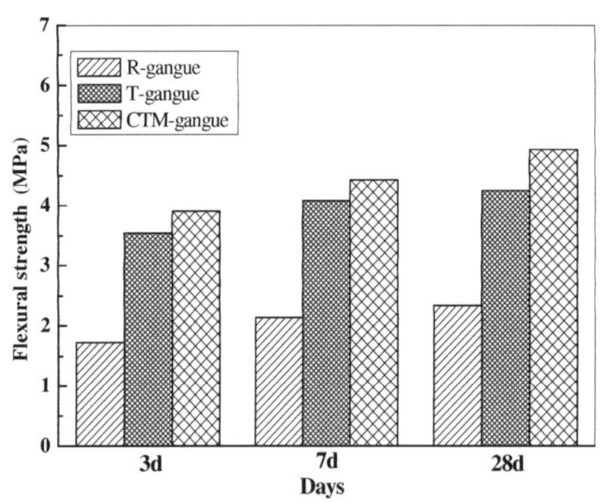

Figure 5. Flexural strength of coal gangue based unburned bricks.

As shown in figure 4 and 5, the compression and flexural strengths of the unburned bricks prepared by the activated coal gangue were higher than those of the unburned bricks prepared by raw coal gangue. In the experiment process, we found that the unburned bricks prepared by raw coal gangue had black powder on the surface, which seriously affected the appearance quality of the test bricks. Moreover, the strength of the test bricks were not up to the requirements. The two kinds of activation methods could effectively improve the cementitious property of coal gangue based unburned bricks. The cementitious property of composite activation was better than that of the single thermal activation. The compressive strength at 3d was 16.62MPa, and the flexural strength was 3.90MPa. At the 28d, the compressive strength was improved by 15.64%, and the flexural strength was improved by 26.41%. The cementitious property could fully satisfy the standard of JC/T422-2007 MU15.

4. Conclusions

i: The activity of coal gangue with high iron and silica contents in Liupanshui was activated. The thermal activation decomposed kaolinite in coal gangue to form the metastable structure with porous state, multiple internal broken bonds and large specific areas. Based on thermal activation, lime added in the composite activation provided the alkaline environment, which reduced the bond energy between the particles in reactant and the degree of crystallinity of quartz in coal gangue.

ii: The thermal activation and composite activation could effectively improve the cementitious property of unburned bricks. The effect of composite activation on the coal gangue was better than single activation. The cementitious property of the unburned bricks prepared by composite activation coal gangue could fully meet the standard of JC/T422-2007 MU15.

5. References

[1] Wang S W and Liang F 2003 *Techniques and Application of Converting Solid Wastes into Resources.* Beijing: Metallurgical Industry Press, 276-7.
[2] Len F G 2000. *Build. Sci. Res. Sichuan.* **26(2)** 44-6.
[3] Jiang A L 2000 *China Coal,* **26(3)** 25-7.
[4] Deng, D H andCen W L 1999 *China Min. Mag.* **8(6)** 87-91.
[5] LI Y F, Wang W X and Yang X Y 2007 *J. Chin. Ceram. Soc.* **35(9)** 1258-69.
[6] Gu B W and Wang P M 2009 N. Build. Mater. **36(5)** 12-5.
[7] Liu Y M 2007 *Application research on activity of activated coal gangue.* Xian: Xi'an University of Architecture and Technology.
[8] Feng Q and Wang P M 2005 *J. Build. Mater.* **8(3)** 215–21.
[9] Yang N R and Yue W H 2000 *The Handbook of Inorganic Metalloid Materials Atlas.* Wuhan: Wuhan University of Technology Press, 230-40.
[10] Song X Y, Gong C C and Li D X 2004 *J. Chin. Ceram. Soc.* **32(3)** 358-63.
[11] Li A S 1995 *Build. Mater. Shanxi.* **(1)** 19-23.

Acknowledgment

The financial support for this work was provided by the Guizhou Provincial Science and Technology Foundation, No. LKLS[2013]29, National Natural Science Foundation of China, No. 50874107,Guizhou Provincial Ordinary College Innovation Team of Coal Solid Waste Recycling Technology, No. [2014]46, Solid Waste Recycling Laboratory of Coal Utilization, No. 2011-278, Guizhou Provincial Educational Innovation Foundation, No. KY[2017]054, Discipline of Guizhou Provence, No. ZDXK[2016]24, 2011 Collaborative Innovation Center of Guizhou Province, [2016] 02.

Heat Transfer Analysis of Thermal Protection Structures for Hypersonic Vehicles

Chen Zhou[1], Zhijin Wang[1*], Tianjiao Hou[2]

[1]Minister Key Discipline Laboratory of Advanced Design Technology of Aircraft, Nanjing University of Aeronautics and Astronautics, Nanjing, 210016, China
[2]College of Aeronautics, Nanjing University of Aeronautics and Astronautics, Nanjing, 210016, China

Abstract. This research aims to develop an analytical approach to study the heat transfer problem of thermal protection systems (TPS) for hypersonic vehicles. Laplace transform and integral method are used to describe the temperature distribution through the TPS subject to aerodynamic heating during flight. Time-dependent incident heat flux is also taken into account. Two different cases with heat flux and radiation boundary conditions are studied and discussed. The results are compared with those obtained by finite element analyses and show a good agreement. Although temperature profiles of such problems can be readily accessed via numerical simulations, analytical solutions give a greater insight into the physical essence of the heat transfer problem. Furthermore, with the analytical approach, rapid thermal analyses and even thermal optimization can be achieved during the preliminary TPS design.

1. Introduction

Space vehicles encounter severe aerodynamic heating when traveling at high speeds in the atmosphere. Thermal protection systems (TPS) are required to maintain the underlying structural temperature within acceptable limits [1]. Since the inner surface temperature is usually taken as a design variable and the temperature distributions are needed to carry out the following linear static analysis as well as bulking analysis, it is of great need and importance to be able to predict the temperature profile through the TPS. Normally, numerical simulations can be adopted to achieve this purpose, however, analytical solutions, whether exact or approximate, are always useful in engineering analysis since they could offer a clear insight into the physical essence of the heat transfer problem [2].

De Monte [3] summarized a few analytical methods that could be used to address transient heat transfer problems, as well as their strengths for specific applications. Miller and Weaver [4, 5] developed analytical models based on techniques of integral transforms and separation of variables, and described the transient temperature distribution through an air-filled box structure and a multi-layered plate, respectively. Finite element (FE) models were also established for comparison purpose and showed a good fit. Ferraiuolo and Manca [6, 7] developed an analytical procedure and a semi-analytical procedure based on the integral method and the Green's function method for layered TPS and hot structures.

In the current work, analytical approaches are used to study the transient heat transfer problem of the TPS. Laplace transform and integral method are adopted to describe the temperature distribution through the TPS subjected to aerodynamic heating. Radiation to ambient environment and time dependent heat flux are taken into account. Comparisons of the results from analytical approaches and one-dimensional (1-D) FE analyses are conducted for validation.

2. Model description

Excessive aerodynamic heating occurs when a space vehicle travels at hypersonic speeds through a planetary atmosphere. Since the temperature gradient along the TPS thickness is much larger than those along the other two directions, the heat transfer problem of a TPS is usually simplified to a 1-D issue [8], as shown in figure 1. In this study, thermal properties are assumed to be constant with respect to time and temperature.

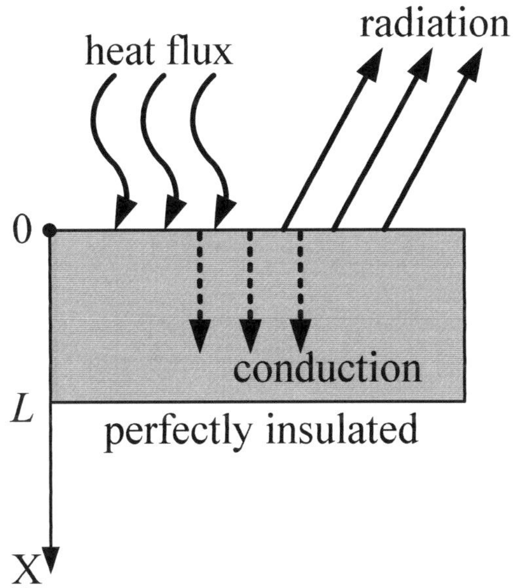

Figure 1. Illustration of boundary conditions.

As shown in figure 1, the TPS is subjected to a heat flux at x=0, while conservatively assumed to be adiabatic at x=L. Some heat is conducted through the TPS, and some heat is radiated away to the ambient environment.

The thermal conduction differential equation to be solved in this study is given as

$$\frac{1}{\alpha}\frac{\partial T}{\partial t} = \frac{\partial^2 T}{\partial x^2} \tag{1}$$

where α is the thermal diffusivity. The boundary conditions to define this problem are

$$-k\left.\frac{\partial T}{\partial x}\right|_{x=0} = q_h - q_r = q_h(t) - \varepsilon\sigma(T_s^4 - T_a^4) \tag{2}$$

$$k\left.\frac{\partial T}{\partial x}\right|_{x=L} = 0 \tag{3}$$

and the initial condition is

$$T\big|_{t=0} = T_i \tag{4}$$

where k is the thermal conductivity, ε is the surface emissivity, σ is the Stefan-Boltzmann constant, q_h and q_r refer to the incident heat flux and radiation, T_s and T_a stand for the surface temperature and ambient temperature, respectively.

3. Analytical model

3.1. Integral method

The heat transfer problem becomes nonlinear once radiation is involved. Exact analytical solutions are always difficult or even impossible to obtain for nonlinear problems. Herein, integral method is adopted to gain approximate solutions. The differential equation of heat conduction is integrated over a phenomenological distance $\delta(t)$, i.e. thermal layer.

The integral method for a 1-D finite region is performed in two stages [2]. For the first stage where $\delta(t)<L$, according to the definition of thermal layer, boundary conditions at $x=\delta(t)$ can be described as

$$\left.\frac{\partial T}{\partial x}\right|_{x=\delta(t)} = 0 \tag{5}$$

$$T\big|_{x=\delta(t)} = T_i . \tag{6}$$

To simplify the mathematical derivation process, following expressions are used

$$Q_1(t) = \frac{q_h(t)}{k} \tag{7}$$

$$Q_2(T_s) = \frac{-\varepsilon\sigma\left(T_s^4 - T_a^4\right)}{k}. \tag{8}$$

Equation (1) is integrated from $x=0$ to $x=\delta(t)$. By introducing the boundary conditions of equations (2) and (5)-(6), together with expressions (7) and (8), the corresponding energy integral equation is obtained

$$Q_1(t) + Q_2(T_s) = \frac{1}{\alpha}\frac{d}{dt}\left[\int_0^{\delta(t)} T dx - T_i\delta(t)\right]. \tag{9}$$

A quadratic polynomial representation is chosen for the temperature profile

$$T_1(x,t) = a_0 + a_1 x + a_2 x^2 . \tag{10}$$

By applying equations (2) and (5)-(6), the coefficients a_0, a_1 and a_2 are determined, and the temperature profile can be further expressed as

$$T_1(x,t) = T_i + \frac{\delta(t)}{2}\left[Q_1(t) + Q_2(T_s)\right]\left(1 - \frac{x}{\delta(t)}\right)^2 . \tag{11}$$

By setting $x=0$, the relation between the surface temperature T_s and the thermal layer $\delta(t)$ is obtained

$$T_s = T_1(0,t) = T_i + \frac{\delta(t)}{2}\left[Q_1(t) + Q_2(T_s)\right] \tag{12}$$

according to which the thermal layer $\delta(t)$ can be expressed as a function of T_s. Then the substitution of equation (11) and $\delta(t)$ into the energy integral equation (9) gives

$$\frac{3}{2}\alpha\left[Q_1(t) + Q_2(T_s)\right] = \frac{d}{dt}\left[\frac{(T_s - T_i)^2}{Q_1(t) + Q_2(T_s)}\right]. \tag{13}$$

The heat flux profile $q_h(t)$ is usually known as an input, and $\varepsilon\sigma T_a^4$ is omitted since it is small compared with the other terms in equation (2). By substituting $Q_1(t)$ and $Q_2(T_s)$, an ordinary differential equation (ODE) for the determination of T_s is obtained

$$2k^2(T_s - T_i)\left[q_h(t) + \varepsilon\sigma T_s^4 - 2\varepsilon\sigma T_i T_s^3\right]\frac{dT_s}{dt} = \frac{3}{2}\alpha\left[q_h(t) - \varepsilon\sigma T_s^4\right]^3 + k^2(T_s - T_i)^2\frac{dq_h(t)}{dt}. \quad (14)$$

Equation (14) can then be integrated numerically with the initial condition (4). Specifically, Maple is employed here in this study to solve the ODE. Once the relation between T_s and t is established, temperature distribution for the range $0<x<\delta(t)$ is therefore derived

$$T_1(x,t) = T_i + (T_s - T_i)\left[1 - \frac{Q_1(t) + Q_2(T_s)}{2(T_s - T_i)}x\right]^2 \quad (15)$$

while temperature for the range $\delta(t)<x<L$ stays at the initial value.

The ending time of the first stage t_L is obtained by setting $\delta(t)=L$. For the second stage, i.e. $t>t_L$, the differential equation (1) is integrated from 0 to L

$$Q_1(t) + Q_2(T_s) = \frac{1}{\alpha}\frac{d}{dt}\int_0^L T dx. \quad (16)$$

It is assumed that the temperature has a quadratic polynomial distribution through the thickness

$$T_2(x,t) = a_0 + a_1 x + a_2 x^2. \quad (17)$$

Then the application of the boundary conditions (2) and (3) solves the coefficients and yields

$$T_2(x,t) = T_s - \left[Q_1(t) + Q_2(T_s)\right]x + \frac{Q_1(t) + Q_2(T_s)}{2L}x^2. \quad (18)$$

The substitution of this expression into the energy integral equation (16) gives the following relation

$$3\alpha\left[Q_1(t) + Q_2(T_s)\right] = \frac{d}{dt}\left[3T_s L - \left(Q_1(t) + Q_2(T_s)\right)L^2\right]. \quad (19)$$

Similarly, $Q_1(t)$ and $Q_2(T_s)$ are introduced into equation (19), and $\varepsilon\sigma T_a^4$ is neglected

$$(3kL + 4\varepsilon\sigma L^2 T_s^3)\frac{dT_s}{dt} = \frac{dq_h(t)}{dt}L^2 + 3\alpha q_h(t) - 3\alpha\varepsilon\sigma T_s^4. \quad (20)$$

By combining the initial condition of the second stage $T_s(t_L)=T_1(0, t_L)$, T_s can be solved numerically and then substituted into equation (18) to determine the temperature profile of the second stage $T_2(x, t)$.

3.2. Laplace transform and integral method

For cases where the incident heat flux is constant and the surface temperature is relatively low before the heat penetration approaches to the inner side of the TPS, radiation is neglected when its contribution is less than 5% of the aerodynamic heating

$$q_r = \varepsilon\sigma(T_s^4 - T_a^4) < 0.05 q_h. \quad (21)$$

In this simplified linear case, the Laplace transform method is used to obtain the temperature profile through the structure thickness, which is as follows

$$T_1(x,t) = T_i + \frac{q_h L}{k}\left[Fo + \frac{x^2}{2L^2} - \frac{x}{L} + \frac{1}{3} - \frac{2}{\pi^2}\sum_{n=1}^{\infty}\frac{1}{n^2}\cos(n\pi\frac{x}{L})\mathrm{e}^{-(n\pi)^2\cdot Fo}\right] \quad (22)$$

where $Fo=\alpha t/L^2$, is the Fourier number.

By setting q_r=0.05q_h, the ending time of the first stage t_r is obtained, after which the radiation must be taken into account as it is comparable to the aerodynamic heating. Therefore, integral method is adopted to gain approximate solutions for the second stage.

To simplify the mathematical derivation process, the following expression are defined

$$Q(T_s) = \frac{q_h - \varepsilon\sigma(T_s^4 - T_a^4)}{k}. \tag{23}$$

An energy integral equation similar to equation (16) is obtained by integrating equation (1) from 0 to L with the boundary conditions (2) and (3)

$$Q(T_s) = \frac{1}{\alpha}\frac{d}{dt}\int_0^L T dx. \tag{24}$$

A quadratic polynomial expression is employed to describe the temperature profile

$$T_2(x,t) = a_0 + a_1 x + a_2 x^2. \tag{25}$$

The coefficients are solved using (2) and (3). The resulting temperature profile becomes

$$T_2(x,t) = T_s - Q(T_s)x + \frac{Q(T_s)}{2L}x^2. \tag{26}$$

Afterwards, the temperature profile is substituted into the equation (24),

$$3\alpha Q(T_s) = \frac{d}{dt}\left[3T_s L - Q(T_s)L^2\right]. \tag{27}$$

Again, $\varepsilon\sigma T_a^4$ is ignored to simplify the derivation. The following ODE to determine the surface temperature T_s is obtained

$$(3kL + 4\varepsilon\sigma L^2 T_s^3)\frac{dT_s}{dt} = 3\alpha q_h - 3\alpha\varepsilon\sigma T_s^4. \tag{28}$$

T_s can be computed numerically with the initial condition $T_s(t_r)=T_1(0, t_r)$. Once T_s is obtained, it is substituted into equation (26) to calculate the temperature distribution.

4. Validation

4.1. Constant heat flux case

The leading-edge radius significantly influences the heat flux. Heat flux is inversely proportional to the square root of the radius. For hypersonic vehicles with sharp leading edges, the maximum temperature might exceed 2000°C, in which case the widely used C/C material in blunt leading edges is not feasible because of the severe oxidization issue. Ultra-high temperature ceramics (UHTCs) [9] have high melting temperature and good chemical and thermal stability. Unlike the working mechanism of traditional insulation materials, UHTCs normally have a high thermal conductivity and transfer heat through the material and reradiate it at cooler locations away from the stagnation region.

Considering a 120mm UHTC leading edge and the 1-D thermal model shown in figure 1 is used. ZrB_2 with SiC of 20% volume fraction is adopted and the material thermal properties are shown in table 1. The aerodynamic heating q_h=200kW/m^2 is applied at x=0. Radiation to the ambient atmosphere is considered with a constant emissivity of 0.86, which is typical for the TPS exterior surface treatments. Both ambient temperature and initial structural temperature are assumed to be 295K.

Table 1. Materials thermal physical properties.

Parameter	ρ (kg/m^3)	k (W/(m·K))	c (J/(kg·K))
ZrB_2/20vol%SiC	5570	99	625

The analytical results of the previously mentioned methods are compared with those obtained by FE analyses. The FE model is established using the 2-node link element (DC1D2) with a refined mesh density. Figure 2 shows the temperature variations with time for x=0, L/2 and L. It is seen that the combination of Laplace transform and integral method can predict the temperature accurately at the beginning while later give slightly higher results than the FE results. The prediction of the integral method has a small deviation at the early stage, after that it approximately coincides with the FE results.

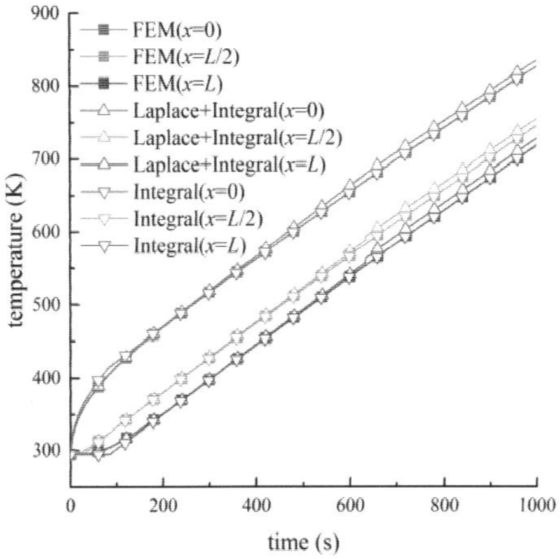

Figure 2. Temperature variations with time at x=0, L/2 and L.

4.2. Variable heat flux case

The integral approximate method can also deal with variable heat flux cases. A transient heat flux shown in figure 3 is applied at x=0. The other parameters are the same as those used in the constant heat flux case in section 4.1.

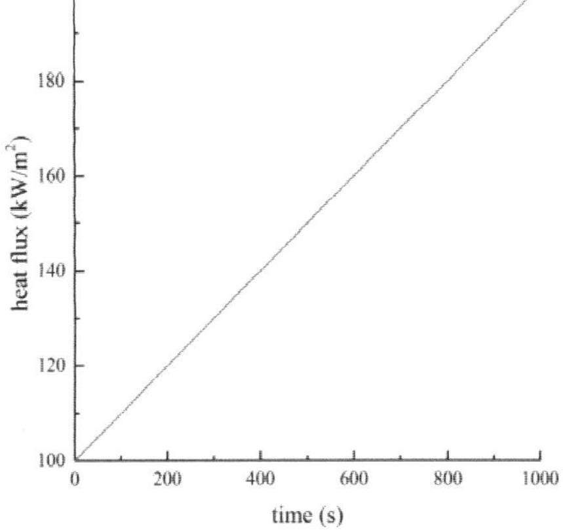

Figure 3. Heat flux profile.

Figure 4 shows the comparison between the results of integral method and FE analyses. Similarly, the maximum error occurs at the quite early stage of the whole heating process. Afterwards, the integral method shows a good fit with the FE results.

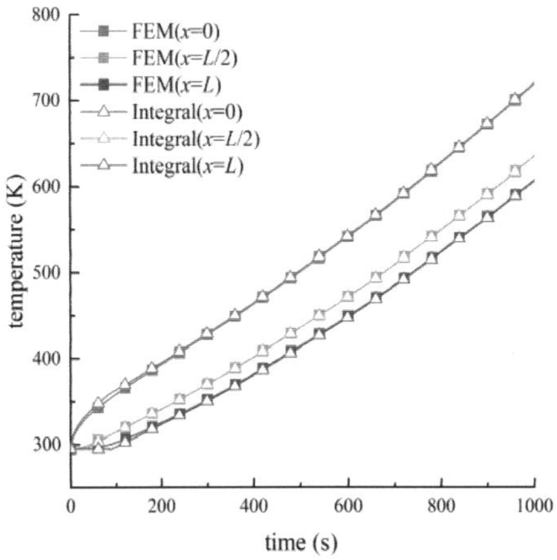

Figure 4. Temperature variations with time at x=0, L/2 and L.

5. Conclusion

The approaches of Laplace transform and integral method are adopted to predict the temperature distribution through the TPS thickness. Radiation boundary conditions are taken into account. Both constant heat flux and variable heat flux cases are studied. Comparisons with the FE results show that both approaches can predict the temperature distribution satisfactorily under constant heat flux and the integral method is also capable of solving variable heat flux cases with a relatively high accuracy.

References

[1] Glass D E 2008 *Ceramic Matrix Composite (CMC) Thermal Protection Systems (TPS) and Hot Structures for Hypersonic Vehicles.* 15th AIAA International Space Planes and Hypersonic Systems and Technologies Conferences, Dayton, Ohio: AIAA 2008-2682.

[2] Hahn D W and Ozisik M N 2012 *Heat Conduction. Third Edition.* John Wiley & Sons.

[3] De Monte F 2000 *Int. J. Heat. Mass. Transf.* **43** 3607-19.

[4] Miller J R and Weaver P M 2003 *J. Thermophys. Heat. Transf.* **17** 269-77.

[5] Miller J R and Weaver P M 2003 *Comp. Struct.* **59** 267-78.

[6] Ferraiuolo M, Manca O and Riccio A. 2009 *ASME Int. Mech. Eng. Congr. Exposit.* 1-9.

[7] Ferraiuolo M and Manca O. 2012 *Int. J. Therm. Sci.* **53** 56-70.

[8] Myers D E, Martin J C J and Blosser M L 2000 *Parametric weight comparison of advanced metallic, ceramic tile, and ceramic blanket thermal protection systems* NASA Technical Memorandum, NASA TM-2000-210289.

[9] Tandon R, Dumm H P, Corral E L, et al. 2006 *Ultra high temperature ceramics for hypersonic vehicle applications.* Livermore: Sandia National Laboratories.

Acknowledgments

This work was supported by Funding of Jiangsu Innovation Program for Graduate Education [Grant No. CXLX13_163], the Fundamental Research Funds for the Central Universities [Grand No. NZ2016101] and a project Funded by the Priority Academic Program Development of Jiangsu Higher Education Institutions (PAPD).

Research Progress in Plasma arc welding of Magnesium Alloys and Magnesium Matrix Composites

Li Hui[1], Zou Yang[2], Li Yongbo[2], Jiao Lei[2], Hou Ruijun[2]

[1]School of Materials Science and Engineering, Jiangsu University of Science and Technology Zhenjiang Jiangsu 212003, China)
[2]School of Materials Science and Engineering, Jiangsu University, Zhenjiang, Jiangsu 212013, China)

Abstract. Magnesium alloys and magnesium matrix composites by means of its excellent performance have wide application prospect in electronics, automotive, biotechnology, aerospace field, and welding technology has become a key of restricting its application. This paper describes the welding characteristics of magnesium, the obvious advantages in the application and the domestic and foreign research advance technology of plasma arc welding of magnesium, and summarizes the existing problems and development trends of plasma arc welding technology of magnesium.

1. Introduction

Magnesium alloy and magnesium matrix composites in some performance and price are better than the aluminum alloy and aluminum matrix composites, and will be more competitive in 21th Century. The density of magnesium at 20℃ is only 1.738g/cm^3, which is 36% lower than the density of aluminum. Compared to the aluminum matrix composite, magnesium matrix composite has higher specific stiffness, specific strength, electromagnetic shielding properties, damping properties, and the lower cost of processing and recycling. In recent years, the price of magnesium alloy keep declining, and it is 10~20% cheaper than aluminum alloy in the same volume, considerably eliminating the restriction of the development of magnesium matrix composites [1, 2]. Magnesium alloys and magnesium matrix composites have a wide application prospect in the development of high-tech industry, biological technology and other fields, due to performance and lower price, but the development and application of magnesium alloy and magnesium matrix composites have been hindered by welding technology. Welding technology of magnesium mainly have resistance spot welding, MIG welding, TIG welding, electron beam welding, friction stir welding, plasma arc welding and laser welding. This paper elaborate the current development of plasma arc welding of magnesium.

2 Magnesium welding characteristics and plasma arc welding

2.1. The welding characteristics of magnesium alloy and magnesium matrix composites

Active chemical properties of magnesium and high melting point of oxide limit the range of the development and application of magnesium, but magnesium matrix composite not only has innate excellent properties of magnesium alloy, and improves the mechanical properties and oxidation resistance, but also has more competitive in development potential of high-tech field than other metal materials. These properties will produce a series of difficulties in the process of welding of magnesium alloys and magnesium matrix composites, and because of the reinforcement phase existing, the welding puddle reaction is more complicated. The main problems are as follows [3]: oxidation and evaporation, coarse grain, thermal stress, weld joints collapse, porosity, heating cracking, reaction between the matrix and the reinforcement, burnthrough of thin workpiece. In addition, due to magnesium active chemical property, welding needs under the protection of inert gas or welding flux to prevent magnesium burning.

2.2. The plasma arc welding characteristics of magnesium alloy and magnesium matrix composites
In 1954, in the American UnionCarbide company RobertGate found after arc compression the plasma arc more concentrated, the arc temperature and jet velocity also increased significantly and the density of arc free energy high, so the welding method known as plasma arc welding had good effect on performance and application in industry [4].

Plasma arc (or compression arc) is cylindrical shape with good straightness and small diffusion angle. Besides it has higher temperature and energy density, more stable properties, lower flame flow velocity, higher rigidity, faster welding speed, and stronger penetration compared to normal arc. According to the difference of the power connection and plasma arc, plasma arc welding includes transfer-plasma arc welding, non-transferred plasma arc welding, twin-body plasma arc welding. The plasma arc welding makes magnesium matrix composite melted through at one time without subplate and the weld joint surface smooth, showing good welding performance.

Compared with plasma arc welding, the resistance spot welding of magnesium has narrow range of application, tendency to adhere and welding deformation due to the small surface tension of molten magnesium when the welding current is too high; the place between electrode wire and the reinforcing phase in molten liquid matrix is bound during MIG welding and spatter in high temperature; Friction-stir welding of magnesium matrix composite with low temperature and small deformation, is a kind of environment-friendly welding method, but the fractional reinforcing phase and the changed organization structure lead to the softening and strength reduction of welding joint. As using laser welding, the laser welding power of magnesium must reach to a certain threshold, and because of the low ionization energy of magnesium alloy, high thermal conductivity and the small absorption rate of laser radiation, it will lead to poor welding stability .

According to the differences of physical properties of matrix and reinforced phase, the welding produce is complicated and not easy to control, therefore plasma arc welding has obvious advantages for magnesium matrix composite compared with other welding methods.

3. Research progress in plasma arc welding of magnesium alloys and magnesium matrix composites

3.1. Research progress in plasma arc welding of magnesium alloys
China's institute comprehensively studied plasma arc welding of magnesium alloys, and published papers in the domestic and foreign periodicals. Early Dalian University of Technology research team deeply studied the influence of variable polarity plasma arc welding (VPPAW) on welding seam and joint of AZ31 magnesium alloy, besides the method had the advantages of good welding structure, small deformation of the magnesium alloy, good performance of joint and simple operation process.

In the variable polarity plasma arc welding of thick plate, Liu Liming, Shen Yong [5] by optimizing the welding parameters such as welding speed, welding current, ion flow number and time ratio, make particle fine and uniform distribution, thus the joint strength was more than 90% of the parent material.

In terms of variable polarity plasma arc welding and keyhole behavior, Jiang Jianbo et al. [6] improved the method of forming protection and beveling, obtaining double sided forming joint with small heat affected zone by single layer welding. The mechanical properties were excellent and the hardness was higher than that of parent metal.

In terms of corrosion behavior, Zhao Hongkai et al [7] put salt spray corrosion test and electrochemical test on the welding joint. The result showed that the corrosion current density of the welding seam is similar to that of the parent metal and the microstructure of the joint was well formed without obvious heat affected zone. Therefore, the corrosion resistance of the welding seam was stronger than that of the parent metal.

In terms of thermal cycle simulation, Yu Xike [8] drew a conclusion high welding temperature time long, fast cooling as well as experimental data optimization through the ANSYS finite element analysis of AZ31 magnesium alloy welding, establishing the double ellipsoid heat source model to realize the numerical calculation of the transient temperature field of the welding seam and better reflect the temperature distribution.

VPPAW was an efficient welding method for non-ferrous metals such as magnesium and aluminum. However, VVPPAW was sensitive to changes in welding process and experimental parameters due to its nature, leading to small reasonable parameter margin, narrow range and poor stability in the case of obtaining a good mechanical joint .Therefore, in the future, the research direction of VPPAW for magnesium alloys trended to comprehend the regularity of parameters, optimize the process, improve the quality of plasma arc welding and enlarge its application range.

W. G. Essers and A. C. Liefkent [9] (of PHILPS company in Holland) first proposed the Plasma-MIG hybrid Welding method with strong arc, the metal filling capacity and the high energy density of plasma arc, realizing the high-efficiency production. According to the position of plasma arc at the position of the gun the method was divided into (a) bias-type plasma welding gun, the plasma arc burns at the tungsten pole, and (b) nozzle-type welding gun, the plasma arc burns under the nozzle.

According to the physical and chemical properties of magnesium and aluminum and the similarity of the welding difficulties of them, the research directions of welding of magnesium alloys can be determined.

Yang Tao [10] studied on the coupling mechanism of Plasma-MIG-Hybrid arc welding process, and research of welding procedure. The outer plasma arc of MIG was compressed, and the internal plasma arc flowed into the outer to reduce the internal arc temperature, contributing to changing the transform of droplet, weakening the thermal effect of MIG arc, and then improving the stability of welding process.

Kohei, Ono et al. [11] used this method to weld aluminum alloy. The result showed that MIG welding wire can be heated not only by resistance and MIG arc, but also by plasma arc. Therefore, it has higher deposition rate than plasma arc welding.

BAI Yan et al. [12] Studied on the influence of plasma-MIG arc welding parameters on welding porosity by orthogonal test, and the paper showed plasma gas flow had the greatest influence on the porosity. Research group of Southwest Jiao Tong University [13] welded AZ31B magnesium alloy by the DC plasma -MIG-Hybrid welding. The result showed the experimental parameters which mainly affects the reinforcement weld and weld width of the workpiece had little effect on the welding penetration, and melted droplet spattered obviously, thus the joint strength is as high as 98.77% of the parent metal.

The active agent is helpful to increase welding speed and improve penetration and welding seam. After adding active agent, plasma arc welding cross section was funnel shape with smaller back width. Research on welding activator types of magnesium alloy and magnesium matrix composites and direction of development are few. Marya (of Colorado mining school) [14] studied chloride activator for magnesium alloy effecting on the workpiece during direct current TIG welding was investigated. Harbin Institute of Technology had also studied three kinds of activator: Fluoride activator, Chloride activator and Oxide activator, and the influence of three kinds of activator used in welding. The study of Welding Research Institute found that the key of plasma arc welding activator of magnesium alloys is to find out the order and maximum multiple of increasing the melting depth.

Claus Emmelmann et al. [15] used common welding equipment by plasma-laser-hybrid welding method, optimizing the optimal parameters by orthogonal experiments and compensating the disadvantages of laser welding and plasma welding. Under the optimum parameters, the production had the lower rejection rate, the more stable welding process, and the easier welding process.

3.2. Research progress in plasma arc welding of magnesium matrix composites

Although the performance of magnesium matrix composite is superior to magnesium alloy, the reinforcement phase size, distribution and interface make magnesium matrix composites more difficult to weld than magnesium alloys. Therefore, there is some research on the welding of composites at home and abroad.

Lei Yucheng et al. [16] carried out plasma arc in-situ welding of aluminum matrix composites. The result showed that ion gas (N_2+Ar) improving, filling material Ti increasing, in-situ welding or arc-ultrasonic method applying, contributing to reducing interface reaction, improving wettability, increasing fine particle. Thus, the defects of structure were reduced, and the joints had good mechanical properties. According to the similarity between magnesium matrix composites and aluminum composites, this method can still be applied to plasma arc welding of magnesium matrix composites and obviously resolve the problems of reinforcement phase in welding

Ruiying Zhang et al. [17] found that compared to the non-transferred plasma arc welding, twin-body plasma arc welding with an arc-end nozzle make the workpiece not connected to the loop and the energy loss small, thus this method provided a better method to control the plasma arc energy and expanded the application of plasma arc.

4. Outlook

(1) In terms of magnesium alloy and magnesium matrix composites, interface between reinforced phase and matrix will become the key of welding technology. The main research directions includes the formation of interface, the relationship between structure and performance, then further researching is consist of the reaction mechanism and synthesis methods between different reinforcement phase and the magnesium matrix. According to the 13th Five-Year national plan attached importance to environmental protection, how to resolve the problem of the environmental adaptability of magnesium alloys and magnesium matrix composites, to realize the recycling, and to achieve the sustainable development of the society, which will draw more attention.

(2) In terms of plasma arc welding technology, the main research direction aims at the relationship between process parameters with the welding joint performance, and the strength loss of weld joint attributed to various macroscopic welding defects in the joint. Thus the following research on molten-pool reaction not only needs to realize metallurgical bonding between the matrix, but also tneeds to get the good interface bonding of the welding joint, strengthening mechanical properties by reinforcement phase. The welding with ultrasonic and magnetic coupled field or friction stir will become the focus of the research on welding of magnesium.

(3) In terms of the existing welding technology, traditional welding is not good to resolve the existing welding problems, so it is necessary to develope an advanced welding method about magnesium alloy and magnesium matrix compositere for solvimg the existing problems .

References

[1] Wan H Y and Xiao X 2011 *J. Price Theory Pract.* **03** 50-1.
[2] Shi X M 2017 *N. China Nonferr. Met. News.* **06** 13007.
[3] Deng C P and Tong Y G 2010 *J. Electr. Weld. Mach.* **03** 10-4.
[4] Dong C L, Wu L and Chen S 2000 *J. Chinese Mech. Eng.* **11(5)** 577-81.
[5] Shen Y, Liu L M and Zhang Z Q 2005 *J. Weld. J.* **06** 1-4.
[6] Jiang J B, Liu L M, Zhu M L and Shen Y 2007 *J. Weld. Jour.* **05** 65-8.
[7] Zhao H K, Xiao F, Yang X D, Lu W and Li M T 2011 *J. Weld. Tech.* **08** 22-4.
[8] Yu X K 2010 *Thermal cycle simulation and structure properties of magnesium alloy by variable polarity,* Chongqing University of Technology.
[9] Jelmorini G 1979 *Method and device foe plasma-MIG-welding.* U.S. Patent 4,147,917.

[10] Yang T 2013 *Research on the technology of Plasma-MIG and coupled mechanism of arc welding 2219 Aluminum Alloy* Harbin Institute of Technology.
[11] Ono K, Liu Z, Era T, et al. 2009 *J. Weld. Nat.* **23(ll)** 805-9.
[12] Yan B, Gao H, Lin W, et al. 2010 *Trans. Nonferr. Met. Soc. China* **20(8)** 1392-6.
[13] Liu Z 2016 *Research on plasma-MIG hybrid welding process of AZ31B magnesium alloy profiles,* South West Jiao Tong University.
[14] Marya M 2002 *J. Weld. World* **46(7/8)** 7-21.
[15] Claus E, Marc K and Nikolai P 2011 *J. Phys. Proc.* **12** 194-200.
[16] Lei Y C, Li X, Chen G, Cheng X N and Zhang J H 2008 *J. Jiangsu Univ.* **(01)** 39-42.
[17] Zhang R Y, Jiang F and Chen S J 2016 *J. Manuf. Proc.* **(24)** 152-60.

Acknowledgement
This research was financially supported by the National Natural Science Foundation of China (51605206), Senior talent project of Jiangsu university (15JDG077) and Jiangsu university industrial centre student innovation project (ZXJG201622 ZXJG201623).

Study and Optimization of Helicopter Subfloor Energy Absorption Structure with Foldcore Sandwich Structures

Zhou HuaZhi[1, *] and Wang ZhiJin[1]
[1]Key Laboratory of Fundamental Science for National Defence -Advanced Design Technology of Flight Vehicles, Nanjing University of Aeronautics and Astronautics, 29 YuDao Street. Nanjing 210016, China

Abstract. The intersection element is an important part of the helicopter subfloor structure. In order to improve the crashworthiness properties, the floor and the skin of the intersection element are replaced with foldcore sandwich structures. Foldcore is a kind of high-energy absorption structure. Compared with original structure, the new intersection element shows better buffering capacity and energy-absorption capacity. To reduce structure's mass while maintaining the crashworthiness requirements satisfied, optimization of the intersection element geometric parameters is conducted. An optimization method using NSGA-Ⅱ and Anisotropic Kriging is used. A significant CPU time saving can be obtained by replacing numerical model with Anisotropic Kriging surrogate model. The operation allows 17.15% reduce of the intersection element mass.

1. Introduction

The subfloor structure is one of the most important structural components in helicopter energy-absorbing structures. So it has to be designed in order to limit the deceleration forces by structural deformation and provide post-crash structural integrity of the cabin floor [1]. Figure 1 shows a typical subfloor structure. In such a structure, the intersection elements play an important role in the overall crash response of the subfloor structure because of its high stiffness and strength. Experimental and numerical studies of intersection element are conducted by Bisagni C [1-4]. A fast reanalysis methodology based on neural network is proposed with the intent to reproduce the crash behaviour of structural intersections. Based on the methodology, size and topological optimization are conducted.

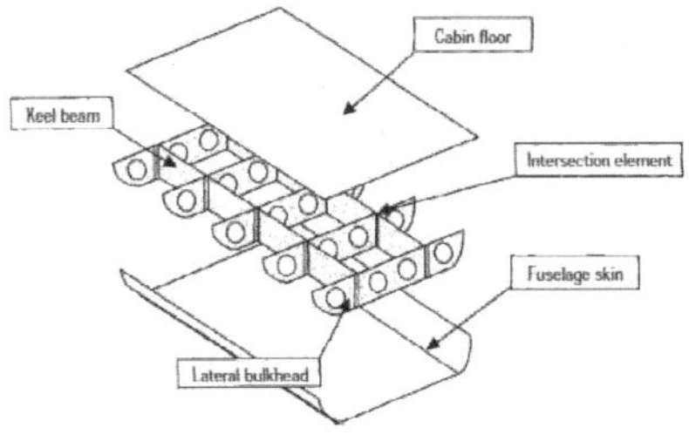

Figure 1. Typical subfloor structure.

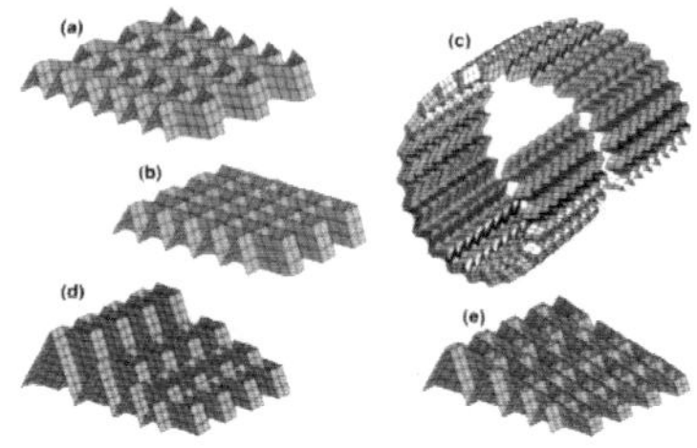

Figure 2. Some kinds of foldcore.

In order to increase the energy-absorbing capacity of the intersection element, structures with high energy-absorption capacity should be used. Foldcore sandwich structure is such a kind of structure. Foldcore, as shown in figure 2, are structures formed by folding plates or foils according to regular repeated lines. Foldcore sandwich structure has been studied a lot. The geometric design method and the mechanics model of foldcore have been studied by Wang Z J [5, 6]. The energy absorption capacity of foldcore sandwich structures is studied by Zhang Y C. Due to their study, foldcore is proved to be a good anti-crash structure [7, 8]. Compression and shear experiments have been conducted on foldcore from CFRP and aramid paper by S. Heimbs [9-15]. The residual strength and the impact property of honeycomb core and foldcore sandwich structures have been studied with the help of experiments and numerical simulations.

Foldcore sandwich structures have been proved to have good energy absorption capacity. In this paper, an intersection element containing foldcore sandwich structures is designed and studied. And in order to reduce the intersection element mass while maintaining the crashworthiness requirements satisfied, optimization of geometric parameters is conducted.

2. Numerical simulation of the intersection element with foldcore sandwich structures

A typical intersection element is shown in figure 3. It is a drop experiment sample in paper [1]. The intersection element with foldcore sandwich structures can be got by replacing the skin and floor of the intersection element with foldcore sandwich structures. The floor is replaced with V-type foldcore sandwich structure and the skin is replaced with M-type foldcore sandwich structure. The new intersection element is shown in figure 4. The two kinds of foldcore are shown in figure 5. The simulation method of the foldcore sandwich structures are introduced in paper [16, 17]. In the simulation method, random geometric errors are introduced. The material used is aluminium alloy 2024 T3. The thickness of the foldcore sandwich structure plate t_c is 0.3mm and the thickness of the foldcore material is 0.15mm. The parameters of the foldcores are set as: $H_M=H_V=20$mm, $A_M=A_V=15$mm, $a_M=a_V=30^{\circ}$, $\lambda_M=\lambda_V=20^{\circ}$, $B_M=5$mm. The thickness of the rest panels are 0.81mm. The distance of the two mid-planes of the sandwich structures is 195mm. The impact velocity is 7.4m/s.

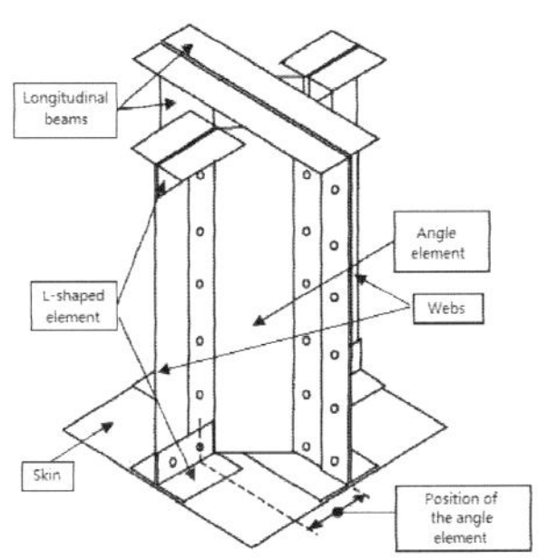

Figure 3. A typical subfloor intersection element.

Figure 4. The intersection element with foldcore sandwich structures.

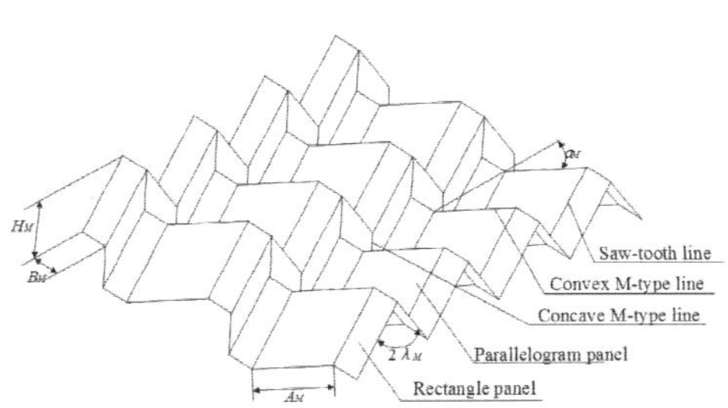

(a) M-type foldcore.

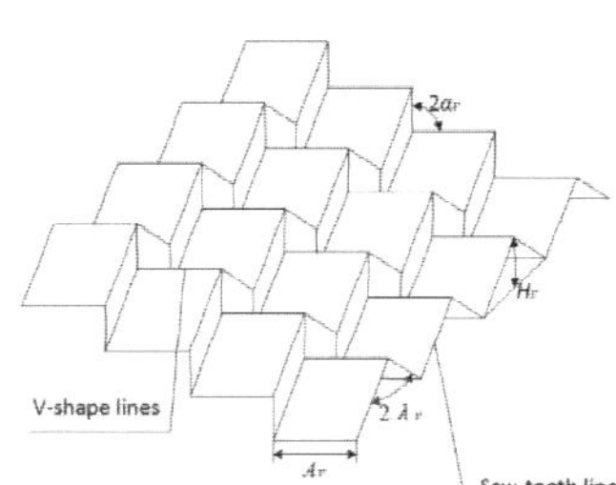

(b) V-type foldcore.

Figure 5. M-type and V-type foldcore and their geometric parameters.

In the numerical simulation model, the element type is chosen as 4-node shell reduced integration element S4R. The impact mass is 110kg and is applied on the middle of the floor top surface. Contact in the structure is modelled using ABAQUS general contact controls. The contact property is defined as "Hard" contact in the normal direction and penalty in the tangential direction with friction coefficient 0.17. The rivets are simulated by connecting nodes around the rivets of two panels using tie constraint.

The deformation process of the intersection element with foldcore sandwich structures is shown in figure 6. The crush analysis results of the new intersection element and the experiment result in paper [1] are shown in table 1. It can be seen that the structure with foldcore sandwich structures has smaller peak force and bigger crush force. It means that the structure with foldcore sandwich structures has better buffering capacity and energy-absorption capacity.

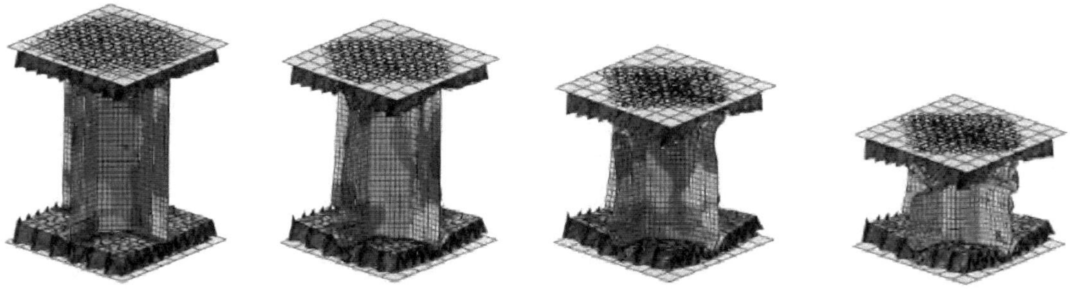

(a) 1.35ms (b) 3.15ms (c) 8.1ms (d) 31.5ms

Figure 6. The deformation process of the intersection element with foldcore sandwich structures.

Table 1. Comparison of the intersection element with foldcore sandwich structures and typical intersection element.

	Typical intersection element	Structure with foldcore sandwich structures	Variable ratio[%]
Peak load[kN]	52.0	41.9	-19.4
Average load[kN]	22.7	26.8	18.1
Mass[kg]	0.481	0.480	-0.2

3. Optimization of the intersection element with foldcore sandwich structures

In order to minimum the total mass while maintaining the crashworthiness requirements of the intersection element, geometric parameters optimization should be conducted, In this paper, an optimization method based on NSGA-II and Anisotropic Kriging is used. NSGA-II is chosen as the optimization algorithm. In each design point, the data needed are calculated through Anisotropic Kriging surrogate model, which greatly reduce the time cost of the calculation.

NSGA-II was improved from NSGA by Deb, Pratap and Agarwal. It can deal with optimization problems with both discrete and continuous variables. Besides, NSGA-II has good global searching ability due to its cross-operation. It means that NSGA-II can deal with optimization with irregular domain and complex constraints, which are just the characteristics of the optimization under crashworthiness requirements.

Anisotropic Kriging surrogate model is a kind of surrogate model which is based on statistical theory. In the numerical simulation model of foldcore, random geometric errors are applied in order to model the crashworthiness property more accurately. Besides, the importance of the geometric variables is different. The effectiveness of Anisotropic Kriging is not depending on the existence of random error. Anisotropic fitting technology is used in Anisotropic Kriging, which enables it to control the importance of the variables. So Anisotropic Kriging surrogate model is suitable.

The mathematical model of the optimization problem can be expressed as:

$$\begin{cases} \min & m(\boldsymbol{x}) \\ \text{s.t.} & \boldsymbol{x}_{\min} \le \boldsymbol{x} \le \boldsymbol{x}_{\max} \\ & A_{\max}(\boldsymbol{x}) \le A_{\text{limit}} \\ & V_{\max}(\boldsymbol{x}) \ge 0 \end{cases} \tag{1}$$

In equation 1, $m(\boldsymbol{x})$ is the total mass of the structure, $\boldsymbol{x}$ is the geometric parameter vector, $A_{max}(\boldsymbol{x})$ is the maximum acceleration of the impact mass in z direction. A_{limit} is 48g, which is set according to MIL-STD-1290A (AV). V_{max} is the maximum velocity of the impact mass. It should be noticed that the initial velocity is a negative value. The forth formula in equation 1 means that the kinetic energy of the impact mass is totally absorbed. The range of variables is shown in table 2.

Table 2. Range of the design parameters in optimization.

Parameter	A_M	a_M	λ_M	B_M	A_V	a_V	λ_V	t_c
Minimum value	3.00mm	15.0°	10.0°	3.00mm	3.00mm	15.0°	10.0°	0.3mm
Maximum value	20.00mm	65.0°	45.0°	10.00mm	20.00mm	65.0°	45.0°	0.5mm
Step	0.02mm	0.1°	0.1°	0.02mm	0.02mm	0.1°	0.1°	0.02mm

Before optimization, Anisotropic Kriging surrogate model must be established. 200 intersection element models are established and analysed. Based on the analysis results, Anisotropic Kriging surrogate models expressing the relations of A_{max}, V_{max} and the geometric parameters are established. The total mass can be calculated as:

$$m = 0.305 + 0.14^2 \rho_{Al} \left[4t_c + \frac{B_M + A_M\sqrt{\sin^2\lambda_M + \cos^2\alpha_M \cos^2\lambda_M}}{\sin\alpha_M\left(B_M + A_M \sin\lambda_M\right)}\delta_M + \frac{\sqrt{1+\cot^2\alpha_V \cos^2\lambda_V}}{\sin\alpha_V}\delta_V \right] \tag{2}$$

25 intersection element models are used to check the accuracy of the surrogate model. The accuracy data of the surrogate models are shown in table 3.

Table 3. Error statistics of the Anisotropic Kriging surrogate modes.

Parameter	A_{max}	V_{max}	m
Maximum error	17%	12.7%	14.3%
Average error	5.6%	4.5%	4.6%

The accuracy of the surrogate models can meet the requirements.

The numerical simulation analysis of different intersection elements need CPU times from 1.5 hour to more than 12 hours. And with the help of Anisotropic Kriging surrogate model, one intersection element need only less than 1 second. Anisotropic Kriging surrogate model can greatly reduce the time cost.

NSGA-II is used as the optimization algorithm. In each generation there are 50 design points. After 60 generations, the optimization converges. The optimization results are shown in table 4.

Table 4. Optimization results.

A_M[mm]	a_M[°]	λ_M[°]	B_M[mm]	A_V[mm]	a_V[°]	λ_V[°]	t_c[mm]	A_{max}[m/s^2]	V_{max}[m/s]	m[kg]
10.10	65.0	35.5	8.10	7.62	64.5	35.5	0.30	443.42	0.7116	0.399

Compared with the original structural mass 0.481kg, mass of the optimized intersection element reduces 17.15%.

4. Conclusions

(1) The intersection element is an important part of the helicopter subfloor structures. The floor and skin of the intersection element are replaced with foldcore sandwich structures. The new intersection element shows good crashworthiness properties.

(2) An optimization method using NSGA-II and Anisotropic Kriging is introduced. NSGA-II can deal with optimization under crashworthiness requirements with irregular domain and complex constraints. Anisotropic Kriging can deal with models with random errors and can greatly reduce optimization time cost.

(3) After optimization, the mass of the intersection element with foldcore sandwich structures reduce 17.15% while all the crashworthiness requirements satisfied.

References

[1] Bisagni C, 1999 *J. Aircraft Eng. Aeros. Tech.* **71(1)** 6-11.

[2] Bisagni C, 1999Int. J. Crashworth. **4(2)** 199-212..

[3] Bisagni C, Lanzi L and Ricci S 2013 *Size and Topological Optimization for Crashworthiness Design of Helicopter Subfloor.* Aiaa/issmo Symposium on Multidisciplinary Analysis and Optimization. Atlanta, 1-11.

[4] Lanzi L, Bisagni C and Ricci S, 2004 *J. Comput. Struct.* **82(1)** 93-108.

[5] Wang Z J and Khaliulin V 2003 *J. Nanjing Univ. Aeronaut. Astronaut.* **34(1)** 6-11.

[6] Wang Z J and Xu Q H 2004 *J. Nanjing Univ. Aeronaut. Astronaut.* **36(4)** 449-53.

[7] Zhang Y C, Yu J M, Zhang S L, et al., *J. Vib. Shock.* **33(1)** 113-8.

[8] Zhang Y C, Wang Z L and ZHang S L, 2010 *J. Ship. Mech.* **14(Z1)** 114-20.

[9] Heimbs S, Middendorf P, Hampf C, et al, 2008 *Aircraft Sandwich Structures with Folded Core under Impact Load* International Conference on Sandwich Structures. Porto 169-380.

[10] Heimbs S, Middendorf P, Kilchert S, et al., 2007 *J. Appl. Comp. Mater.* **14(5-6)** 363-77.

[11] Heimbs S, 2009 *J. Comput. Mater. Sci.* **45** 205-16.

[12] Heimbs S, Mehrens T, Middendorf P, et al., 2007 *Numerical Determination of the Nonlinear Effective Mechanical Properties of Folded Core Structures for Aircraft Sandwich Panels* 6th LS-DYNA Users Conference. Gothenburg, 29-30.

[13] Heimbs S, Cichosz J, Kilchert S, et al, 2009 *Sandwich Panels with Cellular Cores Made of Folded Composite Material: Mechanical Behaviour and Impact Performance* International Conference on Composite Materials. 1485–97.

[14] Heimbs S, Cichosz J, Klaus M, et al., *J. Comp. Struct.* **92(6)** 1485-97.

[15] Klaus M, Reimerdes H G and Gupta N K, 2012 *Int. J. Imp. Eng.* **44(44)** 50–8.

[16] Zhou H Z and Wang Z J. 2016 *Acta Aeronaut. ET Astronaut. Sin.* **37(2)** 579-87.

[17] Zhou H Z and Wang Z J. 2017 *J. Acta Mater. Comp. Sin.* **2017** 1-8.

Investigation on the Acoustic Absorption of Flexible Micro-Perforated Panel with Ultra-Micro Perforations

Guoxin Li[1, 2], Xiaoning Tang[3], Xiaoxiao Zhang[4], Y. J. Qian[5], Deyi Kong[1, 2]

[1]State Key Laboratory of Transducer Technology, Institute of Intelligent Machines, Chinese Academy of Sciences, Hefei 230031, China
[2]Hefei Institutes of Physical Science, University of Science and Technology of China, Hefei 230026, China
[3]Key Laboratory of Textile Science and Technology, Ministry of Education, College of Textiles, Donghua University, Shanghai 201620, China
[4]xQuant Company Ltd, Hangzhou 310012, China
[5]IOT Engineering College, Hohai University, Changzhou 213022, China

Abstract. Flexible micro-perforated panel has unique advantages in noise reduction due to its good flexibility compared with traditional rigid micro-perforated panel. In this paper, flexible micro-perforated panel was prepared by computer numerical control (CNC) milling machine. Three kinds of plastics including polyvinylchloride (PVC), polyethylene terephthalate (PET), and polyimide (PI) were taken as the matrix materials to prepare flexible micro-perforated panel. It has been found that flexible micro-perforated panel made of PET possessing good porosity and proper density, elastic modulus and poisson ratio exhibited the best acoustic absorption properties. The effects of various structural parameters including perforation diameter, perforation ratio, thickness and air gap have also been investigated, which would be helpful to the optimization of acoustic absorption properties.

1. Introduction

Micro-perforated panel (MPP) has been widely used in acoustics design for many years, and it has the advantages of tunable sound absorption peak frequency and robust noise reduction properties. The perforation diameter of micro-perforated panel was reduced to sub-millimeter dimension, thus to provide enough acoustic resistance and low acoustic mass reactance for desired acoustic absorption behavior. The theory model of micro-perforated panel acoustic absorber was firstly proposed by Maa [1], who has developed the general model to predict the acoustic absorption coefficients [2, 3]. The acoustic behavior is related to various structural parameters including perforation diameter, hole distance, panel thickness, and the depth of back air gap. Compared with traditional rigid micro-perforated panel, flexible micro-perforated panel has good flexibility and deformability. Therefore, it could well meet the demand of particular noise reduction of equipment with complicated geometrical shapes. Flexible micro-perforated panel has gradually become the research hot spot in recent years.

To investigate the vibro-acoustic effects of flexible micro-perforated panel, Lee and Swenson [4] have studied the acoustic absorption behavior of compact sound absorber. The results indicated that panel vibration plays an important role in the acoustic properties of the absorber. Hereafter, various models have been proposed to characterize the vibro-acoustic effects on the acoustic absorption of flexible micro-perforated panel, where Maa's transfer impedance definition of aperture in parallel was taken [5-7]. On the other hand, various acoustic absorption structures based on flexible micro-perforated panel have been developed, such as hybrid passive-active noise control system [8], multiple layers of flexible micro-perforated panel for enhanced acoustic absorption [9], and flexible micro-perforated panel absorber made of piezoelectric film [10]. According to Maa's model [11], micro-perforated panel with diminished perforation diameters could effectively increase the energy dissipation of viscous flow and thus improve the acoustic absorption properties. To the best knowledge of us, there is few studies concerning the acoustic behavior of flexible micro-perforated panel with micro-scale perforations (approximate 100 μm). It is feasible for conventional processing methods to make micro-perforated panel with perforation diameters more than 200 μm. However, it is still a challenge to fabricate flexible micro-perforated panel with such small sized perforations of approximate 100 μm. Furthermore, the effects of different structural parameters on the acoustic absorption of flexible micro-perforated panel has not been reported.

In the present work, computer numerical control (CNC) milling machine was taken to prepare flexible micro-perforated panel. Three kinds of plastics including polyvinylchloride (PVC), polyethylene terephthalate (PET), and polyimide (PI) were utilized as the raw materials of flexible micro-perforated panels. The purpose of this study is to investigate the effects of different plastic materials on acoustic absorption behavior. In addition, we have further studied the role of perforation diameter, perforation ratio, thickness and air gap distance in the determination of acoustic absorption properties.

2. Materials and methods

Micro-perforated panel could be considered as a lattice consists of short narrow tubes. The perforations are separated by distances much larger than hole diameters, but rather small compared to the wavelength of acoustic waves. The propagation of acoustic wave in the tube was defined by Rayleigh and Crandall, Maa has further developed an approximate solution for micro-perforated panel with sub-millimeter sized apertures [1]. The diagrammatic sketch and equivalent circuit of micro-perforated panel (MPP) are shown in figure 1. According to Maa's model, the acoustic wave in the system is equivalent to a source of 2p pressure, and micro-perforated panel backed with an air gap space could be seen as a resonant system. The acoustic behavior of such a system was determined by the perforation diameter d, the panel thickness t, the distance between centers of adjacent perforations b and the depth of the air gap D. The acoustic impedance of micro-perforated panel Z and impedance of the air space ZD can be expressed as follows:

$$Z = R + j\omega M \tag{1}$$

$$Z_D = -j\rho c \cdot \cot(\omega D / c) \tag{2}$$

The normal incident acoustic absorption coefficient α and maximum normal incidence acoustic absorption coefficient αm of micro-perforated panel with air gap can be calculated by the equations as follows [11, 12]:

$$\alpha = \frac{4r}{(1+r)^2 + (\omega m - \cot(\omega D / c))^2} \tag{3}$$

$$\alpha_m = \frac{4r}{(1+r)^2} \tag{4}$$

where r is the normalized specific acoustic resistance, m is the normalized acoustic mass respectively. In addition, ρ is the density of air, c is the sound velocity in air, η is dynamic viscosity constant of air, k is the perforated constant, μ = η/ρ is kinematic viscosity constant of air. The equations are as follows:

$$r = \frac{32\mu t}{\rho c d^2} \cdot \left(\sqrt{1 + \frac{k^2}{32}} + \frac{\sqrt{2}k}{8} \cdot \frac{d}{t} \right) \tag{5}$$

$$m = \frac{t}{\rho c} \cdot \left(1 + \frac{1}{\sqrt{1 + k^2 / 2}} + 0.85 \frac{d}{t} \right) \tag{6}$$

$$k = \sqrt{\frac{2\pi f}{\mu}} \cdot \frac{d}{2} \tag{7}$$

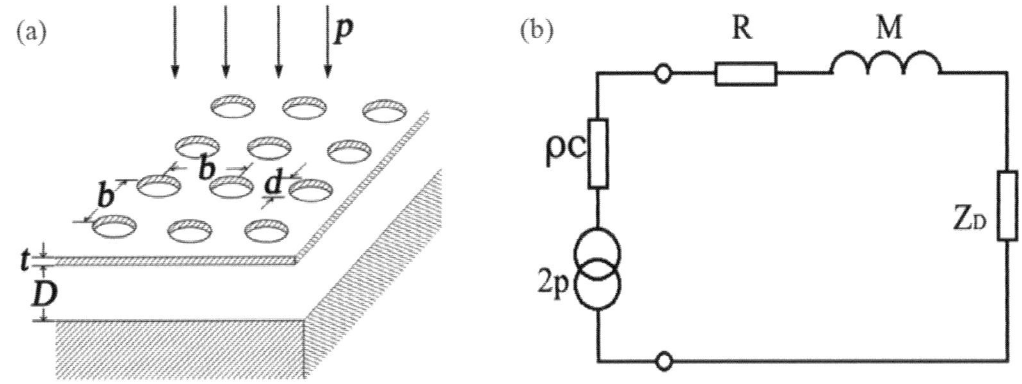

Figure 1. (a) Schematic diagram and (b) equivalent circuit of micro-perforated panel (MPP) absorber.

In this study, the normal incidence absorption coefficient was measured by an impedance tube based on the standing wave method of ISO 10534-2 (Determination of sound absorption coefficients and impedance tubes, 1998). The schematic diagram of test equipment is shown in figure 2. The standard acoustic source of loudspeaker was placed at one end of the tube, and test sample was fixed at the other end. The acoustic field in the tube consists of incident and reflected waves. The maximum and minimum values of standing wave acoustic pressure can be collected by moving the pulley in the orbit. These measured results were repeated at least three times and the average values were taken.

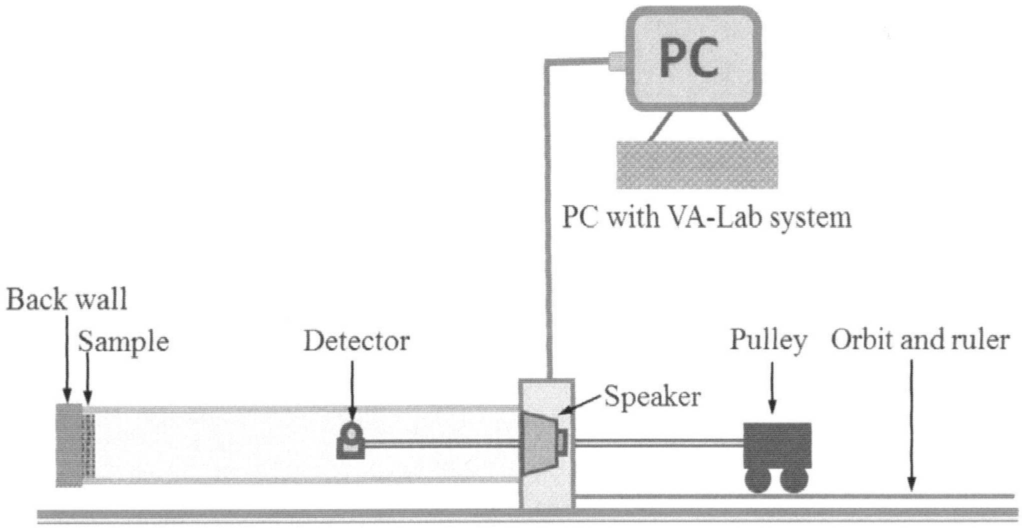

Figure 2. Illustration of impedance tube test system for acoustic absorption coefficients.

Micro-perforated panels were made by computer numerical control (CNC) milling machine. The perforation diameters d of prepared micro-perforated panel was ranged from 0.085 mm to 0.15 mm, which have been measured by stereo microscopes. Three kinds of materials including polyvinylchloride (PVC), polyethylene terephthalate (PET), and polyimide (PI) were taken to fabricate flexible micro-perforated panels.

3. Results and discussions

3.1. Effects of materials

In this section, polyvinylchloride (PVC), polyethylene terephthalate (PET), and polyimide (PI) were used to prepare flexible micro-perforated panels. All the fabricated micro-perforated panels have the consistent structural parameters, where the perforation diameter d is 0.1 mm, thickness t is 0.1 mm, hole spacing b is 0.633 mm, and perforation ratio p is 1.957 %. The acoustic absorption coefficient of obtained flexible micro-perforated panels were tested by impedance tube. The measured results were further compared with Maa's model, as shown in figure 3. It could be observed that the absorption bandwidths of three specimens are moved to low frequency range. Considering that the flexible micro-perforated panel has better flexibility and deformability than the rigid MPP of Maa's, the absorption peaks due to the panel's vibration and the micro-perforation appear at the lower resonant frequencies. The measured results of PI sample exhibited the biggest deviation in comparison with the simulation results of Maa's model, and the acoustic absorption property in high frequency is obviously decreased, which could be by reason of large viscosity and poor machinability. PVC and PET samples have the approximative absorption curves at the frequency range higher than 3000 Hz, while PET has better acoustic absorption properties when the frequency is lower than 3000 Hz. For instance, the acoustic absorption coefficient of PET is 10.6 % higher than that of PVC at 1500 Hz. By comparing the material properties of the two, it has been found that they have similar poisson ratio, but PET has better elastic modulus and bending performance, which account for this consequence. Therefore, it could be concluded that flexible micro-perforated panel made of PET has wider absorption bandwidth and better acoustic properties than PVC and PI samples.

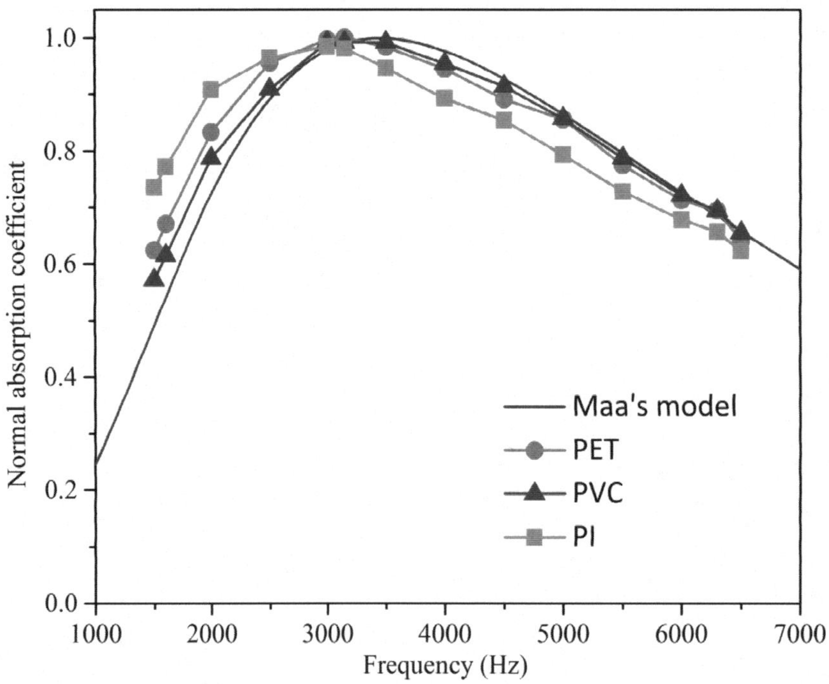

Figure 3. Comparison of Maa's model with the measured normal absorption coefficient of different materials including PET, PVC and PI.

3.2. Effects of structural parameters

We have further studied the effects of different structural parameters on acoustic absorption properties. Firstly, four flexible micro-perforated panels with different perforation diameter were prepared. The structural parameters are listed in table 1. It can be seen that all the samples with the same thickness t of 0.1 mm and the consistent perforation ratio p of 1.957 %, while the diameter d ranged from 0.085 mm to 0.15 mm. The test results of absorption coefficient are shown in figure 4. The absorption peak is higher than 0.95 when the diameter is 0.085, 0.1 or 0.12 mm. However, the acoustic absorption peak is decreased obviously when the diameter is 0.15 mm. The absorption bandwidth is gradually increased with the decrease of diameter, and the flexible micro-perforated panel with 0.085 mm diameter has approximate 3 absorption octaves. Therefore, it could be concluded that flexible micro-perforated panel with reduced aperture diameter is benefit to broaden the absorption bandwidth and increase the absorption peak.

Table 1. Structural parameters of four flexible micro-perforated panels with different perforation Diameter.

Specimen	d (mm)	t (mm)	b (mm)	D (mm)	p (%)
#1	0.085	0.1	0.538	15	1.957
#2	0.1	0.1	0.633	15	1.957
#3	0.12	0.1	0.76	15	1.957
#4	0.15	0.1	0.95	15	1.957

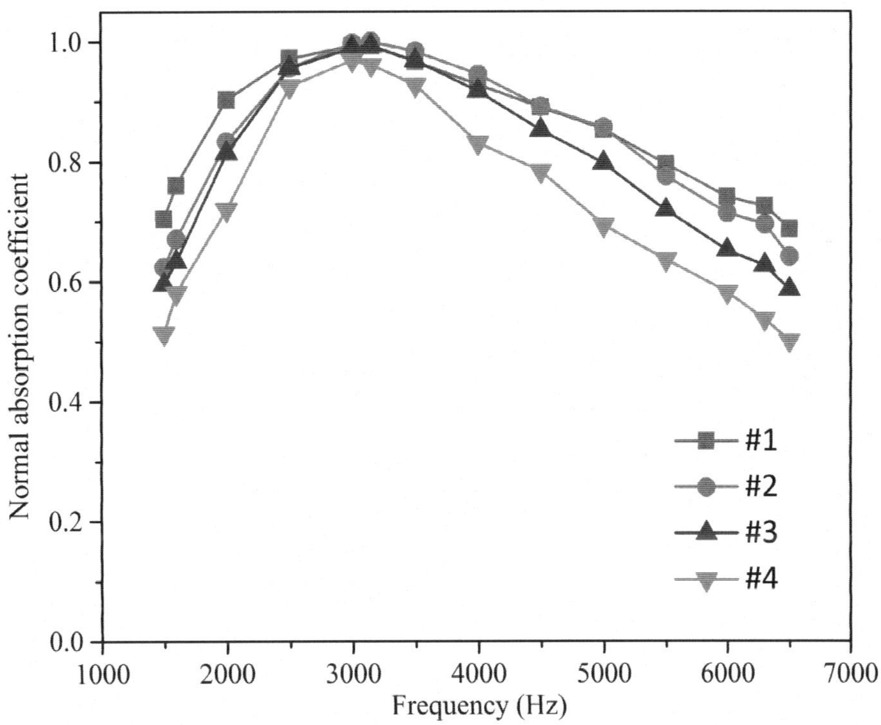

Figure 4. Normal absorption coefficient of flexible micro-perforated panels with different diameters.

Flexible micro-perforated panels with different perforation ratio were also prepared; the structural parameters are listed in table 2. The perforation ratio of four samples ranged from 1.061 % to 4.55 %. As can be seen from figure 5, with the increase of perforation ratio, the maximum absorption coefficient is gradually decreased, and the absorption peak is moving to high frequency range. Therefore, the appropriate perforation ratio is benefit to increase acoustic absorption properties at high frequency range and broaden the absorption bandwidth. In this study, flexible micro-perforated panel with the perforation ratio of 1.957 % has the best acoustic absorption properties.

Table 2. Structural parameters of four micro-perforated panels with different perforation ratio.

Specimen	d (mm)	t (mm)	b (mm	D (mm)	p (%)
#1	0.1	0.1	0.86	15	1.061
#2	0.1	0.1	0.633	15	1.957
#3	0.1	0.1	0.52	15	2.903
#4	0.1	0.1	0.42	15	4.55

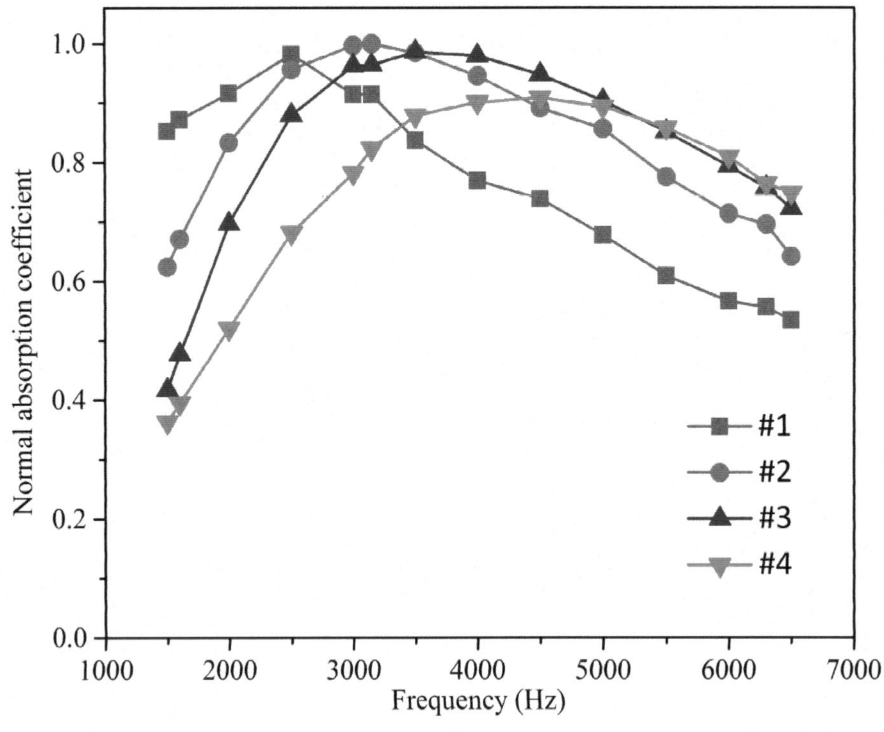

Figure 5. Normal absorption coefficient of flexible micro-perforated panels with different perforation ratio.

The structural parameters of four micro-perforated panels with different thickness are shown in table 3. It can be seen that the thickness of all samples ranged from 0.1 to 0.3 mm. With the increase of thickness, the maximum absorption coefficient is gradually decreased, and absorption peak moved to the low frequency range. In addition, the absorption bandwidth is narrowing, as can be observed from figure 6. Therefore, it can be stated that flexible micro-perforated panel with thinner thickness is benefit to improve acoustic absorption.

Table 3. Structural parameters of four micro-perforated panels with different thickness.

Specimen	d (mm)	t (mm)	b (mm	D (mm)	p (%)
#1	0.1	0.1	0.633	15	1.957
#2	0.1	0.15	0.633	15	1.957
#3	0.1	0.2	0.633	15	1.957
#4	0.1	0.3	0.633	15	1.957

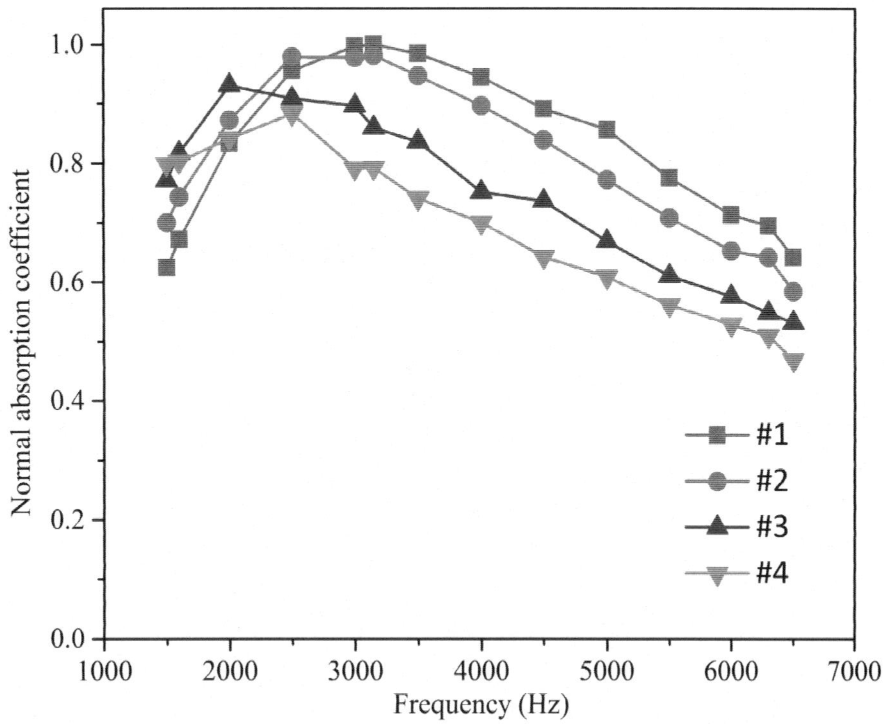

Figure 6. Normal absorption coefficient of flexible micro-perforated panels with different thickness.

Furthermore, we have studied the acoustic absorption properties of flexible micro-perforated panels with different air gap distance, the detailed parameters are listed in table 4. It can be observed from figure 7 that all the four specimens have the similar absorption peak with different air gap distance. However, the increase of air gap distance is benefit to improve the acoustic absorption properties at low frequency range.

Table 4. Structural parameters of flexible micro-perforated panels with different air gap.

Specimen	d (mm)	t (mm)	b (mm	D (mm)	p (%)
#1	0.1	0.1	0.633	10	1.957
#2	0.1	0.1	0.633	15	1.957
#3	0.1	0.1	0.633	20	1.957

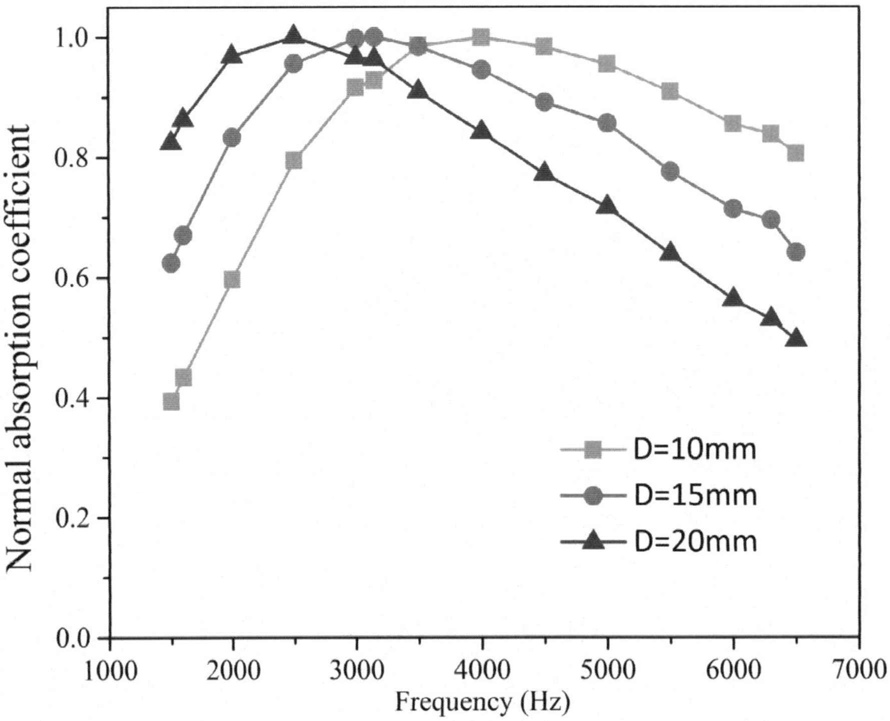

Figure 7. Normal absorption coefficient of flexible micro-perforated panels under different air gap distance.

4. Conclusions

In this work, we have mainly studied the acoustic absorption properties of flexible micro-perforated panel. It has been found that flexible micro-perforated panel appropriately fusing the structural resonances and the micro-perforation exhibited wider absorption bandwidth than the simulation values of Maa's model. Flexible micro-perforated panel made of PET has better mechanical behaviors such as excellent toughness and elasticity, thus show greater acoustic absorption properties than PVC and PI samples. The maximum absorption peak and absorption bandwidth were mainly determined by perforation diameter and perforation ratio. The reduced diameter is benefit to increase the acoustic absorption bandwidth, and flexible micro-perforated panel with 100 μm diameter exhibited approximate 3 octaves. Appropriate perforation ratio is also helpful to improve the acoustic absorption properties, while the air gap depth has little effect on the absorption bandwidth of the flexible. Flexible micro-perforated panel with thinner thickness can also improve the acoustic absorption properties.

References

[1] Maa D Y 1975 *Sci Sin.* **17** 55-71.
[2] Maa D Y 1987 Noise Contr. Eng. J. **29(3)** 77-84.
[3] Dah-You M 1998 J. Acoust. Soc. Am. **104** 2861-6.
[4] Lee J and Swenson G W 1992 *Noise Contr. Eng. J.* **38(3)** 109-17.
[5] Lee Y Y, Lee E W M and Ng C F 2005 J. Sound Vibr. **287(1)** 227-43.
[6] Bravo T, Maury C and Pinhede C 2012 *J. Acoust. Soc. Am.* **131** 3853-63.

[7] Wang Y H, Zhang C C, Ren L Q, Ichchou M, Galland M A and Bareille O 2014 *Comp. Struct.* **108** 400-8.
[8] Zheng W G, Huang Q B, Li S D and Guo Z Y 2011 *J. Low Frequency Noise Vibr. Act. Contr.* **30** 313-28.
[9] Bravo T, Maury C and Pinhede C 2013 *J. Acoust. Soc. Am.* **134** 3663-73.
[10] Duan X H, Wang H Q, Li Z B, Zhu L K, Chen R and Kong D Y 2015 *Appl. Acoust.* **88** 84-9.
[11] Maa D Y 2006 Acta Acust. **31** 181-4.
[12] Maa D Y 1998 J. Acoust. Soc. Am. **104** 2861-6.

Acknowledgements
The authors are grateful to the National Natural Science Foundation of China (No. 11474291, 11604077) and Foundation of State Key Laboratory of Transducer Technology (No. SKT1603) for their support.

Development Directions for Various Types of the Light Wood-Framed Structures

J Malesza[1] and M Baszeń[1] and Cz Miedziałowski[1]

[1]Department of Geotechnics and Structural Mechanics, Bialystok University of Technology, Bialystok, Poland

Abstract. The paper presents current trends in the development of the wood-framed structures. Authors describe the evolution of the technology of implementation, the production process of precast elements of buildings as well as selected realization on the site of these kinds of structures. The attention has been paid to the effect of implementation phases on construction and erecting technology of the wood-framed structures. The paper draws attention to the importance and enhancement of structural analysis of structures in individual phases of building realization.

1. Introduction

The wood-framed technology of building construction is used in many countries around the world. The participation of this type of construction in the entire building market varies depending to the country [1]. These types of buildings are very popular in the Scandinavian countries, England, North America or Japan.

Increasing popularity of the wood-framed structures is a result of many advantages as: low workload, lack of wet processes, short realization time, lightweight, good ecological parameters, use of renewable raw materials.

The untreated solid wood as well as the converted wood is used. The converted technologically modified wood could be made of glued veneers, boards (e.g. LVL, OSB, plywood, chipboard, etc.) or chemically modified (e.g. Accoya®). Recently it could be observed that erected construction are often made in technology of the wood-framed structures and this applies not only to single-family residential houses and one- or two-storey family building but also to multi-storey residential and multifamily buildings. There are also attempts to build in wood-frame technology tall buildings, mainly based on glued or modified wood.

Another tendency is to prepare building components such as walls, floors, roofs in the form of precast in factories. This affects the speeding up of the realization process of construction, but also it is improving the quality of the factory made precast walls, floors or roofs. The lightweight of this type of the structure allows the transport of structural components even at great distances. However it creates the next phase of production, starting from manufacturing in the factory, through the transport, ending with assembly. This fact imposes the importance of design of this type of structure for individual phases [2].

The paper will present current trends in the development of technology and methods of wood-frame buildings.

2. Implementation technology

The technology of implementation evaluates from the execution of all works on the site, by partial prefabrication, through the prefabrication of individual building elements such as walls, floors, roofs, up to the modular spatial parts of buildings [3].

Erecting work on the construction site with numerous construction elements and then mounting the insulation and facade layers is shown in figure 1.

a)

b)

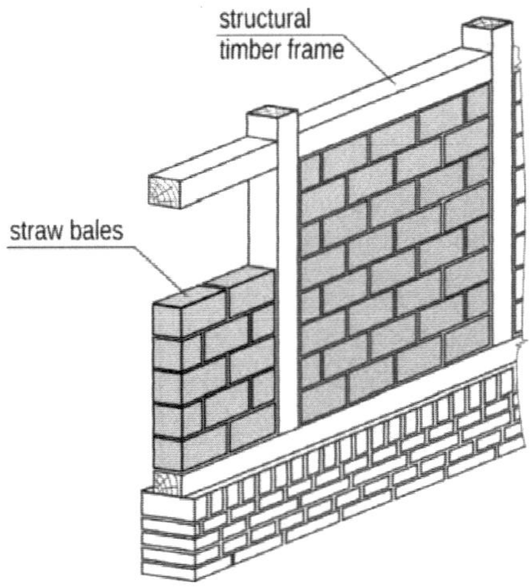

Figure 1. Erecting work on the site: a) construction of beam and stud elements, b) filling the wooden frame by isolation layer of hollow block, sawdust concrete or concrete with sugar cane fibers.

Partial prefabrication consists of the manufacturing of small-dimensions items such as blocks or hollow blocks, which together with the wooden framing make up the construction of buildings. The construction of wall structures based on hollow blocks is presented in figure 2.

a)

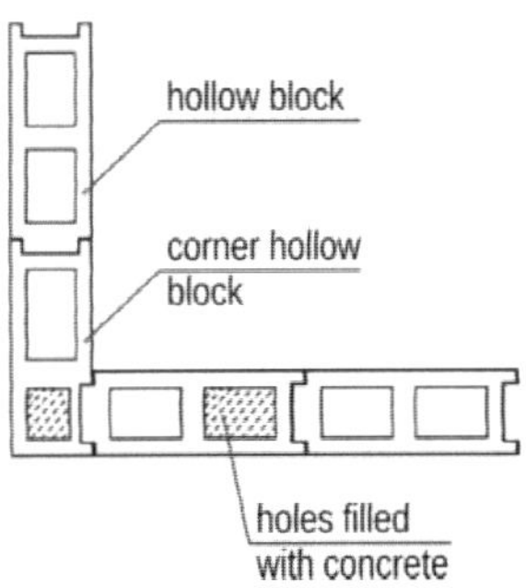

b)

Figure 2. Construction of the wall made of the hollow blocks: a) hollow blocks in the wall layer [4], b) visualization of construction corner [5].

The next step in the development of light wood-frame buildings is the production of large-dimensions panel walls, floors and roofs made of wooden framing and sheathed by the OSB, chipboard or plywood panels, and finally the assemblage of these panels into spatial modules as part of the building, as shown in figure 3.

Modular elements are often pre-finished by window and door frames and facade elevation layers.

IOP Conf. Series: Materials Science and Engineering **269** (2017) 012024 doi:10.1088/1757-899X/269/1/012024

a)

b)

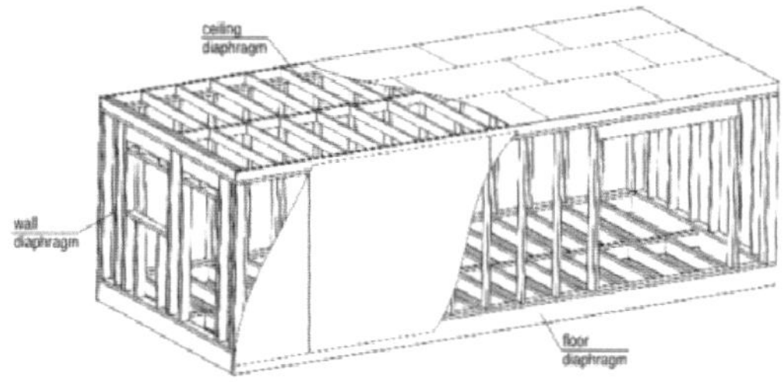

Figure 3. Precast large-dimension elements: a) wall panel element, b) spatial modular element.

3. Factory production of building elements

The production of precast elements of building in manufacturing factories and subsequently the elements transport to the erection site, assemblage and exploitation cause the successive phases of building realization. The phases are shown in figure 4.

a)

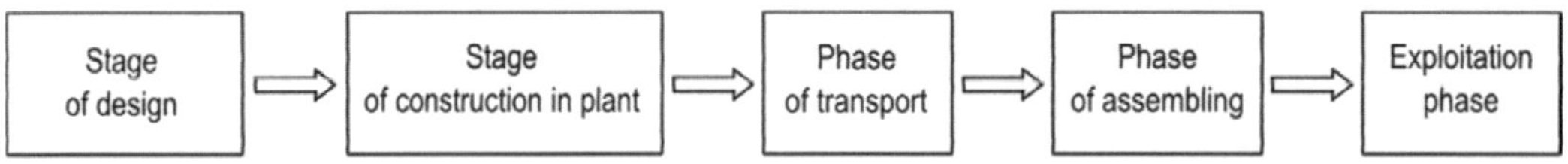

b)

c)

Figure 4. Phases of realization for modular building: a) particular phases of realization, b) production of the module, c) modular elements prepared for transport.

The elements made in precast factories and transported to the erecting site are assembled into entire building, as shown in figure 5.

a)

b)

Figure 5. Building assembly of precast elements: a) panel elements, b) modular elements.

The accuracy of assembly and proper realization of joints is achieved as an important technological process and controlling in this case [6].

4. Design of structural elements and buildings

Prefabrication of panel and modular components and phases of building implementation imply incorporating in the computational analyzes those steps in which the static schemes and the loads are changing. In opposition to previously used 2D calculation models [7, 8] the 3D methods are used to develop the analyzed problem [9, 10] as shown in figure 6.

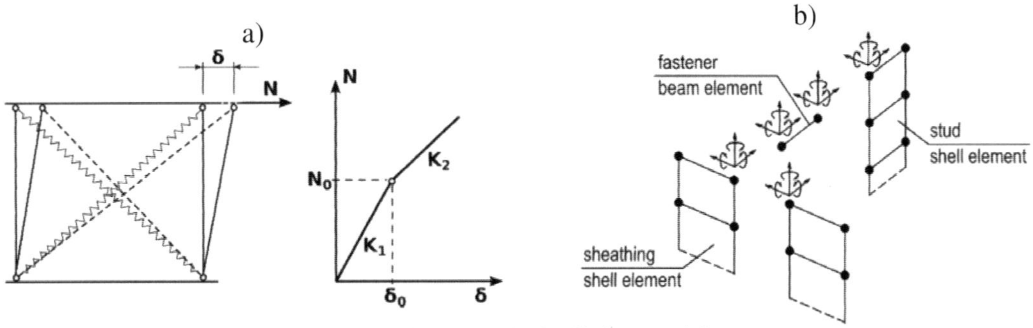

Figure 6. Calculation models for light wood-frame structures:
a) 2-dimensional [7], b) 3-dimensional [9].

Analytical methods like Force Method or Direct Displacement Method allow calculation only in very simplified cases. Development of computer science and faster computer machines caused the calculation of more complex numerical problem stop to be a problematic. Issues with great number of unknowns could be develop using Finite Element Method.

Aggregation of global stiffness matrix is being done by introducing global coordination system, boundary conditions and summation of particular elements of its various types.

$$\boldsymbol{K} = \sum_c \sum_t \sum_e \boldsymbol{K_e} \quad (1)$$

Similar situation is for load vector, which is presented as follows:

$$\boldsymbol{P} = \sum_c \sum_t \sum_t (\boldsymbol{P_h} + \boldsymbol{P_v}) \quad (2)$$

where:

K_e – element stiffness matrix,
c – individual components of the building (walls, floors, etc),
t –types of elements (studs, beams, connectors, sheathing),
e – particular elements of given type,
P_h ,P_v – respectively horizontal and vertical loads.

Displacement vector for beam finite elements is given as:

$$\boldsymbol{q_b} = \{u_1, v_1, w_1, \phi_{x1}, \phi_{y1}, \phi_{z1}, u_2, v_2, w_2, \phi_{x2}, \phi_{y2}, \phi_{z2}\} \quad (3)$$

while for 4-node shell elements is presented as:

$$\boldsymbol{q_s} = \{u_1, v_1, w_1, \phi_{x1}, \phi_{y1}, \phi_{z1}, \dots, u_4, v_4, w_4, \phi_{x4}, \phi_{y4}, \phi_{z4}\} \quad (4)$$

The stiffness matrix is derived from standard finite element method relationships. The final set of equations of the calculation model, which may be nonlinear, has the form:

$$\boldsymbol{K(q)q} = \boldsymbol{P} \quad (5)$$

5. Summary

As mentioned above, the technologies of buildings made of wooden frames evolve from execution directly on the construction site to the full factory production. Recently, the level of technological advancement is the production of ready-made spatial modules covering a part of the completed buildings. Factory production is conducive to the preparation of pre-finished elements with higher quality. However, it introduces to the technological process the implementation phases: production of components, transport of the modules, assembly the modules into an entire building in the erecting site. An important element of this technology is the accuracy of assembly and properly made connections. In addition, prefabrication and phasing of investment process create new, higher demands on structural, technological and organizational design.

References

[1] Shulze H 1996 *Holzbau, Wande – Decken – Dacher* (Stuttgart: B.G. Teubner).
[2] Asiz A, Iranpour M and Smith I 2005 *Analysis Of Structural Stresses During Handling And Transportation Of Factory-built Housing Construction*Proc.of 33rd Ann. Conf. of the Can. Soc. for Civ. Eng. 2.
[3] Lawson M, Ogden R G and Bergin R 2012 *J. Arch. Eng.* **18(2)** 148-54.
[4] Dawdo C 1994 *Sawdust concrete in civil engineering(in Polish)*(Białystok: Bialystok University of Technology).
[5] Krupa A 2016 *Properties of cementitious composites with organic fillers(in Polish)*PhD thesis (Białystok: Bialystok University of Technology).
[6] Mohammad M 2006 *Connection Systems For Prefabricated Wall Panels, Final Report, No.FCC25* (Ottawa: Natural Resources Canada).
[7] Tarabia A M and Itani R Y 1997 *Comput. Struct.* **63(2)** 319-44.
[8] Kasal B, Leichti R J and Itani R Y 1994 *J. Struct. Eng.* **120(12)** 100-19.
[9] Miedziałowski Cz and Malesza M 2006 The wood-framed with sheathing buildings (in Polish) *Studia z zakresu inżynierii* **55** (Warszawa-Białystok).
[10] Pang W and Ziaei E 2012 *Struct. Congr.* **2012** 1767-77.

Acknowledgements

The paper was realized as a part of BUT project No S/WBiIS/5/13 and was financed by MNiSW of Poland.

Flexible Engineering Structures from the Corrugated Metal Sheets - Comparison of Costs of Solutions used in the Road Building

E. Ołdakowska[1]

[1]Department of Building and Road Engineering, Bialystok University ofTechnology

Abstract. The flexible structures from the corrugated metal sheets are used in particular in the road building, especially as passages for animals. Easy and quick assembly, as well as lower realization costs when compared to the traditional solutions increase interest in such structures. Availability and variety of systems allows for searching for solutions which are the best and optimal in the economical range. The article presents the comparison of costs of the basic materials used in various systems of flexible structures from the corrugated metal sheets. In order to determine the costs of the material solutions the data for two systems used in Poland (for construction of the upper passages for animals) since 2008 have been used. The cost estimation for the basic materials required for realization of 1 m^2 of the flexible structure from the corrugated steel sheets have been prepared with use of prices obtained directly from the Polish contractors and manufacturers, as well as process included in the quarterly information (Sekocenbud). The difference of prices of materials available on the market allows the investor for selecting the structure depending on the needs and financial possibilities, as well as for achieving some savings. The savings in case of purchasing sheets of identical parameters (thickness, profile characteristics) are from approx. 4% to 8% per 1 m^2 of sheet. The connectors in form of bolts M20 cl. 8.8 of various lengths are an expense from 3.00 PLN to 3.50 PLN. Those values may seem low, but taking into consideration amounts connected with construction of many square meters of structure they may become very important factor in the total investment costs.

1. Introduction

The first applications of the flexible structures reach the end of the 19th century [1]. Currently the flexible structures from the corrugated metal sheets are commonly used in the road building as passages, bridges, tunnels, pedestrian subways, agricultural passages, passages for animals. In particular the last use is especially desired due to the harmful influence of the communication investments on free migrations of animals, as there are many important environmental areas in Poland inhabited by rare species, like elks, bisons, lynxes, or wolves. On one hand we have barrier influence of roads, fragmentation of animal habitats and crossing of ecological corridors [2, 3], as well as necessity of building the passages for animals as minimization of unfavourable influence of roads. On the other, we have structures from the corrugated metal sheets which are easy in designing due to lower number of details, readily available portfolio of profile drawings, and prepared base of strength calculations for the typical applications, quick in assembly, and less expensive in realization when compared to the traditional solutions. The corrugated metal sheets are used in construction of structures of either lower and upper passages [4]. The passages for animals are divided into the following types [5]:

a) in respect of dimensions and fulfilled functions:

- small passages - mainly for amphibia and reptiles,
- medium passages - mainly for small and medium mammals,
- large passages - mainly for big mammals.

b) in respect of position of road grade line:

- upper passages - located above the road on the footbridges, overpasses, or tunnels in which the road is located,
- lower passages - located below the road in the tunnels or culverts across the road embankment, or under the overpasses.

c) in respect of connection with other functions:

- independent passages,

combined passages (with the dirt roads or water course).

Classification including size, fulfilled functions, and position in respect of the road grade line divides the passages into the following types:

- small lower passages,
- medium lower passages,
- large lower passages,
- medium upper passages,
- large upper passages.

2. Materials used for flexible engineering structures from the corrugated metal sheets

The engineering objects are made from flexible structures consisting of spiral-coiled sheets and sheets connected with bolts [6]. The spiral pipes are manufactured in a continuous way from the galvanized steel sheet, which is then subjected to crimping in the bending machine and connected with use of machine seam depending on designed diameter. In order to protect it from corrosion the sheet is covered with special protective foil [7]. The separate sheets are obtained in the cold rolling process. In the prepared elements the holes are drilled for the bolt connections. The corrugated form of sheet increases the element rigidity and the wave dimensions are strictly connected with desired and required rigidity. The most common waves are: 150 x 50 mm, 200 x 55 mm, 381 x 140 mm, where the values are the wave length and height. The sheet dimensions as well as bolt holes number and spacing depend on the wave profile, structure dimensions, and sheet location in the structure cross section. The sheet length is the multiple of axial distance between the bolt holes located along the longer edge.

The individual elements from the corrugated steel sheets are connected with use of galvanized bolts, and the connection method is designed by the structure manufacturer. The connectors (bolts, nuts, washers) are selected according to the structure type and should fulfil the requirements specified in appropriate standards. The most commonly used bolts are M20 of class 8.8 and of length from 32 to 75 mm. The bolt length depends on the sheet thickness. The common shapes of bolt heads include hexagons, cones, and cylinders. All bolts must be tightened to the specified torque. The tightening torque must be in range of 240 - 450 Nm depending on sheet thickness and structure span. Exceeding the torque limit value may cause incorrect operation of the structure.

The structures made from corrugates steel sheets are placed on the reinforced concrete foundations, and fixing consists in placing the structure with welded flat bar in the hole made in the foundation, and then placing the concrete or fixing with use of bolts to the anchored channel bar.

The structures are protected with dip zinc coating, alu-zinc coating, paint coating, or polymer coating. While selecting the coating thickness for a given structure it is required to assess the environmental conditions, technological possibilities, and required durability. An important element influencing safety during the structure operation is the backfill. It must fulfil the specified requirements during the object construction and during many years of operation.

In case of some objects from the corrugated sheets it is required to use the additional elements, like unloading slab (from corrugated sheets or reinforced concrete), transverse bracing, longitudinal bracing, reinforcing of the object inlet and outlet. They are aimed for increasing the load capacity and reducing the structure deformations.

The article presents the comparison of costs of the basic materials used in various systems of flexible structures from the corrugated metal sheets.

IOP Conf. Series: Materials Science and Engineering **269** (2017) 012025 doi:10.1088/1757-899X/269/1/012025

3. Prices of the basic materials used in the flexible structures from the corrugated metal sheets

In order to determine the costs of the material solutions the data for two systems (A and B) used in Poland (for construction of the upper passages for animals) since 2008 have been used. The cost estimation for the basic materials required for realization of 1 m^2 of the flexible structure from the corrugated steel sheets have been prepared with use of prices obtained directly from the Polish contractors and manufacturers, as well as process included in the quarterly information (Sekocenbud - issue no. 2/2017 [8]).

Table 1 presents prices (1 PLN = 0.26 USD) of 1 m^2 of the corrugated steel sheet of various thicknesses but comparable profile characteristics (wave), and table 2 presents the summary of prices of the basic connectors for the corrugated steel sheet used in the flexible engineering structures.

Table 1. Prices of 1 m^2 of the corrugated steel sheet used in the flexible engineering structures.

No.	Specification	Net price in PLN per 1 m^2
1	Corrugated sheet 7 mm thick - system A	509.48
2	Corrugated sheet 5.5 mm thick - system A	440.52
3	Corrugated sheet 4 mm thick - system A	335.91
4	Corrugated sheet 3 mm thick - system A	272.90
5	Corrugated sheet 7 mm thick - system B	490.00
6	Corrugated sheet 5.5 mm thick - system B	402.27
7	Corrugated sheet 4 mm thick - system B	315.90
8	Corrugated sheet 3 mm thick - system B	256.65

Table 2. Prices of the basic connectors for the corrugated steel sheet used in the flexible engineering structures.

No.	Specification	Net price in PLN per 1 piece
1	Bolt M20 cl. 8.8 70 mm long	3.50
2	Bolt M20 cl. 8.8 63 mm long	3.37
3	Bolt M20 cl. 8.8 50 mm long	3.20
4	Bolt M20 cl. 8.8 37 mm long	3.00
5	Anchor ø 20 365 mm long	9.75
6	Anchor ø 20 225 mm long	7.30

The sheet thickness and the wave characteristics are determined with use of calculations and they depend on the external loads, structure form, structure span, and environment aggressiveness. However, while analysing data from table 1 and comparing sheets of identical parameters (thickness, profile characteristics) but used in different systems the difference in the purchase cost of those sheets is seen. The net price of 1 m^2 of 7 mm thick sheet in system A is higher by 19.48 PLN from the price of 1 m^2 of 7 mm thick sheet in system B. That difference is almost 4%. The analysis of prices of 5.5 mm thick sheets reveals that the purchase cot of 1 m^2 of sheet in system A is440.52 PLN, while in system B it is 402.27 PLN (difference of over 8%). In case of 4 mm thick sheets system A has value higher by almost 6%. Very similar price relation is seen in case of 3 mm thick sheets - 1 m^2 of the corrugated sheet (used in the flexible engineering structures) of system A is more expensive by almost 6% from 1 m^2 of sheet of system B.

A required element for connecting the corrugated steel sheets is bolts. Their length depends on thickness and number of connected sheets. Comparing the prices of those basic connectors (with nut and washer) in table 2 one may see that the price increase is directly proportional to their length. The difference in the purchase price for bolt M20 cl. 8.8 of length of 70 mm and bolt M20 of the same class and length of 63 mm is approximately 4%, for bolt M20 cl. 8.8 of length of 63 mm and bolt M20 cl. 8.8 of length of 50 mm - 5%, while for bolts (cl. 8.8) 50 mm and 37 mm it is 6%.

An important element of correct fixing of the structure from the corrugated sheets are the anchors, channel bars and flat bars, which prices depend on their basic dimensions. Analysing data in table 2 and data included in the quarterly information of "Sekocenbud" one may see that the increase of the geometric parameters is correlated with the increase of prices.

The difference of prices of materials available on the market allows the investor for selecting the structure depending on the needs and financial possibilities, as well as for achieving some savings.

4. Summary

Construction of roads poses a threat for the natural environment. It is necessary to use the passages for animals. They allow the animals to use terrain located on both sides of the road and for moving to distant locations (the basic function of passages for animals, particularly important for protection of rare species having significant spatial requirements), influencing the maintaining of genetic variety of many species of animals. Currently used material and design solutions allow for very large possibilities of technical realization of such objects. The passages for animals are constructed from steel, concrete, and plastics. For over twenty years there are more and more small bridges and culverts from the corrugated steel sheets designed and constructed in Poland. They are easy in designing due to lower number of details, readily available portfolio of profile drawings, and prepared base of strength calculations for the typical applications, quick in assembly, and less expensive in realization when compared to the traditional solutions.

The comparison of costs of the basic materials used in the engineering structures from the corrugated steel sheets allows for determination of some more or less "economical" solutions. However, while selecting the system and materials it is also required to take into consideration the realization potential, experience, and know-how.

The savings in case of purchasing sheets of identical parameters (thickness, profile characteristics) are from approx. 4% to 8% per 1 m^2 of sheet. The connectors in form of bolts M20 cl. 8.8 of various lengths are an expense from 3.00 PLN to 3.50 PLN. Those values may seem low, but taking into consideration amounts connected with construction of many square metres of structure they may become very important factor in the total investment costs.

5. References

[1] General Direction of National Roads and Motorways 2004 *Design and technological recommendations for the flexible engineering structures from the corrugated steel sheets.*

[2] Kurek R 2011 *Passages for animals in the vicinity of roads - optimal solutions, experiences, and problems in range of designing, part 1*Modern engineering building **3**.

[3] Kurek R 2011 *Passages for animals in the vicinity of roads - optimal solutions, experiences, and problems in range of designing, part 2* Modern engineering building **4**.

[4] Styś D 2014 *Passages and culverts for animals in the communication building,*Motorways **4**.
[5] General Direction of National Roads and Motorways 2002 *Catalogue of road devices for environment protection.*
[6] Janusz L and Madaj A 2007 *Engineering objects from the corrugated steel sheets.* Designing and realization. Transport and Communication Publishers.
[7] Wysokowski A and Howis J 2009 *Materials for construction of culverts, part II. Culverts in communication infrastructure – part 6* Modern engineering building **5**.
[8] Sekocenbud - Issue no. 2/2017. Information of prices of the building materials in the 2nd quarter of 2017, Construction Economical and Organizational Implementation Centre "Promocja", Warsaw.

Acknowledgements
The article has been prepared in range of work no. S/WBiIŚ/1/2015 and has been financed from the resources of The Ministry of Science and Higher Education for science.

Process Optimization and Microstructure Characterization of Ti6Al4V Manufactured by Selective Laser Melting

Li junfeng[1], Wei zhengying[1]

[1]State key laboratory for manufacturing systems engineering, Xi'an Jiaotong University, Xi'an, Shaanxi

Abstract. Process optimization and microstructure characterization of Ti6Al4V manufactured by selective laser melting (SLM) were investigated in this article. The relative density of sampled fabricated by SLM is influenced by the main process parameters, including laser power, scan speed and hatch distance. The volume energy density (VED) was defined to account for the combined effect of the main process parameters on the relative density. The results shown that the relative density changed with the change of VED and the optimized process interval is 55~60J/mm^3. Furthermore, compared with laser power, scan speed and hatch distance by taguchi method, it was found that the scan speed had the greatest effect on the relative density. Compared with the microstructure of the cross-section of the specimen at different scanning speeds, it was found that the microstructures at different speeds had similar characteristics, all of them were needle-like martensite distributed in the β matrix, but with the increase of scanning speed, the microstructure is finer and the lower scan speed leads to coarsening of the microstructure.

1. Introduction

Ti6Al4V, a kind of titanium alloy, is a typical two-phase alloy. With a reputation of corrosion resistance, high specific strength and so on. Ti6Al4V is called "space metal" or "ocean metal", due to its excellent performance and a wide application prospect in the military and civilian industry [1]. With the wide application of titanium alloy in biological and medical fields in recent years, it is a general trend to fabricate complex structure of Ti6Al4V alloy device [2-4]. However, conventional manufacturing technology is unable to meet the requirements in the manufacturing of complex structure, such as internal channel, space gradual changing surface and ultrathin wall.

To break the bottleneck, metal additive manufacturing technologies have, therefore, been developed, especially Selective laser melting(SLM) technology. It melts the powder layer by layer, forming a good metallurgical bonding between the layers without any auxiliary tools. Theoretically, it can form a completely dense and any complex components. However, there are many factors during the process of SLM, causing it is difficult to obtain fully dense component [5, 6]. Therefore, improvement of density of samples fabricated by SLM has become a main concern for many scholars. In addition, special microstructure due to extreme non-equilibrium state during process of SLM also has been noticed. Sun J et al. have conducted a optimization of parameters of Ti6Al4V fabricated by SLM using Taguchi method [7]; defect has been discussed via experimentally studied and defect generation mechanisms were different at too high or too low of energy input [8]; Shi X et al.have investigated influence layer thcikness on performance with a high layer thickness (200μm) [9]; Khorasani A M et al. have discussed the mechanisms and critical parameters on solidification, such as laser power, scan speed and strategy, layer thickness, hatch spacing [10]; Kabir M et al. proposed a comprehensive model to help understand the interaction between process parameters, pore morphology and mechanical properties [11]; Song B et al. have investigated process parameters based temperature distribution simulation model [12]; Process parameters of AlSi10Mg fabricated by SLM were conducted by Kempen K et al. [13]. Porosity content control of CoCrMo fabricated by SLM was studied by Joguet D et al. [14]; Thijs Lore et al. have investigated the development of microstructure of Ti6Al4V alloy processed by SLM at different scan strategies [15]; Han J et al have studied microstructure of the SLM processed Ti6Al4V at various laser energy densities [16]. It can be seen that the process parameters are the key factors which affect the defect and microstructure of samples fabricated by SLM, and it is still need to conduct the process optimization of Ti6Al4V fabricated by SLM.this paper aims to investigate the effect of process parameters (laser power, scan speed, and hatch distance) on relative density of Ti6Al4V manufactured by SLM, and obtain a suitable process parameter range; Subsequetly, Taguchi method was used to study the influence degree of three process parameters on relative density. Further, the microstructure under different scan speed was investigated.

2. Experiment and method

Gas atomized Ti6Al4V powders was used in this investigation. The morphology of powder as shown in figure 1. A near spherical shape was present in figure 1 and the surface of powder was not smooth as presented in figure 1(b). Process parameters used in this experiment was shown in table 1. The layer thickness (LT) used in the experiment was kept in 0.03mm. The scan strategy used in this experiment and the sample (6mm×6mm×6mm) processed by SLM were described in figure 2.

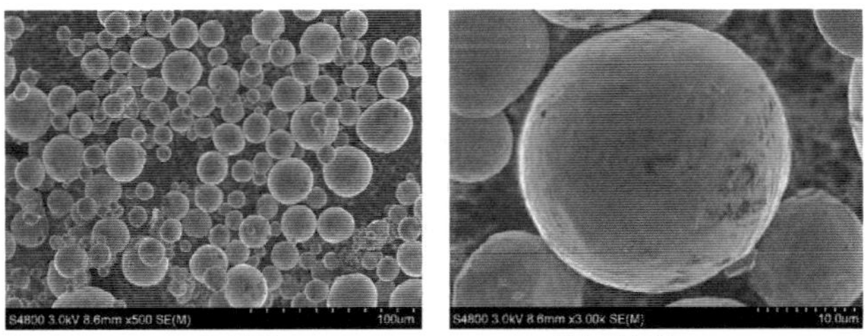

(a) 500× (b) 3000×

Figure 1. Morphology of Ti6Al4V powder.

(a) Scan strategy (b) samples fabricated by SLM

Figure 2. Scan strategy and samples.

Table 1. Process parameters used in the experiment.

	Level 1	Level 2	Level 3
Laser power/(LP,W)	280	300	320
Scan speed/(SS, mm/s)	2800	3000	3200
Hatch distance/(HD, mm)	0.05	0.06	0.07
Layer thickness/(LT, mm)	0.03	0.03	0.03

All parts were fabricated on the self-developed SLM 150 machine. This SLM machine is equipped with an IPG YLR-400 fiber laser, which produces a laser beam with a wavelength of 1070nm and can reach a maximum power of 400W in continuous mode. The scanning velocity can be varied from 10 to 7000mm/s, the maximum diameter of the useful building area is 150mm×150mm. The building chamber is first evacuated and then filled with an inert argon atmosphere to keep an atmosphere with a low oxygen content during building.

The relative density was measured based on Archimedes principle by weighing the samples in the air and subsequently in water. Standard metallographic process was conducted for microstructure, and a solution of 80 ml H_2O, 5 ml HF and 10 ml HNO_3 was used to reveal microstructure. Microstructures were observed under an optical microscope (Nikon MA 200). Morphology of Ti6Al4V powder was characterized by SEM (Hitachi, S4800).

3. Results and discussion

3.1. Process optimization of Ti6Al4V fabricated by SLM

Part density is one of the most important factors to optimise, for it has a direct correlation on the part's mechanical and physical properties. Volume energy density (VED) was defined to describe the relationship between relative density and the combination of three factors. The equation as follows:

$$\text{VED} = LP/(SS \cdot LT \cdot HD)$$

The relationship between relative density and VED was depicted in figure 3. With the increase of VED, the relative density increased first and then decreased. When the VED is near $60\text{J}/mm^3$, the relative density reaches the maximum value, which is close to 100%. There are two reasons for this phenomenon. Figure 4 shows the melting path profile under different LP/SS, when the LP/SS value is too low, as shown in figure 4(a) the depth of the melting channel is lower and spheroidization occurs, resulting in a weak bonding between the layers and un-melted powder particles, therefore the relative density is low; When the LP/SS value is too high, as shown in figure 4.(b) the melting track depth is too large, the height is smaller, it is difficult to form, and the phenomenon of element evaporation and ablation is easy to occur during the forming process, and the density decrease is also caused. Figure 5 shows the melting path profile under different HD. When Hatch distance is too large, the overlap area of molten track is too small, as shown in figure 5(a), which leads to the partial melting of the powder and the inclusion occurred; the phenomenon of porosity results in the appearance of evaporation and ablation of elements after overlap, as shown in figure 5(c). Therefore, only the suitable LP/SS and HD can obtain higher relative density, as shown in figure 4(b) and figure 5(b).

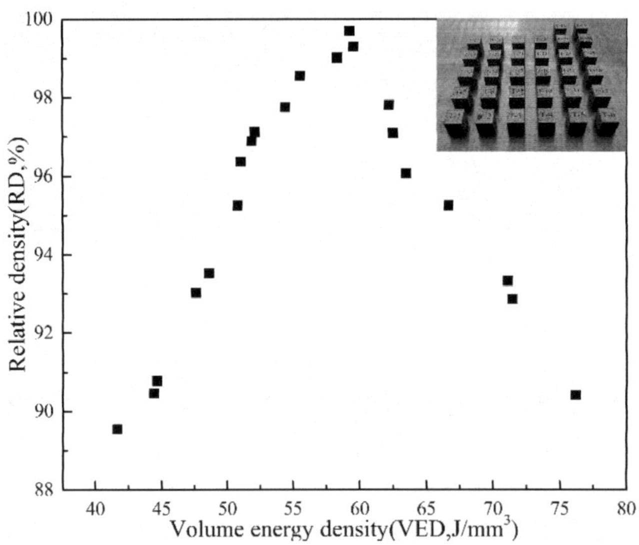

Figure 3. Variation of relative density under different volume energy densities.

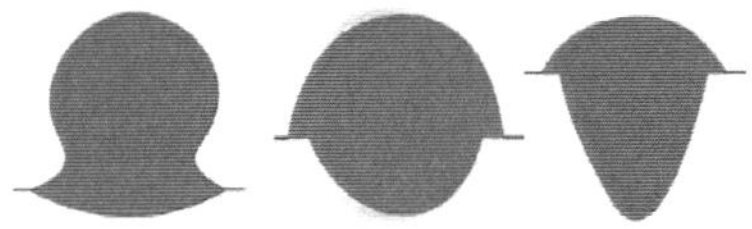

(a) Low (b) Medium (c) High

Figure 4. The melting path profile under different LP/SS.

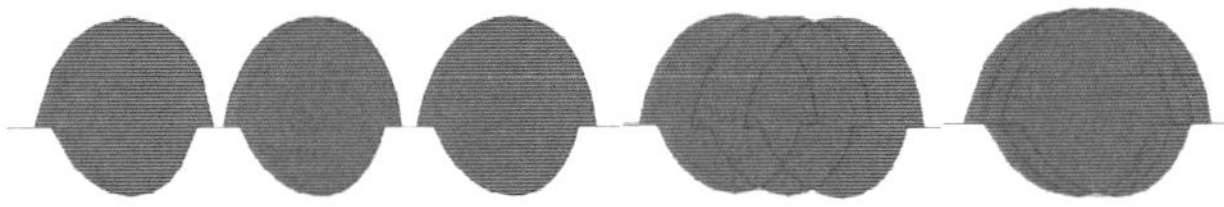

(a) High (b) Medium (c) Low

Figure 5. The melting path profile under different HD.

3.2. The Influence degree of three factors on relative density

After optimizing the manufacturing process, the influence degree of the three factors on the relative density was studied by Taguchi method. Table 2 illustrates the S/N results of the cubic samples produced in the experiment. According to the range analysis from S/N, the influence degree of each factor on the density is Scan speed>Hatch distance>Layer power. Figure 6 shows the change curve of S/N main effect value and the change trend density with various factors.

Table 2. Responses for density and the corresponding S/N.

Std.	Factors			
	A:LP/(W)	B:SS/(mm/s)	C:HD/mm	RD/%
1	280	2800	0.05	95.25
2	280	3000	0.06	96.89
3	280	3200	0.07	89.55
4	300	2800	0.05	92.86
5	300	3000	0.06	98.55
6	300	3200	0.07	90.78
7	320	2800	0.05	90.42
8	320	3000	0.06	99.69
9	320	3200	0.07	93.02
S/N-1	39.45	39.38	39.57	
S/N-2	39.46	39.86	39.48	
S/N-3	39.49	39.19	39.35	
Range	0.04	0.67	0.22	

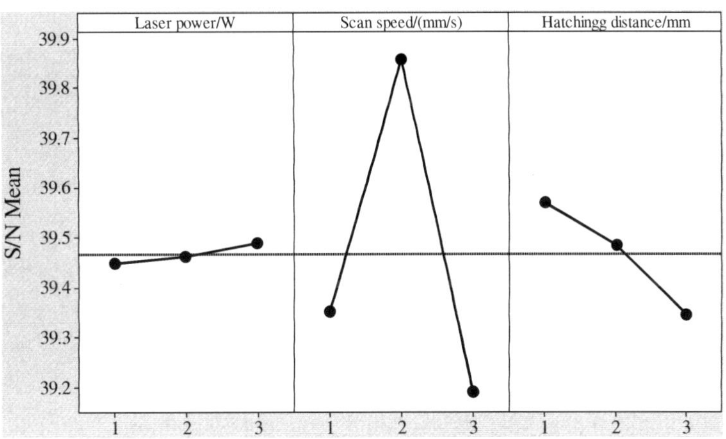

Figure 6. S/N main effect graph.

3.3. Microstructure characterization of Ti6Al4V fabricated by SLM

Microstructure has a direct influence on mechanical properties of components. Therefore, microstructure of samples fabricated by SLM at different scan speed was characterized. Due to the special building process, the microstructure of SLM processed sample is quite different with other traditional manufacturing technology, such as welding, casting, as shown in figure 7. As expected the present microstructure mainly is a very fine acicular martensite, due to the fast cooling rate that occur during the SLM manufacturing process. However, the effect of utilized scan speed on thermal histories during fabrication and, consequently, the resultant microstructure of part had still some differences. With the increase of scan speed, the refined acicular martensite appeared in the matrix of β. This can be due to the fact that the different cooling rates experienced during fabrication at different scan speed may result in different microstructures; That is, the increase of scan speed leads to an increase in the cooling rate, which in turn results in a finer microstructure. The lower scan speed thus results in a coarser microstructure.

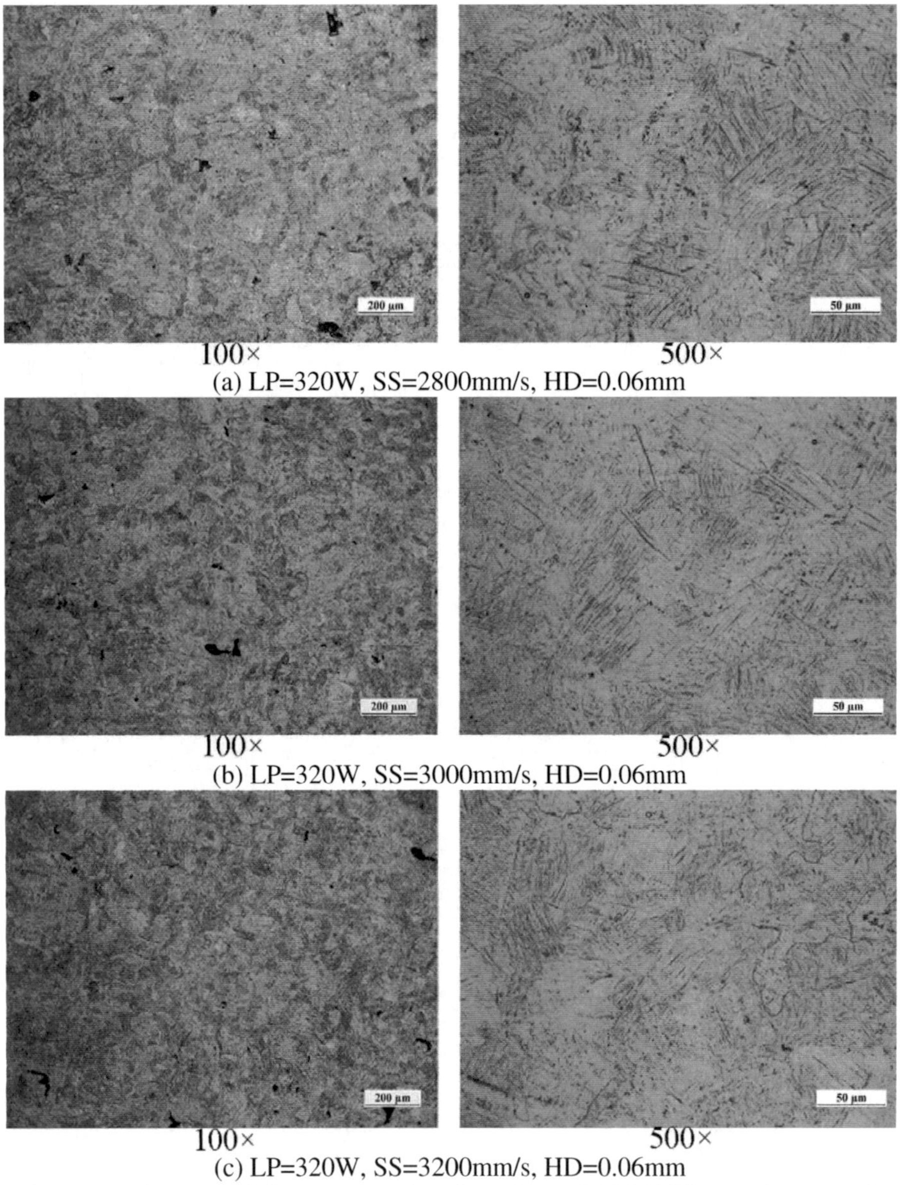

Figure 7. Microscope images of the cross-sectional surface of Ti6Al4V samples manufactured at different scan speed (red line is the boundary of β matrix).

4. Conclusion

In this study, effects of process parameters (i.e. laser power, scan speed and hatch distance) on relative density of Ti6Al4V samples manufactured by selective laser melting (SLM) and microstructure corresponding to different scan speed were investigated. It was found that the relative density of Ti6Al4V samples manufactured via SLM was greatly influenced by process parameters. Volume energy density (VED) was defined to show the combined effect of process parameters, and the optimized VED interval is 55~60J/mm^3. The influence of three process parameters (i.e. laser power, scan speed, hatch distance) on the relative density was studied by Taguchi method. It was found that among the three process parameters, scan speed had a greater effect on relative density. Furthermore, microstructure characterization shown that with the increase of scan speed, the refined acicular martensite appeared in the β matrix. This can be due to the fact that the different cooling rates experienced during fabrication at different scan speed may result in different microstructures. The lower scan speed thus results in a coarser microstructure. In a word, the purpose of this paper is to optimize the process parameters and microstructure characterization of Ti6Al4V fabricated in the self-developed SLM equipment. It is found that the nearly dense Ti6Al4V component can be realized and the microstructure can be controlled by the process parameters. The customization of the microstructure may be obtained by the design of process conditions.

References

[1] Leyens C, Peters M 2003 *Titanium and Titanium Alloys -Fundamentals and Applications*. DLR, **1** 513.
[2] Bartolomeu F, Sampaio M, Carvalho O, et al. 2017 *J. Mech. Behav. Biomed. Mater.* **69** 128.
[3] Liu Y J, Li S J, Wang H L, et al. 2016 *Acta Mater.* **113** 56-67.
[4] Dobrzanski L A, Dobrzanska-Danikiewicz A D, Achtelik-Franczak A, et al. 2017 *Porous Selective Laser Melted Ti and Ti6Al4V Materials for Medical Applications* Powder Metallurgy - Fundamentals and Case Studies.
[5] Wimpenny D I, Pandey P M and Kumar L J 2017 *Advances in 3D Printing & Additive Manufacturing Technologies*. Springer Singapore.
[6] Yadroitsev I, Bertrand P and Smurov I 2007 *Appl. Surf. Sci.* **253(19)** 8064-9.
[7] Sun J, Yang Y and Wang D. 2013 *Opt. Las. Technol.* **49(7)** 118-24.
[8] Gong H, Rafi K, Gu H, et al. 2014 *Addit. Manuf.* **s1–4** 87-98.
[9] Shi X, Ma S, Liu C, et al. 2016 *Materials*, **9(12)** 975.
[10] Khorasani A M, Gibson I, Goldberg M, et al. 2016 *Mater Des*, **103** 348-55.
[11] Kabir M and Richter H 2017 *Materials*, **145**.
[12] Song B, Dong S, Liao H, et al. 2012 *Int. J. Adv. Manuf. Tech.* **61(9-12)** 967-74.
[13] Kempen K, Thijs L, Humbeeck J V, et al. 2015 *Mater. Sci. Technol.* **8**.
[14] Joguet D, Costil S, Liao H, et al. 2016 *Rapid. Prototyping. J.* **22(1)** 20-30.
[15] Thijs L, Verhaeghe F, Craeghs T, et al. 2010 *Acta Mater.* **58(9)** 3303-12.
[16] Han J, Yang J, Yu H, et al. 2017 *Rapid. Prototyping. J.* **2(217)** 217-26.

IOP Conf. Series: Materials Science and Engineering **269** (2017) 012027 doi:10.1088/1757-899X/269/1/012027

The Reliability of the Finite Element Simulation of the Interface Mechanical Effect of Inclusion and Matrix Material in Powder Metallurgy Superalloy

Liu Xin-ling[1, 2, 3, a], Wang Tian-yu[1, 2, 3, b], Zhao Kai[1, 2, 3, c]

[1]AVIC Failure Analysis Center, Beijing Institute of Aeronautical Materials, Beijing 100095, China
[2]Beijing Key Laboratory of Aeronautical Materials Testing and Evaluation, Beijing 100095, China
[3]Aviation Key Laboratory of Science and Technology on Materials Testing and Evaluation, Beijing 100095, China

Abstract. The inclusion defects in powder metallurgy superalloy have the characteristics of randomness, such as the shape, size, location and composition. It is important to establish effective and reliable finite element model and method to study the influence of the different defects. In this paper, the inclusion/matrix interface model was proposed, and on the basic theory of two-dimensional elastic problems of complex variable function, the analytical formula was derived and the analytical solution was calculated for the condition of circular inclusion in matrix. On the other hand, stress distribution of the interface of inclusion and matrix was simulated by finite element method. The finite element model is proved effective, reliable through comparison analysis between the numerical analysis method and finite element simulation.

1. Introduction

Due to the manufacturing process characteristics of powder metallurgy superalloy, the defects are inevitable. At present, inclusion is the main defect category in powder metallurgy superalloy. The inclusion can reduce the fatigue performance, especially the low cycle fatigue property of powder metallurgy superalloy. Different from common particle, fiber reinforced composite materials, inclusions in powder metallurgy superalloy have characteristics of random in aspects of shape, size, location and composition, and cannot establish the overall model using macro homogenization method. Therefore, it is of great significance to study the mechanical effect of inclusions in the matrix.

The main test methods to study the defects are low cycle fatigue tests and analysis from the fracture surfaces, and using statistical method to study the relationship between the test conditions and the defects. Although this method may give the relationship between defects and macroscopic mechanics, the defects influence cannot be studied fully because the randomness of the shape, size, location and composition of the defects. The computer simulation is an important aided method. The interface of defect and matrix improves the simulation complexity. Reliable simulation methods and models are the prerequisite for researching the defects.

Lin Tao [1] analysed the effect of shot peening on powder metallurgy superalloy using ABAQUS software, established the finite element model of a single projectile impacting target material, compared the residual stress and equivalent plastic strain of different velocity and surface friction factor, the collision frequency and the distance of projectiles. A numerical simulation of the isothermal forming process of precasting plate of powder metallurgy superalloy was carried out [2, 3]. The heating process of isothermal forging mould of FGH96 superalloy plate was simulated with FEA by Zhang Mingjie et al. [4]. The stress field and temperature field of the mould during the heating process were studied, the thermal damage caused by thermal fatigue was analysed, and the heating system of the mould was optimized. It is lacking on interface stress analysis of powder metallurgy superalloy by FEA.

The inclusion/matrix interface is a weak joint surface, easy to cause crack initiation and fracture damage, and the mechanical effect produced by inclusion in the matrix is determined by the interface stress distribution of inclusion and matrix. In order to study the reliability of finite element simulation results, the distributions of stress and strain of inclusion interior, matrix nearby inclusion and the interface of inclusion/matrix were comparatively analysed by theoretical derivation and the analytical solution and numerical simulation method. On the other hand, the effect of finite size boundary conditions on the stress distribution was also considered, the validity and reliability of finite element model was studied in this article.

2. Research method

The matrix and inclusions are completely elastic material, where the elastic modulus of the matrix E1 = 193000MPa, the Poisson's ratio v1 = 0.3, the elastic modulus of Al_2O_3E2 = 400000MPa, and Poisson's ratiov2 = 0.3.

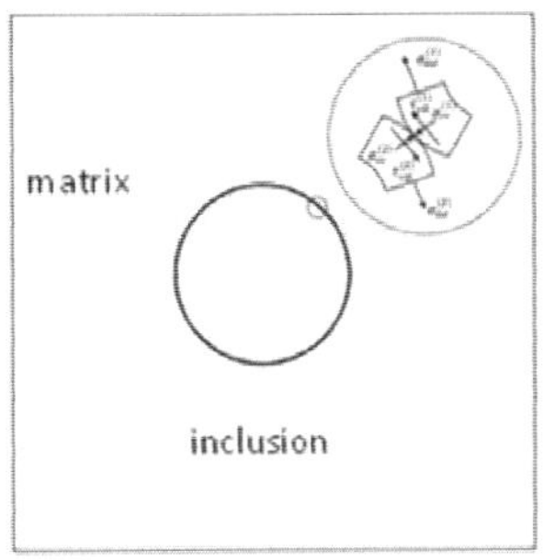

Figure 1. Interface connection model and interface phase model diagram.

Using the inclusion/matrix two-phase model, the interface was considered as the geometric curve or surface of the bonded matrix with the inclusion phase, and the bond strength is certain. In this paper, the circular inclusion and the excellent interface was studied. The boundary conditions of the stress and displacement were as follows.

$$\sigma_{rr}^{(1)} = \sigma_{rr}^{(2)}, \tau_{rr}^{(1)} = \tau_{rr}^{(2)}, u_1^{(1)} = u_2^{(2)}, v_1^{(1)} = v_2^{(2)}$$

According to the boundary conditions of upper formula and the geometrical and physical relations of elastic mechanics, it can be seen that the normal strain and the shear strain perpendicular to the interface are discontinuous. The normal stress parallel to the interface is also discontinuous. Based on these considerations, the interfacial normal stress and the interface shear stress are main destruction parameters, and the maximum main stress and maximum plastic strain are main destruction parameters of matrix.

According to the basic theory of complex function, the stress components can be expressed separately [5].

$$\begin{cases} \sigma_x = 2\operatorname{Re}\Phi(z) - \operatorname{Re}\left[\bar{z}\Phi'(z) + \Psi(z)\right] \\ \sigma_y = 2\operatorname{Re}\Phi(z) + \operatorname{Re}\left[\bar{z}\Phi'(z) + \Psi(z)\right] \\ \tau_{xy} = \operatorname{Im}\left[\bar{z}\Phi'(z) + \Psi(z)\right] \end{cases} \quad (1)$$

The function expression of the displacement component can be obtained according to the geometric equation and the physical equation of the elastic theory.

$$2\mu(u+iv)=\kappa\varphi(z)-z\overline{\varphi'(z)}-\overline{\psi(z)} \tag{2}$$

U and v are the displacement of two directions in a rectangular coordinate system. μ is the shear modulus, $\mu=E/2(1+\nu)$, E is modulus of elasticity, ν is Poisson's ratio. K is the combination parameter of the material, and $\kappa=(3-\nu)/(1+\nu)$ under plane stress condition, $\kappa=3-4\nu$ under plane strain condition.

In the infinite and multiply connected area, the stress function may be represented by the following two analytic functions.

$$\begin{cases} \varphi(z)=-\dfrac{F_x+iF_y}{2\pi(1+\kappa)}\ln z+(B+iC)z+\varphi_0(z) \\ \psi(z)=\dfrac{\kappa(F_x-iF_y)}{2\pi(1+\kappa)}\ln z+(B'+iC')z+\psi_0(z) \end{cases} \tag{3}$$

F_x and F_y are the sum of all the external forces on the edge of the hole, B, B', C, C' are the combined expressions of the far field load, $B=\dfrac{\sigma_y^\infty+\sigma_x^\infty}{4}$, $C=0$, $B'=\dfrac{\sigma_y^\infty-\sigma_x^\infty}{2}$, $C'=\tau_{xy}^\infty$, $\varphi_0(z)$ and $\psi_0(z)$ are just the items of negative power of z.

3. Analytic solution of the stress distribution of circular inclusions in the matrix

The complex function elastic theory was used to obtain the analytic formula. Based on the analytic calculation formula, the interface stress of circular inclusion of infinite flat bearing two-way tensile load was calculated.

3.1. Mechanical model

As shown in figure 2 (a), infinite flat is subjected to a two-way tensile load in the y direction at infinity. A circular inclusion with radius R is in the center of the plate, the plate's material constants are $\mu 1$ and $\kappa 1$, the inclusion's material constants are $\mu 2$ and $\kappa 2$. The matrix and inclusion is perfectly connected and the interface is the boundary between the two different materials. The interface model and its boundary conditions are shown in figure 2 (b). (Note: mark 1 refers to matrix, 2 refers to inclusions)

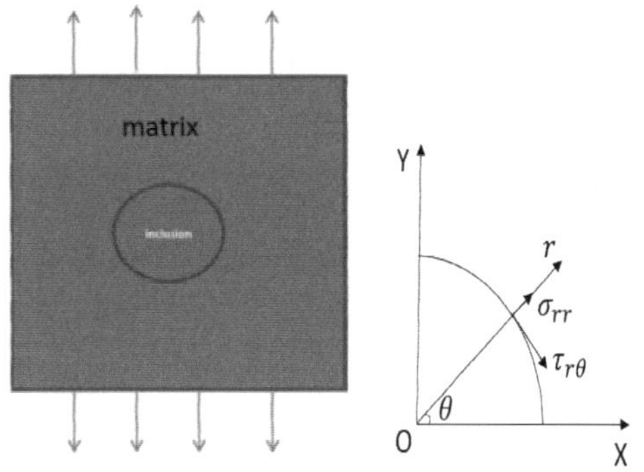

(a) Inclusion in the plate (b) Diagrammatic sketch of angle θ

Figure 2. Inclusion and diagrammatic sketch of angle θ.

3.2. Formula derivation for analytical calculation

The problem may be considered as complex area, which includes infinite plate with a hole in the center and the inclusion in the center of the plate, and the stress components and displacement are continuous. First of all, the analytic function of the solution is expanded into a complex series of the corresponding region. Then, it may be transformed into solving the problem of linear equations using interface and infinity boundary conditions. The last, according to the relationship between analytical function and stress, one-way tensile problem of a single circular inclusion is obtained.

Because the interfacial boundary of inclusion and matrix is arc, boundary conditions of the interface are not easy to represent in rectangular coordinates, thenitis need to be converted to polar coordinates. It can be seen from the geometric knowledge that the displacement (u, v) in the polar coordinates and the displacement in the rectangular coordinate system have the following coordinate changes.

$$\begin{cases} u_r = u\cos\theta + v\sin\theta \\ u_\theta = -u\sin\theta + v\cos\theta \end{cases} \tag{4}$$

Complex function is composed of the upper form:

$$u_r + iu_\theta = (u + iv)e^{-i\theta} \tag{5}$$

Complex expression of the displacement of the rectangular coordinate system is substituted into the formula (2), which is the complex representation of the polar coordinates.

$$2\mu(u_r + iu_\theta) = e^{-i\theta}\left[\kappa\varphi(z) - z\overline{\varphi'(z)} - \overline{\psi(z)}\right] \tag{6}$$

As shown in figure 3, it is necessary to establish the stress equilibrium condition of the interface microelement in order to obtain the normal stress and shear stress of the inclusion and matrix interface of the polar coordinates. The surface area of the microelement cant is dA, the circle angle is θ, and the micro-equilibrium equation in both X and Y directions are as follows.

$$\begin{cases} \sigma_{rr}\cdot dA\cdot\cos\theta + \tau_{r\theta}\cdot dA\cdot\sin\theta - \tau_{yx}dA\cdot\sin\theta - \sigma_x\cdot dA\cdot\cos\theta = 0 \\ \sigma_{rr}\cdot dA\cdot\sin\theta - \tau_{xy}dA\cdot\cos\theta - \tau_{r\theta}\cdot dA\cdot\cos\theta - \sigma_y\cdot dA\cdot\sin\theta = 0 \end{cases} \tag{7}$$

Remove dA, according to the square formula of the trigonometric functions we can obtain the function as follows.

$$\begin{cases} \sigma_{rr} = \frac{1}{2}(\sigma_x + \sigma_y) + \frac{1}{2}(\sigma_x - \sigma_y)\cos 2\theta + \tau_{xy}\sin 2\theta \\ \tau_{r\theta} = \frac{1}{2}(\sigma_y - \sigma_x)\sin 2\theta + \tau_{xy}\cos 2\theta \end{cases} \tag{8}$$

Notice that negative values of σ_{rr}, $\tau_{r\theta}$ represent the opposite direction.

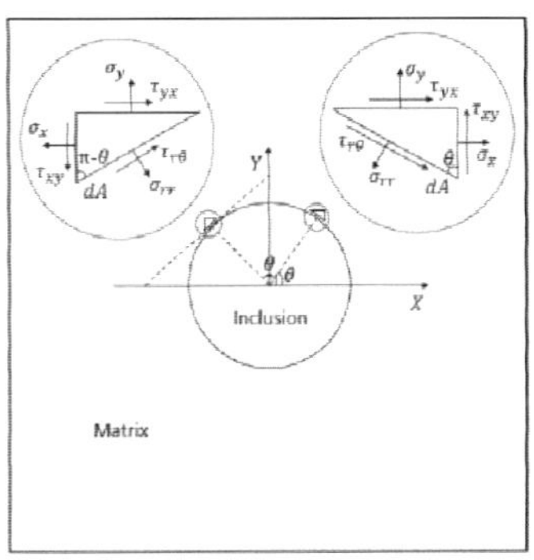

Figure 3. A schematic diagram of the force of the microelement near the interface.

Combine equation (8) with equation (1),

$$\sigma_{rr} - i\tau_{r\theta} = \varphi'(z) + \overline{\varphi'(z)} - e^{i2\theta}\left[\overline{z}\varphi''(z) + \psi'(z)\right] \tag{9}$$

Through the above analysis, we have a complex expression of displacement and stress in polar coordinates.

$$\begin{cases} 2\mu(u+iv) = e^{-i\theta}\left[\kappa\varphi(z) - z\overline{\varphi'(z)} - \overline{\psi(z)}\right] \\ \sigma_{rr} - i\tau_{r\theta} = \varphi'(z) + \overline{\varphi'(z)} - e^{i2\theta}\left[\overline{z}\varphi''(z) + \psi'(z)\right] \end{cases} \tag{10}$$

Substitute the far load conditions $\sigma_x^\infty = 0, \sigma_y^\infty = P, \tau_{xy}^\infty = 0$ into the formula (3), then the infinite plane area with round hole can be expressed as:

$$\begin{cases} \varphi_1(z) = \sum_{n=1} \left[a_n \left(\frac{R}{z}\right)^n \right] + \frac{P}{4} z \\ \psi_1(z) = \sum_{n=1} \left[b_n \left(\frac{R}{z}\right)^n \right] + \frac{P}{2} z \end{cases} \tag{11}$$

According to the series theory in the complex function, the complex expression of the circular area domain is obtained.

$$\begin{cases} \varphi_2(z) = \sum_{n=1} c_n \left(\frac{z}{R}\right)^n \\ \psi_2(z) = \sum_{n=1} d_n \left(\frac{z}{R}\right)^n \end{cases} \tag{12}$$

According to the displacement continuity conditions of the interface, we may obtain:

$$\begin{cases} \varphi_1(z) = \frac{P}{2}\frac{(\mu_2 - \mu_1)R^2}{(\mu_2\kappa_1 + \mu_1)z} + \frac{P}{4}z \\ \psi_1(z) = \frac{P}{2}\frac{(\mu_2\kappa_1 - \mu_1\kappa_2 + \mu_1 - \mu_2)R^2}{(2\mu_2 + \mu_1\kappa_2 - \mu_1)z} \\ \qquad + \frac{P}{2}\frac{(\mu_2 - \mu_1)R^4}{(\mu_2\kappa_1 + \mu_1)z^3} + \frac{P}{2}z \end{cases} \tag{13}$$

$$\begin{cases} \varphi_2(z) = \frac{\mathrm{P}}{4}\frac{\mu_2(\kappa_1 + 1)}{(\kappa_2 - 1)\mu_1 + 2\mu_2} z \\ \psi_2(z) = \frac{P}{2}\frac{\mu_2(1 + \kappa_1)}{\mu_1 + \mu_2\kappa_1} z \end{cases} \tag{14}$$

Substitute the formula (14) into formula (1), then the stress components in the rectangular coordinate system can be obtained.

$$\begin{cases} \sigma_x^{(2)} = \frac{P}{2} \frac{\mu_2(\kappa_1+1)}{(\kappa_2-1)\mu_1+2\mu_2} - \frac{P}{2} \frac{\mu_2(1+\kappa_1)}{\mu_1+\mu_2\kappa_1} \\ \sigma_y^{(2)} = \frac{P}{2} \frac{\mu_2(\kappa_1+1)}{(\kappa_2-1)\mu_1+2\mu_2} + \frac{P}{2} \frac{\mu_2(1+\kappa_1)}{\mu_1+\mu_2\kappa_1} \\ \tau^{(2)}{}_{xy} = 0 \end{cases} \tag{15}$$

Substitute the formula (15) into formula (9), then the stress component in polar coordinates can be obtained.

$$\begin{cases} \sigma_{rr}^{(2)} = \frac{P}{2} \frac{\mu_2(\kappa_1+1)}{(\kappa_2-1)\mu_1+2\mu_2} - \frac{P}{2} \frac{\mu_2(1+\kappa_1)}{\mu_1+\mu_2\kappa_1}\cos 2\theta \\ \tau_{r\theta}^{(2)} = \frac{P}{2} \frac{\mu_2(1+\kappa_1)}{\mu_1+\mu_2\kappa_1}\sin 2\theta \end{cases} \tag{16}$$

Substitute the material parameters under the condition of plane stress $\mu = E/2(1+\nu)$, $\kappa = (3-\nu)/(1+\nu)$ and material parameters under the condition of plane strain $\mu = E/2(1+\nu)$, $\kappa = 3-4\nu$ into formula (15), then under the condition of plane stress,

$$\begin{cases} \sigma_{rr}^{(2)} = \frac{P}{E_1}\left(\frac{1}{\frac{1+\nu_1}{E_1}+\frac{1-\nu_2}{E_2}} - \frac{2\cos 2\theta}{\frac{1+\nu_2}{E_2}+\frac{3-\nu_1}{E_1}} \right) \\ \tau_{r\theta}^{(2)} = \frac{2P E_2 \sin 2\theta}{E_1(1+\nu_2)+E_2(3-\nu_1)} \end{cases} \tag{17}$$

Under the condition of plane strain,

$$\begin{cases} \sigma_{rr}^{(2)} = \frac{P(1-\nu_1)}{E_1}\left[\frac{1}{\frac{(1-2\nu_2)(1+\nu_2)}{E_2(1+\nu_1)}+\frac{1}{E_1}} - \frac{2\cos 2\theta}{\frac{1+\nu_2}{E_2(1+\nu_1)}+\frac{3-4\nu_1}{E_1}} \right] \\ \tau_{r\theta}^{(2)} = \frac{2PE_2(1+\nu_1)(1-\nu_1)}{E_1(1+\nu_2)+E_2(1+\nu_1)(3-4\nu_1)}\sin 2\theta \end{cases} \tag{18}$$

3.3. Analytical calculation and result analysis

From equation (15), we know that the stress components of the circular inclusion in rectangular coordinate systems, σ_x, σ_y, τ_{xy} are independent of the coordinates, and they are constant parameters related to material and load conditions.

The stress components of the interface of the matrix/inclusion are material correlation. The matrix material is FGH95 powder metallurgy superalloy, and the modulus of elasticity E is193000MPa, poisson's ratio is 0.3. The inclusion is Al_2O_3, and the modulus of elasticity E is 400000Mpa, poisson's ratio is 0.3. The load is 100Mpa. All of the parameters are substituted into equation (17) and (18), the relationships between interface stress and angle θ under plane stress and plane strain, as shown in figure 4.

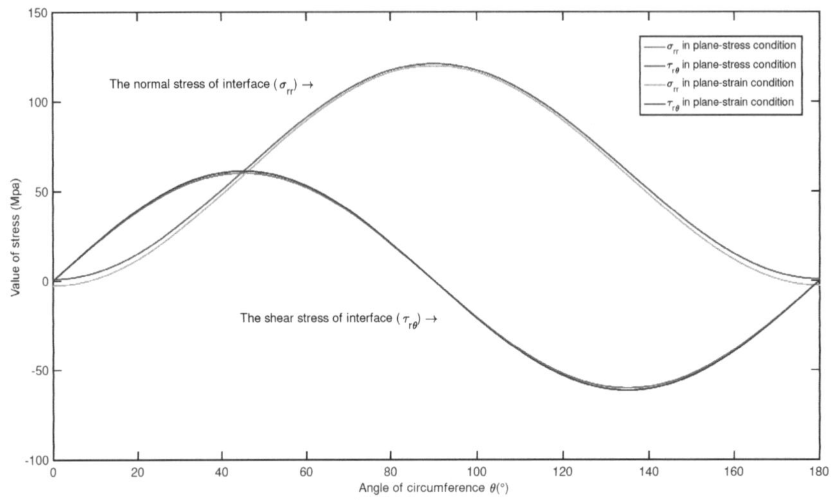

Figure 4. The stress component of the interface varies with angle θ.

It can be seen from figure 4 that the interface stress components have the same trend as the circumferential angle θ under the plane stress condition and the plane strain condition. At the same circumferential angle θ, the interfacial stress component under the plane stress condition is higher than that under the plane strain condition. In addition, from the trend of each curve, the interface normal stress reaches the maximum at the 90° circumference angle θ, and the interface shear stress reaches the maximum at θ 45° and 135°. The interface normal stress and the interface shear stress are the main stress components to measure the interfacial strength. Considering the law of the interface normal stress and shear stress in the range of 0~180°, we consider that the interface area of 45~135° is dangerous.

4. Finite element simulation of stress distribution of circular inclusion in matrix

Using the finite element software Abaqus, the two - dimensional plane problem with circular inclusion in the case of uniaxial stretching is simulated. The reliability of the finite element model is verified by comparing the numerical solution with the analytical solution.

4.1. Finite element model

For simulating the far-off boundary conditions, the inclusion size and the finite large plane size proportion is set to 1 to 10000. The matrix size is 1 m long and 1 m width. The radius R of inclusion is 100 um (equivalent to the size of inclusion in powder metallurgy superalloy). Taking into account the symmetry of the load and geometry, a quarter of the whole was used to build the model.

4.2. Grid division

Considering the difference between the size of inclusion and the size of the matrix, the grids are divided in a hierarchical manner. In addition, in order to obtain the stress value conveniently and accurately at different angle θ, the grid seeds of the interface of the inclusion/matrix are divided into 48 parts within the scope of 0~90°. The angle span of every four grid nodes is 7.5°.

4.3. Results analysis of finite element simulation

Stress component distributions inside of the inclusion under the conditions Plane stress and plane strain are shown in figure 5 and figure 6. S11, S22, S33 represent the stress of the X direction, Y direction and Z direction of the unit cell of the normal stress, and the positive symbol is tensile stress, the negative symbol is pressure stress. The S12 is the shear stress of the unit cell along the Y axis, and the clockwise sign is positive, the counter clockwise sign is negative.

You can see from the figure 5, 6 that each stress component inside of the inclusions is constant under rectangular coordinate system whether it is plane stress or plane strain condition, which is consistent with the result of the analytic solution.

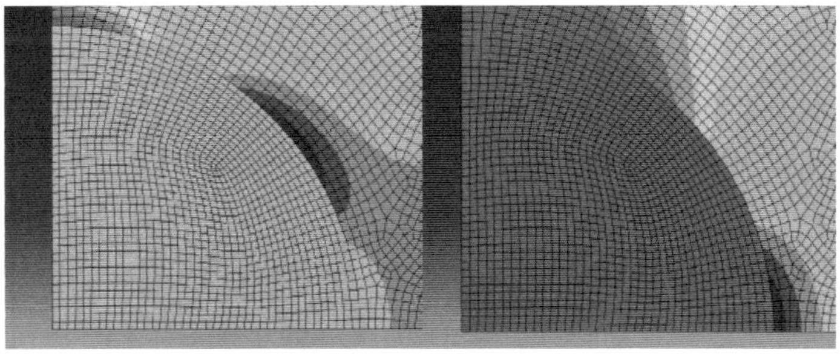

(a)S11Stress component distribution (b)S22 Stress component distribution

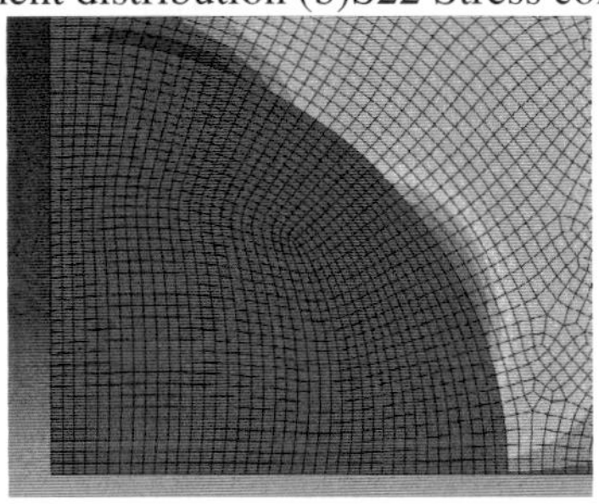

(c)S12 Stress component distribution

Figure 5. Stress component distributions inside of the inclusion under the conditions Plane stress.

(a)S11 Stress component distribution (b) S22 Stress component distribution

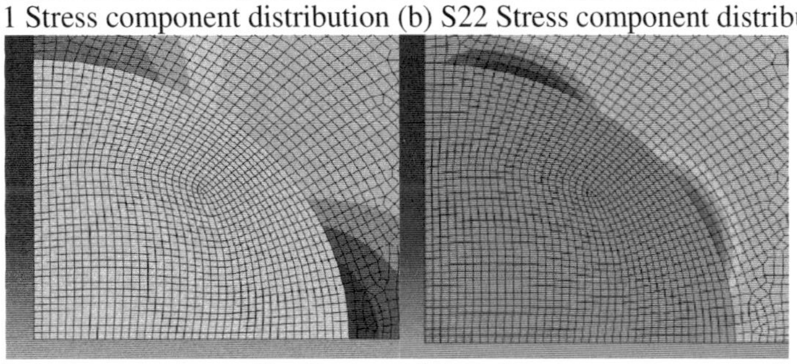

(c)S33 Stress component distribution (d) S12 Stress component distribution

Figure 6. Stress component distributions inside of the inclusion under the conditions Plane strain.

Through formula (9), stress component under rectangular coordinate system can be transferred to normal stress and shear stress of the interface, the calculation results after the transformation are shown in table 1, 2. The analytical solution and the finite element results were compared in figure 7, from figure 7, the finite element calculation result is consistent with the analytic solution, which shows that the finite element calculation model is reliable.

Table 1. Interface stress under the conditions Plane stress.

θ (°)	σ_{rr} (MPa)	$\tau_{r\theta}$ (MPa)	θ (°)	σ_{rr} (MPa)	$\tau_{r\theta}$ (MPa)
0.000	0.948	0.000	52.500	76.617	58.062
7.500	2.996	15.558	60.000	91.115	52.057
15.000	9.001	30.055	67.500	103.564	42.504
22.500	18.554	42.505	75.000	113.117	30.055
30.000	31.003	52.057	82.500	119.122	15.557
37.500	45.501	58.062	90.000	121.170	0.000
45.000	61.059	60.111			

Table 2. Interface stress under the conditions Plane strain.

θ (°)	σ_{rr} (MPa)	$\tau_{r\theta}$ (MPa)	θ (°)	σ_{rr} (MPa)	$\tau_{r\theta}$ (MPa)
0.000	-2.661	0.000	52.500	74.551	59.247
7.500	-0.571	15.875	60.000	89.344	53.119
15.000	5.555	30.669	67.500	102.048	43.372
22.500	15.303	43.372	75.000	111.795	30.668
30.000	28.006	53.120	82.500	117.923	15.875
37.500	42.800	59.247	90.000	120.013	0.000
45.000	58.675	61.337			

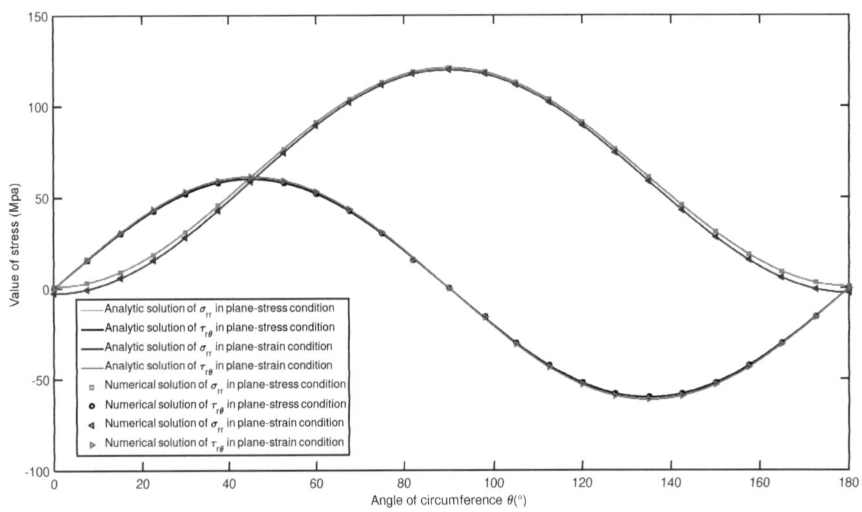

Figure 7. The comparison of analytical solution and the finite element results.

5. Conclusions

(1) The necessary simplification is done for the interface of powder metallurgy superalloy and the inclusion, the interface from the mechanics meaning was defined, which is considered as the set curve or surface of the bonded matrix and the inclusion phase. Two phase model of inclusion and matrix interface in powder metallurgy superalloy is established.

(2) On the basic theory of two-dimensional elastic problems of complex variable function, the analytical formula is derived and the analytical solution is calculated for the condition of circular inclusion in matrix.

(3) The stress and displacement boundary conditions were defined under the condition of good connection interface between the inclusion and matrix in the two phase model. On both sides of the interface, the normal stress perpendicular to the interface is equal, the shear stress parallel to the

interface is equal and interface displacement is continuous. The evaluation parameters of interface cracking are determined as the normal stress and shear stress, and the model of interface stress analysis under different conditions is established.

(4) The validity and reliability of the finite element model are verified through Comparing the numerical analysis method and the finite element simulation results.

References

[1] Lin T and He Y H 2011 *J. Aeronaut. Mater.* **131S1.**
[2] Wang X N, Li F G and Xiao J 2009 *Mater. Sci. Tech.* **17(3)** 315-21.
[3] Zhang R P, Li F G and Xiao J 2007 *Cast. Forg.* **36(9)** 70-4.
[4] Zhang M J, Li F G and Wang S Y. 2012 *J. Aeronaut. Mater.* **32(5)** 18-25.
[5] Xu Z L 2006 *Elastic mechanics.* Beijing: Higher Education Press.

Acknowledgements

The authors would like to acknowledge the financial support of the National High-tech R&D Program (863 Program) (2015AA034401).

IOP Conf. Series: Materials Science and Engineering **269** (2017) 012028 doi:10.1088/1757-899X/269/1/012028

Wear Behaviour of Iron Matrix Composite Reinforced by ZTA Particles in Impact Abrasion

B Qiu[1], S M Xing[1], Q Dong[1]

[1]School of Mechanical, Electronic and Control Engineering, Beijing Jiaotong University, China

Abstract. Zirconia toughened alumina (ZTA) particles reinforced high chromium cast iron composites (ZTA/Iron composites) were prepared by a two-step processing method, i.e. mixing particles by the molten metal and cohering by high pressure, which based on the squeeze casting process. The impact wear resistance under different impact energies were investigated using dynamically loaded abrasive wear tester at room temperature. For comparison, the wear tests of high chromium cast iron were also carried out under the same conditions. Worn surfaces of the samples were observed under scanning electron microscopy equipped with an energy dispersive detector. The results showed that the composites have better impact wear resistance than that of high Cr cast iron regardless of impact energy level, and the wear resistance of the two materials all decrease with the increase of the impact energy. The main wear mechanisms of the high Cr cast iron were micro-cutting and fatigue peeling, while the wear of composites occurred through micro-cutting of the matrix (lower impact energy) and breaking and shedding of the reinforced particles (higher impact energy).

1. Introduction

Wear is one of the three most commonly encountered industrial problems leading to the replacement of components and assemblies in engineering. According to Eyre estimates, abrasive wear comprises roughly about 50% of industrial wear situations [1]. And among various abrasive wear, impact abrasive wear is an extremely bad wear condition in the practical application, which resulting in the materials shows the rapid wear and extremely short life. Compared with the conventional metal wear-resistant materials, particles reinforced iron matrix composites have attracted more and more attentions due to their high wear resistance and excellent comprehensive properties [2, 3]. Meanwhile, these composites can bring the excellent characteristics of reinforcement and matrix into full play, have become one of the development directions of high performance structural materials [4-6]. With more and more composites components were applied into the impact wear conditions in various industries, it would be of great importance to study the impact wear properties of particles reinforced iron matrix composites.

In this paper, high Cr cast iron matrix composites reinforced with ZTA particles were prepared by a two-step processing method [7], i.e. mixing particles by the molten metal and cohering by high pressure, which based on the squeeze casting process. The method belongs to green casting, which eradicate the pollution of moulding sand existed in the traditional casting process, as well as has a lower cost. The wear behaviour of the composites in impact abrasion conditions was investigated. The present work highlights the influence of impact energy on the wear characteristic and the wear mechanism of ZTA/ Iron composites.

IOP Conf. Series: Materials Science and Engineering **269** (2017) 012028 doi:10.1088/1757-899X/269/1/012028

2. Experimental materials and methods

Commercial ZTA particles (containing 60 wt.% Al_2O_3 and 40 wt.% ZrO_2) were used as the reinforcements while micro-alloying high chromium cast iron was chosen as the matrix of the composite material, and the chemical composition of high Cr cast iron matrix are presented in Table 1. The ZTA reinforced particles with an average size of 2-4 mm possessed a micro-hardness of 1500-1700 HV, a density of 4.0 g·cm^{-3} and a melting point of 1890°C.

Table 1. The chemical composition of high Cr cast iron matrix (wt.%).

Element	C	Si	Mn	Mo	W	Ni	Cr	Fe
Content	2.8-3.0	0.8-1.0	1.8-2.0	0.7-1.0	0.9-1.3	0.3-0.5	25-27	Bal.

An intermediate frequency induction furnace was used for the high Cr cast iron melting, and the tapping temperature was higher than 1600°C. The mould, especially the ceramic and metal cavity, was preheated to 220-260°C using a self-made preheater made by electric heating tube, moreover, the particles contained by crucible were preheated to 1100°C in a resistance furnace. The schematic diagram for fabricating ZTA/Iron composites by two-step processing method was shown as figure1, and the experimental procedures are as follows, firstly, added preheated ZTA ceramic particles and poured molten metal into the ceramic and metal cavity to full, respectively. Then making the upper and lower mould closed and locked at the specific pressure of 115MPa, followed by an external pressure of 125MPa was exerted to drive the ceramic and metal pressure head with a filling speed of 150mm/s, thus made the particles and molten iron from ceramic cavity and metal cavity continuous pushed into the mould filling runner. When the molten metal flows over the ceramic cavity, its kinetic energy and viscosity were utilized to achieve the uniform dispersion of ceramic particles, meanwhile, formed the solid-liquid mixture wrapped with particles to fill the workpiece cavity. After then, the mixture in the workpiece cavity achieved solidification and feeding at the supplement pressure of 130MPa offered by workpiece pressure head. Finally, released the pressure of ceramic, metal and workpiece pressure head, opened the mould and removed the composites workpiece.

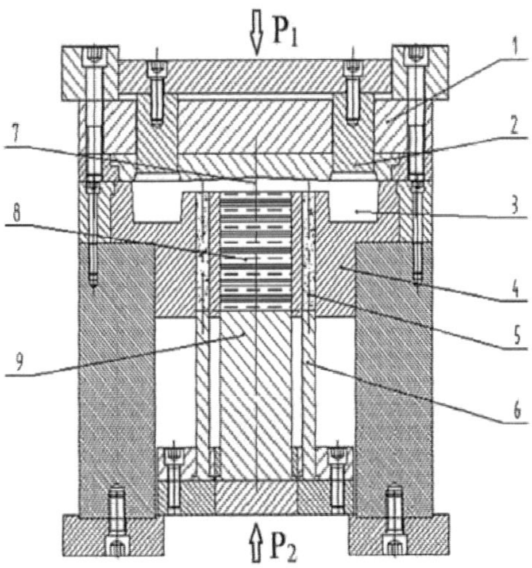

(1) upper mould; (2) workpiece pressure head; (3) workpiece cavity;
(4) lower mould; (5) ceramic particles; (6) particles pressure head;
(7) mould filling runner; (8) molten metal; (9) metal pressure head

Figure 1. Schematic diagram for fabricating ZTA/Iron composites.

The ZTA/Iron composites were designed for industries involving impact loads. Therefore, the impact–abrasion wear tests were carried out to better assess the performance of the composites in service. Meanwhile, the single high chromium cast iron was selected as the reference material for the wear tests to compare with ZTA/Iron composites in the purpose of evaluating the wear properties of composites.

The wear tests were performed on a dynamically loaded abrasive wear tester (MLD-10, produced by Zhangjiakou Cheng Xin Testing Equipment Manufacturing Co. Ltd., China) at room temperature, the schematic representation was exhibited in figure 2(a), the impact–abrasion wear of the samples is realized with the non-stop rotating of the lower sample and the vertical reciprocating motion of the upper sample. The wear (upper) sample were cut into 10 mm × 10 mm × 30 mm size, as shown in figure 2(b), and the surface of 10 mm × 10 mm was mechanically polished for impact-abrasive wear tests. The cylindrical contrast (lower) sample was made of 45# steel (58HRC) with the size of 50 mm in diameter and 20 mm in length, which was quenched at 840°C in water and tempered at 300°C for 1h. The wear surfaces of the tested samples were observed and examined by the scanning electron microscopy (SEM, Hitachi S-4800) equipped with an energy dispersive detector (EDS, Vantage).

During the test, the wear sample is fixed on a clamping chuck which connects with a hammer. The impact frequency of wear sample was 150 times/min and the wear tests were performed for 60 minutes. The rotation speed of contract sample was 200r/min, the impact energy is set at 2 J, 3 J and 4 J, respectively. Refined quartz sands are used as the abrasive material (with particle size of 10-20 mesh, density of 2.6 g/cm^3 and hardness value of 800 HV–1200 HV), the SEM micrograph of abrasive particles as shown in figure 2(c). The wear tests were repeated three times with new samples for each test condition to ensure the reliability, and each sample was subjected to pre-wear for 30min before the formal tests. The wear mass loss was measured every 10 minutes in the tests, before and after each test, the wear sample was cleaned ultrasonically, dried and then weighed by a precision electronic balance with an accuracy of 0.0001g, and take the average value of three measurements as the results.

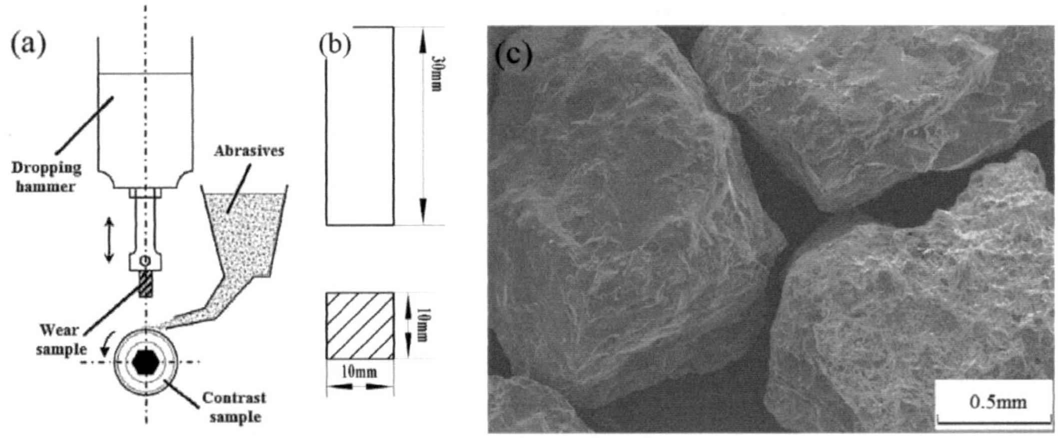

Figure 2. (a) Schematic of MLD-10 wear tester; (b) Size of wear (upper) sample; (c) the SEM micrograph of quartz abrasive particles.

3. Results and discussion

The text of your paper should be formatted as follows: Relative wear resistance β was used to evaluate the impact wear properties of ZTA/Iron composites, which can be obtained through the following process.

First the abrasive volume ΔV was calculated as:

$$\Delta V = \frac{m_1 - m_2}{\rho_m f_m + \rho_c f_c}$$

where ΔV is the abrasive volume; m_1, m_2 is the weight loss before and after abrade, relatively; ρ_m, ρ_c is the density of high chromium cast iron and ceramic particle, relatively; f_m, f_c is the volume fraction of high chromium cast iron and ceramic particles in the composites, relatively.

Then the relative wear resistance β was achieved through the following equation:

$$\beta = \frac{\Delta V_m}{\Delta V_c}$$

where β is the relative wear resistance; ΔV_m is the abrasive volume of reference matrix (high chromium cast iron) sample; ΔV_c is the abrasive volume of ZTA/Iron composites sample.

The test results of wear samples under different impact energies were shown in figure 3. It can be seen that the cumulative volume loss of the single high Cr cast iron and composites all have a greater change in the different impact energies, and the wear properties both decrease with the increase of impact energy. However, the impact abrasive wear property of the composites all superior to that of the high Cr cast iron in various impact energy. Figure 4 shows the relative wear resistance of composites with high Cr cast iron, it has been calculated that the relative wear resistance of composites is 1.84, 1.43, 1.1 times higher than the high Cr cast iron when the impact energy is 2, 3, 4 J, respectively, which indicates that the wear resistance of composites is more excellent in the condition of lower impact energy.

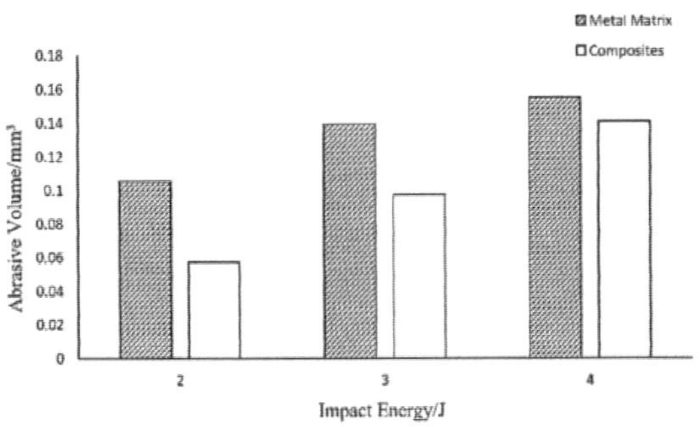

Figure 3. Wear test results of composites and matrix under different impact energies.

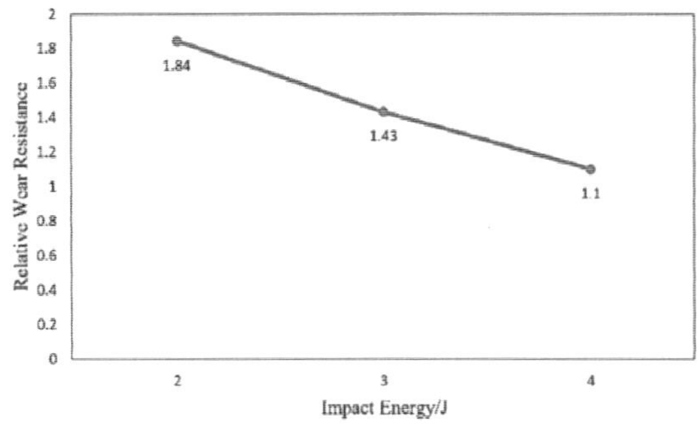

Figure 4. The relative wear resistance of composites under different impact energies.

In the condition of impact abrasive wear, the materials not only bearing tangential sliding wear of the abrasive, but also subject to an impact action in the normal direction. The abrasive piercing the surface of the material in the normal action, then cutting the materials under the action of tangential force.

The quartz sand is hard abrasive compared with the high Cr cast iron, the serious cutting and rolling effect will be produced on high Cr cast iron under the external load. The wear morphology of high Cr cast iron samples were shown in figure 5, it can be seen obvious micro-cutting marks (A in figure 5(a)) from the wear surface when the impact energy is lower (2J), which demonstrates that the wear mechanism is dominated by micro-cutting in the lower energy. The local stress of the material increased with the increase of the impact energy, leading to the appearance of the cracks and developing into the fatigue peeling (B in figure 5(b)). Meanwhile, the abrasive particles will be embedded into the wear surface (figure 5(c)) with the increase of the impact degree.

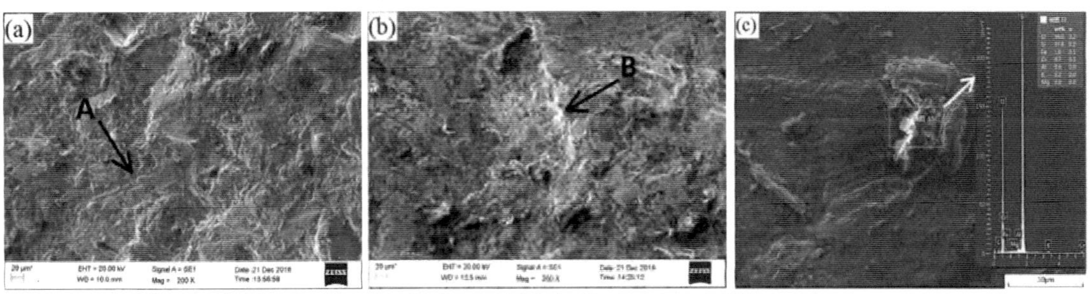

Figure 5. The wear morphology of high Cr cast iron (a) lower impact energy; (b) higher impact energy; (c) The abrasive particles embedded into the wear surface.

When the ZTA ceramic particles were introduced into the high Cr cast iron matrix, the impact wear resistance of the composites depends on the synergistic effect of the metal matrix and reinforced particles. On the one hand, the good supporting role of the metal matrix ensured the ZTA particles hard to fall off in the wear process. On the other hand, the hardness of ZTA particles is much higher than that of the quartz abrasive, which can resist the cutting effect of abrasive on the metal matrix, thus protecting the matrix. However, the wear resistance of the composites will be greatly influenced by whether the reinforced particles break or fall off under the impact action. The wear morphology of composites samples was shown in figure 6, it can be seen that the stress of the particles is small and the fracture is not obvious when the lower impact energy (2.0J) is applied. The ZTA particles are protruded from the matrix during the wear process (figure 6(a)), reduced the contact area of the matrix with abrasive particles and made the matrix hard to cut and wear, so that the wear resistance of the composites is much better than that of the high Cr cast iron. When the impact energy is 4J, the cracks appeared within particles due to the larger stress, and the cracks in different directions further extended, leading to breaking of the particles (C in figure 6(b)), even shedding of the entire particles (D in figure 6(c)). In this case, the ZTA particles lose their protective effect on the matrix, resulting in a greater wear of the composites.

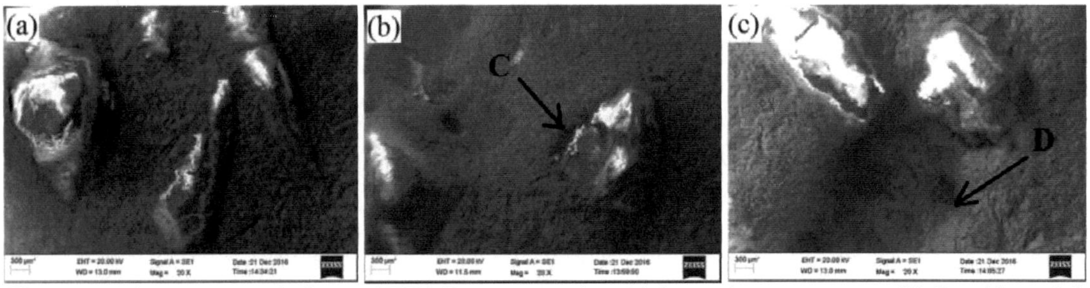

Figure 6. The wear morphology of ZTA/Iron composites (a) lower impact energy; (b) Breaking of the particles in the higher impact energy; (c) shedding of the particles in the higher impact energy.

4. Conclusions

(1) A high Cr cast iron matrix composite reinforced with zirconia toughened alumina (ZTA) particles 2-4mm in size was prepared by a two-step processing method.

(2) The composites have better impact wear resistance than that of high Cr cast iron regardless of impact energy level, and the wear resistance of the two materials all decrease with the increase of the impact energy.

(3) The wear of high Cr cast iron occurred through micro-cutting and fatigue peeling under the lower and higher energy, respectively. While the wear mechanism of ZTA/Iron composites appears as micro-cutting of the matrix in the lower impact energy and breaking and shedding of the reinforced particles in higher impact energy.

5. References

[1] Eyre T S 1976 *Tribol. Int.* **9** 203.
[2] Wiengmoon A and Chairuangsir T 2005 *Acta Mater.* **53** 4143.
[3] Bayraktar E, Ayari F and Katundi D 2011 *Adv. Mater. Res.* **264** 475.
[4] Chen Y and Wang H M 2004 *Surf. Coat. Technol.* **179** 252.
[5] Yang Y F, Wang H Y, Liang Y H and Zhao R Y 2008 *Mater. Sci. Eng. A* **474** 355.
[6] Miracle D B 2005 *Compos. Sci. Technol.* **65** 26.
[7] Xing S M, Qiu B and Bao P W 2015 *China Pat.* No. CN104439185A.

IOP Conf. Series: Materials Science and Engineering **269** (2017) 012029 doi:10.1088/1757-899X/269/1/012029

Finite Element Analysis of Increasing Column Section and CFRP Reinforcement Method under Different Axial Compression Ratio

Zhou Jinghai[1], Kang Tianbei[1], Wang Fengchi[1] and Wang Xindong[1]

[1]School of Civil Engineering, Shenyang Jianzhu University, Shenyang, China, 110168

Abstract. Eight less stirrups in the core area frame joints are simulated by ABAQUS finite element numerical software. The composite reinforcement method is strengthened with carbon fiber and increasing column section, the axial compression ratio of reinforced specimens is 0.3, 0.45 and 0.6 respectively. The results of the load-displacement curve, ductility and stiffness are analyzed, and it is found that the different axial compression ratio has great influence on the bearing capacity of increasing column section strengthening method, and has little influence on carbon fiber reinforcement method. The different strengthening schemes improve the ultimate bearing capacity and ductility of frame joints in a certain extent, composite reinforcement joints strengthening method to improve the most significant, followed by increasing column section, reinforcement method of carbon fiber reinforced joints to increase the minimum.

1. Introduction

Increased column section reinforcement and carbon fiber reinforced plastics (CFRP) method are common structural reinforcement technology in recent years [1]. Both of these methods have advantages and disadvantages, and it's very important to reinforce the beam-column joints by the method of increasing the column section and carbon fiber composite reinforcing.

Many researchers have attempted to ensure the axial compression ratio has a great influence on the seismic performance of the joint [2-4]. In this paper, the cross-section frame structure with fewer stirrups in the core area is taken as the object of this study. The finite element software ABAQUS is used to simulate the various reinforcement schemes under the different axial compression ratio. The ultimate bearing capacity, ductility, stiffness and other effects are considered.

2. Establishment of finite element model

2.1. Fundamental assumption

(1) The concrete is an isotropic elastic material, and the plastic deformation is accompanied by plasticity damage after cracking.

(2) The volume of reinforcement is neglected as linear during the simulation.

(3) It is assumed that there is no relative bond slip between the reinforcement and concrete.

(4) It is reasonable to suppose the carbon fiber fabric is an anisotropic brittle material. The elastic modulus and tensile strength in the direction of the carbon fiber bundle is much larger than the tangent direction and the thickness direction [5].

(5) The CFRP is assumed to be a membrane structure of the concrete without a relative slip.

2.2. Constitutive model

The uniaxial stress-strain curve of concrete and the monotonic tensile stress-strain curve of reinforce are shown in figures 1 and 2. The yield strength f_y and ultimate strength $k_4 f_y$ of ordinary steel in figure 2 are taken as the mean value according to the reference [1], and the curved segment is simplified into a straight line segment. Carbon fiber sheets are anisotropic brittle materials with no yield strength, and the ultimate strength is σ_{cf}, $\sigma_{cf}=E_{cf}\varepsilon_{cf}(\varepsilon<\varepsilon_{cf})$.

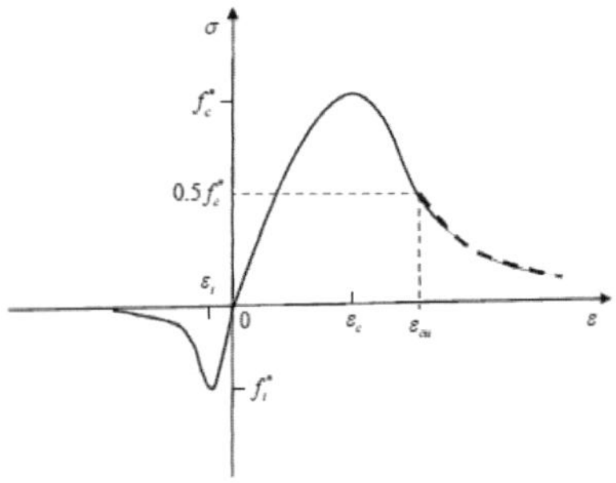

Figure 1. Uniaxial stress-strain curve of concrete.

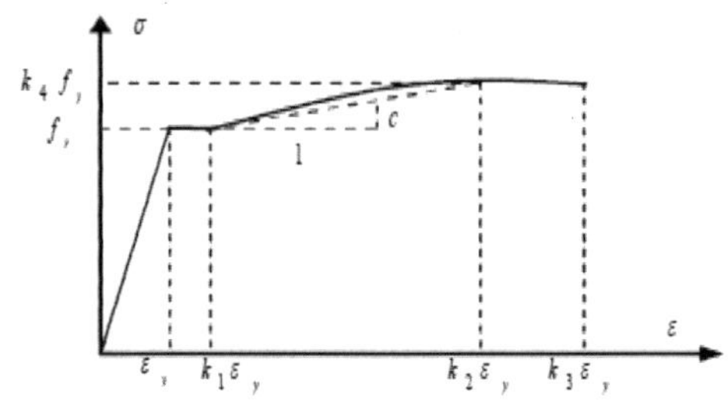

Figure 2. Monotonic stress-strain curve of steel bar.

2.3. Finite element model

Table 1 lists the designing scheme of concrete specimens. The size of beam and column of reference specimen are 150×250×1000mm and 200×200×600mm respectively. The strength of the concrete is C25. The stirrup and longitudinal reinforcements adopt HPB300 and HRB335 respectively.

Table 1. Specimen design.

Series	Increasing column section reinforcement	The layers of CFRP	remarks
KJG-1	reinforced	0	Core area does not match stirrups [6]
KJG-2	reinforced	0	Core area with stirrups
KJG-3	unreinforced	1	Core area does not past carbon fiber sheet
KJG-4	unreinforced	1	Core area paste carbon fiber sheet
KJFG-5	reinforced	1	Core area does not match stirrups, does not past carbon fiber sheet
KJFG-6	reinforced	1	Core area with stirrups, does not past carbon fiber sheet
KJFG-7	reinforced	1	Core area does not match stirrups, paste carbon fiber sheet
KJFG-8	reinforced	1	Core area with stirrups, paste carbon fiber sheet

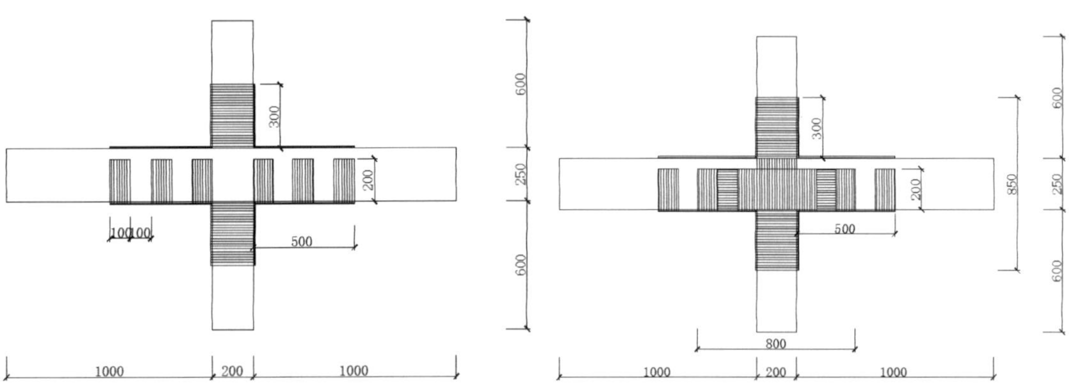

(a) Core area does not past carbon fiber sheet (b) Core area paste carbon fiber sheet

Figure 3. Carbon fiber reinforced plastics (CFRP).

Increasing the column section reinforcement method is to add 60mm thick concrete along the long direction around the column, the thickness of concrete protection layer is 30mm, and the strengthening concrete is C30. Carbon fibre reinforcement is a method of pasting carbon fibre sheets in the 300mm length range between the top and bottom of the cross-shaped core area. Two forms of pasting CFRP as depicted in figure 3.

2.4. Loading method

Monotonic static loading is adopted. The end of the column is hinged seat and the force of 200kN, 300kN and 400kN is applied at the end of the column respectively. The axial compression ratio is 0.3, 0.45 and 0.6. At the same time, the displacement loading mode is applied to the concentrated vertical load at both ends of the beam. In the course of loading, a rigid block is added at the ends of the beam to minimize the effect of stress concentration and make the component in the same plane as a whole loading system.

3. Results and discussion

3.1. Load-displacement curve

The ultimate bearing capacity of the reference specimens are expressed by the load peak of the load-displacement curve. Figures 4-11 indicate the load-displacement curves under different axial compression ratios.

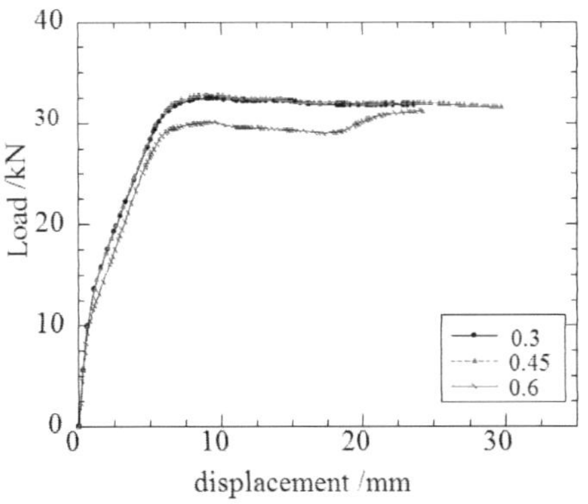

Figure 4. The load-displacement curve of KJG-1.

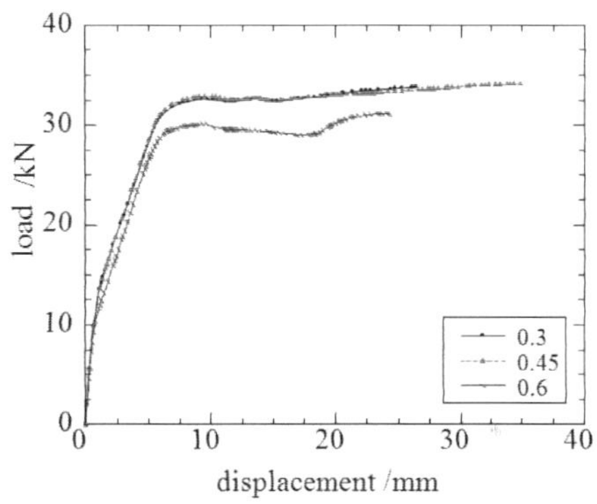

Figure 5. The load-displacement curve of KJG-2.

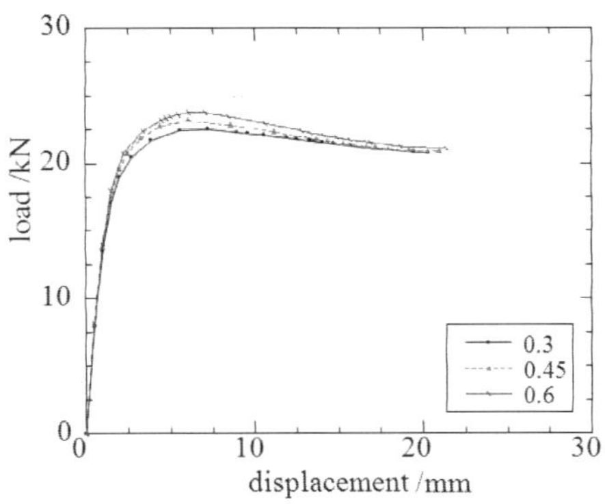

Figure 6. The load-displacement curve of KJG-3.

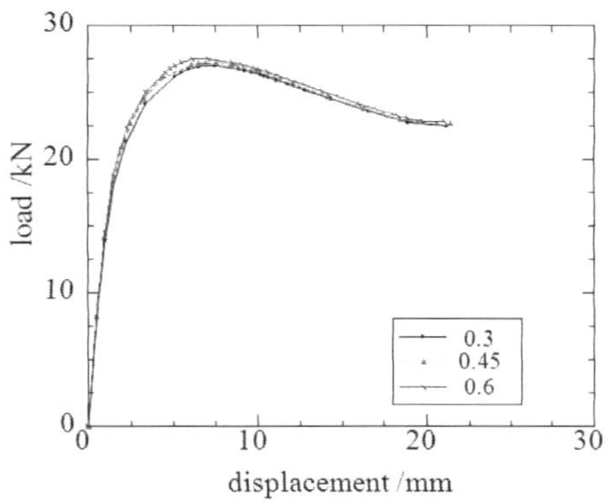

Figure 7. The load-displacement curve of KJG-4.

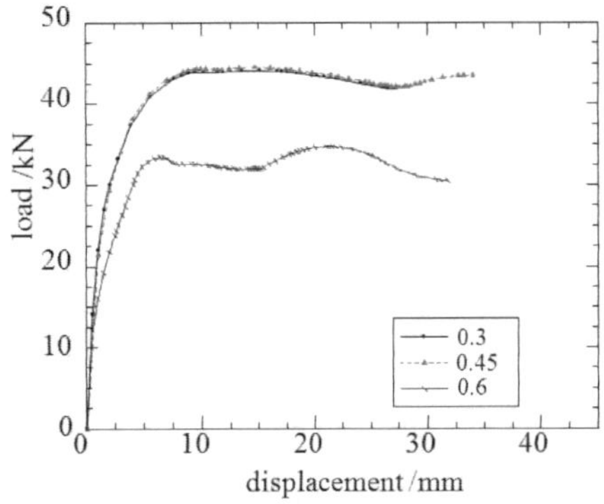

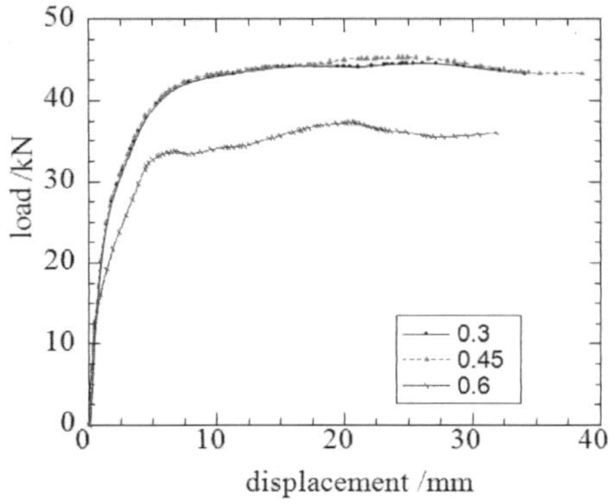

Figure 8. The load-displacement curve of KJFG-5. **Figure 9.** The load-displacement curve of KJFG-6.

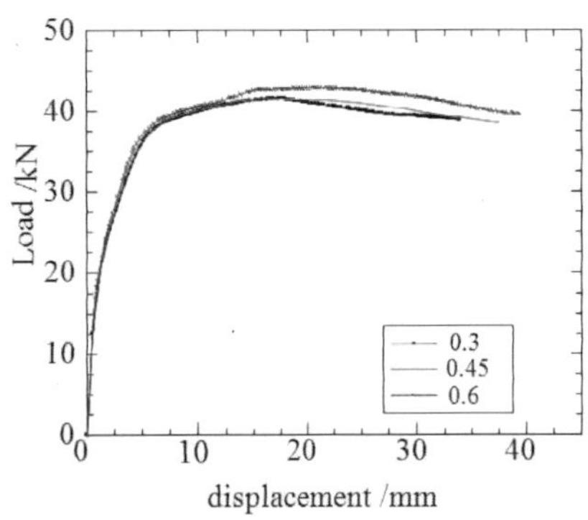

Figure 10. The load-displacement curve of KJFG-7.

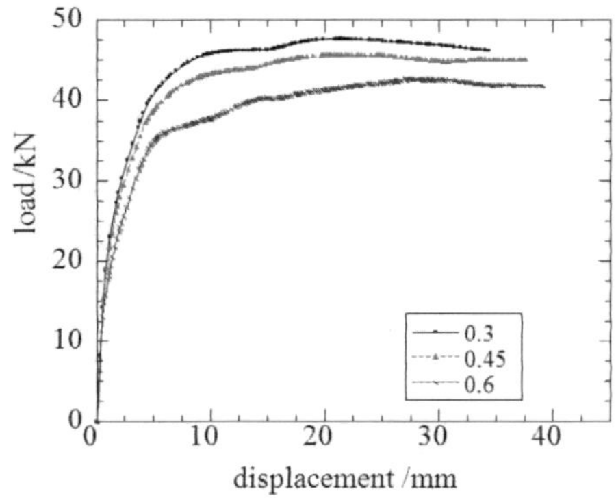

Figure 11. The load-displacement curve of KJFG-8.

Increasing the column section reinforcement method has less influence on the axial compression ratio, as showed in figures 4 and 5. The load-displacement curves of component KJG-1 and KJG-2 almost coincide when the axial compression ratio is 0.3 and 0.45, and the ultimate bearing capacity is less affected and the ultimate displacement is improved. When the axial compression ratio is 0.6, the yield load and ultimate displacement decreased significantly. Meanwhile the ultimate load decreased less.

The load-displacement curves of the component KJG-3 and KJG-4 under the axial compression ratio of 0.3, 0.45 and 0.6 are illustrated in figures 6 and 7. The ultimate bearing capacity of specimens is increased with the increase of axial pressure ratio. The amplitude is small and the limit displacement is almost constant. The CFRP obviously improved the load-carrying capacity when the axial pressure is larger.

The load-displacement curves of the component reinforced by the increasing column section and CFRP are given in figures 8-11. The load-bearing capacity of the reinforced bar in the core of the column section is decreased relatively. The limit displacements of paste carbon fiber sheet (KJFG-7 and KJFG-8) are increased with the increase of axial compression ratio.

3.2. Ductility

The displacement ductility coefficient β_Δ is defined as the ratio of the ultimate displacement of the yield displacement in the case of retaining the basic bearing capacity, as shown in formula (1). Where the Δu is the corresponding displacement when the load-displacement curve has a significant descending section reaching the load peak after 85%, and Δy is yield displacement.

$$\beta_\Delta = \Delta u / \Delta y \tag{1}$$

It can be seen from table 2 that the ductility of the joints strengthened by increasing the column section is much higher, the carbon fiber reinforcement is less, and the composite reinforcement is the greatest. Increasing the column section method, the ductility coefficient increases with the increase of axial compression ratio. The ductility coefficient of the carbon fiber reinforced joints decrease with the increase of the axial compression ratio and the adhesion of the carbon fiber sheet in the core area has an important effect on the ductility.

Table 2. Ductility coefficient.

Series	Δu/mm			Δy/mm			β_Δ		
	0.3	0.45	0.6	0.3	0.45	0.6	0.3	0.45	0.6
KJG-1	23.6	29.7	24.3	4.2	4.3	3.5	5.62	6.91	6.94
KJG-2	26.4	34.8	29.2	4.3	4.5	3.8	6.14	7.73	7.68
KJG-3	20.4	21.0	21.4	3.0	3.1	3.3	6.80	6.77	6.48
KJG-4	21.4	21.2	21.1	3.2	3.2	3.8	6.69	6.63	5.55
KJFG-5	27.3	34.0	31.7	3.9	4.1	3.8	7.00	8.29	8.34
KJFG-6	34.1	37.6	31.9	4.0	4.3	3.9	8.53	8.74	8.18
KJFG-7	33.9	37.5	39.4	4.2	4.3	4.0	8.07	8.72	9.85
KJFG-8	34.4	37.6	39.1	4.0	4.4	4.3	8.60	8.55	9.09

3.3 Rigidity

The ratio of force and deformation at the load displacement curve $0.75P_u$ is used to indicate the resistance to deformation of the member. The rigidity of the specimen is calculated in accordance with formula (2).

$$K_j = 0.75P_u / \Delta w \tag{2}$$

where, P_u is the ultimate load, Δw is the displacement corresponding to $0.75P_u$ in the load-displacement curve. The rigidity of all reference specimens are listed in table 3. No matter what kind of reinforcement method is expected to be adopted, when the axial pressure ratio is 0.6, the rigidity of the specimens decreases, and the stiffness of the composite reinforcement method is the biggest. It is not obvious that the rigidity improved by increasing the column section reinforcement, although the rigidity of the column is improved. The stiffness of the structural member is mainly affected by the beam. The rigidity of the carbon fiber reinforced joint is obviously increased because of the restraint between the beam end and the column end.

Table 3. Rigidity.

Series	$0.75P_u$/ kN			Δw/mm			Rigidity /kN·m^{-1}		
	0.3	0.45	0.6	0.3	0.45	0.6	0.3	0.45	0.6
KJG-1	24.36	24.62	22.59	3.9	3.7	3.8	6.25	6.65	5.94
KJG-2	25.35	25.58	24.3	4.1	3.9	4.3	6.18	6.56	5.65
KJG-3	16.91	17.4	17.79	1.7	1.7	1.9	9.95	10.21	9.36
KJG-4	20.24	20.39	20.63	2	2.1	2.2	10.12	9.71	9.38
KJFG-5	33.1	33.4	26.0	3.0	2.8	2.8	11.04	11.92	9.29
KJFG-6	33.4	34.0	28.0	3.2	3.1	3.1	10.45	10.96	9.02
KJFG-7	31.1	31.2	32.1	3.1	2.9	3.2	10.04	10.74	10.03
KJFG-8	35.7	34.2	32.0	3.4	3.2	3.6	10.51	10.69	8.88

4. Conclusion

(1) The influence of different axial compression ratios on the bearing capacity of the strengthened method of column section is larger, and the bearing capacity of CFRP is less.

(2) The reinforcement scheme can improve the ultimate bearing capacity and ductility of the frame joints in different degree, and the reinforcement method of the composite reinforcement joint is the most notable. The reinforcement of the column section is larger, and the strengthening method of CFRP is the smaller.

(3) The rigidity of the reference specimen with the composite reinforcement scheme increases the maximum and the degradation is stable. Increasing the column section reinforcement method has little effect on the rigidity of the component. The rigidity of the CFRP specimen is improved remarkably, but the stiffness decreases quickly after the peak load.

5. References

[1] Juiru T 1989 *Earthquake resistance of reinforced concrete frame joints,* Nanjing: Southeast University press.
[2] Jianping F, Chuan Z and Tao C 2006 *J. Build. Struct.* **42** 17-26.
[3] Song Y, Fan G 2014 *Mag. Concr. Res.* **66** 1-18.
[4] Jinghai Z, Ting Z and Fengchi W 2016 *J. Shenyang Jianzhu Univ. (Nat. Sci.)* **32** 70-9.
[5] China National Industrial Building Diagnosis and Transformation Engineering Technology Center 2008 *Technical specification for concrete structures strengthened with carbon fiber sheets,* Beijing: China Planning Press.
[6] Lee H S, Kage T and Noguchi T 2003 *Cem. Concr. Res.* **33** 563-70.

Study on the Influence of Grilles on the Flow-Noise Generated by the Cavity

Li Rong-hua[1], Lou Jing-jun[2] and Zhu Shi-jian[2]

[1]College of Power Engineering, Naval Univ. of Engineering, Wuhan 430033, China;
[2]National Key Laboratory on Ship Vibration and Noise, Naval Univ. of Engineering, Wuhan 430033, China

Abstract. The two kinds of cavities, respectively with opening hole and slotted hole, numerically simulated based on the theory of large eddy simulation and acoustic analogy were performed. Firstly, according to the differences between vorticity and pressure magnitude contours of the flow filed around the slotted hole and opening hole, the influences on the flow field of the baffles were illustrated. Secondly, by analysing the power spectrum density of the monitoring points, the main noise source of these two models was found. Lastly, the flow-noise frequency spectrum of slotted hole and opening hole indicated that the grilles play an significant role in reducing the radiation of the flow-noise generated by the cavity with hole.

1. Introduction

Flowing over cavities is often a source of pure tone noise, it has been studied by many investigators. Xu Jun, Li Xiao-dong, Liu Cong-wei and Chen Can studied on the control of cavity flow tone by adding a plate in the cavity, altering the thickness of the boundary layer, changing the bottom slope of the cavity and the length to depth ratio [1-4]. In underwater vehicles, the use of slats is a common control strategy to improve the acoustic radiation. Zhang Nan concluded that the resistance performance caused by the baffles is better than by the longitudinal holes [5]. S Arunajatesan brought forward that the presence of the grilles obstruct the vortex to the cavity, but had little impact on the control of fluctuating pressure inside the cavity [6]. Dai Shao-shi et al pointed out that the fluid pattern is more complex, and the fundamental frequency of the first model of the shear layer increases due to the existence of the grids [7]. Xiao Zhi-xiang concluded that the grilles can decrease the thickness of the shear layer, restrain the fluctuating force and reduce the far field sound radiation intensity [8]. The flow field and acoustic field of the cavity with opening hole and slotted hole will be simulated in this dissertation, then between the two situations the difference of the SPL spectrum and the flow field force will be analyzed to study the mechanism of the grill influence the sound generation and radiation.

2. Model of cavity with two kinds of holes

A schematic of the cavities are shown in figure 1 and figure 2, the length, width and depth of the cavities are L=400mm, W=300mm, H=500mm, while the length and width of the holes are L_{hole}=300mm, W_{hole}=200mm, and the height of the hole neck is H_{hole}=20mm, which can be classified as deep cavity, according to the classification criteria put forward by Sarohia [9]. The computational domain is shown in figure 3, which extends over 5000mm×1800mm×1500mm. The upstream and side boundaries are respectively 2000mm and 900mm far away from the center of the hole.

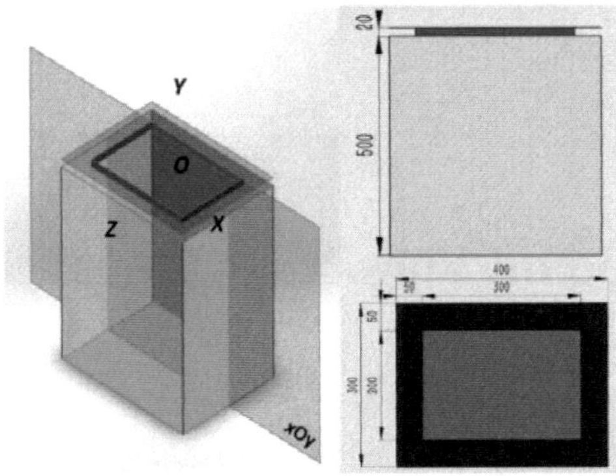

Figure 1. The model of cavity with opening hole.

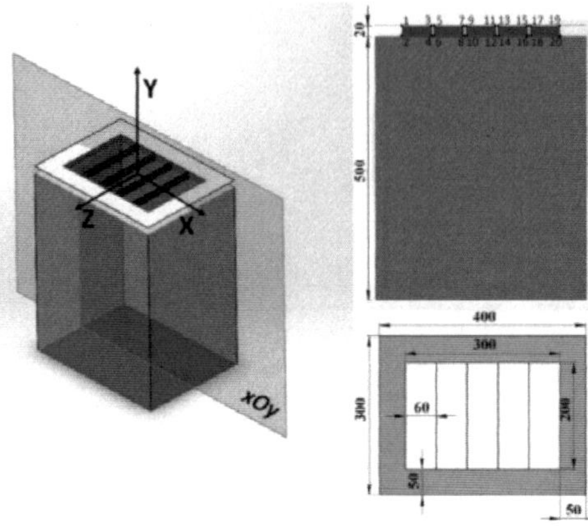

Figure 2. The model of cavity with slotted hole.

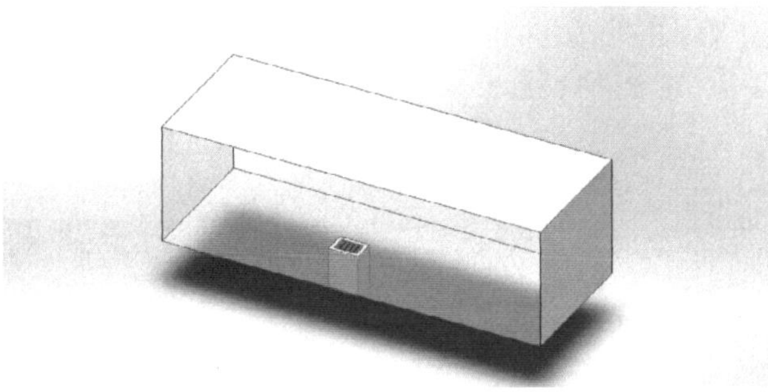

Figure 3. The figure of calculation zone.

The software Solidworks was used to modelling, then the model was imported into ICEM CFD to mesh. The meshes and the boundary conditions of the cavity are shown in figure 4 and figure 5, the CFD meshes include 2307220 meshes and 1676229 meshes of the cavity with opening hole of non-uniform size. The first layer scale near the wall was 0.05mm in order to catch the viscous stress change. In view of the distribution of the flow field parameters, the much coarser mesh size can be used in the places away from the hole. At the inlet boundary, a uniform normal velocity was defined with u=10m/s, whereas the outflow boundary condition was used with a relative pressure $p_{ref,static}$=0 Pa at the outlet. Along the solid walls, the non-slip condition was applied to the cavity wall, and the free-slip condition was imposed to the other walls.

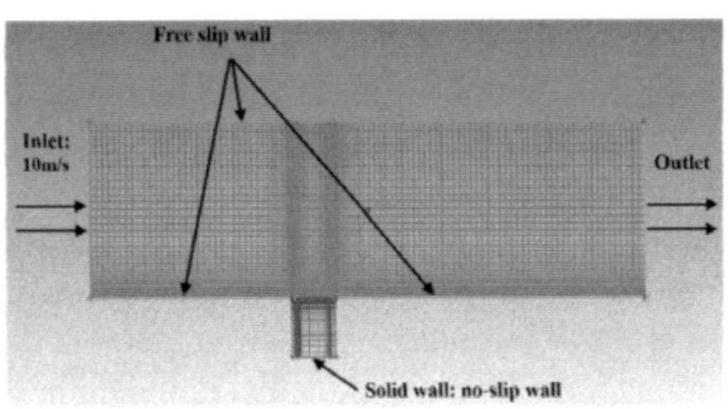

Figure 4. The meshes and boundary conditions of opening hole.

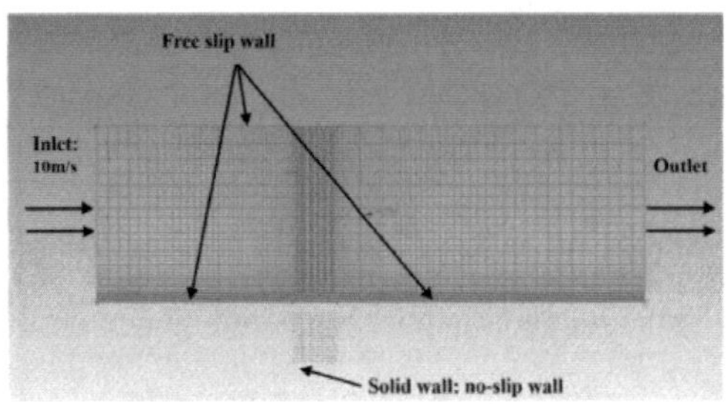

Figure 5. The meshes and boundary conditions of slotted hole.

The calculation was carried out in Fluent, containing the simulation of flow field and acoustic radiation. In the calculation, LES module was firstly used, FW-H acoustic module was set to perform the computation of sound radiation until the flow field is stabilized. In order to facilitate the analysis of the simulation results, serials of monitor points, receiver and slice which were assigned to detect the pressure changes on the wall of cavity and holes, pick up the sound pressure of the sound field and display the vorticity and pressure counters of the field are shown in figure 1 and figure 2.

IOP Conf. Series: Materials Science and Engineering **269** (2017) 012030 doi:10.1088/1757-899X/269/1/012030

3. Results and discussion

3.1. Flow field results

The diagram of three-dimensional steady streamline of the cavity with slotted hole was shown in figure 5. Due to the baffles function, the velocity of the flow was significantly reduced when flowing through the grid, and the fluid outside the grid area was not obviously affected. The diagrams of two-dimensional steady streamline on xOy slice of the two kinds of cavities are shown in figure 7 and figure 8. The velocity of the fluid flowing through the hole into the cavity was lower than the fluid outside the cavity. Only two eddies are found around the opening hole, both are near the rear wall of the opening hole. While the vortices are formed between each of two baffles and the rear wall. It indicates that the grids result in more vortices structure generation, which means that the baffles make the flow field more complex. Points *A~E* are the velocity monitoring points during both the simulation of two cavity models, the velocity of these points are shown in table 1. It obviously shows that the velocity of the slotted hole monitoring points are smaller than the points of the opening hole.

Table 1. The velocity of monitoring points A~E.

	A	B	C	D	E
Opening hole (ms^{-1})	5.7	5.1	3.8	5.3	8.1
Slotted hole (ms^{-1})	5.2	4.1	3.1	4.7	3.6

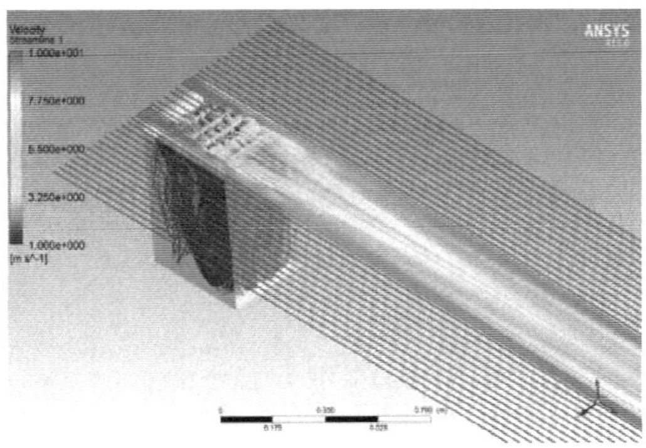

Figure 6. Diagram of 3D steady streamline of the cavity with slotted hole.

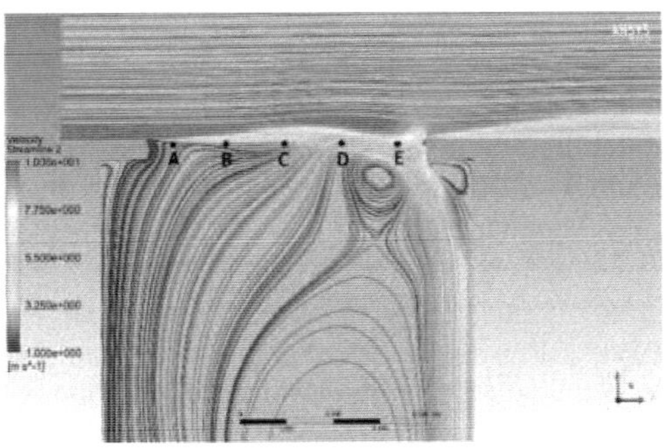

Figure 7. Diagram of two-dimensional steady streamline of the cavity with opening hole on *xOy*.

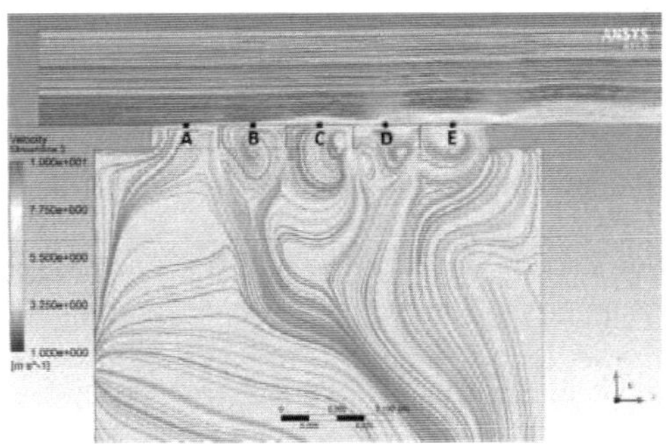

Figure 8. Diagram of two-dimensional steady streamline of the cavity with slotted hole on *xOy*.

One period of velocity invariant Q contours of the slotted hole is shown in figure 9, velocity invariant Q is used to display the vorticity distribution of the slotted cavity. As the fluid flowing through the grilles, the boundary curvature mutation caused by the existence of hole leads to the flow separation, a series of vortices are generated in the first grid hole. As the vortices moving backward and collapsing with the baffles, the vorticity magnitude is increased and the vortices structures continue migrating to the downstream, until the vortices move to the last grid hole. The vortices collide with the rear wall of the hole, and derive from the hole, then gradually annihilate in the flow field. In the upstream, the vortices generated by the flow separating and collision is drove to the downstream to supplement the vortices in the downstream, which results in the repeated process of generation, motivation, collision and annihilation. It can be saw from the figure 9 that the vortices migration process is mainly from the top edge of the baffles, because the low speed fluid in the cavity and the acoustic feedback caused by the collision cannot decelerate the fluid outside the cavity to make the fluid move inside the cavity. Only a handful separated vortices caused by the collision motivates into the cavity and blends with the low speed fluid existed in the cavity.

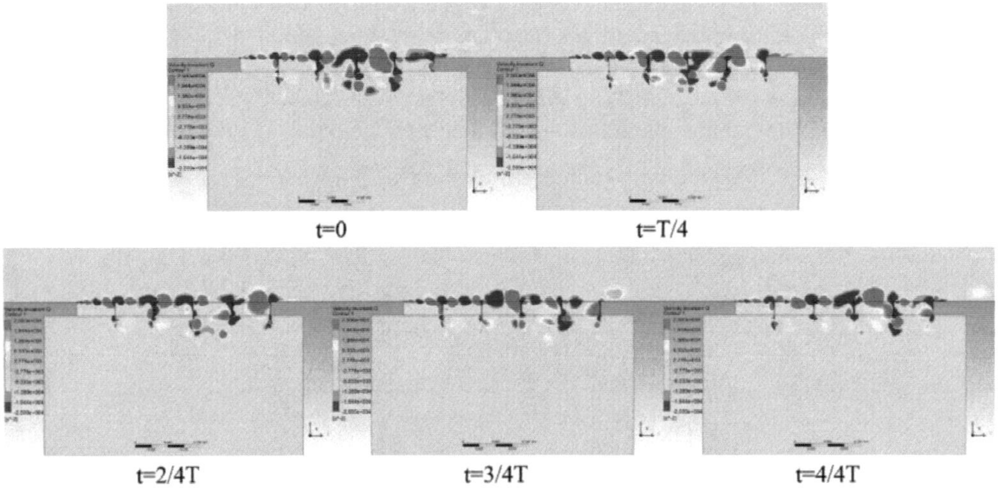

Figure 9. One period of velocity invariant Q contours of the slotted opening.

3.2. Pressure analysis

The pressure distribution contours of the slotted hole in one period are shown in figure 10. The region of high and low pressure varies with time. As shown in figure 11, the pressure time sequence of the monitoring points 1~20 in one period illustrate that the cycle of each point is approximately the same, around 0.033s, while the phase between the points are slightly different.

In the process of the vortices motivation, the different phase between the points may be caused by the concentration of the vortices. On the one hand, the acoustic feedback and vortices collision with the baffle will cause the vortices to move toward the upper or lower edge of the baffle. The pressure of the center of the vortices is descended, due to the centrifugal action of vortex motion, therefore when the vortices concentrates on the different region, then the pressure will be descended. When the vortices collapses with the baffle may cause the vortices separating into two parts, one will flow over the upper edge of the baffle and the other one will be drove into the cavity. The pressure of the monitoring point will be at a low value approach to the upper (lower) edge of the baffle and the pressure of the upper (lower) monitoring point will be at a low value, when the vortices concentrates on the upper (lower) edge of the baffle. On the other hand, as the vortices motivates from the upstream to the downstream, the vortices will concentrate at each grid hole successively as time goes by. Once the vorticity concentrates in a grid hole, the pressure of this region will be decreased, and the pressure of the other region will be at a relatively higher value.

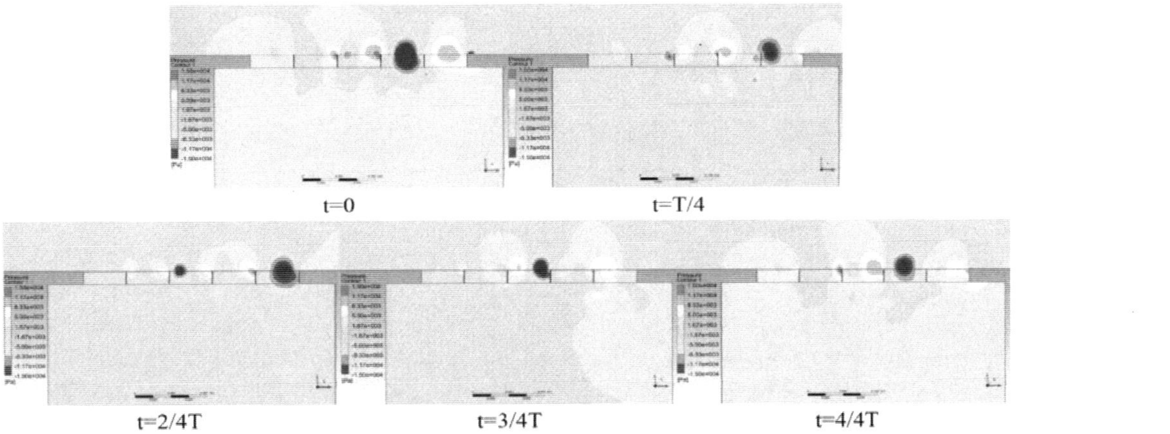

Figure 10. One period of pressure distribution contours of the slotted hole.

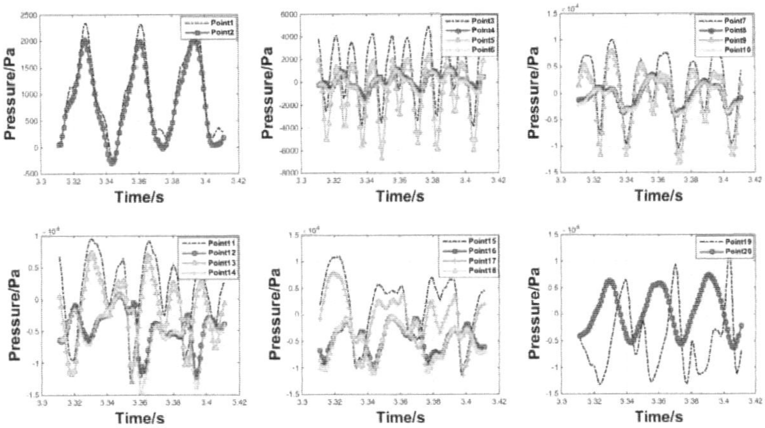

Figure 11. Pressure time sequence of the monitoring points.

The pressure power spectrums density of the monitoring points of the two kinds cavities are shown in figure 12 and figure 13, the values of the opening cavity are significantly higher than the slotted cavity. The main frequencies of the variant points are basically the same, with the fundamental frequency 17.5Hz of the opening cavity, and the fundamental frequency 30Hz and the second order frequency 59.99Hz of the slotted cavity. As shown in the figure 13, the value of the power spectrum density of the variant points is higher, as the location of the point is closing to the downstream, where the value of point 19 on upper edge of the rear wall of the hole is the highest. It is caused by the increasing vorticity magnitude, which resulted from the generation, motivation and collision with more baffles of the vortices.

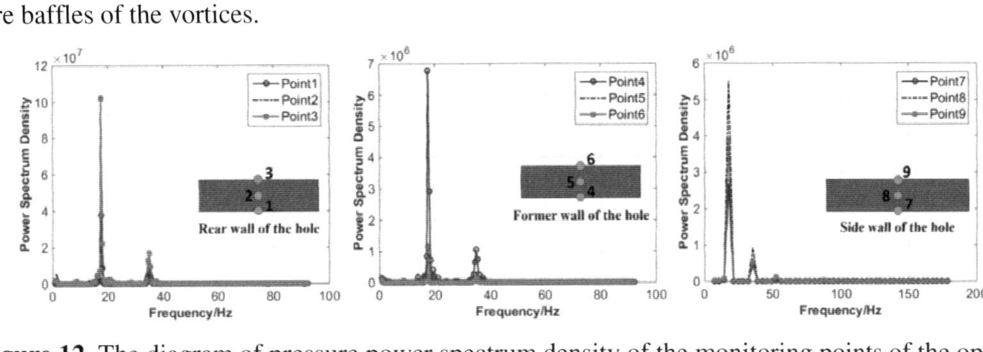

Figure 12. The diagram of pressure power spectrum density of the monitoring points of the opening cavity.

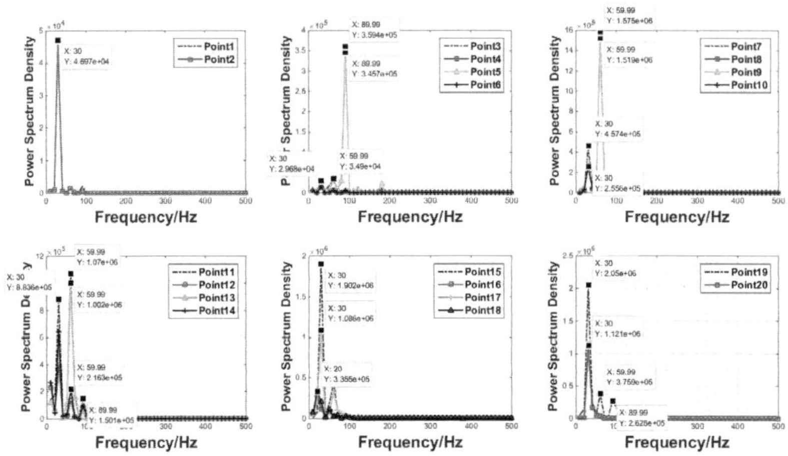

Figure 13. The diagram of pressure power spectrum density of the monitoring points of the slotted cavity.

3.3. Acoustic analogy

The flow-induced noise frequency spectrum of slotted hole and opening hole is shown in figure 14. It indicates that as the baffles installed to the opening hole, the fundamental frequency increases from $f_{os1} = 17.5Hz$ to $f_{ss1} = 30Hz$, and the SPL of the slotted hole model at the fundamental frequency of 30Hz: 124.2dB, is lower than the SPL of the opening hole model at the fundamental frequency of 17.5Hz. Due to the existence of the baffles, the stroke between fluid and the cavity is cut from 300mm to 60 mm, therefore the period of the vortices motivation is shortened, and the fundamental frequency is increased. The velocities of the five points of the slotted cavity are significantly decreased by the baffles compared to the model of cavity with opening hole, which results in the excited force between fluid and baffles reducing, and furthermore makes the SPL at the fundamental frequency reduced. For the problem of the flow over the open cavity of an underwater structure, a semi-empirical formula by Rossiter can be used, where the frequency is obtained as [10].

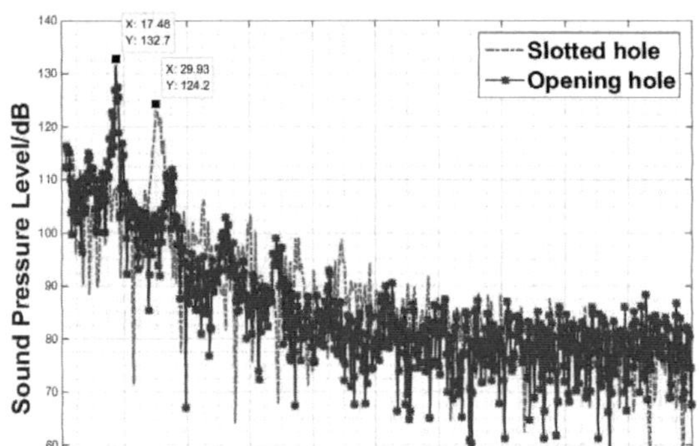

Figure 14. The flow-noise frequency spectrum of the cavity with slotted hole and opening hole.

$$f_n = \frac{U_\infty}{L_{hole}} \frac{n-\alpha}{M_\infty + 1/k_c}, \quad n = 1,2,3\cdots \tag{1}$$

where U_∞ is the velocity of the far field fluid, L_{hole} is the stream direction scale of the cavity hole, M_∞ is the Mach number, $\alpha = 0.25$ and $k_c = 0.57$. Using the above equation, the fundamental frequency of the opening hole is f_{o1}=14.29Hz, at $L_{hole} = 0.3m$, $U_\infty = 10m/s$ the result of the calculation

$f_{os1}=17.5Hz$ is closed to f_{o1}=14.29Hz, while the fundamental frequency of the slotted hole is f_{s1}=70.98Hz, $L_{slotted}=0.3m$, $U_{\infty}=10m/s$, the result of the calculation $f_{ss1}=30Hz$ is far different from f_{s1}=70.98Hz. The reason for the case of the Rossiter formula does not apply to the slotted hole model is that the presence of the baffles destroy the effect of acoustic feedback.

4. Conclusion

(1) The presence of the baffles can decrease the velocity of the fluid, can be reduce the fluctuating force between the fluid and the wall, and play a salient role in controlling the flow noise radiation generated by the cavity with hole.

(2) The upper edge of the rear hole wall and the up-water surfaces of the baffles are the main source of the flow noise generation, therefore, the chamfer or fillet should be performed in engineering.

(3) The flow field near the slotted hole is much more complex than the flow filed near the opening hole. Rossiter formula is only suitable to the prediction of the simple flow filed. Rossiter formula can be used to predict the fundamental frequency of the model of the cavity with opening hole, but can not be applied to the model of the cavity with slotted hole.

Reference

[1] Xu J, Tang K F and Zhang X 2014 *Chinese J. Hydrodyn.* **29(05)** 618-29.
[2] Li X D, Liu J D and Gao J H 2006 *Chinese J. Theoretic. Appl. Mech.* **38(5)** 599-604.
[3] Liu C W, Wu F L, Li H, Peng Y L and Li W P 2014 *Chinese J. Hydrodyn.* **29(2)** 218-24.
[4] Chen C, Wu F L, Li H, Zhang Z G and Liu C W 2015 *Chinese J. Hydrodyn.* **3** 272-8.
[5] Zhang N, Shen H C, Yao H Z, Gao Q X and Gu M 2004 *J. Ship Mech.* **8(01)** 1-11.
[6] Arunajatesan S and Sinha N 2005 *Modeling Approach for Reducing Helmholtz Resonance in Submarine Structure.* 11th AIAA/CEAS Aeroacoustics Conference.
[7] Dai S S, Yao X L, Yang G J and Li Z. 2010 *J. Vibr. Shock*, **29(04)** 207-12.
[8] Xiao Z G and Zhu J G 2015 *Study on the influence of grilles on cavity noise.* The Chinese Congress of Theoretical and Applied Mechanics.
[9] Sarohia V 1977 *AIAA J.* **15(7)** 984-91.
[10] Rossiter J E. 1964 *Aeronaut. Res. Council Rep. Memoranda*, **3438** 8-12.

IOP Conf. Series: Materials Science and Engineering **269** (2017) 012031 doi:10.1088/1757-899X/269/1/012031

Study on Ballistic Absorbing Energy Character of High Performance Polyethylene Needle Felt

Zhu Kailiang[1], Fu Jianqiao[1]

[1]Fire Research Institute of Shanghai, Shanghai 000021

Abstract. The ballistic performance of polyethylene needle felt is tested and the failure morphology after test is also observed. The results showed that when the non-dimensionally non-stressed fibers in polyethylene needles are subjected to high-speed projectile, secondary movement such as stretching and twisting occurs first. This secondary movement is very full, it is the main way of ballistic absorbing energy of the polyethylene needle felt which can avoid the polyethylene fiber short-term rapid heating-up and destroyed. Analysis results show that under normal temperature and humidity conditions, the V50 of 6-layer forded polyethylene needle felt sample is 250m/s. At (450 ± 50) m/s speed range of the target missile, the mean value of the penetrative specific energy absorption for 3-layer forded polyethylene needle felt anti-1.1g simulated projectiles (tapered column) reaches 24.1J·m^2/kg.

1. Introduction

Since the 1960s, the war has been constantly around the world, the military equipment industry has been developed rapidly, the development and research of bullet-proof materials have got more and more attention and a series of bullet-proof new materials are developed and applied [1 2]. The software bullet-proof material is highly valued due to its excellent performance such as low quality, high performance and good wearing comfort [3 4].

High-performance polyethylene fibre as a thermoplastic organic fibre has excellent physical properties [5]: low density (only 0.97kg/m^3), linear straight chain molecular structures, orientation degree approximate 100%, high strength (3.0~3.6GPa). But the disadvantage is that the melting point is low. When woven fabric made of high-performance polyethylene fiber is impacted by the high-speed shrapnel, the strain rate of the polymer can reach $10^5 s^{-1}$ or more, at this point the polymer chain is too late to move in a short time. Some secondary sports: vibration, rotation, etc. due to the effect of woven weaving force cannot deform to absorb energy too. So when polyethylene fiber woven fabric bullet proofs bullets, it is extremely easy to friction heat and burn which results in not very satisfactory bullet-proof performance [6].

The state of the fibers in the polyethylene needle felt is completely in a non-dimensionally non-stressed state, which is completely different with the fibers in coven fabric. In this study, the bullet-proof energy absorption performance of needle felt made of high performance polyethylene fiber was investigated·

2. Experimental

The raw materials of the experiment were polyethylene monofilament with diameter of 1.5D and mechanical properties ≥30CN / dtex and length of 100mm. A needle punching process was used to produce a polyethylene needle felt with a thickness of about 2 mm and an area density of (200 ± 15) g. The sample size of ballistic impact test was (400×400) mm. Samples forded of six layers and three layers of the above size of polyethylene needle felt were used respectively to test the ballistic limit speed V50 and penetration specific energy absorption.

Ballistic experiment was carried out at room temperature and humidity conditions using 1.1g tapered column fragment simulated projectile which were made of 45 steel with hardness of 30HRC. And the size of simulated projectile was shown in figure 1. The trajectory limit speed V50 of test target was measured with reference to the public safety industry standard GA950-2011. A 12.7mm ballistic speed gun was used, the speed forward and behind target of simulated bullet penetration test were measured by the light curtain speed target. The impact incident speed, penetration residual speed and ballistic limit speed V50 of simulated projectile were got by correcting the results of light screen target speed. During the test, the target plate is fixed on the target frame, and the periphery of the target is bound and clamped by the frame. The firing distance used in the experiment (ie, the distance from the launch device port to the fiber target) is 5m, and the shooting direction is vertical to target plate. The shooting test was shown in figure 2 [7]. The Lecia-S6E type Leica microscope and JSM-6000 SEM were used to observe the damage morphology of the polyethylene needle felt after the ballistic experiment.

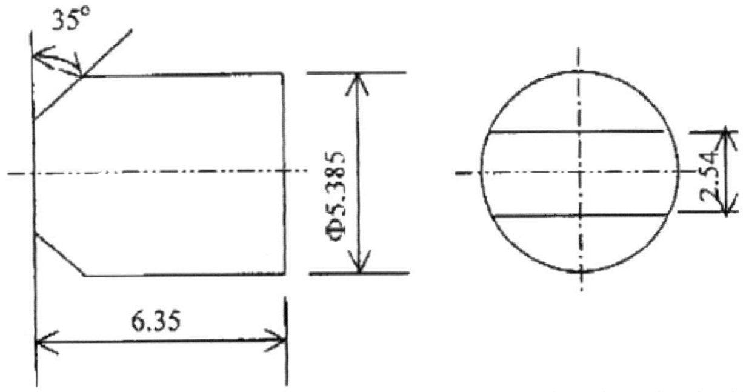

Figure 1. The shape and size of tapered column simulated missile (mm).

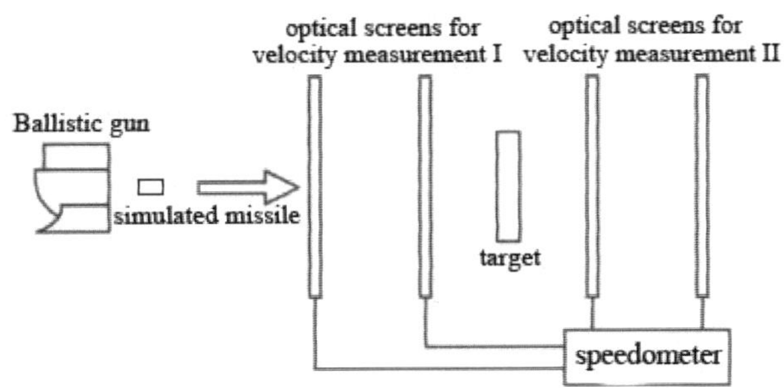

Figure 2. The schematic of shooting experiment.

3. Experimental results and analysis

3.1 Morphological analysis

Figure 3(a) is the original topography of the polyethylene needle felt. It can be seen from the figure that the polyethylene fiber is a non-dimensionally non-stressed state in the needle felt. The movement of fibers in the polyethylene needle felt is different from woven fabric under the impact of high-speed fragment. The stress of the fibers in the polyethylene needle felt is only positive stress which stretches the fiber along impact point, as the show of figure 3(b). The tensile deformation of the fiber in the polyethylene needle felt is more adequacy than woven fabric. As the show of figure 3(c), the fibers in the polyethylene needle felt are stretched by 13mm along the direction of projectile movement at the impact speed of 300m/s. So when the polyethylene needle felt is impacted, the first occurrence of fiber is the tensile deformation of the secondary movement which is the main means of energy absorption.

Followed by the secondary movement of polyethylene fibers there is plenty time for the polymer chains in the fiber to give the reaction to absorb the energy of the fragment.

The figure 3(d) is the SEM microstructure of the polyethylene felt after the impact of the fragment, and the splitting and tensile fracture of fibers did not appear at the impact point. But there is a significant melting and softening phenomenon, the occurrence of high temperature damage is mainly due to the friction between the fibers and the high-speed fragment. When the local temperature exceeds the glass transition temperature of the polyethylene fiber, the mechanical properties of the fibers are rapidly reduction which makes the polyethylene felt loss the protective effect. Therefore, the main failure mode is the softening or melting of the fibers caused by the high temperature when the polyethylene needle felt is impacted by the fragment. Secondly, from the torsion zone in figure 3(d) can be seen that the polyethylene fiber has obvious differences in axial stretching and compression, the anti-shear ability and anti-axial compression capacity is relatively weak.

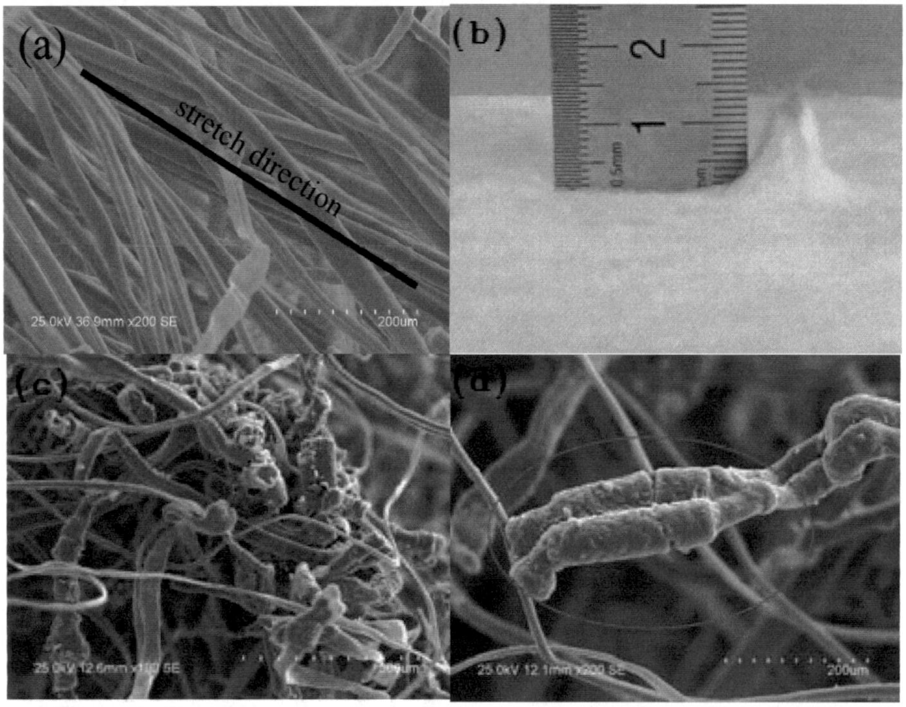

Figure 3. The topography of polyethylene needle felt after ballistic impact.

3.2 Ballistic performance analysis

According to the GA 950-2011 (V50 test method for ballistic materials and products), the V50 test was carried out on 6-layers of the above-mentioned polyethylene needle-felt samples using 1.1g columnar wedge-shaped fragment simulated projectiles and the results show in table 1. The standard deviation is 13.2m/s and the $\bar{V}$ (average velocity of the test points) is 260.5 m/s. According to the conversion formula $V50 = \bar{V}e^{-ax}$ (a is the flying attenuation coefficient of the missile, and x is the flight distance from the measuring point to the target), the V50 of the 6-layer polyethylene needle felt was calculated to be 250 m/s. Table 2 is the V50 values of polyethylene needle felt and other fabric target samples, through the comparison can be found that polyethylene needle felt has good bulletproof energy absorption performance.

Table 1. The V50 test results of polyethylene needle felt.

Numbers	Projectile velocity (m/s)	Test results
1	271.1	unpenetrated
2	429.2	penetration
3	549.2	penetration
4	211.4	unpenetrated
5	295.1	penetration
6	299.9	penetration
7	304.9	penetration
8	252.1	penetration
9	332.1	penetration
10	269.0	penetration
11	239.0	unpenetrated
12	258.8	unpenetrated
13	272.8	penetration

Table 2. The polyethylene needle felt ballistic performance comparison with others [8].

Target sample	Single-layer surface density (g/m^2)	Layer number	Target density (kg/m^2)	V50 (m/s)
CT736 (resin content20%)	510	17	8.71	645
CT709 multi-layer fabrics	220	26	5.24	583
Gold Flex	230	23	5.34	510
polyethylene needle felt	200	6	1.2	250

At (450±50) m/s speed range of the target missile, the perforation specific energy absorption of the three-layer polyethylene needle felt composite samples were tested and the results are shown in table 3. The results show shat polyethylene needle felt has a good penetration of energy absorption performance, the mean value of the penetrative specific energy absorption reaches for 3-layer forded polyethylene needle felt anti-1.1g simulated projectiles (tapered column) reaches 24.1J·m^2/kg. The main means of energy absorption is the tensile deformation of fiber as show in figure 3.

Table 3. The test results of penetrative specific energy absorption.

Sample number	Shooting number	Impact velocity (m/s)	Residual velocity (m/s)	Perforation specific energy absorption (J·m^2/kg)
1#	1	406.6	371.4	23.9
	2	450.6	424.5	19.9
	3	413.9	376.9	25.5
	4	450.8	419.7	23.6
	5	490.9	468.2	19.0
	6	429.6	390.2	28.2
2#	7	446.5	424.7	16.5
	8	430.1	385.5	31.8
	9	454.5	424.5	23.0
	10	435.1	397.4	27.4
	11	424.7	376.3	33.9
	12	454.2	430.9	18.0
Mean value of perforation specific energy absorption (J·m^2/kg)				24.1

Conclusions

Polyethylene needle felt has a good energy absorption performance. The V50 of the 6-layer polyethylene needle felt is calculated to be 250 m/s. At (450 ± 50) m/s speed range of the target missile, the mean value of the penetrative specific energy absorption for 3-layer forded polyethylene needle felt anti-1.1g simulated projectiles (tapered column) reaches $24.1J\cdot m^2/kg$. The main way of ballistic absorbing energy is the secondary movement. When the non-dimensionally non-stressed fibers in polyethylene needles are subjected to high-speed projectile, secondary movement such as stretching and twisting occurs first and this secondary movement is very full, so it is the main way of ballistic absorbing energy. High temperature melting is the main way for the fragments to break the polyethylene needle felt.

References

[1] OGrujicic M, Bell W C, et al. 2012 *Mater. Des.* **34** 808.
[2] Appleby-Thomas G J, Hazell P J, et al. 2009 *J. Appl. Phys.* **105(6)** 064916.
[3] Ogorkiewcz R M 1991 *Int. Def. Rev.* **24** 349-55.
[4] Barwood M J, Newton P S and Tipton M J 2009 *Aviat. Space Envir. Med.* **80(4)** 353-9.
[5] Hearle J W S 2001 *High-performance fibers,* Woodhead Pub-lishing.
[6] Kim H and Nam I 2012 *J. Appl. Polys. Sci.* **123(5)** 2733.
[7] Peng G, Feng J C and Liu Y D 2007 *J. Ballist.* **19(03)** 10-4.
[8] Zhang H P 2002 *Failure Mechanisms of Ballistic Resistant Material and Decay of Its Fibers* Shanghai: Dong Hua University, 2002.

Acknowledgements

The authors are grateful to the support of Research Program of Science and Technology Commission Foundation of Shanghai (No.14DZ1206800).

IOP Conf. Series: Materials Science and Engineering **269** (2017) 012032 doi:10.1088/1757-899X/269/1/012032

Finite Element Analysis and Experimental Study on Elbow Vibration Transmission Characteristics

Dai Qing-shan[1, 2], Zhang Zhen-hai[2] and Zhu Shi-jian[2]

[1]College of Power Engineering, Naval Univ. of Engineering, Wuhan 430033, China
[2]National Key Laboratory on Ship Vibration & Noise, Wuhan 430033, China

Abstract. Pipeline system vibration is one of the significant factors leading to the vibration and noise of vessel. Elbow is widely used in the pipeline system. However, the researches about vibration of elbow are little, and there is no systematic study. In this research, we firstly analysed the relationship between elbow vibration transmission characteristics and bending radius by ABAQUS finite element simulation. Then, we conducted the further vibration test to observe the vibration transmission characteristics of different elbows which have the same diameter and different bending radius under different flow velocity. The results of simulation calculation and experiment both showed that the vibration acceleration levels of the pipeline system decreased with the increase of bending radius of the elbow, which was beneficial to reduce the transmission of vibration in the pipeline system. The results could be used as reference for further studies and designs for the low noise installation of pipeline system.

1. Introduction

The elbow not only has the function of changing the flow direction of the transmission medium, but also plays an important role in adjusting the installation position. When the fluid flows past through the elbow, the velocity and pressure will be changed greatly, resulting in separation and reflux. The flow will definitely become very heterogeneous, leading to the production of vibration. Pipeline system vibration transmission characteristic is affected by many factors, such as elbow bending radius, fluid flow, Mach number, the movement direction of the fluid, etc. In this research, we mainly aimed at observing the changes of elbow vibration transmission characteristics following the variation of bending radius. Furthermore, flow velocity, as a major factor, was also taken into consideration in the test part.

In the aspect of elbow structural parameter, Liu et al. [1] analysed the variation of the strain and amplitude in the straight tube through simulation and experiments, and summarized the vibration characteristics of the straight pipe section of the single-bend system. Although they considered some influence of elbow structure parameters in this system, the keystone of their research was the vibration characteristics of the straight pipe section. Zhang et al. [2] studied the effect of elbow wall thickness on the natural frequency of pipeline system, and concluded that the amplitude of the system decreased with the increase of wall thickness. However, the function of elbow bending radius was not considered. On the basis of the research above, researchers took the bending radius and fluid-structure coupling into account, and the results showed that bending radius had a large effect on high order natural frequency [3]. Another study illustrated the elbow bending radius from 60°- 90° bend had major influence on amplitude through the finite element method [4]. This study was of some theoretical value. The finite element model of elbow was established by Lu et al. [5], and the elbow vibration characteristics under three different flow velocities were analysed. The influence of fluid excitation was considered, but the influence of elbow bending radius was not taken into account.

IOP Conf. Series: Materials Science and Engineering **269** (2017) 012032 doi:10.1088/1757-899X/269/1/012032

In the aspects of hydrodynamics, researchers has established a fluid-solid coupled mathematical model of the 2-d single elbow pipeline system with cavitation phenomenon, and the model has been verified by experiment [6, 7]. The theoretical calculation is in good agreement with the experimental results. Yang et al. [8] simulated elbow vibration that caused by different flow velocities fluid, and illustrated the relation between fluid velocity and elbow vibration through the simulation results. But the function of bending radius was neglected. From another perspective, researchers found out the change rule and influence factors of elbow pressure field and velocity field by adopting the method of experimental contrast, and provided a theoretical basis for the future research of the elbow [9, 10].

At present, there is lack of research on the influence of elbow parameters on the vibration transmission characteristics. In this paper, the changes of bending radius elbow vibration transmission characteristics under different working conditions are analysed through simulation calculation and experiment, which will provide reference for the low noise installation and design of pipeline system.

2. Simulation

Take DN80 type pipe into consideration and establish finite element model of the elbow. The diameter was set to 88.4 mm and wall thickness was set to 4 mm. The required elbow model was simplified to shell structure and created in the part module of ABAQUS. Then the material and cross section properties of the model were defined in property module. The material parameters of the model were shown in table 1. And then Load module and boundary conditions were defined in the load module. In the left side line, there was a harmonic exciting force. The finite element model of the pipeline system was shown in figurc 1. From left to right, there are different elbow finite element models of which the bending radius was 0, 1, 2, 3 times the size of the pipe diameter. These models were shown in figure 2: (a)-(d).

Table 1. Parameters of pipe model material.

Material parameters	Elastic modulus(pa)	Density (kg×m^3)	Poisson ratio
	1.47×10^{11}	7300	0.25

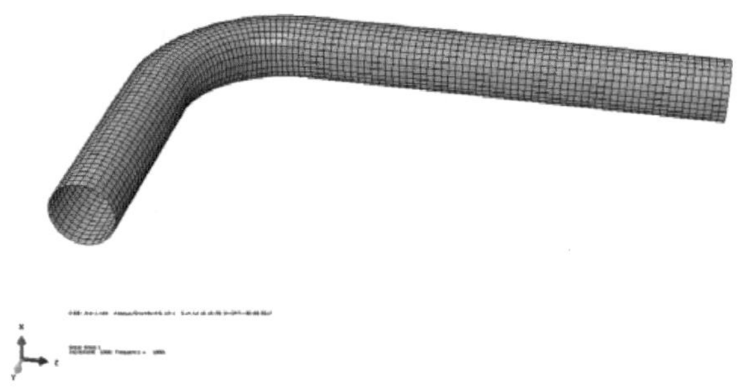

Figure 1. Pipeline system finite element model.

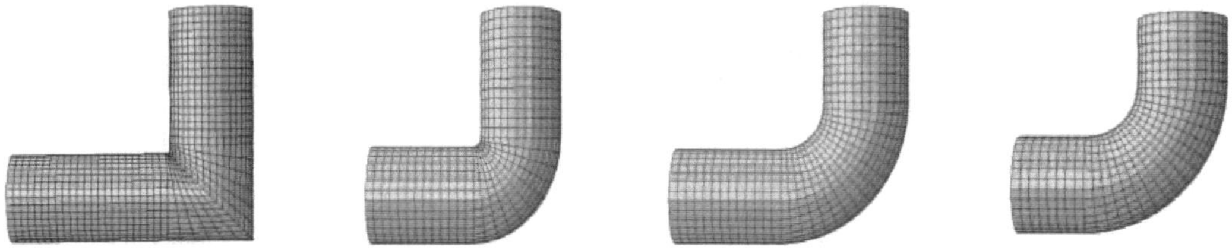

(a) (b) (c) (d)

Figure 2. Different bending radius elbows finite element model.

Figure 3, 4 and 5 presented the vibration acceleration level spectrums of four different bending radius elbows under the *x*, *y* and *z* axial unit excitation force that applied to the left end of the pipeline system finite element model. The frequency range was 0-500 Hz, and the unit of vibration acceleration level was dB (the reference level was 10^{-6}).

IOP Conf. Series: Materials Science and Engineering **269** (2017) 012032 doi:10.1088/1757-899X/269/1/012032

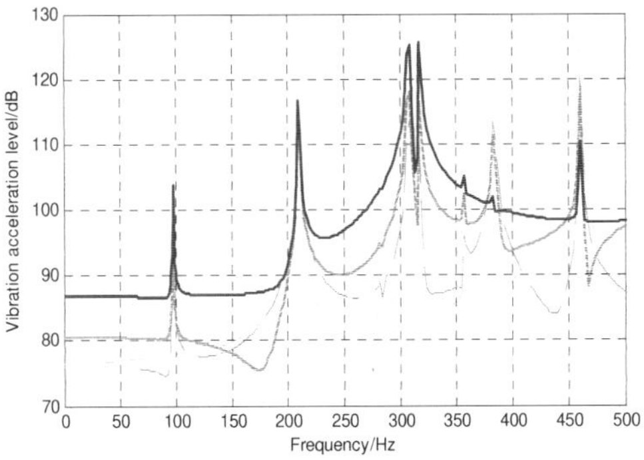

Figure 3. Vibration acceleration level of four elbows under the X-axis excitation.

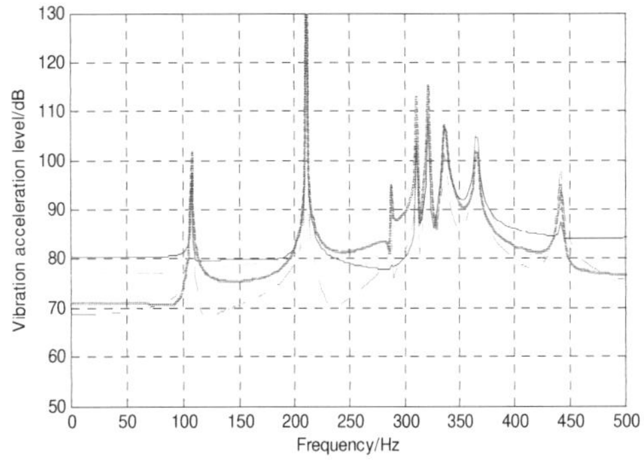

Figure 4. Vibration acceleration level of the four elbows under the Y-axis excitation.

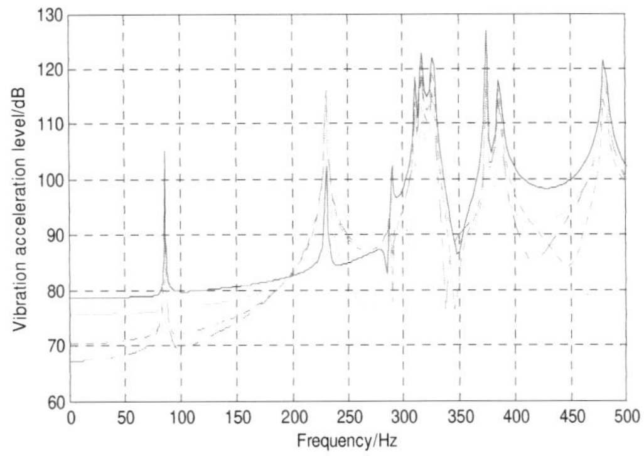

Figure 5. Vibration acceleration level of four elbows under the Z-axis excitation.

As seen in figures above, the four vibration acceleration level curves of elbow shown a trend of decreasing at the 0-250 and 400-500 Hz frequency bands, so we could say that in both frequency bands, elbow bending radius had a great influence on the vibration transmission characteristics. However, in 300-400 Hz frequency band, the peaks of four curves were similar. Consequently, elbow bending radius made little difference for the vibration transmission characteristics in that frequency range. From figure 3, we could draw the conclusion that the bending radius had the greatest influence on the vibration transmission characteristics of the pipeline system under the X-axis excitation. 3 times bending radius vibration acceleration level was lowest than other elbows. Therefore we draw the conclusion: the increase of bending radius contributed significantly to reducing the vibration transmission in the pipeline system to some extent. Though the vibration acceleration levels under mechanical excitation in different directions have been proved by simulation analysis, further researches conducted in the experimental part were necessary.

3. Experiment research

Based on the results of finite element analysis above, further experimental research was taken at the last part of the paper. We took the flow velocity into consideration and explored the vibration of different bending radius elbow under different flow velocity. From left to right, there were four kinds of elbow which bending radius were 0, 1, 2, 3 times the size of the pipe diameter and shown in figure 6.

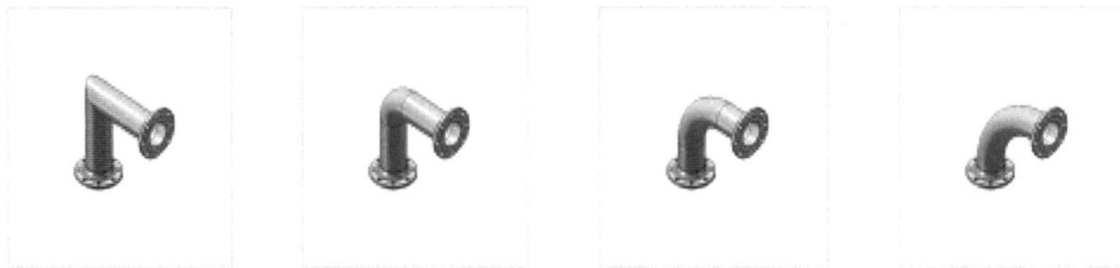

Figure 6. Four types of bend pipes.

The bend pipe vibration was measured with a three-way accelerometer mounted on both ends of the flange shown in figure 7.

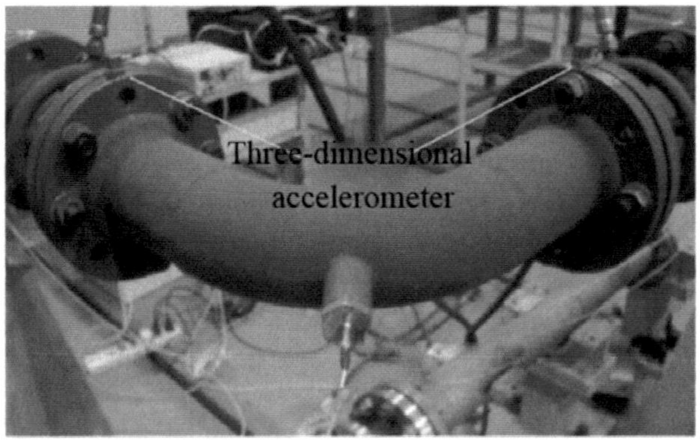

Figure 7. The bend pipe to be tested.

The comparison results of the horizontal, axial and vertical vibration acceleration level were shown in figure 8(a)-(f).

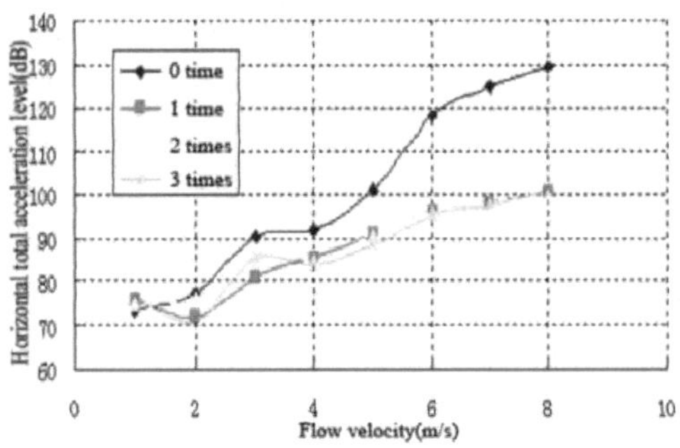

(a)

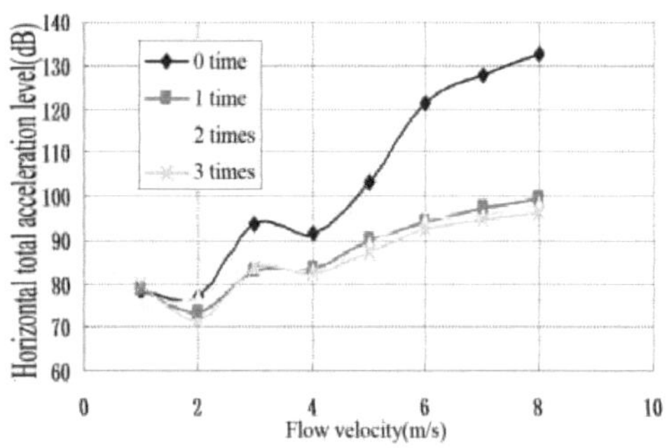

(b)

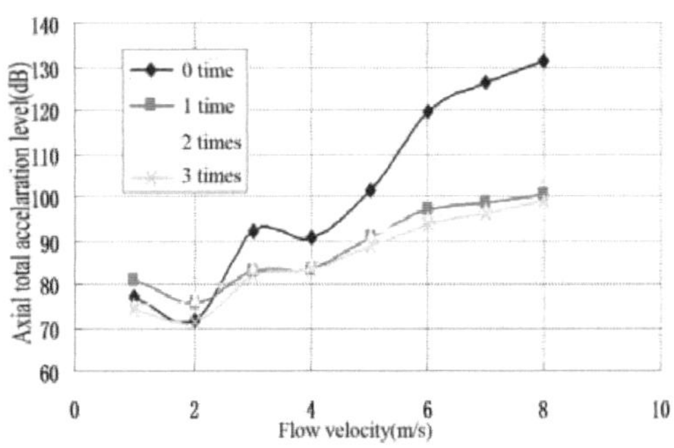

(c)

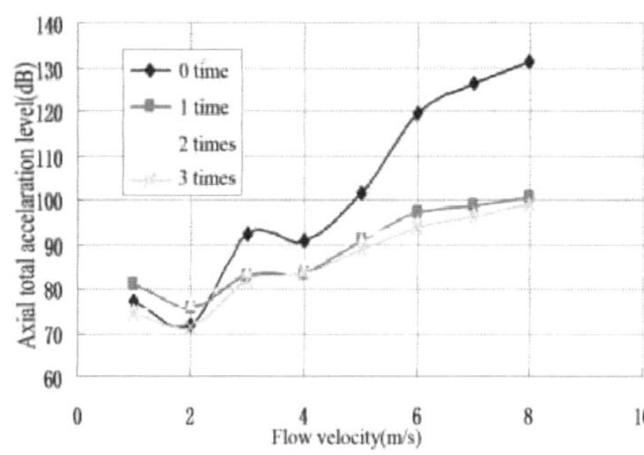

(d)

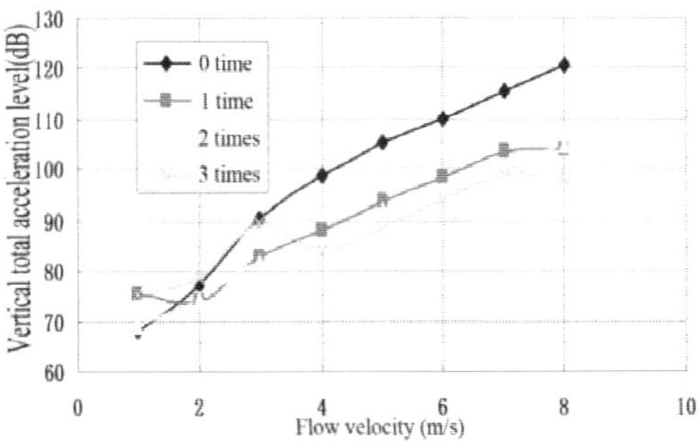

(e)

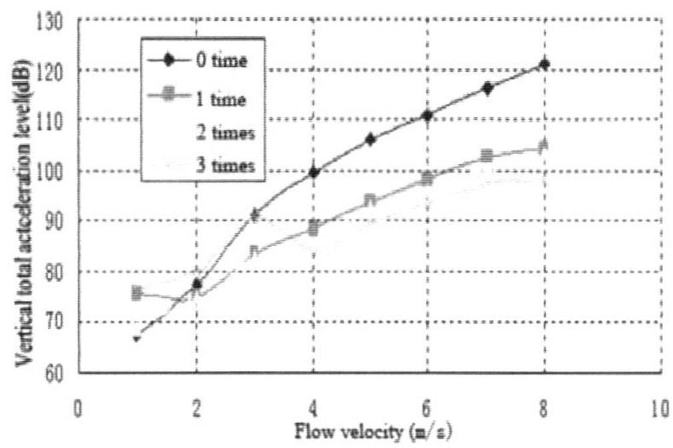

(f)

Figure 8. The corporation curve of the acceleration at the downstream of the elbow is with the change of velocity.

It is shown by these figures that the total vibration acceleration level in any direction of the bending tube increases with the increase of velocity when the diameter ratio is certain. When the bending radius is 0 time the size of the pipe diameter, the total vibration level of the bend in any direction, upstream region or downstream region, is almost to the maximum relative to other diameter ratios. Moreover, the difference becomes larger and larger with the increase of flow velocity. When the bending radius increases from 1 to 3 times the size of the pipe diameter, the total vibration level in any direction of the elbow is almost the same, only small gradually decreasing. For the upstream and downstream of the elbow, the acceleration of the vertical vibration is generally the maximum, the axial second and the horizontal minimum. The vertical vibration of the elbow is consistent with different diameter ratio. The total axial vibration is more consistent when the diameter ratio is larger than 0. But only when the flow velocity is greater than 5m/s, it is slightly higher in the downstream region than in the upstream region. When the diameter ratio is 0, the result is opposite. For the horizontal vibration acceleration level, it is a little bit higher in the upstream region than in the downstream region. On the contrary, it is slightly lower in the upstream region than in the downstream region in other elbow bending radius.

4. Result

Finite element method in this paper can simulate the vibration response of the elbow under different excitation characteristics, but greatly depends on the accuracy of finite element model. Sometimes the model is not accurate enough which will have an impact on the simulation results. In the calculation process, the size of the mesh will also affect the accuracy of the calculation. In the finite element software, how to establish accurate elbow finite element model still needs further complement, improve and verify.

The results of the test show that the replacements of large bending radius have certain effect on tube fluid in irregular flow. However, the modification of the single bend does not affect the vibration of the whole piping system. Therefore, in practical engineering applications, it is possible to consider minimizing the number of elbows and optimizing the radius of the elbow composition in the piping system. At the same time, it is suggested that the bending radius of metal pipe shall be no less than 2D (D is 0 of outer diameter of pipe). If there are some difficulties to install the components, the bending radius shall be less than 2D, but the minimum shall not be less than 1.5D.

5. Reference

[1] Liu Q Y Lu J and Tao S Y 2017 *J. Sci. Tech. Eng.* **17(3)** 321-7.
[2] Zhang J, Liang Z and Han C J 2014 *J. Saf. Sci. Tech.* **10(8)** 6-10.
[3] Ye H L, Shao P Z, Chen N and Liu Z M 2014 *Modal analysis with fluid-structure interaction and study on Parameters' Influences of Pipes conveying fluid* Chinese Congress on Computational Mechanics 1-5.
[4] Fu Y L and Jing H Q 2013 *J. Vibr. Shock* **32(13)** 165-8.

[5] Lu C Y, Zeng Q P and Meng X 2013 *Mach. Tool Hydraul.* **41(19)** 141-3.
[6] Wang Y M 2015 *Locomot. Veh. Tech.* **2(1)** 35-6.
[7] Tijsseling A S, Vardy A E and Fan D 1996 *J. Fluid. Struct.* **10** 395-420.
[8] Yang D W, Xie J H and Tian K 2010 *Mod. Manuf. Eng.* **8** 144-8.
[9] Wang J P, Shen Y L and Cao K Q 2006 *Mach. Tool Hydraul.* **5** 113-5.
[10] Qiu L J, Zhang G F and Hao M 2013 *J. Liaoning Shihua University* 48-52.

The Three-Dimensional Elemental Distribution of 3D Printing Stainless Steel Gear via Confocal 3D–XRF Analysis

Min Qin[1], Longtao Yi[2], Jingbang Wang[1], Yue Han[1], Tianxi Sun[1], Zhiguo Liu[1]

[1] The Key Laboratory of Beam Technology and Material Modification of the Ministry of Education, Beijing Normal University, Beijing 100875, China
[2] The Institute of Nuclear Physics and Chemistry, China Academy of Engineering Physics, Mianyang 621000, China

Abstract. The macroscopic mechanical properties of 3D printing product are closely related to their microstructure, it has significant importance to accurately characterize the micro-structure of 3D printing products. Confocal three-dimensional micro-X-ray fluorescence (3D-XRF) is a good surface analysis technology widely used to analyse elements and elemental distributions. Therefore, this technique is also very suitable for element distribution measurement of 3D printing product which is printed layer by layer. In this paper the 3D-XRF technique was used to study the spatial elemental distribution of a micro zone from the 3D printing stainless steel gear. An elemental mapping of two orthogonal sections in the depth direction and three dimensional elemental rendering of one micro-region were obtained. The result shows that elemental distribution of the sample is not uniform, the elemental layer structure is formed in the depth direction, the content of the element in measured area vary smoothly, and with no elemental mutation region. This indicates that the 3D printing sample are fused well between layers and layers, with no large pores or bubbles inside the sample. This study demonstrates that it is feasible to make assessment for micro-structure of 3D printing metal product by using confocal 3D-XRF.

1. Introduction

3D printing is one of rapid prototyping technologies with the characteristic of manufacturing layer by layer [1]. Because the macroscopic mechanical properties of 3D printing products are closely related to their microstructure, it has significant importance to accurately characterize the micro-structure of 3D printing products in order to improve the performance and quality of 3D printing products. Currently, many research institutes has carried out many study works about characterization of micro-structure of 3D printing product [2-4], mainly using some of the traditional characterization methods, such as density detection, morphology analysis and phase analysis [5, 6]. The latest report concerning the characterization of 3D printing metal sample is that ROSS CUNNINGHAM et al. [7] obtained the synchrotron XCT rendering of 3D printing Ti-6Al-4V alloy sample by using the synchrotron X-ray microtomography in 2016. The traditional characterization methods e.g. Electron spectroscopy can only get the surface elemental distribution, and the sample need to be destructively prepared. Although the synchrotron XCT technique has higher resolution and analysis depth, but it is too expensive and hard to apply for test time. The 3D-XRF is a non-destructive technique, easy to implement in general laboratory. What's more, the spatial elemental distribution can be obtained non-destructively by the confocal 3D-XRF.

Confocal three–dimensional micro X-ray fluorescence (3D-XRF) based on the polycapillary X-ray lenses is an excellent surface analysis technology. It has been widely used in the biological science, material science, environmental science, archaeology etc. [8-13], since Gibson and Kumakhov [14] proposed the principle of 3D-XRF combined with the use of two individual polycapillary X-ray lenses in 1993. By using polycapillary X-ray lenses in both the excitation and detection channels, a confocal volume could be obtained. Because the polycapillary X-ray lens limits the viewing area of the detector, radiation were shielded from the detector. Therefore only radiation from the confocal volume could be detected. If the sample is moved through this micro-volume, the space information about elementary composition of the sample can be obtained. Based on the above principles, it is possible to perform 3D elemental mapping non-destructively using this method. It is also an ideal method to analyse samples with lamellar structure. So it is suitable to make spatial elemental distribution analysis of 3D printing sample by confocal 3D-XRF. By analyzing the distribution of elements in the micro-region of the sample, it is possible to make an assessment to fusion situation of interlayer inside 3D printing products, such as whether exists pores or bubbles.

In this work the 3D-XRF technique was used to study the spatial elemental distribution of a micro region from the 3D printing stainless steel gear. An elemental mapping of two orthogonal sections in the depth direction and three dimensional elemental rendering of a micro-region were obtained by using a tabletop confocal 3D-XRF set-up in laboratory. Based on the results, the microstructure of the 3D printing stainless steel gear can be evaluated.

2. Experimental

2.1. Confocal 3D-XRF set-up

Figure 1 (b) shows the confocal 3D XRF setup. An X-ray tube with a Mo target (XTG UltraBright Microfocus X-ray Source, Oxford, USA) was operated at 20kV and 0.5mA. The focal spot size of the X-ray tube was 16.45μm at 20kV. The polycapillary full X-ray lens (PFXRL) for X-ray irradiation was attached to the X-ray tube, and a polycapillary parallel X-ray lens (PPXRL) was attached to a silicon drift detector (X-123, AMPtek, USA). Each polycapillary lens was designed and manufactured by the Key Laboratory of Beam Technology and Material Modification of the Ministry of Education, Beijing Normal University. The input focal distance and output focal distance of the full lens was 52.9mm and 11.5mm, respectively. The focal distance of the half lens was 14.9mm. The sizes of the focal spot of the half lens and full lens were experimentally determined to be 33μm and 32.4 μm at 17.4keV (Mo Ka), respectively. Both polycapillary X-ray lenses were positioned in the confocal geometry, as shown in figure 1 (a). The micro volume overlaps the output focal spot of the PFXRL and the input focal spot of the PPXRL. The angle between the incident and detection beams was set to 90°.

The depth resolution is 47μm at 7.4 keV (Ni Ka). The sample stage was installed on an X-Y-Z stage [DS102 Series, SURUGA SEIKI, Japan], and the corresponding step resolution was 0.5μm of each axis. A CCD camera was also used to confirm the relative position between the sample and confocal volume.

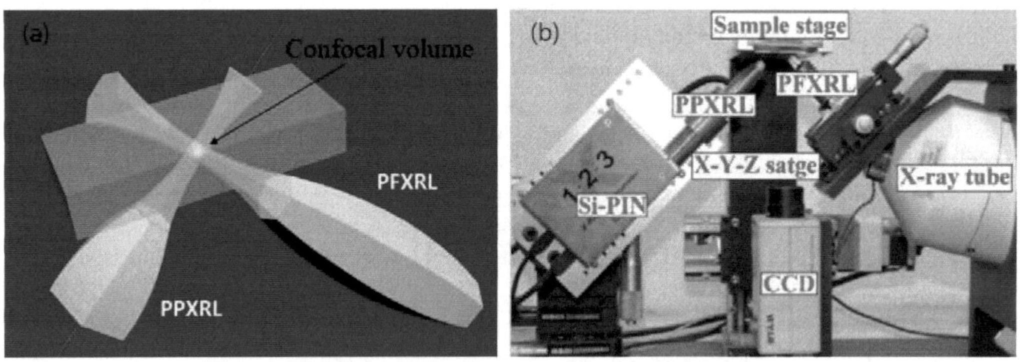

Figure 1 (a) Schematic diagram of the confocal geometry
(b) Experimental setup of the confocal 3D-XRF.

2.2. Sample

Figure 2 shows the photographs of the 3D printing metal gear whose material is 316L stainless steel. It was printed by 3D systems Company, with the thickness of printing layer is about 20μm. To study the spatial distribution of the sample, a series of elemental mapping analysis based on the confocal 3D-XRF technique were applied to the region (R) and region (C). The micro-region (R) is a 1000×700×40 μm small cuboid, as shown in figure 2 (c). As shown in figure 2 (d), the micro-region (C) is two orthogonal sections in the depth direction, with the area dimension is 600×184 μm.

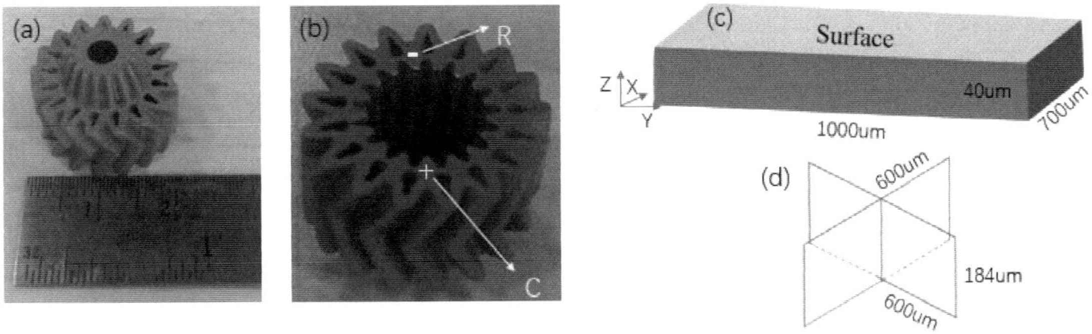

Figure 2. Photographs of the 3D printing stainless steel gear (a) the size of sample (b) schematic diagram of measurement region (c) the size of measurement R-region (d) the size of measurement C-region.

3. Result and discussion

3.1. Elemental components

To ensure the main elemental composition of the sample in this study, a confocal 3D-XRF was applied to this 3D printing stainless steel gear, with the measurement time is 45s. As shown in figure 3, Cr, Mn, Fe, and Ni were detected. This detection results are consistent with the main metal elements contained in 316L stainless steel.

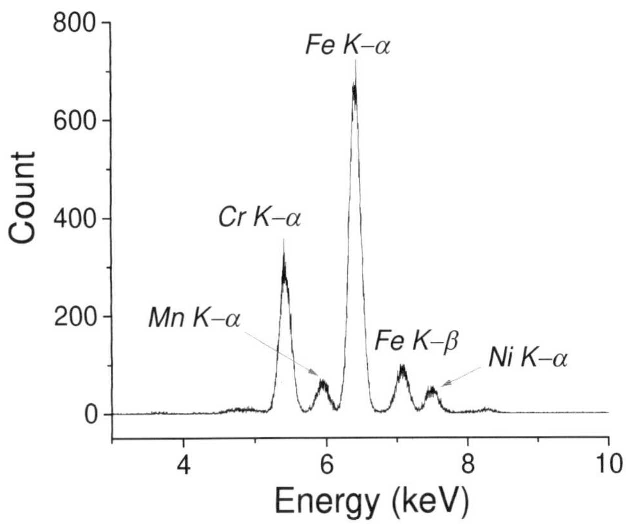

Figure 3. Fluorescence spectra of the 3D printing stainless steel gear.

3.2. Elemental mapping in depth direction

Figure 4 shows the elemental mapping of two orthogonal sections in depth direction of region C with an analysis depth of 184μm. The dimensions of two sections are the same. As shown in figure 5 that the content of 4 elements in measured area vary smoothly, and with no elemental mutation region. It can be observed from the enlarge figures that the fluorescence counting strength of three elements in terms of Cr, Fe and Mn increase first, then decrease, and then increase again. These trends indicates that the elemental distribution isn't uniform and the elemental layer structure is formed in the depth direction inside the 3D printing sample.

It can be inferred from figure 4 that the 3D printing sample is fused well between layers and layers, with no large pores or bubbles. But the elemental layer structure is formed along the depth direction, perhaps because of the layer and layer forming method.

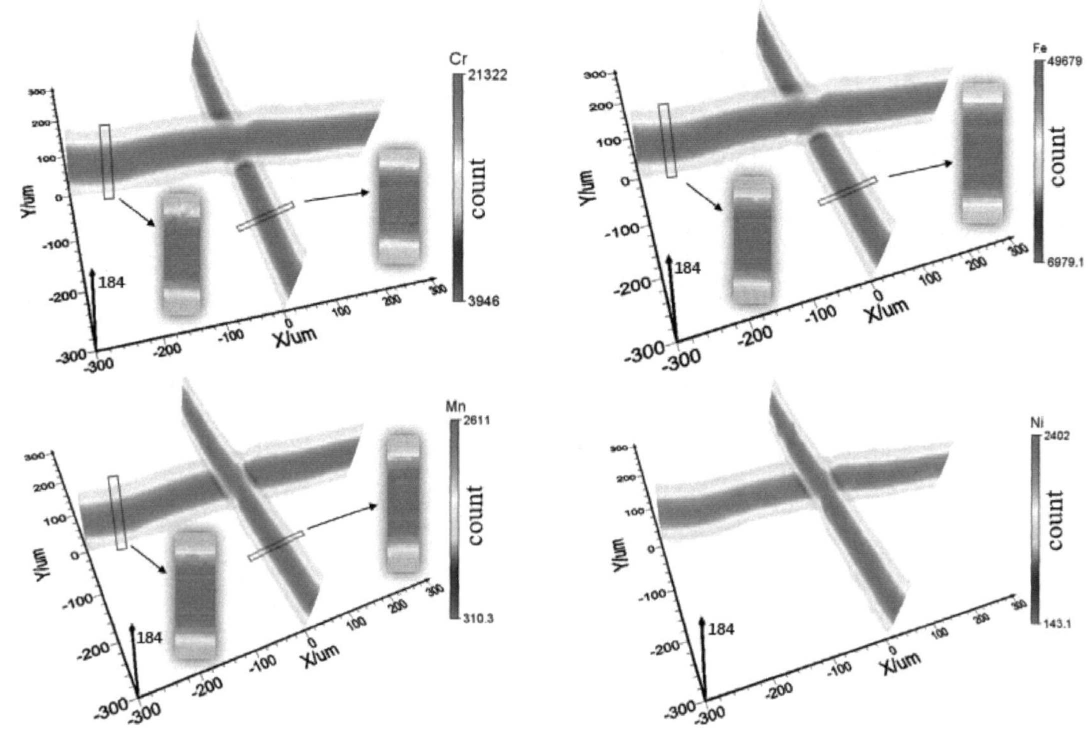

Figure 4. Two orthogonal sections' elemental mapping of Cr, Fe, Mn, Ni in the region of C measured by confocal 3D-XRF set-up.

3.3. Three-dimensional elemental rendering
To confirm the 3D elemental distributions in the region R, a series of 3D elemental mapping analyses were performed at different depths. The scan depth range was from 0 to 40μm, and the step was 10μm. To more vividly show the spatial distributions of the elements in the sample, background intensity was wiped out from the 3D rendering.

As shown in figure 5, the spatial distributions of 4 elements are not uniform, respectively. The elemental distribution trends of Cr and Fe are consistent with the distribution trends of Ni and Mn. The content of Cr and Fe element was higher than the content of Ni and Mn in sample. It can be inferred that this elemental nonuniformity is caused by the uneven mixing of printing powder material or the insufficient elements diffusion during the printing process. Additionally, the content of 4 elements in measured volume vary smoothly, and with no elemental mutation region. So it indicates that the 3D printing sample is fused well between layers and layers inside that measured micro region, with no large pores or bubbles.

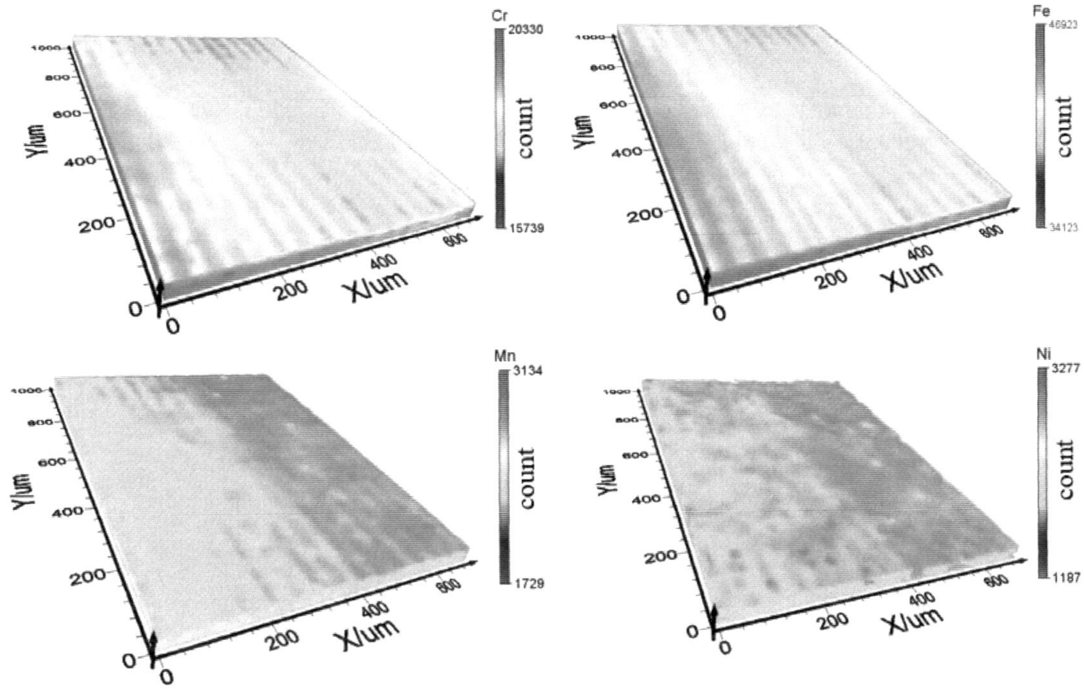

Figure 5. Three dimensional rendering of Cr, Fe, Mn, Ni in region of R measured by confocal 3D-XRF set-up.

4. Conclusion

We have firstly applied the confocal 3D-XRF technique to study the spatial elemental distribution of a 3D printing stainless steel gear. An elemental mapping of two orthogonal sections and three dimensional elemental rendering of one micro-region were obtained.

The results show that elemental distribution of the sample is not uniform, the elemental layer structure is formed in the depth direction, the content of the elements in measured region changes smoothly, and with no elemental mutation region. It can be inferred that this elemental non-uniformity is caused by the uneven mixing of printing powder material or the insufficient elements diffusion during the printing process. The results also indicate the 3D printing sample are fused well between layers and layers inside measured micro region, with no large pores or bubbles.

Hence, it is feasible to make assessment for micro-structure of 3D printing metal product by using confocal 3D-XRF. This method is a good supplement to the traditional characterization methods of 3D printing product.

References

[1] Parthasarathya J and Starly B, 2010 *J. Mech. Behav. Biomed. Mater.* **3** 249-59.
[2] Amato K N and Gaytan S M 2012 *Acta Mater.* **60** 2229–39.
[3] Heinl P and Muller L 2008 *Acta Biomater.* **4** 1536–44.
[4] Ramirez D A and Murr L E 2011 *Acta Mater.* **59** 4088–99.
[5] Yadroitsev I and Bertrand Ph 2007 *Appl. Surf. Sci.* **253** 8064–9.
[6] Gu D D and Shen Y F 2008 *Appl. Surf. Sci.* **255** 1880–7.
[7] Cunningham R and Narra S P 2016 *Min. Met. Mater. Soc.* **68** 765-70.
[8] West M and Ellis A T 2013 *J. Anal. At. Spectrom.* **28** 1544–90.
[9] Šmit Z, Janssens K, Proost K and Langus I 2004 *Nucl. Instrum. Meth. B* **219** 35-40.
[10] Woll A R, *Appl. Phys. A Mater. Sci. Process.* **83** 235-8.
[11] Sun T and Liu H 2014 *Nucl. Instrum. Meth. B* **323** 25–9.
[12] Kanngießer B and Malzer W 2012 *Appl. Phys. A Mater. Sci. Proc.* **106** 325–38.
[13] Yi L T and Qin M 2016 *Appl. Phys. A.* **122** 856.

[14] Gibson W M and Kumakhov M A, *Int. Soc. Opt. Photonics* **9**.

Acknowledgments
This research was supported by National Natural Science Foundation of China (11675019). The 3D Print stainless steel gear was provided by 3D systems Company.

IOP Conf. Series: Materials Science and Engineering **269** (2017) 012034 doi:10.1088/1757-899X/269/1/012034

The Design of Software for Three-Phase Induction Motor Test System

Xu Haixiang[1], Wu Fengqi[1], Xue Jiai[1]

[1]Shanghai special equipment supervision and Inspection Institute of Technology, No. 915, Jinshajiang Road, Shanghai, China

Abstract. The design and development of control system software is important to three-phase induction motor test equipment, which needs to be completely familiar with the test process and the control procedure of test equipment. In this paper, the software is developed according to the national standard (GB/T1032-2005) about three-phase induction motor test method by VB language. The control system and data analysis software and the implement about motor test system are described individually, which has the advantages of high automation and high accuracy.

1. Introduction

The three-phase induction motor is an important mechanical and electrical production, which is widely used as the prime mover in machinery, steel and other industry. In the traditional motor test system, the Oscilloscope or other wave recorder are used to measure the stator current, voltage and the power input current, voltage, then the recorded waveforms are offline processed to calculate the relevant parameters of each test item. These test system has some disadvantages, such as low test accuracy, low automation, manual recording of test results, and difficult to carry out statistical analysis, which is unable to meet the requirement of the modern testing.

With the rapidly development on industry and information technologies in recent years, the control of all equipment has been performed through the use of computers. Most equipment uses intelligent instruments and PLC to connect with computers to monitor each load and electricity consuming devices. The waveforms of motor parameters are collected by the intelligent instrument through the sensors, and then transferred to the computer. The analysis of motor running status is realized on the computer by the control system software, at the same time the instructions are sent to the PLC to control the motor running. In this way, the close-loop control of the motor test system is established. In addition, there are also other ways to automate the motor test, such as virtual instrument technology, adaptive fuzzy PID controller. Therefore, the test system would be more simple, flexible and convenient with the powerful computer software replacing the traditional instrument.

In the paper, we design and implement the three-phase induction motor test system with high accuracy and highly automation, which has great practical significance for improving the production quality and reducing the labour intensity.

2. The composition of three-phase induction motor test system

There are two variable frequency power supplies and a control board in the test system, one provides power to the test motor, and the other provides power to the load motor. The system control block diagram is shown in figure 1.

IOP Conf. Series: Materials Science and Engineering **269** (2017) 012034 doi:10.1088/1757-899X/269/1/012034

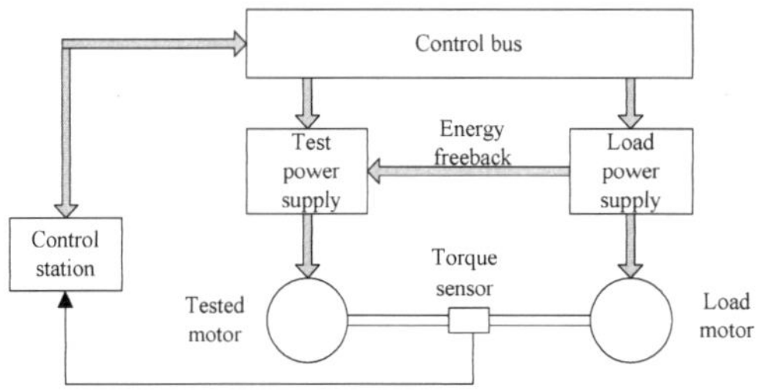

Figure 1. The system control block diagram.

The motor test system consists of a control station, an electric cabinet and a motor test platform, the control station concludes industrial computer, intelligent instrument; and the electric cabinet concludes the variable frequency power supply, PLC, the capacity, the transformer and relay; the motor test concludes T-slot platform, the tested motor, the load motor and the torque sensor. The test power supply and the load power supply are controlled by PLC, meanwhile, PLC's instructions are sent by the industrial computer. The tested motor, the load motor and the torque sensor are rigidly connected. When the test system works, the speed, torque, voltage and current of tested motor are collected by the intelligent instrument, then these test data is sent to the industrial computer, and saved as data file. When the load motor is in the status of power generation, the generated energy could be fed back to the test power supply. The motor is controlled by adjusting of voltage and frequency, which is based on the data of sensors, the motor is in closed-loop control.

3. The design of the test software

The software of test system is written by Visual Basic 6.0, which concludes two parts, one is the control system software; the other is the test data processing software.

3.1. The design of control system software

3.1.1. The design of the operator interface. As shown in figure 2, the operator interface concludes the experiment number, the nameplate of tested motor, environmental parameter, display instrument, test items, the system status area, and test record.

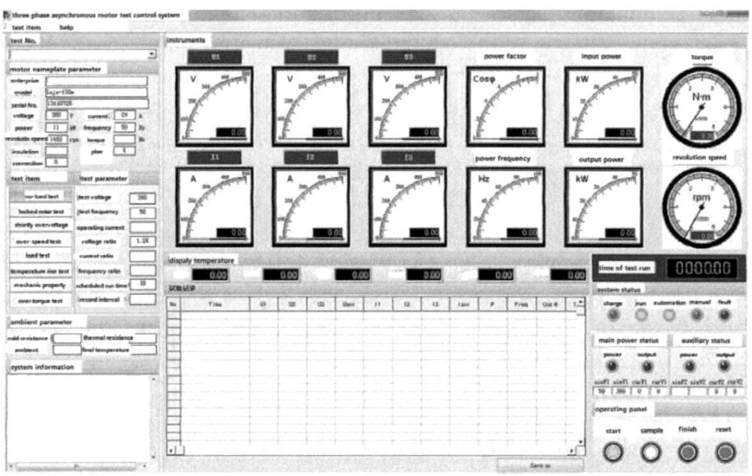

Figure 2. The operator interface of control system

3.1.2. The design of the test items. According to the national standards GB/T 1032-2005《Test procedures of three-phase induction motors》, there are eight test items such as no load test, locked rotor test, load test, mechanical characteristic test, over torque test, over speed test and heating test. The control flow of each test item is determined according to the test procedure, at the same time the data of tested motor is also sampled and saved, which is processed as the input data of the processing test data software.

3.1.3. The flow of test item. The flow of each test item is shown in figure 3, and test items are corresponding to the eight choice buttons in figure 2. The main parameters and the target control parameters values of tested motor are inputted to the table on the operator interface. Among these test items, the test procedure of heating test is the same as that of load test except the collection of data.

3.1.4. The control step of equipment.

3.1.4.1 The starting and stopping of tested and load motor. In order to prevent the starting current from too large, the tested and load motor are started (unloaded) with the frequency and voltage raised (decreased) step by step. The terminal voltage and frequency are applied to the motor according the VF control, thus the load motor can be smoothly connected to the tested motor without impacting.

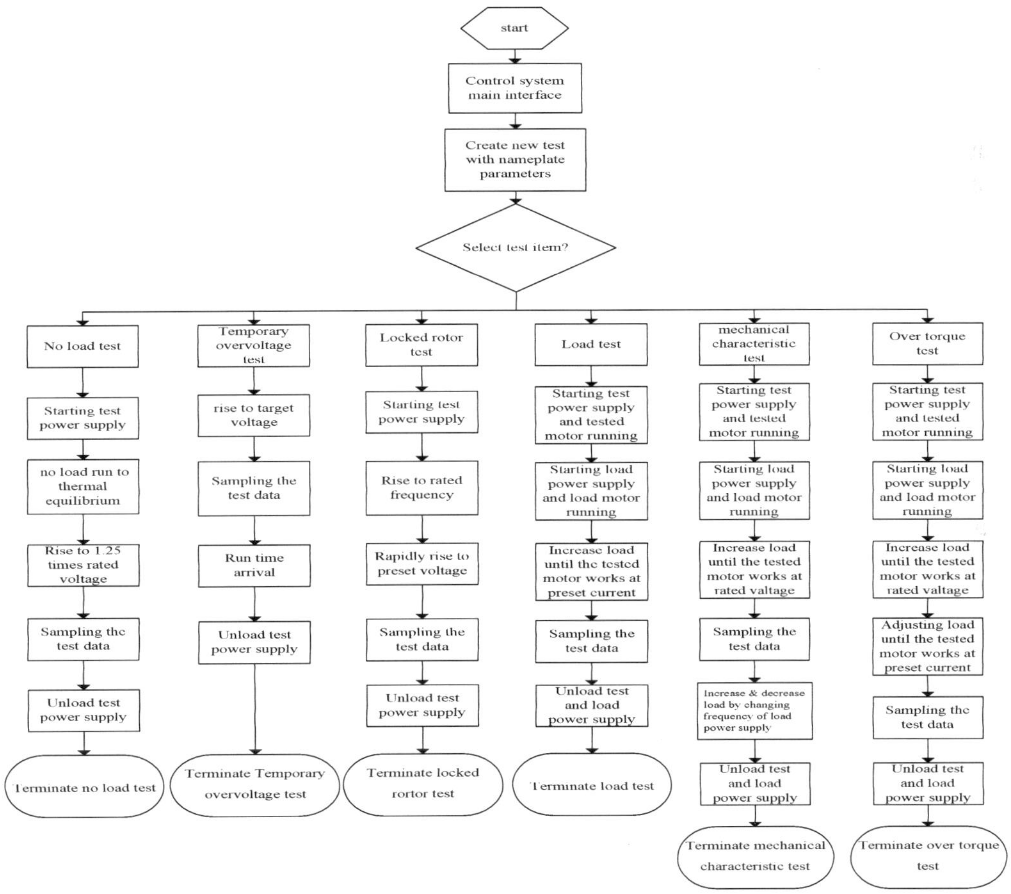

Figure 3. The procedure of test items.

3.1.4.2 The control of mechanical characteristic test. Firstly, the test voltage is preset, the tested and load motor are started according the relevant step. Secondly, the test data is sampling and recorded, and the frequency of load motor is decreased with 2Hz each cycle, it is descended from 50Hz to 0.5Hz in 20s. Thirdly, the ascending frequency process begins with the target frequency is 50Hz, at the end of ascending frequency process the tested motor and load motor are all at no-load working condition.

3.2. The design of test data processing software

3.2.1. The design of the operator interface. The operator interface is designed according to test items, which concludes test number, nameplate parameter, data analysis options and instructions, the calculation interfaces of every test items. In each test item there are the raw data area, the processing data area, the environment and auxiliary data area, the analyzing data area, which is shown in figure 4, the default is no-load interface.

3.2.2. The classification and flow chart of processing test data software.
3.2.2.1. The calculation of no-load test data. When the tested motor runs in thermal stable status, the no-load test data is sampled. The no-load test data processing flow chart is shown as figure 5.

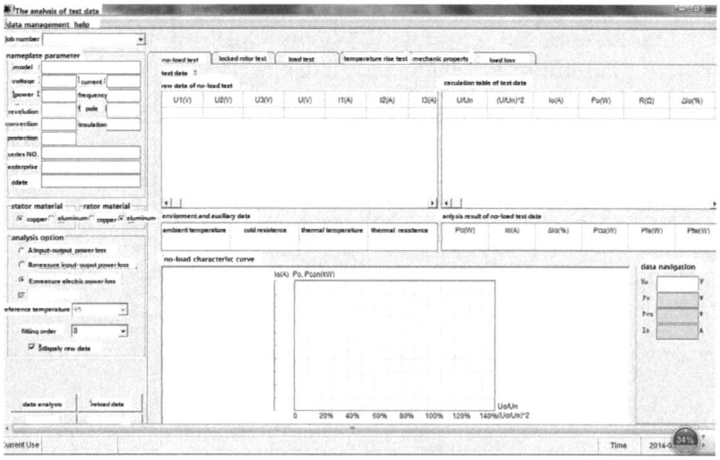

Figure 4. The operate interface of processing test data software.

The raw data of no-load test concludes three-phase terminal voltage, stator line current and input power, the terminal voltage is normalized as U/U_N and $(U/U_N)^2$ (U_N is the rated voltage 380V) so that the tested data curve could be drawn conventionally. The figure 6 is the curves of no-load test data, which is the least square fitting curves of the existing data points, these curves include $P_0 = f(U_0/U_N)$, $I_0 = f(U_0/U_N)$, $P_{con} = f(U_0/U_N)^2$, $P_{fe} = f(U_0/U_N)$. The first two curves are drawn from the raw test data, the third curve is P_{con} versus $(U_0/U_N)^2$, and the wind friction loss is obtained when this curve intersects the vertical axis, thus the fourth curve is equal to the third curve minus the wind friction loss.

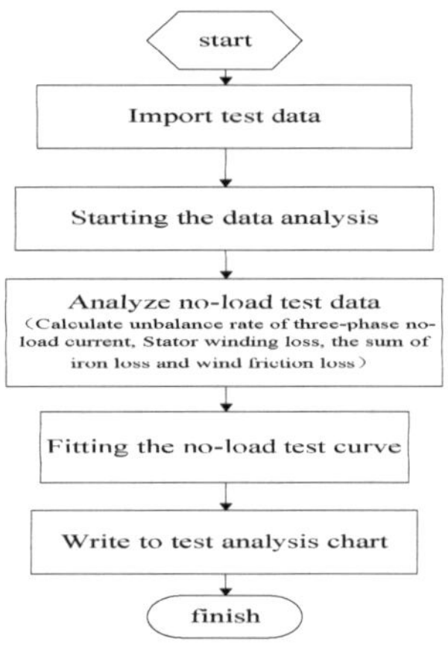

Figure 5. The flow chart of no-load test data.

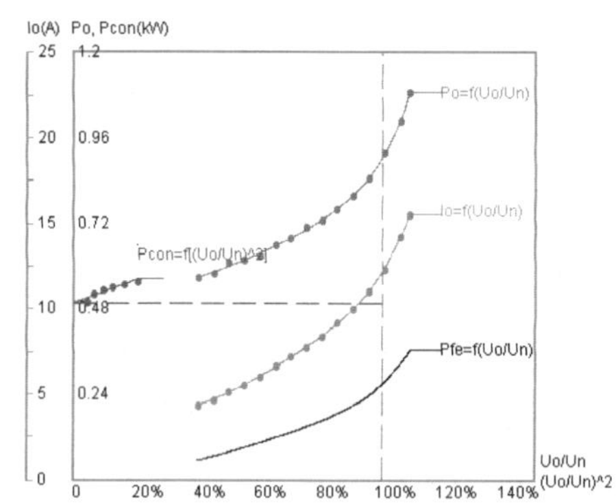

Figure 6. The curve of no-load test.

3.2.2.2. The calculation of locked-rotor test data. The locked current and locked torque are measured in the locked-rotor test data. The raw data concludes the locked voltage U_K, locked current I_K, locked torque T_K and locked input power P_K. The locked-rotor test data processing flow chart is shown in figure 7, when $U_K \geq 0.9U_N$, the curves of $I_k = f(\mathrm{U}_k/\mathrm{U}_N)$, $T_k = f(\mathrm{U}_k/\mathrm{U}_N)$ and $P_k = f(\mathrm{U}_k/\mathrm{U}_N)$ are fitted by the least square method based on some locked-rotor test data; when $U_K < 0.9U_N$, the parameters are calculated by linear regression method, that is, linear $Lg(I_K) = F(Lg(U_K))$ is fitted at first, then the locked current at rated voltage is obtained from the extended curve of this line, just as the dashed line in figure 8, the locked torque is calculated according $T_{kn} = T_K(I_{kn}/I_k)^2$, and the locked power is calculated according $P_{kn} = P_K(I_{kn}/I_k)^2$, where I_{kn} & T_{kn} is the locked current and locked torque at rated voltage, I_k & T_K are the locked current and locked torque at the maximum locked voltage.

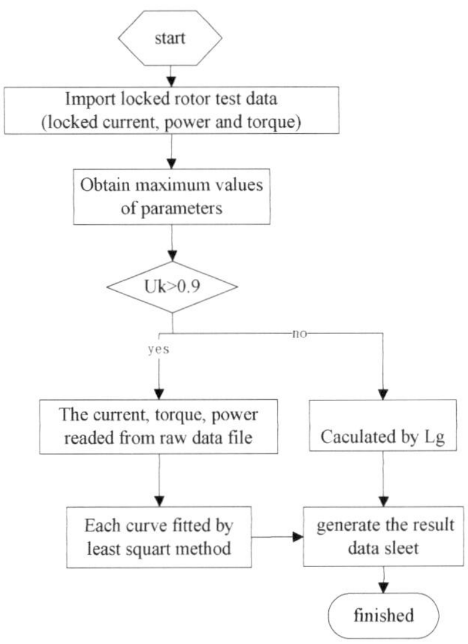

Figure 7. The flow chart of locked-rotor test data.

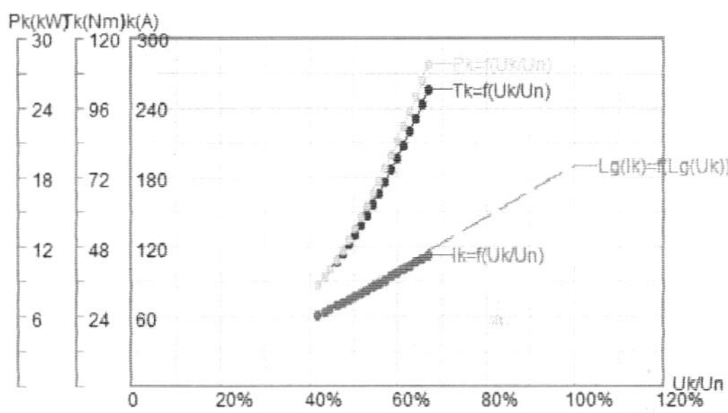

Figure 8. The curve of locked-rotor test.

4. Conclusion

The test system software discussed in the paper has advantages of having fine interface, manipulating conveniently and having reliable operation. The operator interface of control system is designed as entering some test parameters and clicking the starting buttons, then the test system could complete the requirements of main test items; similarly, by using the processing test data software the raw test data is transformed to the needed data, then the export of related parameter calculated results is obtained with drawing the curve diagram, which is dramatically reduced the test labor, and the accuracy of test data meets the requirement of the test.

On the other hand, the test system also has some shortcomings, such as the parameters of tested motor are not be coved those of the test system, the modification of the control software is a complex thing, therefore, the perfections and stability system software are greatly depended on the user's feedback to faults .

Up to now, more than ten types of motor tests have been carried out on the test system. The system software has also been slightly modified in accordance with different motors, and satisfactory results have been obtained. With the improvement of computer, we will consider the use of MATLAB to optimize the processing software for experimental data analysis.

References

[1] GB/T 1032-2005, 2005 *Test procedure for three-phase induction motors* China Standard Press.
[2] Cai J G and Wu Y Q 2011 *Handbook of motor test technology and equipment* China Machine Press, Beijing.
[3] Sun X M and Gong J H 2011 *1200 cases developed by Visual Basic*, Tsinghua University press, Beijing.

Acknowledgment

This work is supported by the science and technology plan project of general administration of quality supervision, inspection and quarantine of P.R.C (the development of brake motor automatic test bench) & (Research on key technology of crane safety supervision based on big data analysis).

IOP Conf. Series: Materials Science and Engineering **269** (2017) 012035 doi:10.1088/1757-899X/269/1/012035

Simplified Model and Response Analysis for Crankshaft of Air Compressor

Li Chao-bo[1], Lou Jing-jun[2] and ZhangZhen-hai[2]

[1]College of Power Engineering, Naval Univ. of Engineering, Wuhan 430033, China
[2]National Key Laboratory on Ship Vibration and Noise, Naval Univ. of Engineering, Wuhan 430033, China)

Abstract. The original model of crankshaft is simplified to the appropriateness to balance the calculation precision and calculation speed, and then the finite element method is used to analyse the vibration response of the structure. In order to study the simplification and stress concentration for crankshaft of air compressor, this paper compares calculative mode frequency and experimental mode frequency of the air compressor crankshaft before and after the simplification, the vibration response of reference point constraint conditions is calculated by using the simplified model, and the stress distribution of the original model is calculated. The results show that the error between calculative mode frequency and experimental mode frequency is controlled in less than 7%, the constraint will change the model density of the system, the position between the crank arm and the shaft appeared stress concentration, so the part of the crankshaft should be treated in the process of manufacture.

1. Introduction

CAD-CAE-CAM is a complete set of industrial production process. The finite element method in CAE has strong ability of analysis and calculation, and is widely used in different field. CAE analysis needs preprocessing the original model established by CAD. The progress mainly includes assigning material attributes, meshing, loading stress and so on, then converts it to finite element model. Many details are included in original CAD model. When the detail size is less than the finite element mesh size, dense grid or even meshes error is generated. Then contribute to long calculation time or even calculation error halfway. Therefore, it is necessary to simplify the geometric features of the original model under the premise of guaranteeing original CAD model characteristics. The main function of CAE is to realize the dynamic design of structure and to verify the rationality of design. Modal analysis is an important part of structural dynamic design, and it is the basis of vibration and noise prediction. Combining the theoretical modal analysis method with the experimental modal analysis method can solve the problems related to the dynamic design of complex structures to a great extent. In this paper, an air compressor crankshaft is taken as the object of study. Firstly, the frequencies and shapes for original modal of the crankshaft between 0~2000Hz are calculated by the finite element model, and compared with the experimental modal at the same time. Then, the CAD model is simplified and the mesh is retiling according to the previous dimensions. Natural frequency variation of the crankshaft in simplified process is compared. The response of the crankshaft reference point under certain constraint conditions is investigated, and the stress distribution of the original model is analysed.

IOP Conf. Series: Materials Science and Engineering **269** (2017) 012035 doi:10.1088/1757-899X/269/1/012035

2. Modal analysis

2.1. Vibration equation

The differential equation of crankshaft vibration is described as

$$[M]\{\ddot{X}\}+[C]\{\dot{X}\}+[K]\{X\}=\{F\} \tag{1}$$

where $[M]$, $[C]$ and $[K]$ are the mass matrix, damping matrix and stiffness matrix respectively, $\{\ddot{X}\}$, $\{\dot{X}\}$ and $\{X\}$ are the acceleration matrix, velocity matrix and displacement matrix respectively, $\{F\}$ is the load matrix.

The influence of damping $[C]$ on modal shapes and modal frequencies can be neglected. In a free vibration status, we ignore the external fore in equation (1), namely $\{F\}=\{0\}$. So the equation (1) can be written as

$$[M]\{\ddot{X}\}+[K]\{X\}=\{0\} \tag{2}$$

When the structure at free vibration status is under natural frequency, substituting $\{X\}=\{A\}\sin(\omega t+\varphi)$, we can get the basic equation of the modal analysis without damping as

$$\{[K]-\omega^2[M]\}\{A\}=\{0\} \tag{3}$$

When the equation has nonzero solution, the condition $\left|[K]-\omega^2[M]\right|=0$ must be satisfied. Characteristic root ω_i is the natural frequency of the structure, characteristic vector $\{A_i\}$ is the corresponding mode. Because of the irregular shape of the crankshaft, it is difficult to solve the differential equation. Therefore, the finite element method is used to solve the problem, and the experiment is used to verify it.

2.2. Original model modality

The air compressor is four stage compression, four connecting rods are connected side by side on a crank pin, the crankshaft axis is perpendicular to the ground, and the four cylinders are 90 degrees around the crank pin. The free suspension diagram of the crankshaft is shown in figure 1. The knocking point and the layout of the accelerometer is shown in figure 2. The modal test adopts the method of multi point knocking and single point response. The red arrow is the direction of sensor installation, the black small hammer indicates hammer point and direction, and knocking 27 points altogether. According to the data provided by the manufacturer, the density is $7820Kg/m^3$, the elastic modulus is $2.1\times10^{11}N/m^3$, and the Poisson's ratio is 0.266. The mesh is divided into tetrahedral units, and the grid is refined at the corners and round holes. Analysis bandwidth is 0-2000Hz, and the mode shape contrast is shown in figure 3. The bending direction of the vibration mode is right, the error of first order frequency is 2.5%, and the error of two order frequency is 3.1%. Because of the influence of casting process, and residual oil in the oil hole, the error is less than 5%. Basically, it can meet the requirement of calculation.

Figure 1. Free suspension diagram of crankshaft.

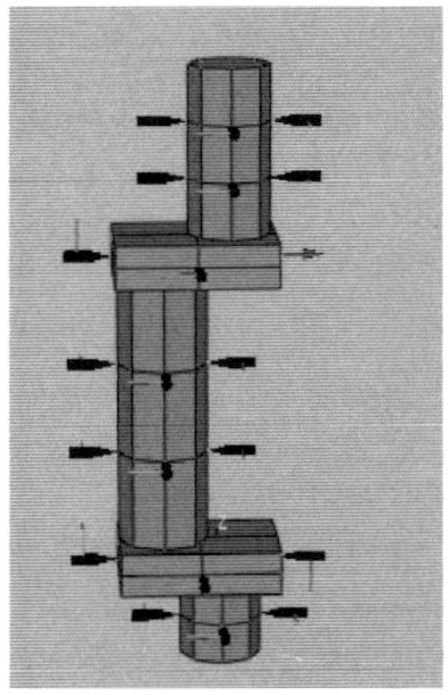

Figure 2. Knocking point and layout of accelerometer.

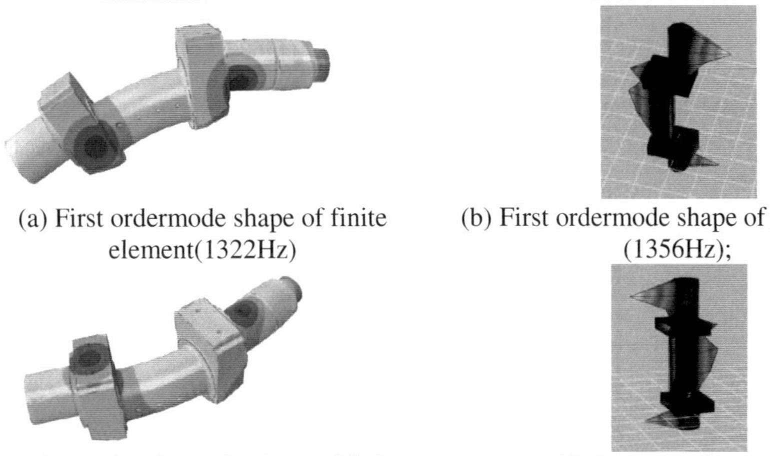

(a) First ordermode shape of finite element(1322Hz)

(b) First ordermode shape of experiment (1356Hz);

(c) Second ordermode shape of finite element (1442Hz)

(d) Second ordermode shape of experiment (1489Hz);

Figure 3. Comparison of mode shapes.

2.3. Model simplification

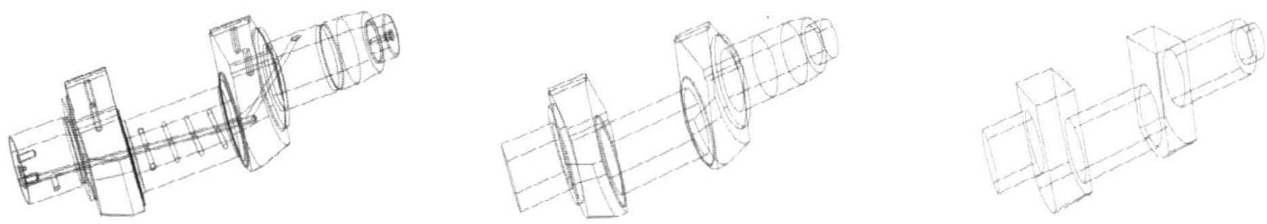

Original model Delete hole and chamfer model Modify peripheral detail

Figure 4. Simplified process of crankshaft.

In order to facilitate mesh division and save computer resources, we should simplify the model in the process of finite element analysis. To discuss the effect of simplification on modal frequencies, the simplified procedure of the model is shown in figure 4. The finite element modal analysis of the three cases is carried out, and the results are shown in table 1. The error of first order natural frequency between finite element model and the experimental is 6.6%, and the error of second order natural frequency between finite element model and the experimental is 5.6%. It can meet the computational requirements basically. For the simplified finite element, the higher order natural frequencies can be controlled less than 7% compared with the original finite element model.

Table 1. Modal frequency under different degree of simplification (Hz).

	Order 1	Order 2	Order 3	Order 4	Order 5	Order 6
Original model	1322	1442	2198	2852	3211	4443
Delete hole and chamfer model	1321	1461	2205	2904	3263	4499
Modify peripheral detail model	1266	1405	2081	2770	3142	4364

3. Harmonic response analysis

Constraint and load addition is shown in figure 5, the short end of the shaft neck is fixed, the semi-circular surface connected to the coupling on the other end is applied 1Pa pressure, the direction point to the crankshaft axis, and the crank end labeled as a response point. The sine sweep frequency is set between 0~5000Hz, and the load step is set to 500, then the frequency response of the reference point is shown in figure 6. Each peak of the frequency correspond is the modal frequency under the constraint condition, which is understood as the installation state for the moment. It is obvious that the modal frequencies between 0Hz and 2000Hz are two more than those of free modes and the constraint condition will change the modal density. When the excitation frequency of the system is very low, the response is close to the static displacement under constant force, and belongs to the stiffness control zone. As the excitation frequency increases, the response frequency is shifted from one modal frequency to another.

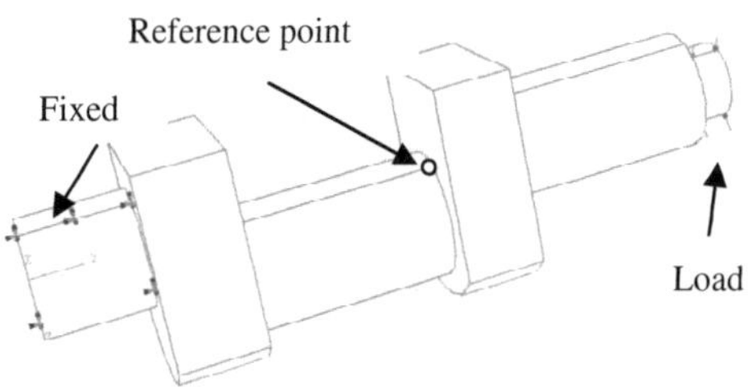

Figure 5. Constrain and load addition.

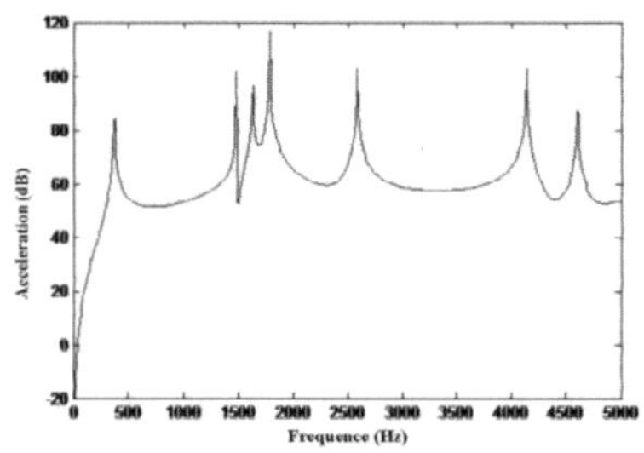

Figure 6. Frequency response of the reference point.

4. Stress analysis

According to the corresponding formula for the exhaust pressure of the air compressor, the piston force variation of the four cylinders with the crank angle is obtained, as is shown in figure 7. The order of the gas pressure magnitude is 10^4. With the two cylindrical surfaces of the crankshaft journal completely fixed, take the force is 10^4, divided by the semi cylindrical area of the crank pin, then the pressure is acquired as $5.7\times10^5\,Pa$, loaded on the crank pin, then get the stress distribution shown in figure 8. Generally, the stress concentration is easy to occur at the two sides of the crank arm. It can be seen that the stress concentration occurs in the part between the crank arm and the shaft neck, but the stress concentration is not very obvious besides crank pin. During the running of the crankshaft, cracks are easy to occur in the stress concentration area and rolling or nitriding can be carried out on the stress concentration part in the process of manufacture, and the transition fillet can also be increased.

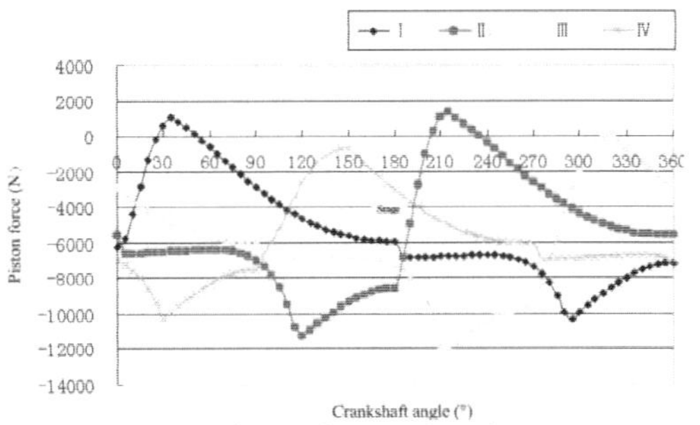

Figure 7. Piston force of each stage.

Figure 8. Stress distribution of the crankshaft.

5. Conclusion

(1) Original finite element modal calculation results are closest to the experiment, but the calculation efficiency is the lowest. Simplified model greatly improve calculation efficiency, although the error is about two times of the original finite element model, compared with the experimental modal, the error can be controlled within 7%.

(2) Certain constraint conditions will change the modal density of the system, resulting in an increase in the modal density in some frequency bands.

(3) The stress analysis of the crankshaft shows that the stress concentration occurs at the place where the crank arm and the shaft neck are connected. It can roll, nitriding or increase the transition fillet in the stress concentration of the crankshaft during the manufacture.

6. Reference

[1] Wu M 2010 *Research on model automatic simplification based on feature in polymorphic theory*HeFei University of Technology.
[2] Liu C C, Chen J Y, Li Q P, Qiu C and Luo X L 2012 *Fluid. Mach.* **40(08)** 17-21.
[3] Xu Z J and Wang S J 2010 *J. Xi'an Jiao Tong Univ.* **44(03)** 100-4.
[4] Tang H N, Wang S J, Xu Z J and Zhao J 2014 *Mech. Str.* **36(04)** 492-8.
[5] Tang B, Liu G B, Xu H P, Yang Q C, Wang Y, Zhao Y Y and Li L S 2014 China Mech. Eng. **25(07)** 873-7.
[6] He Y C, Hu Z H, Dai X X and Pan S L 2005 *Fluid. Mach.* **32(04)** 39-41.
[7] Bae Y J, Kim J B, et al. 2008 *J. Mech. Sci. Tech.* **25(22)** 1391-9.
[8] Zhu S J and He L 2006 *Vibration control of ship machinery* National Defence Industry Press.

Influence of Torsion Effect on the Mechanical Characteristics of Reinforced Concrete Column

Debin Wang[1], Guoxi Fan[*2]

[1]School of Civil and Safety Engineering, Dalian Jiaotong University, Dalian, Liaoning, 116028, China;
[2]School of Engineering, Ocean University of China, Qingdao, Shandong, 266100, China)

Abstract. The purpose of this paper is to study the effect of torsional effect and loading rate on the flexural capacity of RC members. Based on the fiber model of finite element software ABAQUS, a model has been established with the consideration of the strain rate sensitivity of steel and concrete. The model is used to reflect the influence of the rotational component of ground motion by applying the initial angular displacement. The mechanical properties of RC columns under monotonic loads are simulated. The simulation results show that there has been a decrease in the carrying capacity and initial stiffness of RC columns for high initial torsion angle. With the increase of initial torsion angle, the influence of loading rate on RC columns gradually increases.

1. Introduction

Reinforced concrete (RC) column is the main bearing member in building. So a large number of experimental studies on the mechanical properties of RC columns under bending moment and shear force have been performed. At present, domestic and foreign scholars have carried out a lot of research on the mechanical properties of RC members under dynamic loads. Some conclusions can be summarized as follows: 1) the carrying capacity of RC member increases with the increase of loading rate; 2) the ductility of RC member decreases with the increase of loading rate; 3) the initial stiffness of RC member increases with the increase of loading rate, but the rate of stiffness degradation accelerates [1-3]. The similar conclusions can be draw for other studies reported by Wang [4], Fan [5], Fang [6], and Xiao [7].

However, due to the multi-dimensional and non-static characteristics of ground motion, RC member is often subjected to complex dynamic force, accompanied by torsion effect. Tirasit [8] studied the mechanical properties of RC members under combined bending moment and torsion force. Test results show that the carrying capacity of RC member decreases while the damage area of RC member changes. The bending and torsion tests of RC members were carried out by Belarbi [9, 10], it was found that the failure mode, deformation capacity, and other mechanical behavior of the member changed significantly. Among them, due to the effect of the torsion, the bearing capacity of specimen decreased and the stirrup is yielded in advance. Ray [11] has also carried out tests on RC columns subjected to a combined torsion and bending. Thus, the mechanical properties of RC members are significantly changed after considering the torsion effect.

Based on the above considerations, the load-displacement curves, the carrying capacity and other mechanical behavior of reinforced concrete columns under different loading rate and torsion displacement are studied in this paper.

2. Finite element model and dynamic constitutive model

2.1. Finite element model

The numerical simulation of the RC column is performed by employing finite element software ABAQUS. The 8-node solid reduced integration element (C3D8R) is selected for concrete, while the 2-node truss element (T3D2) is selected for longitudinal reinforcements or stirrup reinforcements. The embedded element is used to represent the interaction between steel reinforcement and concrete. During the process of the simulation, the load at column end was applied with displacement control mode. In order to avoid the stress concentration, the rigid block is built at the loading point corresponding to the bottom and top of RC column, and the rigid connection between the reference point and the rigid block is adopted. Then the displacement and initial torsion angle are applied at the reference point. The influence of loading rate on the mechanical properties of RC column is mainly considered in some cases. Therefore, "Dynamic Explicit solver" is used to calculate the dynamic response of RC column in this paper.

2.2. Dynamic constitutive model of materials

Only compressive skeleton curves and hysteretic rulers are introduced for the dynamic constitutive model of concrete. The CEB model is adopted for concrete, and the detailed introduction of the concrete constitutive model is shown in Reference [12]. Besides, the ideal elastic-plastic model is used as the calculation model for steel reinforcement. The relationship of dynamic strength and quasi-static strength of steel reinforcement is shown as follows [13].

$$\frac{f_{yd}}{f_{ys}} = 1 + c_f \lg \frac{\dot{\varepsilon}}{\dot{\varepsilon}_0} \tag{1}$$

$$\frac{f_{ud}}{f_{us}} = 1 + c_u \lg \frac{\dot{\varepsilon}}{\dot{\varepsilon}_0} \tag{2}$$

$$c_f = 0.1709 - 3.289 \times 10^{-4} f_{ys} \tag{3}$$

$$c_u = 0.02738 - 2.982 \times 10^{-5} f_{ys} \tag{4}$$

where $\dot{\varepsilon}$ is the strain rate of the steel reinforcement under dynamic loading; $\dot{\varepsilon}_0$ is the strain rate of the steel reinforcement under quasi-static loading, which is selected as 2.5×10^{-4}/s.

3. Influence of torsion effect

Due to the multi-dimensional and non-static characteristics of ground motion, RC member is often subjected to complex dynamic force, accompanied by torsion effect. Therefore, based on the finite element model mentioned above, the initial angular displacement is applied at rigid region of the top of the member to reflect the torsion effect. The torsion displacements are given in the form of an initial angular displacement, with the value of 0°, 5°, 10° and 15°, respectively. At the same time, the finite element software ABAQUS can consider the rate-dependent of the material through dynamic constitutive model. Moreover, in order to consider the influence of loading rate on the mechanical properties of the member, a method to calculate the strain rate level of the material under near-field ground motion proposed by Asprone [14] is adopted. The strain rate levels of the material corresponding to the loading rates of 0.1mm/s, 1.0mm/s, 10mm/s and 50mm/s are 10^{-5}/s, 10^{-4}/s, 10^{-3}/s, and 10^{-2}/s, respectively.

As shown in figure 1, the load-displacement curves of the member with the initial angular displacement of 0°, 5°, 10° and 15° are obtained under different loading rates. It can be seen from figure 1 that the carrying capacity of the member decreases with the increase of initial angular displacement. Table 1 shows the carrying capacity of RC members under different loading rates and initial angular displacements. By calculating the data in table 1, it is found that the reduction extent of the carrying capacity are 33.0%, 28.2%, 29.4% and 30.2% corresponding to the loading rate equalling to 0.1mm/s, 1.0mm/s, 10mm/s, and 50mm/s, respectively, while the initial angular displacement increases from 0° to 15°. It can be seen that the carrying capacity of the member is significantly reduced while the initial angular displacement increases.

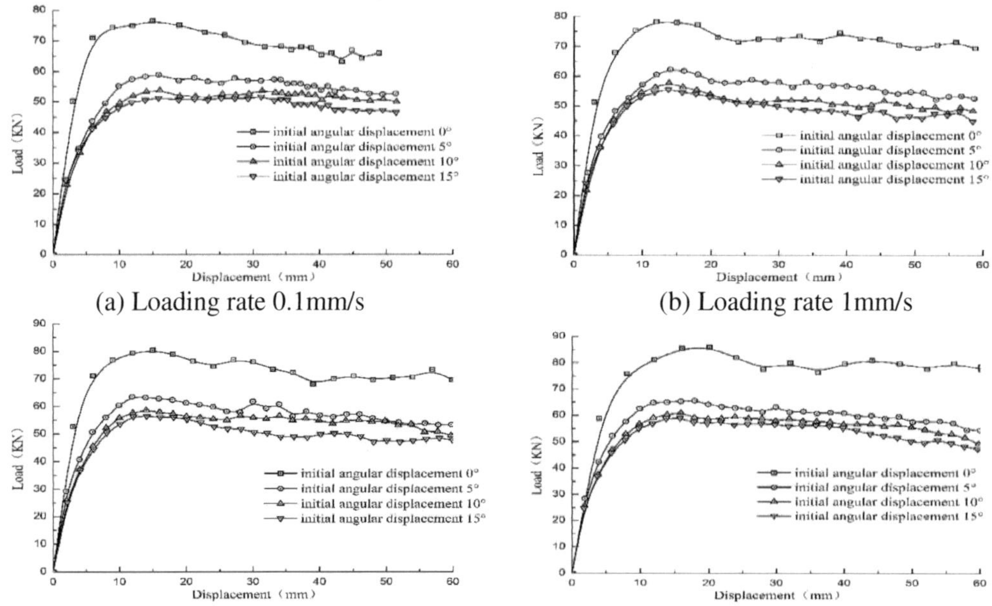

(a) Loading rate 0.1mm/s (b) Loading rate 1mm/s

(c) Loading rate 10mm/s (d) Loading rate 50mm/s

Figure 1. Load-displacement curves under different loading rates and initial torsion angles.

Table 1. Carrying capacity under different loading rates and initial angular displacements.

Initial angular displacement \ Loading rate	0.1mm/s	1mm/s	10 mm/s	50 mm/s
0°	76.6	78.3	80.5	84.5
5°	59.0	62.4	63.5	65.7
10°	53.9	57.8	58.6	61.0
15°	51.3	56.2	56.8	58.9

Note: The unit in table 1 is kN.

Besides, variation of the initial stiffness can be estimated by the calculation results. It can be found that the initial stiffness of the member decreases with the increase of initial angular displacement. Take calculation results of the loading rate equalling to 0.1mm/s for an example, the initial stiffness of members with the initial angular displacement equalling to 0°, 5°, 10°, and 15° are 11.8kN/mm, 7.68kN/mm, 6.98kN/mm and 6.76kN/mm, respectively. As the initial angular displacement increases from 0° to 15°, the initial stiffness of the member decreases by as much as 42.7%. It can be found that the influence of torsion effect on the mechanical properties of the member is significant.

4. Dynamic mechanical properties of members under torsion action

Seismic load has random and dynamic characteristics, so it is very important to analyze the mechanical properties of members under fast loading conditions. Based on the consideration of the rotation component of the ground motion, this paper analyzes the changes of the mechanical properties of members under different initial angular displacements and loading rates. In the same way, load-displacement curves of each member are re-adjusted by using the above calculation results. Figure 2 shows load-displacement curves of each member under different initial angular displacements.

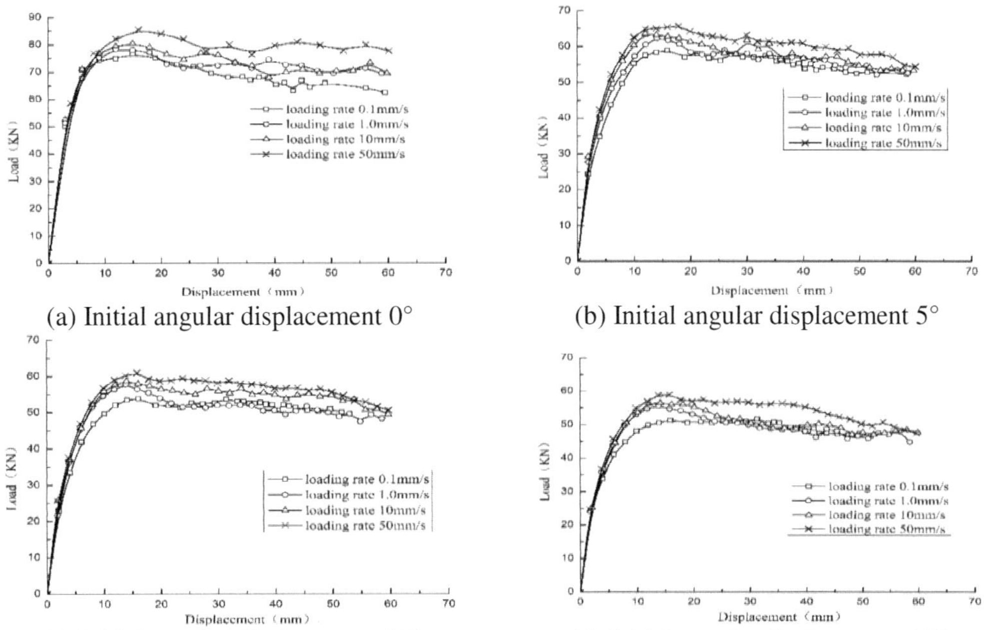

(a) Initial angular displacement 0°
(b) Initial angular displacement 5°
(c) Initial angular displacement 10°
(d) Initial angular displacement 15°

Figure 2. Load-displacement curves under different initial torsion angles and loading rates.

By comparing four groups of load-displacement curves, regardless of changes of initial angular displacement, the carrying capacity of the member always increases with the increase of loading rate. The carrying capacity of the member with a loading rate of 0.1mm/s is selected as the reference value, and the carrying capacity growth percentage of members with a loading rate of 1.0mm/s, 10mm/s and 50mm/s are calculated, as shown in table 2. It can be found that as the initial angular displacement increases, the increasing amplitude of carrying capacity of different members is obvious for different loading rates. For example, when the loading rate increases from 0.1 mm/s to 50 mm/s, the carrying capacity of the member is increased by 10.3% for the initial angular displacement equalling to 0°, while that is increased by 14.8% for the initial angular displacement equalling to 15°.

Figure 3 shows the change trend of carrying capacity under different initial angular displacements and loading rates. With the initial angular displacements of 0°, 5°, 10° and 15°, the carrying capacity of the member increases as the loading rate increases, and the overall growth trend is consistent. Based on the above analysis, it can be seen that the increase of initial angular displacement will lead to the increase of strain rate sensitivity of the member. Therefore, it is very important to consider the torsion and dynamic effects of ground motion in seismic design.

Table 2. Growth rate of carrying capacity under different initial torsion angles.

Initial angular displacement / Increase percentage of bearing capacity	0°	5°	10°	15°
$\frac{F_1 - F_{0.1}}{F_{0.1}}$ (%)	2.2	5.8	7.2	9.6
$\frac{F_{10} - F_{0.1}}{F_{0.1}}$ (%)	5.1	7.6	8.7	10.7
$\frac{F_{50} - F_{0.1}}{F_{0.1}}$ (%)	10.3	11.4	13.2	14.8

Note: $F_{0.1}$, F_1, F_{10}, F_{50} are the carrying capacity of members with loading rate of 0.1mm/s, 1.0mm/s, 10mm/s and 50mm/s at different initial angular displacements, respectively.

IOP Conf. Series: Materials Science and Engineering **269** (2017) 012036 doi:10.1088/1757-899X/269/1/012036

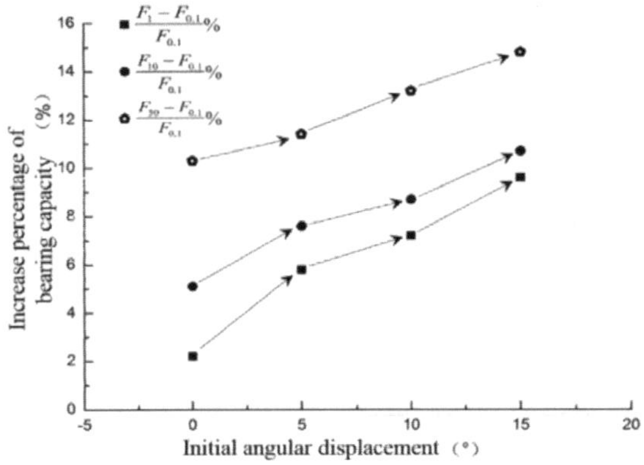

Figure 3. Influence of loading rate on the carrying capacity under different initial torsion angles.

5. Summary

Based on the above analysis, the following conclusions can be drawn:

1) With the increase of initial angular displacement, the flexural carrying capacity and the initial stiffness of the member decrease.

2) The stiffness degradation amplitude with larger initial angular is more obvious. In addition, as the initial torsion angle changing from 10° to 15°, the initial stiffness degradation is the most significant.

3) The influence of loading rate on reinforced concrete column is more obvious with larger initial angular displacement. Under the condition of higher initial displacement angle, the carrying capacity of the member decreases obviously.

References

[1] Lamarche C P and Tremblay R 2011 J. Constr. Steel Res. **67(9)** 1401-10.
[2] Adhikary S D, Li B and Fujikake K 2012 *Int. J. Imp. Eng.* **47** 24-38.
[3] Kulkarni S M and Shah S P 1998 *ACI Struct. J.* **95(6)** 705-15.
[4] Wang D B, Li H N and Li G 2013 *Constr. Build. Mater.* **47** 1167-81.
[5] Fan G X, Song Y P, Wang L C, et al. 2014 *China Civil Eng. J.* **47(11)** 11-8.
[6] Fang Q, Liu J C, Zhang Y D, et al. 2001 *Eng. Mech.* **18(2)** 1-8.
[7] Xiao S Y, Cao W B and Pan H H 2012 *J. Build. Struct.* **33(12)** 142-6.
[8] Tirasit P and Kawashima K 2007 J. Earthq. Eng. **11(3)** 425-52.
[9] Belarbi A, Prakash S and You Y M 2009 *Struct. Eng. Mech.* **33(2)** 137-58.
[10] Belarbi A, Ayoub A, Greene G, et al 2007 *Struct Eng Res Frontiers.* 1-16.
[11] Ray M B and Mallick S K 1980 Indian Concr J. 54(3), 80-83.
[12] Eibl J 1988 *Concrete structures under impact and impulsive loading (CEB-bulletin d'information, NO.187).* Eubrovnik: Comite Euro-International du Beton.
[13] Li M and Li H N. 2010 *China Civil Eng. J.* **43(4)** 70-5.
[14] Asprone D, Frascadore R, Ludovico M D, et al. 2012 *Eng. Struct.* **35** 29-36.

Acknowledgements

The authors would like to thank the National Natural Science Foundation of China (Grant No. 51408093); the Scientific Research Project of Education Bureau of Liaoning Province, China (Grant No. L2015098); the Liaoning Provincial Natural Science Foundation Project (Grant No. 20170540134); the Key Project of State Key Laboratory of Coastal and Offshore Engineering for Dalian University of Technology (Grant No. LP1617); the Fundamental Research Funds for the Central Universities (Ocean University of China, Grant No.201713032); the Postdoctoral Application Research Program of Qingdao City (Grant No.861605040024); the Outstanding Young Teacher Cultivation Program. The fundings and supports from the above projects are greatly acknowledged.

Application of COMSOL Multiphysics in Thermal Effect Analysis of Electromagnetic Active Vibration Absorber

Su Hang[1], WengXue-tao[2]

[1]College of Power Engineering, Naval Univ. of Engineering, Wuhan 430033, China
[2]National Key Laboratory on Ship Vibration and Noise, Naval Univ. of Engineering, Wuhan 430033, China)

Abstract. At present, there are some researches in the thermal analysis of electromagnetic absorbers. The heating principle of electromagnetic absorber magnetic circuit is analysed, and the finite element method is used to numerically solve the temperature field in the working process of electromagnetic vibration absorber. The magnetic circuit simulation model of electromagnetic vibration absorber is established in Comsol Multiphysics finite element analysis software. And the grid Division, simulation analysis of the vibration absorber magnetic circuit structure of the internal temperature distribution, you can get the vibration absorber magnetic circuit in the working process of the temperature field of two-dimensional distribution graphics and magnetic circuit structure of different parts of the temperature rise contrast chart. The conclusion provides some theoretical reference for the design and research of electromagnetic active vibration absorber.

1. Introduction

With the improvement of warship technology, the warship and civilian have been developing rapidly, and they have vibration characteristics due to the presence of power equipment. For warships, reducing hull vibration noise can reduce the possibility detected by the other ship sonar [1], and for civilian, mechanical vibration will affect the comfort of passengers, and further more excessive vibration can even cause machine and instrument equipment disordered, hull structure damage and so on. Dynamic Vibration Absorber is a good method to solve these problems. With the rapid development of computer and the application of fast Fourier transform algorithm, the absorber gradually turns to active vibration absorption. Compared with passive vibration absorption technology, active vibration absorption technology has greater flexibility and adaptability.

According to the different actuators, the active vibration absorber can be divided into the following common types: Electromagnetic power absorbers [2], magnetic levitation type power absorbers [3] and piezoelectric power absorbers [4]. The electromagnetic actuator is a kind of more traditional actuator, the frequency range is wide, and it can control the complex periodic vibration better, and its volume and output force characteristic can satisfy the actual engineering demand of the active vibration absorber.

But the modern vibration problem is more and more demanding for the vibration absorber, which requires the dynamic range of absorbers and the power to withstand more, the frequency response is wider and flatter, and the transient response is better. The vibration absorber is a kind of power exchange energy system, its electric energy converts to the output force, but during the conversion process, most of the electric energy in the current flows through the copper coil when the Joule effect is converted into heat, and the rotor and assembly body generate eddy heat. The coil and the moving subassembly are the heat source of the absorber, which transmits heat through different heat transfer modes to the surrounding area. It will cause magnetic clearance, magnetic circuit, copper counterweight and shell components such as temperature rise even caused by high-frequency strong circuit coil and magnet overheating damage and other consequences, and usually, the main cause of damage to the absorber is high temperature that caused the magnet failure. So the research on the thermal effect of the absorber becomes very meaningful, and the conclusion provides a theoretical basis for the design and improvement of the electromagnetic active vibration absorber structure.

2. Introduction of Eddy Current effect

The coil is turned into alternating current, then the coil produces alternating magnetic field. As the conductor in the middle of the coil in the circumferential direction can be equivalent to a circle of circuit, as shown in figure 1, **오류! 참조 원본을 찾을 수 없습니다.**Circuit in the magnetic flux in the constant change, so in the direction of the conductor's circumferential can produce induction electromotive force and induced current, and the direction of electric currents along the circumference of the conductor circle is like a circle of the vortex. Soin the entire conductor of electromagnetic induction and the occurrence of induction current phenomenon is known as eddy current phenomenon

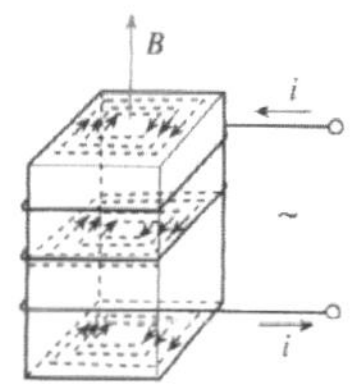

Figure 1. Eddy current effect.

The longer the outer perimeter of the conductor is, the higher the frequency of the alternating magnetic field is, the larger the eddy is.

The eddy currents inside the conductor also generate heat, in line with Joule's law. If the conductor's resistivity is small, then the eddy current is very strong, the heat generated is very big. The thermal effect of the current can increase the temperature of the iron core and the insulating material will be aged easily.

In the process of the electromagnetic active vibration absorber designed in this paper, the alternating current in the stator coil which is the main body of the coil assembly is used to generate alternating magnetic field, and the movement of the core of the actuator in the electromagnetic actuator in the transformation magnetic field inevitably has the problem of electric eddy current heating. These transient temperature fields will have a great influence on the operation of the equipment, what's worse, the safety of the equipment will have unfavourable effects.

The purpose of the thermal field analysis is to calculate the temperature distributions of the model at different times under a certain frequency of current.

3. Establishment of thermal field model of electromagnetic absorber

3.1. The required theoretical part

While the vortices are formed between each of two baffles and the rear wall, it indicates that the grids result more vortices structure generation, meaning that the baffles make the flow field more complex. Points *A~E* are the velocity monitoring points during both the simulation of two kinds cavity models, and the velocity of these points are shown in table 1. It obviously shows that the velocity of the slotted hole monitoring points is smaller than the points of the opening hole. The system is solved by the equation of heat transfer equations, and the heat transfer equation is a parabolic partial differential equation, which describes the temperature distributions of specific regions in a given time:

$$\rho C_p \frac{\partial T}{\partial t} - \nabla \cdot k \nabla T = Q(T) \tag{1}$$

Among it, ρ is the density, C_p is the specific heat capacity, k is the thermal conductivity, and Q is induction heating.

The conductivity of copper σ is given by an expression:

$$\sigma = \frac{1}{\rho_0(1+a(T-T_{ref}))} \tag{2}$$

Wherein, ρ_0 is a reference temperature of 293K resistivity, a is the temperature coefficient of resistivity, and T is the actual temperature in the field. The average time of induction heating is given by the following formula:

$$Q = \frac{1}{2}\sigma\left|E\right|^2 \tag{3}$$

Wherein, σ is electrical conductivity and E is electric field strength.

3.2. The establishment of simulation

(1) COMSOL Multiphysics is a finite element analysis software, the so-called finite element method is to divide the continuous model into a finite number of small modules, set up each module of the formula and solve them, then combine them. These modules represent the original model, so we can obtain the approximate solution of the continuous model. The following simulations are performed in Comsol Multiphysics.

Determine the physical field of the simulation: select "Heat Transfer-electromagnetic heat-Induction heating" module. The "Induction heating" interface combines all characteristics of the time harmonic "magnetic field" interface with the "heat Transfer" interface for induction and eddy current heating modeling. This predefined multi-physical field coupling adds electromagnetic power loss as a heat source, and the properties of the electromagnetic material vary with the temperature. This physical field interface is based on the assumption that the magnetic cycle time is shorter than the thermal time scale (adiabatic assumption).

The induction heating interface, a multiple physical field interface, is used to simulate induction heating and eddy current heating, which comprises a magnetic field interface and a solid heat transfer interface. The electromagnetic power loss is defined as a heat source by the preset coupling of multiple physical fields, and the properties of the electromagnetic material vary with the temperature.

Based on different licenses, the two-dimensional support for steady-state and transient simulations. In addition, the Frequency Domain field interface and the steady state solid heat transfer interface coupling, called the frequency domain-steady state, similar also has the frequency domain-instantaneous state and so on.

Constitutive physical Field Interface: The magnetic field interface is used to calculate the magnetic field and the current distribution in the coils, conductors and magnetic bodies. Depending on the license, the two-and three-dimensional support for steady-state, frequency-domain, and time-domain formats. When approaching the steady-state boundary, the frequency and time domain formats become increasingly uncomfortable, and you can extend the frequency range by adding a lower conductivity. The magnetic field interface solves the Maxwell equation using the magnetic vector potential format, and the scalar potential is used as the dependent variable for the coil.

Solid heat transfer is used to simulate heat conduction, convection and radiation. By default, solid heat transfer is activated in all domains, and it can contain other domain types, such as fluid fields. The temperature equation defined in the solid field corresponds to the differential form of the Fourier law and can include other contributions such as heat sources.

(2) Select the preset study of the Physical interface: "Frequency domain-instantaneous state". The frequency domain-transient study is used to calculate temperature variation over time, including the distribution of electromagnetic field in frequency domain. It is used in the model of physical field interface including induction heat, microwave heating, laser heating, inductively coupled plasma and

microwave plasma. For the plasma physical field interface, the temperature represents the electron temperature.

(3) Model establishment: In the research, focusing on the magnetic circuit part of the heat transfer, the rest of the part is not involved in the analysis, so the model only gives the magnetic circuit part. In Comsol Multiphysics, a two-dimensional axisymmetric geometrical model of the magnetic Circuit section shown in figure 3 is established, and the dimensions and specifications are determined according to the kind.

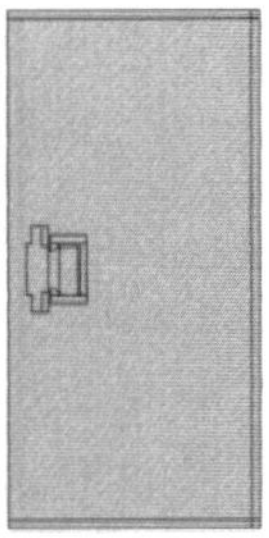

Figure 2. Two-dimensional axisymmetric model of magnetic circuit of electromagnetic absorber.

(4) Add material: According to the physical model of the addition of materials, the main materials are: the addition of circuit board materials to the model, copper coil, air, iron core and coil assembly body. The relative permeability, conductivity and relative permittivity are determined.

(5) Increase physical Field

1) In the “Field” field, add the “Ampere’s Law”: Maxwell-Ampere's law formula is as follows:

$$\nabla \times H - \frac{\partial D}{\partial t} = J \tag{4}$$

The amount in the left side of the equation is the physical quantity describing the electromagnetic field, where D is the electric displacement vector $(C \cdot m^{-2})$, H is the magnetic field strength $(A \cdot m^{-1})$, and the right is the source of the excitation electromagnetic field, where J is the current strength $(A \cdot m^{-2})$.

2) Add "multi-turn coil" in “Field” field: Multi-turn coils satisfy the following hypothetical equations

$$J_e = \frac{NI_{coil}}{A} e_{coil} \tag{5}$$

wherein, J_e is the external current, N is the coil turn number, I_{coil} is the coil current, A is the coil conductor cross-sectional area, and e_{coil} is the planar outside unit vector.

(6) Grid division: In the model, due to the different boundary conditions, the process and complexity of each component are different, and in the interior of the magnetic circuit due to different parts, different locations of the structural characteristics and various needs of the solution (such as in small size and the critical position of two regions), the tiny monocular divided is also different. As shown in figure 2, where the structure is simpler and the local change scope is small, the grid is larger, and the grid is smaller where the structure is more complex and the adjacent position varies greatly more meticulous [5].

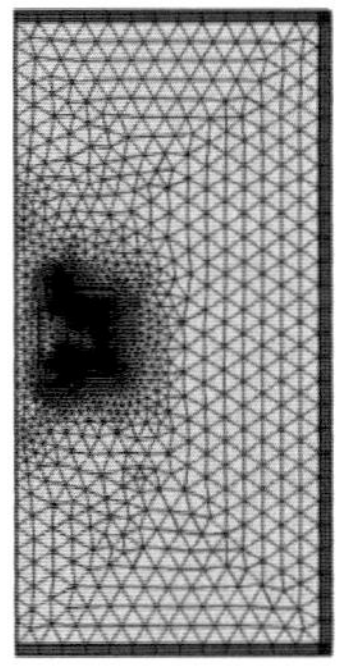

Figure 3. Part Grid division of active vibration absorber magnetic circuit.

4. Solution of partial thermal analysis of magnetic circuit

The principle of magnetic circuit analysis has been described in front of the electromagnetic absorber a constant power, heat generation in the iron core caused by eddy current effect in alternating magnetic field, which immediately diffuses through the shell, air and other heat conduction. In the study, the thermal distribution of the electromagnetic actuator is effectively analyzed in the magnetic circuit structure.

In the analysis process, the iron core is the only heat source, heat transfer process contains solid heat transfer and radiation heat transfer. Over time, the temperature of the core gradually increased, heat was diffused into the magnetic circuit system by thermal radiation. Then heat transfer in solid is achieved in the magnetic circuit system. So temperature of the whole system gradually increased. The analysis method of this paper chooses transient thermodynamic analysis, and the general equation of analysis is:

$$[C]\{\dot{T}\}+[K]\{T\}=\{Q\} \tag{6}$$

In the formula: $[K]$ is the coefficient matrix, including thermal conductivity coefficient, thermal convection coefficient and thermal radiation coefficient and shape coefficient, $[C]$ is the specific heat matrix, considering the increase of energy in the system, $\{T\}$ represents the temperature vector of each node, $\{\dot{T}\}$ is the derivative of the temperature to time of the node, and $\{Q\}$ is the heat flux vector of the node, including thermal generation.

Table of material parameters used in the study:

Table 1. Material parameters.

Part	Relative permeability	Conductivity (S/m)	Atmospheric heat Capacity (J/(kg*K))	Density (kg/m^3)	Coefficient of thermal Conductivity (W/(m*k))
Copper coil	1	5.998e7	385	8960	400

The transient solution is done by setting the required research time, time step and frequency.

The results of the analysis are as follows: The transient analysis by the coupling of multiple physical fields. Figure 4 and figure 5 are the three-dimensional temperature distribution chart of the magnetic circuit structure in 1800s and 3600s respectively, which intuitively indicates that the internal temperature of electromagnetic active vibration absorber is different, as the temperature rises. According to figure 6 and figure 7respectively, 1800s, 3600s magnetic circuit structure of the two-dimensional contour distribution of the temperature, the temperature of iron core is the highest, and the outer surface is the room temperature 293K (20$^\circ C$). From the graph and the legend, it can be seen that there are obvious differences in the electromagnetic active vibration absorber's different parts. Figure 8 is the temperature rise curve of iron core over time.

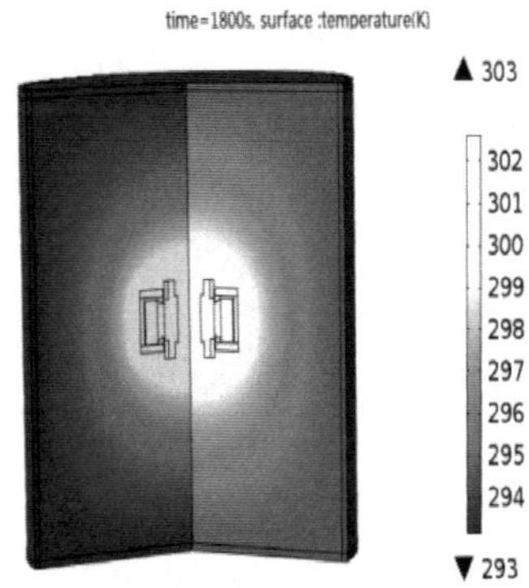

Figure 4. Three-dimensional temperature diagram of posterior magnetic circuit structure after 1800s.

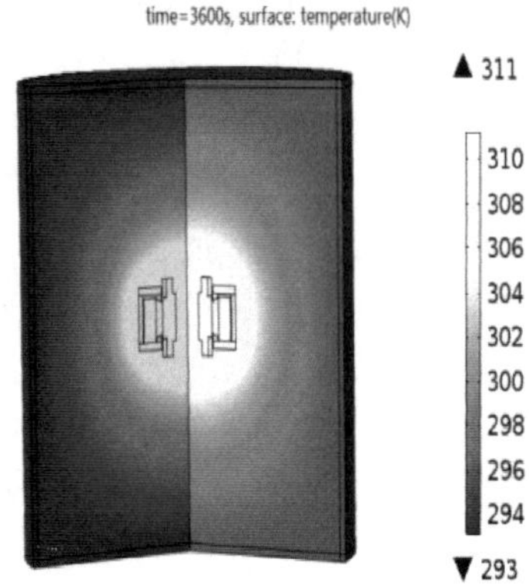

Figure 5. Three-dimensional temperature diagram of posterior magnetic circuit structure after 3600s.

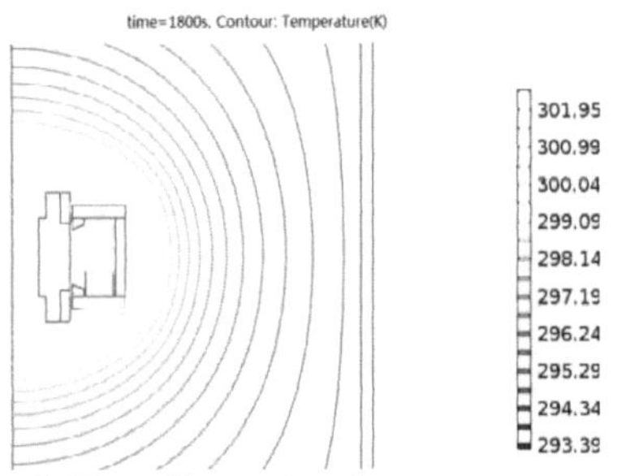

Figure 6. Two-dimensional temperature posterior magnetic circuit structure after 1800s.

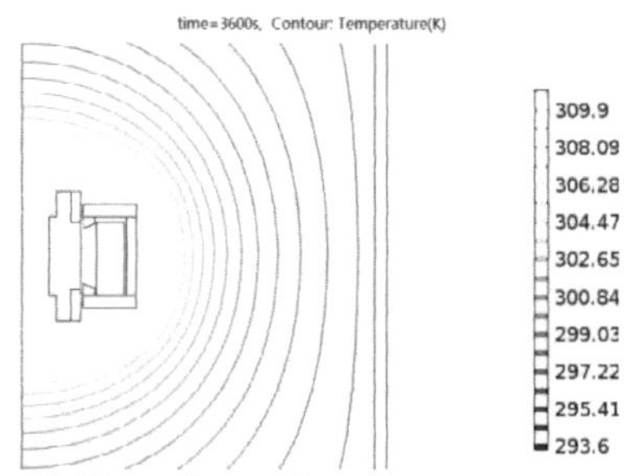

Figure 7. Two-dimensional temperature posterior magnetic circuit structure after 3600s.

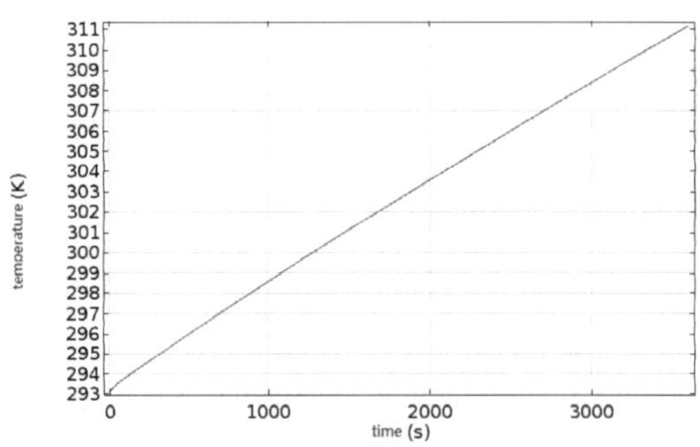

Figure 8. Iron Core Heating diagram.

From figure 4 to figure 8, the current in the simulation is 2 A and the frequency is 40Hz. It can be seen from the graph that the temperature increases gradually as the working time increases, and the temperature grows slowly and tends to stability. When the active vibration absorber continues to work up to 3600s, the electromagnetic part of the heating temperature up to 311K (38°C), when the temperature exceeds 150 degrees, permanent magnet will be demagnetized, and coil insulation layer will melt, so the vibration absorber continued to work 3600s still can maintain a very good working condition.

IOP Conf. Series: Materials Science and Engineering **269** (2017) 012037 doi:10.1088/1757-899X/269/1/012037

5. Conclusion

By using the induction heating module in the heat transfer of COMSOL Multiphysics, the thermal effect of the magnetic circuit structure is analyzed and the temperature distribution of different components inside the magnetic circuit structure is obtained. Iron core has the highest temperature, followed by the copper coil. The results require us to consider temperature range when selecting the active vibration absorber material. So that it can get rid of adverse effects on the working state of components caused by too high temperature and burns and other hazards on the operators. The longer the working time, the higher the temperature, so it is necessary to consider the process of cooling the active vibration absorber in the subsequent design process, and create a good working environment. The study of the thermal effect of the magnetic structure of the active vibration absorber provides some reference for the material selection in the production, and has some reference value for the research and development of the active vibration absorber.

Reference

[1] Duan X S 2010 *Dynamic modeling and self-convergent control of magnetic levitation isolator* University of Science and Technology of China.
[2] Zhang H T, Li W Y and Liu Z G 2001 J. Vibr. Eng. **14(1)** 113-7.
[3] Liu G D, Ma G L and She L H 2003 *Nois. Vibr. Contr.* **23(6)** 18-20.
[4] Jalili N and Knowles W D 2004 *Smart Mater. Struct.* **13** 998-1005.
[5] Chen K Y, Li B Q and Yan M X 2014 *Light Ind. Tech.* **9** 59-60.

IOP Conf. Series: Materials Science and Engineering **269** (2017) 012038 doi:10.1088/1757-899X/269/1/012038

The Viscous Dissipation Effect in a Regular Polygonal Duct for H2 Boundary Conditions

T J Teleszewski[1]

[1]Department of HVAC Engineering, Faculty of Civil and Environmental Engineering, Bialystok University of Technology, Wiejska 45A, 15-351 Bialystok, Poland

Abstract. A numerical investigation based on the Boundary Element Method (BEM) was carried out to solve laminar forced convection with viscous dissipation in a straight regular polygonal duct with H2 boundary conditions. The axial heat conduction in the fluid is neglected. The effects of the Brinkman number and Nusselt number for both the wall heating case and the wall cooling case are considered. Nusselt numbers are obtained as a function of the number of sides of a regular polygonal duct and of the Brinkman number. The singular points of the Brinkman number and the Brinkman number for a limited Nusselt number when the total heat flux reached the heat flux that is generated internally by viscous dissipation processes are obtained.

1. Introduction

The flow of Newtonian fluids and heat transfer through a straight non-circular duct has been studied by many researchers in recent years. Laminar forced convection in a duct with constant axial and peripheral wall heat flux (H2 boundary conditions) has been found in many different systems and applications for very low conductive materials, such as electrical resistance heating, nuclear heating and flow through micro-channels [1]. In particular, laminar forced convection heat transfer in ducts with H2 boundary conditions has been studied by neglecting the effects of viscous dissipation [1]. Wang [2] presented results for H2 forced convection in rounded rectangular ducts. Etemad et al. [3], using the Galerkin Finite Element Method solved the steady laminar flow and H2 heat transfer of non-Newtonian fluids in equilateral triangular ducts. Ray and Misra [4] analysed the evaluation of pressure drop and heat transfer characteristics of laminar forced convection through square and equilateral triangular ducts with rounded corners, for H2 boundary conditions. Gao and Hartnett [5] using a Finite Difference Method presented a numerical solution for fully developed laminar flow forced convection of a power law non-Newtonian fluid in a rectangular duct under different combinations of H2 boundary conditions with adiabatic walls. Morini and Spiga [6] studied the velocity and temperature distribution for Newtonian fluid laminar flow and forced convection heat transfer in rectangular ducts for modified H2 thermal boundary conditions for eight versions involving different combinations of heated and adiabatic walls. Shahmardan et al. [7] proposed an analytical solution for convective heat transfer in rectangular ducts for H2 boundary conditions. Dharaiya and Kandlikar [8] performed a numerical analysis of laminar forced convection in rectangular microchannels subject to H2 boundary conditions. Wang [9], using an analytical method, calculated the Nusselt number for H2 heat transfer in rectangular ducts with large aspect ratios. Wang [9] found the limiting value of the Nusselt number for H2 boundary conditions approaches 2.9162 as the aspect ratio tends to infinity. Chung and Zhang [10] numerically determined the Nusselt number from the thermally developing flow of non-Newtonian fluids in rectangular ducts under H2 boundary conditions. Turgut [11] numerically investigated laminar flow and heat transfer in smooth hexagonal ducts subject to H2 boundary conditions. Shahsavari et al. [12] presented an analytical solution for H2 heat transfer in microchannels of hyperelliptical and regular polygonal cross sections for flow with a negligible viscous dissipation effect.

The effects of viscous dissipation cannot by neglect for some systems, such as flow through ducts, where the viscosity is large or flow in microchannels. In the literature, some analyses of the effects of viscous dissipation of laminar forced convection in channels with H2 boundary conditions are available. Barletta et al. [13] studied the effect of viscous dissipation for laminar flow in stadium-shaped ducts with H2 boundary conditions. Sheikhalipour and Abbassi [14] investigated the viscous dissipation effects in trapezoidal microchannels under H2 boundary conditions using the Finite Difference Method. Rij et al. [15] studied the frictional and convective heat transfer characteristics of rarified flows in rectangular microchannels with H2 boundary conditions. Sayed-Ahmed et al. [16] analysed the Graetz problem for a Bingham plastic fluid in laminar tube flow for H2 boundary conditions by taking account of viscous dissipation.

The aim of this study is analysis of the viscous dissipation effects on the laminar forced convection in regular polygonal channels subject to H2 thermal boundary conditions. Teleszewski and Sorko [17] provide a compact relationship for the Nusselt and Brinkman numbers for different values of sides in regular polygonal channels under H1 boundary conditions with laminar viscous dissipation. No compact solutions were found for the Nusselt number in the H2 case with viscous dissipation effects. The influence of the Brinkman number on the Nusselt number for different values of sides of a regular polygonal duct is obtained for the H2 thermal boundary conditions.

2. Mathematical model

The present analysis is based on the fully developed, laminar, steady, incompressible flow with a constant dynamic viscosity μ and constant thermal conductivity k. Under the aforementioned assumptions, the continuity (1), momentum (2) and energy (3) equations of a Newtonian fluid in a straight duct with uniform heat flux are given as the form:

$$\frac{\partial \mathrm{u}_z}{\partial z} = 0 \tag{1}$$

$$\mu\left(\frac{\partial^2 \mathrm{u}_z}{\partial x^2} + \frac{\partial^2 \mathrm{u}_z}{\partial y^2}\right) = \frac{\partial p}{\partial z}, \qquad \frac{\partial p}{\partial x} = \frac{\partial p}{\partial y} = 0 \tag{2}$$

$$k\nabla^2 T = \rho c_p \frac{\partial T}{\partial z}\mathrm{u}_z(x,y) - \Phi(x,y), \qquad \Phi(x,y) = -\mu\left[\left(\frac{\partial \mathrm{u}_z}{\partial x}\right)^2 + \left(\frac{\partial \mathrm{u}_z}{\partial y}\right)^2\right] \tag{3}$$

where u_z is the axial velocity, p is the pressure and Φ is the viscous dissipation therm. Figure 1 shows the duct configuration and coordinate system.

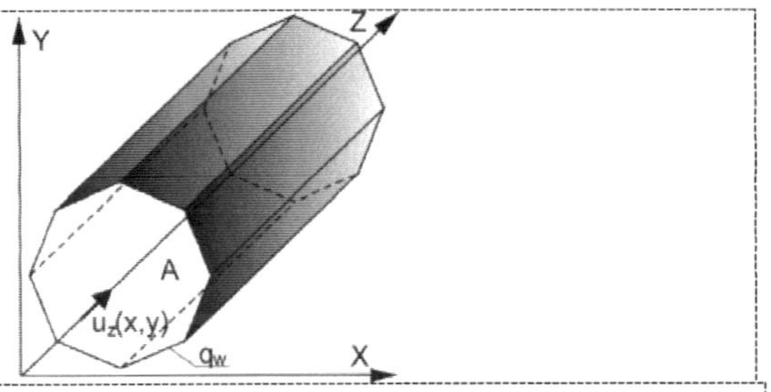

Figure 1. Geometry of the regular polygonal duct.

In the momentum equation (2), the fully developed velocity profile $u_z(x,y)$ is obtained from the boundary element method (BEM) with no-slip conditions on the duct wall u_z=0. The BEM method was also employed for the energy equation (3) with H2 boundary conditions. The H2 thermal boundary conditions are defined as uniform peripheral wall heat flux (q_w=const) with constant axial wall heat flux. The BEM numerical method is used to solve the momentum (2) and energy equations (3) with 1000 constant elements and 40602 triangular cells. To determine the relative differences between the numerical and analytical values, the numerical results obtained by BEM were compared to the analytical values of the Nusselt number with the Brinkman number equal to zero for a circular tube and square duct. For 1000 constant boundary elements and 40602 triangular cells in the duct cross-section, the maximum relative differences in the Nusselt number were found to be ΔNu=0.005%.

3. Results and discussion

The Brinkman number is commonly used to describe the ratio between the heats generated by viscous forces to the heat exchanged by conduction at the walls. The Brinkman referred to the wall heat flux density q_w is given by

$$\mathrm{Br}_q = \frac{\mu u_m^2}{D_h q_w}, \qquad \mathrm{u}_m = \frac{1}{A}\iint_A u_z(x,y)dA, \qquad D_h = \frac{4A}{L} \tag{4}$$

where u_m is the mean velocity, L is the wetted perimeter of the cross-section, D_h is the hydraulic diameter and A is the cross-sectional area. The wall heat transfer q_w is positive wall heat transfer, indicating that the fluid is heated by the wall and q_w is negative, showing that the fluid is cooled by the wall. The Nusselt number is obtained as

$$Nu = \frac{q_w D_h}{k\left(T_w - T_b\right)}, \qquad T_b = \frac{1}{A\mathrm{u}_m}\iint_A \mathrm{u}_z(x,y)T(x,y)dA \tag{5}$$

where T_b is a bulk temperature.

In the literature [18-21], the relationship between the Brinkman number and the Nusselt number in flows through a straight duct is described by the formula:

$$Nu = \frac{a_1}{a_2 + a_3 Br_q} \tag{6}$$

where a_1, a_2, a_3 are coefficients that depend on the geometrical shape of the cross sections. Similarly, the Nusselt's relation to the Brinkman number is proposed in regular polygonal channels subject to H2 conditions

$$Nu = \frac{Nu_0}{1 + \sigma_n Nu_0 Br_q} \tag{7}$$

where Nu_0 is the Nusselt number with H2 boundary conditions on the assumption of negligible effects due to viscous dissipation (Br_q=0) and the shape factor σ_n. Nu_0 and σ_n depend on the number of n-sided regular polygonal ducts. For a circular pipe the analytical value of σ_n is equal to 1 [21] and the fully developed average Nusselt number Nu_0 is equal to 48/11 [21]. The Nusselt number Nu_0 for H2 boundary conditions, the shape factor σ_n for regular polygonal ducts with the correlation coefficient $R^2 \approx 1$, is proposed:

$$Nu_0(n), \sigma_n(n) = \frac{a + bn + cn^2 + dn^3}{e + fn + gn^2 + hn^3} \tag{8}$$

where a, b, c, d, e, f, g, h coefficients are reported in Table 1. Note that the coefficients of shape factor σ_n are calculated in recent paper [17].

Table 1. Coefficients of the Eq. (8) to determine the Nusselt Nu_0, and shape coefficients in regular polygonal ducts.

Coefficients of the Eq. (45)	a	b	c	d	e	f	g	h
$\mathbf{Nu_{0\_H2}}$	1.076	-3.7656	1.36446	0.03714	1.000	-0.8052	0.31154	0.00851
σ_n**[17]**	0.9193	-0.6928	0.12666	-0.010517	1.000	-0.7082	0.1277	-0.010518

The total wall heat flux q_w contains the positive thermal heat flux transferred to the walls q_f and negative heat flux generated by viscous dissipation q_{vd}. In fact, when the total heat flux q_w is positive, then its direction is from the wall to the fluid and the fluid is heated by the wall, otherwise negative total heat flux goes from the fluid to the wall and the fluid is cooled by the wall. Figure 2 shows the variation of the Nusselt with the Brinkman number for different n side numbers of polygonal ducts: n=3 (triangular shape), n=4 (square shape), n=5 (pentagon shape) and n=1000 (circular tube) for H2 boundary conditions. The trend of the Nusselt number and the Brinkman number with the number of sides of a regular polygonal duct in the case of the boundary condition H2 is similar to that of the H1 condition [17]. When the heat flux generated by viscous dissipation and the Brinkman number equal zero ($q_f/q_w=1$), then the Nusselt number corresponds to the fully developed flow in ducts without viscous dissipation and $Nu=Nu_0$. When the Brinkman number goes from zero to infinity, then the total wall heat flux, the Brinkman number, the Nusselt number and the temperature difference between the wall temperature and the fluid bulk temperature are positive. In this case, the value of the Nusselt number decreases with the increase of the Brinkman number and the Nusselt number goes to zero because the total wall heat flux goes to zero. Zanchini [21] proved that, whenever the wall heat flux tends to zero, the asymptotic Nusselt number is zero. After this point total heat flux changes its sign from positive to negative and the Nusselt number decreases with the increase in the Brinkman number. When the temperature difference between wall temperature and fluid bulk temperature changes its sign from positive to negative, then the singularity point is observed as shown in figure 2. In fact, at the singularity point the power generated by the shear rate is balanced with the power transferred to the wall [19, 20]. The singular points of the Brinkman number for H2 boundary conditions can be obtained by the general solutions (7): $Br_{qs}=-1/(\sigma_n Nu_0)$. The function of the singular points of the Brinkman number versus the number of sides of a regular polygonal duct ($3 \leq n \leq 16$) with H2 boundary conditions are reported in figure 3. A similar increasing trend is observed for singular points of the Brinkman number versus number of sides of a regular polygonal duct in the case of the boundary condition H2, as well as for the H1 boundary conditions [17].

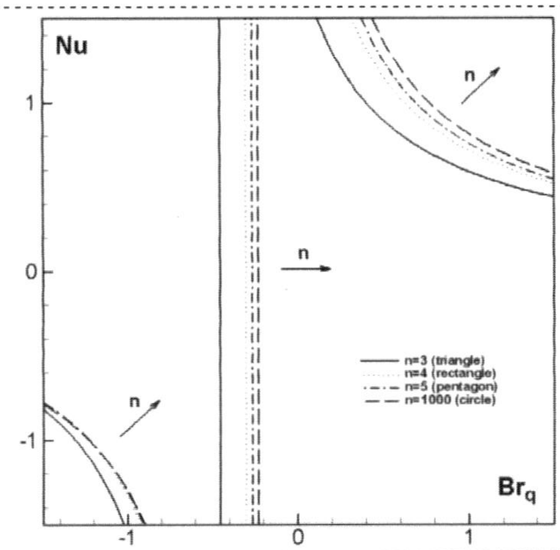

Figure 2. Nusselt numbers versus the Brinkman for different n side numbers of polygonal ducts for H2 boundary conditions.

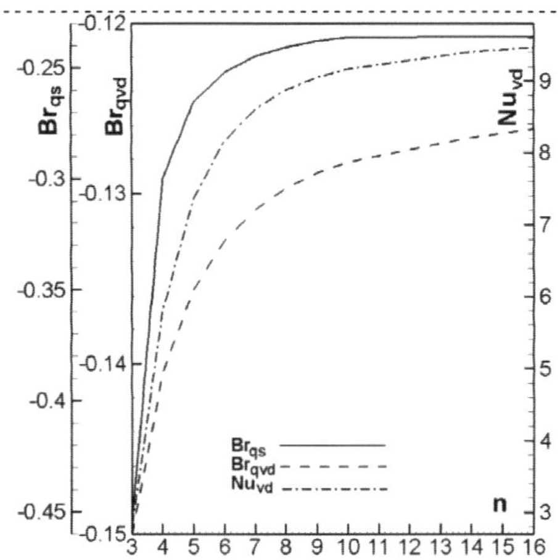

Figure 3. Variation of Br_{qs}, Br_{qvd}, Nu_{vd} with n for H2 boundary conditions.

After the singularity point, the Nusselt number decreases with the increase in the Brinkman number until the total heat flux reaches the heat flux generated internally by viscous dissipation processes and the heat flux supplied by the wall into the fluid approaches zero. For a circular pipe with H1 boundary conditions the analytical value of a limited Brinkman number Br_{qvd} is equal to 1/8 [21] and the limited Nusselt number Nu_{qvd} is equal to 9.6 [22]. For a circular tube with H2 boundary conditions, this value is also the same. When $Br=Br_{qvd}$ and $Nu=Nu_{qvd}$ then the ratio of heat flux generated by viscous dissipation to the total wall heat flux is equal to one ($q_{vd}/q_w=1$). In figure 3, the limited Brinkman number Br_{qvd} and the limited Nusselt number Nu_{vd} versus the number of sides of a regular polygonal duct ($3 \leq n \leq 16$) are given for H2 boundary conditions. The Brinkman number in singularity points, the limited Brinkman and Nusselt number increase in the range of $3 \leq n < \infty$ for polygonal ducts, with the increasing value of n.

4. Conclusions

The BEM method has been used to numerically solve the momentum and the energy equations for the laminar viscous dissipation of flow through a straight regular polygonal duct with H2 boundary conditions under the assumption that the axial condition in the fluid is neglected. The effects of the Brinkman number, the number of sides of a regular polygonal duct and the Nusselt number are studied. A compact correlation between the Nusselt number and the Brinkman number has been proposed for regular polygons with a different number of sides. The singular points of the Brinkman number and the Brinkman number for a limited Nusselt number when the total heat flux reached heat flux is generated by internally by viscous dissipation processes and the heat flux supplied by wall into the fluid approaching zero is obtained. The value of the Brinkman number in a singularity, and the Brinkman number for a limited Nusselt number increased with the number of sides increases for H2 boundary conditions.

References

[1] Shah R K and London A L 1978 *Laminar Flow Forced Convection in Ducts A Source Book for Compact Heat Exchanger Analytical Data* (New York: Academic Press).
[2] Wang C Y. 2014 *J. Thermophys. Heat Transf.* **28(4)** 811.
[3] Etemad S Gh, Mujumdar A S and Nassef R 1996 *Appl. Math. Model.* **20(12)** 898.
[4] Ray S and Misra D 2010 *Int. J. Therm. Sci.* **49(9)** 1763.
[5] Gao S X and Hartnett J P 1992 *Int. Commun. Heat Mass. Transf.* **19(5)** 673.
[6] Spiga M and Morini G L 1996 *Int. J. Heat Mass Transf.* **39(6)** 1165.
[7] Shahmardan M M, Norouzi M, Kayhani M H and Amiri Delouei 2012 *J. Zhejiang Univ. Sci. A* **13(10)** 768.
[8] Dharaiya V V and Kandlikar S G 2012 *Trans. ASME, J. Heat Transf.* **134.**
[9] Wang C Y 2014 *J. Heat Transf.* **136(7)**.
[10] Chung B T F and Zhang Z J 1994 *Trans. Eng. Sci.* **5** 183.
[11] Turgut O 2014 *Trans. Mech. Eng.* **38(M1)** 45.
[12] Shahsavari S, Tamayol A, Kjeang E and Bahrami M 2012 *ASME J. Heat Transf.* **104** 091701-10.
[13] Barletta A, Rossie di Schio E, Hahne E and Spindler K 2001 *Int. Commun. Heat Mass Transf.* **28(4)** 449.
[14] Sheikhalipour T and Abbassi A 2009 *2nd Micro and Nano Flows Conference* (West London UK Brunel University Research Archive) p 1.
[15] Rij van J, Ameel T and Harman T 2009 *Int. J. Heat Mass Transf.* **52** 2792.
[16] Sayed-Ahmed M E, Saleh H K and Hamdy I 2013 *Appl. Math.* **3(1)** 27.
[17] Teleszewski T J and Sorko S A 2016 *Proc. Eng.* **157** 171.
[18] Morini G L and Spiga M 1999 *J. Heat Transf.* **121** 1083.
[19] Aydin O 2005 *Energ. Convers. Manag.* **46** 757.
[20] Avci M and Aydin O A 2006 *C. R. Mecaniq.* **324** 164.
[21] Zanchini E 1997 *Int. J. Heat Mass Transf.* **40(1)** 169.
[22] Basu T and Roy D N 1985 *Int. J. Heat Mass Transf.* **28(3)** 699.

Flexural Vibration Band Gap Characteristic of Double Periodic Structure Beams

Ning Ronghui[1], **Zhang Zhenhai**[2]

[1]College of Power Engineering, Naval Univ. of Engineering, Wuhan 430033, China
[2]National Key Laboratory on Ship Vibration and Noise, Wuhan 430033, China

Abstract. A double-periodic structural beam is designed. The complex band structure of the structure is deduced by spectral element method. The influence of the length of the third element on the band gap characteristics of the double-periodic structural beam is studied. The results show that the double-periodic structure beams can reduce the band gap frequency, but it will affect the bandwidth of the band gap while maintaining the structural characteristics of the mechanical structure. The length of the third element has little effect on the centre frequency of each band gap, but it will affect the band gap bandwidth.

1. Introduction

The periodic structure has the characteristic of band gap, which has wide application prospect in vibration reduction, sound filtering and new sensor [1]. Therefore, the band gap characteristics in periodic structures have been a hotspot and difficult. There are many researches on One-dimensional periodic structure [2], since the preparation is simple, and the effect is obvious. However, most of the researches are on the general binary cycle structure [3]. There are also ternary periodic structure [4] researches. These researches have great significance to improve the periodic structure of the band gap characteristics and apply the band gap to vibration and noise reduction [5]. However, since the composition of the binary periodic structure is simple, change the band gap characteristics mainly by changing the length of the cell, the use of new materials [6]. Although these methods are very effective, but in changing the length of the cell, it will obviously have a greater impact on the mechanical properties of the structure [7]. And the use of new materials, on the one hand will make the design cost increases, on the other hand subject to material constraints [8]. Many designs still need to further change the length of the cell so that the structure of the band gap to meet the design requirements, so these two design methods are very effective, but the actual use of relatively difficult.

On the basis of the binary periodic structure beams, the five-element double-periodic structural beam is designed in this paper. On the one hand, it can ensure that the mechanical properties of the double-periodic structural beams are similar to those of the binary periodic structure. On the other hand, the length of the cell of the five-element periodic structure beam changes will have an impact on the structure of the band gap. And because of the increase in the third component, the mechanical properties and band gap characteristics of the beam are enriched, and there is more tunability in vibration and noise reduction.

2. Band gap calculation theory of double periodic structures

As shown in figure 1, the double periodic structural beam [9] is a periodic structure composed of AB-AB-C cells. There is an internal period AB-AB structure in the cell, so it is called a double-periodic structure. Compared with the single cycle, the double periodic structure introduces the third component C, which makes the structure have more adjustable space and makes it easy to adjust the band gap of the periodic structure. In a single cell, the length of the beams of component A is L_1, component B is L_2 and component C is L_3. The length of the beam of each component is much greater than the width and height of the beam. The beam is satisfied with the condition of Bernoulli-Euler beam. The structural model of the beam is established by the spectral element method. Each component beam can be regarded as a spectral unit, and the single cell can be regarded as the five spectral elements.

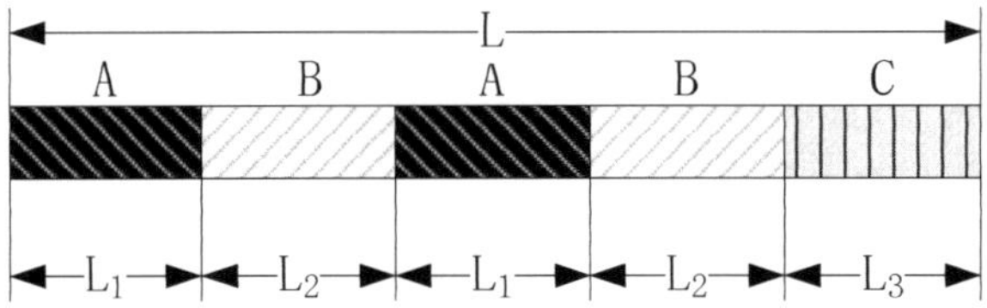

Figure 1. Cell schematic diagram of five-element double periodic structure.

2.1. Connection of spectral unit

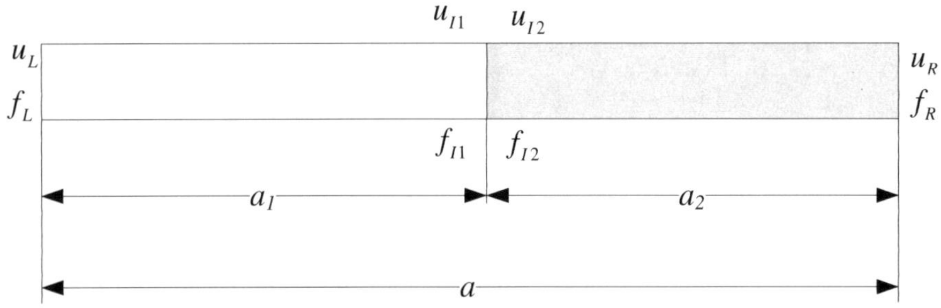

Figure 2. single cell diagram of periodic beam structure.

As shown in figure 2, the beam a is composed of beam a_1 and beam a_2, which are connected Head-to-tail. Each beam can be represented by a spectral unit. The force at the left and right end of beam a_1 are f_L and f_{I1}. The displacement at the left and right end of beam a_1 are u_L and u_{I1}. The force at the left and right end of beam a_2 are f_{I2} and f_R. The displacement at the left and right end of beam a_2 are u_{I2} and u_R. The dynamic equation of the beam a_1 is:

$$\begin{Bmatrix} \boldsymbol{f}_L \\ \boldsymbol{f}_{I1} \end{Bmatrix} = \begin{bmatrix} \hat{\boldsymbol{k}}_{11}(a_1) & \hat{\boldsymbol{k}}_{12}(a_1) \\ \hat{\boldsymbol{k}}_{21}(a_1) & \hat{\boldsymbol{k}}_{22}(a_1) \end{bmatrix} \begin{Bmatrix} \boldsymbol{u}_L \\ \boldsymbol{u}_{I1} \end{Bmatrix} \tag{1}$$

For the flexural vibration of the beams, the force vector contains two components of shear force and bending moment, and the displacement vector includes two components of lateral displacement and section angle. As a result the component of the stiffness matrix $\hat{\boldsymbol{k}}_{11}(a_1)$ in the kinetic equation is a second order matrix.

Similarly, the dynamic equation of beam a_2 is:

$$\begin{Bmatrix} \boldsymbol{f}_{I2} \\ \boldsymbol{f}_R \end{Bmatrix} = \begin{bmatrix} \hat{\boldsymbol{k}}_{11}(a_2) & \hat{\boldsymbol{k}}_{12}(a_2) \\ \hat{\boldsymbol{k}}_{21}(a_2) & \hat{\boldsymbol{k}}_{22}(a_2) \end{bmatrix} \begin{Bmatrix} \boldsymbol{u}_{I2} \\ \boldsymbol{u}_R \end{Bmatrix} \tag{2}$$

By the continuous condition of displacement, it is known that $\boldsymbol{u}_{I1} = \boldsymbol{u}_{I2} = \boldsymbol{u}_I$ and $\boldsymbol{f}_{I1} = -\boldsymbol{f}_{I2}$.

Then, the dynamic equation of the cell can be written as follows:

$$\begin{Bmatrix} f_L \\ 0 \\ f_R \end{Bmatrix} = \begin{bmatrix} \hat{k}_{11}(a_1) & \hat{k}_{12}(a_1) & 0 \\ \hat{k}_{21}(a_1) & \hat{k}_{22}(a_1)+\hat{k}_{11}(a_2) & \hat{k}_{12}(a_2) \\ 0 & \hat{k}_{21}(a_2) & \hat{k}_{22}(a_2) \end{bmatrix} \begin{Bmatrix} u_L \\ u_1 \\ u_R \end{Bmatrix} \tag{3}$$

Equation (3) can be simplified to get only the left and right ends of the displacement and force of the expression as follows:

$$\begin{Bmatrix} f_L \\ f_R \end{Bmatrix} = \begin{bmatrix} \hat{k}_{11} & \hat{k}_{12} \\ \hat{k}_{21} & \hat{k}_{22} \end{bmatrix} \begin{Bmatrix} u_L \\ u_R \end{Bmatrix} \tag{4}$$

where:

$$\begin{aligned} \hat{k}_{11} &= \hat{k}_{11}(a_1) - \hat{k}_{12}(a_1)\left(\hat{k}_{22}(a_1)+\hat{k}_{11}(a_2)\right)^{-1}\hat{k}_{21}(a_1) \\ \hat{k}_{12} &= -\hat{k}_{12}(a_1)\left(\hat{k}_{22}(a_1)+\hat{k}_{11}(a_2)\right)^{-1}\hat{k}_{12}(a_2) \\ \hat{k}_{21} &= -\hat{k}_{21}(a_2)\left(\hat{k}_{22}(a_1)+\hat{k}_{11}(a_2)\right)^{-1}\hat{k}_{21}(a_1) \\ \hat{k}_{22} &= \hat{k}_{22}(a_2) - \hat{k}_{21}(a_2)\left(\hat{k}_{22}(a_1)+\hat{k}_{11}(a_2)\right)^{-1}\hat{k}_{12}(a_2) \end{aligned} \tag{5}$$

Contrary to equations (1) and (4), it is clear that the dynamic equation after the connection of the two spectral elements is exactly the same as the form of the dynamic equation for a single spectral element. Similarly, the dynamic stiffness equations of a single spectral unit can be obtained by satisfying any of the spectral unit equations connected to each successive sequence under continuous conditions, and the dynamic stiffness matrix after connection can be obtained from the dynamic stiffness of the connected spectral units.

2.2. Band gap characteristics calculation of double periodic structures

According to the above reasoning, we can see that although the cell of double periodic structure contains three components of five components. But by using spectral unit analysis method, the final dynamic equation is as shown:

$$\begin{Bmatrix} f_L \\ f_R \end{Bmatrix} = \begin{bmatrix} \hat{k}_{11} & \hat{k}_{12} \\ \hat{k}_{21} & \hat{k}_{22} \end{bmatrix} \begin{Bmatrix} u_L \\ u_R \end{Bmatrix} \tag{6}$$

According to the periodic structure of the Bloch theorem [10], the boundary displacement and force vectors of the cells satisfy the following relationship:

$$u_R = e^{-iqL} u_L, \quad f_R = -e^{-iqL} f_L \tag{7}$$

Combing equation (6) with (7), we can get the characteristic equation with displacement only:

$$\left(\begin{bmatrix} \hat{k}_{21} & \hat{k}_{22} \\ 0 & I \end{bmatrix} - e^{-iqL} \begin{bmatrix} -\hat{k}_{11} & -\hat{k}_{12} \\ I & 0 \end{bmatrix} \right) \begin{Bmatrix} u_L \\ u_R \end{Bmatrix} = \begin{Bmatrix} 0 \\ 0 \end{Bmatrix} \tag{8}$$

For each frequency point, solving the eigenvalues of equation (11), we can get two pairs of identical Bloch wave solutions $\pm q_1$, $\pm q_2$. The solution of each q represents a kind of intrinsic wave. When the imaginary part of q is not 0, the eigen wave propagates in the process of propagation, so all the frequency points that make the imaginary part of q are not zero is called the band gap of the structure.

3 .Band gap calculation and result analysis

Aiming at the five-element double periodic structure beam, the band gap of the double period structure and the simple binary period structure is compared and the influence of the third component on the double periodic structure band gap is considered.

3.1. Comparison between double periodic and binary period structure

The double periodic structure model is shown in figure 1, the parameters are as follows. The length $L_1 = L_2 = L_3 = 0.1m$. The shape of each component is the same. Section size: $b = 0.01m$, $h = 0.01m$. The material parameters of each component are shown in table 1.

Table 1. Material parameters of component.

Code	Material	Density (kg/m^3)	Elastic modulus (GPa)
A	Aluminum	2799	72
B	Plexiglass	1142	2.0
C	Epoxy resin	1180	4.3

In order to compare with the binary periodic structure, two binary periodic structures are designed. The cell length of the first binary periodic structure is equal to the cell length of the double periodic structure. The length of each element is 0.25 meters, The cross-sectional size is consistent with the bending of the double-periodic structure. The cell composition is aluminum and plexiglass. Because the structure is consistent with the characteristic parameters of the double-periodic structure, it is called the characteristic binary periodic structure in this paper. For the second binary periodic structure, the element length is consistent with that of the double-periodic structure. The geometrical parameters of the double periodic structure are close to each other, and this is called geometric binary periodic structure.

In this paper, the complex band structure is calculated. The complex band structure of the double periodic structure, the characteristics binary periodic structure and the geometric binary periodic structure are shows as figure 3, figure 4and figure 5, respectively. The figure 3 shows that the first three band gap frequency of the double periodic structure ranges are 24Hz-30Hz, 102Hz-140Hz, 262Hz-340Hz, and the figure 4 shows that the first three band gap frequency of the characteristics binary periodic structure are 24Hz-38Hz, 108Hz- 212Hz, 308Hz-464Hz. The figure 5 shows that the first two band gap frequency of the geometric binary periodic structure of the first two band gap frequency range of 150Hz-230Hz, 672Hz-1316Hz.

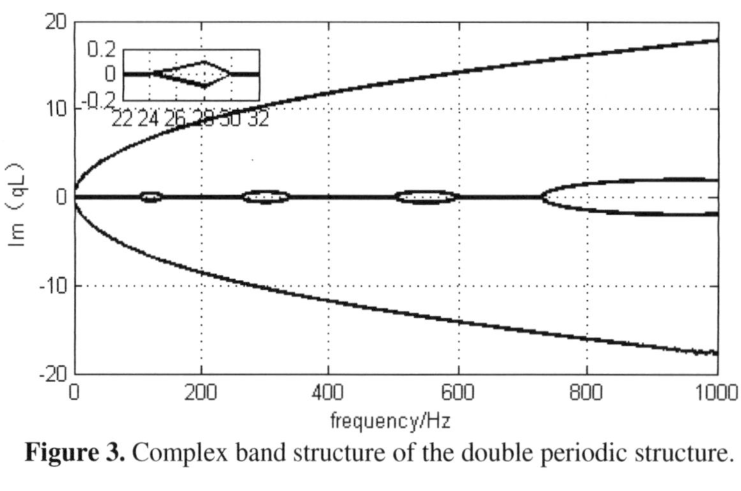

Figure 3. Complex band structure of the double periodic structure.

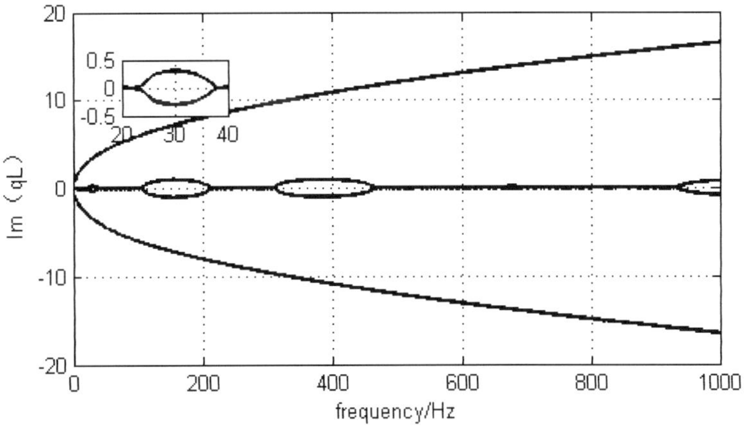

Figure 4. Complex band structure of the characteristics binary periodic structure.

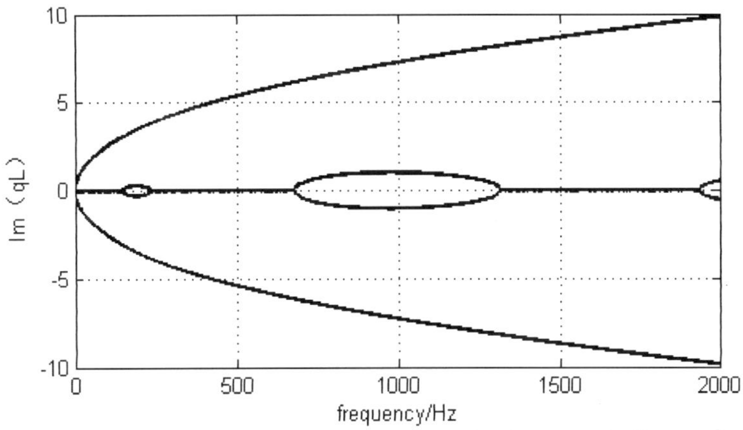

Figure 5. Complex band structure of the geometric binary periodic structure.

Compared with the double periodic structure, the starting frequency of the band gap is slightly lower than the starting frequency of the characteristic binary structure, but the frequency range is narrower. Compared with the double periodic structure and the geometric binary structure, the band gap start frequency of the double periodic structure is obviously reduced in the case of the geometrical feature size of the structure, and the change of the strength and stiffness of the structure due to the small geometric parameters is not large, thus realizing the band gap frequency of the structure without affecting the use function of the structure.

3.2. The influence on the band gap of the third group

For the double periodic structural beam, due to the presence of the third component, the element obviously affects the band gap characteristics of the structure. In this paper, we study the influence of the length of the third element on the gap in the case of a constant cell length. The composition of the calculated model and the cross-section of the structure are consistent with the double periodic structural model in table 2. The length of the cell is 0.5m, and the length of each element of aluminum and plexiglass is equal, the length of the epoxy resin is changed, the first three frequencies obtained band gap in table 1 under different epoxy length is shown.

As it shown in table 2, the frequency band of the first band gap of the double periodic structure is slightly widened with the increase of length of the third element, but the change is not obvious. The starting frequency of the second band gap rises, the termination frequency decreases. The frequency of the third band gap band is reduced, the termination frequency is increased, and the frequency band is widened. For the whole, the center frequency of each band gap does not change much, but the effect of each band gap is changed due to the change in length Therefore, in the actual engineering application, should be based on the actual needs, determine the target attenuation band, select the optimal length of the third component, making the structure to produce the greatest attenuation effect.

Table 2. The first three frequencies obtained band gap under different epoxy length.

Length of epoxy (m)	The first band gap	The second band gap	The third band gap
0.18	24Hz-32Hz	112Hz-128Hz	246Hz-336Hz
0.14	24Hz-32Hz	106Hz-134Hz	250Hz -344Hz
0.1	24Hz-30Hz	104Hz-140Hz	262Hz-340Hz
0.06	24Hz-30Hz	98Hz-148Hz	278Hz -328Hz
0.02	24Hz-28Hz	96Hz-150Hz	292Hz -308Hz

4. Conclusions

Based on the combination of spectral unit theory and Bloch theory, the band gap calculation formula of bi-periodic structural beam is introduced, and the band gap characteristics of the double periodic structural beam are calculated. The periodic structure beam with the same characteristic parameters and the periodic structure beam with several parameters And the effect of the third component on the band gap characteristics is studied. The main conclusions are as follows:

(1)In the case where the cell length is constant, the starting frequency of the band gap is slightly lower than the starting frequency of the characteristic binary structure, but the band is narrower.

(2)In the case of the geometrical feature size of the structure, the band gap start frequency of the double periodic structure is obviously reduced, and the geometrical parameters change little, so that the strength and stiffness of the structure are not changed greatly, the use of the function of the case, reducing the structure of the band gap frequency.

(3) The change of the length of the third element almost does not affect the center frequency of each band gap of the double periodic structural beam, but with the increase of the length, the frequency band of the first band gap of the bi-periodic structure is slightly widened, but the change is not obvious; The starting frequency of the two band gap increases, the termination frequency decreases, the frequency band becomes narrow, the starting frequency of the third band gap band decreases, the termination frequency increases, and the frequency band becomes wider.

References

[1] Cicek A, Salman A, Kaya O A, et al. 2016 *Ultransonics*, **65** 78-86.
[2] Yu D L, Liu Y Z, Qiu J, et al. 2005 *J. Vibr. Shock.* **24(2)** 92-4+110.
[3] Liu J W, Yu D L, Wen J H, et al. 2016 *J. Vibr. Shock,* **35(6)** 141-5.
[4] Hua J, Zhang S and Cheng J C. 2005 *Acta Phys. Sin.* **54(3)** 1261-6.
[5] Richards D and Pines D J. 2003 *J. Sound Vibr.* **264(2)** 317-42.
[6] Qi P S, Du J, Jiang J L, et al. 2016 *Acoust. Tech.* **35(4)** 303-7.
[7] Wen J H, Yu D L, Wang G, et al. 2005 *Chinese J. Mech. Eng.* **41(4)** 1-6.
[8] Chen B and Huang X C. 2005 *Noise Vibr. Control.***5** 37-41.
[9] Li Y F, Shen H J, Zhang L, et al. 2017 *Chinese J. Acoust.* **42(3)** 335-40.
[10] Xiao Y, Wen J H, Yu D L, et al. 2013 *J. Sound Vibr.* **332** 867-93.

A New Type of Tea Baking Machine Based on Pro/E Design

Xin-Ying Lin[1], Wei Wang[1]

[1]Industrial robot measurement and control and mold rapid manufacturing of key laboratories of Fujian province, Minnan University of Science and Technology, Shishi, Fujian, China, 362700

Abstract. In this paper, the production process of wulong tea was discussed, mainly the effect of baking on the quality of tea. The suitable baking temperature of different tea was introduced. Based on Pro/E, a new type of baking machine suitable for wulong tea baking was designed. The working principle, mechanical structure and constant temperature timing intelligent control system of baking machine were expounded. Finally, the characteristics and innovation of new baking machine were discussed.The mechanical structure of this baking machine is more simple and reasonable, and can use the heat of the inlet and outlet, more energy saving and environmental protection. The temperature control part adopts fuzzy PID control, which can improve the accuracy and response speed of temperature control and reduce the dependence of baking operation on skilled experience.

1. Introduction

In the whole process of wulong tea, baking is a very important process after fixing, twisting. The quality of tea baking has a great impact on the quality of tea. Before tea baking, the water content in the tea is relatively large, Chen flavor and Oxidation of polyphenols increased. Through baking, inclusions in the tea for thermal physicochemical reaction under the heat can enhance the quality of tea, such as Aromatic substances spread from the inside out to strengthen the tea. At the same time, the reduction of moisture content can slow down the tea aging rate, easy to store. Baking can tighten the strip, evaporate moisture, develop aroma, and promote the transformation of ingredients in tea. In short, baking has played an important role in promoting the quality of wulong tea. In the finishing process of wulong tea, baking can enhance the fragrance and ripeness, changing the quality of tea to meet the needs of different taste. Tea drying is generally carried out by using hot air as a medium, and the formation of excellent quality of tea is closely related to the drying temperature. According to the type of wulong tea, variety, season, shape and different needs of consumer tastes, the operator should take different baking time and baking temperature.

In the tea baking, baking temperature is usually changed to meet the quality requirements of tea by fixing the baking time and the amount of put in order to facilitate the operation. In general, the baking temperature of high-grade tea is 80-100 ℃, the baking temperature of mid-range tea is 100-110 ℃, the baking temperature of low-grade tea is 110-120 ℃, the baking temperature of Tea sticks etc is preferably 110-130 °C.

2. The design principle of baking machine

The baking machine is the electric type, using circulating hot air to bake. Every layer has access to the outlet, making the box more uniform temperature, dehydration better. Baking machine is mainly composed of mechanical parts and intelligent temperature control part. Mechanical parts are mainly composed of the fan, heater, box, motor, belt drive and other components. The Intelligent control part is mainly composed of industrial computer, micro-controller, heater and sensor detection department. The main structure of the baking machine schematic diagram is shown in figure 1.

The main structure of the baking machine is circular box, the fan and heating electric wire are placed in the lower part of the baking box, and the electric motor is arranged at the bottom. The electric motor drives the fan through the belt drive, sucks the air flow from the middle air inlet, enters the hot air through the electric furnace wire, enters the inner oven through the tuyere, So that the layers of tea evenly heated, evaporated hot and humid wind from the exhaust port, so continue to cycle. Tea is left on the sieve frame the in middle of the baking box. In the process of tea baking, setting up air purification mechanism in the circulating air duct so that the volatile green and other miscellaneous substances in time to be filtered out in the baking and will not eventually remain in the tea inside when the wind from the air purification mechanism through the tea .It can be a good way to improve the aroma and quality of tea. Also insulation cotton is placed in the baking box inside and outside the wall in order to reduce heat loss and improve thermal efficiency.

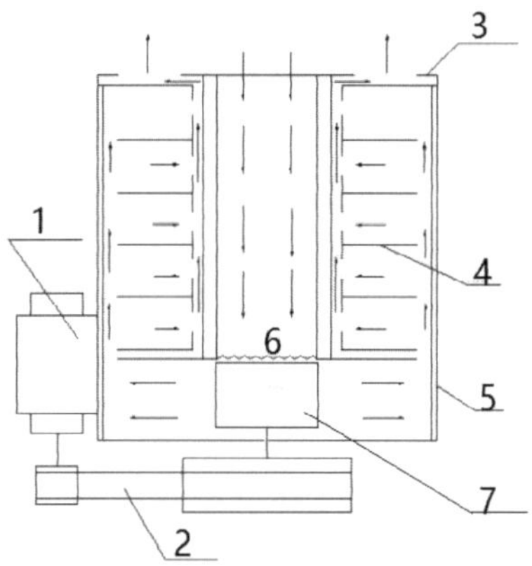

1. Motor, 2. Belt drive, 3. Cover, 4. Sieve frame, 5. Box, 6. Electric wire, 7. Fan

Figure 1. Schematic diagram.

The main structure of the existing rotary baking machine on the market structure is shown in figure 2. The main structure of the baking machine is a rectangular box. The fan and the heating electric wire are placed in the upper part of the baking box and the bottom is equipped with a low speed rotating electric motor. Baking boxes are inside the air inlet and exhaust. The upper part of the baking box is placed with an electric control box for temperature control. The tea leaves on the sieves inside the baking box. In the process of tea baking, the control flow generated by the fan becomes into hot air after through the electric furnace wire, and enters the oven through the outlet. The motor in the bottom of box drives sieve frame in low-speed rotation, making the layers of tea evenly heated, evaporated hot and humid wind from the exhaust mouth to discharge, so continue to cycle.

IOP Conf. Series: Materials Science and Engineering **269** (2017) 012040 doi:10.1088/1757-899X/269/1/012040

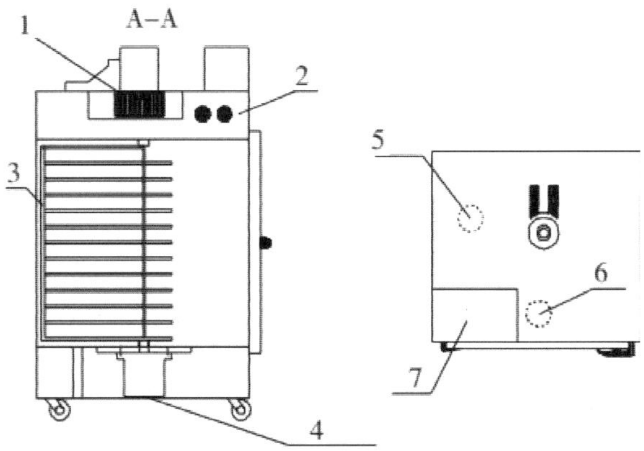

1. Fan, 2. Electric wire, 3. Sieve frame, 4. Rotating motor,
5. Exhaust hole, 6. Intake hole, 7. Electric control box

Figure 2. Schematic diagram of the structure of an electric baking machine.

The above two kinds of structures are compared with each other, the structure of the new type of tea baking machine has the following characteristics and innovation:

(1) The main structure of the baking machine is cylindrical, making the heat more uniform. At the same time baking machine can compensate the air volume and heat. And the rotating motor at the bottom is not required so that the cost can be reduced.

(2) The motor is placed externally and does not require a high temperature resistant motor. At the same time the centrifugal fan speed can be easily adjusted by changing the pulley combination.

(3) The outlet and the inlet have a large heat transfer contact area. The air inlet can be effectively recycled to the heat of the outlet to achieve energy efficiency.

(4) The structure of baking machine is more compact, better to achieve miniaturization, easy to apply to drying tea.

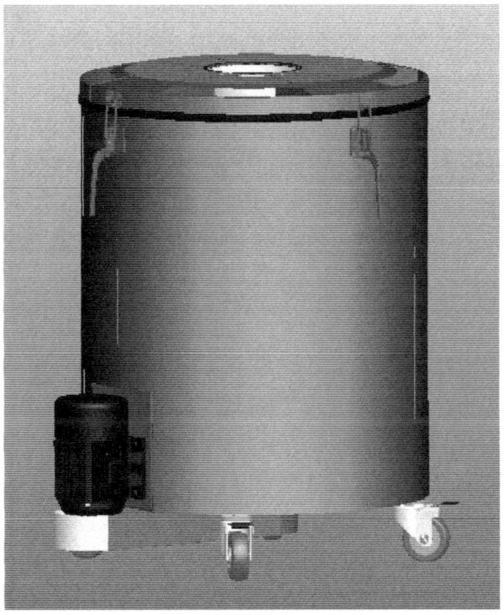

Figure 3. External structure of the baking machine.

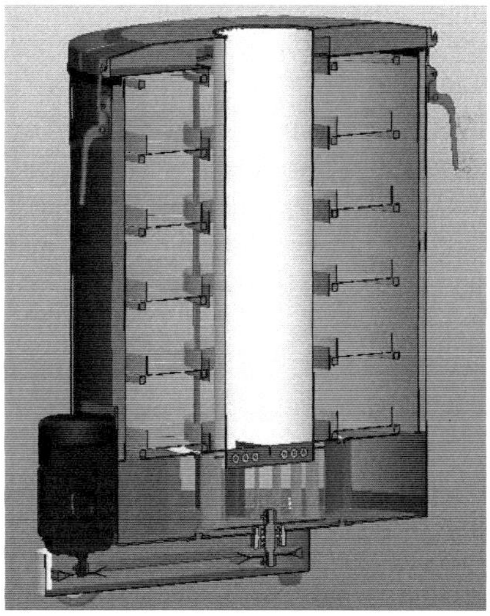

Figure 4. Internal structure of the baking machine.

Figure 5. Structure of the baking machine.

3. The mechanical structure design of the baking machine

Pro/Engineer is a parametric design, based on the characteristics of the entity model of the design software. Engineering designers can generate models by using feature-based features with intelligent features, and models can be easily changed. At the same time, Pro/Engineer is built on the unified grassroots on the database. As long as the entire design process changes anywhere, can be reflected in the entire design process related to the link. Pro/E using modular design, designers can achieve sketching, parts design, assembly design, sheet metal design, processing and so on. Designers can choose to use as needed.

With the Pro/E three-dimensional design software platform, the author has designed the working principle, created the parts, assembled the components into components, and then assembled components into the whole machine, and finally completed the whole baking machine design and improvement process.

The external structure of the baking machine is shown in figure 3, and the internal structure is shown in figure 4, and the decomposed structure is shown in figure 5.

4. The control system design of baking machine

This machine has designed the temperature control system based on fuzzy PID control and the humidity control system. For the temperature control, the system takes for PC as the host computer (VB platform for the design interface software), single-chip for the next machine. PC mainly completes the complex data processing and the control of the next bit machine, single-chip mainly finish data acquisition and the control of the heating device. In the next bit machine module, the temperature and humidity of baking box is collected real-time by using temperature and humidity sensors, and temperature and humidity data is sent by serial communication to the VB software interface. In VB software interface we can achieve fuzzy controller real-time correction PID controller parameters by temperature difference and rate of temperature difference through using VB and Matlab mixed programming. And we can obtain the control amount by PID algorithm, and send the control commands and control amount through the serial communication to the micro-controller, then the micro-controller will achieve the heating device power and duty cycle control by the relay on-off controlling to adjust the baking temperature.

For humidity control, thresholds can be set. When the humidity reaching the threshold, the humidity must be reduced by getting rid of moisture through the exhaust fan to ensure the aroma and quality of tea. At the same time, the tuning parameters have been adjusted by using Matlab/Simulink simulation experiments on the temperature and humidity control system.

5. Conclusion

In this paper, the effects of tea baking on the quality of tea were described in detail. Through the analysis and optimization of the existing baking machine structure, a new type of baking machine suitable for wulong tea baking is designed, and the overall structure and thermostat intelligent control system of baking machine are introduced in detail. The mechanical structure of this baking machine is more simple and reasonable, and can use the heat of the inlet and outlet, more energy saving and environmental protection. The temperature control part adopts fuzzy PID control, which can improve the accuracy and response speed of temperature control and reduce the dependence of baking operation on skilled experience.

References

[1] Yang J and Li M R 2011 *J. Yichun College* **8** 70-2.
[2] Su X M 2012 *J. Tea Fujian* **3** 27-8.
[3] Lin D Y and Li Z C 2012 *J. Food & Mach.* **1** 122-3.
[4] Li B 2011 *J. Agric. Mech. Res.* **8** 149-52.
[5] Han Q Y and Zhao C M. 2013 *J. Tech. Innov. Appl.* **14** 52.
[6] Liu H H and Wang j H 2009 *J. Cereal & Feed Ind.* **5** 10-1.
[7] Song Z Y 2010 *Energy-saving automation control of tea roller fixing machine* Hefei: Anhui Agricultural University.

Acknowledgement

This work is sponsored by Fujian Provincial Key Laboratory Project - Industrial Robot Measurement and Control and Mold Rapid Manufacturing Laboratory (Fujian Education [2017] No. 8), Fujian Province Young Teachers Education and Research Project (Project No: JAT170736)

IOP Conf. Series: Materials Science and Engineering **269** (2017) 012041 doi:10.1088/1757-899X/269/1/012041

Effect of Activation Function and Post Synaptic Potential on Response of Artificial Neural Network to Predict Frictional Resistance of Aluminium Alloy Sheets

T Trzepiecinski[1] and H G Lemu[2]*

[1]Rzeszow University of Technology, Department of Materials Forming and Processing, al. Powst. Warszawy 12, 35-959 Rzeszów, Poland
[2]University of Stavanger, Department of Mechanical and Structural Engineering, N-4036 Stavanger, Norway

Abstract. Many technological factors affect the friction phenomenon in sheet metal forming process. As a result, the determination of the analytical model describing the frictional resistance is very difficult. In this paper, a friction model was built based on the experimental results of strip drawing tests. Friction tests were carried out in order to determine the effect of surface and tool roughness parameters, the pressure force and mechanical parameters of the sheets on the value of coefficient of friction. The strip drawing friction tests were conducted on aluminium alloy sheets: AA5251-H14, AA5754-H14, AA5754-H18, AA5754-H24. The surface topography of the sheets was measured using Taylor Hobson Surtronic 3+ instrument. In order to describe complex relations between friction and factors influencing tribological conditions of sheet metal forming, the multilayer artificial network was built in Statistica Neural Network program. The effect of activation function and post synaptic potential function on the sensitivity of multilayer neural network to predict the friction coefficient value is presented. It has been found that the difference in the prediction of error of neural network for different approaches can reach 400%. So, the proper selection of activation and post synaptic potential functions is crucial in neural network modelling.

1. Introduction

Friction in metal forming process is an important parameter affecting the possibility to plastically deform different materials. In sheet metal forming, in particular, frictional resistance is a complicated function of mechanical properties of the sheet metal, temperature, lubrication conditions, and topography of both the tools and the workpiece [1-4]. Friction occurred at high pressures may essentially differ from friction in kinematic pairs due to a great influence of plastic deformations on dynamic character of contact [5]. Friction occurs due to some mechanisms which are simultaneously taking place with frictional contact, including the ploughing effect between asperities, the adhesion between contacting asperities, the flattening due to stretching, flattening due to sliding, and the appearance of hydrodynamic friction stresses [6, 7]. Complex interactions between the parameters influencing the frictional resistance and coefficient of friction value make it difficult to analyse and understand this relation. In the last few decades, many investigations were conducted to find real friction models. Challen and Oxley [8], for instance, developed a friction model taking the combined effect of ploughing and adhesion on the coefficient of friction into account. Wilson and Sheu [9] developed an analytical upper bound model to describe the flattening behaviour using wedge-shaped asperities with a constant angle. Trzepiecinski and Lemu [10] noticed that the proper selection of variables that influence the coefficient of friction value can improve the sensitivity of neural network friction model.

Advanced systems of data processing allow for automatic analysis of complex data set and generation of response on interrogated questions. Due to a large number of parameters and phenomena affecting the tribological phenomena, developing analytical relations between input variables and coefficient of friction value is practically impossible. This task is successfully realised by artificial neural networks (ANN) that belong to the artificial intelligence technology. In this paper, the ANN model of friction is built in Statistica Neural Networks program based on the results of strip drawing friction test of aluminium alloy sheets. In the sheet metal forming process, the strip-drawing tests simulate the friction phenomenon that exists between the punch and the wall of the drawpiece. The effect of learning algorithm, network architecture and the method of data selection are topics of recent papers of the authors [10, 11]. In addition, the authors aim to report, in this paper, about the effect of activation function (AF) and post synaptic potential function (PSPF) on the sensitivity of neural network to predict the friction coefficient value.

2. Material and method

2.1. Experiments

To determine the value of the coefficient of friction, the strip drawing tests were performed using a friction simulator (figure 1) mounted in standard tensile machine. During the tests, a strip of the aluminium alloy sheets (AA5251-H14, AA5754-H14 AA5754-H18, and AA5754-H24) was clamped with specified forces (400, 800, 1200, 1600, and 2000 N) between two cylindrical rolls of equal radii r = 10 mm. Next, when the displacement of moving crosshead of the test machine was engaged, the pulling force F_p and clamping force F_c were recorded continuously using a load cell and a computer programme. The value of coefficient of friction has been determined as a ratio of clamping force and pressure (normal) force (figure 1).

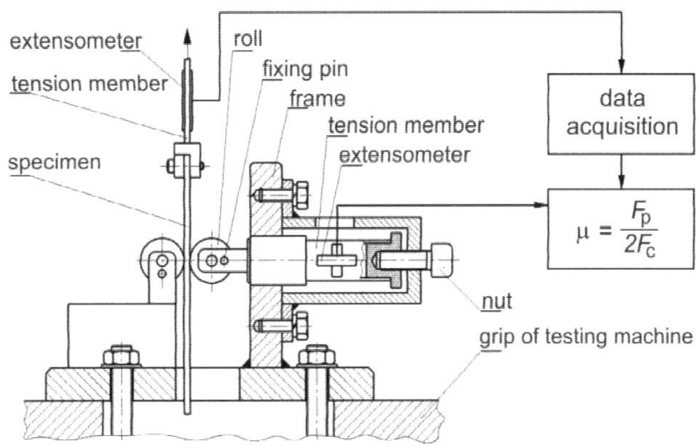

Figure 1. The schematic of friction simulator.

The specimens for the friction tests were prepared as the strips with 20 mm width and about 200 mm length, which were cut along the rolling direction of the sheet. Four sets of rolls (counter samples) with different surface roughness values (0.32, 0.63, 1.25 and 2.5 μm) were used. The tests were conducted using machine oil LAN-46 as lubricant. Surface roughness parameter measurements were carried out using Taylor Hobson Surtronic 3+ instrument to determine arithmetic average height along the rolling (Ra^0) and transverse directions (Ra^{90}) of the sheet metal. The value of yield the stress σ_y of the tested sheets was determined in the uniaxial tensile test.

2.2. ANN modelling

The friction model, based on the multilayer perceptron (MLP) network, was built in Statistica Neural Networks program. The methodology of creating computations in this program was described in the recent works of authors [10, 11]. Based on the training data, the network architecture MLP 5:5-11-1:1 was found as the optimal structure which ensures the minimal prediction of error of the ANN [11]. The following parameters were selected as input set of variables: yield stress of the sheets materials σ_y, arithmetical mean deviation of the assessed profile of sheet surface Ra^0 and Ra^{90} measured along and transverse to sheet rolling direction respectively, clamping force F_c and average surface roughness of rolls Ra^r.

The significance of the influence of input parameters on the output variable (coefficient of friction) is confirmed by genetic algorithm [10]. The total number of experiments was 80. The training set (64 training pairs) consists of values of input variables and corresponds to value of the friction coefficient. For the purpose of keeping an independent check on the progress of the learning algorithm from all observations, a validation set containing 8 training pairs (i.e. 10% of the all training samples) was randomly separated. Furthermore, the independent set of 10% training samples was selected and assigned to the test set. This set was used for assessment of the quality of ANN prediction.

To study the effect of activation function and post synaptic potential functions on response of MLP 5:5-11-1:1 network, seven activation functions (table 1) and two post synaptic potential functions (linear and radial) were considered. In the case of MLP network, the highest performance is received for the network trained using back propagation (BP) algorithm. This BP algorithm was used with the following settings: learning rate value was 0.1 and momentum value was set to 0.3 [11].

Comparison of the ANNs with different activation functions and post synaptic potentials was carried out in reference to MLP in which the logistic activation function and linear post synaptic potential were used. As a stop criterion of the training process, the moment at which the value of Root-Mean-Square (RMS) error [12] of the validation set tends to decrease was assumed. Based on this criterion, learning process of the MLP with AF and PSPF was stopped after about 1200 epochs. So, the analysis of effect of the type of AF and PSPF on the quality of learning process was performed after 1200 epochs.

Table 1. The analysed activation functions of the neurons.

Function	Logistic	Hiperbolic	Softmax	Unitsum	Sine	Ramp	Step
Formula	$x = \frac{1}{1-e^{-1}}$	$\frac{e^x - e^{-x}}{e^x + e^{-x}}$	$\frac{e^x}{\sum_i e^{x_i}}$	$\frac{x}{\sum_i x_i}$	sin(x)	-1 $x \leq 1$ x -1< x <+1 +1 $x \geq 0$	0 $x < 0$ +1 $x \geq 0$
Range	(0, +1)	(-1, +1)	(0, +1)	(0, +1)	[0, +1]	[-1, 1]	[0, +1]

3. Results and discussion

Application of *step* AF ensures the highest rate of convergence of training process of the MLP in conjugation with *linear* PSPF. However, it was observed that use of the *unitsum* function considerably increases the time of learning process and the network error decreased just after 450 epochs. The network error for the validation set, for all the used activation functions, except for the *unitsum* AF, is stabilized after about 300 epochs. The highest network training error after 1200 epochs is observed for *unitsum* AF, and the lowest for *step* AF function.

The effect of number of learning epochs on the training and validation sets of an MLP network error for the cases of linear and radial PSPF assumptions are plotted in figure 2 and 3 respectively. As depicted in the figures, the MLPs errors for validation set are characterized by high noise (figures 2b, 3b), while normal effect is observed when the number of training pairs in the validation set is significantly smaller than the number of training pairs in the training set. The *radial* PSPF (figure 3a) eliminates the initial lack of convergence in *linear* PSPF (figure 2a). In this case, the *step* AF also ensures the lowest training error after 1200 epochs.

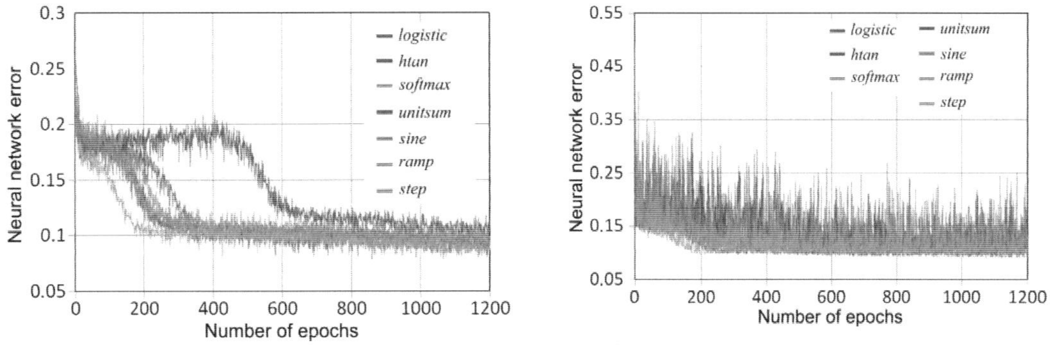

Figure 2. The effect of number of learning epochs on MLP's error of (a) training set and (b) validation set, in the case of assumption *linear* PSPF.

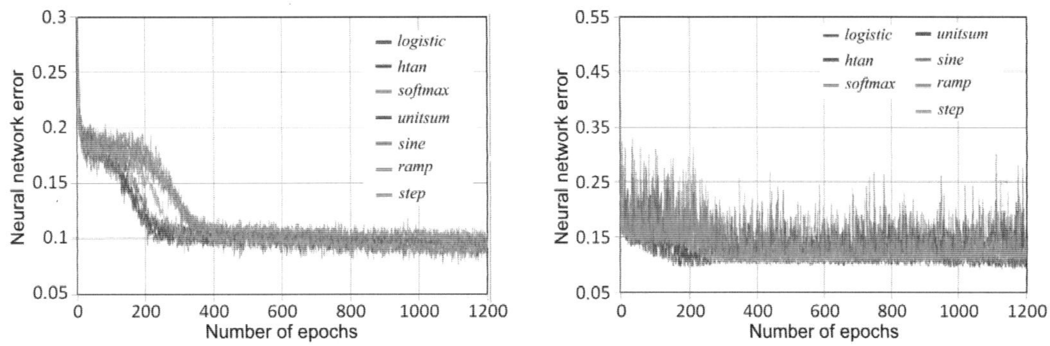

Figure 3. The effect of number of learning epochs on MLP's error of (a) training set and (b) validation set (b), in the case of assumption *radial* PSPF.

The activation function is used for transformation of the activation level of the neuron in output signal. Typically, activation functions include a "flattening" effect. Together with the PSPF, it defines a neuron type. Similar to the logistic function, the hyperbolic tangent function (*tanh*) is an S-shaped curve, with the difference that the output values are in the range (-1, +1). Because of the symmetry of the *tanh* function, it often works better than the *logistic* function. The *tanh* function is preferred for activation of neurons in multilayer neural networks. The trigonometric AF activation functions (i.e., *sine*) is useful for recognizing radially distributed data. *Ramp* AF is a piecewise linear version of a sigmoid function which ensures relatively low efficiency of learning process, but enabling a very fast operation of the neural network.

When using *linear* PSPF, the highest correlation for training set is observed for *tanh* AF. When using *radial* PSPF, the highest correlation coefficient assures *sine* activation function (figure 4a). The highest difference in the correlation coefficient, when using both considered PSPFs, is noticed in the case of *tanh* and *ramp* AF. The quality of the network prediction is described by the correlation coefficient for the test set, which is the highest for conjugation of *unitsum* AF and *radial* PSPF, and in conjugation of *logistic* AF and *linear* PSPF (figure 4b). It is visible that selection of the PSPF function is a crucial element of ANN modelling. PSPF is a function which evaluates the value of neuron activation based on the input values, weights and threshold value. In the linear PSPF activation, the activation is expressed as a difference between a sum of weighted inputs minus and threshold value. In radial PSPF, the activation is a scaled square of an input vector and vector of weights.

The importance of proper selection of neuron activation function and post synaptic potential function is clearly visible in figure 5. In the case of *softmax* AF the difference in mean error of prediction between *linear* and *radial* PSPFs differs in about 400%.

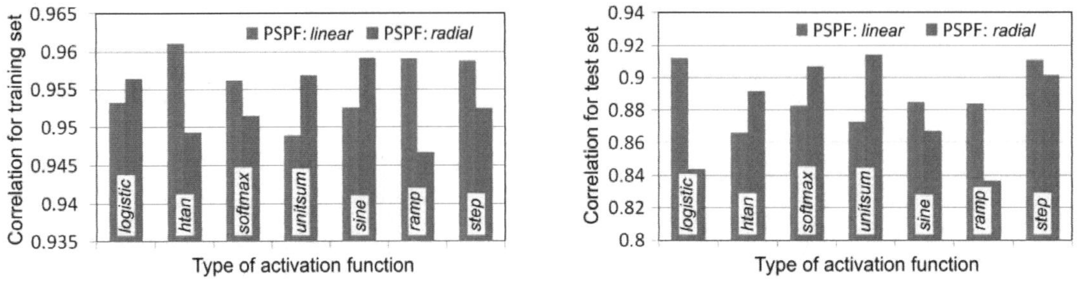

Figure 4. The effect of activation function on correlation coefficient of (a) training set and (b) test set.

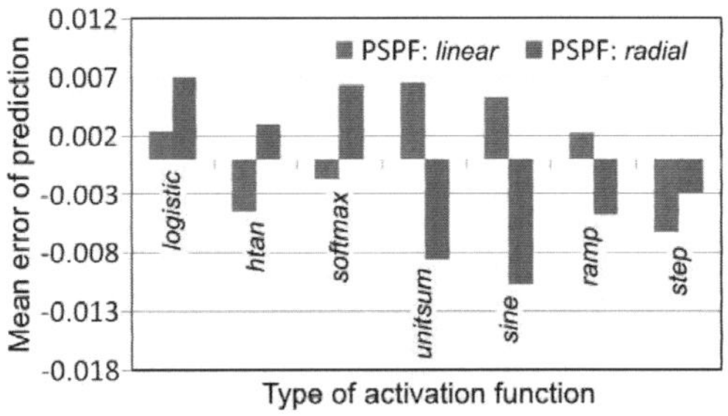

Figure 5. The effect of activation function on value of mean error of MLP prediction.

4. Conclusions

The results of modelling the frictional phenomena in sheet metal forming using artificial neural networks showed that the appropriate choice of the parameters of neuron activations is an important parameter, and this determines the rate of learning process and ability of prediction of ANN. Although, in our research, the conjugation of *softmax* AF and *linear* PSPF assured the minimal error of prediction, both types of the activation and post synaptic potential functions should be selected separately for individual set of training data. The analyses of MLP 5:5-11-1:1 network simulated with seven types of neuron activation functions showed that the difference in prediction error between all approaches may reach about 400%.

5. References

[1] Figueiredo L, Ramalho A L, Oliveira M C and Menezes L F 2011 *Wear* **271** 1651.
[2] Gil I, Galdos L, Mugarra E, Mendiguren J and de Argandona E S 2016 *IOP Conf. Series: Mater. Sci. Eng.* **159** 012019.
[3] Sigvant M, Pilthammar J, Hol J, Wiebenga J H, Chezan T, Carleer B and van den Boogard A H 2016 *IOP Conf. Series: Mater. Sci. Eng.* **159** 012021.
[4] Wang C, Ma R, Zhao J and Zhao J 2017 *J. Mater. Proc.* **27** 126.
[5] Trzepieciński T, Bazan A and Lemu H G 2015 *Int. J. Automot. Technol.* **16** 849.
[6] Wilson W and Chang D 1996 *J. Tribol.* **118** 83.
[7] Hol, J, Cid Alfaro M V, de Rooij M B and Meinders T 2012 *Wear* **286-287** 66.
[8] Challen J and Oxley P 1979 *Wear* **53** 229.
[9] Wilson W and Sheu S 1988 *Int. J. Mech. Sci.* **30** 475.
[10] Trzepieciński T and Lemu H G 2012 *J. Mech. Eng. Autom.* **2** 69.
[11] Lemu H G and Trzepieciński T 2014 *Adv. Mater. Res.* **1051** 204.
[12] STATISTICA Neural Networks. Addendum for version 4. Tulsa: StatSoft, 1999.

IOP Conf. Series: Materials Science and Engineering **269** (2017) 012042 doi:10.1088/1757-899X/269/1/012042

Experimental Investigation of Frictional Resistances in the Drawbead Region of the Sheet Metal Forming Processes

T Trzepiecinski[1], R Fejkiel[2] and H G Lemu[3]*

[1]Rzeszow University of Technology, Department of Materials Forming and Processing, al. Powst. Warszawy 12, 35-959 Rzeszów, Poland
[2]State School of Higher Vocational Education in Krosno, Rynek 1, 38-400 Krosno, Poland
[3]University of Stavanger, Department of Mechanical and Structural Engineering, N-4036 Stavanger, Norway

Abstract. Drawbeads are used in sheet metal forming to restrain the sheet from flowing freely into die cavity, especially in the case of forming unsymmetrical drawpieces. This process is necessary to produce an optimal stamped part without wrinkles and cracks. In this paper, a special tribological simulator is used to evaluate the frictional resistances during flowing the sheet through the circular shape bead. The tests were conducted on DC04 carbon steel sheets with a sheet thickness of 0.8 mm. Experiments were carried out at different process parameters: friction conditions, specimen widths, heights and surface roughness of drawbead. The results obtained in the drawbead friction test show that the value of friction coefficient depends on the width of the sample. The character of sheet deformation during bending and reverse bending on the sheet thickness over the drawbead changes the surface topography and real contact area of sheet and tool.

1. Introduction

Sheet metal forming processes are influenced by many parameters such as friction and lubrication conditions, die geometry, shape of drawpiece, blank holding force and draw-bead force. In the case of forming drawpieces with complex shape, the material flow into die cavity must be controlled to prevent a drawpiece against defects such as wrinkling, galling, and tearing. Prevention of wrinkling of large complicated shapes like automobile body panels is based on the use of drawbeads in suitable regions of the drawpiece flange. The other means of applying a restraining force to control the material flow is application of higher blankholder pressure. However, this process may cause excessive wear of the tools and galling the sheet metal, especially in the case of aluminium alloys. The drawbeads play an important role in sheet metal forming and cause an increase in radial tensile stresses and a decrease in compressive stresses in drawpiece material. The number and arrangement of the drawbeads in stamping tool depends on the shape of the workpiece, and the drawing depth. They are placed in the locations subjected to small deformations, in which the hamper of the material flow and an increase in tensile stress are required. High blankholder pressure forces and drawbeads, however, increase the risk of sheet fracture. So, the determination of the frictional resistances that arise in drawbead is a crucial task in the process of producing an optimal stamped part with minimum material usage.

An analytical, experimental and numerical investigations of material flow controlled by the restraining force of the drawbead have been the subject of extensive studies in the past few decades. Firstly, Nine [1] developed a drawbead simulator which has been used by several researchers for further investigations. The draw-bead test allowed the simulation of the bending and unbending in a sheet metal forming process and measurement of the friction coefficient during the sliding of the sheet against a drawbead [2]. Cao and Boyce [3] analyzed the restraining force with respect to the depth of drawbead. Triantafyllidis et al. [4] considered the effects of the drawbead in sheet forming using experiments and numerical modelling. The effects of friction coefficient, bead geometry, and material properties were investigated. Smith et al. [5] developed a novel device for obtaining pulling and holding forces for drawbead tooling on inclined binder surfaces.

Keum et al. [6] developed an expert database system that computes the drawing characteristics of different kinds of drawbead: the circular drawbead, stepped drawbead, and squared drawbead. The predictions obtained from this expert system agreed well with the experiments. The experimental and numerical research on the effect of sheet metal surface roughness, sample orientation according to sheet rolling direction and lubricant conditions on the value of friction coefficient in the drawbead region conducted by Trzepieciński and Lemu [7] have shown that the sample orientation and the lubrication conditions are crucial parameters influencing the value of the coefficient of friction. Firat and Cicek [8] applied numerical modeling of drawbead to improve the accuracy of finite element simulations in terms of part draw-in and thickness distribution predictions. The results of the modeling have been validated with channel drawing experiments of high-strength low-alloy steel.

The use of expert systems and numerical simulations to predict frictional resistances in mass production is limited due to the high costs. Furthermore, accurate prediction using these systems requires the knowledge of many parameters that must be determined and often have to be verified experimentally. So, better understanding of the evolution of friction coefficient at drawbead still demands experimental testing. In this paper, the drawbead apparatus is used to evaluate the frictional resistances in drawbead region in forming of DC04 carbon steel sheets that are widely used in automotive industry.

2. Material and method

The experimental works were conducted at different process parameters, including specimen widths, drawbead penetrations (figure 1a) and drawbead surface roughness. Drawbead simulator (figure 1b) that was mounted on the Zwick/Roell Z030 tensile testing machine was employed. In this setup, the wrap angle of middle roll may be changed by a nut. During the test, the pulling (vertical) force and clamping (horizontal) force were measured and registered simultaneously using tension gauges and computer program respectively. The friction test consisted of two stages: (1) pulling the specimen between freely rotating cylindrical rolls, and (2) measuring the pulling force and the clamping force that gave the bending and unbending resistance of the sheet under “frictionless” conditions, respectively. The sheet was displaced between the rotating rolls in order to minimize the friction between the sheet and rolls, while the second specimen was pulled between the fixed rolls [7]. The coefficient of friction (COF) value is then calculated according to the following expression:

$$\mu = \frac{\sin\Theta}{2\Theta} \cdot \frac{F_P^{fix} - F_P^{free}}{F_C^{fix}} \tag{1}$$

where θ is the quarter contact angle of actual engagement of the strip over the bead, and F_C^{fix} is the normal force or clamping force obtained with the fixed beads, F_P^{fix} is the pulling force obtained with the fixed rolls, and F_P^{free} is the pulling force obtained with the freely rotating rolls.

The test material was a DC04 steel sheet ($R_{p0.2}$ = 187 MPa, R_m = 307 MPa, A = 22.5%) with a thickness of 0.8 mm. The basic roughness parameters of the sheet determined using the Talysurf CCI Lite 3D instrument are as follows: *Sa* = 1.178 μm, *Sq* = 1.467 μm, *Sp* = 8.628 μm, *Sz* = 17.902 μm, *Ssk* = -0.128, *Sdq* = 2.97, *Sdr* = 0.265%. Three sets of rolls with different surface roughness values (*Ra* = 0.32, 0.64 and 1.25 μm) were used. Machine oil lubrication L-AN46 was used for the friction tests.

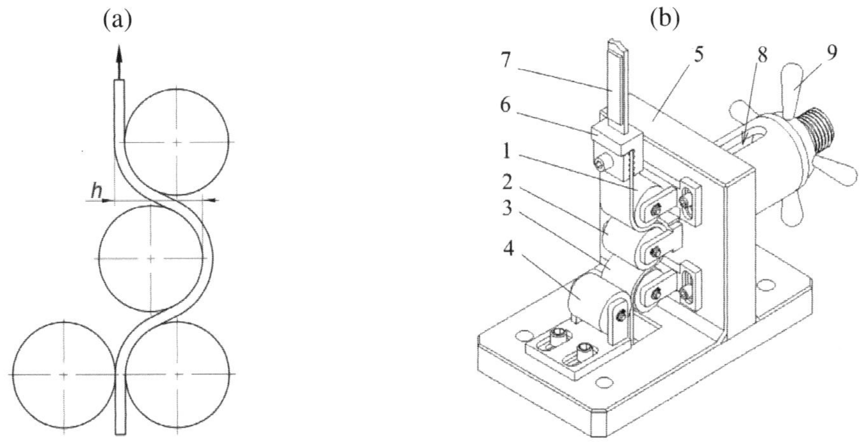

Figure 1. (a) The model of drawbead and (b) measurement system used for friction testing; 1 – front roll; 2 - middle roll; 3 – back roll; 4 – supporting roll; 5 – frame; 6 – tension member; 7 and 8 – extensometers; 9 – nut.

3. Results and discussion

As depicted in figure 2, the highest values of COF, for all sample widths, are observed at the penetration depth of middle roll h = 12 mm. Results of the statistical analysis showed that standard deviation of the registered experimental data did not exceed ±0.016. The effect of rolls surface roughness value on the friction coefficient value is also clearly visible. In the case of the smallest sample width (figure 2a) the highest COF is observed for surface roughness of Ra = 1.25 μm. The lowest friction coefficient value is observed for the rolls with surface roughness of Ra = 0.63 μm. However, for the highest tested sample width w = 20 mm (figure 2c) the situation is the contrary. This can be explained by the effect of sample width on the character of sheet deformation. The bending and reverse bending of the sheet during pulling the sheet through the drawbead caused that the initial rectangular section of the sample is changed into concave-convex section after leaving the drawbead. The contact of the sheet and front roll exists only at the edges of the sample (figure 3), where the roughness asperities are more flattened than in the middle area of the sample width. This effect is not indicated by previous researches that were conducted to investigate frictional resistances based on the idea of Nine's simulator [1].

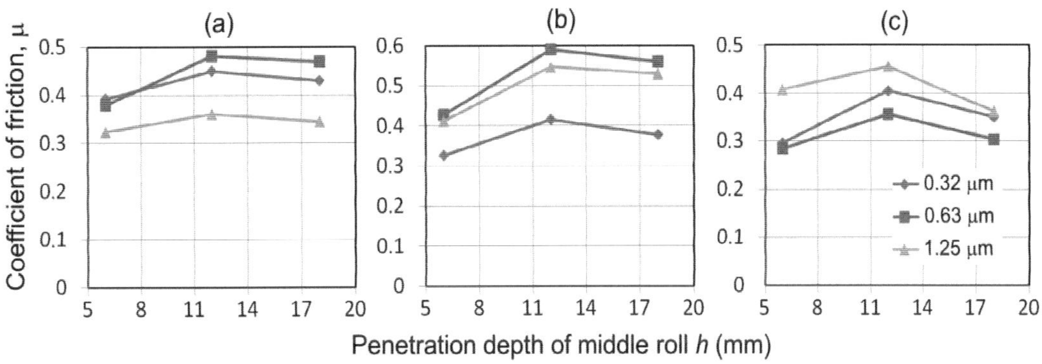

Figure 2. The effect of penetration depth of middle roll on the value of COF for sample widths: (a) w = 7 mm, (b) w = 14 mm and (c) w = 20 mm.

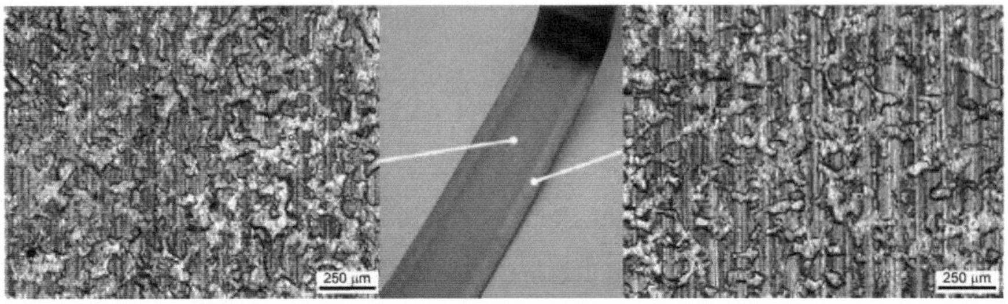

Figure 3. The view of the sample surface after the friction test.

The values of COF determined at lubricated conditions seem to be too high. However, it should be noted that the values determined using drawbead simulator are conventional parameters resulted from the difference of plastic and friction resistance of pulling the sheet through the drawbead. Furthermore, the pulling direction of the sheet is not parallel to the direction of sheet movement around the middle roll. The sheet is subjected to strain hardening phenomenon during friction testing, so the mechanical properties of the sheet greatly influence the COF value than the contact conditions.

Increased sample width leads to decreased COF value only in the case of the roll with surface roughness of *Ra* = 0.32 μm. Furthermore, the values of COF for tested sample widths (figure 4a) are the most similar among all used rolls (Figure 4a-c). The most unfavourable friction conditions are observed in the interaction of the roll roughness *Ra* = 0.63 μm with the sample width *w* = 14 mm (figure 4b). Further, two-fold increase in roll roughness causes slight decrease in the COF value (figure 4c). To understand the relation between the sample width and COF for roll roughness values of *Ra* = 0.63 μm and *Ra* = 1.25 μm (figures 4a and 4b), it must be taken into account that lower surface roughness leads to an increased COF value due to increased contact area. In materials forming, where there exists high contact plastic deformations, according to the friction law of Amontons-Coulomb, the COF, in many cases, does not depend on the area of contact, but this is not satisfied. However, the high surface roughness of the tool causes flattening of the softer sheet material and the frictional resistances increase.

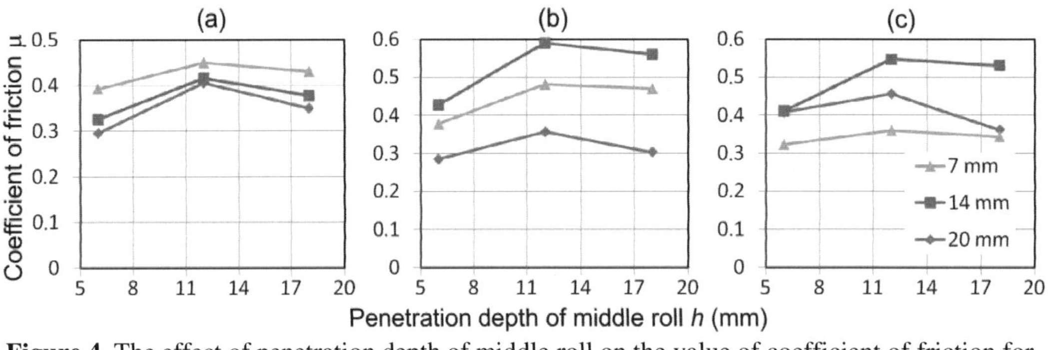

Figure 4. The effect of penetration depth of middle roll on the value of coefficient of friction for surface roughness of rolls: (a) *Ra* = 0.32 μm, (b) *Ra* = 0.63 μm and (c) *Ra* = 1.25 μm.

The values of the friction coefficient determined for the samples cut according to the rolling direction and cut transverse to this direction (figure 5) are close to the range of the standard deviation of data. So, it can be noted that sample orientation slightly influences the friction coefficient. Considering the sample width *w* = 20 mm, the optimal surface roughness ensured the minimal value of COF that is equal to *Ra* = 0.63 mm (figure 6). As mentioned above, the high surface roughness corresponds to the intensification of flattening mechanism, so the COF value increases. In the case of low surface roughness, the lubricant is squeezed out from the contact interface, so the load is carried out by plastically deformed roughness asperities.

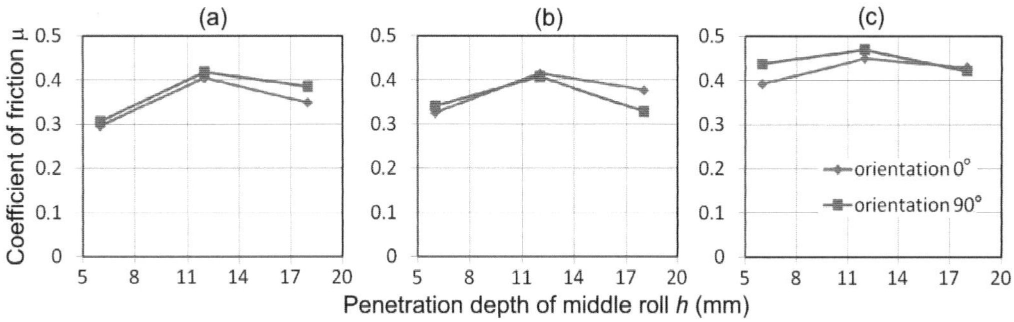

Figure 5. The effect of penetration depth of the middle roll on the value of the coefficient of friction for sample width w = 20 mm at: (a) Ra = 0.32 μm, (b) Ra = 0.63 μm and (c) Ra = 1.25 μm.

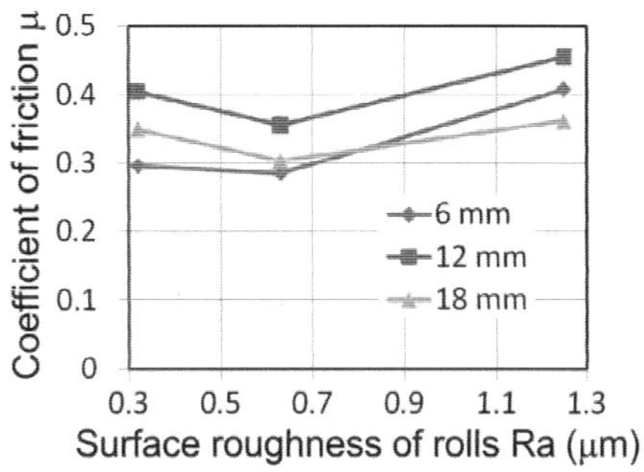

Figure 6. The effect of surface roughness of rolls on the value of the coefficient of friction for sample width w = 20 mm.

4. Conclusions

The action of the drawbead changes the stress-strain state in the drawpiece, and this is because of strain hardening phenomenon of the mechanical properties of sheet material. The plastic deformation of the sheet material causes the change of the topography of sheet surface, and as a consequence, the conditions of frictional contact change. Experimental tests using the drawbead simulator showed that the sample width is an essential parameter in analysis of frictional resistance while pulling the sheet through the drawbead. During a series of local bending and reverse bending, the sheet width determines the character of sheet deformation and real area of contact between sheet and rolls. The high surface roughness ensures better lubrication due to high volume of voids. However, high profile height causes the intensification of flattening of roughness asperities. Then, the friction coefficient value increases. In contrast, the low surface roughness of tool ensures worse lubrication due to low volume of open voids, which result in increased COF value.

5. References

[1] Nine H D 1978 *Drawbead forces in sheet metal forming*, Mechanics of Sheet Metal Forming eds D P Koistinen, N M Wang (New York: Plenum Press) pp 179–211.
[2] Figueiredo L, Ramalho A, Oliveira M C and Menezes L F 2011 *Wear* **271** 1651.
[3] Cao J and Boyce M C 1993 *Drawbead penetration as a control element of material flow*, Proc. Sheet-Metal and Stamping Symposium (Detroit, USA, 02–04 March 1993) pp 145–153.
[4] Triantafyllidis N, Maker B and Samanta S K 1986 *J. Eng. Mater. Technol.* **108** 321.
[5] Smith L M, Zhou Y J, Zhou D J, Du C and Wanintrudal C 2009 *J. Mater. Proc. Technol.* **209** 4942.
[6] Keum Y T, Kim J H and Ghoo B Y 2001 *Int. J. Solids Struct.* **38** 5335.
[7] Trzepieciński T and Lemu H G 2014 *Stroj. Vestn. J. Mech. Eng.* **60** 51.
[8] Firat M and Cicek O 2011 *Int. J. Adv. Manuf. Technol.* **55** 107.

IOP Conf. Series: Materials Science and Engineering **269** (2017) 012043 doi:10.1088/1757-899X/269/1/012043

The Synthesis of a Chemical Stimuli Nanowhiskers of Cellulose (CNWs) Composite Hydrogel

Jiufang Duan[1]*, Yuxue Gao[1]

[1]MOE Engineering Research Center of Forestry Biomass Materials and Bioenergy, Beijing Forestry University, Beijing 100083, China

Abstract. Long aspect-ratio CNWs derived from non-toxicity, sustainable, biocompatibility cellulose has high mechanical properties and often be applicated as strengthening agents in composites and hydrogels. N, N'-bis (propionyl) cystamine and acrylamide can form chemical stimuli hydrogel through copolymerization in water solution. Disulfide bond in N, N'-bis (propionyl) cystamine may perform link/ break transition in response to chemical stimuli. CNWs were performed as reinforce agent of acrylamide- N, N'-bis (propionyl) cystamine hydrogel. A series of CNWs composite hydrogels were prepared by acrylamide, N, N'-bis (propionyl) cystamine solution with CNWs suspensions homogeneously dispersed. The structure of the composite hydrogel was characterized by fourier transform infrared spectroscopy (FTIR).

1. Introduction

Hydrogel is a kind of three-dimensional crosslinked hydrophilic polymer networks that can swellen in water and can hold on a large volume of water in swollen state. Hydrogels have received considerable attention since 1954 [1], for it has wide applications, such as water treatment, agriculture products [2], drug control and delivery systems [3, 4], tissue engineering [5], pharmaceuticals [6], wound dressing [7] and so on.

Stimuli-responsive hydrogels [8] have recently gained attention owing to its smart nature in potential use in control and release of micro-fluidics. This kind of hydrogels has the ability of change their volume when meet certain environmental parameters, during volume change the absorption and release of aqueous solution could be realized. Disulfide bond is an ideal reversible connection chemical bond that has the ability of chemical stimuli properity. Acrylamide copolymerize with N, N'-bis (propionyl) cystamine can introduce disulfide bond into the hydrogel system and got a chemical stimuli hydrogel. At the same time, biodegradable CNWs are attractive as reinforce agent to make environmentally friendly bio-based composites hydrogels [9]. In sequence, CNWs were used to reinforce this hydrogel. The structure, crystal properties and swelling behavior of the hydrogel composite were examined.

2. Experimental

2.1. Material

Acrylamide, N, N'-bis (propionyl) cystamine, ethanol, ammonium persulfate, N, N, N', N'-tetramethylethylenediamine (TEMED) were of analytical grade and used without further purification.

2.2. Preparation of CNWs nanocomposite hydrogels

CNWs (CNWs here used were isolated from cotton) [10] is uniformly dispersed in an aqueous solution and prepared into solutions of concentration 0.09 g/ml, 0.067 g/ml, 0.045 g/ml, 0.03 g/ml, 0.011g/ml, 0.006 g/ml. 19g anhydrous ethanol, 0.4368g N, N'-bis (propionyl) cystamine, 3.724 g acrylamide and 4500 μL TEMED were added into 57g CNWs solution. The solution was stirred until fully dissolved. 9100 μL freshly prepared 10% w / v ammonium persulfate were added to initiate the polymerization reaction. This reaction was maintained at room temperature for 15 h to get the hydrogel.

2.3. Methods

IR spectra were recorded by FTIR (Nicolet iN10Thermo Fisher Scientific China) in the region of 400–4000 cm−1. The gravimetric method was employed to measure the swelling ratios of the hydrogels in distilled water at 25°C. After immersion in distilled water for about 48 hr to reach swelling equilibrium, the gel samples were taken out and weighed after removing the excess water on the surfaces. Each data was measured three samples, and the average value of three measurements was taken. The equilibrium swelling ratio (SR) was calculated as SR = Ws/Wd, where Ws is the weight of the swollen gel and Wd is the weight of the hydrogel at the dry state.

3. Results and discussion

3.1. The synthesis of the CNWs composite hydrogels

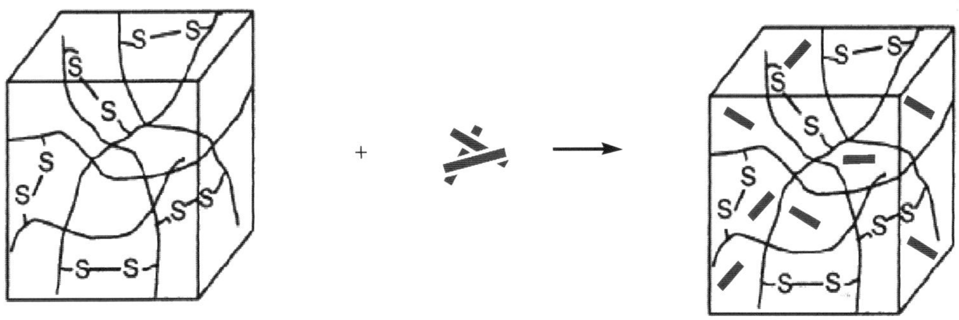

Figure 1. Acrylamide- N,N'-bis(acryloyl)-cystamine CNWs composite hydrogels equation.

CNWs here used were isolated from cotton. As seen in figure 1, CNWs dispersed in the hydrogel, except for the covalent bonds inacrylamide / N, N'-bis (propionyl) cystamine hydrogel, the disulfide bonds act as multifunctional cross-links and bridges at the adjacent molecular chain. Thus, the above disulfide bonds in hydrogels network contribute to the chemical stimuli behaviors.

3.2. Characterization of the CNWS composite hydrogels

Figure 2 shows the FTIR spectra of the CNWS composite hydrogels. As shown in figure 2, absorption peaks at 3430cm-1 is attributable to OH, peak at 2890 cm-1 corresponding to the band CH_2, peak at 1440 cm-1 due to the amide C-N and N-H stretching, peak at 1720 cm-1 is assigned to the C=O stretching vibration of the amide group.

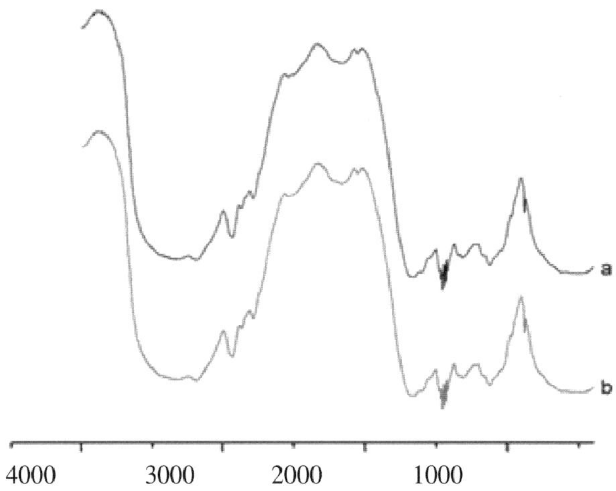

Figure 2. IR spectra of Acrylamide- N,N'-bis(acryloyl)-cystamine CNWs composite hydrogels (nanocellulose concentration a:0.09 g/ml, b:0.006 g/ml).

4. Conclusion

Hydrogels mechanical propertie is important to its application. CNW is an excellent reinforcing agent. Disulfide bond is an ideal reversible connection chemical bond that has the ability of chemical stimuli properity. In this study, CNWs were performed as reinforce agent of acrylamide- N, N'-bis (propionyl) cystamine hydrogel, and the same time disulfide bond has been designed into this hydrogel. A series of CNWs composite hydrogels were prepared by acrylamide, N, N'-bis (propionyl) cystamine solution with CNWs suspensions homogeneously dispersed. Acrylamide- N,N'-bis(acryloyl)-cystamine composite hydrogels with CNWS were successfully prepared in aqueous solution at 25°C by simple method. The structure of the composite hydrogel was characterized by infrared Spectroscopy (IR). The CNWs concentration affects the water absorption of the hydrogel.

References

[1] Wichterle O, Lim, D 1960 *Nature* **185(4706)** 117-8.
[2] Amulya K. 2010 *J. Thoracic. Cardiovasc. Surg.* **139(2)** 496-7.
[3] Ashley G W, Henise J, Reid R and Santi D V 2013 *Proc. Nat. Acad. Sci. U.S.A.* **110(6)** 2318-23.
[4] Brandl F, Kastner F, Gschwind R M, et al. 2010 *J. Contr. Releas. Offic. J. Contr. Releas. Soc.* **142(2)** 221-8.
[5] Zhang L, Li K, Xiao W, Zheng L, Xiao Y, Fan H, et al. 2011 *Carbohydr. Polym.* **84(1)** 118-25.
[6] Kashyap N, Kumar N and Kumar M 2005 *Crit. Rev. Ther. Drug. Carr. Syst.* **22(2)** 107-49.
[7] Sikareepaisan P, Ruktanonchai U andSupaphol P 2011 *Carbohydr. Polym.* **83(4)** 1457-69.
[8] Andreas R, Georgi P, Stephan K, Jens L, Karl-Friedrich A and Hans-Jürgen P A. 2008 *Sensors* **8(1)** 561-81.
[9] Abdul Khali H P S l, Bhat A H and Ireana Yusra A F 2012 *Carbohydr. Polym.* **87(2)** 963-79.
[10] Li Y, Guan C h, Yang P Y, Liu A F, Li J and Duan J 2015 *Paripex-Indian J. Res.* **4(6)** 33-6.

Acknowledgments

This research was financially supported by the National Natural Science Foundation of China (31600464).

IOP Conf. Series: Materials Science and Engineering **269** (2017) 012044 doi:10.1088/1757-899X/269/1/012044

The Principle and the Application of Self-cleaning Anti-pollution Coating in Power System

Y J Zhao[1, 2]*, Z B Zhang[3], Y Liu[1, 2], J H Wang[1, 2], J L Teng[1, 2], L S Wu[1, 2], Y L Zhang[1, 2]

[1]Nari group corporation (state grid electric power search stitute), Naijing, 211106, China
[2]Beijing Guodian Futong science and technology development corporation, Ltd, Beijing, 100070, China
[3]Institute of Chemistry Chinese Academy of Sciences, Beijing, 100190, China

Abstract. The common problem existed in power system is analyzed in this paper. The main reason for the affection of the safe and stable operation to power equipment is flash-over caused by dirt and discharge. Using the self-cleaning anti-pollution coating in the power equipment surface is the key to solve the problem. In the work, the research progress and design principle about the self-cleaning anti-pollution coating was summarized. Furthermore, the preparation technology was also studied. Finally, the application prospect of hard self-cleaning anti-pollution coating in power system was forecast.

1. Introduction

With the development of economy, the emission of industrial pollutants is increased year by year. And the air pollution is more and more serious. All kinds of electrical equipment and devices in outdoor power, such as insulator, reactor, electrical cabinets and solar photovoltaic panels, are covered by pollution easily in the electric field or natural environment. The pollution layer on the electric equipment surface, which maybe contain metal particles, oil pollution, dirt, and acid, alkali, salt that can be affected with damp, not only affect the appearance and efficiency of the power equipment, but also cause the accidents, such as leakage, flashing, flash-over and so on in the conditions of fog, dew and rain.

The common used method of preventing and controlling flash over is using anti-pollution coating on the surface of the electric equipment. And the coating system based on polymer silicone rubber is the most widely used. Based on the principle, compared with other types of elastomer coatings, advantage of the room temperature vulcanized silicone rubber anti-pollution flash-over coating is hydrophobicity property and hydrophobicity transfer property, which can ensure that the water is not easy to stay on the surface of the coating, thus preventing the phenomenon of flash-over. The hydrophobicity property is due to the low degree of cross-linking of silicone rubber, and the unresponsive siloxane with the low surface tends to migrate to the surface. However, it also has a negative effect, such as, the hardness of silicone rubber coating is low, and the hydrophobicity property is directly affected by the rate of the migration of siloxane polymer. In addition, although Si-CH_3 in the silicon rubber structure has excellent hydrophobicity, the Si-OH has oil absorption property, which makes the coating surface to stain easily. Based on the above reasons, the equipment surface with the anti-pollution coating accumulates pollution easier. This not only affects the appearance, but also the hydrophobicity transfer property will be losing [1]. With the pollution accumulating, even the flash-over accident will happen.

The self-cleaning anti-pollution coatings used on the surface of the electrical equipment can reduce the pollution accumulation. The pollution will be washed down under the wind, rain, washing even if there is a little pollution. The electric power equipment surface can keep clean. This can reduce creepage distance and reduce electrical equipment investment for electrical equipment insulation. Furthermore, using the self-cleaning anti-pollution coating with strong hydrophobic property can effectively prevent moisture condensation and the accident for indoor distribution equipment located in wet regions.

2. Research progress of self-cleaning anti-pollution coating

Self-cleaning coating can achieve self-cleaning effect by changing the surface property of the base material such as glass, ceramics, metal or plastic. With the advantages of water saving, energy saving and environmental protection, it has important application prospects in construction, transportation, new energy and other industries [2, 3]. According to the principle of function, there are two kinds of self-cleaning coatings [3], super hydrophobic surface and super hydrophilic surface. Super hydrophobic surface, also named the "lotus effect", achieves self-cleaning effect by the water droplets scrolling down to wash away pollutants. The reason for the super hydrophilic surface keeping the surface clean is that when water droplets contact with the coating surface spreading rapidly form a uniform water film, because of the dirty gravity, the stain can be effectively taken away.

Considering the actual requirements for the power equipment, water cannot form a continuous film in insulation parts surface layer. The super hydrophobic coating technology is most likely to be applied in the electric power equipment insulation configuration. Organic silicon materials, fluorine materials, and silazane are the commonly used hydrophobic materials.

2.1. Organic silicon materials

Organic silicon has the property of low surface energy due to Si-O and Si-CH_3 in the structure. But the biggest disadvantage of this kind of material is that it is soft, low to the mechanical strength and easy to be smudge after curing. Silicone resin or silica sol with branched structure has high hardness after curing, but the coating is crisp and the hydrophobic property is weak because of the reduction of Si-CH_3 structure. At present, the superior organic silicon low-surface energy material products mainly come from foreign companies, such as Silres brand silicon resin of Wacker and Hydrozo silica sol of BASF SE. For the above two kinds of materials, the common problem is its low surface energy, which causes the weak adhesion properties to the base material. It greatly increased the construction difficulty and cost if base material processing and surface precoating are needed.

2.2. Fluorine materials

Fluorine polymer materials occupy a large portion in the field of low surface energy materials because of its good chemical stability, corrosion resistance, aging resistance, chemical inertness. The development time of fluorine polymer materials is the longest and the market share of fluorine polymer materials is the highest.

In 1943, DuPont, a foreign company, developed Teflon material, applied for patent, and applied it to high temperature resistant equipment such as non-stick pan, which obtained the ideal application effect. In 1965 ~ 1970, Atofina company and the Elf Altochem company successfully developed poly (vinylidene fluoride) (PVDF) and partial fluorine ethylene-co- tetrafluoroethylene-co-six fluoride resin. But the poor solubility and strong role in water-and oil-repellent made the two kinds of products can only be used in the form of solid powder. The application is limited greatly. Three fluoride vinyl chloride-co-perfluorinated alkyl vinyl ether copolymer (FEVE) resins, which were developed by Japanese Asahi Glass, opened up a new era in 1982. The product not only overcome the defects of PTFE and PVDF sintering to form film at high temperatures, and can be made after dissolved in organic solvent at room temperature curing solvent coatings. DuPont company developed a kind of water-based coating based on poly-perfluorinated alkyl surfactant, which can form a layer of excellent viscosity transparent film after baking at 100 ℃. This product has good applications at the car shell protective coating, interior wall coating and ship coating. 3M has developed new fluorine-containing materials, including Scotchgard, which is the most famous brand, which has been widely used. In addition, Solvay has developed a variety of special fluorinated monomers for low surface coating development for downstream manufacturers.

But general perfluoroalkane resins can be dissolubility and need to be heated and melted to apply. Hybridization with the other organic polymer, can solve the problem of the solubility and room temperature curing. But the low surface energy property, the aging resistance and the service life of perfluorinated resin will be reduced by adding the polyurethane, isocyanate resin into the original materials.

In addition, the fluorine carbon as the main structure of the whole fluorine alkane polymer materials is difficult to degrade, which has certain harm to the environment. Releasing small molecules with no reaction will pollute the environment, and has potential harm to human and biology. Moreover, the fluorine alkane resin material has low surface friction coefficient, good wear resistance. But its adhesion to the base material is poor because of the characteristics of low surface energy. When the fluorine alkane resin material is used as coating, it will affect its reliability. Formation of the fluorocarbon resin after hybridization with other organic resin, improves adhesion. But low coefficient of friction properties of resin is lost for the whole fluorine alkane. Because of the low coating hardness, general hardness under 2 h, and poor wear-resisting performance, the coating has the long-term shortcomings.

2.3. Silazane materials

In recent years, the organic-inorganic hybrid (hybrid) low-surface coating materials have attracted more and more attention, especially the Si-N based silazane resin material has been widely used. The characteristic of this type of material is that the structure contains an inorganic Si-N bond, while the side chain can be organic or H. The Si-N bond in silazane resin is polar, which is easy to combine with all kinds of substrate. Therefore, silazane resin has high adhesion. In addition, silazane can be cured with moisture, which can be single-component with high curing density and a maximum hardness of 9H. Inorganic coatings also have obvious advantages in durability, corrosion resistance and resistance to dielectric erosion. Compared with the traditional organic silicon materials, the curing process does not exist with Si-OH group, which does not cause contamination. In addition, the Si-N polar group in its chemical structure is oriented towards the base material, and the organic Si-CH_3 is out. Therefore, the coating can still have the characteristics of low surface energy under the precondition of excellent adhesion. At the same time, the Si-H key in its structure can also make the material have further modification, such as the hydrophobic fluorine structure can be introduced into the material system, which will further improve the self-cleaning performance of the material. The hybrid material has the characteristics of single unit and room temperature curing. It has high density, high hardness and excellent wear resistance. Moreover, the non-degradable materials in the hybrid material have no potential harm to the environment.

At present, the material has been used in high-speed trains in Europe and India, as well as in Japanese cars, which can be used as a water-repellent cleaner to protect the car's paint. There is a series of research results to poly silazane to the Institute of Chemistry Chinese Academy of Sciences, such as the application in military industry and other special areas. In recent years, many researches are development in poly silazane coating to the Institute of Chemistry Chinese Academy of Sciences [4-6]. It has also made significant breakthrough based on the low surface energy of silazane hydrophobic coating.

3. Design principle of self-cleaning anti-pollution coating

Generally speaking, the insulation protection layer of power equipment can be decreased when the power equipment often runs in high temperature, high humidity, and strong light and electrostatic field. The main reasons include: (1) The aging of the membrane materials under the action of wet heat transfer, light, acid and alkali corrosion. (2) The coating cannot have the characteristics of hydrophobic and hate oily. And for a long time, the oil in the environment is deposited on the surface of the coating, which destroys the original micro-nano structure and coating surface. (3) Sand erosion or rain erosion caused by coating in wind, sand and rainstorm.

Based on the above analysis, film-forming material must have aging resistance, hydrophobic, hate oily, high self-cleaning, high mechanical strength and wear-resisting.

A new type of coating material with high hardness, high self-cleaning and excellent insulation performance is developed from silazane polymer. The silicone self-cleaning coating developed has the following characteristics.

3.1. Excellent hydrophobic property

Silazane resin can form the silicon oxide as the nuclear material after curing with Si-N chain structure of low surface energy silicon nitrogen resin as film forming material. The formation process is shown in figure 1. If perfluorinated polyether structure is introduced on Si-N structure, the surface energy will be further reduced, which can ensure coating has excellent hydrophobic, water contact angle of 120°, rolling angle of less than 5°.

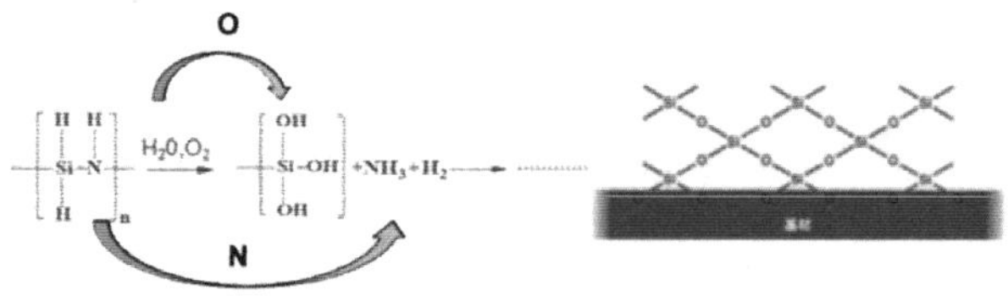

Figure 1. Curing process of silazane resin.

3.2. High self-cleaning

A suitable TiO_2 with photolysis can be added to the micronanometer filler in the design of coating formula with the silazane resin as membrane material. TiO_2 is a kind of n-type semiconductor material, and its forbidden band width is 3.2 eV.

When it is irradiated by UV-light with the wavelength less than or equal to 387.5 nm, the electron transition occurs. It will oxidize the surface -OH and H_2O to OH radicals. The free radicals, has high activity, can break down most of the organic matter into inorganic small molecules, CO_2 and H_2O. This also provides guarantee for the self-cleaning of the coating. However, due to the large amount of TiO_2 added, the surface hydrophilicity of the coating will be increased. Therefore, the photodecomposition performance of this project only serves as a supplement to oil and gas performance, to achieve high self-cleaning characteristics and not as the main function of coating.

3.3. High adhesion

The structure of silazane resin material as film-forming substances contains an inorganic Si-N bond while the lateral base can be organic or H (figure 2). The Si-N bond in silazane hybrid resin material is polar, easy to combine with all kinds of substrate. Therefore, the adhesion property is high.

3.4. High hardness

The silazane resin is cured with moisture, with high curing density and a maximum hardness of 9H, with good sand resistance.

In addition, the coating also has obvious advantage in durability, corrosion resistance and resistance to medium erosion ability, which will ensure the excellent aging resistances, anti-corrosion, prevent ultraviolet, heat resistance performance.

4. Preparation and construction technology of hard self-cleaning anti-pollution coating

Coating preparation technology is through the choice and cooperation of micro/nano particles to build micro/nano structure of the coating and through design and select the appropriate low surface energy materials for coating film-forming substances. The coating was formed by mixing and grinding micro/nano filler with low surface energy materials. Further through spraying, brush, dip coating and other conventional coating preparation method for the preparation of super hydrophobic coating.

The preparation process of the silazane self-cleaning coating is discussed in this work. Firstly, the micro/nano particles were choiced. Meantime, the surface of nano silica, titanium dioxide, alumina and other inorganic filler were treated by fluoride silane coupling agent in order to obtain the coating with super hydrophobic characteristics and hate oil characteristics. It also can improve the

IOP Conf. Series: Materials Science and Engineering **269** (2017) 012044 doi:10.1088/1757-899X/269/1/012044

compatibility of nano particles and nitrogen silicon resin and enhance the storage stability of coatings. Secondly, the prefabricated base glue was obtained by the initial mixing of the inorganic fillers with fluoride silicone resin and surface modification of the special design by the mixer. Thirdly, the prefabricated base glue was treating for a period of time by vacuum kneading machine. Base gum was produced by keeping the vacuum degree and temperature to a certain value. Through vacuum kneading polymer and inorganic filler can be more fully dispersed. Next, the materials produced by vacuum kneading machine will be further porphyries to avoid the possibility of powdery reunite through three roller grinding machines. Finally, mix the base gum with a certain amount of dispersants and a certain amount of dispersants through the two-planar mixer, and make high self-cleaning and anti-fouling flash paint and filling and sealing.

5. The prospect analysis of self-cleaning anti-pollution coating applied in power system

Through the optimization of micro/nano particles and the surface modification, and with low surface energy silicon nitrogen as the membrane material for resin material, the self-cleaning anti-pollution coatings with excellent hydrophobic, excellent self-cleaning, high adhesion and high corrosion resistance, were developed. The coating can satisfy the operation requirements of some electrical equipment. And it should have broad prospect of application in the power system.

5.1. Self-cleaning for Power equipment

The power equipment, which runs in the outdoor, can be accumulated by pollution easily in electric or natural conditions. Especially there are large factories produced dust around the power equipment, the original color of the surface almost cannot be seen, as shown in figure 2.

The surface of power equipment can keep cleaning with coating silicone self-cleaning coating. Even if there is brief surface area pollution, under the irrigation of heavy rain, the surface will remain clean.

Therefore, the silazane hard self-cleaning coatings can be widely used to maintain electrical equipment surfaces cleaning.

Figure 2. Dirty insulators.

5.2. Anti pollution flash-over for power equipment

The reason for the power equipment flash-over is following. First, the pollution layer was formed because of the pollution accumulation on the equipment surface. In dry environment, resistance of pollution layer is very high and the insulation performance will not be reduced. However, in wet environment, such as fog, dew, rain and snow, surface conductivity can be increased and insulation performance can be decreased after the pollution layer is wetting. This will cause flash-over discharge.

Because of excellent hydrophobic and self-cleaning properties of the silazane hard self-cleaning coating, the dust on the surface of the electric power equipment can be effectively removed. Even in the fog, dew, rain and snow weather, the water on the surface of the electrical equipment will be rolled into water droplets. The conductive path can't be formed. Therefore, silazane hard self-cleaning coatings can be widely used in electric power equipment surface to prevent pollution flashover.

5.3. Prevent condensation for power equipment

The environment temperature changes, condensate maybe occur in the surface of the electrical equipment. With condensate formed for long time on the surface of electrical equipment, the cabinet and operating mechanism will be rusted, which will cause the equipment operating mechanism unsmooth, parts corrosion damage and other problems. Therefore, the service life of the equipment cannot meet the design needs [7]. At the same time, the condensate can cause the insulation strength of the insulation material to be reduced, which can cause the accident, such as creepage, flash-over and so on [8].

Because of silazane self-cleaning coating has excellent hydrophobic, water contact Angle of 120°, rolling Angle of less than 5°, can make the water droplets under the action of gravity from the coating surface, thus achieve the goal of surface prevent condensation.

6. Conclusion

The development, design principles and possible applications in power system of silazane self-cleaning anti-pollution coatings were introduced in this paper. Because of the usage of low surface energy silazane as film-forming material make the coating developed has high hydrophobic property, high self-cleaning, high resistance, high adhesion, high hardness and high corrosion resistance. Therefore, the potential application value is existed for anti pollution, prevent flashover and prevent condensation of the power equipment.

References

[1] Wu L S, Zhao Y J, Wang G G, et al, 2015 *Silicon. Mater.* **29(5)** 353-9.
[2] Ganesh V A , Raut H K, Nair A S and Ramakrishna S R, 2011 *J. Mater. Chem.* **21** 16304-22.
[3] Sun X D and Zhang Z Y, 2010 *Paint. Coat. Ind.* **40(12)** 65-71.
[4] Wang D, Zhang Z B, Wang X F, et al, 2017 *Micronanoelectr. Tech.* **54(8)** 514-21.
[5] Zhang Z B, Xiao F Y, Luo Y M and Xu C H, 2013 *Paint. Coat. Ind.* **43(4)** 74-9.
[6] Zhang Z B, Xiao F Y, Luo Y M and Xu C H, 2013 *Fine. Spec. Chem.* **21(7)** 25-8.
[7] He X N and Zhang X, 2013 *Sci. Tech. Innov. Herald.* **11** 27-30.
[8] Yang R and Zhao F, 2014 *Major Sci. Tech.* 283-4.

Acknowledgements

This work was supported by Beijing Guodian Futong Science and Technology Development CO., LTD (FTZY201706).

Fatigue Reliability Analysis of 6156-T6 Riveted Joints based on Detail Fatigue Rating

An Chen[1], Jun Yang[1], Kan Zhang[1] and Haiying Zhang[1]

[1]Aircraft Strength Research Institute of China, Xi'an, China

Abstract. The fatigue behavior of two kinds of riveted joints 6156-T6 aluminum alloy structures was experimentally investigated. The detail fatigue rating (DFR) values of 6156-T6 is calculated based on statistical analysis of fatigue tests. The fatigue crack initiation and propagation behavior are examined using scanning electron microscopy. The results show that DFR value of 6156-T6 aluminum alloy with riveted lap joints is 97.81MPa, and the DFR value of rivet-filled countersink hole structure is 168.39MPa. The crack initiation sites occurred at the vicinity of the fastener hole in all specimens. The fatigue striations were uniformly spaced in the region of stable crack growth.

1. Introduction

6156 aluminum alloy was developed for higher damage tolerance capability than its predecessors (6xxx series aluminum alloy, e.g. 6056) by France Pechiney Company. In addition, the low density, excellent weldability and corrosion resistance made 6156 aluminum alloy more attractive for widespread used in aircraft structures [1, 2]. She studied the fatigue crack propagation rate of 6156 aluminum alloy in various aging states and different stress ratios [3]. Zhang investigated the effects of two-step aging treatment on the mechanical properties, electrical conductivity, intergranular corrosion and microstructure of 6156 aluminum alloy by means of tensile, conductivity measurement, intergranular corrosion experiments and transmission electron microscopy [4]. Lin researched the effects of aging treatments on the mechanical properties and corrosion behavior of 6156 A1-alloy [5].

However, the fatigue behavior of riveted joints in 6156 aluminum alloy remains rather limited. In the present work, fatigue tests on two kinds of riveted joints specimens from 6156-T6 A1-alloy were carried out. The DFR value of riveted joints was obtained by statistical analysis of experimental data. The fractography of fatigue fracture surface was examined also using scanning electron microscopy (SEM).

2. Experiment

Two categories of specimen were manufactured for fatigue tests: (i) riveted lap joints (RLJ); (ii) rivet-filled countersink hole (RCH). The dimensions of riveted lap joints specimens were shown in figure 1. The specimens were 300 mm-long, 40 mm-wide sheets, with 80 mm-long lap joints made of three two-rivet rows. The dimensions of rivet-filled countersink hole specimens were shown in figure 2. The rivets were NAS1097AD6, spaced 20 mm along the width of the specimen.

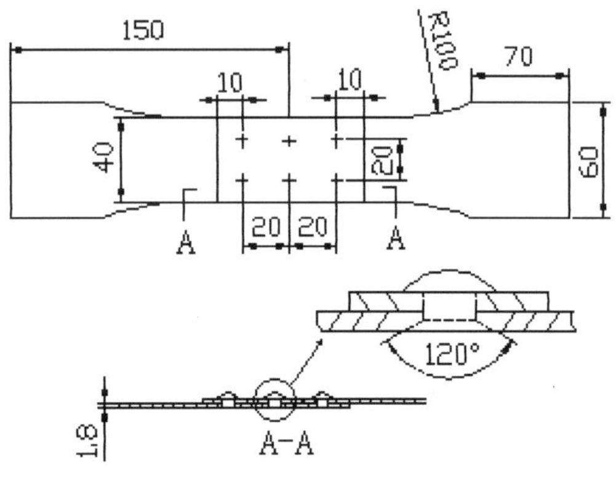

Figure 1. Dimension of riveted lap joints specimen.

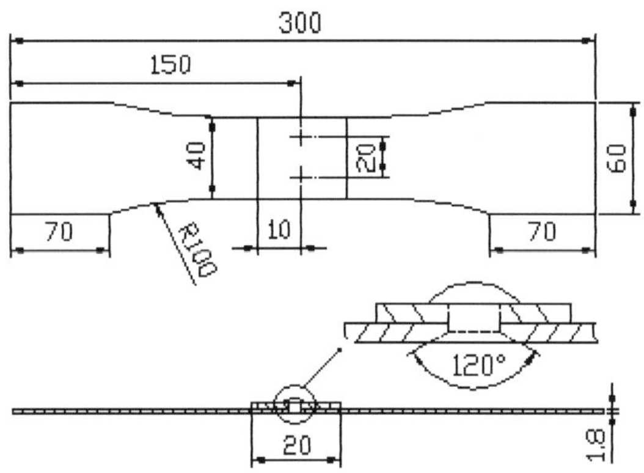

Figure 2. Dimension of rivet-filled countersink hole specimen.

The specimens was assembled from two aluminum alloy 6156-T6 plates with nominal thicknesses of 1.8 mm. Table 1 listed the chemical composition of 6156-T6 aluminum alloy.

Table 1. Chemical composition of 6156-T6 (mass fraction, %).

Si	Fe	Cu	Mn	Mg	Zn	Cr	Al
0.83	0.13	0.84	0.51	0.71	0.39	0.10	other

Fatigue tests were performed on INSTRON-8801 servo-hydraulic material testing machine (figure 3) in ambient air. The specimens were loaded in uniaxial tension with an R ratio of 0.06. The fatigue tests have been performed under a frequency of 20 Hz until crack occurred. Fatigue tests were conducted according to HB5287-96 specification.

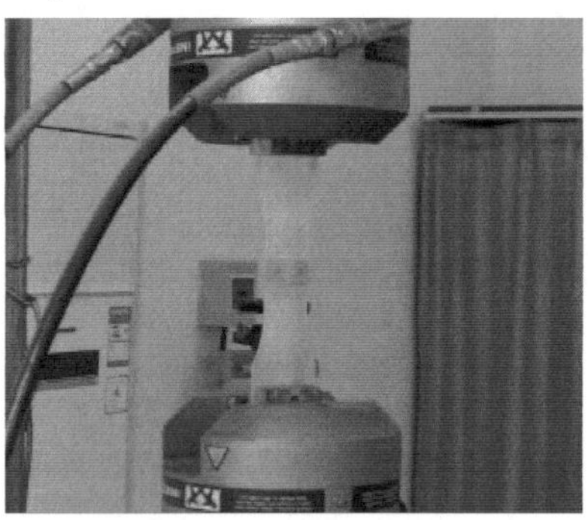

Figure 3. Riveted joints test.

3. Results and Analysis

3.1. Test Results

Table 2 shows the experimental fatigue results of two categories of 6156-T6 aluminum alloy riveted structures.

Table 2. Test results of fatigue test.

type	σ_{max}(MPa)	fatigue life (cycle)					
RLJ	95	246585	342293	276885	437748	405612	272703
RCH	180	292738	198533	160906	228576	176038	238931

3.2. Fatigue reliability assessment

The detail fatigue rating (DFR) means the maximum nominal of 10^5 cycles under constant amplitude loading with stress ratio 0.06, 95% confidence level and 95% confidence [6]. A structural joints having a higher DFR value leads to a better fatigue performance and a longer fatigue life. Existing uncertainties related to fatigue damage assessment are also accounted in DFR method. The relationship between the DFR and other fatigue parameters is shown as follows [7].

It is assumed that the fatigue life follows a two-parameter Weibull distribution, which can be expressed as:

$$F(N) = 1 - \exp[-\left(\frac{N}{\beta}\right)^{\alpha}] \tag{1}$$

where β is characteristic fatigue life, α is the scale parameter, which determines the shape of the distribution density curve of Weibull distribution.

According to maximum likelihood method, the point estimation of characteristic life can be expressed as:

$$\hat{\beta} = \left(\frac{1}{n}\sum_{i=1}^{n} N_i^{\alpha}\right)^{\frac{1}{\alpha}} \tag{2}$$

where n is number of valid data, N_i is experiment life, The fatigue material parameters used here is α = 4 for aluminum alloy.

The reliable fatigue life $N_{95/95}$ can be determined by:

$$N_{95/95} = \frac{\hat{\beta}}{S_C \cdot S_R \cdot S_T} \tag{3}$$

where $\hat{\beta}$ is the point estimation of characteristic fatigue life, S_C is the confidence level factor, S_R is the reliability level factor and S_T is the specimen factor. S_C with a 95% confidence level for n=6 is 1.15. S_C with a 95% confidence level is calculated as S_R=2.1. Specimen factor S_T is determined as 1.3.

Finally, the DFR of a structural detail can be calculated as:

$$DFR = \frac{(1-R)\sigma_{m0}}{0.94\sigma_{m0}/\sigma_{\max} S^{(5-\lg N)} - (0.47S^{(5-\lg N)} - 0.53) - R(0.47S^{(5-\lg N)} + 0.53)} \tag{4}$$

The fatigue life reliability assessment method is applied here to a structure subjected to a constant amplitude fatigue loading. The fatigue material parameters used here are S=2 and σ_{m0} =310MPa for aluminum alloys.

The calculated results of fatigue reliability assessment of 6156-T6 aluminum alloy are shown in table 3. The DFR value of 6156-T6 aluminum alloy with riveted lap joints is 97.81MPa, and the DFR value of rivet-filled countersink hole structure is 168.39MPa.

Table 3. Statistical analysis of test results.

Type	$\hat{\beta}$	$N_{95/95}$	DFR(MPa)
RLJ	352483	112273	97.81
RCH	229057	72960	168.39

3.3. Fractography

Figure 4 is the SEM image of fatigue fracture surface. Fatigue fracture can be divided into three areas: fatigue source area, fatigue crack growth area and instantaneous fracture area, as seen in figure 4(a). The crack initiation sites occurred at the vicinity of the fastener hole in all specimens. Fracture surfaces near the crack initiation sites show a cleavage-like microstructure caused by transcrystalline sliding fracture, as seen in figure 4(b). This is frequently observed in aluminum alloys when fatigue tests are conducted under ambient air. The fatigue crack growth area (figure 4(c)) shows very fine fatigue striations. The fatigue striations were uniformly spaced in the region of stable crack growth providing evidence of the nature of crack propagation characteristics of the alloy in the T6 microstructural condition. The instantaneous fracture area (figure 4(d)) shows a ductile fracture feature resulting from void growth and coalescence. Dispersion of microscopic voids and a number of shallow dimples are present at the grain facets.

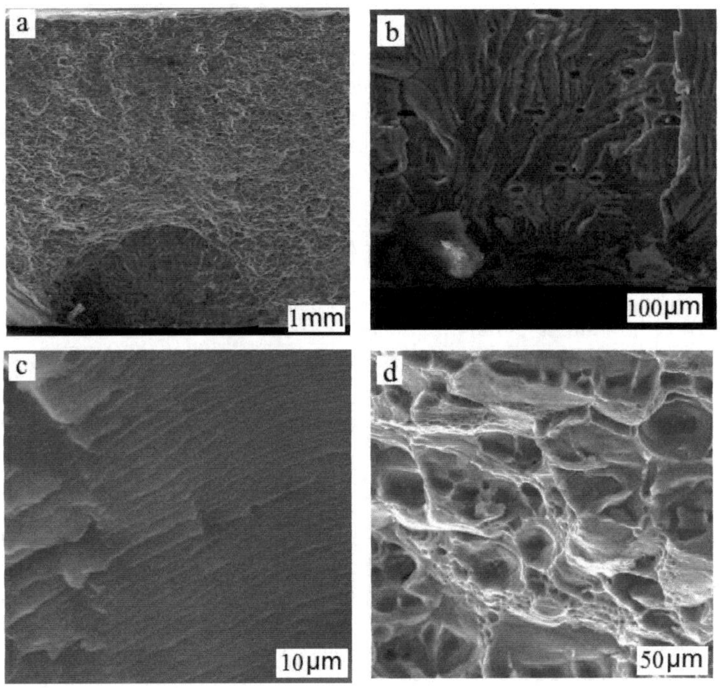

Figure 4. SEM image of fatigue fracture surface.

4. Conclusions

Fatigue tests on two kinds of riveted joints 6156-T6 aluminum alloy specimens are carried out. The DFR values of riveted joints are calculated by statistical analysis of experimental data.The DFR value of 6156-T6 aluminum alloy with riveted lap joints is 97.81MPa, and the DFR value of rivet-filled countersink hole structure is 168.39MPa. Fatigue fractography can be divided into three typical areas:

fatigue source area, fatigue crack growth area and instantaneous fracture area. The crack initiation sites occurred at the vicinity of the fastener hole in all specimens. Fatigue source area shows a cleavage-like microstructure. The fatigue striations were uniformly spaced in the region of stable crack growth, which is the crack growth characteristics of the alloy in the T6 microstructural condition. The instantaneous fracture area shows a ductile fracture feature resulting from a number of shallow dimples and microscopic voids.

5. References

[1] Manabu N and Takehiko E 2000 *Mater. Sci. Eng. A* **285** 62–8.
[2] Morgeneyer T F, Starink M J and Sinclair I 2006 *Mater. Sci. Forum.* **519** 934–42.
[3] She L J, Zheng Z Q, Zhong J and Lin L 2011 *Mater. Mech. Eng.* **35** 19–23.
[4] Zhang H F, Zheng Z Q, Zhong S, Luo X F and Zhong J 2012 *Chin. J. Nonferrous. Met.* 22 1025–32.
[5] Lin L, Zheng Z Q and Li J F 2012 *Rare. Metal. Mat. Eng.* **41** 2163–68.
[6] Chen Y J 1989 *Hand Book of Durability and Damage Tolerance Design for Flight Structure, Fatigue Analysis for Flight Structure* (Xi'an: Aeronautics and Astronautics Ministry Scientific Design Academy) pp 361–3.
[7] Chen A, Li S S, Dong D K and Xu F 2016 *Mech. Sci. Technol. Aerosp. Eng.* **35** 489–92.

IOP Conf. Series: Materials Science and Engineering **269** (2017) 012046 doi:10.1088/1757-899X/269/1/012046

Study on Macro-fiber Composite Coupled-Plate Structures

Jiarui Zhang[1], Jianwei Tu[1], Fangpeng Lai[1], Wei Luo[1], Chenfei Zhu[1]

[1]Hubei Key Laboratory of Roadway Bridge and Structure Engineering, Wuhan University of Technology, Wuhan, Hongshan District, Luoshi Road No 122

Abstract. In this paper, the dynamic and vibration control effect of macro-fiber composite (MFC) were studied on the plate and shell structure, proposing a new calculation method of MFC coupled-plate structure actuating equation. Based on the first kind of piezoelectric equation, this paper deduces the equation of P1-type MFC composite plate structure, considering the influence of the Poisson's ratio of the controlled structure and the effect of the MFC plane strain on MFC's actuating force and actuating moment. In this paper, the vibration control experiment of the MFC coupled-plate structure is completed. The deviation between the experimental results and the simulation results is less than 7.5%, indicating that the P1-type MFC actuating equation is correct and can be used in the simulation calculation of vibration control for the MFC coupled-plate structure.

1. Introduction

As the most common piezoelectric materials so far, piezoelectric ceramics are characterized by high sensitivity and temperature stability. Some deficiencies, such as poor flexibility and piezoelectric effect, restrict its application into practical projects, though. Macro Fiber Composite (MFC), a new type of piezoelectric composite, is compounded of piezoelectric ceramic fibers, polymer matrix and interdigitated electrodes. Compared with traditional piezoelectric ceramics, polymer matrix in MFC enhances the material flexibility [1]. By making piezoelectric ceramics fibrous phase and coordinating them with finger electrodes, it tremendously improves piezoelectric effect of piezoelectric materials. Therefore, MFC is of the best overall performance at present.

With its “inverse piezoelectric effect”, a main application of MFC is to change its actuating force through adjusting input voltage, so as to implement active control over the structural vibration. Therein, a crucial step is establishing an accurate MFC actuating equation which many scholars have studied in recent years. Considering the influence of polymer matrix on the composite, S. Sreenivasa Prasath et al. have made further researches on the influence of the volume fraction represented by piezoelectric ceramic fibers on MFC elasticity modulus, piezoelectric coefficients, stress strain, etc. by establishing the MFC mechanical model with equivalent stratification [2]. Kateřina Steiger et al. have established the MFC finite element model based on the hypothesis of approximating MFC into clintheriform well-distributed piezoelectric material [3]. Only with the mechanical analysis of MFC material, those researches have not obtained the relationship between the actuating force and input voltage, remaining a certain gap to MFC implementing the active control. Guo Yi et al. have simplified the constitutive equation of MFC composite to be two-dimensional, ignored the transformation of through-thickness direction, and deduced the one-dimensional stress and electricity domino effect formula of MFC [4]. It is considered that the thickness direction is the polarization direction of the piezoelectric fiber, not the longitudinal direction. And he has not considered the Poisson’s ratio influence of the controlled structure when calculating the activation force and activation bending moment, which produces a great error in the result.

Hereby, this paper not only considers the influence of Poisson's ratio of the controlled structure on MFC activation force and activation bending moment, but the place strain produced by MFC actuation, deduces the MFC two-dimensional actuating equation for plate structures, and obtains the accurate MFC strength and electricity relationship, on the basis of which the active control simulation is performed on the vibration of a plate structure and the active control experiment is conducted.

2. The actuating equation of MFC coupled-plate structures

The actuating equation of MFC coupled-plate structures is deduced on the basis of the first kind of piezoelectric basic equations. The stress in MFC is obtained by transformation of the first kind of piezoelectric basic equation; the actuating equation of MFC coupled-plate structures is deduced by the stress equation; and then the MFC activation force and activation bending moment are obtained for plate structures.

2.1. *MFC piezoelectric equation*

The active layer of MFC is treated as homogeneous piezoelectric fibers, as is shown in figure 1. The constitutive equation of MFC linear-elastic piezoelectric material with orientation polarization on x is:

$$\begin{bmatrix} \varepsilon_z \\ \varepsilon_y \\ \varepsilon_x \\ \gamma_{yx} \\ \gamma_{xz} \\ \gamma_{zy} \end{bmatrix} = \begin{bmatrix} s_{11}^E & s_{12}^E & s_{13}^E & 0 & 0 & 0 \\ s_{21}^E & s_{22}^E & s_{23}^E & 0 & 0 & 0 \\ s_{31}^E & s_{32}^E & s_{33}^E & 0 & 0 & 0 \\ 0 & 0 & 0 & s_{44}^E & 0 & 0 \\ 0 & 0 & 0 & 0 & s_{55}^E & 0 \\ 0 & 0 & 0 & 0 & 0 & s_{66}^E \end{bmatrix} \begin{bmatrix} \sigma_z \\ \sigma_y \\ \sigma_x \\ \tau_{yx} \\ \tau_{xz} \\ \tau_{zy} \end{bmatrix} + \begin{bmatrix} 0 & 0 & d_{31} \\ 0 & 0 & d_{32} \\ 0 & 0 & d_{33} \\ 0 & d_{24} & 0 \\ d_{15} & 0 & 0 \\ 0 & 0 & 0 \end{bmatrix} \begin{bmatrix} 0 \\ 0 \\ E_x \end{bmatrix} \tag{1}$$

In the equation, s_{ij}^E is the flexibility coefficient, in which *i* represents the MFC strain direction, *j* the MFC stress direction; d_{ii} is the piezoelectric strain coefficient, in which *i* represents the electric field action direction, *j* the strain direction of piezoelectric material under electric field action, the numbers 1, 2, 3 respectively correspond to *z*, *y*, *x* directions in Cartesian coordinate system, and 4, 5, 6 the shear deformation directions *yx*, *xz*, *zy*; E_x is the electric field intensity in *x* direction.

MFC is lamelliferous with the thickness of only 0.3mm. According to the vibration theory of thin shells, stress and strain in the through-thickness direction can be ignored. Since MFC shear strain γ_{xz} is irrelevant to the electric field intensity E_x and its own shear strain τ_{xz} is rather small, the influence of shear strain can be ignored. Therefore, equation (1) can be simplified as:

$$\begin{bmatrix} \varepsilon_z \\ \varepsilon_x \end{bmatrix} = \begin{bmatrix} s_{11}^E & s_{13}^E \\ s_{13}^E & s_{33}^E \end{bmatrix} \begin{bmatrix} \sigma_z \\ \sigma_x \end{bmatrix} + \begin{bmatrix} d_{31} \\ d_{33} \end{bmatrix} E_x \tag{2}$$

The electric field intensity E_x in MFC is:

$$E_x = \frac{U}{t} \tag{3}$$

in which U is the applied voltage amplitude, and t is the spacing of interdigital electrodes.

2.2. *The actuating equation of MFC coupled-plate structures*

MFC coupled-plate structure is shown in figure 2, in which M_x is the MFC equivalent moment for the controlled structure; F_x is the MFC equivalent acting force for the controlled structure; H is the plate thickness; a is the effective length of MFC, b is the width, and h is the thickness; L is the position of neutral axis. Because its thickness is extremely small, L=0.5H.

IOP Conf. Series: Materials Science and Engineering **269** (2017) 012046 doi:10.1088/1757-899X/269/1/012046

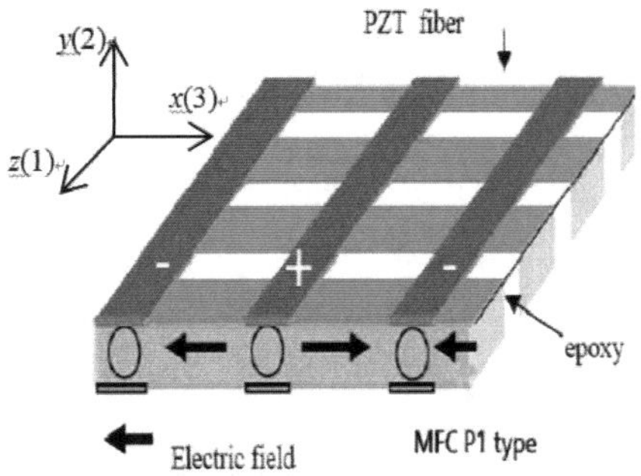

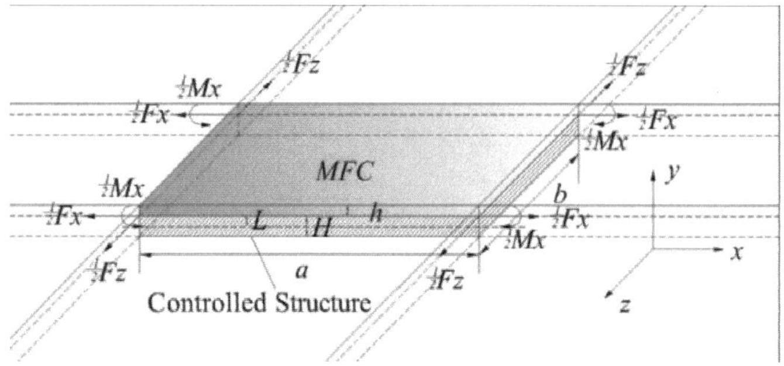

Figure 1. The construction of MFC. **Figure 2.** The construction of plate equipped with MFC.

The two-dimensional MFC stress is obtained through formula (2):

$$\begin{bmatrix}\sigma_z\\ \sigma_x\end{bmatrix}=\begin{bmatrix}\dfrac{s_{33}^E}{s_{11}^E s_{33}^E-s_{13}^E s_{31}^E} & \dfrac{-s_{13}^E}{s_{11}^E s_{33}^E-s_{13}^E s_{31}^E}\\ \dfrac{-s_{31}^E}{s_{11}^E s_{33}^E-s_{13}^E s_{31}^E} & \dfrac{s_{11}^E}{s_{11}^E s_{33}^E-s_{13}^E s_{31}^E}\end{bmatrix}\begin{bmatrix}\varepsilon_z-d_{31}E_x\\ \varepsilon_x-d_{33}E_x\end{bmatrix} \tag{4}$$

The two-dimensional flexibility matrix can be represented as:

$$\begin{bmatrix}s_{11}^E & s_{13}^E\\ s_{31}^E & s_{33}^E\end{bmatrix}=\begin{bmatrix}\dfrac{1}{E_{fz}} & \dfrac{-v_f}{E_{fx}}\\ \dfrac{-v_f}{E_{fz}} & \dfrac{1}{E_{fx}}\end{bmatrix} \tag{5}$$

in which E_{fx}, E_{fz} are elasticity modulus of MFC in x direction and in y direction respectively; v_f the Poisson's ratio of MFC.

$$\varepsilon_{fx}=\frac{d_{33}U}{t},\varepsilon_{fz}=\frac{d_{31}U}{t} \tag{6}$$

in which ε_{fz}, ε_{fx} are the strain of MFC created by voltage in z direction and in x direction respectively;

The actuating force and actuating moment of MFC for the plate structure are:

$$\begin{cases}F_z=-a\int_L^{L+h}\sigma_{fz}dy=a\int_L^{L+h}\left[Q_{11}^E(\varepsilon_{fz}-\varepsilon_z)+Q_{13}^E(\varepsilon_{fx}-\varepsilon_x)\right]dy\\ F_x=-b\int_L^{L+h}\sigma_{fx}dy=b\int_L^{L+h}\left[Q_{31}^E(\varepsilon_{fz}-\varepsilon_z)+Q_{33}^E(\varepsilon_{fx}-\varepsilon_x)\right]dy\end{cases} \tag{7}$$

$$\begin{cases}M_z=a\int_L^{L+h}(-\sigma_{fz})ydy=a\int_L^{L+h}\left[Q_{11}^E(\varepsilon_{fz}-\varepsilon_z)+Q_{13}^E(\varepsilon_{fx}-\varepsilon_x)\right]ydy\\ M_x=b\int_L^{L+h}(-\sigma_{fx})ydy=b\int_L^{L+h}\left[Q_{31}^E(\varepsilon_{fz}-\varepsilon_z)+Q_{33}^E(\varepsilon_{fx}-\varepsilon_x)\right]ydy\end{cases} \tag{8}$$

Since ε_x, ε_z in those equations are unknown, ε_x, ε_z are assumed to be in linear distribution along the through-thickness direction of the structure and the constitutive relation is obtained between the plate and MFC, as is shown below [5].

$$\varepsilon_x=\alpha_x y+\varepsilon_{x0},\varepsilon_z=\alpha_z y+\varepsilon_{z0} \tag{9}$$

$$\begin{cases} \int_{L}^{L+h} \frac{E_{fz} y}{1-v_f^2}\left(\varepsilon_z + v_f \varepsilon_x - \varepsilon_{fz} - v_f \varepsilon_{fx}\right) dy + \int_{-(H-L)}^{L} \frac{E_p y}{1-v_p^2}\left(\varepsilon_z + v_p \varepsilon_x\right) dy = 0 \\ \int_{L}^{L+h} \frac{E_{fz}}{1-v_f^2}\left(\varepsilon_z + v_f \varepsilon_x - \varepsilon_{fz} - v_f \varepsilon_{fx}\right) dy + \int_{-(H-L)}^{L} \frac{E_p}{1-v_p^2}\left(\varepsilon_z + v_p \varepsilon_x\right) dy = 0 \\ \int_{L}^{L+h} \frac{E_{fx} y}{1-v_f^2}\left(\varepsilon_x + v_f \varepsilon_z - \varepsilon_{fx} - v_f \varepsilon_{fz}\right) dy + \int_{-(H-L)}^{L} \frac{E_p y}{1-v_p^2}\left(\varepsilon_x + v_p \varepsilon_z\right) dy = 0 \\ \int_{L}^{L+h} \frac{E_{fx}}{1-v_f^2}\left(\varepsilon_x + v_f \varepsilon_z - \varepsilon_{fx} - v_f \varepsilon_{fz}\right) dy + \int_{-(H-L)}^{L} \frac{E_p}{1-v_p^2}\left(\varepsilon_x + v_p \varepsilon_z\right) dy = 0 \end{cases} \tag{10}$$

α_z, α_x are slopes of strain distribution; ε_{x0}, ε_{z0} are the membrane strain in x direction and in z direction respectively; σ_{fx}, σ_{fz} are the stress in MFC, and E_p is the elasticity modulus of the plate; ε_{pz}, ε_{px} are the strain of the plate; v_p is the Poisson's ratio of MFC. Then the equilibrium equation is written to solve α_x, α_z, ε_{x0}, ε_{z0}.

To sum up

$$\begin{cases} F_x = \frac{bE_p H\left[2CT - 2DS + (H-2L)(CU - DT)\right]\left(\varepsilon_{fx} + v\varepsilon_{fz}\right)}{2\left(T^2 - SU\right)\left(1-v^2\right)} \\ F_z = \frac{aE_p H\left[2QA - 2BP + (H-2L)(AR - BQ)\right]\left(\varepsilon_{fz} + v\varepsilon_{fx}\right)}{2\left(Q^2 - PR\right)\left(1-v^2\right)} \end{cases} \tag{11}$$

$$\begin{cases} M_x = \frac{bE_p H\left(\varepsilon_{fx} + v\varepsilon_{fz}\right)}{6\left(T^2 - SU\right)\left(1-v^2\right)}\left[2\left(H^2 - 3HL + 3L^2\right)(DT - CU) + 3(H-2L)(DS - CT)\right] \\ M_z = \frac{aE_p H\left(\varepsilon_{fz} + v\varepsilon_{fx}\right)}{6\left(Q^2 - PR\right)\left(1-v^2\right)}\left[2\left(H^2 - 3HL + 3L^2\right)(QB - RA) + 3(H-2L)(BP - AQ)\right] \end{cases} \tag{12}$$

in which

$$A = \frac{E_{fz}\left[(L+h)^2 - L^2\right]}{2\left(1-v^2\right)}, B = \frac{E_{fz} h}{1-v^2}, C = \frac{AE_{fx}}{E_{fz}}, D = \frac{BE_{fx}}{E_{fz}}, P = \frac{E_{fz}\left[(L+h)^3 - L^3\right]}{3\left(1-v^2\right)} + \frac{E_p\left[(H-L)^3 + L^3\right]}{3\left(1-v^2\right)}$$

$$Q = \frac{E_{fz}\left[(L+h)^2 - L^2\right]}{2\left(1-v^2\right)} + \frac{E_p\left[L^2 - (H-L)^2\right]}{2\left(1-v^2\right)}, R = \frac{E_{fz} h}{1-v^2} + \frac{E_p H}{1-v^2}, S = \frac{E_{fx}\left[(L+h)^3 - L^3\right]}{3\left(1-v^2\right)} + \frac{E_p\left[(H-L)^3 + L^3\right]}{3\left(1-v^2\right)}$$

$$T = \frac{E_{fx}\left[(L+h)^2 - L^2\right]}{2\left(1-v^2\right)} + \frac{E_p\left[L^2 - (H-L)^2\right]}{2\left(1-v^2\right)}, U = \frac{E_{fx} h}{1-v^2} + \frac{E_p H}{1-v^2}$$

Because the Poisson's ratios hardly differ among regular materials, it is assumed that the Poisson's ratios of the plate and MFC are equal and the minimum value is taken to calculate the action of MFC over the controlled structure and solve the simultaneous the mechanical equilibrium equation.

3. The actuating equation of MFC coupled-plate structures

To verify the correctness of the equation, the model is verified through vibration control experiment and analogue simulation.

In the analogue simulation, modeling is conducted through ANSYS, and the modeling parameters of the plate model are illustrated in table 1. Then the exciting force and the actuating force of MFC are

IOP Conf. Series: Materials Science and Engineering **269** (2017) 012046 doi:10.1088/1757-899X/269/1/012046

defined, with exciting position of the exciting force in the center of the plate and the paste position of MFC shown by the red lines in figure 3. The actuating force and actuating bending moment provided by MFC and the plate structure are obtained by the actuating equation deduced in the first section of the paper, which are exerted into the node located at the red lines in figure 3. In simulation, the exciting force and the MFC working voltage are both in simple harmonic form, and the exciting force are made opposite to the displacement phase created by MFC actuating force and actuating bending moment, so that the MFC actuating force and actuating bending moment can decrease vibration. The exciting forces in simulation analysis are in two working conditions, 20Hz, 10N and 50Hz, 10N respectively, and MFC working voltage is 300V.

Table 1. Material Properties.

Material	Elasticity Modulus (GPa)	Density (kg m^3)	Poisson's Ratio	Length (mm)	Width (mm)	Thickness (mm)
Plate (aluminum)	70.5	2700	0.31	400	400	2

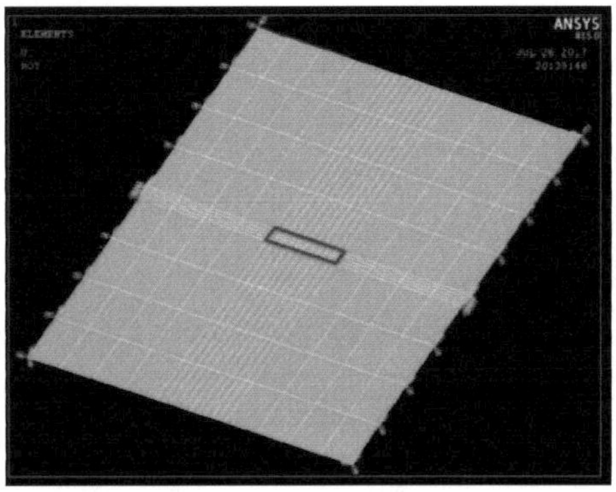

Figure 3. The ANSYS model of plate.

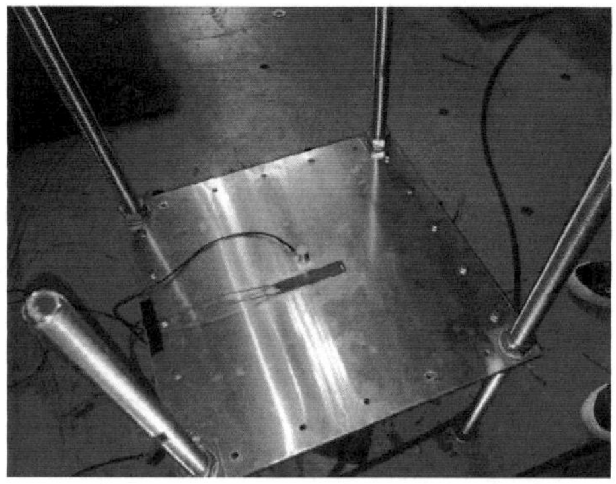

Figure 4. The model of plate experiment.

In the vibration control experiment of MFC coupled-plate, the plate parameters are the same as in table 1. The experimental system is composed of sensing, controlling, and exciting, in which the acceleration sensor is adopted as the sensor part, MFC the control part, and the exciter the exciter part, shown in figure 4. In the experiment, two sinusoidal signals, 20Hz and 50Hz, output by dSPACE provide exciting signals for the exciter and MFC. The figure 5 shows the acceleration comparison of simulation and experiment before and after the control with 20Hz and 50Hz excitation frequency and 300V working voltage. It is evident that the deviation of simulation and experiment under two working conditions is less than 7.5%, proving that the actuating equation deduced in this paper is accurate and applicative. Furthermore, MFC can decrease vibration of the plate structure effectively. When the exciting force is 20Hz, 10N, and the working voltage is 300V, MFC in the experiment can decrease the acceleration amplitude by 7%; when the exciting force is 50Hz, 10N, and the working voltage is 300V, MFC in the experiment can decrease the acceleration amplitude by 13%.

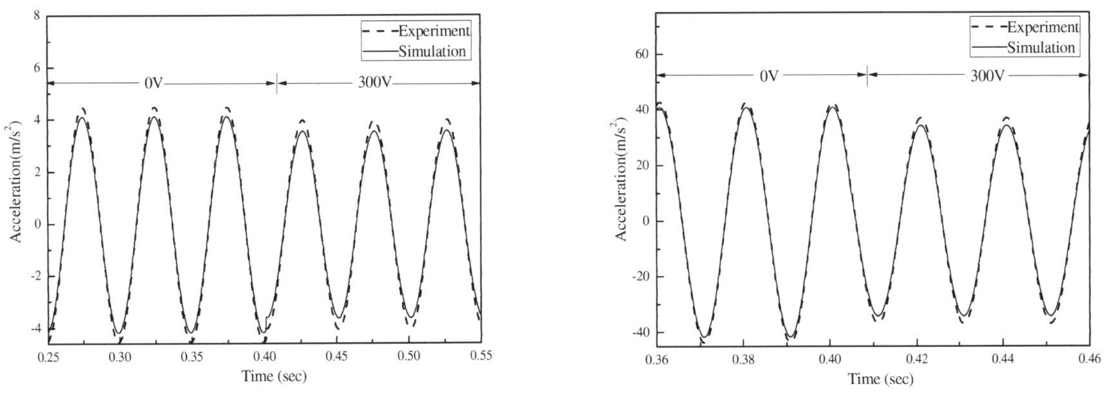

Figure 5. The comparison of the acceleration of simulation and experiment before and after control under 20Hz (left) and 50Hz (right) excitation.

4. Conclusion

This paper studies plate structures, deduces a new actuating force and actuating bending moment equation of MFC coupled-plate structures, and conducts the comparison validation about simulation and experiment, reaching conclusions as follows:

The deduced actuating equation of two-dimensional MFC coupled-plate structures considers the influence of membrane strain produced by actuation between plate-structural Poisson's ratio and MFC on MFC actuating force. This paper obtain MFC's actuating force and actuating moment on the plate structure.

The plate vibration control simulation and experiment is accomplished. Through analogue simulation and experimental verification, the comparison results are obtained before and after the control under 20Hz, 10N and 50Hz, 10N excitation. Under the working conditions of 20Hz, 10N and 50Hz, 10N, the deviation of simulation and experiment is less than 7.5%, which proves that the MFC actuating equation in this paper is correct and applicative to the plate structure, and that MFC produces good effect on vibration control of plate structures.

References

[1] Hou Z W, Chen R W, Xu Z W and Liu Q 2010 *J. Vibr. Meas. Diagn.* **30** 51-2.
[2] Sreenivasa Prasath S and Arockiarajan A 2013 *Int. J. Mech. Sci.* **77** 98-106.
[3] Steiger K and Mokrý P 2015 *Smart. Mater. Struct.* **24** 25-51.
[4] Yi G, Liu Y J and Leng J S 2011 *Proc. SPIE* **79772C** 1-7.
[5] RezaMoheimani S O and Fleming A J 2006 *Piezoelectric transducers for vibration control and damping* (London:Springer-Verlag) 23-9.

Acknowledgments

This research work is supported by the Chinese National Natural Science Foundation, project No 51478372, Hubei Province Innovation Group Project, project No 2016CFA020 and the Fundamental Research Funds for the Central Universities, project No 2017-zy-055.

IOP Conf. Series: Materials Science and Engineering **269** (2017) 012047 doi:10.1088/1757-899X/269/1/012047

Dynamic Response of Two Rigid Foundations on Multi-Layered Poroelastic Medium

Suraparb Keawsawasvong[1], Teerapong Senjuntichai[1,*]

[1]Applied Mechanics and Structures Research Unit, Department of Civil Engineering, Faculty of Engineering, Chulalongkorn University, Bangkok 10330

Abstract. This paper presents dynamic response of two rigid foundations resting on a multi-layered poroelastic medium under time-harmonic vertical loading. The contact surface between the foundations and the layered medium is assumed to be smooth and fully permeable. This dynamic interaction problem is studied by employing a discretization technique and an exact stiffness matrix scheme. The accuracy of the present solution is verified and selected numerical results for impedances of two square foundations are also presented.

1. Introduction

The study of dynamic interaction between foundations and soils is of considerable importance in civil engineering. Several researchers investigated dynamic response of vertically loaded rigid foundations resting on an elastic half-space by employing a variety of analytical and numerical techniques. For example, Wong and Luco [1] presented dynamic response of rigid foundations of arbitrary shape resting on an elastic half-space by using a discretization technique. Subsequently, Wong and Luco [2] extended their technique to study interaction between two rigid square foundations. In addition, the case of two rigid rectangular foundations was also studied by Triantafyilidis and Prange [3]. Geomaterials are often two-phase materials consisting of an elastic solid skeleton with voids filled with water, commonly known as poroelastic materials. In the past, dynamic response of vertically loaded foundations involving poroelastic materials was investigated extensively by adopting Biot's poroelastodynamics theory [4]. For example, the case of a rigid rectangular foundation on a homogeneous poroelastic half-space was studied by Halpern and Christiano [5]. Vertical vibrations of a circular plate in a multi-layered poroelastic medium were also presented by employing the exact stiffness matrix method [6]. In practice, there exists a situation of a closely spaced foundation system where the interaction between adjacent foundations needs to be taken into consideration. A review of literature indicates that studies on dynamic interaction between two rectangular foundations and a multi-layered poroelastic medium have never been reported in the past.

In this paper, dynamic interaction between two rigid rectangular foundations and a multi-layered poroelastic medium subjected to time-harmonic vertical loading is presented. The geometry of foundations-layered system is shown in Figure 1. Each layer of the multi-layered half-space is governed by Biot's poroelastodynamics theory [4]. The contact surface between the foundation and the layered medium is assumed to be smooth and fully permeable. To solve this interaction problem, the contact surface is discretized into a number of square elements. The unknown contract traction within each discretized element is then solved from the flexibility equations based on the influence functions, which are determined by employing the exact stiffness matrix method [6]. The accuracy of the proposed solution scheme has been verified, and selected numerical results for impedances of two identical square foundations resting on a multi-layered poroelastic half-space are also presented.

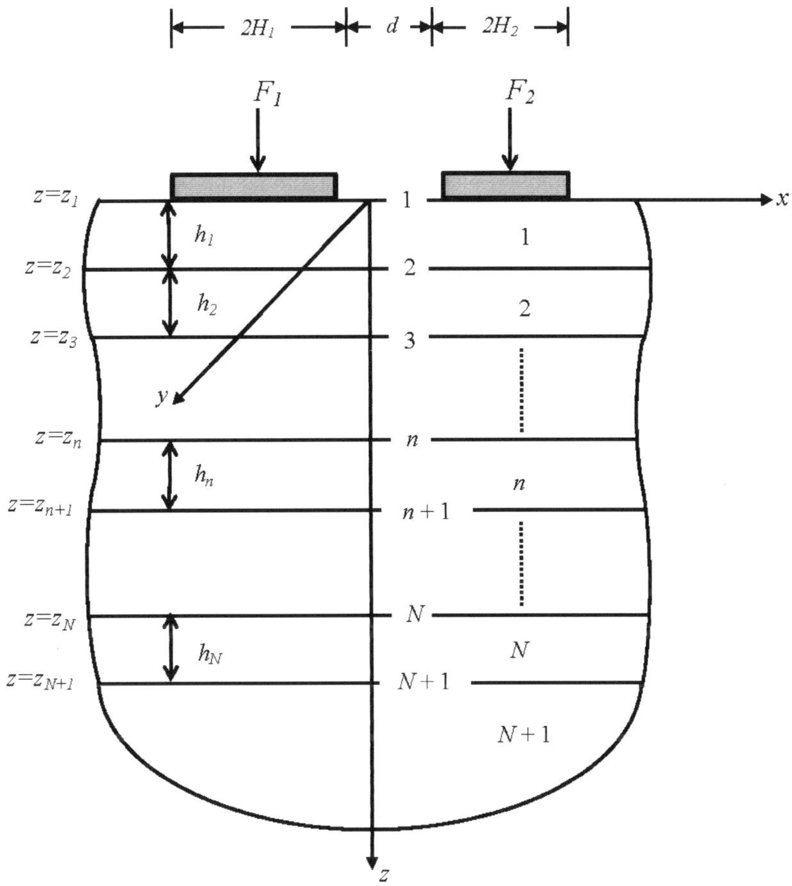

Figure 1. Geometry of rigid foundations on a multi-layered poroelastic medium.

2. Basic equations

Consider a poroelastic half-space with a Cartesian coordinate system (x, y, z) defined such that the z-axis is perpendicular to the free surface as shown in figure 1. Let $u_i(x, y, z, t)$ and $w_i(x, y, z, t)$ denote the average displacement of the solid matrix and the fluid displacement relative to the solid matrix in the i-direction ($i = x, y, z$) respectively. The constitutive relation can be expressed following Biot's theory for a poroelastic material by using the standard indicial notation as

$$\sigma_{ij} = 2\mu\varepsilon_{ij} + \lambda\delta_{ij}e - \alpha\delta_{ij}p \quad , \qquad i, j = x, y, z \tag{1a}$$

$$p = -\alpha M \varepsilon_{kk} + M\zeta \tag{1b}$$

In equations (1a) and (1b), σ_{ij} is the total stress component of the bulk material; ε_{ij} is the strain component of the solid matrix; μ and λ are Lame' constants of the bulk material; p is the excess pore fluid pressure; and ζ is the variation of fluid content per unit reference volume. In addition, α and M are Biot's parameters accounting for compressibility of the two-phased material. In this study, the motion is assumed to be time-harmonic with the factor of $e^{i\omega t}$, where ω is the frequency of excitation.

Lu and Jeng [7] showed that the general solutions to Biot's equations can be obtained by using a Helmholtz representation and the Fourier integral transforms. The general solutions [7] can be expressed in the frequency-wave number domain in the following matrix form:

$$\mathbf{u}(k_x,k_y,z) = \mathbf{R}(k_x,k_y,z)\mathbf{c}(k_x,k_y) \tag{2a}$$

$$\boldsymbol{\sigma}(k_x,k_y,z) = \mathbf{S}(k_x,k_y,z)\mathbf{c}(k_x,k_y) \tag{2b}$$

where

$$\mathbf{u}(k_x,k_y,z) = \left[\mathrm{i}\bar{u}_x \ \mathrm{i}\bar{u}_y \ \bar{u}_z \ \bar{p}\right]^T \tag{3a}$$

$$\boldsymbol{\sigma}(k_x,k_y,z) = \left[\mathrm{i}\bar{\sigma}_{xz} \ \mathrm{i}\bar{\sigma}_{yz} \ \bar{\sigma}_{zz} \ \bar{w}_z\right]^T \tag{3b}$$

$$\mathbf{c}(k_x,k_y) = \left[A \ B \ C \ D \ E \ F \ G \ H\right]^T \tag{3c}$$

and the arbitrary functions $A(k_x,k_y)$ to $H(k_x,k_y)$ can be determined by employing appropriate boundary and continuity conditions.

3. Analysis of interaction problem

Consider a soil-structure interaction problem involving two rectangular foundations, namely Foundation 1 and Foundation 2, separated by a distance d and resting on a multi-layered poroelastic half-space as shown in figure 1. The contact areas under Foundations 1 and 2 are denoted by A_1 and A_2, respectively. The two foundations are assumed to be rigid, fully permeable and in smooth contact with the layered half-space. When Foundation 1 is subjected to time-harmonic vertical loading, Foundation 2 will also experience a part of radiated energy dissipated from Foundation 1. Consequently, Foundation 2 starts vibrating accordingly, and also generates some radiated energy that could affect Foundation 1. It is noted that the interaction between foundations would be significant if the two foundations are relatively close. To investigate this interaction problem, Foundation 1 is assumed to undergo time-harmonic vertical displacement of amplitude Δ_z^1 whereas the displacement of the second foundation is restrained to be zero. The following relationship can then be established:

$$\int G_{zz}^{11}(\mathbf{r},\hat{\mathbf{r}})T_z^1(\mathbf{r})dA_1 + \int G_{zz}^{12}(\mathbf{r},\hat{\mathbf{r}})T_z^2(\mathbf{r})dA_2 = \Delta_z^1 \tag{6a}$$

$$\int G_{zz}^{21}(\mathbf{r},\hat{\mathbf{r}})T_z^1(\mathbf{r})dA_1 + \int G_{zz}^{22}(\mathbf{r},\hat{\mathbf{r}})T_z^2(\mathbf{r})dA_2 = 0 \tag{6b}$$

In equations (6a) and (6b), $T_z^i(\hat{\mathbf{r}})$ (i = 1, 2) denote the normal contact traction generated at the contact surface of Foundation i; $G_{zz}^{ij}(\mathbf{r},\hat{\mathbf{r}})$ ($i = j$ = 1, 2) denotes the influence functions, which are the vertical displacement of Foundation i at any point with the position vector $\mathbf{r}$ due to a unit vertical point load applied at a point on Foundation j with the position vector $\hat{\mathbf{r}}$. Note that the influence functions $G_{zz}^{ij}(\mathbf{r},\hat{\mathbf{r}})$ are computed by employing an exact stiffness matrix scheme [6]. A number of $N_x^i \times N_y^i$ sub-elements are employed for Foundation i (i = 1, 2) where N_x^i and N_y^i denote the number of sub-elements in the x and y directions, respectively. In the numerical study, only the case of identical square foundations is investigated. The size and the number of sub-elements for both foundations are then identical. The computed impedance functions of dynamic interaction between both foundations denoted by K_{ij} ($i = j$ = 1, 2) can then be obtained as follows:

$$K_{11} = K_{22} = \int T_z^1(\hat{\mathbf{r}})dA_1 \tag{7a}$$

$$K_{12} = K_{21} = \int T_z^2(\hat{\mathbf{r}})dA_2 \tag{7a}$$

4. Numerical results and discussion

The accuracy of the present solution scheme is first verified. Wong and Luco [2] presented dynamic response of two identical rigid square foundations resting on a homogeneous elastic half-space with the Poisson's ratio of 0.333. The size of both square foundations is $2H \times 2H$ and the distance between the two foundations is denoted by d. In addition, $\delta = \omega H\sqrt{\rho\mu}$. The homogeneous elastic half-space is modelled in the present study by using 10 homogeneous layers with a thickness of $h/a = 0.2$ and an underlying half-space of identical material properties of $\nu = 0.333$ and negligibly small values of poroelastic parameters. In addition, each foundation is discretized into 64 or 8 × 8 sub-elements (see figure 2(a)). Figure 2(b) shows the comparison of non-dimensional impedance functions of two identical rigid square foundations between the present solutions and those given by Wong and Luco [2] for $d/H = 1$. Since the two foundations are identical, $K_{11} = K_{22}$ represents the total force under Foundation 1 and $K_{12} = K_{21}$ represents the total force under Foundation 2. It is evident that a good agreement between the two solutions is observed from the comparison shown in figure 2(b).

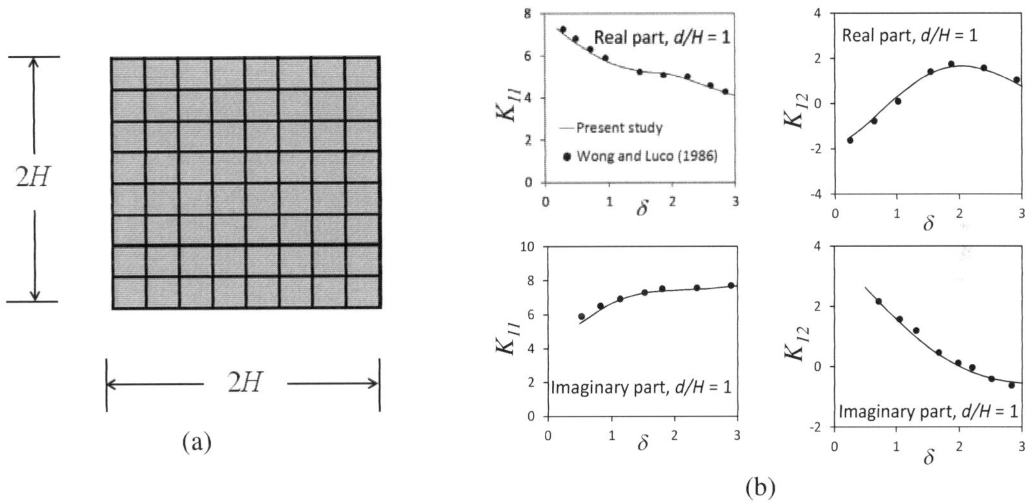

Figure 2. (a) Discretization of contact area; (b) Comparison of impedance functions K_{11} and K_{12} for two identical rigid square foundations with $d/H = 1$ resting on a homogeneous elastic half-space.

The dynamic interaction between two rigid foundations and a multi-layered poroelastic half-space is presented next. Figure 3 shows impedance functions K_{11} and K_{12} of two identical square foundations of size $2H \times 2H$ with various distances between the foundation, i.e., $d/H = 0.5$, 1 and 2, resting on a multi-layered poroelastic half-space. The layered half-space under consideration consists of two poroelastic layers with the same thickness H overlying a poroelastic half-space. The material properties of the two layers and the half-space are given in table 1. Both real and imaginary parts of the impedance functions K_{11} and K_{12} are shown in figure 3. Both foundations are fully permeable and the discretization shown in figure 2(a) is employed. It can be seen from figure 3 that the impedance functions of the two foundations depend significantly on the distance d. Both real and imaginary parts of K_{11} varies smoothly over the frequency range $0 < \delta < 3$ for all values of d/H. As expected, the maximum response is found in the case of the shortest distance between two foundations, i.e., $d/H = 0.5$ followed by the $d/H = 1$ and 2 respectively. The case of single foundation is also shown in figure 3 for comparison, and it is found that the impedance of two foundations converges to that of single foundation when $d/H > 2$ implying that K_{11} of two-foundation system can be obtained from the analysis of single-foundation system when the distance between them is greater than their width. It is also found that impedance function K_{12} shows complicated variations with frequency and its values do not diminish as the distance d/H increases. This indicates that an unloaded foundation would experience a radiated energy dissipated from a loaded foundation even though the distance between them is greater than their width. Additional parametric studies are currently underway and the results will be discussed in a future publication.

Table 1. Material properties of the layered system considered in figure 3.

Location	$\mu^{\dagger}$	$\lambda^{\dagger}$	$M^{\dagger}$	$\rho^{\dagger\dagger}$	$\rho_f^{\dagger\dagger}$	$m^{\dagger\dagger}$	α	$b^{\dagger\dagger\dagger}$
First layer	2.5	5	25	2	1	3	0.95	1.5
Second layer	1.25	1.88	18.8	1.6	1	1.8	0.98	0.75
Half-space	10	10	20	2.4	1	4.8	0.9	4.5

$^{\dagger}\times 10^8$ N/m^2; $^{\dagger\dagger}\times 10^3$ N/m^2; $^{\dagger\dagger\dagger}\times 10^6$ N/m^2.

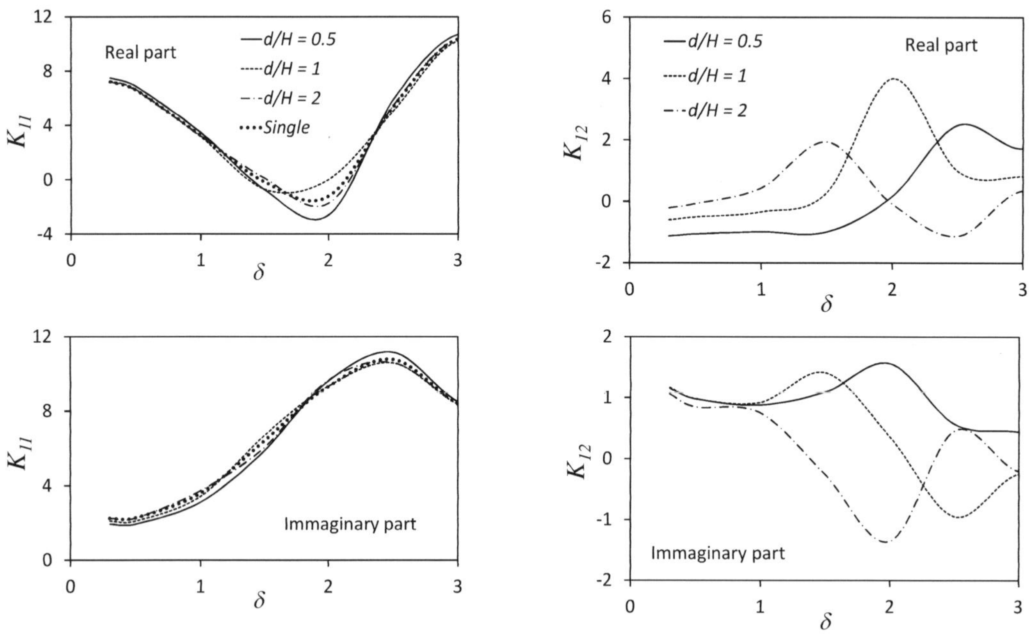

Figure 3. Impedance functions K_{11} and K_{12} of two identical rigid square foundations.

5. Conclusions

Dynamic response of two rigid foundations resting on a multi-layered poroelastic half-space is successfully studied in this paper by employing the discretization technique and the exact stiffness matrix scheme. Selected numerical results for impedance functions K_{11} and K_{12} for two identical rigid square foundations are presented, and it is found that the impedance functions are clearly influenced by the frequency of excitation and the distance between the two foundations. Extensive parametric studies on foundations of other shapes will be presented in a future publication.

6. References

[1] Wong H L and Luco J E 1976 *Earthq. Eng. Struct. Dyn.* **4** 579-87.
[2] Wong H L and Luco J E 1986 *Soil Dyn. Earthq. Eng.* **5** 149-58.
[3] Triantafyilidis T and Prange B 1987 *Soil Dyn. Earthq. Eng.* **6** 164-79.
[4] Biot M A 1956 *J. Acoust. Soc. Am.* **28** 168-78.
[5] Halpern M R and Christiano P 1986 *Earthq. Eng. Struct. Dyn.* **14** 439-54.
[6] Senjuntichai T and Sapsathiarn Y 2003 *J. Eng. Mech.* **129** 1330-41.
[7] Lu JF and Jeng DS 2007 *Int. J. Solid. Struct.* **44** 573-86.

Acknowledgement
The work was supported by the Thailand Research Fund (grant numbers PHD/0064/2558 and BRG 5880017).

IOP Conf. Series: Materials Science and Engineering **269** (2017) 012048 doi:10.1088/1757-899X/269/1/012048

Automatic Acquisition and Utilization of Stress Characteristic Knowledge for Excavator Boom Structural Optimization

Hua Haiyan[1], Shen Zhenhui[2]

[1]FuJian University of Technology, Fuzhou, China
[2]Fujian Jiangxia University, Fuzhou, China

Abstract. To improve the efficiency of coordinating the lightweight goal and structural performance for excavator boom, a knowledge-based optimization mechanism is proposed. By this mechanism, the stress characteristics in excavator boom can be extracted automatically at first. After that, stress characteristics knowledge can be acquired based on Monte Carlo simulation and processed under multistate constraint. Depending on the cycle operation between knowledge utilization module and optimization module, the strategy of utilizing characteristic knowledge to guide the structural optimization can be realized. Finally, different testing cases are taken in the structural optimization of excavator boom as examples to demonstrate the feasibility and effectiveness of proposed method.

1. Introduction

Excavator is one kind of important engineering machinery, which often works under multiple working conditions. How to improve its comprehensive performance is still a research focus [1, 2]. In particular, boom is the largest load-bearing part in the working equipment of excavator and affects the performance of excavator directly. But the contradictory between lightweight goal and structural performance is usually difficult to be coordinated during the structural optimization of boom. This difficulty is more serious when considering multiple working conditions. It is necessary to seek effective methods to overcome above trouble. In recent years, many researches are focused on the method of combining knowledge engineering with intelligent design to improve the efficiency of design, which has been utilized in the design of aircraft, ships and other fields [3, 4]. However, coordinating the lightweight goal and structural performance is still a bottleneck. Therefore, a new strategy of automatic acquiring and utilizing stress characteristic knowledge for excavator boom structural optimization is proposed in this paper.

2. A new knowledge-based intelligent optimization mechanism

In order to realize the automatic knowledge acquisition and utilize knowledge to guide the structural optimization, a knowledge-based intelligent optimization mechanism is constructed, which is shown in figure 1. This mechanism includes nine modules, some of which are illustrated as follows:

Solid modeling module is constructed to realize batch modeling by connecting VC++ with Pro/Engineering. Boom samples required for characteristic extraction are generated by this module.

Kinematics and kinetostatics analysis module is constructed to calculate the forces of boom, which are required to the loading of finite element analysis.

Characteristic extraction module is constructed to extract characteristic automatically by connecting VC++ with ANSYS. This module includes two sub-modules, i.e., the stress survey sub-module and the stress characteristic extraction sub-module. Firstly, using the stress survey sub-module, the stress characteristic (SC) regions for boom samples can be identified by the method of finite element analysis and danger evaluation. And then, characteristic stress, i.e., the maximum stress in every SC region, can be extracted by the stress characteristic extraction sub-module.

Data processing module is constructed for analyzing the data generated by characteristic extraction module.

Knowledge processing module can be used for acquiring stress characteristic knowledge, which includes the evaluation knowledge of characteristic regions, the influence evaluation knowledge and the main influence factors (MIF) knowledge. All knowledge is saved in knowledge base.

Knowledge utilization module and knowledge-based intelligent optimization module are constructed to form a loop. By these two modules, the strategy of utilizing characteristic knowledge to guide the structural optimization can be realized.

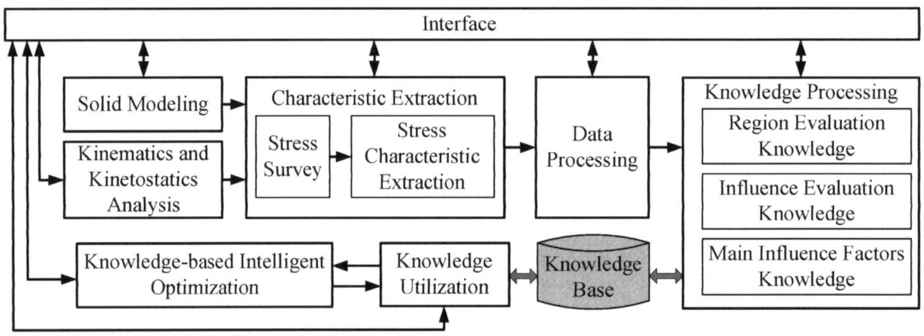

Figure 1. Knowledge-based intelligent optimization mechanism.

3. Automatic acquisition and utilization of stress characteristic knowledge

3.1. The method of automatic knowledge acquisition

Knowledge acquisition is carried out before optimization, which includes two steps shown in figure 2.

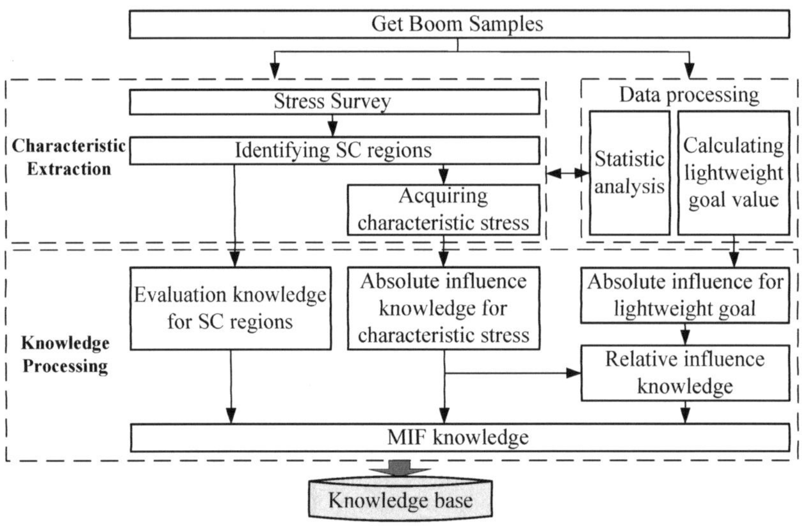

Figure 2. The process of knowledge acquisition before optimization.

The first step is characteristic extraction. In order to coordinate stress distribution, the SC regions need to be identified as the controlled targets. But most types of excavator boom have continuous geometric shape, which makes it more difficult to determine the SC regions. Therefore, the double-step stress survey [5] is carried out to explore the dangerous regions where the maximum stress often occurred. Those dangerous regions are taken as SC regions, and the maximum stress in SC region is called characteristic stress.

The second step is knowledge processing. In this step, the evaluation knowledge for SC regions, the influence evaluation knowledge and the MIF knowledge can be acquired. The first part is the evaluation knowledge for SC region. After SC regions identified, the times of occurring danger for every region is counted and the danger grade is evaluated. The second part is influence evaluation knowledge, which includes absolute influence knowledge and relative influence knowledge. In order to extract absolute influence knowledge for characteristic stress, the Monte Carlo simulation technique and the Spearman rank correlation analysis method are used. The Spearman rank correlation models between structure variables and characteristic stresses are constructed. And then, the Spearman correlation coefficients are taken as the absolute influence knowledge for characteristic stresses. Similarly, the absolute influence knowledge for lightweight goal is acquired. According to absolute influence knowledge, the relative influence knowledge of the structure variables for characteristic stress and lightweight goal are acquired.

After that, MIF knowledge is reasoned under multi-state constraints. The general expression of rule is shown as Rule 1, and the meaning of the symbols is shown in table 1.

Rule1: IF c_1 and c_2 and (c_3 or c_4) and c_5, THEN r_1 and r_2;

Table 1. Meaning of the symbols in Rule1.

Symbols	Meaning
c_1	Adjusting expectation for corresponding state; $c_1 = 1$: In state1, lightweight priority. Stress is better to be reduced, but allowed to increase greatly. $c_1 = 2$: In state2, lightweight priority. Stress is better to be reduced, but allowed to increase slightly. $c_1 = 3$: In state3, stress reduction must be ensured.
c_2	The danger grade of SC regions.
c_3	The absolute influence knowledge for characteristic stress.
c_4	The absolute influence knowledge for lightweight goal.
c_5	The relative influence knowledge.
r_1	The influence grade for characteristic stress.
r_2	Adjusting weighting

3.2. The strategy of knowledge utilization.

Knowledge utilization module is connected with knowledge-based intelligent optimization module to form a loop. According to the evolution population, the corresponding stress characteristic knowledge saved in knowledge base can be acquired to guide the optimal searching of next generation. The cycle of utilizing knowledge to guide optimization is shown in figure 3, which includes three important parts.

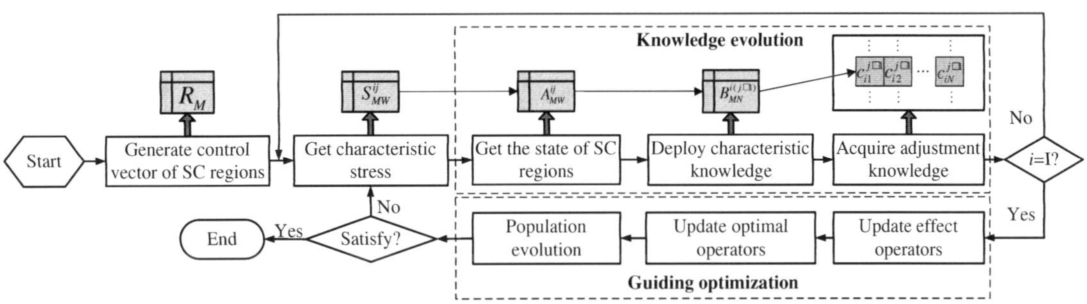

Figure 3. The cycle of utilizing knowledge to guide optimization.

Generate control vector of SC regions. Because of the different danger situation occurring in different SC regions, the control strategy for SC regions should be identified before optimization. According to the evaluation knowledge, SC regions can be divided into three classes. The first class called main control region has the highest adjustment priority. The second class called conditional control region is often depended on geometry of boom. The third class called auxiliary control region has the lowest adjustment priority. The control vector of SC regions is generated to guide knowledge deployment and knowledge integration, as well as coordinate knowledge conflict.

3.2.1. Knowledge evolution. In order to match corresponding knowledge from knowledge base, it is necessary to get the information of individuals firstly. For each individual, because excavator often works under multiple working conditions, the characteristic stresses of individual in every condition should be extracted and saved in a matrix. And then, the states of SC regions for such individual can be identified by the rules. According to the state of each SC region in every condition, the corresponding MIF knowledge can be matched, which is integrated and combined to the adjustment knowledge vector of such individual.

3.2.2. Guiding optimization. In order to utilize adjustment knowledge to guide the optimal searching, the effect operators are designed according the optimization algorithm. In this paper, the genetic algorithm is utilized, so that some effect operators, such as elitist preservation influence operators, selection influence operators, crossover influence operator and mutation influence operators [6], are designed. After acquiring adjustment knowledge, the effect operators are updated. And then, the genetic operators are also updated to realize knowledge-based population evolution.

4. Application

The structural optimization of gooseneck type boom for medium-sized excavator is taken as an example, in which the lightweight goal of minimizing the volume of boom under four working conditions [7] is proposed. By using the mechanism shown in figure 1, the SC regions are identified and shown in figure 4. And then, the stress characteristic knowledge is acquired and saved in knowledge base.

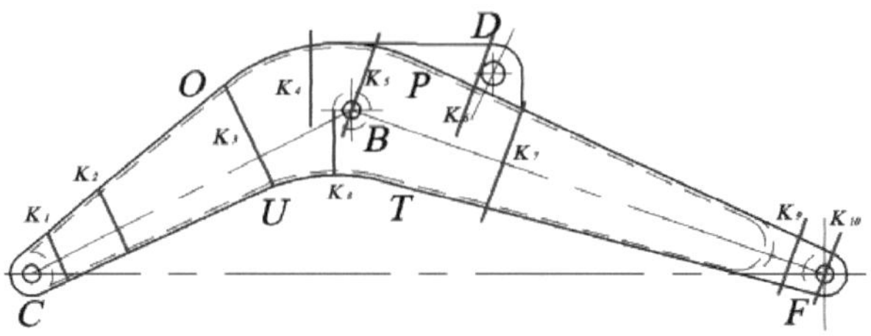

Figure 4. The SC regions of gooseneck type boom.

In order to verify the feasibility and effectiveness of proposed method, four testing cases with different control strategies for SC regions are shown in table 2.

Table 2. Control strategies for SC regions in testing cases.

Case Number	Main control region	Conditional control region	Auxiliary control region
Case 1	Maximum stress in boom	—	—
Case 2	Maximum stress in boom	—	—
Case 3	K_1, K_2, K_7, $K_{9\text{-}10}$	$K_{3\text{-}4\text{-}8}$	—
Case 4	K_1, K_2, K_5, K_6, K_9, K_{10}	$K_{3\text{-}4\text{-}8}$	K_7

Case 1 is the traditional experience design case and only the maximum stress in the boom is controlled. Case 2 uses the existing genetic algorithm and also only controls the maximum stress. Case 3 includes four main control regions, i.e., K_1, K_2, K_7 and the region combining K_9 with K_{10}. The conditional control region in this case is the combined control of K_3, K_4 and K_8, but the auxiliary control region is not considered. Case 4 includes six main control regions, i.e., K_1, K_2, K_5, K_6, K_9 and K_{10}. The conditional control region in Case 4 is the same to Case 3, but the auxiliary control region is K_7. In Case 3 and Case 4, knowledge is utilized to guide the genetic operators, which are illustrated in our preliminary work [6]. The results of testing cases are shown in table 3.

In table 3, it is shown that the volume of boom in Case 3 is decreased by 40.97% and 4.76% comparing with Case 1 and Case 2 respectively. Comparing with Case 3, the volume is further reduced in Case4, but the maximum stress is not increased. D*M* represents the maximum difference value of stress among four digging conditions. Comparing all of testing cases, the value of D*M* in Case 3 and Case 4 are much less than that in Case 1 and Case 2.

Table 3. The results of testing cases.

Case	Volume (m^3)	Maximum Stress under Four Digging Conditions (MPa)				D*M* (MPa)
		Condition 1	Condition 2	Condition 3	Condition 4	
Case 1	0.1662	166.25	180.05	243.27	145.59	97.68
Case 2	0.1030	186.44	165.37	236.17	180.92	70.80
Case 3	0.0981	192.25	188.14	234.98	189.23	46.84
Case 4	0.0967	201.68	176.57	233.57	187.59	57.00

In addition, the evolution processes of Case 2, Case 3 and Case 4 are shown in figure 5. The evolutionary rate of Case 4 is the fastest and Case 3 is secondly, which indicates that the searching capacity can be improved greatly under the guidance of stress characteristic knowledge.

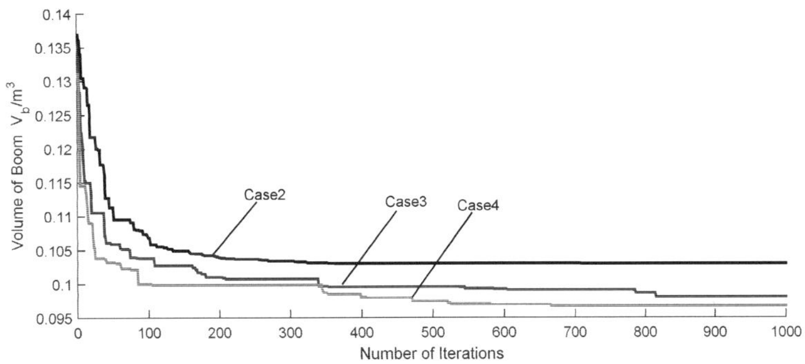

Figure 5. The evolution processes of Case 2, Case 3 and Case 4.

The comparing of characteristic stresses for Case 3 and Case 4 is shown in table 4, where M_{max} represents the value of characteristic stresses in every control region; E_{rr} represents the difference between characteristic stress and allowable stress.

Table 4. The comparing of characteristic stresses for Case 3 and Case 4. MPa.

Name	Item	K_1	K_2	K_5	K_6	K_7	K_9	K_{10}	$K_{3\text{-}4\text{-}8}$
Case 3	M_{max}	214.58	205.85	144.41	158.72	167.53	234.98	181.23	164.91
	E_{rr}	30.42	39.15	100.59	86.28	77.47	10.02	63.77	80.09
Case 4	M_{max}	233.57	223.62	164.94	186.81	145.43	222.56	186.50	162.61
	E_{rr}	11.43	21.38	88.06	58.19	99.57	22.44	58.50	82.40

In table 4, although the genetic operators in Case 3 and Case 4 are the same, the knowledge guiding to the optimization are different. In Case 3, the values of E_{rr} in K_5 and K_6 are larger than other regions. Similarly in Case 4, the maximum value of E_{rr} is in the auxiliary control region K_7. It is indicated that the priority of control regions may influence the stress distribution. The result also shows that the average value of E_{rr} is 60.97MPa in Case 3, but this value is decreased by 55.25MPa in Case 4, which means the situation of equal strength in Case 4 is better than in Case 3.

5. Conclusion

By verifying the proposed strategy of automatic acquiring and utilizing stress characteristic knowledge, the conclusions are shown as follows:

The proposed knowledge-based intelligent optimization mechanism can acquire knowledge from RC regions automatically.

The priority of control RC regions as well as the strategy of utilizing knowledge may influence the efficiency and effectiveness of optimization greatly.

Utilizing stress characteristic knowledge to guide optimization, the effectiveness of coordinating the conflict between structure performance and lightweight goal can be improved remarkably.

6. References

[1] Cevdet C U 2008 *Shape optimization of an excavator boom by using genetic algorithm* TURKEY: Middle East Technical University.

[2] Solazzi L 2010 *J. Terramech.* **47** 201-7.

[3] La Rocca G and Van Tooren M J L. 2010 *J. Aerospc. Eng.* **224(9)** 1041-55.

[4] Curran R, Verhagen W J C and Van Tooren M J L 2010 *Exp. Syst. Appl.* **37** 7336-50.

[5] Hua H Y and Lin S W 2014 *The Extracting of Structural Performance Characteristics for Excavator Boom Based on Double-Step Stress Survey.* 2014 International Conference on Mechanical Design, Manufacture and Automation Engineering (MDMAE2014). Pennsylvania, USA: DEStech Publications, Inc. 27-31.

[6] Hua H Y, Lin S W 2014 Chinese J. Mech. Eng. **27(2)** 392-401.

[7] Tianjin Mechanical Engineering Research Institute. Hydraulic excavators-testing method of structure strength. GB9141-88. Beijing: Standardization Administration of the People's Republic of China, 1988.

Acknowledgements

This work is supported by the Education Scientific Research Project for the Young Teacher of Fujian Province, China (Grant No. JAT170374), and the Scientific Research Foundation Project of Fujian University of Technology, China (Grant No. GY-Z14075).

IOP Conf. Series: Materials Science and Engineering **269** (2017) 012049 doi:10.1088/1757-899X/269/1/012049

Study on Dynamic Simulation of New Cosine Gear

Yu-yu Cai[1], Wei Wang[1]

[1]Industrial robot measurement and control and mold rapid manufacturing of key laboratories of Fujian province, Minnan University of Science and Technology, Shishi, Fujian, China, 362700

Abstract. Based on the gear meshing theory, after the process of construction is discussed, the mathematical model about profile curve equation, conjugate equation and action line equation is built. It provides some theoretical proofs for research of performance and manufacture of cosine-gear. In this article, the profile curve equation of cosine gear is introduced, and the conjugate equation is deduced. The solid model of cosine-gear is built by example with Pro/E software, and the meshing process of a pair of cosine-gears is simulated. From the result of simulating, it can detect the interference and the dynamic characteristics of a pair of cosine gears quickly, and during the meshing process, there is no meshing interference. Moreover, the cosine gear is line contact. It is significant to improve the efficiency of design and manufacturing of cosine gear.

1. Introduction

Inspired by Logix gear [1], double involute gear [2] and other new gears, in order to overcome some defects in the application of gears in gear pumps [3]. Such as low volumetric efficiency, low flow rate and oil retention, a new type of gear drive, cosine gear drive is put forward. The profile of the cosine gear is a cosine curve, it consists of two parts: convex curve and concave curve. The radius of curvature in middle of cosine tooth profile is very large, it is beneficial to improve contact strength, the normal meshing performance of conjugate tooth profile can be guaranteed, the utility model can be widely used in gear pumps and other fields, it is positive to relieve the oil loss, improve the volumetric efficiency and increase the flow and pressure of the gear pump.

Before the real cosine gear is machined, using Pro/ENGINEER software, the 3D solid model can be established accurately, it is transformed into dynamic simulation environment, add complex forces and constraints, then the movement simulation analysis is done. Through the simulation results, gear profile interference and continuous transmission during gear transmission can be detected, then the model can be modified quickly and accurately. So as to shorten the modification time and improve the design efficiency.

2. Derivation of cosine profile curve and conjugate tooth profile equation

2.1. The equation of profile curve

The profile of cosine gear is similar to the double circular arc gear, they are all concave at the top of the tooth and protruding at the root of the tooth, The tooth profile curve is shown in figure 1. In polar coordinates, the polar coordinate equation of the gear profile is:

$$\rho = r + A\cos(z\varphi) \tag{1}$$

A is the amplitude, just the radius of the circle, and Z is the number of teeth, $z = 2\pi r / p$

The parameter equation is:

$$\begin{cases} x = [r + A\cos(z\varphi)]\sin(\varphi) \\ y = [r + A\cos(z\varphi)]\cos(\varphi) \end{cases} \tag{2}$$

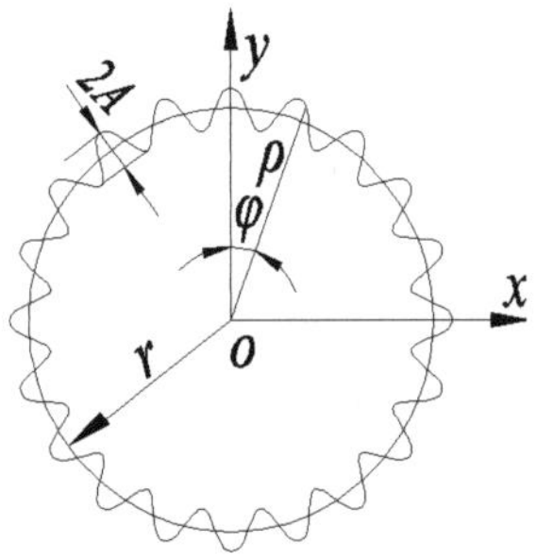

Figure 1. Equation of tooth profile curve.

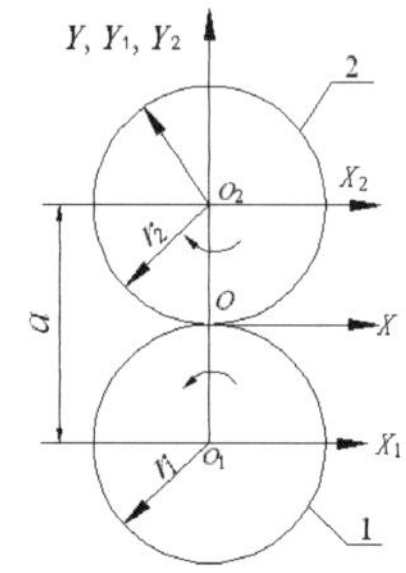

Figure 2. Gear meshing coordinate system.

2.2. Conjugate tooth profile equation

Just as figure 2, set the coordinate system $\Sigma_0(O: x, y)$ as the static coordinate system, its origin O coincides with the meshing point P, the coordinate system $\Sigma_1(O_1: x_1, y_1)$ is connected with the tooth profile C_1, The coordinate system $\Sigma_2(O_2: x_2, y_2)$ is connected with the tooth profile C_2, O_1 and O_2 are the origin of Σ_1 and Σ_2 respectively. In the coordinate system Σ_1, the equation C_1 of the tooth profile is:

$$\begin{cases} x_1 = [r_1 + A\cos(z\varphi)]\sin(\varphi) \\ y_1 = [r_1 + A\cos(z\varphi)]\cos(\varphi) \end{cases} \tag{3}$$

According to the meshing principle of tooth profile, the equation of meshing motion of tooth profile C_1 is expressed in coordinate system Σ_1:

$$\begin{cases} \overrightarrow{r^{(1)}} = x_1(u)\vec{i_1} + y_1(u)\vec{j_1} \\ \phi(u,t) = \overrightarrow{n^{(1)}} \cdot \overrightarrow{V^{(12)}} = 0 \end{cases} \tag{4}$$

In order to deduce the parametric equation of tooth profile C2, the radius of the meshing point in the coordinate system Σ_1 is changed directly to the coordinate system Σ_2 by the coordinate system Σ_1:

$$\begin{cases} \overrightarrow{r^{(2)}} = M_{21}\overrightarrow{r^{(1)}} + \overrightarrow{r_{01}^{(2)}} \\ \overrightarrow{r_{01}^{(2)}} = M_{20}\overrightarrow{r_{o1}} + \overrightarrow{r_0^{(2)}} \end{cases} \tag{5}$$

In style, $r_{01} = \begin{pmatrix} 0 \\ -r_1 \end{pmatrix}$, $r_0^{(2)} = \begin{pmatrix} r_2 \sin\varphi_2 \\ -r_2\cos\varphi_2 \end{pmatrix}$

$$M_{20} = \begin{pmatrix} \cos\varphi_2 & -\sin\varphi_2 \\ \sin\varphi_2 & \cos\varphi_2 \end{pmatrix}$$

$$M_{21} = \begin{pmatrix} \cos(\varphi_1 + \varphi_2) & -\sin(\varphi_1 + \varphi_2) \\ \sin(\varphi_1 + \varphi_2) & \cos(\varphi_1 + \varphi_2) \end{pmatrix}$$

Thus obtained

$$r_{01}^{(2)} = \begin{pmatrix} \cos\varphi_2 & -\sin\varphi_2 \\ \sin\varphi_2 & \cos\varphi_2 \end{pmatrix}\begin{pmatrix} 0 \\ -r_1 \end{pmatrix} + \begin{pmatrix} r_2\sin\varphi_2 \\ -r_2\cos\varphi_2 \end{pmatrix} = \begin{pmatrix} a\sin\varphi_2 \\ -a\cos\varphi_2 \end{pmatrix} \quad (6)$$

$$r^{(2)} = \begin{pmatrix} \cos(\varphi_1+\varphi_2) & -\sin(\varphi_1+\varphi_2) \\ \sin(\varphi_1+\varphi_2) & \cos(\varphi_1+\varphi_2) \end{pmatrix}\begin{pmatrix} x_1 \\ y_1 \end{pmatrix} + \begin{pmatrix} a\sin\varphi_2 \\ -a\cos\varphi_2 \end{pmatrix} \quad (7)$$

That is

$$\begin{cases} x_2 = x_1\cos(\varphi_1+\varphi_2) - y_1\sin(\varphi_1+\varphi_2) + a\sin\varphi_2 \\ y_2 = x_1\sin(\varphi_1+\varphi_2) + y_1\cos(\varphi_1+\varphi_2) - a\cos\varphi_2 \end{cases} \quad (8)$$

Among them, φ_1, φ_2 respectively stands for the angle of gear 1 and gear 2 is turning, $\varphi_2 = i * \varphi_1$, i stands for the ratio of gear transmission, a stands for the distance of gear center.

As shown in figure 3, the instantaneous center line of gear 1 is a circle with a radius r_1, P is the instantaneous center point, its position in the certain coordinate system $\sum_1$ is fixed, the known tooth profile a-a is tied together with the coordinate system $\sum_1$, the angle between the tangent and the axis x_1 at any point $m_1(x_1, y_1)$ on the tooth setting a-a is β, draw the perpendicular line of $m_1 p_1$ through o_1, and the angle between it and $o_1 p_1$ is ψ, then we can see by the diagram:

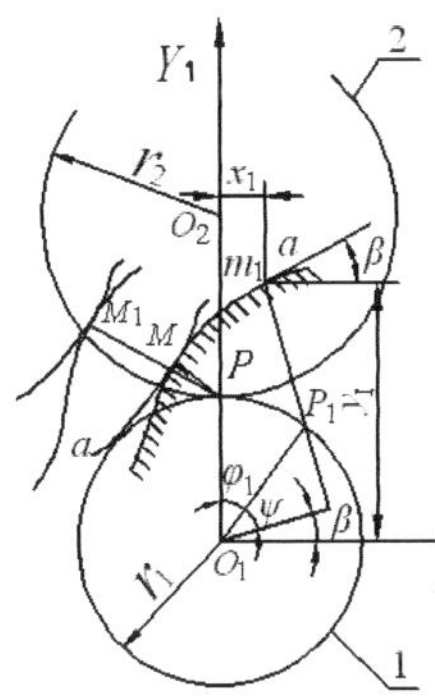

Figure 3. Calculation parameters of conjugate tooth profile.

Equation (8): p -he normal of the point on a-a and the intersection of Y_1, p_1 - The intersection of the normal point m_1 of the tooth profile and the node.

Simultaneous equations (3), (7) and (8) can obtain the equation of tooth profile C_2. The shape of the conjugate tooth profile is also a curve of approximately cosine shape.

3. Establishment of gear solid model

The part of gear teeth is the most complicated in the gear, the cosine gear is the same as other gears, in order to meet the needs of kinematics simulation analysis and finite element analysis, the 3D solid model is required to be mapped accurately. In this paper, the number of teeth $z_1 = 13$, modulus $m = 3mm$, pressure angle $\alpha = 20^0$, width of the cosine gear $b = 30mm$ (as specified in gear 1) as an example, mating gear 2 $z_2 = 28$, according to the method of establishing 3D model of involute gear [6], describe the construction of the solid model of gear 1 and its conjugate gear (set as gear 2).

3.1. The solid model of gear 1

Use the command of Curve/From equation to enter the following equation in the Cartesian coordinate system, and obtain a cosine curve equation of gear 1.

$$\begin{cases} \varphi = t * 9 \\ x = [20 + 2 * \cos(20 * \varphi)]\sin(\varphi) \\ y = [20 + 2 * \cos(20 * \varphi)]\cos(\varphi) \\ z = 0 \end{cases} \tag{9}$$

Obtain another cosine curve by the Mirror command, in order to ensure that the gear pair does not appear meshing interference in the meshing process, The teeth of the gear 1 and 2 are moved by a circular arc that is tangent to the cosine curve, considering the oil film thickness [7] between the gear pairs, the radius of the arc is 1.

Generate a single gear with the Extrude command, and the Pattern command generates the blank model of gear 1, the Cut command generates the key groove, and a 3D solid model of gear 1 is obtained, as shown in figure 4.

Figure 4. Solid model of gear 1.

Figure 5. Solid model of gear 2.

3.2. The solid model of gear 2

Because the curve equation of conjugate tooth profile is complex, it is difficult to generate cosine curve according to gear 1. Using Matlab software to take 20 points on the conjugate tooth profile, the tooth profile of gear 2 is obtained by connecting the points. Finally, according to the same method as gear 1, a three-dimensional solid model of gear 2 is established, as shown in figure 5.

4. Meshing simulation analysis of cosine gear pair

Taking a pair of gear pairs ($z_1 = 13$, $z_2 = 30$) as an example, to mesh simulation analysis of gear pair is carried out. Figure 6 shows the virtual assembly of the gear pair. Figure 7 shows the shape of the contact area of a single tooth profile.

Figure 6. Assembly drawing of gear pair.

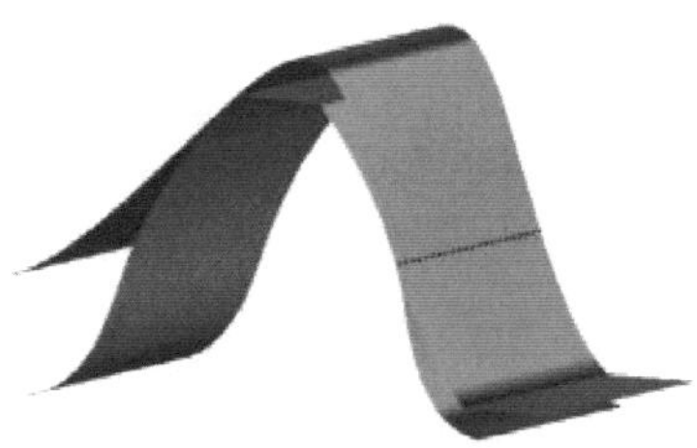

Figure 7. Shape of contact area of gear pair.

We can know in figure 6-7:

1. The simulation results show that the gear pair has no mesh interference and can continuously drive.

2. The contact shape of gear pair is straight line, so it can be deduced that the form of cosine gear transmission is line contact.

5. Conclusion

The equation of tooth profile curve of cosine gear is established, using the gear meshing principle and differential geometry method, the conjugate tooth profile equation is derived, by using the modeling function of Pro/E, a pair of cosine gear pairs is generated, and the meshing process is simulated. Meshing simulation of cosine gear pair, the actual assembly effect of gear pair can be detected, the assembly relation and motion relation among the parts are tested. It is very convenient to observe whether the cosine tooth profile formed by the selected parameters can cause tooth profile interference and whether the gear pair can be continuously transmitted, and the shape of the contact area can also be observed dynamically, so that the design, assembly and transmission of the cosine gear are reasonable. It is of great value to instruct designers to modify design models directly and quickly, to improve design efficiency, and to shorten the design and manufacture time of gears. During the meshing process, there is no meshing interference. Moreover, the cosine gear is line contact.

6. References

[1] Komori T, Ariga R and Nagata S. 1990 *Trans. ASME,* **12** 430-6.
[2] Zhang G H, Li F J and Wei J 2008 *J. Mech. Transmis.* **32** 1-5.
[3] Dong H L, Yuan S H, Hu J B and Li X Y 2013 *J. Tribol.* **33** 436-41.
[4] Daniel T Y, Yang C H and Tong S H 2001 *Mech. Mach. Theor.* 105-7.
[5] Barone S 2001 *Eng. Comput.* 66-74.
[6] Ji B, Wang C Y, Lu L J, Sun T H 2016 J. Taiyuan Univ. Sci. Tech. **37** 50-3.
[7] Zhao N, Jia Q J, Ouyang B and Wu H M 2015 J. Mech. Sci. Tech. Aerospc. Eng. **34** 329-33.

Acknowledgement

This work is sponsored by Fujian Provincial Key Laboratory Project - Industrial Robot Measurement and Control and Mold Rapid Manufacturing Laboratory (Fujian Education [2017] No. 8), science and technology plan projects of Quanzhou (2014Z140), science and technology plan projects of Shishi (2014SK12).

IOP Conf. Series: Materials Science and Engineering **269** (2017) 012050 doi:10.1088/1757-899X/269/1/012050

FEM Simulation of the Effect of Coefficient of Thermal Expansion and Heat Capacity on Prediction of Residual Stresses of Compression Molded Glass Lenses

Bo Tao[1], **Ye Yuan**[1]

[1]Key Laboratory of Metallurgical Equipment and Control Technology of Ministry of Education, Wuhan University of Science and Technology, Wuhan, Hubei 430081, China

Abstract. In this research, the effects of the coefficient of thermal expansion (CTE) and heat capacity on the prediction of residual stresses in BK7 compression molded glass lenses were studied. Three different groups of CTE and two different kinds of heat capacity, which are constant and proportional to temperature, were chosen to investigate the impacts of residual stresses. The simulation results show a big difference and suggest that the properties of glass materials determine the residual stresses and should be measured carefully.

1. Introduction

Compression molding as a hot pressing method is ideal for fabricating precision glass optical elements for its net-shape and high volume fabrication capability [1-3]. However, during cooling, residual stresses are eventually frozen in the molded lenses due to the structural relaxation phenomena when glass material goes through its transition temperature (T_g) region. Residual stresses, inside glass optical elements, can introduce refractive index variation, resulting unwanted light path deviation that may cause image quality deterioration. Therefore, residual stresses are important criteria for evaluating glass optical elements [4].

Finite element method (FEM) has been employed to study and optimize the compression molding process parameters [5-7]. The Tool-Narayanaswamy-Moynihan (TNM) model was applied to model the structural relaxation behavior of glass material during cooling [8, 9]. In this research, the structural relaxation behavior of an optical glass material was simulated by the FEM software MSC/MARC. Therefore, the residual stresses inside molded glass lenses can be predicted. In order to get realistic predictions of the residual stresses, the parameters used in FEM should be accurate. However, during the simulation, the CTE and heat capacity were measured with uncertainty. Different values may resulting variant prediction of residual stresses. Therefore, the effects of CTE and heat capacity on residual stresses were carefully studied.

2. Numerical simulation theory of glass molding

The compression molding process can be grouped into two major stages: heating and hot forming of the glass material, cooling. In the heating stage, the glass is heated to the molding temperature T_{molding}, which is slightly higher than the glass's T_g. In T_{molding}, the viscosity of glass will decrease and the glass can be modeled as a Newtonian fluid. During forming, the behavior of glass deformation is governed by a function of the stress σ_{ij} and strain rate $\dot{\varepsilon}_{ij}$, viscosity $\eta(T)$, shown in equation (1):

$$\sigma_{ij} = 2\eta(T)\dot{\varepsilon}_{ij} \tag{1}$$

During cooling, the change of glass material's property is non-linear because it depends on current temperature and temperature history. A response function $M_v(t)$ is adopted to describe the change of a glass property, presented in equation (2) [9]:

$$M_v(t) = \sum_{i=1}^{n} (w_g)_i \exp\left(-\frac{t}{\tau_{vi}}\right) \tag{2}$$

where, $(w_g)_i$ are called weighting factors and τ_{vi} are the associated structural relaxation times. The structural relaxation times at a given temperature can be calculated using the TNM model, expressed in equation (3) [8]:

$$\tau_v = \tau_{v,ref} \exp\left(-\frac{H}{R}\left(\frac{1}{T_{ref}} - \frac{\mu}{T} - \frac{1-\mu}{T_f}\right)\right) \tag{3}$$

where $\tau_{v,ref}$ is structural relaxation time at reference temperature T_{ref}. H is activation energy, R is the ideal gas constant and μ is a fraction parameter. The value of μ lies between 0 and 1. Once the fictive temperature T_f is known, the property of glass material $p(T)$ can be calculated using equation (4) [9]:

$$\frac{1}{p(0)}\frac{dp(T)}{dT} = \alpha_g(T) + \left(\alpha_l(T_f) - \alpha_g(T_f)\right)\left(\frac{dT_f}{dT}\right) \tag{4}$$

where α_l and α_g are thermal expansion coefficients (CTE) of liquid and solid glass material, respectively. $p(0)$ represents the instantaneous value of the property of the glass material.

3. Numerical simulation

The numerical simulation was based on an aspherical glass lens. Figure 1(a) shows the schematic of the aspherical glass lens. The aspherical surface described by equation (5),

$$Z = \frac{-Cx^2}{1+\sqrt{1-(1+K)C^2x^2}} \tag{5}$$

where, x is the coordinate of the optical surface, K (4.3666) is the conic constant. C is surface curvature ($C = 1/R_1$, R_1 is the vertex radius of the aspherical surface).

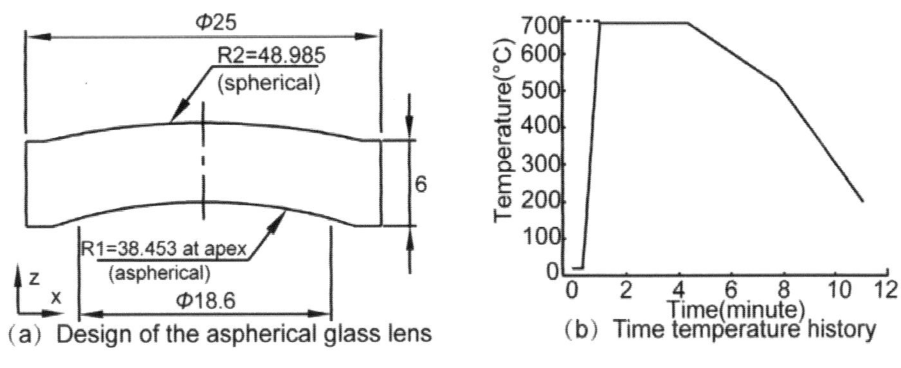

Figure 1. (a) Schematic of the aspherical lens (Units: mm),
(b) Time-temperature history of glass molding process.

Figure 1(b) shows the time-temperature history of the molding process. Molding was performed at 684 °C, and the temperature was kept until stresses due to forming were released. A two-steps cooling process was followed. First, the temperature was decreased to 520 °C at a rate of 0.8 °C/s, then to

200 °C at a rate of 1.6°C/s. When the temperature was decreased to 200 °C, the molded lens was taken out and cooled naturally.

In simulation, two dimensional (2D) axisymmetric models were applied because of the rotational symmetrical geometry of experiment settings. The glass and molds were meshed by four-node isoparametric quadrilateral element. Important mechanical and thermal properties of the glass (BK7) and molds (Tungsten carbide), are summarized in table 1. The parameters used in TNM model are summarized in table 2.

Table 1. Mechanical and thermal properties of BK7 glass [6] and Tungsten carbide [10].

Material Properties	BK7	Tungsten carbide
Elastic modulus, E [MPa]	82,500	570,000
Poisson's ratio, ν	0.206	0.22
Density, ρ [kg/m^3]	2,510	14,650
Thermal conductivity, κ_c [W/m °C]	1.1	63
Specific heat, C_p [J/kg °C]	858	314
Transition temperature, T_g [°C]	557	
Solid CTE, α_g [/°C]	5.6×10^{-6}	4.9×10^{-6}
Liquid CTE, α_l [/°C]	1.68×10^{-5}	--
Viscosity, η [MPa-sec] (at 685°C)	60	--

Table 2. Structural relaxation parameters used in numerical simulation [6].

Material properties	Value
Reference temperature, T_{ref} [°C]	685
Activation energy/gas constant, H/R [K]	47,750
Fraction parameter, μ	0.45
Weighting factor, w_g	1
Structural relaxation time, τ_v [sec] (at 685 °C)	0.019
Stress relaxation time, τ_s [sec] at 685 °C	0.0018

Figure 2 shows the model of deformed aspherical lens at the end of molding. The lens was meshed into 3,500 elements. Upper and lower molds were meshed into 381 and 365 elements, respectively.

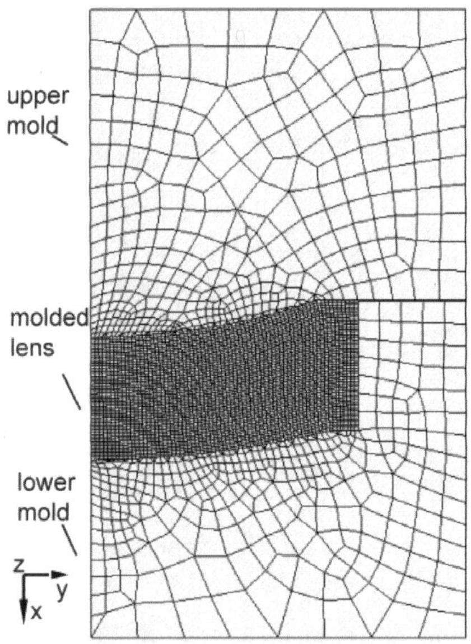

Figure 2. Deformed model of the aspherical glass lens.

Figure 3 shows the simulated distribution of the residual stresses inside the molded glass lens in cylindrical coordinates. The stress distribution displays half of the cross section of the aspherical lens viewed in the normal direction. In figure 3, negative value means compressive stress and positive value means tensile stress.

According to figure 3, the lens is mainly under tensile stress in the center and compressive stress at the top and bottom surfaces. Because the glass lost heat from its surfaces and the temperature of its center part was always higher than its surfaces. Therefore the center part will be under tensile stress as compared to the surfaces.

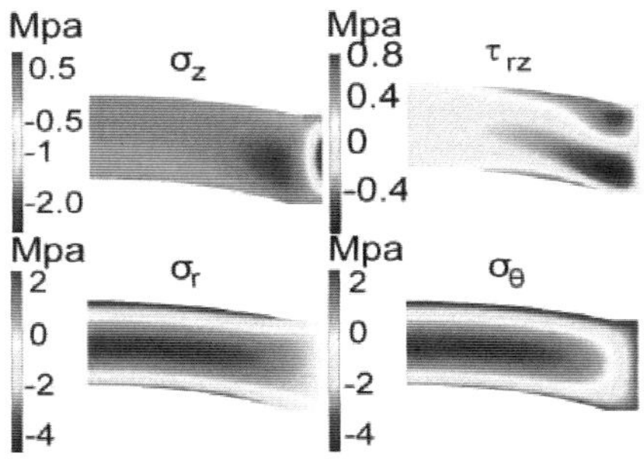

Figure 3. Numerical simulated residual stresses distribution in the aspherical glass lens in cylindrical coordinates: axial stress σ_z, shear stress τ_{rz}, radial stress σ_r, circumferential stress σ_θ.

4. Simulation results and discussion

4.1. Effect of coefficient of thermal expansion

Generally, the solid CTE of BK7 glass, in table 1, is an average CTE from 20 °C to 300 °C. And the CTE from 300 °C to T_g is unknown. Reference [6] measured the solid CTE of BK7 using Orton Dilatometer (Model 1000D), $\alpha_g = 8.3\times10^{-6}$ /°C. Then the liquid CTE α_l was calculated by using two different methods based on α_g and experimental data: $\alpha_l = 3.77 \times 10^{-5}$ /°C and 3.16×10^{-5} /°C. Reference [11], also mentioned that $\alpha_g = 5.6\times10^{-6}$/°C, $\alpha_l = 1.68\times10^{-5}$/°C by using $\alpha_l/\alpha_g=3$. Table 3 shows these three kinds of CTE.

Table 3. Coefficient of thermal expansion.

Solid CTE, α_g(/°C)	5.6×10^{-6}	8.3×10^{-6}	8.3×10^{-6}
Liquid CTE, α_l(/°C)	1.68×10^{-5}	3.77×10^{-5}	3.16×10^{-5}
$\alpha_l/\ \alpha_g$	3	4.54	3.81

Figure 4 shows the simulated radial stress σ_r based on CTE shown in table 3. The other parameters used in the simulation are shown in table 1 and 2. In figure 4, the radial stresses show the same trend, but with different magnitude. The maximum tensile stress shown in figure 4(b) is almost as twice as the stress shown in figure 4(a).

IOP Conf. Series: Materials Science and Engineering **269** (2017) 012050 doi:10.1088/1757-899X/269/1/012050

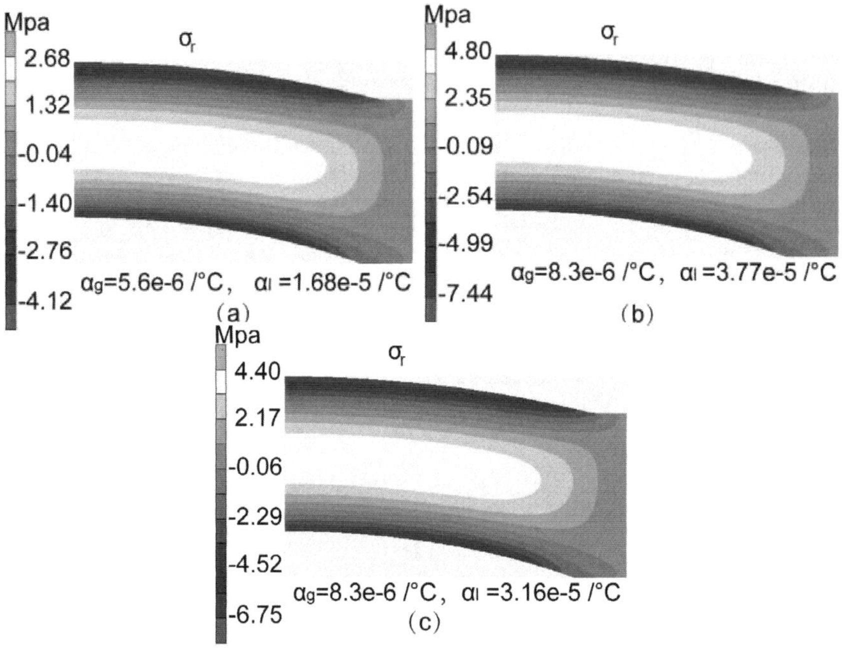

Figure 4. Numerical simulated radial stress σ_r under different CTE, (a) α_g=5.6×10^{-6} /°C, α_l=1.68×10^{-5} /°C, (b) α_g=8.3×10^{-6} /°C,α_l= 3.77×10^{-5} /°C, (c) α_g=8.3 ×10^{-6} /°C, α_l= 3.16×10^{-5} /°C. .

4.2. Effect of heat capacity

The heat capacity of the simulations shown in section 3 is set as a constant. However, in the region of T_g, due to structural relaxation, the volume of glass, elastic modulus, viscosity, heat capacity, etc. are related to temperature changes and temperature history. For poly-oxide, the relationship between liquid heat capacity C_{pl} and solid heat capacity C_{pg} is $C_{pl} / C_{pg} \approx 1.3$ [11] .

Figure 5 shows temperature-related heat capacity. The heat capacity is proportional to temperature until C_{pl}= 1.3 C_{pg} at T_g.

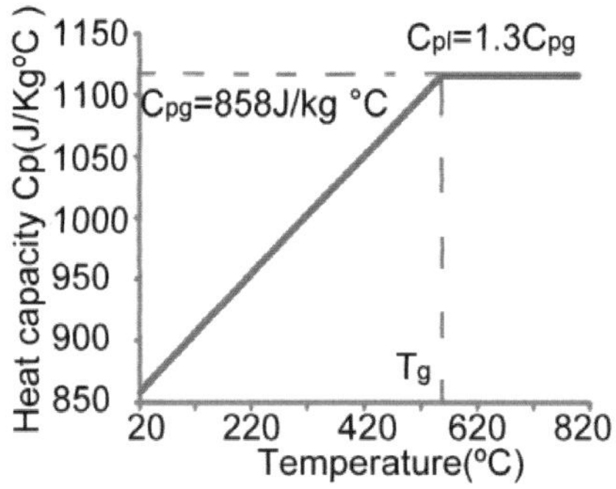

Figure 5. Heat capacity of BK7 changes with temperature.

Figure 6 (a) and (b) are the simulated radial stress of the fixed heat capacity and the temperature-related heat capacity, respectively. The corresponding maximum tensile stress and compressive stress are: (a) 2.68 Mpa, 4.12 Mpa; (b) 3.41 Mpa, 5.33 Mpa. It can be seen that the heat capacity has a certain effect on the residual stresses and can not be neglected.

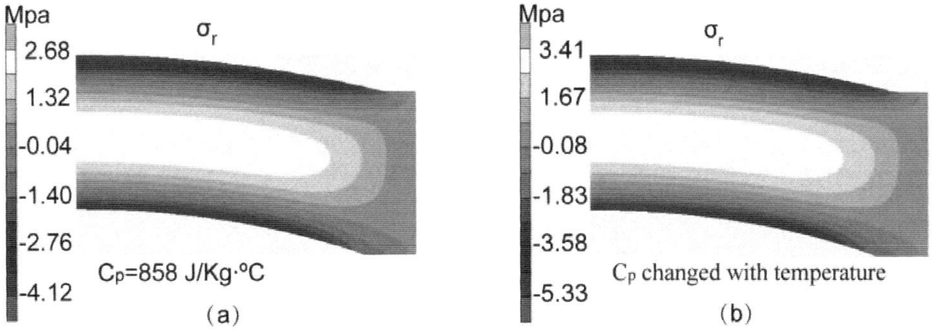

Figure 6. Finite element simulation of radial stress σ_r in the aspherical glass lens (a) C_p=858 J/Kg·°C, (b) C_p changed with temperature shown in figure 5

5. Conclusions

By using FEM with a structural relaxation model, it is able to predict the residual stresses inside compression molded glass lenses. Three types of CTE and two kinds of heat capacity were applied to investigate the influences on prediction of residual stresses. The results show that the simulated residual stresses have the same distribution. However the magnitude of residual stresses shows a big difference of the effect of CTE and heat capacity. Under different CTE (α_g, α_l), the maximum tensile stress of the radial stress at (8.3x10^{-6} /°C, 3.77x10^{-5} /°C) is almost as twice as the maximum tensile stress at (5.6x10^{-6} /°C, 1.68x10^{-5} /°C). For heat capacity, the heat capacity changed with temperature causes 20% larger than constant heat capacity in radial stress.Therefore, in numerical simulation, the parameters should be conform to reality in order to obtain accurate residual stresses.

References

[1] Allen Y Y and Jain A 2005 *J. Am. Ceram. Soc.* **88(3)** 579-86.
[2] He P, Wang F, Li L, Georgiadis K, Dambon O, Klocke F and Allen Y Y 2011 *J. Opt.* **13(8)** 085703.
[3] Li L, He P, Wang F, Georgiadis K, Dambon O, Klocke F and Allen Y Y 2011 *J. Opt.* **13(5)** 055407.
[4] Aben H and Guillemet C 1993 *Photoelasticity of Glass* (Springer Science & Business Media).
[5] Chen Y, Allen Y Y, Su L, Klocke F, and Pongs, G 2008 *J. Manuf. Sci. E.* **130(5)** 051012.
[6] Jain A, Firestone G C, and Allen Y Y 2005 *J. Am. Ceram. Soc.* **88(9)** 2409-14.
[7] Jain A 2006 *Experimental study and numerical analysis of compression molding process for manufacturing precision aspherical glass lenses* (Doctoral dissertation, The Ohio State University).
[8] Narayanaswamy O 1971 *J. Am. Ceram. Soc.* **54(10)** 491-8.
[9] Soules T F, Busbey R F, Rekhson S M, Markovsky A and Burkem A 1987 *J. Am. Ceram. Soc.* **70(2)** 90-5.
[10] Fujidie Co 2003 *Material: Cemented Carbide Alloys* (http://www.fujidie.co.jp).
[11] Scherer G W 1986 *Relaxation in glass and composites.*

Acknowledgment

The work is supported by National Natural Science Foundation of China (Grant No. 51505349, 51578430), and Key Laboratory of Metallurgical Equipment and Control of Ministry of Education in Wuhan University of Science and Technology (2015B06).

Dimensional Precision Research of Wax Molding Rapid Prototyping based on Droplet Injection

Huang Mingji[1], Wu Geng[1], Shan yan[2]

[1]University of Science and Technology Beijing, 100083
[2]Chinese people's liberation army army armored military academy, 130017

Abstract. The traditional casting process is complex, the mold is essential products, mold quality directly affect the quality of the product. With the method of rapid prototyping 3D printing to produce mold prototype. The utility wax model has the advantages of high speed, low cost and complex structure. Using the orthogonal experiment as the main method, analysis each factors of size precision. The purpose is to obtain the optimal process parameters, to improve the dimensional accuracy of production based on droplet injection molding.

1. Introduction

With the continuous development of manufacturing products in various industries, the shape of the casting requirements are more and more complex, the production cycle requires more compact, the traditional process has been unable to meet the actual production requirements [1, 2]. In view of the shortcomings of the traditional pressure-type manufacturing process, in the production of small and medium-sized scale, the application of rapid prototyping technology to solve a lot of traditional process defects [3-5].

Researchers at home and abroad to do a lot of exploration. Zhang Yu [6] that SLS, FDM, SLA technology can be used for the formation of wax mold, but the wax mold strength is low. Nicholas [7] have published a method for casting and 3D printing wax molds, including paraffin-based resin materials. University of Munich, Thomas Ottnad [8] found that piezoelectric actuators based on piezoelectric elements can produce drops of droplets. Shi Yusheng [9] suggested that rapid welding technology could solve the problems existing in the production of complex thin-walled castings, and reduce the production and development costs.

However, the rapid development of wax molds based on droplet jet molding is not yet mature, and there is great room for improvement in materials, equipment and technology [10]. This article use of traditional casting wax, compared with PS powder, casting wax has low-shrinkage, small-deformation, environmental-friendly non-toxic, recyclable characteristics. In addition, this process is suitable for a variety of solid cast wax on the market, not just limited to PLA-like wire. In this paper, dimensional accuracy is studied from the aspects of dimensional compensation and shrinkage characteristics. Improve the dimensional accuracy of the wax mold by optimizing the rapid prototyping process.

2. Equipment preparation

2.1. Equipment and materials

The main structure includes a set of three-dimensional motion console, a heating device with a heat preservation pipeline, a high-precision piezoelectric injection valve, as shown in figure 1. The main parameters are shown in the following configuration table 1. The main components of the material used in the study are 316-2B medium temperature precision casting wax, forming surface high precision, low shrinkage, high strength, less ash, suitable for 3D printing.

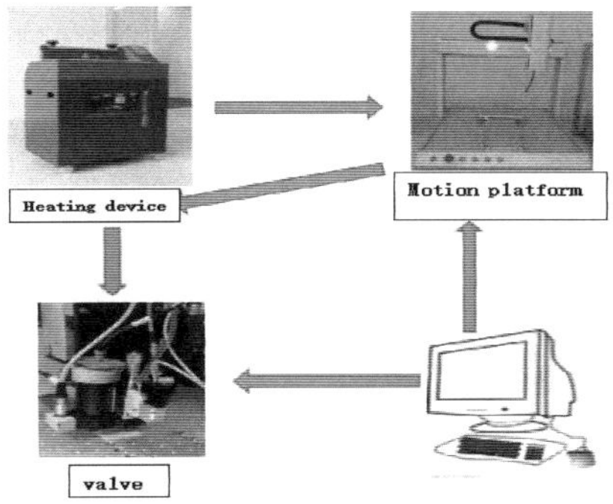

Figure 1. Equipment schematic diagram.

Table 1. Parameters of moving platform.

Project	Parameter
Spraying range	500/500/100mm
Plateform speed	0--150 $mm \cdot s^{-1}$
Frequency range	20-200 HZ
Minimum delivery	2 nl
Operate temperature	0--130 ℃

2.2. Mechanical error compensation

Place the precision cast wax in the heating tube to melt. Start the wax injection molding until the equipment is stable. Form a set of rectangle with a radius of 40 mm and a height of 2 mm.

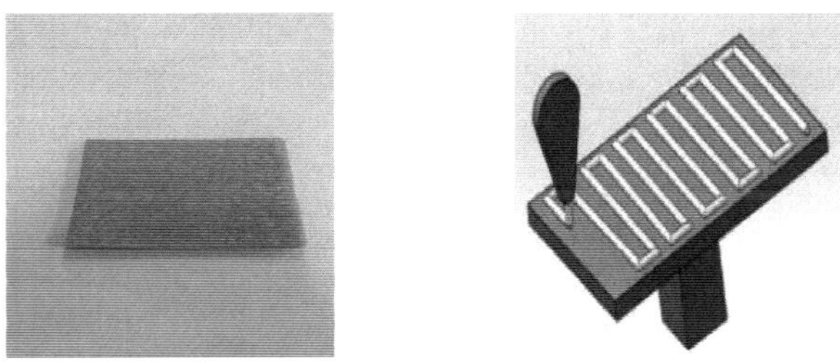

Figure 2. Machining methods and test sample pieces.

The dimensions are measured immediately after processing is complete. Respectively, take the average of each group as the measured size, calculate the X, Y, Z direction of the machining error compensation factor.

The mechanical error compensation factor in the X direction is 0.99, the mechanical error compensation factor in the Y direction is 0.98, and the mechanical error compensation factor in the Z direction is 1.04.

3. Orthogonal test

Select the spray molding sample shown in figure 2, the test parameters include: spray wax temperature (T), molding speed (V), layered thickness (D). The orthogonal test was designed to measure the shrinkage of X, Y and Z dimensions, and the variance analysis was carried out to quantitatively obtain the influence of each process parameter on the experimental results.

The reasonable range of the three factors: spray wax temperature range of 75 - 95°C, processing speed range of 0-100 mm*s-1, the thickness of the layer is 0.2-1mm. Based on the above data, select the three-factor four-level orthogonal table, the establishment of test factors group table, as shown in table 2.

Table 2. Experimental factors and levels.

	factors		
Lvl	Wax spraying temperature °C (T)	Forming speed mm/s (V)	Thickness height mm (D)
1	75	20	0.2
2	80	40	0.4
3	85	60	0.6
4	90	80	0.8

The sample size is 90mm (X) * 90mm (Y) * 60mm (Z), the middle cylinder diameter is 40mm, the test sample is shown in figure 2.

According to the design principle of orthogonal experiment, each factor is arranged in different ways according to the orthogonal way, forming 16 groups of tests, each group made three samples, processed immediately after the completion of measuring its size, and then put 20°C constant temperature container for 24 hours after the insulation, and then measure the sample X, Y, Z direction size, and find the amount of each measurement direction of the amount of expansion and contraction. DAta in table 3.

Table 3. Shrinkage in all directions.

	Average Shrinkage (mm)															
x	1.16	0.97	1.23	1.42	1.20	1.26	1.44	1.27	1.36	1.41	1.39	1.51	1.48	1.46	1.53	1.66
y	1.23	1.04	1.25	1.45	1.13	1.21	1.37	1.32	1.46	1.53	1.46	1.48	1.61	1.48	1.69	1.62
z	1.07	0.95	0.99	1.13	1.21	1.09	1.20	1.13	1.28	1.29	1.16	1.41	1.46	1.27	1.39	1.42

The analysis of the measured data (in the X direction as an example), analysis of the impact of various factors significant degree.

Table 4. Effect of four factors on shrinkage.

	Quadratic sum	Free degree	Square deviation	F-value	P-value	Significance
T	9.18	3	3.06	50.51	0.0010	Significantly
V	1.62	3	0.54	8.92	0.0125	
D	0.35	3	0.11	1.93	0.2253	
error	0.36	6	0.06			
sum	11.52	15				

According to this table can be seen in the X direction, the three factors affect the ability of the order: spray wax temperature> molding speed> layer thickness. The following clear in the multi-factor role and single factor, the three factors on the specific impact of specimen contraction trend.

4. Result analysis

4.1. Influence of size accuracy on multi - factor interaction

The figure above shows the effect of the three factors on the amount of contraction, where the three coordinates represent the grouping in the orthogonal level table, and the color change represents the change in shrinkage. It can be seen that the range of the optimal parameters is consistent with the previous three-factor variables.

At the molding speed of 60 mm / s, the effects of spray wax temperature and layer thickness were studied. According to the above study, the first three horizontal ranges of wax spray temperature and stratified thickness in table 2 were divided into five Group test level. Each level of manufacture of three samples, measuring shrinkage method above.

It can be seen from the figure 6 that the optimum temperature of the spray wax is the second orthogonal level, about 80°C; the optimum stratification thickness parameter is the second orthogonal level, about 0.4 mm.

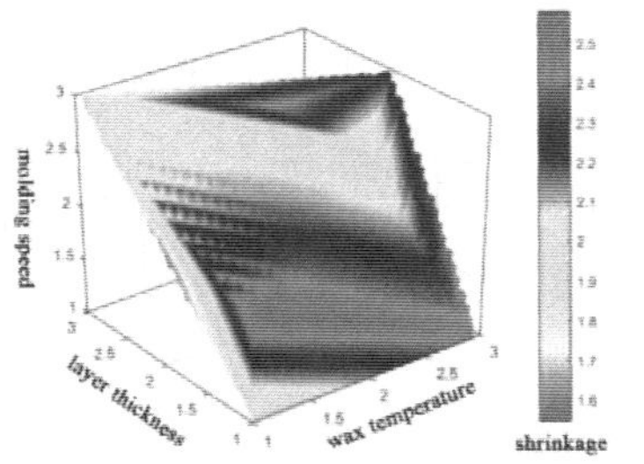

Figure 3. Effect of three factors on shrinkage.

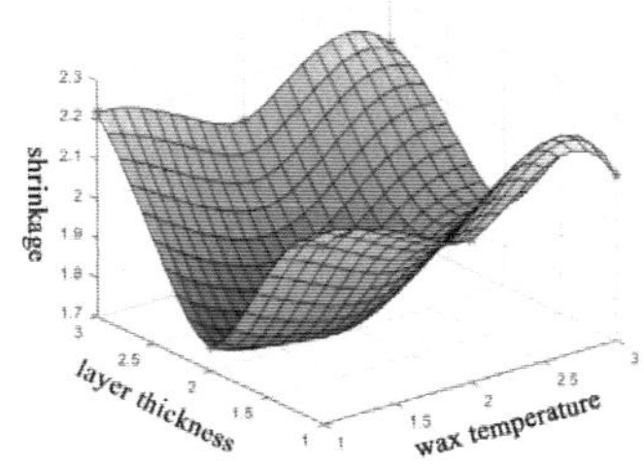

Figure 4. Effects of two significant factors on shrinkage.

4.2. Effect of single factor on dimensional accuracy

The influence of various factors on the forming precision was studied. Get a set of superior parameters, the following single factor control variable method to verify the above parameters.

4.2.1. Effect of spraying temperature on sample forming accuracy. Orthogonal experiments show that the wax spray temperature has the greatest influence on the precision of wax mold forming. Spray the temperature is too low, prone to broken wax or plug the nozzle. If the temperature of the spray is too high, the sprayed wax can not be rapidly solidified, resulting in a significant reduction in machining accuracy and a tendency to produce bubbles. It can be seen that when the wax spray temperature is about 84°C, the sample forming precision is the best.

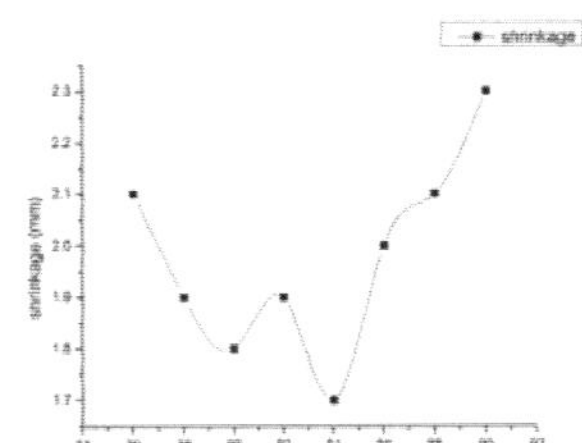

Figure 5. Temperature on shrinkage.

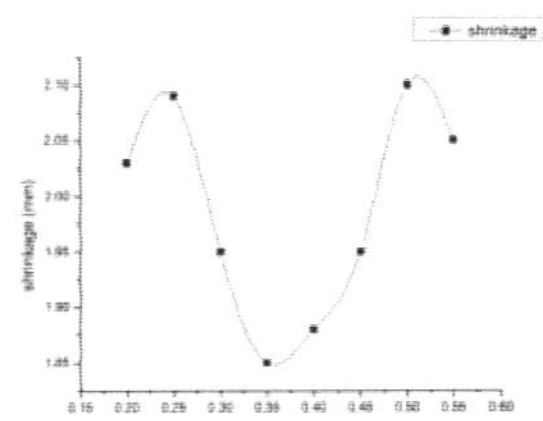

Figure 6. Thickness on shrinkage.

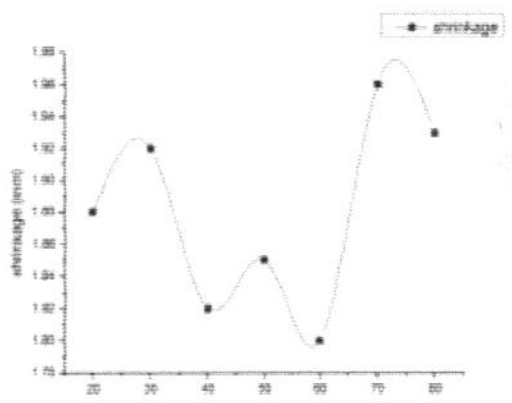

Figure 7. Speed on shrinkage.

4.2.2. Influence of Layered Thickness on Sample Forming Accuracy. The effect of stratified thickness on specimen shrinkage is obvious, especially for Z direction. The higher layers caused by the interlayer fusion, the higher the accuracy of the molding. When the wax spray temperature 80°C, forming speed 60 mm / s case, only to change the thickness of the layer, the data shown in the following figure. As can be seen from figure 6, when the stratified thickness is 0.35 mm, the contraction is shown to be minimal.

4.2.3. Effect of Molding Speed on Sample Forming Accuracy. In a certain range, with the molding speed increases, the specimen of the molding accuracy should be improved. As the speed of the platform increases, the problem of material spillage caused by the long injection time is prevented, and the temperature distribution of each molding is made more uniform. Analysis shows that when the platform moving speed in a large range of changes, the impact on the specimen shrinkage is minimal. The optimum value is 60mm / s.

5. Conclusion

(1) Droplet injection wax mold 3D rapid prototyping, the three process parameters of the significant ability of the order of the order is: spray wax temperature> layer thickness> forming speed.

(2) The best process parameters: molding speed 60 mm / s, spray wax temperature 80°C, layer thickness 0.35 mm, forming speed 60mm/s.

(3) The shrinkage rate in the x direction is 1.6%, the shrinkage in the y direction is 1.9%, the shrinkage of the z direction is 2.3%, and the compensation coefficient in the x direction is 1.006, the y direction. The compensation coefficient is 0.998 and the compensation coefficient in the z direction is 1.064. After the size is corrected, the size of the specimen in the size of the contraction accuracy control within 1mm

References

[1] Fan Z T 2012 *Foundry* **06** 583-91.
[2] Charbonnier B et al. 2016 *J. European Ceram. Soc.* **36** 4269-79.
[3] Sun d 2016 Spec. Cast. Nonferr. Alloy. **11** 1172-4.
[4] Jiang T L 2017 *Electr. Mech. Eng.* **01** 48-51.
[5] Xiao K 2005 *Found. Tech.* **08** 712-3 + 6.
[6] Zhang Y 2016 *Found. Tech.* **04** 759-64.
[7] Nicholas W B et al. 2015 *Science* **6244** 161-5.
[8] Thomas O, Markus K and Franz I 2012 *IEEE* **978** 117-121.
[9] Schumacher C M, et al., 2014 *RSC Adv.* **4(31)** 1603-9.
[10] Shi Y S et al. *Foundry* **10** 749-52.

Study on Transverse Load Distribution of Hinged Hollow Beam

Weiyue Wang[1], Chao Zhang[2] and Shui Wan[1]

[1]School of Transportation, Southeast University, Nanjing 210096, China;
[2]School of highway engineering, Chang'an University, Xian, 710000, China

Abstract. The bridge is a kind of space structure, when the car load on a part of the bridge, the impact of its load will be transmitted to the surrounding. In this paper, the hinge plate method is used to calculate and analyze the simply supported hollow slab of a certain arch bridge. Considering the hinge plate mounting method is suitable for pouring concrete bridge connecting the longitudinal tongue and groove joints, horizontal beams fabricated separate beam only in the middle between the free flaps or reinforced with steel connection. Therefore, the transverse analysis and calculation of the superstructure of box girder are carried out by using hinge plate method. And mechanical analysis of the transverse beam with finite element software MIDAS Civil grillage method.

1. Introduction

The horizontal distribution of loads refers to how the vehicle load acting on the bridge is distributed among the main girders, or how the girders share the vehicle live load. For a beam composed of multiple beams, and through the transverse beams, bridges and other horizontal connections to form a whole beam bridge [1]. When the bridge has a load, because the lateral stiffness of the structure will inevitably make the main beam of different load to share the load. For the determination of the internal force value S of a certain section of a main beam, the concept of influencing line is introduced in the vertical and horizontal directions of the bridge, simplify the problem of space as a plane problem. Which is:

$$S = P \cdot \eta(x, y) \approx P \cdot \eta_2(y) \cdot \eta_1(x) \tag{1}$$

In the formula: $\eta(x, y)$ —— Internal force influence surface of a beam in space calculation.

$\eta_1(x)$ —— Single beam in the x-axis direction of a section of the internal force of the line.

$\eta_2(y)$ ——When the unit load acts in different positions along the transverse direction (y-axis direction) of the deck, the variation ratio of the load ratio assigned by a beam is also referred to as the influence of the lateral distribution of load on a beam.

"$P \cdot \eta_2(y)$"is the load distributed to a beam in the transverse direction when the load "P" acts on point A(x,y). Temporarily expressed as "P". The maximum load "P'_{max}" can be obtained according to the loading at the most unfavorable position. We define that "P" is the axle weight, then m is called the lateral distribution of the load, which means that the maximum load assumed by a root beam is a multiple of the axle (usually less than 1).

2. Articulated plate method

For the cast-in-place concrete longitudinal seam joints of the assembly-type bridge and a welded bridge with no intermediate beams connected between wings or welded crossheads. As the block between the horizontal with a certain connection structure, its connection is very rigid and weak. The kind of structure is practically closing to a plurality of slit plates that parallels to each other. For this type of force structure, the lateral hinge plate theory can be used to calculate the lateral distribution of the load.

The bridge consists of 9 hollow slab beams, the cross-section of the bridge and the typical cross-section of the hollow plate as shown in figure 1 and figure 2.

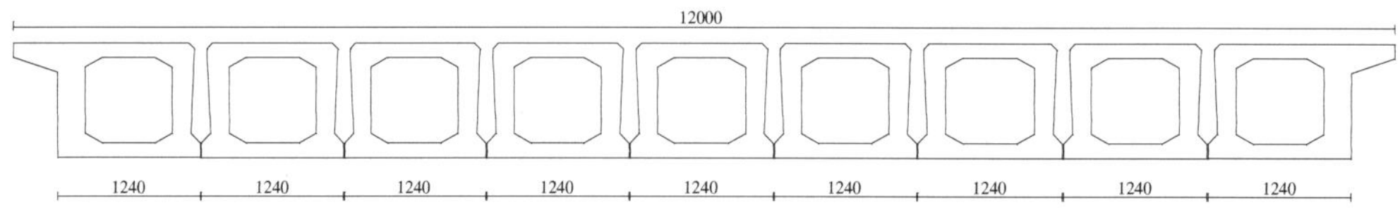

Figure 1. Full bridge cross section (unit: mm).

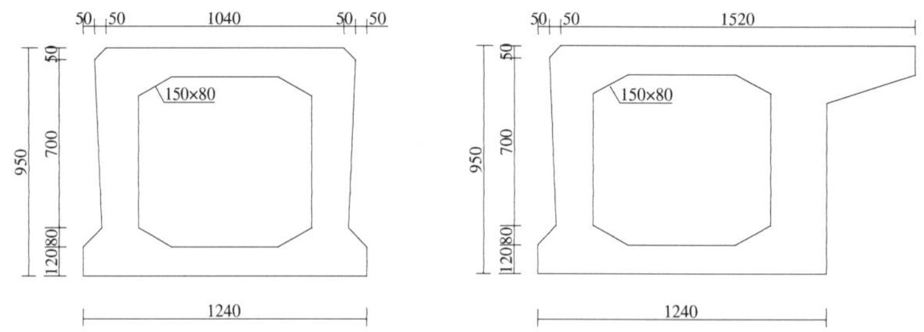

(a) Medium plate cross section (b) Side plate cross section (unit: mm)

Figure 2. Hollow plate typical cross section (unit: mm).

2.1. The calculation process

Calculate the stiffness factor.

$$\gamma = \frac{b\varphi}{2\omega} = \frac{\pi^2 EI}{4GI_T}\left(\frac{b}{l}\right)^2 \approx 5.8\frac{I}{I_T}\left(\frac{b}{l}\right)^2 = 0.012427$$

$$(\omega_1 + a_1\varphi_1)/\omega \approx \frac{I}{I_1} + 23.2\frac{I}{I_{T1}}\left(\frac{a_1}{l}\right)^2 = 0.840786$$

When the unit load is applied to any beam, the vertical load assigned to each beam is solved. The vertical load of a beam calculated by the different load on the board is connected into a curve. And the horizontal impact line of the load of a sheet is formed.

2.2. The calculation result

According to the calculation, the hinge force values of each hinge point and the vertical load values of each beam are obtained under the unit force. The hinge values of the nodes are shown in table 1, and the vertical load values for each plate are listed in table 2.

Table 1. Hinge of hinge points.

Hinge number	g_1	g_2	g_3	g_4	g_5	g_6	g_7	g_8	g_9
1	0.7755	-0.1974	-0.1606	-0.1318	-0.1004	-0.0930	-0.0810	-0.0731	0.0696
2	0.6108	0.6421	-0.3013	-0.2473	-0.1943	-0.1745	-0.1520	-0.1372	0.1306
3	0.4769	0.5014	0.5554	-0.3752	-0.3021	-0.2648	-0.2307	-0.2083	0.1981
4	0.3671	0.3858	0.4274	0.4905	-0.4316	-0.3684	-0.3210	-0.2898	0.2757
5	0.2757	0.2898	0.3210	0.3684	0.4316	-0.4905	-0.4274	-0.3858	0.3671
6	0.1981	0.2083	0.2307	0.2648	0.3021	0.3752	-0.5554	-0.5014	0.4769
7	0.1306	0.1372	0.1520	0.1745	0.1943	0.2473	0.3013	-0.6421	0.6108
8	0.0696	0.0731	0.0810	0.0930	0.1004	0.1318	0.1606	0.1974	0.7755

Table 2. The vertical load values assigned to each beam.

Beam number	p_{i1}	p_{i2}	p_{i3}	p_{i4}	p_{i5}	p_{i6}	p_{i7}	p_{i8}	p_{i9}
1	0.2245	0.1974	0.1606	0.1318	0.1004	0.0930	0.0810	0.0731	0.0696
2	0.1646	0.1605	0.1407	0.1155	0.0939	0.0815	0.0710	0.0641	0.0610
3	0.1339	0.1407	0.1433	0.1280	0.1078	0.0903	0.0787	0.0710	0.0676
4	0.1099	0.1155	0.1280	0.1343	0.1295	0.1036	0.0903	0.0815	0.0775
5	0.0914	0.0961	0.1064	0.1221	0.1368	0.1221	0.1064	0.0961	0.0914
6	0.0775	0.0815	0.0903	0.1036	0.1295	0.1343	0.1280	0.1155	0.1099
7	0.0676	0.0710	0.0787	0.0903	0.1078	0.1280	0.1433	0.1407	0.1339
8	0.0610	0.0641	0.0710	0.0815	0.0939	0.1155	0.1407	0.1605	0.1646
9	0.0696	0.0731	0.0810	0.0930	0.1004	0.1318	0.1606	0.1974	0.2245

3. Finite element space calculation

The beam stiffness method is used to replace the bridge superstructure with the equivalent beam grid. The bending stiffness and torsional rigidity of each section distributed on the space board (beam) are concentrated in the nearest equivalent grating. The longitudinal stiffness of the actual structure is concentrated in the longitudinal beam element, and the lateral stiffness is concentrated in the transverse beam.

3.1. Modeling introduction

The finite element program Midas Civil was used to simulate the bridge, and the model was established by using the beam grid method. Because the hinge only transmits the shear force and does not transmit the bending moment. The two ends of the beam are released when the beam is simulated. At the same time, the bending moment of the beam is similar to the bending moment of the hollow plate beam for the accuracy of the calculation [2]. When the beam grid method is used, the simulation of the lateral connection is very important, which is directly related to the accuracy of the result. In this paper, the stiffness of the virtual beam is less than the rigidity of the stringer [3-5].

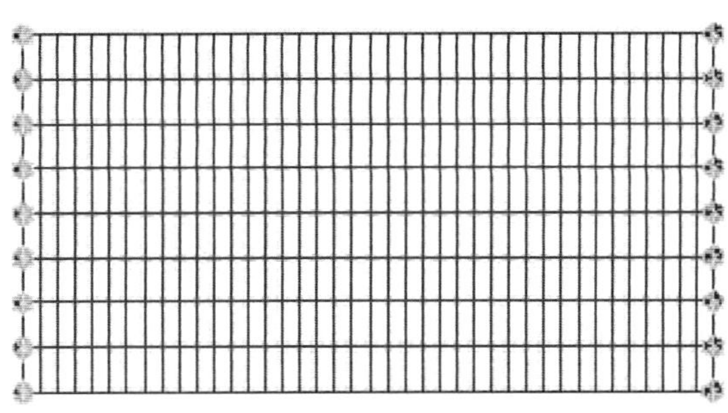

Figure 3. Finite element analysis model.

3.2. Calculation results

According to the concept of horizontal distribution of load, the following relationship should be established in any section of the bridge across the longitudinal direction:

$$\frac{\omega_i(x)}{\omega_j(x)}=\frac{M_i(x)}{M_j(x)}=\frac{Q_i(x)}{Q_j(x)}=\frac{P_i(x)}{P_j(x)}=C \tag{2}$$

In the formula, x (x), M (x), Q (x), P (x) are the deflection, bending moment, shear force and the assigned force and load of section i and j, respectively, and C is constant.

According to equation (2), the vertical distribution of the influence of the horizontal distribution of the load can be expressed as

$$\eta_\omega=\frac{\omega_j}{\sum_{i=1}^{n}\omega_i},\ \eta_M=\frac{M_j}{\sum_{i=1}^{n}M_i},\ \eta_Q=\frac{Q_j}{\sum_{i=1}^{n}Q_i} \tag{3}$$

In the finite element model, the amplitude of the beam is obtained by loading the unit sine load, and the influence of the horizontal distribution of the bridge load and the variation law of the beam are obtained by the formula (3). The figure 4 is the magnitude of the moment value of the load when the sine load is applied to the beam No. 1. The influence line coordinates are summarized in table 3.

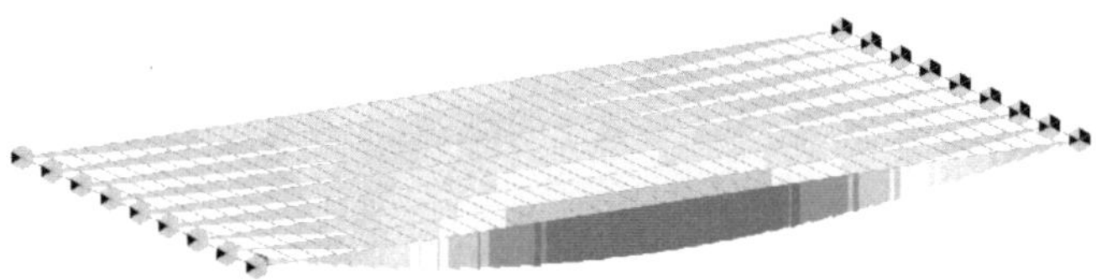

Figure 4. The load acts on the bending moment of beam 1.

Table 3. Finite element method to calculate the vertical load value.

Beam number	p_{i1}	p_{i2}	p_{i3}	p_{i4}	p_{i5}	p_{i6}	p_{i7}	p_{i8}	p_{i9}
1	0.1722	0.1518	0.1297	0.1133	0.1004	0.0905	0.0833	0.0785	0.0761
2	0.1520	0.1508	0.1356	0.1171	0.1037	0.0935	0.0861	0.0812	0.0788
3	0.1305	0.1359	0.1380	0.1259	0.1102	0.0993	0.0914	0.0863	0.0838
4	0.1141	0.1175	0.1261	0.1310	0.1214	0.1080	0.0994	0.0939	0.0912
5	0.1012	0.1042	0.1104	0.1215	0.1287	0.1215	0.1104	0.1042	0.1012
6	0.0912	0.0939	0.0994	0.1080	0.1214	0.1310	0.1261	0.1175	0.1141
7	0.0838	0.0863	0.0914	0.0993	0.1102	0.1259	0.1380	0.1359	0.1305
8	0.0788	0.0812	0.0861	0.0935	0.1037	0.1171	0.1356	0.1508	0.1520
9	0.0761	0.0785	0.0833	0.0905	0.1004	0.1133	0.1297	0.1518	0.1722

3.3. Compare results

The results of the hinge plate method are compared with the finite element results in the same figure.

(a) The load acts on beam 1.

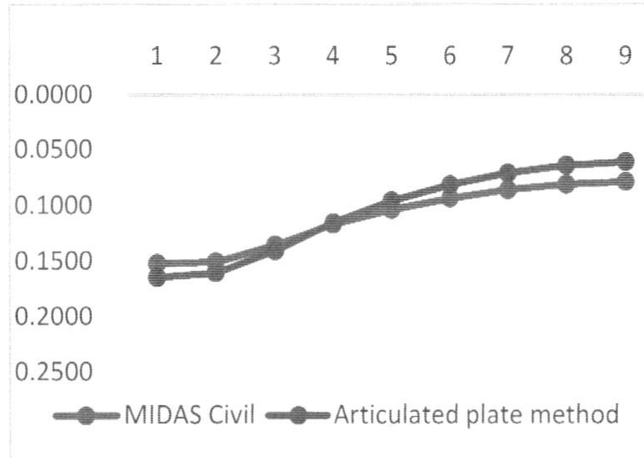

(b) The load acts on beam 1.

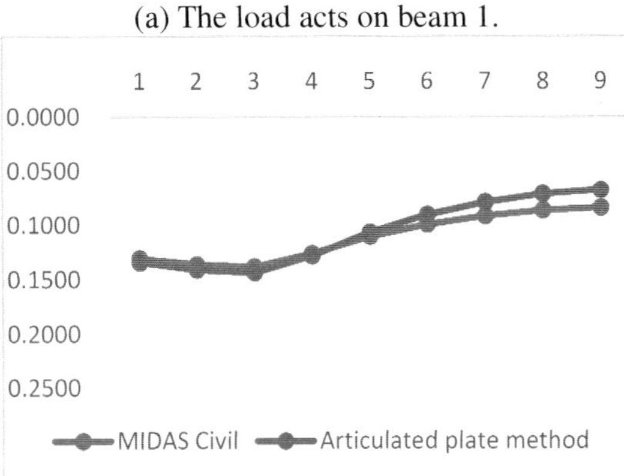

(c) The load acts on beam 3.

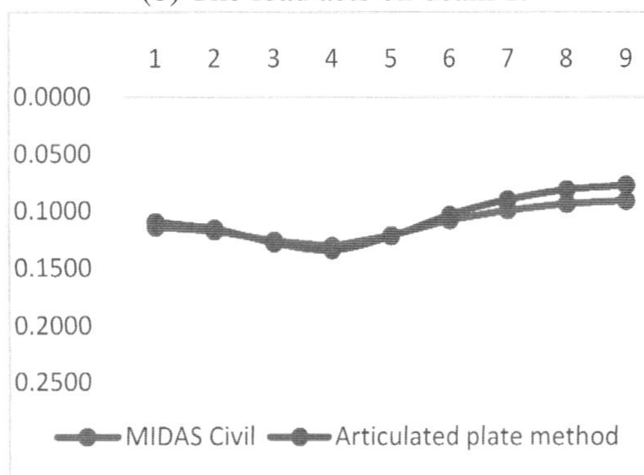

(d) The load acts on beam 4.

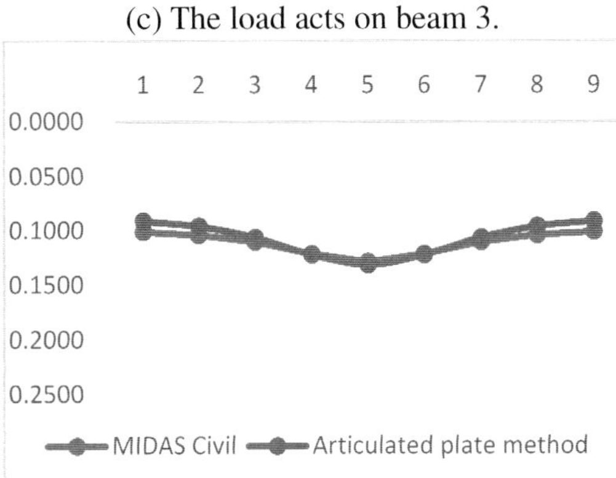

(e) The load acts on beam 5.

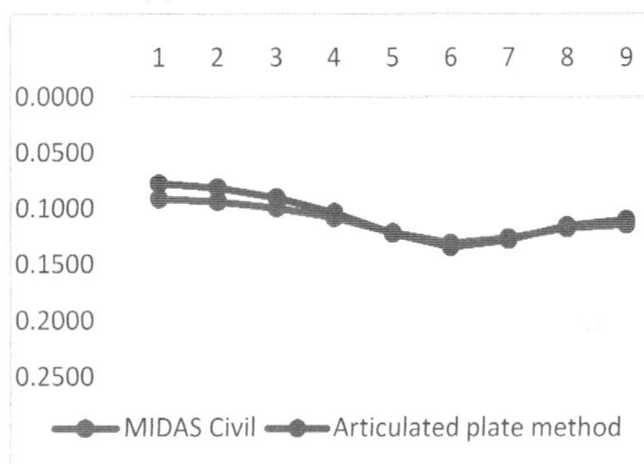

(f) The load acts on beam 6.

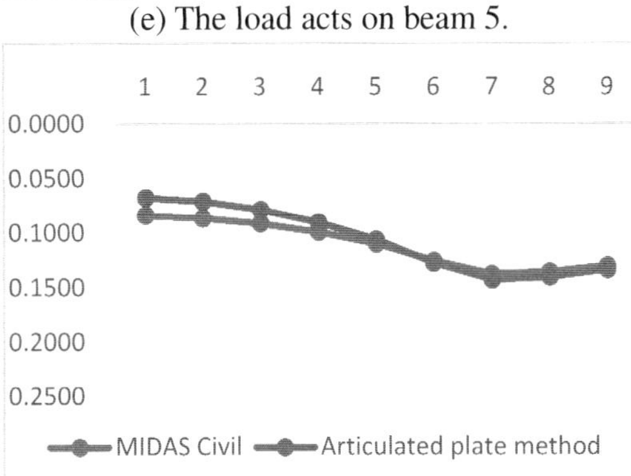

(g) The load acts on beam 7.

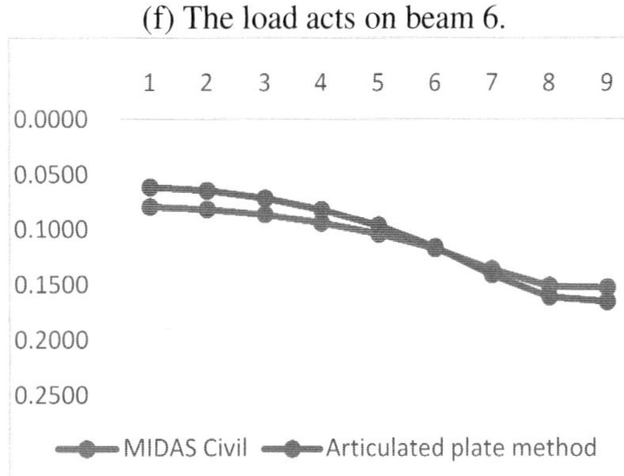

(h) The load acts on beam 8.

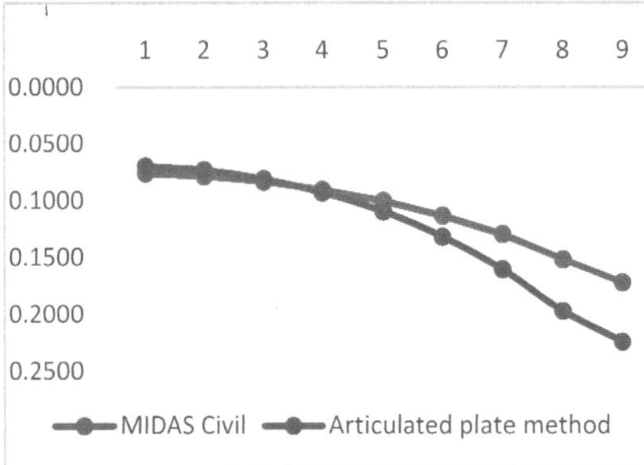

(i) The load acts on beam 9.

Figure 5. Results contrast curve.

4. Conclusion

The results show that the stiffness of the simulated beam is close to that of the longitudinal beam, and the calculated results are more accurate. In this paper, the stiffness of the simulated beam is slightly smaller than that of the rigid beam. In the beam stiffness, did not consider the side beam and the beam of the bending stiffness of the gap, so the edge beam effect line calculation results are low.

The lateral load distribution than the load transverse distribution of beam calculated on the hinge plate is calculated by the method is more uniform, this is because that along the length of plate can transfer shear hinged plate, but only a finite element model of shear transfer in virtual beam position. The beam grid method can calculate the bending moment and shear transverse distribution of the full length of the bridge, and the hinge plate method can only reflect the horizontal distribution of the bending moment.

5. Reference

[1] Shao X D 2016 *Bridge Engineering* China People's Communications Press.
[2] Hambly E C 1982 *Bridge upper structural performance* Beijing: People's Communications Press.
[3] Ge J Y and Ding X Y 2010 *Highway* **04** 103-6.
[4] Mu Y J 2010 *Study on Transverse Load Distribution Characteristics of Simply Supported Hollow Slab Bridge.*
[5] Lin P Z 2011 *Analysis Theory and Application Research on Shear Lag Effect of Concrete Box Girder.*

IOP Conf. Series: Materials Science and Engineering **269** (2017) 012053 doi:10.1088/1757-899X/269/1/012053

Rolling Force Prediction of Hot Strip based on Combined Friction

Xingdong Li[1]

[1]College of Mechanical Engineering, Yanshan University, Qinhuangdao, Hebei 066004, China

Abstract. Rolling force of hot strip was predicted in which slipping and sticking friction were adopted on the contact arc between work roll and hot strip. Karman equation was applied to slipping friction and Orowan equation was applied to sticking friction. Predicted rolling forces of hot strip agreed with measured ones from a seven-stand hot strip mill. Friction stress distributions of one classical hot rolled strip on contact arc were analysed at each pass. From pass 1 to 3, sticking friction dominated while from pass 4 to 7 slipping and sticking friction coexisted. It proved that the applicability and superiority of rolling force model based on combined friction for rolling force prediction of hot strip.

1. Introduction

Rolling force is an important parameter that influences thickness precision and strip shape quality in the control of hot strip rolling. A lot of researchers [1-5] have been involved with simulation of rolling force in hot strip rolling with different methods. Chai X J [1] et al. computed transverse distribution of rolling force of hot strip by finite element and artificial neural network. Considering deformation rate on deformation resistance, Li Z J [2] et al. developed a rolling force predicting model on the basis of SIMS model. Chun S Y [5] calculated rolling force distribution based on improved Karman equation for hot strip rolling.

Karman equation had been widely used to predicting rolling force of cold strip where friction stress is not large and smaller than shearing yield limit of rolled strip. But in computing rolling force of hot strip when friction was large and nearly equal to shearing yield limit, Karman equation had a deviation from measured rolling force in reason that friction was ignored in plastic condition equation. Orowan equation had been widely used to predicting rolling force of hot strip. However, in the past passes of hot tandem rolling mill, rolling force predicted by Orowan equation had a deviation from measured rolling force.

In this paper, friction on the contact arc between work roll and hot strip was adopted as slipping and sticking friction. The models for computing unit rolling force with slipping and sticking friction conditions were developed based on Karman and Orowan unit pressure differential equation respectively, and then the model were developed for predicting the total rolling force. A lot of total rolling force simulations were conducted with collected data from a seven-stand rolling mill to assess the validity of the roll force models based on combined friction. The applicability and superiority of rolling force model in this paper was shown by analyzing the friction stress distribution of one classical hot rolled strip on contact arc.

2. Predicting model of rolling force of hot strip

In this paper, contact arc surface between hot rolled strip and work roll were divided into backward slipping, sticking and forward slipping zones, and point n was neutral plane, as shown in figure 1. From point a to b, friction stress increased and was less than shearing yield limit of rolled strip. Within this zone, it was named backward slipping zone. The zone from point d to c was similar to that from point a to b, and it was named forward slipping zone. From point b to c, absolute value of friction stress was equal to shearing yield limit of rolled strip and remains unchanged. Within this zone, it was named sticking zone.

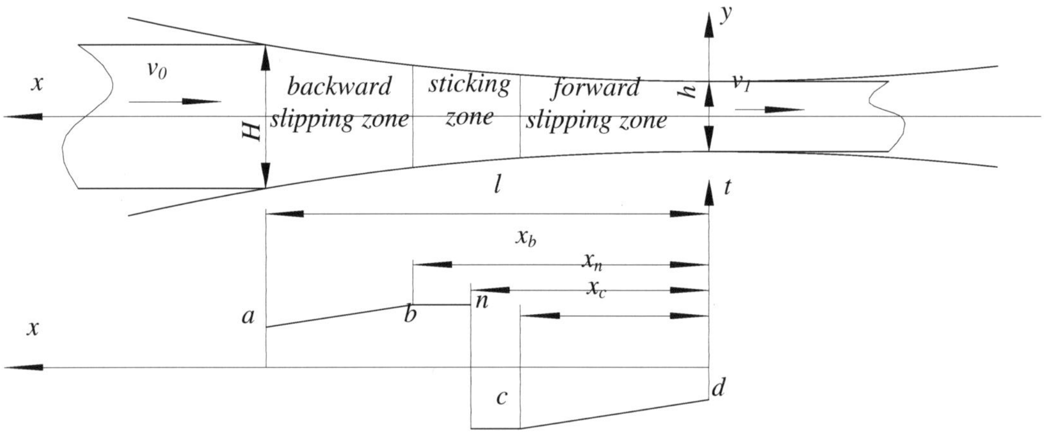

Figure 1. Friction condition in contact arc surface.

2.1. Derivation of rolling forces per unit area p_x

In backward and forward slipping zone, contact arc was adopted as straight line and friction was adopted as dry friction. Friction coefficient μ is unchanged, and friction stress t_x is

$$t_x = \mu p_x \tag{1}$$

Based on Karman differential equation of unit pressure, rolling forces per unit area p_{dc} in forward slipping zone is

$$p_{dc} = 2k\left(1-\sigma_1(2k)^{-1} + tg\theta_{dc}\mu^{-1}\right)\left(h_x h^{-1}\right)^{\mu/tg\theta_{dc}} - 2k\mu^{-1}tg\theta_{dc} \tag{2}$$

where k, σ_1 denote shear yield limit of rolled strip, exit tension respectively; h_x denotes strip thickness at point x and θ_{dc} denotes angle between chord dc and center line of the rolling.

Similar to forward slipping zone, rolling forces per unit area p_{ba} in backward slipping zone is

$$p_{ba} = 2k\left(1-\sigma_0(2k)^{-1} - tg\theta_{ba}\mu^{-1}\right)\left(H\left(h_x\right)^{-1}\right)^{\mu/tg\theta_{ba}} + 2k\mu^{-1}tg\theta_{ba} \tag{3}$$

where σ_0 denotes entrance tension and θ_{ba} denotes angle between chord ba and center line of the rolling.

In sticking zone, friction stress was equal to shearing yield limit of rolled strip. Contact arc was adopted as

$$h_x = h + R\theta_x^{\ 2} \tag{4}$$

where R, θ_x denote radius of work roll and central angle at point x respectively.

Based on Orowan differential equation of unit pressure, rolling forces per unit area p_{cn} in forward sticking zone is

$$p_{cn} = 2k\left\{\frac{\pi}{4}\ln\left(\frac{h_x}{h_c}\right)+\frac{\pi}{4}+\left(\frac{R}{h}\right)^{1/2}\arctan\left[\left(\frac{R}{h}\right)^{1/2}\theta_x\right]+\frac{1-2\mu}{2\mu}-\left(\frac{R}{h}\right)^{1/2}\arctan\left[\left(\frac{R}{h}\right)^{1/2}\theta_c\right]\right\} \quad (5)$$

where h_c, θ_c denote strip thickness and central angle at point c respectively.

Similar to forward sticking zone, rolling forces per unit p_{nb} in backward sticking zone is

$$p_{nb} = 2k\left\{\frac{\pi}{4}\ln\left(\frac{h_x}{h_b}\right)+\frac{\pi}{4}+\left(\frac{R}{h}\right)^{1/2}\arctan\left[\left(\frac{R}{h}\right)^{1/2}\theta_b\right]-\left(\frac{R}{h}\right)^{1/2}\arctan\left[\left(\frac{R}{h}\right)^{1/2}\theta_x\right]+\frac{1-2\mu}{2\mu}\right\} \quad (6)$$

where h_b, θ_b denote strip thickness and central angle at point b respectively.

2.2. Derivation of total rolling forces

It was assumed that total rolling force acted on the roll by the hot strip was equal to its vertical component. Friction was ignored, total rolling force P is

$$P = B\left(\int_0^{x_c} p_{dc}dx + \int_{x_b}^{l} p_{ba}dx\right) + BR\left(\int_{\theta_c}^{\theta_n} p_{cn}d\theta + \int_{\theta_n}^{\theta_b} p_{nb}d\theta\right) \quad (7)$$

where θ_n denotes central angle at point n.

To solve P with equation (7), x_b, θ_n and x_c must be determined beforehand. x_c and x_b is calculated by iterative method. Total contact arc was assumed straight line. When deviation between the friction stress and shear yield limit of rolled strip at point b was smaller than 0.000001 times the shear yield limit, it was considered that this point was x_b. The same method can be used to determine x_c. θ_n was calculated by the intersect method. In the forward and backward sticking zone, both equations (5) and equation (6) can be used to compute rolling forces per unit area at point n. When rolling forces per unit area at point n in two equations was equal, the position was neutral point n. The central angle at the neutral point n was the neutral angle θ_n.

3. Validation and discussion

In order to verify the validity of rolling force model in this paper, many simulations were made with the experimental data collected from a seven-stand finished hot rolling mill. Errors, this is equal to the difference between rolling force of predicted and measured divided by measured rolling force, are shown in figure 2.

As shown in figure 2, at stands 1, 2, and 3, errors were within 2%, and at stands 4, 5, 6 and 7 the errors which were bigger than those at stands 1, 2 and 3 were within 5%. As a whole, errors of rolling force predicted by combined friction model in this paper were within 5%. It can be used to predict rolling force of hot strip for hot strip plant.

Taking one classical hot rolled strip in this plant as example, strip thickness at entrance of rolling mill is 40 *mm* and strip thicknesses at exit of each pass successively are 21.03, 12.97, 8.45, 5.48, 3.72, 2.74, and 2.36 *mm*. Strip width is 720 *mm* and work roll diameters of each pass successively are 824.3, 757.1, 781.8, 617.1, 588.3, 597.1, 638.2 *mm*. Rolling forces measured and predicted by combined friction model and SIMS model are shown in figure 3.

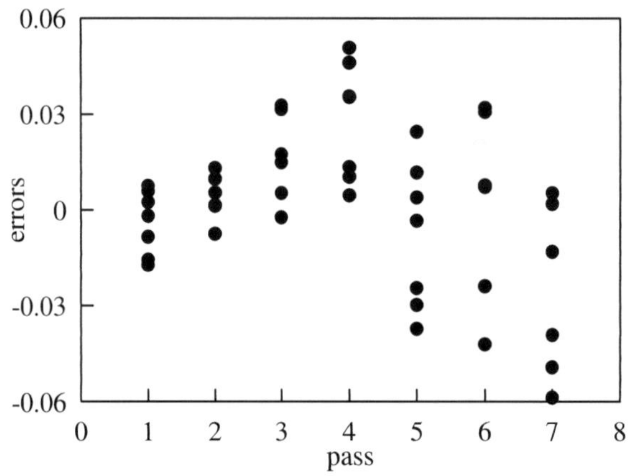

Figure 2. Errors between rolling force of predicted and measured.

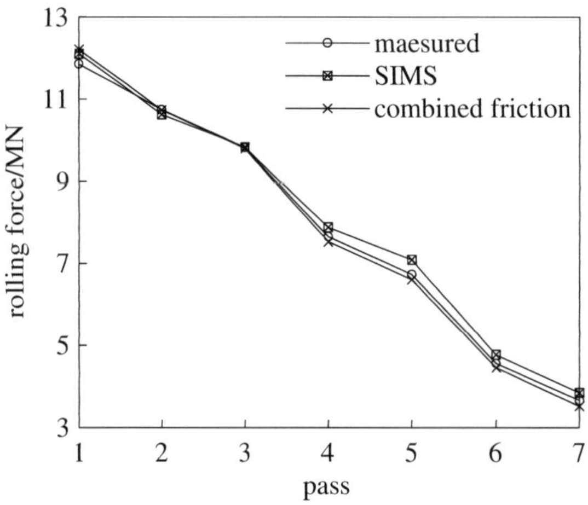

Figure 3. Comparison of rolling force among measured, combined friction and SIMS model.

As shown in figure 3, rolling forces predicted by combined friction model and SIMS model were a little larger than measured ones at pass 1 and rolling forces predicted by combined friction model and SIMS model were nearly equal to measured ones at pass 2 and 3. At pass 4, 5, 6 and 7, rolling forces predicted by combined friction model were smaller than measured ones, while rolling forces predicted by SIMS model were larger than measured ones. This is because of the ratio of slipping zone on the contact arc. Friction stress distribution of the above rolled strip on contact arc between hot strip and work roll at each pass are shown in figure 4.

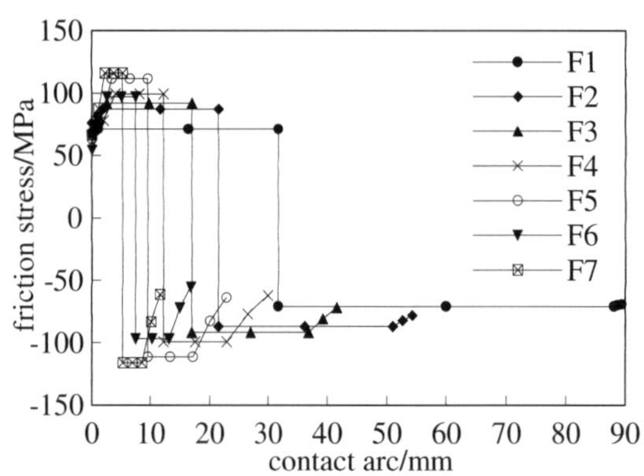

Figure 4. Friction stress on contact arc of each pass.

As shown in figure 4, at pass 1 the length of slipping zone that is small can be ignored because of large friction coefficient and rolling reduction. With the decease of rolling reduction at pass 2 and 3, the length of slipping zone increased, but against the length of contact arc, it is relatively small. Therefore, at pass 1, 2 and 3, sticking friction dominated the contact arc, SIMS model was fitted to predicting rolling forces and rolling forces predicted by combined friction model are close to ones by SIMS model. At pass 4, 5, 6 and 7, with the decease of friction coefficient and rolling reduction, the length of slipping zone visibly increased. The contact arc consisted of slipping and sticking friction. Rolling forces predicted by SIMS model were bigger than measured ones because it assumed the entire contact arc was sticking friction. Rolling force predicted by combined friction model was fitted and errors were smaller than those by SIMS.

4. Conclusions

It was assumed that friction on contact arc between work roll and hot strip consisted of slipping and sticking friction. Within slipping friction zone, friction stress is equal to the coefficient of friction multiplied by the unit rolling force and smaller than or equal to absolute value of shearing yield limit of rolled strip. Within sticking friction zone, friction stress was equal to shearing yield limit of rolled strip. The unit rolling force equation within slipping friction zone was derived based on the principle of Karman equation, and equation within sticking friction zone was derived based on the principle of SIMS equation. Then, total rolling force model was constructed and made to simulate rolling force with the process data collected from a seven-stand hot strip rolling mill. The simulated rolling force agreed with the measured ones. By analyzing the friction stress distribution on contact arc, it showed that from pass 1 to 3 it was close to SIMS friction assumption, while from pass 4 to 7 it consisted of slipping and sticking friction on contact arc. It provides a better applicability method for predicting the rolling force in hot strip rolling.

5. References

[1] Chai X J, Li H B, Zhang J, Zhou Y Z, Ma Y H and Zhang P W 2017 *Iron. Steel.* **52** 52-60.
[2] Li Z J, Peng Y, Liu H M, Wang S F and Xiao L Z 2013 *J. China Mech. Eng.* **24** 2114-18.
[3] Hou L G, Zhang G M and Xiao H 2006 *J. Iron. Steel. Res.* **18** 26-28, 62.
[4] Li Y H, Wang T, Zhang L, Gao L, Guo Q, Xiao H, Ma L L and Bian J 2012 *Heavy Mach.* 37-40.
[5] Chun S Y, Li W G and Liu X H 2014 *Int. J. Mech. Sci.* **89** 256-63.

IOP Conf. Series: Materials Science and Engineering **269** (2017) 012054 doi:10.1088/1757-899X/269/1/012054

Study on the Pure Torsion Performance of I-Type Steel-Concrete Composite Bridge with Double Girders

H C Liu[1], S Wan[1] and Q J Liu[1]

[1]Department of Transportation, Si Pailou campus of Southeast University, Nanjing, Jiangsu, China

Abstract. At present, the study of the torsion performance of the steel-concrete composite bridge with open sections are usually simplified to study the torsion performance of single-girder steel-concrete composite bridge added the influence of transverse vehicle distribution. Due to the existence of transverse connections between the bridge girders, the overall torsion performance of double-girder composite bridge is different from that of single-girder composite bridge. Therefore, it is necessary to study the effect of different forms of the connections on the torsional properties of the double-girder composite bridge through experiments. In this paper, three double-girder steel-concrete composite model bridges are designed to study the effect of different forms of transverse connections and the effect of different shapes of steel webs on torsional properties of composite bridges.

1. Designing of the model bridges

This research contains three model bridges, and is respectively numbered SP1, SP2 and SP3. SP1 is a double-girder bridge with small beams connected to the vertical webs, SP2 is a double-girder bridge with trusses connected to the vertical web, and SP3 is a double-girder bridge with small beams connected to corrugated webs. The standard span of each model bridge is 2.4m, and the calculation span is 2.2m, the height is 0.374m, and the cross section size is shown in figure 1-figure 5.

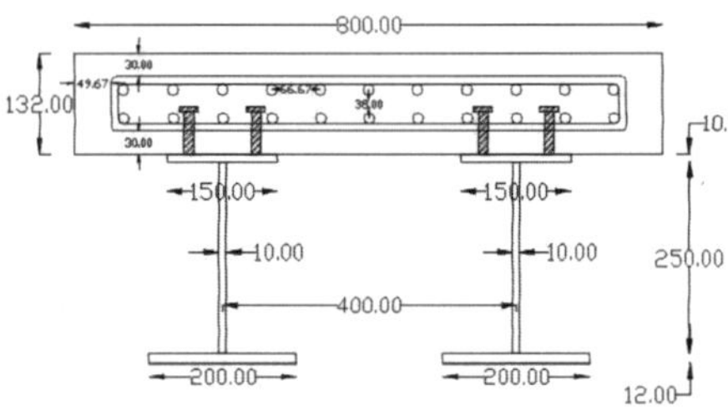

Figure 1. The cross-section of SP1 and SP2.

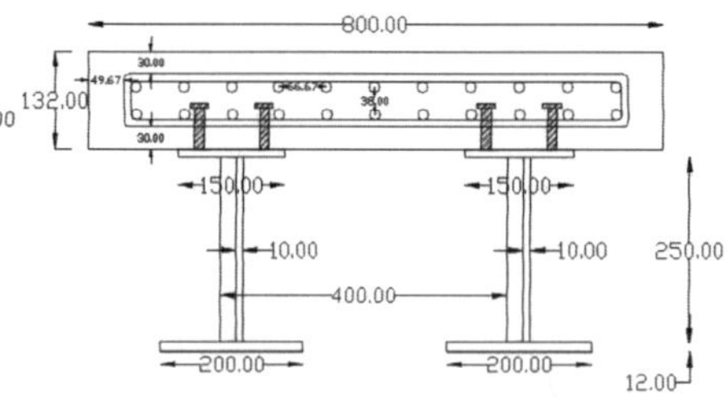

Figure 2. The cross-section of SP3.

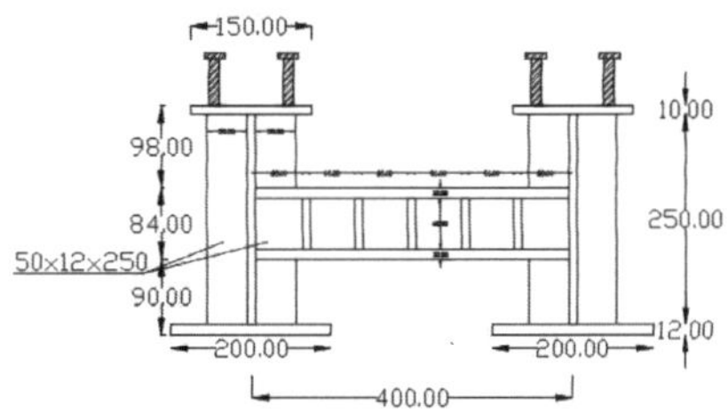

Figure 3. Connections between the webs of SP1.

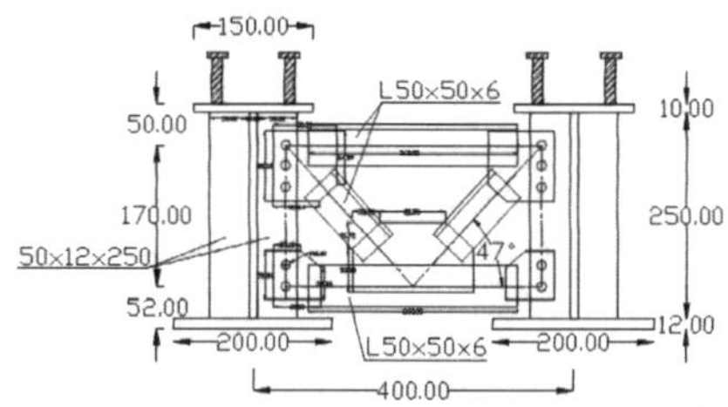

Figure 4. Connections between the webs of SP2.

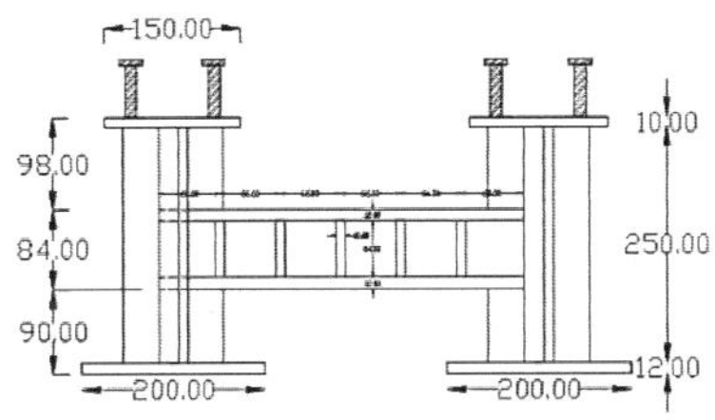

Figure 5. Connections between the webs of SP3.

The transverse connections are set at the end region and the mid-span position, as is shown in figure 6-8. The grade of the concrete used in this research is C50. The type of longitudinal steel and horizontal reinforcement steel used in the girder is R235. The type of steel used in corrugate webs is Q235. Studs are used to connect the steel webs and the concrete. The reinforcement of the model bridge and the fabrication of the corrugated steel web are shown in figure 9-10.

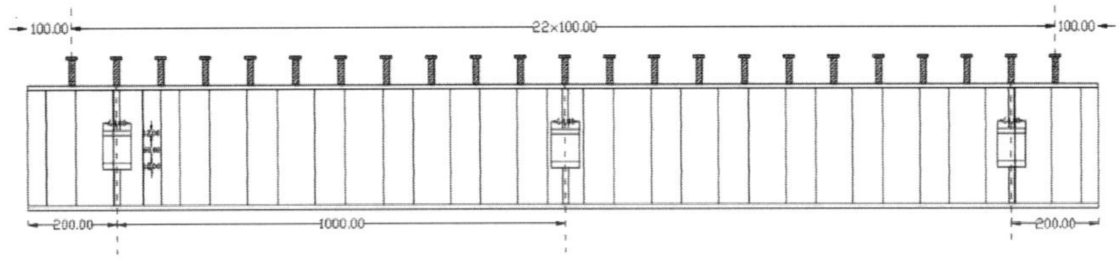

Figure 6. Elevation view of the beam connections of the girder with corrugated steel webs (mm).

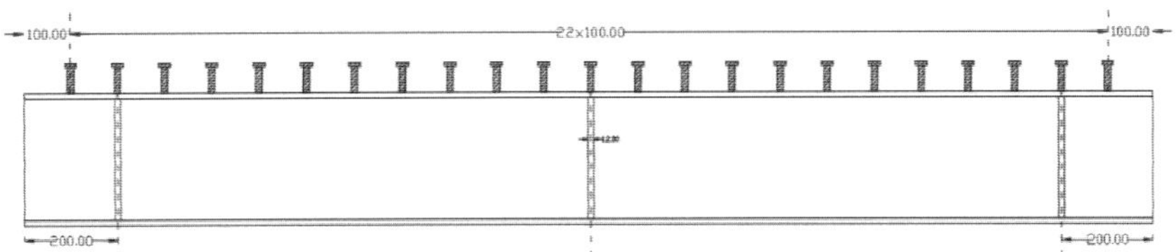

Figure 7. Elevation view of the truss connections of the girder with vertical steel webs (mm).

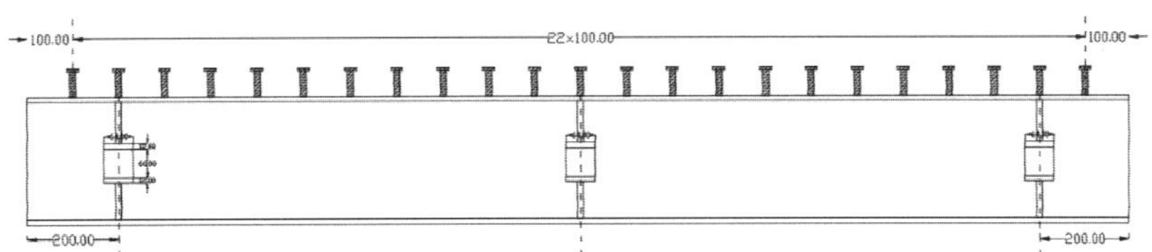

Figure 8. Elevation view of the beam connections of the girder with vertical steel webs (mm).

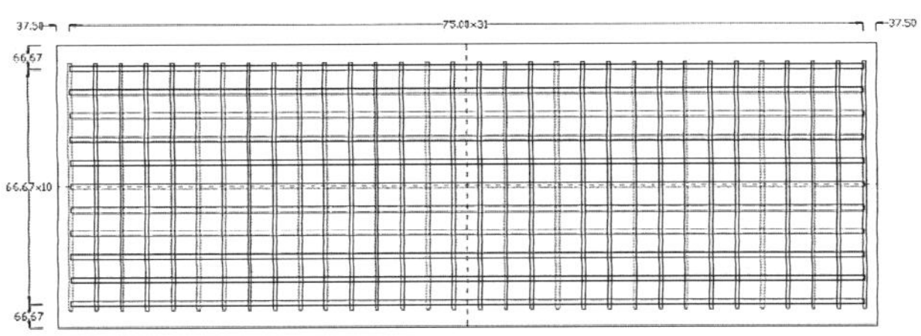

Figure 9. Top view of the reinforcement in the concrete plate.

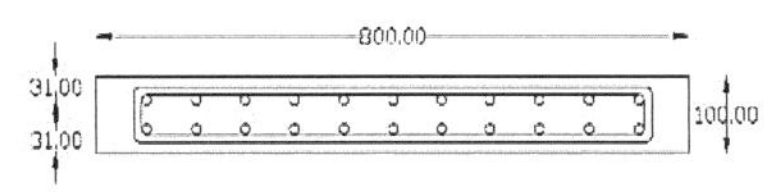

Figure 10. Side view of the reinforcement in the concrete plate.

2. Research Scheme of pure torsion performance of I-type steel-concrete composite bridge with double girders

2.1. Loading device of the research

The specimens are under pure torsion situation in this research. Eccentric loading is applied to the impose pure torque on the specimens. One end of the specimens is a fixed end and the other is a rotating end, which is rotated by the eccentric loading of the rotary hinge and the loading beam so that the specimens can be subjected to a pure torque load. The loaded arm of the torque is 1.35 m as shown in figure 11, and the eccentric load is applied by the hydraulic servo control jack.

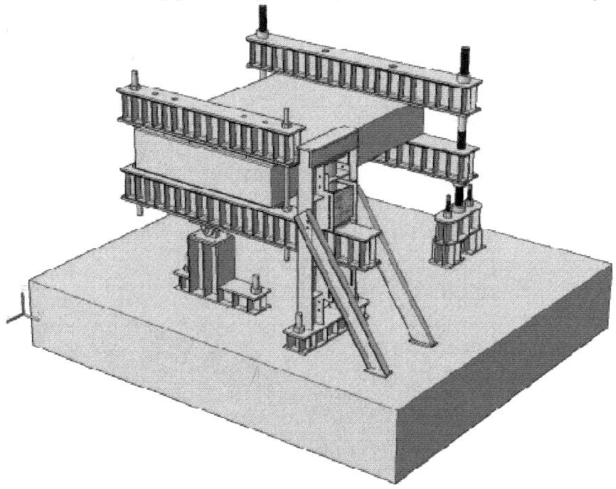

Figure 11. Loading equipment.

2.2. Research contents of the specimens and plant arrangement

The experimenting area of the research is shown in the picture below. The main contents of the experiment are the relative torsion angle of concrete flange and steel girder, the distribution and width of concrete crack in the area that is ± 0.75m away from the middle.

The specific measurement scheme is shown in figure 12-13. The research data is collected automatically using IMP data acquisition system, and the collected data is directly saved into the computer, and is made into a text file. During the whole process of the experiment, the curve of load-inclination and load-midspan deflection of the specimen can be monitored on the monitor screen to observe and adjust the loading speed of the experiment.

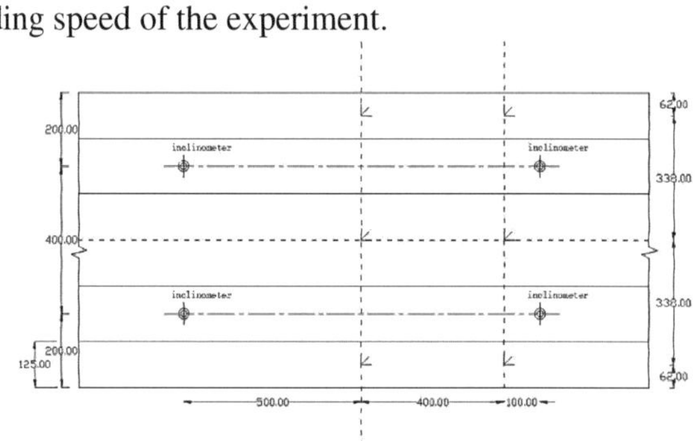

Figure 12. The strain gages and inclinometers in the upper surface of the concrete flange.

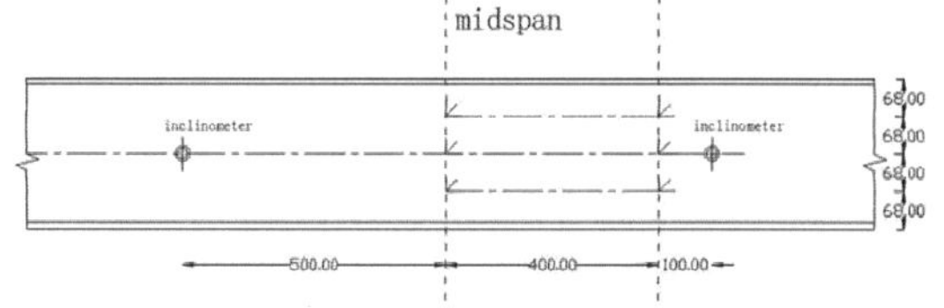

Figure 13. The strain gages and inclinometers in the steel webs.

2.3. The loading process of the experiment

At the beginning of loading, 1.0kN is set as a segment, and graded loading is applied in order to observe and depict cracks and obtain the experimental data. When the applied torque is able to make

IOP Conf. Series: Materials Science and Engineering **269** (2017) 012054 doi:10.1088/1757-899X/269/1/012054

the first crack near the center of the concrete wing near the bearing happen, the cracking torque is recorded and the loading speed is slowed down to observe the degree of crack development. When penetrating cracks appear, the torque value is recorded and the change of the crack width is recorded. The loading continues until the value in the pressure sensor screen no longer increases, at this moment, the composite bridge has reached the torsion limit, and the limit torque is recorded. In the end, the loading is unloaded smoothly.

3. Results and analysis of the experiment

3.1. Failure characteristics of the composite bridges under pure torsion

The Experiment shows that the three specimens share similar damage patterns. The crack pattern of the pure torsion of composite girders is shown in figure 14-16. As can be clearly seen from the figure, almost every spiral crack goes through the entire concrete flange thoroughly, and the cracks are evenly distributed. Due to the restraint of the steel beam, the I-shape steel webs are still in the elastic stage when the concrete flange is damaged. The whole specimen can still bear some torque load. The development of the crack width is restricted and the reduction in bearing capacity is not obvious. By comparison, it is found that with the increase of stiffness of transverse connection, the number of diagonal cracks increases, and the width of cracks gets smaller.

Figure 14. The cracks on the side face of SP1.

Figure 15. The cracks on the end face of SP1.

Figure 16. The cracks on the top face of SP1.

Although there is a considerable torsion angle in the end, the torsional bearing capacity did not significantly decline. Due to the fact that these specimens are double-girder bridges, and are connected by different forms of transverse connections, the torsional bearing capacity of the concrete flange has been greatly improved. Compared with the single-girder bridge with same cross section, the ultimate torque resistance increases significantly. Finally, due to the emergence of one or several large-width cracks in the concrete flange, the whole composite specimen is damaged by torsion.

Comparing the final damage patterns of the double-girder specimens with different shapes of steel webs, the ultimate torsion angle and the ultimate torque of the corrugated-web specimen with small beam connections are larger than that of the vertical-web specimen with small beam connections. The torsional bearing capacity can be improved with the help of corrugated webs. But the corrugated webs cannot effectively restrict the deformation of the composite specimen. Comparing the final failure patterns of the double-girder vertical-web specimens with different transverse connections, it is found that the specimen with small beam connections has a small torsion angle, but a large ultimate torque.

Contrary to this, the specimen with truss connections has a large torsion angle, but a small ultimate torque. However, by comparing the torsion angle and the ultimate torque value of the two specimens, it is found that the difference between the specimens is minimal.

3.2. Main results of the experiment

In table 1, T_{cr} represents the fracture torque of the specimen, which is characterized by the occurrence of the inclined crack in the concrete flange. T_u represents the ultimate torque of the specimen, which is characterized by the failure of applying higher forces. The $\theta_{cr}^{'}$ and $\theta_u^{'}$ respectively represents the ultimate torsional curvature and the cracking torsional curvature.

Table 1. Measured values of feature load and torsional curvature of the specimens under pure torsion.

Specimens	SP1	SP2	SP3
T_{cr} (kN·m)	10.95	10.9	10.80
T_u (kN·m)	82.20	83.59	83.97
T_{cr} / T_u	0.133	0.130	0.129
$\theta_{cr}^{'}$ (10^{-3}rad/m)	2.26	2.09	2.79
$\theta_u^{'}$ (10^{-3}rad/m)	83.0	91.6	46.3
$\theta_{cr}^{'} / \theta_u^{'}$	2.72%	2.28%	6.03%

3.3. Analysis on torsional deformation

3.3.1. Relationship between the torsional deformation of concrete flange and the steel web. As can be seen from the relationship curve of T—θ' shown in figure 17, the torsion of steel webs and concrete flange is well coordinated and basically synchronized in the elastic stage. After the cracking torque is reached, the torsional curvature of steel webs and concrete flange begins to have dissimilarities. When the ultimate torque is reached, the torsional curvature of the concrete is larger than that of the steel webs, but the difference between them is not big. At the same time, it is considered that the loading device has a great effect on the concrete flange. During the loading process, there are no signs of destruction of connectors. And after the concrete is smashed, no signs of significant deformation of the pegs are found, so it can be assumed that in the entire process of torsion loading, the concrete flange and steel webs are well coordinated and basically synchronized.

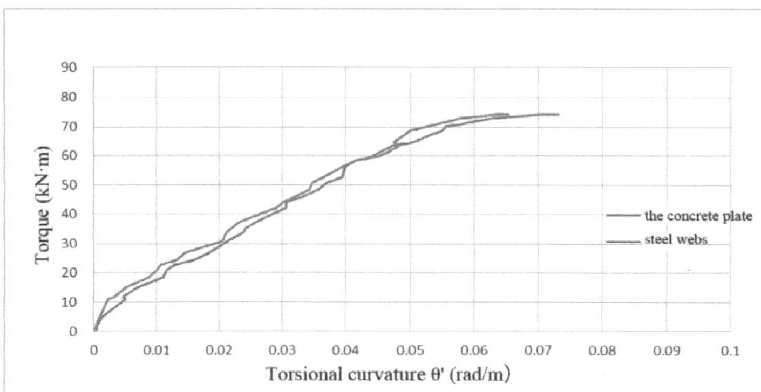

(a) SP1

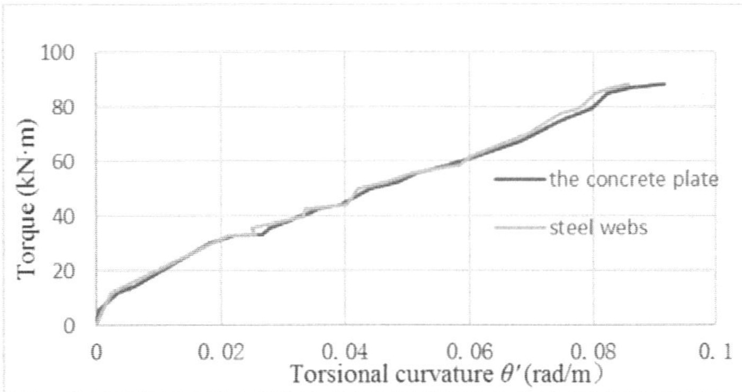

(b) SP2

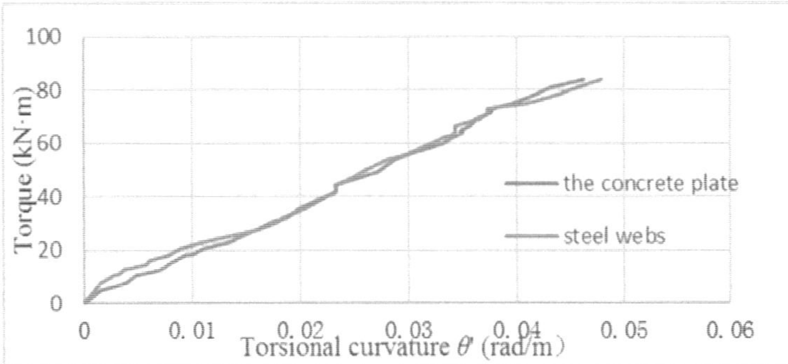

(c) SP3

Figure 17. Relationship between torque and torsional curvature of concrete flange and steel webs.

This result provides a basis for the assumption in the torsional analysis that the torsional curvature of concrete flange and the steel web are equal in values. After the specimens are cracked, the torsional curvature of the concrete flange is slightly larger than that of the steel webs, and the difference rate between them is about 10%, which may be caused by the decrease of the torsional rigidity after the cracking of the concrete happens. So the torsional curvature of concrete flange can be adopted as the torsional curvature of the composite girder for safety purpose.

4. Conclusion

(1) In the entire process of the torsion loading, the concrete flange and steel webs are well coordinated and basically synchronized. This result provides a basis for the assumption in the torsional analysis that the torsional curvature of concrete flange and the steel web are equal in values. And the torsional curvature of concrete flange can be adopted as the torsional curvature of the composite girder for safety purpose.

(2) The effect of the shape of steel webs on the torsional resistance of composite girders is that the torsional bearing capacity can be improved by the usage of corrugated webs, but the influence of the usage of corrugated webs on constraint of deformation is not obvious.

(3) The effect of the forms of transverse connection between the double girders on the torsional resistance of composite girders is that the contribution of the small beams connections to the torsional stiffness of the composite girders is greater than that of the truss connections, and the ultimate torsion angle is smaller with small beam connections, but the improvement on torsional bearing capacity is not obvious.

References

[1] Rahal K N and Collins M P 1995 *ACI Struct. J.* **92(4)** 459-69.
[2] Slutter R Cx and Driseoll G C 1965 *ASCE J. Struct. Div.* **91(ST2)** 71-99.
[3] Singh R K and Mallick S K 1977 *The India Concr. J.* **51(1)** 24-30.
[4] Ghosh B and Mallick S K 1979 *The India Concr. J.* **53(2)** 48-53.

[5] Basu Ray M and Mallick S K 1980 *The India Concr. J.* **54(3)** 80-83.
[6] Dinno K S and Merchant W 1965 *Struct. Eng.* **43(7)** 219-21.
[7] Trahair N S 1999 *J. Struct. Eng.* **125(2)** 175-182.

Arch-Axis Coefficient Optimization of Long-Span Deck-Type Concrete-Filled Steel Tubular Arch Bridge

Q J Liu[1], S Wan[1] and H C Liu[1]

[1]Department of Transportation, Si Pailou campus of Southeast University, Nanjing, Jiangsu, China

Abstract. This paper is based on Nanpuxi super major bridge which is under construction and starts from Wencheng Zhejiang province to Taishun highway. A finite element model of the whole bridge is constructed using Midas Civil finite element software. The most adverse load combination in the specification is taken into consideration to determine the method of calculating the arch-axis coefficient of long-span deck-type concrete-filled steel tubular arch bridge. By doing this, this paper aims at providing references for similar engineering projects.

1. Project overview

Nanpuxi super major bridge is located in Taishun county, Zhejiang province. It spans the cliffs of the two sides of the canyon, and the elevation of the mountaintop is 271m, the elevation of the ditch is 115m, and the relative elevation is 156m.The main bridge is a deck-type concrete-filled steel tubular arch bridge with a net span of 258m, and the net rise-span ratio is 1/4.6. The axis of the arch is catenary. Based on previous engineering experience, the paper uses six arch axis coefficients-1.5, 1.6, 1.7, 1.756, 1.8, 1.9 to carry out the research.

The cross section type of the bridge is integral, and the width of the bridge is 25m. The whole bridge consists of two pieces of arch rib, rib spacing is 17m, and each arch rib consists of four steel pipes with 1200 mm diameter, and is 3m wide, 5.5m high. The steel pipe has a thickness of 22mm, and is filled with C50 expansion concrete. The column above the arch adopts stiffener steel box section. And the bent cap above the column adopts steel box structure. The bridge deck is composed of continuous 20m steel-concrete composite beams. The composite beams are composed of 85cm high I-shape beam and 23cm thick precast and cast-in-place concrete plates, and the nominal beam height of the composite beams is 1.2m. The bridge deck consists of 10cm-thick asphalt concrete and 10cm-thick cast-in place concrete. The layout of Nanpuxi super major bridge is shown in figure 1. The design load is: vehicle highway-class I.

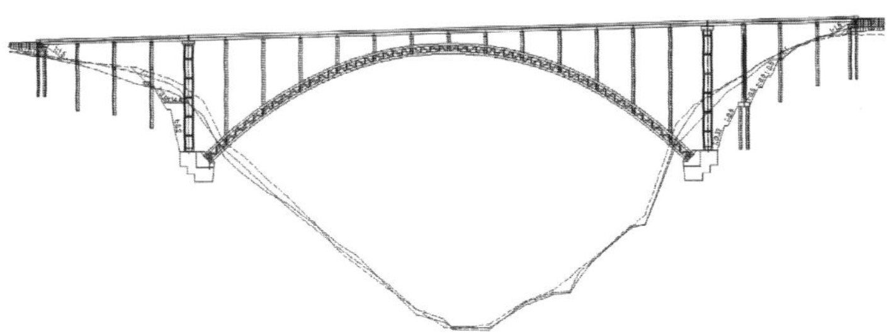

Figure 1. The layout of Nanpuxi super major bridge.

IOP Conf. Series: Materials Science and Engineering **269** (2017) 012055 doi:10.1088/1757-899X/269/1/012055

2. Overall calculation

The bridge integral calculation is carried out by using finite element software Midas civil 8.3.2 to construct a whole bridge finite element model. Space truss elements are used in the model to get stress calculation results. The model is constructed according to the actual construction process of construction stage. The first period dead load and second period dead load and basic variable load such as vehicle load, and temperature load, deformation and additional load are combined in different ways to get the stress results in arch bridge under the most unfavorable load combination. The finite element model of arch bridge is shown in figure 2.

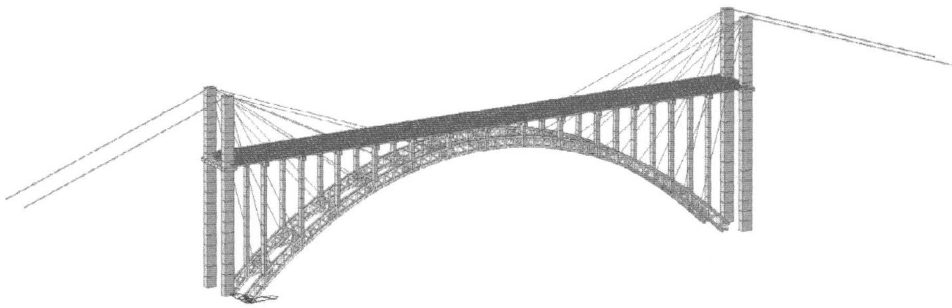

Figure 2. The finite element model of the concrete-filled steel tubular arch bridge.

2.1. Load condition and load combination

2.1.1. Main load conditions. (1) Dead load: the structural self-weight and phase ii dead load including the guardrail and the paving of the bridge deck.

(2) Live load: highway I class load, its fabric can be divided into two different modes according to the specification: symmetrical loading and partial loading.

(3) Bearing settlement:

1) Each bearing moves 1cm backward

2) One bearing sinks 1cm downward

(4) Temperature effect

1) Uniform heating and cooling, the overall rise and drop of the temperature is 25°C.

2) Gradient heating and cooling

The gradient heating and cooling of bridge panel is implemented according to article 4.3.12 in the highway bridge design general specification (JTG d60-2015).

The gradient heating of arch ribs is implemented according to article 4.2.5 in the highway concrete-filled steel tubular arch bridge deign specification (JTG/T D65-06-2015) and the rise of the temperature is 5°C.

(5) Wind load

The calculation results are mainly used to determine the arch axis coefficient of the arch rib, and vertical load is mainly considered in this case, so the wind load is not considered.

(6) Shrinkage creep

The shrinkage creep is determined according to article 4.2.4 of the highway bridge design general specification (JTG d60-2015) and the relevant provisions in appendix F in Code for Design of Highway Reinforced Concrete and Prestressed Concrete Bridges and Culverts (JTG D62-2004).

The shrinkage and creep of the concrete used in the composite beam as well as the bridge panel and the steel tube of the arch rings are considered in the model. The relative humidity is 90%.

2.1.2. Load combination. The load combination is considered according to article 4.1.5 and 4.1.6 of the general specification of highway bridge design (JTG d60-2015).

2.2. Construction stage division

The construction stage is divided as the table 1 shows.

Table 1. Construction stage division.

Construction stage	Construction content
1	constructing arch abutment, constructing transition pier, installing cable tower
2	hoisting the first segment of the arch ribs, installing buckle cable, back stay and k-shape support
3	hoisting the second segment of the arch ribs, installing buckle cable, back stay and k-shape support
4	hoisting the third segment of the arch ribs, installing buckle cable, back stay and k-shape support
5	hoisting the fourth segment of the arch ribs, installing buckle cable, back stay and k-shape support
6	hoisting the fifth segment of the arch ribs, installing buckle cable, back stay and k-shape support
7	hoisting the sixth segment of the arch ribs, installing buckle cable, back stay and k-shape support
8	hoisting the seventh segment of the arch ribs(closure segment), installing H-shape support
9	demolishing buckle cable and placing concrete in skewback to seal the hinges after the closure
10	placing concrete in the first steel tube of the bottom chord
11	placing concrete in the first steel tube of the top chord after concrete in the former phase reaches requirements
12	placing concrete in the second steel tube of the bottom chord after concrete in the former phase reaches requirements
13	placing concrete in the second steel tube of the top chord after concrete in the former phase reaches requirements
14	the concrete in the second steel tube of the top chord reaches enough strength and starts to get into function
15	placing concrete in batten plate
16	installing columns in 3# and 4# and 10# and 11# after the batten plate concrete reaches the required strength
17	installing columns in 5#~9#
18	installing columns in 1# and 2# and 12# and 13#
19	installing copings on columns in 3# and 4# and 10# and 11#
20	installing copings on columns in 5#~9#
21	installing copings on columns in 1# and 2# and 12# and 13#
22	installing bridge deck on columns in 2#~4# and 10#~12#
23	installing bridge deck on columns in 4#~10#
24	installing bridge deck on columns in 0#(transition pier)~2# and 12#~14#(transition pier)
25	secondary dead load (guardrail load and bridge deck pavement load)
26	shrinkage creep in 3 years after the completion and support displacement

3. Optimization of arch-axis coefficient(m)

The arch-axis line adopts the catenary line, combined with the previous engineering experience, this paper intends to take six values of arch-axis coefficient to make comparisons: m = 1.5, m = 1.6, m = 1.7, m = 1.756, m = 1.8, m = 1.9, the static analysis is carried out with the help of Midas civil 8.3.2 program. The whole ridge finite element model is constructed for calculation comparison.

3.1. Comparison of dead load stress under long-term effect after the completion

From the comparison results of the different arch-axis coefficient, the conclusion that difference between axial stress under dead load with different coefficient is minimal. Under dead load situation, there are great changes in the stress of the arch ribs. With the increase of the m coefficient, the absolute value of the negative moment in skewback has gradually reduced, the positive moment in skewback has gradually increased, and the compressive stress of the bottom chord bar has reduced.

Table 2. Comparison of stress in arch rib steel tubes with different arch-axis coefficient under dead load (N/mm^2)[a].

Arch-axis coefficient	Stress of top chord in skewback (N/mm^2)	Stress of bottom chord in skewback (N/mm^2)	Stress of bottom chord in 1/4 midspan (N/mm^2)	Stress of top chord in vault (N/mm^2)	Stress of bottom chord in vault (N/mm^2)
m=1.9	-1.33e+002	-9.98e+001	-9.45e+001	-1.37e+002	-8.18e+001
m=1.8	-1.23e+002	-9.88e+001	-9.28e+001	-1.34e+002	-8.32e+001
m=1.756	-1.19e+002	-1.03e+002	-9.20e+001	-1.32e+002	-8.39e+001
m=1.7	-1.13e+002	-1.08e+002	-9.10e+001	-1.30e+002	-8.46e+001
m=1.6	-1.03e+002	-1.18e+002	-8.90e+001	-1.25e+002	-8.61e+001
m=1.5	-9.22e+001	-1.29e+002	-8.66e+001	-1.23e+002	-8.77e+001

[a] The negative values in the table above represent compressive stress, and positive values represent tensile stress.

Table 3. Comparison of stress in arch rib concrete with different arch-axis coefficient under dead load (N/mm^2)[a].

Arch-axis coefficient	Stress of top chord in skewback (N/mm^2)	Stress of bottom chord in skewback (N/mm^2)	Stress of bottom chord in 1/4 midspan (N/mm^2)	Stress of top chord in vault (N/mm^2)	Stress of bottom chord in vault (N/mm^2)
m=1.9	-1.15e+001	-1.14e+001	-9.80e+000	-1.14e+001	-6.61e+000
m=1.8	-1.05e+001	-1.10e+001	-9.62e+000	-1.12e+001	-6.72e+000
m=1.756	-1.02e+001	-1.09e+001	-9.54e+000	-1.11e+001	-6.77e+000
m=1.7	-9.60e+000	-1.12e+001	-9.44e+000	-1.10e+001	-6.83e+000
m=1.6	-8.65e+000	-1.23e+001	-9.23e+000	-1.05e+001	-6.95e+000
m=1.5	-7.69e+000	-1.36e+001	-9.03e+000	-1.06e+001	-7.08e+000

[a] The negative values in the table above represent compressive stress, and positive values represent tensile stress.

3.2. The comparison of stress under the most unfavorable fundamental combination effects of persistence condition

The table below shows the comparison of stress under the most unfavourable fundamental combination effects of persistence condition with different arch-axis coefficient.

Table 4. Comparison of stress in arch rib steel tubes with different arch-axis coefficient under the most unfavourable fundamental combination effects of persistence condition (N/mm^2)[a].

Arch-axis coefficient	Stress of top chord in skewback (N/mm^2)	Stress of bottom chord in skewback (N/mm^2)	Stress of bottom chord in 1/4 mid span (N/mm^2)	Stress of top chord in vault (N/mm^2)	Stress of bottom chord in vault (N/mm^2)
m=1.9	-2.71e+002	-2.00e+002	-1.60e+002	-2.48e+002	-1.39e+002
m=1.8	-2.48e+002	-1.99e+002	-1.57e+002	-2.38e+002	-1.41e+002
m=1.756	-2.43e+002	-2.05e+002	-1.56e+002	-2.36e+002	-1.42e+002
m=1.7	-2.43e+002	-2.13e+002	-1.54e+002	-2.38e+002	-1.44e+002
m=1.6	-2.28e+002	-2.27e+002	-1.51e+002	-2.33e+002	-1.46e+002
m=1.5	-2.04e+002	-2.43e+002	-1.48e+002	-2.23e+002	-1.48e+002

[a] The negative values in the table above represent compressive stress, and positive values represent tensile stress.

Table 5. Comparison of stress in arch rib concrete with different arch-axis coefficient under the most unfavourable fundamental combination effects of persistence condition (N/mm^2)[a].

Arch-axis coefficient	Stress of top chord in skewback (N/mm^2)	Stress of bottom chord in skewback (N/mm^2)	Stress of bottom chord in 1/4 mid span (N/mm^2)	Stress of top chord in vault (N/mm^2)	Stress of bottom chord in vault (N/mm^2)
m=1.9	-2.24e+001	-1.78e+001	-1.12e+001	-1.76e+001	-8.27e+000
m=1.8	-2.00e+001	-1.75e+001	-1.10e+001	-1.65e+001	-8.39e+000
m=1.756	-1.97e+001	-1.74e+001	-1.09e+001	-1.64e+001	-8.45e+000
m=1.7	-2.06e+001	-1.73e+001	-1.08e+001	-1.71e+001	-8.51e+000
m=1.6	-1.97e+001	-1.80e+001	-1.06e+001	-1.69e+001	-8.64e+000
m=1.5	-1.73e+001	-1.91e+001	-1.04e+001	-1.59e+001	-8.79e+000

[a] The negative values in the table above represent compressive stress, and positive values represent tensile stress.

3.3. Comparison of stress under the most unfavorable standard combination effects of persistence condition

The table below shows the comparison of stress under the most unfavourable standard combination effects of persistence condition with different arch-axis coefficient.

Table 6. Comparison of stress in arch rib steel tubes with different arch-axis coefficient under the most unfavourable standard combination effects of persistence condition (N/mm^2)[a].

Arch-axis coefficient	Stress of top chord in skewback (N/mm^2)	Stress of bottom chord in skewback (N/mm^2)	Stress of bottom chord in 1/4 midspan (N/mm^2)	Stress of top chord in vault (N/mm^2)	Stress of bottom chord in vault (N/mm^2)
m=1.9	-2.28e+002	-1.72e+002	-1.36e+002	-2.11e+002	-1.19e+002
m=1.8	-2.08e+002	-1.70e+002	-1.33e+002	-2.03e+002	-1.21e+002
m=1.756	-2.04e+002	-1.74e+002	-1.32e+002	-2.01e+002	-1.22e+002
m=1.7	-2.04e+002	-1.81e+002	-1.31e+002	-2.03e+002	-1.23e+002
m=1.6	-1.92e+002	-1.94e+002	-1.29e+002	-1.99e+002	-1.25e+002
m=1.5	-1.71e+002	-2.07e+002	-1.26e+002	-1.90e+002	-1.27e+002

[a] The negative values in the table above represent compressive stress, and positive values represent tensile stress.

Table 7. Comparison of stress in arch rib concrete with different arch-axis coefficient under the most unfavourable standard combination effects of persistence condition (N/mm^2)[a].

Arch-axis coefficient	Stress of top chord in skewback (N/mm^2)	Stress of bottom chord in skewback (N/mm^2)	Stress of bottom chord in 1/4 midspan (N/mm^2)	Stress of top chord in vault (N/mm^2)	Stress of bottom chord in vault (N/mm^2)
m=1.9	-1.74e+001	-1.42e+001	-8.36e+000	-1.38e+001	-6.32e+000
m=1.8	-1.54e+001	-1.40e+001	-8.23e+000	-1.29e+001	-6.41e+000
m=1.756	-1.52e+001	-1.40e+001	-8.17e+000	-1.28e+001	-6.46e+000
m=1.7	-1.60e+001	-1.38e+001	-8.10e+000	-1.34e+001	-6.51e+000
m=1.6	-1.54e+001	-1.39e+001	-7.95e+000	-1.33e+001	-6.62e+000
m=1.5	-1.34e+001	-1.47e+001	-7.79e+000	-1.24e+001	-6.73e+000

[a] The negative values in the table above represent compressive stress, and positive values represent tensile stress.

From the results of the stress comparison results above, the conclusion that the absolute value of the cross-section stress in skewback and vault is the smallest when the arch-axis coefficient is 1.5 ~

1.7 under the effect of the most unfavorable fundamental combination effect and the most unfavorable standard combination effect. So the optimal arch-axis coefficient is 1.6.

4. Conclusion

Generally, dead weight of the arch and dead load on the arch are taken into consideration to determine the optimal arch-axis coefficient, and the variable load such as shrinkage and creep of the arch rib concrete and the vehicle load hasn't been taken into account. However, for long-span deck-type concrete-filled steel tubular arch bridge, variable load accounts for a large proportion of the total load, and the shrinkage and creep of the arch rib concrete is unneglectable and exist permanently, so these loads should be taken into account in arch-axis coefficient optimization. By comparing the stress in skewback and vault, the optimal arch-axis coefficient that makes the least differences in these two parts can be obtained.

5. References

[1] Hajjar J F, Molodan A and Schillerr P H 1998 *Eng. Struct.* **20** 398-412.
[2] Nakanishi K, Kitada T and Nakai H 1999 *J. Constr. Steel. Res.* **51** 297-319.
[3] Kitada T 1998 *Eng. Struct.* **20** 347-54.
[4] Elchlakani M 2001 *J. Constr. Steel. Res.* **57** 1141-68.
[5] Pi Y L and Trahair N S 2000 *Eng. Struct.* **20** 993-1005.

IOP Conf. Series: Materials Science and Engineering **269** (2017) 012056 doi:10.1088/1757-899X/269/1/012056

The Application of Radial Basis Function (RBF) Neural Network for Mechanical Fault Diagnosis of Gearbox

Pengbo Wang[1]

[1]Institute of Solid Mechanics, Beihang University (Beijing University of Aeronautics and Astronautics), Beijing, 100191, P.R. China

Abstract. In this paper, the radial basis function (RBF) neural network is used for the mechanical fault diagnosis of a gearbox. We introduce the basic principles of the RBF neural network which is used for pattern classification and features a fast learning pace and strong nonlinear mapping capability; thus, it can be employed for fault diagnosis. The gearbox is a widely-used piece of equipment in engineering, and diagnosing mechanical faults is of great significance for engineers. A numerical example is presented to demonstrate the capability of the proposed method. The results indicate that the mechanical faults of a gearbox can be correctly diagnosed with a trained RBF neural network.

1. Introduction

Gearboxes are important pieces of equipment in industry. Gears are often damaged by collision or erosion, which leads to the gearbox malfunctioning. Mechanical faults in gearboxes may result in serious accidents; hence, the mechanical fault diagnosis of gearboxes is of great significance in engineering. Traditional mechanical fault diagnosis is mainly based on the examination of vibration features related to the faults. However, the workload involved with this approach is usually very large. Sometimes the vibration features are contaminated by noise, and the approach becomes inapplicable. Therefore, scholars have developed a more intelligent approach, the neural network, to diagnose mechanical faults.

Collected vibration features can be integrated into a neural network that can represent the complicated nonlinear relationship mapping from the vibration features to the mechanical faults [1]. Moreover, the neural network is robust to noise 0. The most widely-used neural network is the error back propagation (BP) neural network. However, an obvious disadvantage of BP neural networks is the slow learning speed 0. Thus, researchers developed the radial basis function (RBF) neural network, which possesses a high learning speed 0. In this study, we introduce the RBF neural network to the problem of mechanical fault diagnosis.

The remainder of this paper is organised as follows. In Section 2, the basics of the RBF neural network are briefly introduced. In Section 3, the procedure of using RBF neural network to diagnose the mechanical faults of a gearbox is presented. In Section 4, the different types of mechanical faults in the gearbox are discussed. In Section 5, a numerical example is provided to illustrate the feasibility of the proposed method, and Section 6 presents the conclusions of the study.

2. Basic principles of the radial basis function (RBF) neural network

A typical RBF neural network has three layers: the input layer, the hidden layer, and the output layer, as shown in figure 1. The relationship between the input and output can be described by the node number of the hidden layer, J, and the network weights, w_{jk}, between the hidden layer and the output layer. The mapping between the input and output is thereby established through the neural network.

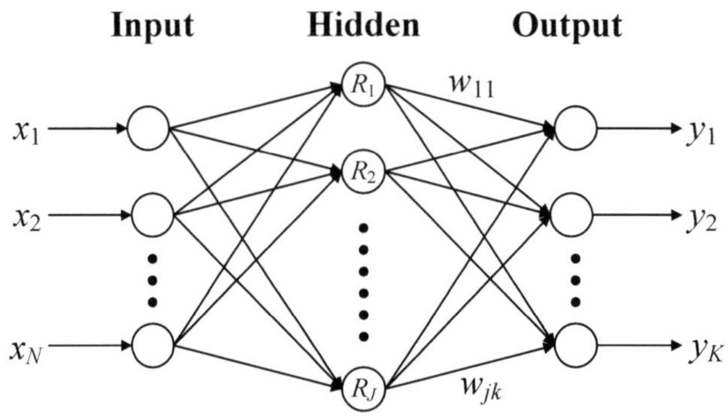

Figure 1. Topology structure of the RBF neural network.

The commonly-used radial basis functions are as follows:

$$f(\boldsymbol{x}) = \exp[-(\frac{\boldsymbol{x}}{\sigma})^2], \tag{1}$$

$$f(\boldsymbol{x}) = \frac{1}{(\boldsymbol{x}^2 + \sigma^2)^\alpha}, \quad (\alpha > 0), \tag{2}$$

$$f(\boldsymbol{x}) = (\boldsymbol{x}^2 + \alpha^2)^\beta, \quad (\alpha < \beta < 1). \tag{3}$$

All these functions are radially symmetric, and the most commonly-used function is Gaussian:

$$R_j(\boldsymbol{x}) = \exp(-\frac{\| \boldsymbol{x} - \boldsymbol{c}_j \|^2}{2\sigma_j^2}), \quad j = 1, 2, \cdots\cdots, J, \tag{4}$$

where $\boldsymbol{x}$ is the n-dimensional input vector; $\boldsymbol{c}_j$ is the center of the j-th RBF; σ_j is the j-th RBF width, which can be adjusted; J is the number of nodes in the hidden layer. The symbol $\| \boldsymbol{x} - \boldsymbol{c}_j \|$ represents the norm of the vectors $\boldsymbol{x}$ and $\boldsymbol{c}_j$, which can also be regarded as the distance between the vectors $\boldsymbol{x}$ and $\boldsymbol{c}_j$.

As shown in Figure 1, the input layer represents the nonlinear mapping from $\boldsymbol{x}$ to $R_j(\boldsymbol{x})$, and the output layer represents the linear mapping from $R_j(\boldsymbol{x})$ to y_k. The k-th output is a weighted summation of all RBFs, i.e.

$$y_k = \sum_{j=1}^{J} w_{jk} \cdot R_j(\boldsymbol{x}), \quad k = 1, 2, \cdots\cdots, K, \tag{5}$$

where K is the number of outputs; w_{jk} is the weight between the j-th node of hidden layer and the k-th node of the output layer; J is the number of nodes in the hidden layer.

3. RBF neural network for mechanical fault diagnosis

The procedure of using an RBF neural network to diagnose the mechanical faults of a gearbox is illustrated in figure 2. The dynamic data is recorded from the gearbox, and then processed by signal-processing technologies, such as the wavelet transform or filtering. Thus, vibration features can be obtained. With these features, the RBF neural network can be established and trained, and after the weights of the network have converged the network can be used. With given test features, the established network can diagnose the mechanical faults of the gearbox.

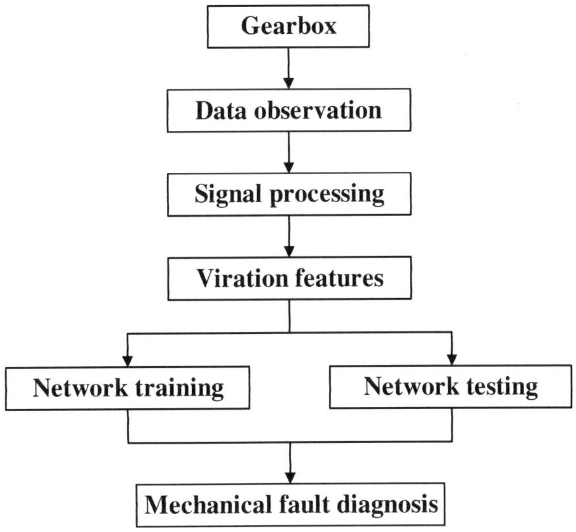

Figure 2. Flowchart of the RBF neural network for mechanical fault diagnosis.

4. Mechanical faults of gearboxes

In this study, it is assumed that gearbox faults are all caused by the gears. We consider three states of the gear: 'normal', 'root crack', and 'tooth broken'. The three states are illustrated in figures 3, 4, and 5, respectively.

For simplicity, we use three row vectors to represent the three states. The row vector [1, 0, 0] represents the 'normal state', the row vector [0, 1, 0] represents the 'root crack' state, and the row vector [0, 0, 1] represents the 'tooth broken' state. The relationships between the states and row vectors are listed in Table 1.

Figure 3. The 'normal' state of the gear.

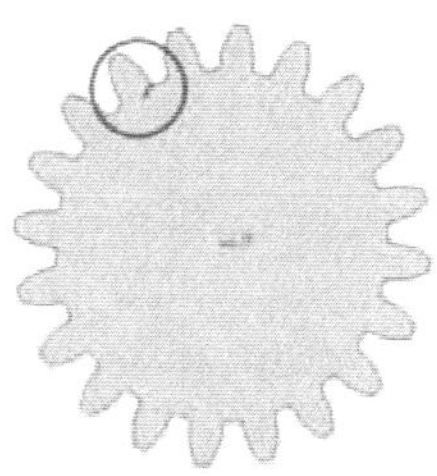

Figure 4. The 'root crack' state of the gear.

Figure 5. The 'tooth broken' state of the gear.

Table 1. Relationships between the gear states and row vectors.

No.	Gear state	Row vector
1	Normal	[1, 0, 0]
2	Root crack	[0, 1, 0]
3	Tooth broken	[0, 0, 1]

5. Numerical example

In this example, we use data from an automobile gearbox to demonstrate our proposed method. We have twelve groups of data in total, and within each group, eight parameters numbered from P_1 to P_8 represent the ten vibration features of the gearbox. Our aim is to use the RBF neural network to realize the mapping between the ten vibration features and the three gear states. In other words, we use the ten inputs to determine the three outputs.

We use nine groups of data as the sample data and three groups of data as the test data. The sample data listed in Table 2 is used to train the RBF neural network. The test data listed in Table 3 is then used to verify the correctness of the RBF neural network.

Table 2. Sample data for RBF neural network training.

No.	P_1	P_2	P_3	P_4	P_5	P_6	P_7	P_8	Gear state	Row vector
S1	0.6858	0.3876	0.2160	0.4776	0.4005	0.2199	0.3477	0.2821	Normal	[1, 0, 0]
S2	0.6271	0.2841	0.4179	0.4161	0.7674	0.2702	0.2313	0.2646	Normal	[1, 0, 0]
S3	0.1326	0.2640	0.3441	0.1689	0.9041	0.3449	0.4359	0.1287	Normal	[1, 0, 0]
S4	0.7809	0.5145	0.2105	0.8133	0.4473	0.3989	0.2904	0.5733	Root crack	[0, 1, 0]
S5	0.9670	0.6667	0.1686	0.9314	0.5616	0.4842	0.4275	0.4518	Root crack	[0, 1, 0]
S6	0.1077	0.3447	0.3692	0.9981	0.5931	0.3744	0.1872	0.2496	Root crack	[0, 1, 0]
S7	0.5277	0.7041	0.5487	0.5432	0.8766	0.1965	0.2322	0.6819	Tooth broken	[0, 0, 1]
S8	0.2172	0.5727	0.4019	0.7227	0.8526	0.1350	0.2472	0.3192	Tooth broken	[0, 0, 1]
S9	0.7902	0.6774	0.3495	0.3462	0.3223	0.1971	0.1831	0.7869	Tooth broken	[0, 0, 1]

Table 3. Test data for RBF neural network check.

No.	P_1	P_2	P_3	P_4	P_5	P_6	P_7	P_8	Gear state	Row vector
T1	0.6303	0.2851	0.3894	0.4077	0.7803	0.3001	0.2259	0.2670	Normal	[1, 0, 0]
T2	0.7779	0.5401	0.2133	0.8403	0.4503	0.3894	0.3004	0.5673	Root crack	[0, 1, 0]
T3	0.7798	0.6705	0.3602	0.3513	0.3306	0.2049	0.1862	0.7791	Tooth broken	[0, 0, 1]

The RBF neural network is built with the sample data, with 25 neurons in the hidden layer. The performance of the network is plotted in Figure 6. The results produced by the network are listed in Table 4, which indicates that faults can be effectively diagnosed with the RBF neural network.

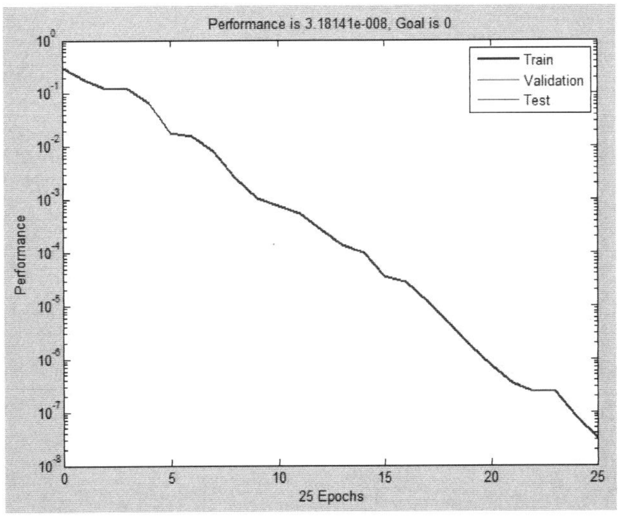

Figure 6. Performance of the RBF neural network.

Table 4. Correctness check of the RBF neural network.

No.	Gear state	Row vector	Result of the RBF neural network	Correct or not
T1	Normal	[1, 0, 0]	[0.9814, 0.0031, 0.0007]	Correct
T2	Root crack	[0, 1, 0]	[0.0008, 0.9541, 0.0013]	Correct
T3	Tooth broken	[0, 0, 1]	[0.0010, 0.0021, 0.9669]	Correct

6. Conclusion

In this paper, the RBF neural network and its basic principles were introduced for the mechanical fault diagnosis of gearboxes. The procedure of using the RBF neural network to diagnose the mechanical faults of a gearbox was put forward. A numerical example was provided to illustrate the proposed method, using twelve groups of data in total: nine groups were used as the sample data and three groups were used as the test data. We used three row vectors to represent the three states of the gearbox, and the RBF neural network was built with the sample data. After the network was trained, it was capable of predicting the state of the test data. The row vectors obtained using the RBF neural network on the test data were compared with the real states of the test data. The results indicate that the states predicted by the RBF neural network are consistent with the real states. This suggests that the mechanical faults of a gearbox can be correctly diagnosed with the trained RBF neural network.

7. References

[1] Zhang Y, Ding X and Liu Y, 2002 *IEEE. T. Power. Deliver.* **11** 1836-41.
[2] Chen Z, Li C and Sánchez R V, 2015 *J. Vibroeng.* **17** 2379-92.
[3] Liu X M, Yao X H and Zhao Q, 2012 *Adv. Mater. Res.* **433** 7563-8.
[4] Guo Z H, Wu J and Lu H Y, 2011 *Knowl-based. Syst.* **24** 1048-56.
[5] Chen J and Liao C M, 2002 *J. Process. Contr.* **12** 277-89.

IOP Conf. Series: Materials Science and Engineering **269** (2017) 012057 doi:10.1088/1757-899X/269/1/012057

Synthesis and Characterization of N-(2-azidoethyl) Alkylnitramines

Fu-lei Gao[1], Ying-lei Wang[1,] *, Yue-ping Ji[1], Wei-xiao Liu[1], Bin Chen[1], Wei Wang[1]

[1]Xi'an Modern Chemistry Research Institute, Xi'an, Shanxi 710065, China

Abstract New azido nitromine energetic plasticizers, methyl-(2-azidoethyl)nitramine (MeAEAN), ethyl-(2-azidoethyl)nitramine (EtAEAN), Buthyl-(2-azidoethyl)nitramine (BuAEAN), have been synthesized starting from the readily available *N*-alkyl-2-ethanolamine. The compounds were structurally characterized by spectroscopic (1HNMR,IR) and elemental analysis. Some physicochemical properties of N-(2-azidoethyl) alkylnitramines were obtained by the tests. Results showed that the denisity of MeAEAN, EtAEAN and BuAEAN are 1.343, 1.320, 1.211 g/cm^3 respectively, and have good heat-resistance ability and insensitive to impact and friction.

1 Introduction

In the last few decades, the azido nitromine [1, 2] compounds have received a great deal of attention as these azido nitromine plasticizers impart additional energy to propellants and explosives [3, 4]. It has been showed that the azido nitromine commonly applied, such as 1,5-diazido-3-nitraza-pentane (DIANP) [5-7], can be formulated into gun and rocket motor propellants as an energetic plasticizers. When DIANP is combined with nitrocellulose, the combination has a higher buring rate and low flame temperature, and it also can lead to high energy outputs when used with solid and liquid energetic compounds as well as providing low gas molecular weights and flame temperatures. In order to search for novel the azido nitromine compounds, methyl-(2-azidoethyl)nitramine (MeAEAN), ethyl-(2-azidoethyl)nitramine (EtAEAN), buthyl-(2-azidoethyl)nitramine (BuAEAN), were designed and synthesized for the first time, using *N*-alkyl-2-ethanolamine as raw materials (Scheme 1), confirmed their structures and measured their physicochemical properties.

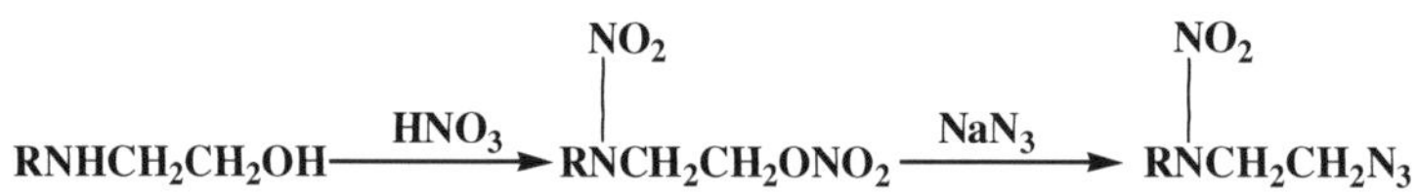

R : CH_3, CH_3CH_2, $CH_3CH_2CH_2CH_2$

Scheme 1. The synthesis of alkyl-AEANs.

2. Experimental

2.1. Synthesis of MeAEAN

2.1.1. Methyl-(2-nitroxyethyl)nitramine(MeNEAN)[8-11]. In a three-necked flask, provided stirrer, thermometer, and dropping funel there was placed 26 g (0.25 mol) of acetic anhydride. While maintaining the temperature of the 10℃ by means of external cooling, 12.6 g (0.2 mol) of nitric acid

(98 percent) was added with vigorous string. After addition of the acid, 6 g (0.044 mol) of anhydride zinc chloride was added. To the nitrating mixture there was added 6 g (0.1 mol) of N-methyl-2-ethanolamine through a dropping funel which required 30 minutes, and the temperature was maintained at 15℃ by means of intermittent cooling with an ice and water bath. The reaction mixture was held at 35-40℃ for 50 minutes the reactants was completed and then poured into 200 ml ice and water. Vigorous string was used during the quenching operation and in a short time MeNEAN precipitated as a white solid. The substance was filtered, washed thoroughly with 50 ml water, air dried for 24 hours. There was obtained 11.1g of crude of MeNEAN melting at 38-40℃. ^{1}HNMR($CDCl_3$/TMS) δ: 3.47(s, 3H, CH_3), 4.11(t, 2H, CH_2), 4.75(t, 2H, CH_2); IR(KBr, cm^{-1}) ν: 1644 (ν_{as}-O-NO_2), 1257 (ν_s-O-NO_2), 1511 (ν_{as}-N-NO_2), 1283 (ν_s-N-NO_2), 850 (ν_s-N-O); Anal. Calcd for $C_3H_7N_3O_5$: C 27.82, N 25.45, H 4.24; Found (%) C 27.97, N 25.01 H, 4.26.

2.1.2. Methyl(2-azidoethyl)nitramine (MeAEAN) [5, 12]. MeNEAN (16.5g, 0.1 mol) was added to 50 mL dimethyl sulfoxide at 45℃, and the solution was heated to 65℃ with good stirring. After MeNEAN was completely dissolved in dimethyl sulfoxide, NaN_3 (16.5g, 0.1 mol) was added in 30 min, then the mixture was held at 80-85℃ for 7 h. After cooling an equal quantity of methylene chloride solvent was added and the entire organic portion was washed four times with water to remove DMSO and inorganic salts. The methylene chloride solution was concentrated to give 12.6 g of light yellow oil with a purity of 98.90% (HPLC). The product was MeAEAN. ^{1}HNMR($CDCl_3$/TMS)δ: 2.00 (t, 2H, -CH_2N_3), 2.01 (t, 2H, -CH_2NNO_2) ; IR (KBr, cm^{-1}) ν: 2948, 1450, 1420 (-CH_2); 2108(-N_3); 1521, 1275 (-N-NO_2).Anal. Calcd for $C_3H_7N_5O_2$: C 24.83, H 4.82, N 48.27; Found (%) C 24.74, H 4.91, N 48.52.

2.2 Synthesis of EtAEAN

2.2.1. Ethyl -(2-nitroxyethyl)nitramine(EtNEAN). EtNEAN was prepared according to the method of MeNEAN. 12.53 g light yellow oil was obtained, Its purity was 98.07%. ^{1}HNMR($CDCl_3$/TMS) δ: 1.3(t, 3H, CH_3), 3.84(m, 2H, CH_2), 4.14(t, 2H ,CH_2), 4.72(t, 2H, CH_2). Calcd for $C_4N_3O_5H_9$: C 26.82, N 23.46, H 5.03; Found (%)C 26.59, N24.14, H 5.21. IR(KBr, cm^{-1}) ν:1644(ν_{as}-O-NO_2); 1270(ν_s-O-NO_2); 1514(ν_{as}-N-NO_2); 1290(ν_s-N-NO_2); 850(ν_s-N-O).

2.2.2. Ethyl -(2-azidoethyl)nitramine(EtAEAN). EtAEAN was prepared according to the method of MeAEAN. 12.78 g yellow oil was obtained, Its purity was 96%. ^{1}HNMR($CDCl_3$/TMS) δ: 3.95 (t, 2H, -$CH_2CH_2N_3$), 3.80 (q, 2H, -CH_2CH_3), 3.63(t, 2H, -CH_2N_3), 1.18 (t, 3H, -CH_3). Calcd for$C_4H_9N_5O_2$:C 31.18, N 44.03, H 5.66; C 30.14, N 43.37, H 5.43. IR(KBr, cm^{-1}) ν:2984(ν_{as}-CH_3); 1450(ν_{as}-CH_2); 2106(ν_{as}-N_3); 1576(ν_s-N-NO_2).

2.3 Synthesis of BuAEAN

2.3.1. Buthyl -(2-nitroxyethyl)nitramine(BuNEAN). BuNEAN was prepared according to the method of MeNEAN. 15.58 g yellow oil was obtained, Its purity was 98.2%. ^{1}HNMR($CDCl_3$/TMS) δ: 0. 82(s, 3H, CH_3), 1. 47(s, 2H, CH_2). Calcd for$C_6H_{13}N_3O_5$: C 34. 78, N 20. 28, H 6. 28; Found (%)C 35. 12, N 20. 07, H 6. 11. IR(KBr, cm^{-1}) ν:1654 (ν_{as}-O-NO_2); 1536 (ν_{as}-N-NO_2); 1574(ν_s-N-NO_2); 833(ν_s-N-O).

2. 3. 2. Buthyl -(2-azidoethyl)nitramine(BuAEAN). BuAEAN was prepared according to the method of MeAEAN. 14. 21g yellow oil was obtained, Its purity was 97. 8%. ^{1}HNMR($CDCl_3$/TMS) δ:0. 76(s, 3H, CH_3), 1. 24(s, 2H, CH_2). Calcd for$C_6H_{13}N_5O_2$:C 38. 50, N 37. 43, H 6. 95; C 38. 45, N 37. 01, H 6. 33. IR(KBr, cm^{-1}) ν:2943 (ν_{as}-CH_3); 1435 (ν_{as}-CH_2); 2105 (ν_{as}-N_3); 1520 (ν_s-N-NO_2).

2. 4. Properties of alkyl-AEANs

Some physicochemical properties of alkyl-AEANs obtained by the tests are all listed in table 1, including the appearance, density, decomposition temperature, friction sensitivity and impact sensitivity.

Table 1. The properties of N-(2-azido) alkyl nitramines (alkyl-AENAs).

Properties	Me-AENA	Et-AENA	Bu-AENA
Formula	$C_3H_7N_5O_2$	$C_4H_9N_5O_2$	$C_6H_{13}N_5O_2$
Appearance	light yellow oil	yellow oil	yellow oil
Relative molecular weighty	145	159	187
Density /g·cm^{-3}	1. 343	1. 320	1. 211
Decomposition temperature	180°C	178°C	167°C
Friction sensitivity /%	0	0	0
Impact sensitivity H_{50}/cm	>320	>320	>320

The results show that the compounds have good heat-resistance ability and insensitive to impact and friction.

3. Result

Methyl-(2-azidoethyl)nitramine (MeAEAN), ethyl-(2-azidoethyl)nitramine (EtAEAN), buthyl-(2-azidoethyl)nitramine (BuAEAN), were synthesized for the first time. Their decomposition temperatures are 180°C, 178°C and 167°Crespectively. They are insensitive to impact and friction, revealing that they can be potentially used as insensitive energetic plasticizers for gun and rocket motor propellants compositions. In order to make these energetic plasticizers applying in propellants compositions, in depth study of the synthesis, stability, compatibility and mechanical properties were required.

4. References

[1] Simmons R L1984US4450110.
[2] Flanagan J E1991 US5013856.
[3] Ji Y P, Lan Y and Li P R2008*Chinese J. Expl.Propell.***8** 11.
[4] Raman K V1988*Propell. Expl. Pyrotech.***13**149.
[5] Gao F L, Ji Y P and Wang W2011*Chinese J. Expl. Propell.***34**12.
[6] Ji Y P, Gao F Land Han R 2013*Chinese J. Energ. Mater.***21**612.
[7] Wang Y L, Ji Y P and Li P R 2010*Chinese J. Energ. Mater.***18** 11.
[8] Gao F L, Ji Y P and Li P R2011*Chinese J. Energ. Mater.***19** 497.
[9] George F W and Walter J C 1949 US2461582.
[10] Kala P C 2004 *Propell. Expl. Pyrotech.***2** 93.
[11] Alfred T B and Fred T F 1954 US2678946.
[12] Wang W, Ding F and Liang Y 2010 *Chinese J. Energ. Mater.* **33** 52.

IOP Conf. Series: Materials Science and Engineering **269** (2017) 012058 doi:10.1088/1757-899X/269/1/012058

Synthesis and Thermal behaviors of 1, 8-Dihydroxy-4, 5-Dinitroanthraquinone Nickel Salt

Ying-lei Wang[1], Feng-qi Zhao[1], Jian-hua Yi[1], Yue-ping Ji[1]*, Wei Wang[1], Si-yu Xu[1], Ting An[1], Fu-lei Gao[1]

[1]Xi'an Modern Chemistry Research Institute, Xi'an, Shanxi 710065, China

Abstract. A novel energetic combustion catalyst, 1, 8-dihydroxy-4, 5-dinitroanthraquinone nickel salt (DHDNENi), was firstly synthesized by the process of metathesis reaction in a yield of 91%, and its structure was characterized by IR, element analysis, thermogravimetric analysis (TG) and differential scanning calorimetry (DSC). The thermal behavior and non-isothermal decomposition reaction kinetics of DHDNENi were studied by means of different heating rate differential scanning calorimetry (DSC). The kinetic equation of major exothermic decomposition reaction for DHDNENi is obtained. The self-accelerating decomposition temperature (TSADT) and critical temperature of thermal explosion (Tb) are 574.4K and 593.4K, respectively. The thermal stability of DHDNENi is good.

1. Introduction

Combustion catalyst as a ballistic modifier is an important ingredient in solid rocket propellant [1-3]. These are generally metal salts of carboxylic acids, and salts of aromatic carboxylic acids are found effective as combustion catalysts [4-8]. However, the carboxylic acid moiety in these salts is inert and does not contribute to the energetic [9-11]. In other hand, most combustion catalyst was organic lead compounds, the toxicities of lead compounds have attracted people's much more attention in these decades, and new kinds of combustion catalysts replacing lead compounds need to be found [12-14]. Nickel compounds, with their catalytic effect close to lead compounds, have been used as eco-friendly combustion catalysts documented in many literatures.

1, 8-dihydroxy-4, 5-dinitroanthraquinone (DHDNE) was an energetic compound with acidity. The structure contains acid group, energetic group and more carbon content. So it could be used as the intermediate to synthesize energetic combustion catalyst. In this paper, the structure of nickel salt of 1, 8-dihydroxy-4, 5-dinitroanthraquinone (DHDNENi) was designed and synthesized, a kind of new metal salt of 1, 8-dihydroxy-4, 5-dinitroanthraquinone, can be used as a green energetic combustion catalyst to modify combustion performance of solid propellants. The studies on the thermal behaviors and catalysis are a very important starting point for the selection, application and exploitation of DHDNENi. To the best of our knowledge, there is no report about it.

The aim of this work is to use the acidity of DHDNE synthesizing DHDNENi (scheme 1), and gain the thermolysis kinetics of DHDNENi under static state by means of differential scanning calorimetry (DSC) at different heating rate [15-17].

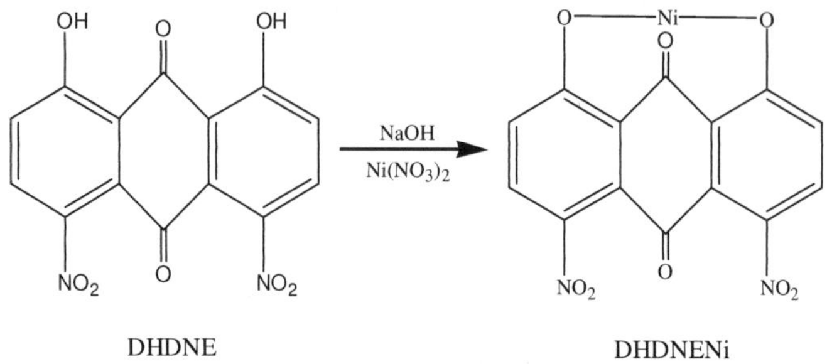

Scheme 1. Synthetic routes of DHDNENi.

2. Results and discussion

2.1. Structural analysis of DHDNENi

2.1.1. The RSFTIR spectrum of DHDNENi at room temperature. The RSFTIR spectrum of DHDNENi at room temperature is shown in figure 1. The antisymmetric and symmetric stretching vibration absorptions of C-NO_2 bonds ($\upsilon_{as(C\text{-}NO2)}$ and $\upsilon_{s(C\text{-}NO2)}$) are at 1548 and 1311 cm^{-1}, the $\upsilon_{as(\text{-}C=O)}$ peak at 1684cm^{-1}.

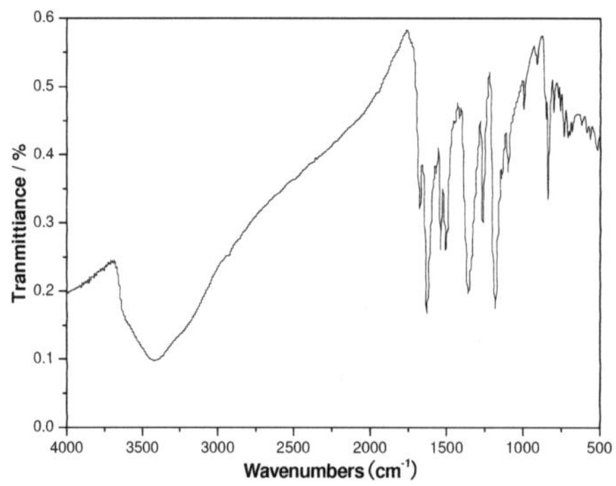

Figure 1. RSFTIR spectrum of DHDNENi at room temperature.

2.1.2. The TG curve of DHDNENi. The TG curve for DHDNENi is shown in figure 2. TG curve shows the mass loss process of DHDNENi is divided into two stages. The first stage begins at about 360K and completes at 553K with a mass loss of 12.8%, and the second stage begins at 586K and completes at 606K with a mass loss of 71.4%. The mass residue radio is 15.8%. We can deduce that the structure of DHDNENi may contain three crystals water, and the main component of the residue is NiO and some hydrocarbon polymer.

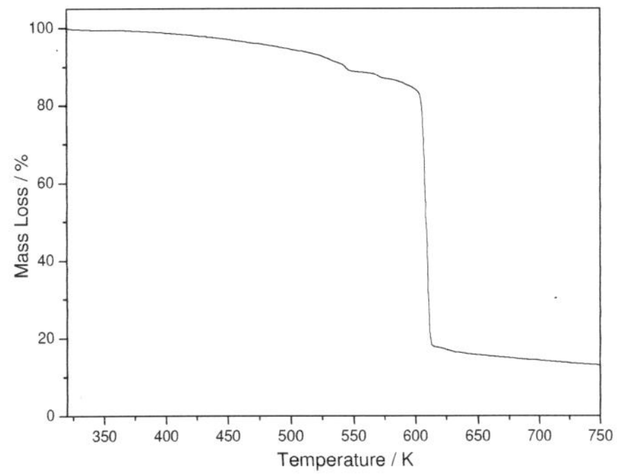

Figure 2. TG curve for the title compound at a heating rate of 10K min^{-1}.

2.1.3. Elemental analysis and X-ray fluorescence of DHDNENi. The elemental analysis can test the content of organic elements (C, H, N, O. et al), and the X-ray fluorescence tests the content of metal. The results of elemental analysis and X-ray fluorescence are C 38.10, H 2.275, N 6.33, O 39.99 and Ni 13.30, respectively.

According to the results of RSFTIR, TG, elemental analysis and X-ray fluorescence, the molecular formula of DHDNENi is can be deduced, and it is $C_{14}H_4N_2O_8Ni \cdot 3H_2O$.

2.2. The thermal decomposition behavior of DHDNENi under static N2 gas condition. The typical DSC curve (figure 3) indicates that thermal behavior of DHDNENi present one stage. The stage is an obvious exothermic decomposition process. The extrapolated onset temperature and peak temperature of the process are 575.1K and 659.7K at the heating rate 10.0K$\cdot$$min^{-1}$, respectively.

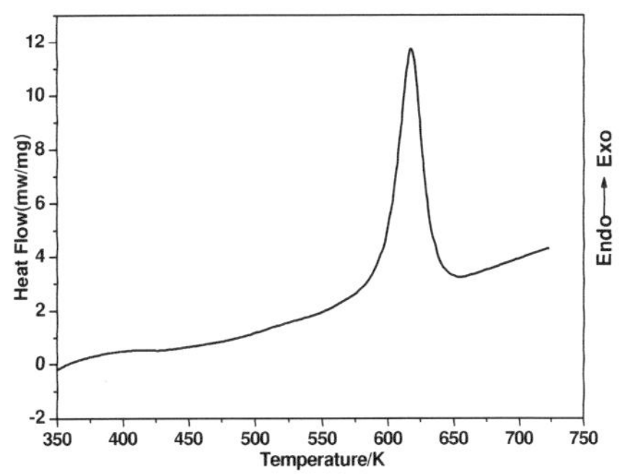

Figure 3. DSC curve for the title compound at a heating rate of 10K min^{-1}.

2.3 Non-isothermal decomposition kinetics of DHDNENi. In order to obtain the kinetic parameters (the apparent activation energy (E) and pre-exponential constant (A)) of the two main exothermic decomposition reactions for the title compound, a multiple heating method (Kissinger's method [18] and Ozawa's method [19]) was employed as follows:

Kissinger's equations:

$$\ln\left(\frac{\beta}{T_{\mathrm{p}}^{2}}\right)=\ln\frac{AR}{E}-\frac{E}{R}\frac{1}{T_{\mathrm{p}}} \tag{1}$$

Ozawa's equations:

$$\log \beta + \frac{0.4567E}{RT_p} = C \tag{2}$$

where T_p is the peak temperature (K), β is the linear heating rate (K min^{-1}), E is the apparent activation energy (kJ mol^{-1}), A is the pre-exponential constant (s^{-1}), R is the gas constant (J mol^{-1} K^{-1}) and C is a constant.

The measured values of the extrapolated onset temperature (T_e), peak temperature (T_p) and enthalpy (ΔH_d) of the exothermic decomposition reaction were listed in table 1. The values of T_{e0} and T_{p0} in the stage corresponding to $\beta \rightarrow 0$ obtained by equation (3) were also listed in table 1 [19].

$$T_{(0,\,\mathrm{e\ or\ p})i} = T_{(00,e0\ \mathrm{or\ p0})} + n\beta_i + m\beta_i^2 \qquad i = 1 \sim 4 \tag{3}$$

where n and m are coefficients.

Table 1. The values of T_e, T_p, T_{e0} and T_{p0} of the exothermic decomposition stage for DHDNENi determined from the DSC curves at various heating rates (β).

β (K min^{-1})	T_e (K)	T_p (K)	T_{e0} (K)	T_{p0}(K)
5.0	586.3	605.2		
10.0	598.7	618.0	574.4	592.7
15.0	611.6	631.1		

The kinetic parameter values (E and A) determined by Kissinger method and Ozawa method and linear correlation coefficients (r) were listed in table 2. We can see that the apparent activation energy (E) obtained by Kissinger method agrees well with that obtained by Ozawa method, and the linear correlation coefficients are all very close to 1. So, the result is credible.

Table 2. The kinetic parameters obtained by the data in table 1[a].

E_k (kJ mol^{-1})	log(A /s^{-1})	r_k	E_O (kJ mol^{-1})	r_O
156.1	11.1	0.9998	153.7	0.9999

[a] Subscript k, data obtained by Kissinger's method; subscript o, data obtained by Ozawa's method.

By substituting corresponding data (β_i, T_i and α_i, i=1, 2, 3,...) into equation (2), the values of E for any given value of α (the conversion degree) were obtained and shown in figure 4. We can see that the values of E steadily distribute from 148.1 kJ mol^{-1} to 1158.1 kJ mol^{-1} in the α range of 0.3-0.9, and the average value of E is 152.8 kJ mol^{-1}, which is in agreement with that obtained by Kissinger's method and Ozawa's method from only peak-temperature values. So, the values were used to check the validity of E by other methods.

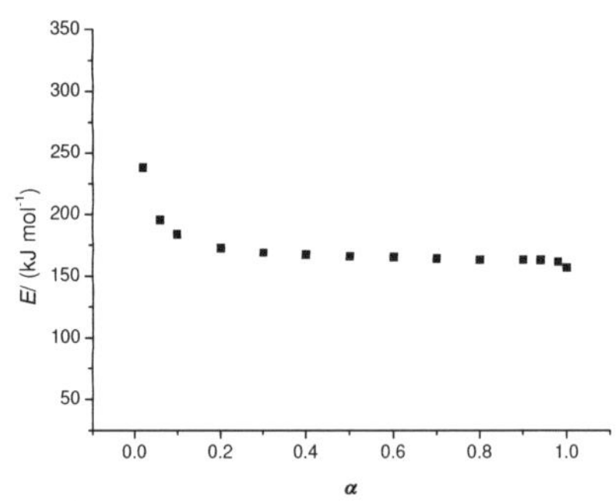

Figure 4. Ea vs α curve for the decomposition reaction of DHDNENi by Ozawa's method.

The four integral equations (Ordinary-integral, MacCallum-Tanner, Šatava-Šesták, Agrawal) were cited to obtained the values of *E*, *A* and the most probable kinetic model function ($f(\alpha)$) from each single DSC curve [20].

Forty-one types of kinetic model functions in reference [20] and corresponding experimental data form DSC curve at different heating rates were put into the four equations (Ordinary-integral, MacCallum-Tanner, Šatava-Šesták, Agrawal) for calculating, respectively. The kinetic parameters (E and A), probable kinetic model function and linear correlation coefficient (r) are presented in table 3. We can see that their values of *E* and lg*A* obtained by the five equations agree well with each other, and the mean value is close to that obtained by Kissinger method and Ozawa method. So, we can conclude that the most probable kinetic model function of the exothermic decomposition reaction is classified as

$f(\alpha)=2(1-\alpha)[-In(1-\alpha)]^{1/2}$, Jander equation (n=1/2) (No.13), and

$G(\alpha)=[-In(1-\alpha)]^{1/2}$, according to the unanimity rule of calculation results to each model equation. So, the kinetic equation of the exothermic decomposition reaction can be described as:

$$\frac{\mathrm{d}\alpha}{\mathrm{d}T}=\frac{10^{11.1}}{\beta}2(1-\alpha)[-In(1-\alpha)]^{1/2}\exp\left(-1.561\times10^{5}/RT\right) \quad (4)$$

Table 3. Calculated values of kinetic parameters of decomposition reaction.

β/K min^{-1}	Eq.	E /kJ mol^{-1}	lg (A/s^{-1})	r
5.0	Ordinary-integral	156.0	11.1	0.9967
	Mac Callum-Tanner	158.7	11.2	0.9971
	Šatava-Šesták	157.9	11.2	0.9971
	Agrawal	156.0	11.1	0.9967
10.0	Ordinary-integral	150.3	10.6	0.9978
	Mac Callum-Tanner	153.1	10.8	0.9981
	Šatava-Šesták	152.7	10.8	0.9981
	Agrawal	150.3	10.6	0.9978
20.0	Ordinary-integral	148.1	10.4	0.9975
	Mac Callum-Tanner	151.1	10.7	0.9979
	Šatava-Šesták	150.9	10.7	0.9979
	Agrawal	148.1	10.4	0.9975
	mean	152.8	10.8	

2.4. Self-accelerating Decomposition Temperature (TSADT) and critical temperature of thermal explosion (Tb). The self-accelerating decomposition temperature (T_{SADT}) and critical temperature of thermal explosion (T_b) are important parameters required to insure safe storage and process operations for the energetic materials and then to evaluate the thermal stability. T_{SADT} and T_b can be obtained by equation (5) and equation (6), respectively. So, T_{SADT} and T_b of DHDNENi are 574.4K and 593.4K, respectively. They are all high, indicating that the thermal stability of the title compound is good.

$$T_{\mathrm{SADT}}=T_{e0} \quad (5)$$

where *a* and *b* are coefficients.

$$T_b = \frac{E_O - \sqrt{E_O^2 - 4E_O RT_{e0}}}{2R} \tag{6}$$

where E_O is the apparent activation energy obtained by Ozawa method.

3. Experimental

3.1. Synthesis

DHDNENi used in this work was prepared according to the following method: 1,8-dihydroxy-4,5-dinitroanthraquinone (3.30g ,0.01mol) and NaOH (0.84g,0.021mol) were dissolved in 100mL H_2O and the pH was about 8. The reaction mixture was stirred at 60°C for 1 hour, then $Ni(NO_3)_2$ (2.91g ,0.01mol) dissolving in 20mL H_2O was added into the above reaction solution. The resulting solids were filtered and dried by air. The product obtained is 3.61g, and the yield of 93.3%.

3.2 Experimental instruments and conditions of thermal analysis

DSC measurements were carried out on a Model TA-910 USA instruments. The operation conditions were as follows: heating rates, 2.5K·min^{-1}, 5K·min^{-1}, 10K·min^{-1} and 20K·min^{-1}; sample mass, 0.5 mg ~1mg; aluminium sample cell; atmosphere, static nitrogen with 0.1MPa.

4. Conclusion

(1) Nickel salt of 1, 8-dihydroxy-4, 5-dinitroanthraquinone (DHDNENi) was synthesized by the process of metathesis reaction, and its structure was characterized by IR, element analysis, TG and DSC.

(2) The apparent activation energy and pre-exponential factor of the exothermic decomposition reaction of DHDNENi obtained by Kissinger's method are 156.1 kJ mol^{-1}and $10^{11.1}$ s^{-1}, respectively. The self-accelerating decomposition temperature (T_{SADT}) and critical temperature of thermal explosion (T_b) are 574.4K and 593.4K, respectively.

(3) The thermal behavior and non-isothermal decomposition reaction kinetics of DHDNENi were studied by means of different heating rate differential scanning calorimetry (DSC). The kinetic equation of major exothermic decomposition is

$$\frac{d\alpha}{dT} = \frac{10^{11.1}}{\beta} 2(1-\alpha)[-In(1-\alpha)]^{1/2} \exp\left(-1.561\times10^5/RT\right)$$

References

[1] Yi J H, Zhao F Q, Hong W L, Xu S Y, Hu R Z, Chen Z Q and Zhan L Y 2010 *J. Hazard. Mater.* **76** 257.

[2] Yi J H, Zhao F Q, Wang B Z, Liu Q, Zhou C, Hu R Z, Ren Y H, Xu S Y, Xu K Z and Ren X N 2010 *J. Hazard. Mater.* **181** 432.

[3] Wang Y L, Zhao F Q and Yi J H 2012 *Chinese J. Explos. Proprl.* **35** 1.

[4] Kulkarni P B, Purandare G N, Nair J K, Talawar M B, Mukundan T and Asthana S N 2005 *J. Hazard. Mater.* **119** 53.

[5] Qiu Q Q, Xu K Z, Yang S H, Gao Z, Zhang H, Song J R and Zhao F Q 2013 *J. Solid State Chem.* **205** 205.

[6] Nair J K, Talawar B and Mukundan T 2001 *J. Energ. Mater.* **19** 155.

[7] Thomas M, Klapotke, Stierstorfer J and Weber B 2009 *Inorg. Chim. Acta.* **362** 2311.

[8] Ashmore C I, J C S and Stephents W D 1978 US04108696A.

[9] Nakagawa T W and Rudy T P 1979 US4166045A.

[10] Ishiguro H and Shimizu H 1991 JP03097687.

[11] Kawasaki H and Kai T 1990 JP11092263.

[12] Sammour M H and Nomp C, Egypt 2007 Proceedings of the NATAS Annual Conference on Thermal Analysis and Applications.

[13] Liu R, Zhou Z N, Qis Y, Yang L, Wu B D, Huang H S and Zhang T L 2013 *Central European J. Energ. Mater.* **10** 17.
[14] Thompson and Stephen B 1996 International Annual Conference of ICT 27th 101.
[15] Hou C H, Geng X H, An C W, Wang J Y, Xu W Z, and Li X D 2013 *Central European J. Energ. Mater.* **10** 123.
[16] Wang Y L, Zhao F Q, Xu K Z, Ji Y P, Pan Q, Yi J H, Chen B and An T 2013 *Inorg. Chim. Acta.* **405** 505.
[17] Lu L, Xu K Z, Zhang H, Wang G, Huang J, Wang B Z and Zhao F Q 2012 B*ull. Korean Chem. Soc.* **33** 2352.
[18] Kissinger H E 1957 *Anal. Chem.* **29** 1702.
[19] Ozawa T 1957 *Bull. Chem. Soc. Japan* **38** 1881.
[20] Hu R Z, Gao S L, Zhao F Q, Shi Q Z, Zhang T L and Zhang J J 2008 Thermal Analysis Kinetics, 2nd ed (Beijing: Science Press).

Acknowledgments
This investigation received financial assistance from the National Science Foundation of China (Grant No. 21173163 and NO. 21303133).

The Properties of N,N'-bis(propionyl) cystamine- acrylamide /Nanowhiskers of Cellulose (CNWs) Composite Hydrogel

Jiufang Duan[1]*,Yuxue Gao[1]

[1]MOE Engineering Research Center of Forestry Biomass Materials and Bioenergy, Beijing Forestry University, Beijing 100083,China

Abstract. Dithiothreitol was employed as a responsive agent for disulfide bond in N,N'-bis (propionyl) cystamine in N,N'-bis(propionyl) cystamine – acrylamide/CNWs composite hydrogel. This hydrogel was obtained by mixing a queous solutions of CNWs, N,N'-bis(propionyl) cystamine and acrylamide at room temperature. The internal morphology, and crystalline performance of the hydrogels were characterized by scanning electronmicroscopy(SEM) and X-ray diffraction. The scanning electronmicroscopy(SEM) showed that the hydrogel has aporous structure. X–ray diffraction patterns confirmed that N,N'-bis(propionyl) cystamine – acrylamide /CNWs composite hydrogel has amorphous structure. The water swelling ratio reaches to the highest value at 0.067g/ml CNWs concentration.

1. Introduction

Hydrogels is a kind of water-swollen networks, which were most intensively studied for the application in wound healing application, biomedical materials, drug delivery, tissue engineering, wastewater treatment, food chemistry and so on [1, 2]. Responsive hydrogels have the function of response to environmental stimuli such as temperature, pH, electric or magnetic fields, light, and enzymes [4-6]. These materials have great potential applications in the controlled drug delivery and release.

CNWs was widely used as reinforcing agents in conventional hydrogel composites materials, owing to it has the properties of high strength, high surface area, tunable surface chemistry, and other desirable properties [7, 8].

In this paper, we describe a hydrogel based on the chemical responding bond of dithiothreitol and incorporating CNWs to this system. It is expected to obtain a chemically responsive gel with certain strength. The preparation and properties of N,N'-bis(propionyl) cystamine –acrylamide /CNWs composite hydrogel were studied.

2. Experimental

2.1. Material

Acrylamide, N, N'-bis (propionyl) cystamine, ethanol, ammonium persulfate, N, N, N ', N'-tetramethylethylenediamine (TEMED) were of analytical grade and used without further purification.

2.2. Preparation of CNWs nanocomposite hydrogels

CNWs (CNWs here used were isolated from cotton) [8] is uniformly dispersed in an aqueous solution and prepared into solutions of concentration 0.09 g/ml, 0.067 g/ml, 0.045 g/ml, 0.03 g/ml, 0.011g/ml, 0.006 g/ml. 19g anhydrous ethanol , 0.4368g N, N'-bis (propionyl) cystamine, 3.724 g acrylamide and 4500 μL TEMED were added into 57g CNWs solution. The solution was stirred until fully dissolved. 9100 μL freshly prepared 10% w / v ammonium persulfate were added to initiate the polymerization reaction. This reaction was maintained at room temperature for 15 h to get the hydrogel.

2.3. Methods

The morphological characterization of gel was performed with scanning electron microscopy (S-3400N, Hitachi, Japan). The X-ray diffraction was performed with X - ray diffractometer (Advance 8, Bruker, Germany). The gravimetric method was employed to measure the swelling ratios of the hydrogels in distilled water at 25°C. After immersion in distilled water for about 48 hr to reach swelling equilibrium, the gel samples were taken out and weighed after removing the excess water on the surfaces. Each data was measured three samples, and the average value of three measurements was taken. The equilibrium swelling ratio (SR) was calculated as SR = Ws/Wd, where Ws is the weight of the swollen gel and Wd is the weight of the hydrogel at the dry state.

3.Results and discussion

3.1. Water swelling ratios of the hydrogels

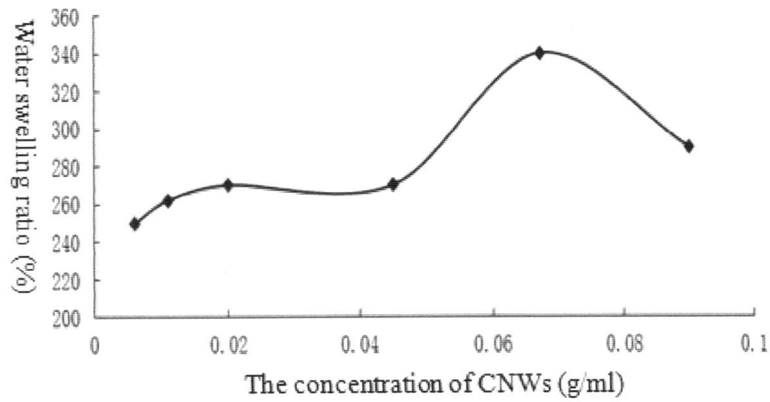

Figure 1. Water swelling ratio of N, N'-bis (propionyl) cystamine – acrylamide/ CNWs composite hydrogel

CNWs concentration affects the water absorption of the hydrogel. When the concentration of increased from 0.0067 g / ml to 0.045 g / ml, the water absorption ratio of hydrogels is increased. The water absorption ratio reaches to the highest value at 0.067 g / ml CNWS concentration.

3.2Hydrogel morphology

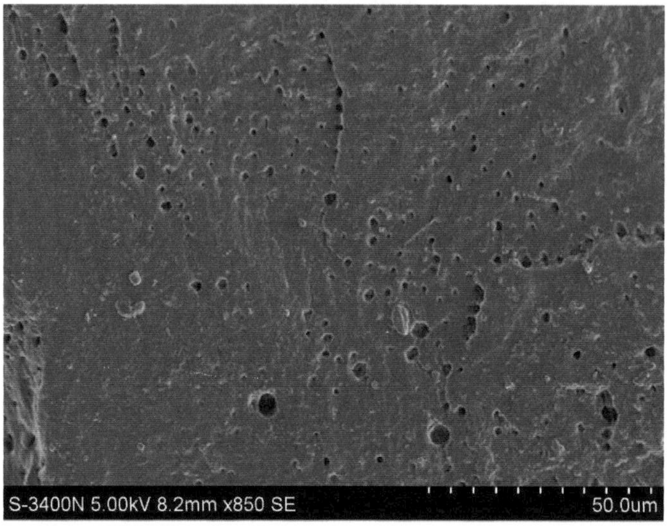

Figure 2. The SEM image of N, N'-bis (propionyl) cystamine – acrylamide/ CNWs composite hydrogel.

As shown in figure 2, the SEM image of N, N'-bis (propionyl) cystamine – acrylamide/ CNWs composite hydrogel, the hydrogel has a porous structure.

3.3. Crystalline performance

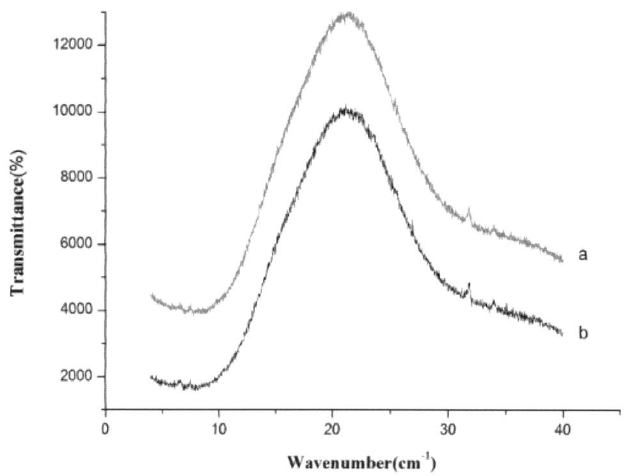

Figure 3. The X-ray diffraction of N, N'-bis (propionyl) cystamine – acrylamide/ CNWs composite hydrogel (CNWs concentration a:0.09g/ml， b:0.006g/ml).

The X-ray diffraction patterns confirmed that N,N'-bis(propionyl) cystamine –acrylamide/CNWs composite hydrogel has amorphous structure.

4. Conclusion

We reported the synthesis, the morphology, the swelling ratios and crystalline performance of the N,N'-bis(propionyl) cystamine –acrylamide/CNWs composite hydrogels. The N,N'-bis(propionyl) cystamine –acrylamide /CNWs composite hydrogel has a porous internal morphology and amorphous structure, which based on the test of SEM and X–ray diffraction of the sample of N,N'-bis(propionyl) cystamine –acrylamide /CNWs composite hydrogels. The water absorption ratio of the composite hydrogel reaches to the highest value at 0.067g/ml CNWs concentration. Based on the effect of chemical responsive on the disulfidebond though contact with dithiothreitol, the obtained N,N'-bis(propionyl) cystamine –acrylamide/CNWs composite hydrogel have the potential application in intelligent response field.

5. References

[1] Bai R, Yang Q, Tang J, Morelle X P, Vlassak J and Suo Z, 2017 *Extr. Mech. Lett.* **15** 91–6.
[2] Casolaro M, Casolaro I and Lamponi S 2012 *European J. Pharmaceut. Biopharmaceut.* **80** 553–61.
[3] Ricka J and Tanaka T 1984 *Macromolec.* **17** 2916–21.
[4] Bae Y H, Okano T and KimS W 1990 *J. Polym. Sci. Part B: Polym. Phys.* **28** 923–36.
[5] Zhao Q, Sun J Z, Chen S W and Zhou Q Y 2010 *J. Appl. Polym. Sci.* **115** 2940–5.
[6] Reinicke S, Dohler S, Tea S, Krekhova M, Messing R, Schmidt A M and Schmalz H, 2010 *Soft Matt.* **6** 2760–73.
[7] De France k J, Hoare T and Cranston E D 2017 *Chem. Mater.* **29(11)** 4609-31.
[8] Tummala G K, Joffre T, Rojas R, Persson C and Mihranyan A *2017Soft Matt.* **13(21)** 3936-45.

Acknowledgments

This research was financially supported by the National Natural Science Foundation of China (31600464).

IOP Conf. Series: Materials Science and Engineering **269** (2017) 012060 doi:10.1088/1757-899X/269/1/012060

Fatigue Characteristics of 3D Printed Acrylonitrile Butadiene Styrene (ABS)

M. M. Padzi[1*], M. M. Bazin[1], W. M. W. Muhamad[1]

[1]Department of Mechanical Engineering, Malaysia France Institute, University Kuala Lumpur, 43650, Bangi, Selangor, Malaysia

mahfodzah@unikl.edu.my

Abstract. Recently, the use of 3D printer technology has become significant to industries, especially when involving the new product development. 3D printing is a technology, which produces the 3D product or prototype using a layer-by-layer technique. However, there becomes less research on the mechanical performance of the 3D printed component. In the present work, fatigue characteristics of 3D printed specimen have been studied. Acrylonitrile butadiene styrene (ABS) has been chosen as a material research due to its wide applications. Two types of specimen used, which is the 3D printing and moulding specimens. Fused deposition modelling (FDM) technique was used to produce the specimens. The dog bone shape part was produced based on ASTM D638 standard and the tensile test has been carried out to get the mechanical properties. Fatigue test was carried out at 40%, 60% and 80% of the tensile strength. The moulded part shows higher fatigue cycles compared to 3D printed part for all loading percentages. Fatigue lives for 40%, 60% and 80%, were 911, 2645 and 26948 cycles, respectively. The results indicated that 3D printed part has a lower fatigue life, which may not suitable for industrial applications. However, the 3D printed part could be improved by using various parameters and may be introduced in low strength application.

1. Introduction

Rapid prototyping is the prior additive manufacturing (AM) process that develops in the 1980's in creating prototypes and model [1]. In addition, according to Karina M.C et. al [2], a 3D printer is an associate innovative technology and was initiated accustomed earlier produce prototypes. 3D printing is one of the additive layer fabrication processes [3]. The primary 3D printer was created by means of Charles Hull in 1984. That enables the tangible 3D object to be made of digital information [4]. During the 90's, the primary SLA (Stereolithographic) machine was made by 3D System. It proved that complex and advanced components could be manufactured. Moreover, this technology has been engaged within the medical space. After getting into the 21st Century, 3D printing technology has been utilized in the various fields, such as medical, manufacturing, automotive, aviation, jewelry, and others. Within the last decades, the world's 1st 3D written robotic craft was inbuilt seven days in 2011. Also, the world's first 3D written automobile was in-built an equivalent year [5].

3D printing enhanced the productivity. 3D printing technology is a saving time process than conventional techniques of making objects which include prosthetics and implants, which require milling, forging, and a protracted transport time [6]. Besides, 3D printer is able to manufacture an object at low cost [7]. The cost to produce a 3D object is least, with the first object being as cheap as the final. That is mainly high quality for organizations, which have lower manufacturing volumes or that, produce components or merchandise, which can be complicated or require frequent adjustments. 3D printing could reduce costs because it can efficiently use raw materials and produce less waste [8]. For example, a pharmaceutical tablet weighing 10mg may want to doubtlessly be custom-fabricated on demand as a 1-mg pill. Some capsules will also be revealed in dosage bureaucracy, which can be less complicated and greater price-effective to deliver to patients. By 3D printing enables to design with novel geometry that might be troublesome or impossible to produce by using the conventional method. So, this can improve engineering component performance in term of creating an innovation and creativity in design for 3D printing.

There are many types of filler that can be used in a 3D printing machine. One of them is Acrylonitrile Butadiene Styrene (ABS). ABS is a group of thermoplastic of "terpolymers" [9]. Typically, ABS contains 15% to 35 % acrylonitrile, 5% to 30 % butadiene and 40% to 60% styrene. However, this proportion could be a change in a quite wide variety. Monomer is an atom or small molecule, which will be secured to alternative identical molecules to create a polymer [10]. This combination of materials makes it become an outstanding impact strength and high mechanical strength [11]. Furthermore, it gives a stiff thermoplastic polymer property which compatible with rough customer products. ABS yields plastically instead of tearing and shows ductile failure. ABS includes a sensible long-term load carrying capability with stresses higher than their tensile strength. However, ABS has a greater modulus of elasticity and hardness. It has comparatively low creep nature and high heat resistance which makes it become a decent dimensionally stability and has good electrical insulation properties. ABS is utilized in the wide selection of an application in trade these days. Its uses among several others producing ways, such as injection molding, blow molding, extrusion and as a filament for 3D printing merchandise [9].

Fatigue is chargeable for up to 90 % of the in-service part failure that occur in industry area [12]. In the 19th century, fatigue was a mysterious phenomenon due to its no warning fracture and did not give any plastic deformation visible. This lead to an erroneous believes that fatigue was merely an engineering matter. In the 20th century, fatigue seems not to be as engineering issues, but also as material and design development [12]. Fatigue on 3D printed component is branches of a part in studying of materials and manufacturing area. There are two cases that have found where the cases of fatigue on 3D printing component. According to [13], 3D printed component has produced dog-bone specimen by using rapid prototyping that fulfills the UNI EN ISO 527-1 (1997) standard, which designates a 10 mm by 4 mm cross-section with a gauge length of 8mm. The maximum number of cycles for running test respectively is set to 10,000 to prevent overflow data in the equipment. The materials that have been used are ABS and ABSplus. For rapidly prototyped component, usually, the printing orientation and pattern will influence the material properties [14]. Besides, there was a case of 3D printed fatigue for elastomer material, which is Tango Black Plus material. The comparison specimen is between interface specimen and specimen with no interface. The ratio of elongation that is conducted on each specimen is about 20 %, 30 %, 40 %, 50 % and 60 %. The maximum 60 % was chosen because it indicates the maximum elongation before it breaks.

2. Experimental

2.1. Sample preparation

The material used was ABS due to its availability and its wide application in most research field. The dimension of the specimen is based on the standard ASTM D638 as shown in figure 1. Sample thickness is 3 mm (figure 2) and was produced by using a 3D printer shown in figure 3. For moulding specimen, the specimen was prepared from ABS plate, cutting in the same shape by using CNC machines, as shown in figure 4. Figure 5 shows the specimen cutting from CNC machine.

2.2. Tensile testing

The tensile testing was conducted using Universal Testing Machine (UTM) 250 kN as shown in figure 6. This test was done to determine the mechanical properties of ABS materials such as yield strength, ultimate tensile strength (UTS) and modulus of elasticity.

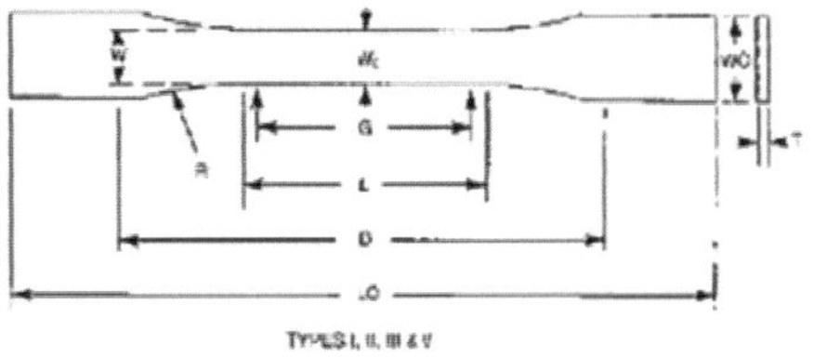

Figure 1. Dimension of specimen.

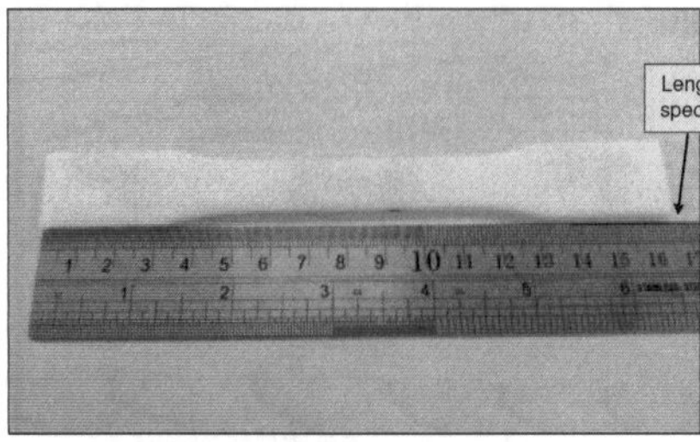

Figure 2. 3D printed Specimen.

Figure 3. 3D printed machine.

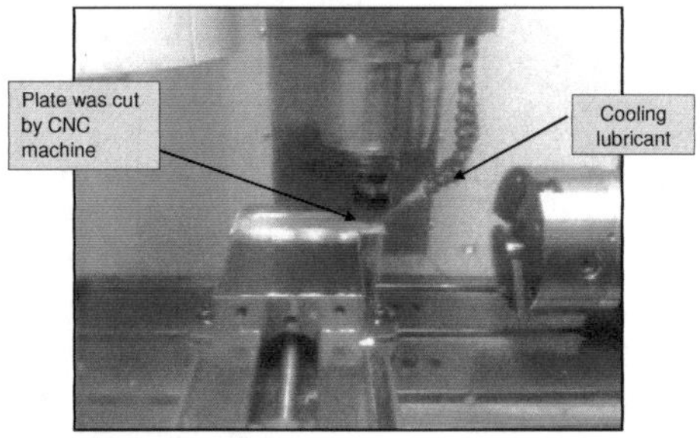

Figure 4. CNC machine used to cut specimen.

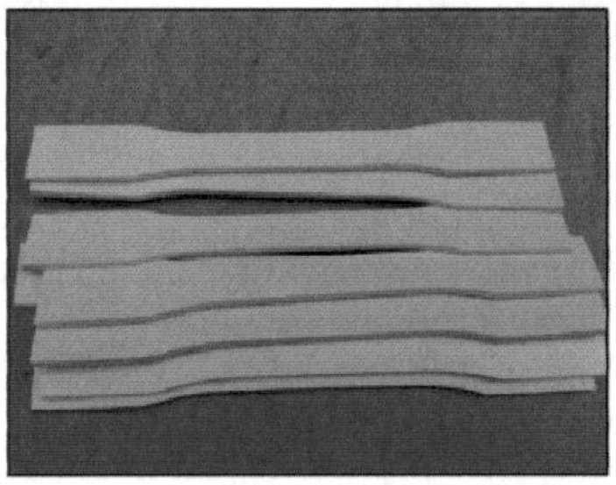

Figure 5. Specimens obtained from cutting process.

2.3. Fatigue testing

Fatigue testing was performed using machine SERVOPULSER model EHF- LV020K1-020 shown in figure 7. The test was done at frequency 1 Hz with tension-tension cyclic loading, R=0. Two different applied loads were used, which are 40%, 60% and 80% from UTS value. Fatigue test was run until the specimen breaks.

Figure 6. Tensile testing machine.

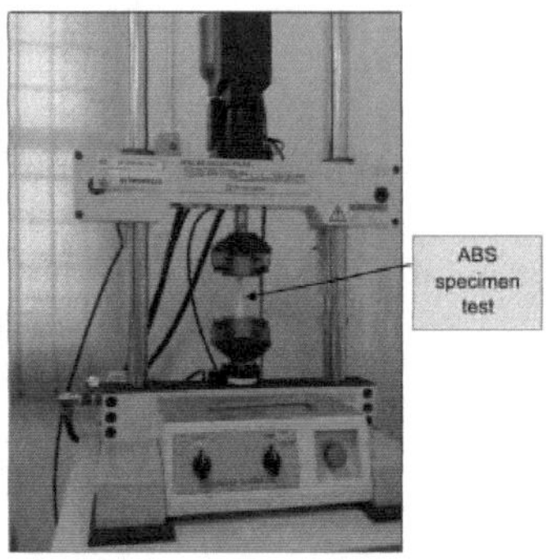

Figure 7. Fatigue testing machine.

3. Results and discussion

3.1. Mechanical properties

Table 1 shows the mechanical properties for both 3D printed and moulding specimens. Ultimate tensile stress and young's modulus of moulding contribute the higher value compared to 3D printed specimen. Young modulus values obtained from both specimens do not have much difference. The percentage difference is shown in table 2 and figure 8. However, the difference was higher for UTS, which was influenced by several factors. The regions in the material, which do not seem to be absolutely dense, have a lower mass/volume quantitative relation resulting in lower absolute strength and stiffness values [15]. Furthermore, the fabrication process technique also influenced the high differences of UTS [16]. This is because moulding is a process where the material was compacted by force into a mould. So, it makes the strength of moulding is higher than 3D printer.

Table 1. Mechanical properties of 3D printing and moulding specimen.

Properties	3D Printed Specimen	Moulding Specimen
UTS (MPa)	17.8	48.5
Modulus Young (GPa)	12.5	17
Breaking point (kN)	0.63	1.27

Table 2. Percentage differences of UTS and Young's Modulus.

Types of specimen	UTS	Young's Modulus	Percentage difference	
			UTS	Young's Modulus
3D printer	17.8	12.5	30.7 %	4.5 %
Moulding	48.5	17		

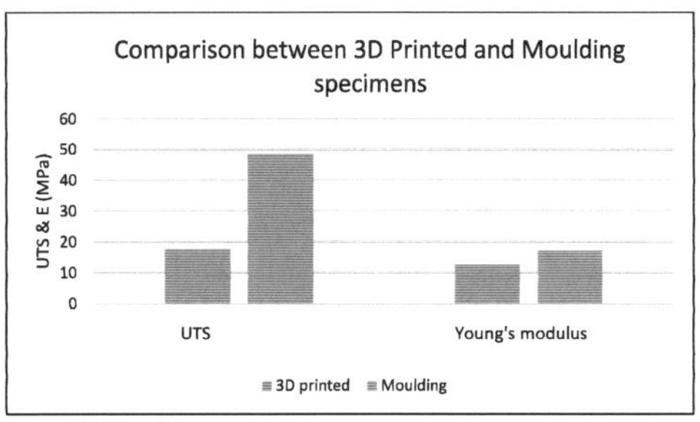

Figure 8. Comparison of percentage differences for printed and moulding specimen.

3.2. Fatigue Properties

Table 3 shows the load applied for fatigue testing. The result of fatigue test under different loading values for 3D printed and moulding specimen is presented in table 4. The fatigue cycle of moulding specimen shows the highest cycle for each load compared to 3D printing specimen. This may due to the fill of material [17]. This result's conceivable in sight of the actual fact that injection moulding involves the applying of high pressures that compact the material upon injection into the die cavity. Moreover, applied holding pressure throughout the injection moulding in addition compensates material shrinkage, leading to additional compact samples compared to FDM components, wherever the individual raster's area unit simply deposited next to every alternative [17].

From table 4, 40 % of load test shows the highest number of cycles for both 3D printed and moulding specimens. This coincides with the fatigue theory which is the higher the magnitude of the stress, the smaller the number of cycles the material is capable to sustain before failure [18]. The stress at where the location of failure happens in a given variety of cycles is that the fatigue strength. The number of cycles needed for a material to fail at an explicit stress is fatigue life [19]. The percentages of differences in fatigue life between 3D printed specimen and moulding specimen for 40 %, 60 % and 80 % are 30 %, 39 %

and 55 % respectively. This fatigue behaviour for 3D printed specimen seems to be accepted to compete with moulding for certain application in industries.

Figure 9 present a comparison of fatigue life between 3D printed and moulding specimens. At 40% and 60%, the number of cycles for moulding specimen was higher than 3D printed. Hence, less differences between both specimens. This is because, at 40 % and 60 %, the amount of load applied is moderate, make the specimen able to withstand longer for fatigue life. While, at 80 %, the percentage differences for 3D printed was half of mouldings specimens. Thus, it shows bigger differences among the specimens. This is because, at 80 %, the amount of load applied was high, thus, make specimen to fail short and low fatigue life.

Table 3. Load applied for fatigue testing.

Types of specimen	σ_u (MPa)	Load test(kN)		
		40 % σ_u	60 % σ_u	80 % σ_u
3D printed specimen	17.8	0.3	0.4	0.5
Moulding specimen	48.5	0.7	1.1	1.5

Table 4. Fatigue life of 3D printed and moulding specimen.

Types of specimen	σ_u (MPa)	No. of cycle		
		40 %	60 %	80 %
3D printed	17.8	1.9 x 10^4	1.62 x 10^3	4.09 x 10^2
Moulding	48.5	2.6 x 10^4	2.65 x 10^3	9.11 x 10^2
Percentages difference (%)	63	30	39	55

The stress - cycle (S-N) curves are shown in figure 10 for 3D printed specimen and moulding specimen, comparing with previous research. S-N curves show that 3D printed experiments exhibit similar fatigue life to the 3D printed specimen from previous research. The fatigue life between 3D printer and previous research shows the fatigue life from previous research is a little bit higher compared to 3D printing for loading 80 % and 60 %. The trend was changed for 40 % fatigue loading which 3D printed contributes a higher fatigue life compared to previous research. Thus, this result can be accepted.

Besides, for fatigue life of moulding specimen demonstrated the higher cycle compared to others' cycles. Shown in figure 9, fatigue life for these three different load values is below 10^5 of the number of cycles. Thus, it means that ABS material is considered as low cycle fatigue materials. The trend of curves between 3D printed and previous research shows that the results obtained were validated. Besides, the trend of curves for moulding specimen was steeper compared to others and was considered in the same range (10^2 to 10^3 for 40 % loading test) with 3D printed and previous research results. The moulding specimen has the highest trend is due to its ability to withstand high stress applied.

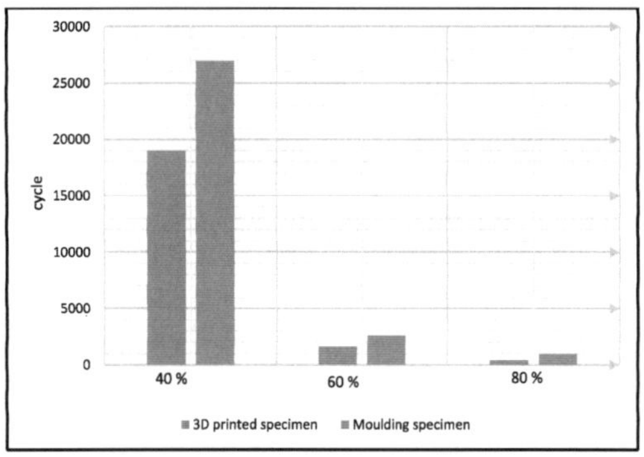

Figure 9. Comparison of fatigue life between 3D printed and moulding specimen.

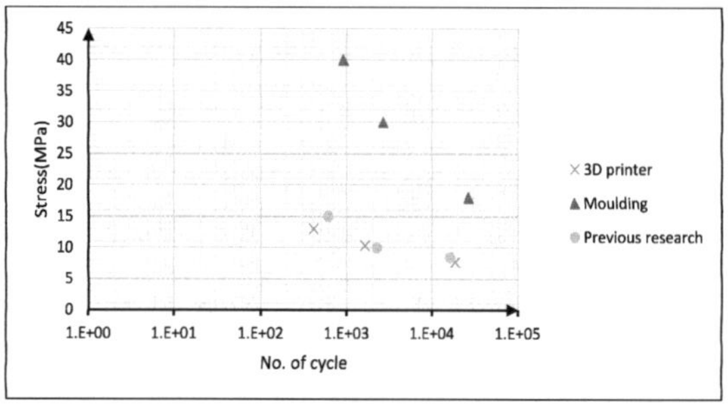

Figure 10. S-N curve for 3D printed and moulding specimen, comparing with previous research data.

4. Conclusion

From this work, it can be concluded that fatigue life was determined by the tensile strength value. Ultimate tensile strength (UTS) value of the moulding specimen (48.5 MPa) was higher compared to 3D printing specimen (17.8 MPa) where it causes the fatigue life of moulding specimen was higher than 3D printed specimen. Fatigue life for 40% σ_u, 60% σ_u and 80% σ_u is 2.6 x 10^4, 2.65 x 10^3 and 9.11 x 10^2 for moulding specimen, respectively, while for a printed specimen is 1.9 x 10^4, 1.62 x 10^3 and 4.09 x 10^2, respectively. This gives the percentage differences in fatigue life between 3D printed specimens and moulding specimens is 30 %, 39 % and 55 %. This is due to moulding process and the fabrication technique that produce more ductile properties [16]. Therefore, 3D printed component might not suitable for high strength industrial application. However, within the context of production, 3D printed components may be grown in low volume, customized design components and for low strength application. Moreover, the technology of 3D printing could produce a complex design of a product that could not be brought about by the moulding method.

5. References

[1] Wong K V and Hernandez A 2012 *A Rev. Addit. Manuf.*
[2] Morales C K and Wisalchai J E 3D Printing for end use products - A case study of the Industry its Capabilities and Value Chain
[3] Vojislav et al 2010 Int. J. Prod. Res. Taylor and Francis 1
[4] Zeyu L (2014) Exploration of 3D Printing 3–41
[5] Maxey K Report Adds 3D Printing to 12 Disruptive Techs for 2025 May 2013
[6] Ventola C L (2014) Medical Applications for 3D Printing: Current and Projected Uses Pharmacy and Therapeutics 39(10) 704–711
[7] Winnan CR (2014) Adventures in 3D Printing Limitless Possibilities and Profit22-25
[8] Fall (2012) 3D printing and The Future of Manufacturing3-5
[9] Hamod H (2014) Suitability of recycled HDPE for 3D printing filament Haruna Hamod Degree Thesis : Plastics Technology
[10] Jadhav KG (2012) Synthesis of Monomers for New Conjugated Polymers
[11] Chaturvedi J (2009) Parametric Optimization of Fused Deposition Modeling Using Response Surface Methodology20-24
[12] Azeez A A (2013) Fatigue failure and testing methods
[13] Lee J and Huang A (2013) Fatigue analysis of FDM materials 4(March 2012) 291–299
[14] Ahn S et al (2011) Anisotropic material properties of fused deposition modelling ABS
[15] Hooreweder BVMoens D Boonen R Kruth JP and Sas P(2013) On The Difference in Material Structure and Fatigue Properties of Nylon Specimens Produced by Injection Moulding and Selective Laser Sintering
[16] Fischer (2016) Thermoplastics: The Strongest Choice for 3d Printing
[17] Dawoud M (2016) Mechanical behaviour of ABS : An experimental study using FDM and injection moulding techniques
[18] Gulshan F (2009) Fatigue Phenomenon in Materials Fatigue Phenomenon in Materials
[19] Önem (2003) Effect of Temperature On Fatigue Properties Of Din 35 NiCrMoV 12 5 Steel

IOP Conf. Series: Materials Science and Engineering **269** (2017) 012061 doi:10.1088/1757-899X/269/1/012061

Preparation of Magnetic Molecularly Imprinted Polymer for Chlorpyrifos Adsorption and Enrichment

M Chen[1], X Ma[2*] and J Sheng[2]

[1]School of Environmental Science and Engineering, Guangdong University of Petrochemical Technology, Maoming, China
[2]School of Environmental Science and Engineering, Guangdong University of Technology, Guangzhou, China

Abstract. Magnetic molecularly imprinted polymer (MMIP) for chlorpyrifos was prepared and characterized. The adsorption performance of MMIP for chlorpyrifos was evaluated under various conditions. The results showed that the adsorption equilibrium was achieved within 1 h, the adsorption capacity was 16.8 mg/g, and the adsorption process could be well described by Langmuir isotherm model and pseudo-second-order kinetic model. The MMIP was used as the selective sorbent for solid–phase extraction of chlorpyrifos from environmental water and vegetable samples. Combined with gas chromatography–mass spectrometry, a LOD of 30 ng/L, spiked recovery of 89.6%–107.3% and RSD of 1.9%–3.8% for chlorpyrifos were obtained.

1. Introduction

Chlorpyrifos is the most widely used herbicide in agriculture and other applications. However, chlorpyrifos is associated with chromosomal damage and bladder cancer [1]. Nowadays, chlorpyrifos has been found in various environmental matrices and food samples. Consequently, it is necessary to develop suitable analytical methods for chlorpyrifos. Solid-phase extraction (SPE) is the most often used method for separation and enrichment of pollutants from various matrices [2-6].

Common adsorbents for SPE such as C18 lack adsorption selectivity. Molecularly imprinted polymer is a tailor-made material designed to present valuable molecular recognition properties for a target compound (template). Thus, MIP can selectively recognize the target molecules in a complex system [7].

Traditional SPE is always a tedious process. However, magnetically separation technology can provide an easy and rapid way for removal of magnetic particles from solution by applying a magnetic field [8, 9].

In this work, magnetic molecularly imprinted polymer (MMIP) for chlorpyrifos was prepared and characterized, and the adsorption behaviors of chlorpyrifos on the MMIP were investigated. Finally, the MMIP was used for preconcentration of chlorpyrifos from water and vegetable samples before instrumental determination.

2. Materials and methods

2.1. Chemicals

Chlorpyrifos was obtained from Dr. Ehrenstorfer GmbH (Germany), p-aminophenyltrimethoxysilane (APTMS) was purchased from Sinopharm Chemical Reagent Co., Ltd (Shanghai, China). Other chemicals and materials were purchased from Tianjin Damao Chemical Reagent Factory (Tianjin, China).

2.2. Apparatus
A 7890A-5975C GC–MS system (Agilent, USA) using a HP-5MS capillary column (30 m × 0.25 mm i.d. × 0.25 μm) was utilized for separation and analysis. The injection volume was 1 μL without splitting. The quantification ion (m/z) for chlorpyrifos was 314.

2.3. Preparation of MMIP
Magnetic mesoporous silica (Fe_3O_4@$mSiO_2$) was synthesized according to the literature [10]. In the preparation of MMIP by surface imprinting technology with a sol-gel process [7], Fe_3O_4@$mSiO_2$ was used as the carrier, chlorpyrifos as the template and APTMS as the functional monomer.

2.4. Sample analysis
Water samples were collected from the Peal River (Guangzhou, China) and the Center Lake (Higher Education Mega Center, Guangzhou, China), and stored at 4 °C until analysis. Vegetable samples were purchased from a market (Maoming, China). 2.00 g vegetable was extracted using 20 mL ethyl acetate.

20 mg of MMIP was added to sample solution. After shaking the mixture at 300 rpm for 20 min at room temperature, the MMIP were isolated from solution with a magnet. The chlorpyrifos adsorbed on the MMIP were eluted with 5 mL methanol/acetic acid (97:3, v/v) for 10 min. After drying with N_2, the chlorpyrifos was redissolved in 1.0 mL dichloromethane. Then, 1 μL of which was injected into the GC–MS system for analysis.

3. Results and discussion

3.1. Characterization of MMIP
The FT-IR spectrum of the prepared MMIP was recorded. In the spectrum, the peak at 576 cm^{-1} was from the Fe-O stretching vibration, the peak at 1093 cm^{-1} was due to the Si-O-Si stretching vibration. The peaks at 1261 cm^{-1} and 1720 cm^{-1} were corresponded to the C-O vibration and C=O stretching vibration, respectively. These results indicated the successful preparation of MMIP.

The surface morphology of the MMIP was provided by SEM. From the SEM observation (figure 1), it was found that the MMIP were almost regular in shape with an average diameter of around 80 nm.

Figure 1. SEM image of MMIP.

The magnetic property of the MMIP was measured using a vibrating sample magnetometer. It was observed that its greatest saturation magnetization was 18.7emu/g, and both its remanence and coercivity were near zero, indicating the MMIP was superparamagnetic.

IOP Conf. Series: Materials Science and Engineering **269** (2017) 012061 doi:10.1088/1757-899X/269/1/012061

3.2. Adsorption properties of chlorpyrifos on MMIP

The adsorption of chlorpyrifos on MMIP was examined using batch experiments and conducted in duplicate. The adsorption capacities varied with initial chlorpyrifos concentrations (figure 2). Langmuir and Freundlich isotherms were used to describe the adsorption behaviors. Langmuir isotherm mode fitted the experimental data better (R^2=0.9963) than Freundlich model (R^2=0.9118), suggesting the adsorption of chlorpyrifos onto the MMIP followed the Langmuir model. The maximum adsorption capacity was 16.8 mg/g at an initial chlorpyrifos concentration of 20 mg/L.

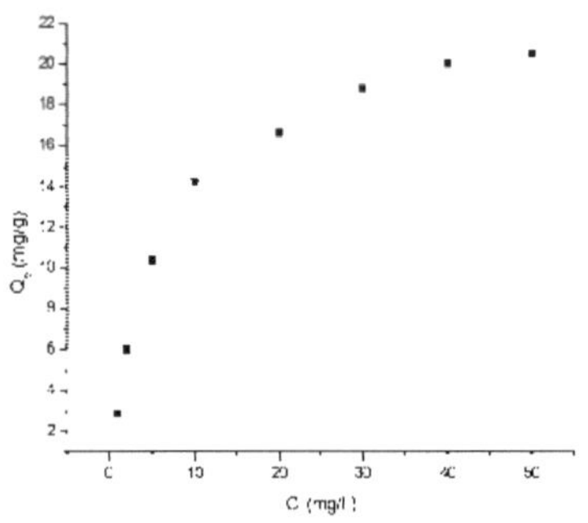

Figure 2. Chlorpyrifos adsorption on MMIP at different initial concentrations.

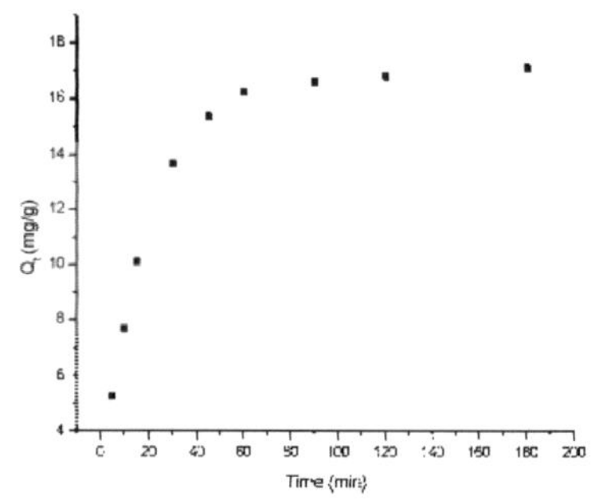

Figure 3. Effect of contact time on chlorpyrifos adsorption on MMIP.

The effect of contact time on chlorpyrifos adsorption with the MMIP was investigated using an initial concentration of 20 mg/ L. The results (figure 3) showed that the adsorption occurred rapidly within the first 0.5 h and then gradually slowed down until reaching equilibrium (1 h). The kinetics of chlorpyrifos adsorption was studied by using pseudo-first-order and pseudo-second-order kinetic models. It was found that the pseudo-second order kinetic equation could preferably describe the adsorption process.

The selectivity coefficient (α) and imprinting coefficient (β) were used to estimate the selectivity properties of MMIP and MNIP toward chlorpyrifos and structurally similar compounds omethoate and dichlorphos.

$$\alpha = K_d(\text{chlorpyrifos})/\ K_d(\text{M}) \tag{1}$$

$$\beta = \alpha_{\text{imprinted}} / \alpha_{\text{non-imprinted}} \tag{2}$$

where K_d is the distribution coefficient and M represents omethoate or dichlorphos.

As shown in table 1, the K_{MNIP} of chlorpyrifos for the MNIP was only slightly larger than those of omethoate and dichlorphos. However, the K_{MMIP} of chlorpyrifos for the MMIP was much higher than others, indicating that the template molecule chlorpyrifos had a higher affinity for the imprinted polymer than omethoate and dichlorphos.

Table 1. The selectivity parameters of MMIP and MNIP for chlorpyrifos.

Compounds	K_{MMIP} (L g^{-1})	K_{MNIP} (L g^{-1})	α_{MMIP}	α_{MNIP}	β
Chlorpyrifos	5.38	1.65	–	–	–
Omethoate	1.56	1.32	3.45	1.25	2.76
Dichlorphos	1.17	1.26	4.60	1.31	3.51

3.3. Determination of chlorpyrifos in water and vegetable samples

The MMIP was used as the sorbent for solid– phase extraction of chlorpyrifos from environmental water and vegetable samples. It was found that the proposed method had excellent linearity between the peak area and concentration over the range of 0.1–50 μg/ L for chlorpyrifos, with a correlation coefficient of 0.9925. The LOD calculated on the basis of signal to noise ratio of 3 was 30 ng/ L.

These water samples and vegetable extracting solution were analyzed, and the results were listed in table 2. Then, these samples were spiked with chlorpyrifos to assess the matrix effects and the results were also shown in table 2. It is obvious that the spiked recoveries of these samples were in the range of 89.6%–107.3%, with precision (RSD) of 1.9%–3.8%. Therefore, the developed method was sensitive, precise and rapid.

Table 2. Analytical results of chlorpyrifos in water and vegetable samples.

Sample	Added (ppb)	Found (ppb)	Recovery (%)	RSD (%)
River water	0	3.26	–	2.9
	5.0	8.50	107.3	2.4
Lake water	0	0.87	–	3.2
	1.0	1.92	105.7	1.9
Chinese cabbage	0	6.35	–	3.8
	10.0	15.69	89.6	2.7
Tomato	0	18.56	–	2.8
	20.0	36.05	94.83	2.4

4. Conclusions

Chlorpyrifos imprinted polymer on magnetic mesoporous silica was successfully synthesized and applied to adsorption of chlorpyrifos from solutions. The results suggested that the prepared MMIP could act as a selective SPE adsorbent for efficient separation and preconcentration of trace chlorpyrifos from environmental water and vegetable extract, and could be removed conveniently from sample solutions with the help of an external magnet. A new method for determination of trace chlorpyrifos in these samples was developed by using the MMIP based SPE coupled with gas chromatography–mass spectrometry. Results showed that the established method possessed good performance in terms of limit of detection, linearity, accuracy and reproducibility. Therefore, the synthesized MMIP can be a powerful alternative to common SPE adsorbents for chlorpyrifos, and the proposed method is a useful tool for routine analysis of chlorpyrifos due to its sensitivity, rapidity, selectivity and reliability.

5. References

[1] Lee W, Blair A, Hoppin J and Alavanza M 2004 *J. Natl. Cancer. Inst.* **96** 1781.
[2] Ma X, Huang B and Chen M 2007 *Rare Metal.* **26** 541.
[3] Zou J, Ma X, Dang Y and Chen Y 2014 *J. Anal. At. Spectrom.* **29** 1692.
[4] Chen J, Ma X and Wu Y 2014 *Food Anal. Methods* **7** 1083.
[5] Li D, Ma X, Wang R and Yu Y 2017 *Anal. Bioanal. Chem.* **409** 1165.
[6] Nong Y, Ma X., Fan S. and Yu Y 2014 *Anal. Methods* **6** 4124.
[7] Li C, Ma X, Zhang X, Wang R, Li X and Liu Q 2017 *J. Sep. Sci.* **40** 1621.
[8] Chen Y, Ma X, Huang M, Peng J, Li C 2016 *Anal. Methods* **8** 824.
[9] Li C, Ma X, Zhang X., Wang R, ChenY and Li Z 2016 *Anal. Bioanal. Chem.* **408** 7857.
[10] Zhao B, He M, Chen, B and Hu B 2015 *Spectrochim. Acta* B **107** 115.

Acknowledgements

This work was financially supported by the Science and Technology Planning Project of Guangdong Province, China (No. 2014A040401077), and the Major Projects (Natural Science) of Education Department of Guangdong Province, China (No.261555101).

IOP Conf. Series: Materials Science and Engineering **269** (2017) 012062 doi:10.1088/1757-899X/269/1/012062

On Heat Transfer Analysis for a Sphere of Combustible Material of Variable Thermal Conductivity

Ramoshweu Solomon Lebelo[1, a*], Radley Kebarapetse Mahlobo[1, b], Samuel Olumide Adesanya[1, 2, c], Mohana Sundaram Muthuvalu[3, d]

[1]Vaal University of Technology, Mathematics Department, Andries Potgieter Blvd, Vanderbijlpark, 1911, South Africa
[2]Redeemer's University, Mathematical Sciences Department, PMB 230 Ede, Osun State, Nigeria.
[3]Universiti Teknologi PETRONAS, Fundamental and Applied Sciences Department, 32610 Seri Iskandar, Perak Darul Ridzuan, Malaysia.

[a*]sollyl@vut.ac.za, [b]radley@vut.ac.za, [c]adesanyas@run.edu.ng, [d]mohana.muthuvalu@petronas.com.my

Abstract. This article considers an exothermic chemical reaction taking place in a stockpile of reactive material with a thermal conductivity that is temperature dependent. The study is modeled in a spherical domain whose carbon-containing material reacts spontaneously with the oxygen trapped within the system. The combustion process results with a complicated process that is nonlinear in nature, and the energy equation used to govern the problem is tackled numerically with the semi-implicit finite difference method (FDM). The results are depicted graphically and discussed to give a theoretical understanding of the heat transfer analysis during combustion.

1. Introduction

Exothermic chemical reaction in combustible materials has drawn a lot of research because of spontaneous reaction that takes place within a stockpile of such materials. Most of the veld fires that are hazardous to live species are caused by spontaneous combustion taking place within stockpiles of combustible materials such as coal, hay, wood, wool or any dumbed rubbish in the open space [1, 2]. Spontaneous ignition is due to the readiness of the carbon-containing material to react with the oxygen trapped within the material when the exothermic chemical reaction takes place [3]. The process of combustion gives heat as one of the products, and this generated heat raises the temperature of the system. Progressive generation of heat rate should not exceed its rate of release to the surrounding environment, because this situation leads to thermal runaway that breeds self-ignition [4-6]. It is known that heat transfer involves energy transport due to conduction, convection and radiation, but in this study, we assume convection as the only means of energy transport. The analysis of heat transfer in this case is done theoretically using mathematical approach that involves computer programing. This approach, requires thorough understanding of the thermo-physical parameters embedded on the partial differential equation governing the problem. The studies of heat transfer in a spherical domain were done in [7, 8] but in these cases, constant thermal conductivity was assumed. In this case, we consider thermal conductivity that is temperature depended. Variable thermal conductivity's influence on thermal ignition was studied in [9, 10], where it is clarified that the thermal conductivity varies with temperature, exponentially, as according the following equation

$$k = \mu e^{b(T-T_w)}, \quad (1)$$

where μ is the thermal conductivity of the material at the ambient temperature T_w, b is the parameter for the variation of the thermal conductivity, and T is the absolute temperature of the sphere.

The combustion process is simplified by assuming a one-step irreversible chemical kinetics expressed as [2, 8]

$$C_iH_j + \left(i + \frac{j}{4}\right)O_2 \rightarrow iCO_2 + \left(\frac{j}{2}\right)H_2O + \text{heat}. \quad (2)$$

In this analysis, it is assumed that heat generation within the combusting material follows Arrhenius law of kinetics and that this generated heat does not depend on the diffusion of the reactant, in other words, the reactant consumption is neglected. The results are graphically expressed and thoroughly discussed.

2. Mathematical modeling

The spherical domain of combustible material is assumed to undergo an nth order chemical kinetics. The convective heat loss to the environment obeys Newton's law of cooling, and it is described by $-\frac{h}{k}[T - T_w]$. h, in this case, represents the heat transfer coefficient and, k the variable thermal conductivity of the material described in equation 1. The negative sign indicates the heat loss from the sphere to the surrounding environment. Figure 1 below illustrates the geometry of the problem.

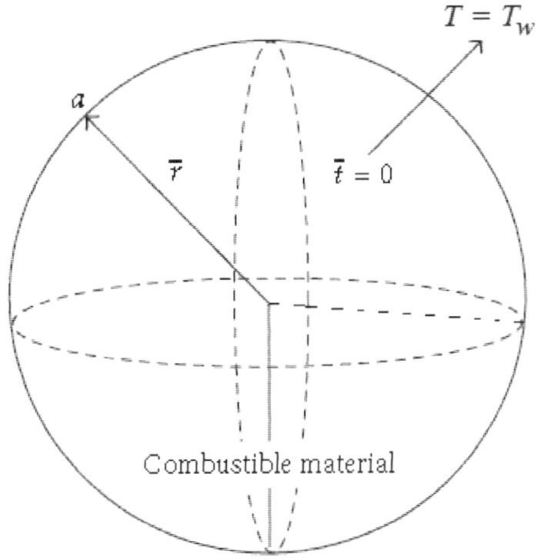

Figure 1. The geometry of the problem.

The following nonlinear partial differential equation in one dimension, governing the problem is given as [7, 8].

$$pc_p \frac{\partial T}{\partial \bar{t}} = k\left(\frac{\partial^2 T}{\partial \bar{r}^2} + \frac{2}{\bar{r}}\frac{\partial T}{\partial \bar{r}}\right) + QAC\left(\frac{KT}{vl}\right)^m e^{-E/RT} - \sigma(T - T_w). \quad (3)$$

Convective boundary conditions are as follows:

$$\frac{\partial T}{\partial \bar{r}}(0, \bar{t}) = 0, \quad (4)$$

$$\frac{\partial T}{\partial \bar{r}}(a, \bar{t}) = -\frac{h}{k}[T(a, \bar{t}) - T_w]. \quad (5)$$

The initial condition is

$$T(\bar{r}, 0) = T_0. \quad (6)$$

Here, T_0 is the initial temperature of the sphere, ρ is the density, c_p is the specific heat at constant pressure, Q is the heat of reaction, A is the rate constant and C is the reactant concentration. We have K as the Boltzmann's constant, v as the vibration frequency, l as the Planck's number, E as the activation energy and R as the universal gas number. Moreover, m is the numerical exponent and takes the following values, -2 for sensitized, 0 for Arrhenius and 0.5 for bimolecular kinetics. Lastly, $\bar{r}$ is the radial length of the sphere, and σ is the heat loss parameter. The following dimensionless parameters were introduced to equation 3-6,

$$\left.\begin{aligned} \theta &= \frac{E(T-T_w)}{RT_w^2}, \theta_0 = \frac{E(T_0-T_w)}{RT_w^2}, \; Bi = \frac{ah}{k}, \\ \beta &= \frac{\mathrm{b}RT_w^2}{E}, \; \alpha = \frac{\sigma a^2}{\mu}, \\ \delta &= \left(\frac{KT_\infty}{vl}\right)^m \frac{QAEa^2C}{\tau RT_\infty^2} exp\left(-\frac{E}{RT_\infty}\right), \\ r &= \frac{\bar{r}}{a}, \; \varepsilon = \frac{RT_w}{E}, \; t = \frac{\mu\bar{t}}{\rho c_p a^2}, \end{aligned}\right\} \tag{7}$$

to have the following dimensionless governing equation:

$$\frac{\partial\theta}{\partial t}e^{-\beta\theta} = \frac{\partial^2\theta}{\partial r^2} + \frac{2}{r}\frac{\partial\theta}{\partial r} + \beta\left(\frac{\partial\theta}{\partial r}\right)^2 + \delta(1+\varepsilon\theta)^m e^{\theta/(1+\varepsilon\theta)} e^{-\beta\theta} - \alpha\theta e^{-\beta\theta}. \tag{8}$$

The initial condition is

$$\theta(r,0) = \theta_0, \tag{9}$$

and boundary conditions are

$$\frac{\partial\theta}{\partial r}(0,t) = 0; \tag{10}$$

$$\frac{\partial\theta}{\partial r}(1,t) = -Bi\theta e^{-\beta\theta}. \tag{11}$$

Here, θ is the dimensionless temperature, θ_0 is the dimensionless initial temperature, δ is the Frank-Kamenetskii parameter also called the rate of reaction parameter, ε is the activation energy parameter, r is the dimensionless radial length of the sphere, and β represents the dimensionless variable thermal conductivity parameter. Bi is the thermal Biot number at the surface of the sphere, and α is the dimensionless heat loss parameter. The dimensionless heat transfer rate at the surface of the sphere is described by $Nu = -\frac{d\theta}{dy}$, called the Nusselt number (N).

3. Numerical approach

The application of the semi-implicit finite difference scheme involves taking implicit terms at intermediate time level ($N + \xi$) for $0 \le \xi \le 1$. For this study, $\xi = 1$ was used to enable use of larger time steps for convenience. The approximation of the partial differential equation that governs the problem, was done at the second-order central differences. The boundary conditions were obtained from the modification of the equations that corresponded to the first and the last grid points, and the finite differences for the discretization of the governing equation are taken from the linear Cartesian mesh and uniform grids. The semi-implicit FDM for the governing equation is ultimately expressed as:

$$\begin{aligned} \left(\frac{\theta^{N+1}-\theta^N}{\Delta t}\right)e^{-\beta\theta} &= \frac{\partial^2}{\partial r^2}\theta^{N+\xi} + \frac{2}{r}\frac{\partial}{\partial r}\theta^N + \beta\left(\frac{\partial}{\partial r}\theta^N\right)^2 - \alpha\left(\theta e^{-\beta\theta}\right)^N + \\ &\delta\left[(1+\varepsilon\theta)^m \Phi^n e^{\left(\frac{\theta}{1+\varepsilon\theta}\right)} e^{-\beta\theta}\right]^N \end{aligned} \tag{12}$$

The unknown new time ($N + 1$) terms are placed on the left, whilst the known terms at the present time (N) are placed on the right side of the equation. The following expression is attained,

$$
\begin{aligned}
-\xi\tau\theta_{j+1}^{N+1} + \left(e^{-\beta\theta} + 2\xi\tau\right)\theta_j^{N+1} - \xi\tau\theta_{j-1}^{N+1} &= -\tau(1-\xi)\theta_{j+1}^{N} - \alpha\Delta t\theta_{j+1}^{N} \\
+\left[e^{-\beta\theta} - 2\tau(1-\xi)\right]\theta_j^N - \frac{\Delta t\beta}{4}\left[\left(\theta_{j+1}^N - \theta_{j-1}^N\right)\right]^2 &+ \tau(1-\xi)\theta_{j-1}^N + \frac{1}{2r_j^2}\tau\left(\theta_{j+1}^N - \theta_{j-1}^N\right) \\
+ \delta\Delta t\left[(1+\varepsilon\theta)^m \Phi^n e^{\left(\frac{\theta}{1+\varepsilon\theta}\right)} e^{-\beta\theta}\right]^N
\end{aligned}
\qquad (13)
$$

where, j represents the position, N the time, Δr the mesh spacing and $\tau = \frac{\Delta t}{\Delta r^2}$ equation 13 forms a tri-diagonal system that was solved using MAPLE in this study. Graphical solutions to understand the temperature behavior during combustion are presented in the following section.

4. Results and discussion.

In this section, results are graphically presented and discussed. Unless or otherwise stated, the following parameters were applied

$$\beta = 1, \delta = 0.1, \alpha = 1, \varepsilon = 0.1, m = 0.5, Bi = 1, t = 10.$$

4.1. Steady state temperature profiles

The steady state is attained when the temperature of the system is no more dependent on the time. This is illustrated in figure 2 and figure 3, where at some time $t > 0$, the temperature of the system remains constant. It is observed that the temperature profiles are high at the center of the sphere, due to continued exothermic chemical reaction within the sphere. The exothermic chemical reaction produces heat as one of the products of combustion and this results with high profiles of temperature. The profiles of temperature drop toward the surface of the sphere because of progressive heat loss due to convection at the surface of the sphere.

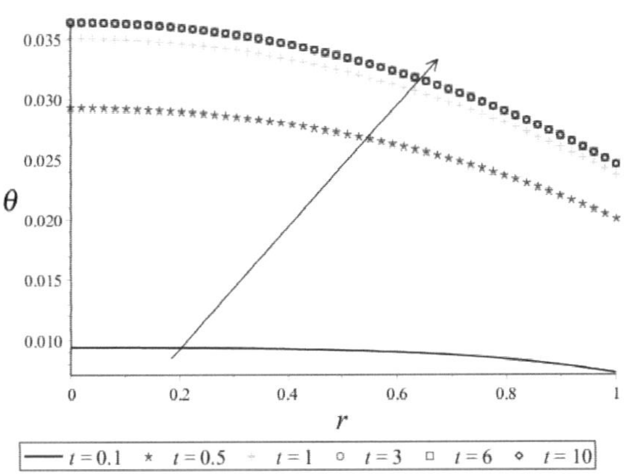

Figure 2. 2-D Steady state.

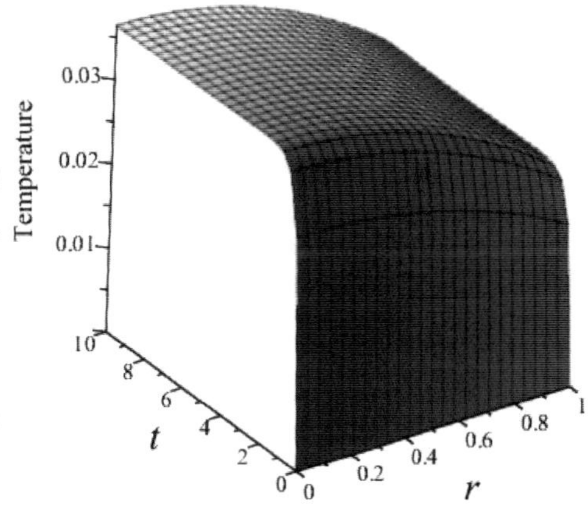

Figure 3. 3-D Steady state.

4.2. Unsteady state temperature profiles

In this subsection, we consider the influence of the variation of selected kinetic parameters on the heat transfer process, hence the temperature behavior of the system. The following parameters $\delta, \varepsilon, \beta, m, \alpha$ and Bi were each varied, while others kept constant, to investigate the heat transfer. The figure 4-9 show the behavior of the temperature due the influence of all the stated parameters, as they are respectively presented. We see from figure 4-7 that a combined increase in the parameters δ (reaction rate), ε (activation energy), β (variable thermal conductivity) and m (chemical kinetics index) gives an increase in the temperature profiles. The increase in δ accelerates the rate of the exothermic chemical reaction to generate more heat. The same is observed with an increase in ε, to indicate that at higher levels of the activation energy, the exothermic chemical reaction is accelerated. The increase in β confirms the dependence of the thermal conductivity on the temperature of the system. It is of interest to note that the exothermic chemical reaction is fastest, during bimolecular kinetics ($m = 0.5$), hence highest temperature profile, compared to Arrhenius kinetics ($m = 0$). The exothermic chemical reaction is very slow during sensitized kinetics ($m = -2$), as indicated by the lowest temperature profile.

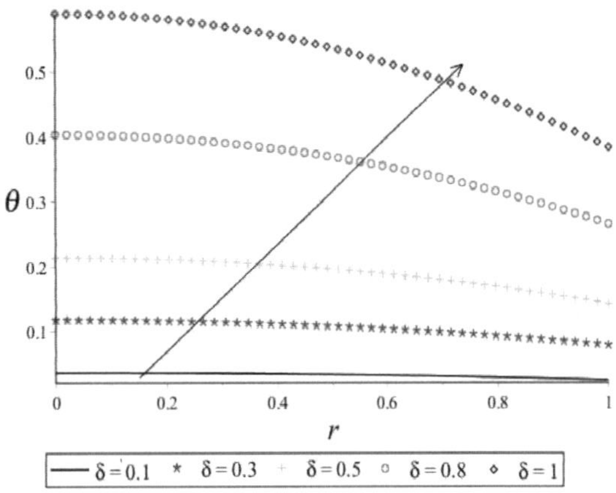

Figure 4. δ on temperature.

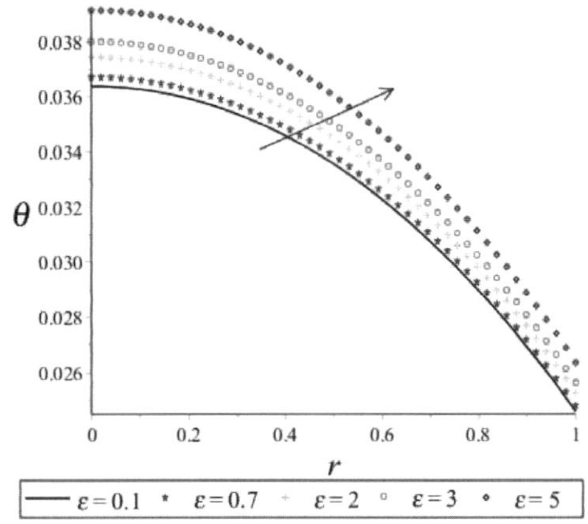

Figure 5. ε on temperature.

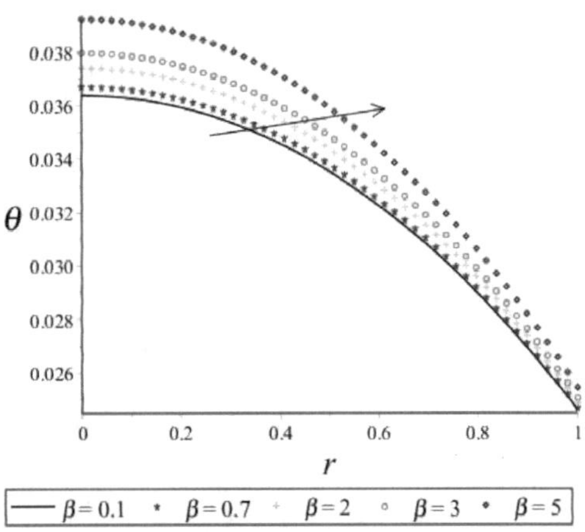

Figure 6. β on temperature.

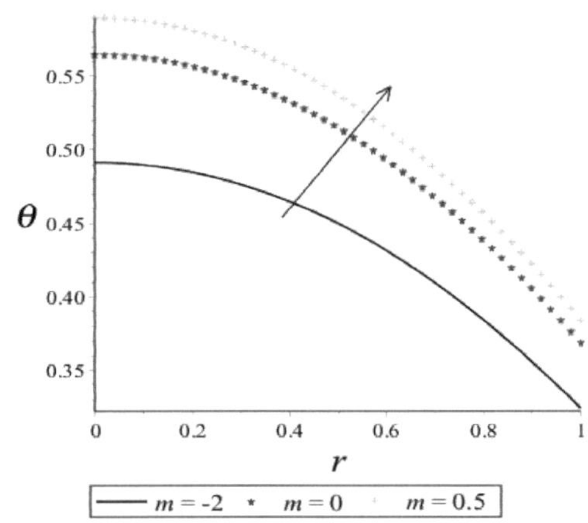

Figure 7. m on temperature.

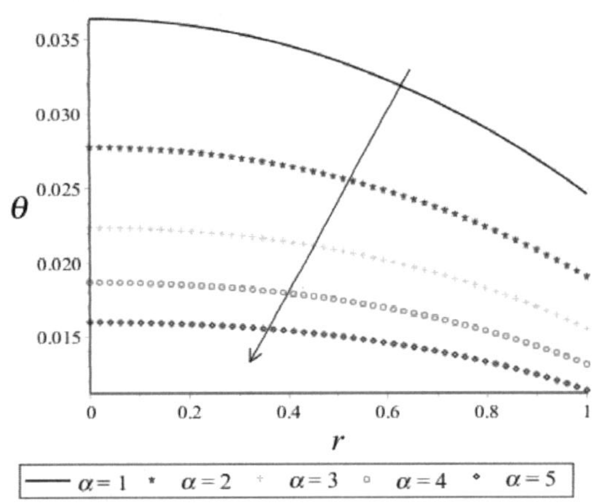

Figure 8. α on temperature.

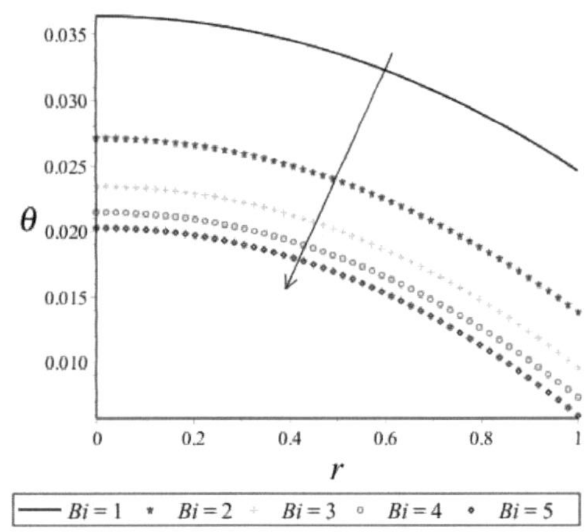

Figure 9. Bi on temperature.

Contrary to what has happened, we observe that a combined increment in the parameters α (heat loss) and Bi (temperature biot number) results with a decline in the temperature profiles, as illustrated in figure 8, 9. An increase in α means that the rate of heat loss at the surface of the sphere is increased, hence the temperature of the system's reduction. The same is observed with the increase in Bi, that there is a decline in the temperature of the system. This is due to increased rate of heat loss due to convection at the surface of the sphere. The last two parameters are good for discouraging the exothermic chemical reaction from taking place to control self-ignited fires.

5. Conclusion

The analysis of heat transfer in a combusting stockpile of reactive materials modeled in a spherical domain of variable thermal conductivity was presented in this article. It was discovered that some parameters encourage exothermic chemical reaction, whilst others discourage the reaction from taking place in stockpiles of combustible materials. The parameters $\delta, \varepsilon, \beta, \alpha$ and, m were found to favor the taking place of the exothermic chemical reaction. The occurrence of the exothermic chemical reaction releases more heat into the combusting sphere, and the accumulation of heat causes the raising up of the temperature gradient during combustion. It is better to keep these parameters very low to reduce self-ignition process in stockpiles of combustible materials. On the other hand, it was found out that the parameters α and Bi favor the reduction of the temperature during the combustion. These parameters should be kept high to avoid self-ignited fires from explosions. This study was done theoretically to make the understanding of the combustion process in a cheaper and a quicker way compared to the experimental one. The understanding of the imbedded parameters on heat transfer can help in the control of self-ignition process in the storage of reactive materials. This study can be extended to two-step exothermic chemical reaction processes.

6. References

[1] Makinde O D 2005 *Mech. Res. Commun.* **32** 191-5.
[2] Lebelo R S 2014 *AIP Conf. Proc.* **1621, 60** 60-8.
[3] Arisoy A, Beamish B B and Cetezen E 2006 *Turkish J. Eng. Env. Sci.* **30** 193–201.
[4] Chinyoka T and Makinde O D 2013 *Commun. Nonlin. Sci. Num. Sim.* **18** 2448–61.
[5] Bowes P C 1984 *Self-Heating: Evaluating and Controlling the Hazards*, Elsevier, Amsterdam.
[6] Balakrishnan E, Swift A and Wake G C 1996 *Math. Comp. Model.* **24** 1–10.
[7] Lebelo R S 2016 *J. Commun. Compt.* **13** 159-63.
[8] Lebelo R S, Makinde O D and Chinyoka T 2017 *Adv. Mech. Eng.* **9** 1–14.
[9] Lacey A A and Wake G C 1982 *J. Appl. Math.* **28** 23-39.
[10] Moise A and Pritchard H O1981 *C. R. Exp. S. Sci. Canada, M3J,* **1P3** 165-7.

IOP Conf. Series: Materials Science and Engineering **269** (2017) 012063 doi:10.1088/1757-899X/269/1/012063

Geological and Geochemical Characteristics of Skarn Type Lead-Zinc Deposit in Baoshan Block, Yunnan Province

Xue Yao[1] and Peng Wang[2]

[1]Faculty of Land Resource Engineering, Kunming University of Science and Technology, Kunming 650093, China
[2]Yunnan Yunling Highway Construction Group CO., LTD., Kunming 650093, China

Abstract. Baoshan block is an important Pb-Zn-Fe-Cu polymetallic ore-concentration area which is located in southern of the Sanjiang metallogenic belt in western Yunnan. The article is studying about the geological and geochemical characteristics of the skarn type lead-zinc deposit in Baoshan block. The skarn-type lead-zinc deposit Baoshan block is characterized by skarn and skarn marble, and the orebodies are layered, or bedded along the interlayer fault, which are significantly controlled by structure. The research about Stable isotope S, H and O indicates that the ore-forming fluids are mainly derived from magmatic water, partly mixed with parts of metamorphic water and atmospheric precipitation. The initial Sr isotopic Sr87/Sr86 ratio suggests that the ore-forming materials derived from deep concealed magmatic rock, age of Rb-Sr mineralization is similar to that of Yanshanian granite. In conclusion, the Yanshanian tectonic-magmatic-fluid coupling mineralization of Yanshan formation is the main reason for the skarn-type lead-zinc deposit in the Baoshan block.

1. Introduction

Baoshan block is located in the southern of the complex orogenic belt of Sanjiang (Nujiang, Lantsang and Jinshajiang) in Southwest China, belonging to the northern part of the Yunnan, Myanmar, Thailand, Malaysia block, which is an important ore-concentrated area of Pb, Zn, Fe, Cu, Au, Ag and Hg polymetallic, and a natural laboratory for the study of skarn-type Pb-Zn-Fe polymetallic deposits. In recent years, with the continuous investment in mineral exploration, Baoshan Hetaoping large-scale lead-zinc-copper mine, Jinchanghe large-scale iron-lead-zinc-copper mine, Zhenkang Luziyuan super-large lead-zinc-iron mine and Shuitoushan lead-zinc mine as well as over dozens of lead-zinc-polymetallic deposits have found in the study area, at the same time, Zhenkang Luziyuan is listed as one of the lead-zinc mine reconnaissance area in the national-level prospecting strategic breakthrough. Geologists have paid great attention to the prospecting breakthrough of this area, and a large number of scholars have studied the metallogenic geological background, regional tectonic evolution, geological and geochemical characteristics of ore deposits and metallogenic chronology, but the genesis of their mineral deposits still remained a controversy. In this paper, the genesis skarn-type Pb Zinc polymetallic deposit study combined with the geological characteristics of three Pb-Zn-polymetallic deposits in the Baoshan block, the typical Hetaoping, the Luziyuan and the Jinchanghe, and combined with the fluid geochemistry, stable isotope and metallogenic chronology studied the causes about the genesis of skarn type Pb Zn polymetallic deposits, in order to provide the basis for the establishment of regional metallogenic and ore-prospecting models.

2. Regional geological background

The eastern boundary of the Baoshan block is adjacent to Lancangjiang fault, the western eastern boundary is adjacent to Nujiang fault, and the northern boundary is lost due to the convergence of the Lantsang and Nujiang faults, besides, the southern part extends to Myanmar, forming the northern extension of the Yunnan Myanmar, Thailand and Malaysia terrain. Before the Early Permian Epoch, Baoshan block and Tengchong, Qiangtang, Lhasa blocks were attached to the northeast part of Gondwanaland, at the end of Sakamarian Stage of the early Permian Epoch, Baoshan and Tengchong blocks began to split and drift from the Gondwanaland to the north, resulting in the entering of meso-tethys; during the late Triassic epoch to the Early Cretaceous Epoch, Lhasa block drifted to the north, and meso-tethys entered into the closing stage, resulting in the collision between Tengchong block and Baoshan block to form the Gaoligong collision structure belt [1, 2]. The Baoshan block experienced the early Paleozoic and Mesozoic-Cenozoic granite intrusion activities: the Paleozoic granitoid intrusion activity is mainly represented by the Pinghe granite base in the south of the block, and the zircon U-Pb is aged to be 502-466 Ma [2-4]. The Mesozoic Indosinian granitoids are mainly Muchang A-type granites, which are complex massif that composed of soda ash syenite, sodicum granite and aegy granite. The age of cyrtolite and nepentinite U-Pb is 253-241Ma, with Sn, Zn, $\sum$Ce comprehensive mineralization, but the ore-forming process is still not clear [5]. The rock bodies formed during the Mesozoic-Cenozoic magmatic activity mainly include Zhibenshan, Kejie and Caojian acid granites, of which, the Zhibenshan and Kejie granites are mainly formed in the late Yanshanian magmatic activity. The granite is composed of biotite granite, two-mica granite and light-colored granite. The age of zircon U-Pb is126.7±1.6 Ma and 93±13 Ma, respectively, and the relevant mineralization include Pb, Zn, Fe, Cu that form polymetallic deposits, such as Baoshan Hetaoping and Zhenkang Luziyuan Pb-Zn-Fe(Cu); the rock bodies formed during the magmatic activity of the Himalayan period is mainly Caojian two-mica granite, and the age zircon U-Pb is 73 ± 0.19 Ma, which may be related to Yunlong Tiechang tin deposit [4, 6].

3. Geological and geochemical characteristics of typical deposits

3.1. Geological characteristics of the deposit

The geologic characteristics of the skarn-type lead-zinc deposits in the Baoshan block are very similar. The ore deposits in Hetaoping, Jinchanghe and Luziyuan are all produced in the Upper Cambrian strata. The ore bodies are produced in layers, strata, veins and stripes along the interlayer fracture zone. Skarn and skarn marble are the main ore-bearing lithology. Minerals mainly include sphalerite, galena, chalcopyrite and magnetite, which is semi-self-shaped and Hypidiomorphic granular structure that produced in the massive, disseminated and vein structure. The skarn minerals are mainly composed of (Manganese) garnet, (Manganese) pyroxene, diopside, ilvaite, epidote, (Manganese) actinolite, quartz and calcite that form a typical Ca-Mn skarn structure, which is mostly represented by the rose pyroxene of Luziyuan deposit, and is considered an important symbol of searching for skarn-type lead-zinc deposits in Baoshan block. The skarn-type Pb-Zn deposits in the study area are characterized by obvious mineralized zonation, and the deep to shallow parts are garnet diopside skarns → magnetite, hematite → manganese skarn-type magnetite → pomegranate garnet skarn → manganese skarn type lead-zinc ore → epidote and actinolite from skarn → actinolite lead-zinc ore → actinolite and chlorite marble Pb-Zn ore, of which, chlorite can be observed in the whole area. The alteration mineralization features are controlled by the degree of metasomatism of deep hydrothermal fluids, the carbonate rocks and tectonic activities.

3.2. Geochemical characteristics of the deposit

The contents of major elements CaO and Fe2O3 in skarn lead zinc deposits in Baoshan block are the highest, followed by MgO, Al2O3, MnO, K2O and Na2O. The omega (CaO) shows the characteristics of marble > skarn ore rocks > >, the other major elements MgO, Al2O3, MnO, K2O, Na2O, TiO2 and Pb, Zn content of ore-forming elements as marble skarn rocks <tectonic rocks < ore occurrence law. CaO and other major elements showed negative correlation, this characteristic shows that the alteration process is a process of silicon and iron, magnesium, aluminum, manganese, calcium and other elements to carry out, so that the content of CaO decreased gradually, formed a set of high manganese skarn composite rock mineral.

On the aspect of sulfur source, Ohmoto [9] studied the δ34S composition of some well-known deposits in the world, and concluded that there are four kinds of sulfur source in the hydrothermal deposit: (1) mantle sulfur, the δ34S value is close to 0, and its composition is close to sulfur isotope, and the variation range is relatively small; (2) from the ocean water and seawater sulfate, featured by enrichment of ^{34}S, and the value of $\delta^{34}S$ = about +5 ‰; (3) mixed sulfur, the source is relatively complex, $\delta^{34}S$ = +5 ‰~+15‰, and the variation range is relatively large; (4) biogenic sulfur, biological reduction is characterized by enrichment of ^{32}S, $\delta^{34}S$ is mostly negative value, and the variation range is large. As shown in table 1 and figure 1, the ^{34}S value of the Hetaoping deposit is 0.99 ~7.20, and the range of variation is large. The 34S value of the Luziyuan deposit is 9.23~10.17, and the variation range is small. Both of the ^{34}S values of the two deposits are in the range of mixed sulfur, which is consistent with magmatic sulfur, but it also mixed with metamorphic sulfur.

Table 1. Isotopic geochemical correlation of typical deposits.

Deposit	**Hetaoping lead-zinc- copper deposit**	**Luziyuan lead - zinc - iron deposit**
^{34}S isotope composition (‰)	0.99~7.20	9.23~10.17
H-O isotope composition (‰)	$\delta D_{V\text{-}SMOW}$: -108~-100, average104; δO^{18}_{H2O}: 6.1~7.6, average 6.7 [7]	$\delta D_{V\text{-}SMOW}$: -81.7~-73.3, average 75.4; δO^{18}_{H2O}: 5.4~6.1‰, average 5.7
Sr^{87}/Sr^{86}	0.7116~0.2720 [6]	0.714 486~0.714 512 [8]
Rb-Sr metallogenic age	116.1±3.1 [6]	141.9± 2.6 Ma [8]

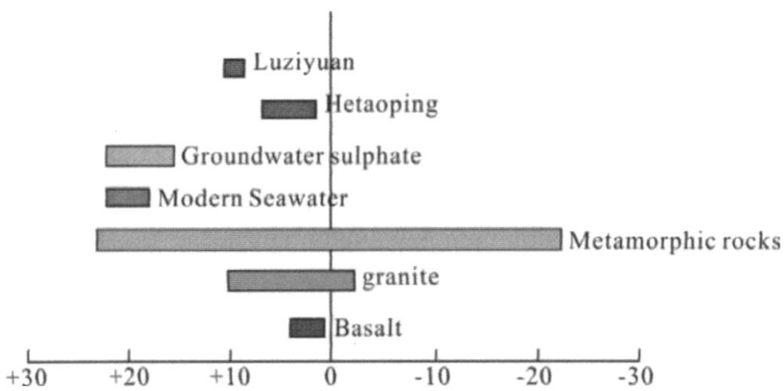

Figure 1. Sulfur isotope composition comparison of typical deposits and natural sulfur-containing materials.

The H and O isotopic composition in fluid inclusions of Augite at Luziyuan deposit: δ DV-SMOW is -81.7 ‰ ~ -73.3 ‰, $\delta^{18}O_{H2O}$ is 5.4 ‰ ~ 6.1 ‰, the variation ranges of $\delta^{18}O_{H2O}$ is narrow and δ DV-SMOW is relatively large; The H and O isotopic composition in fluid inclusions of marble at Hetaoping deposit: δ DV-SMOW ranges from -108‰ to -100 ‰, with wide variation range, and $\delta^{18}O_{H2O}$ ranges from 6.1‰ to 7.6‰, and the variation range is narrow (table 1). In the H-O diagram (figure 2), the throwing-dots at Luziyuan deposit are located in the lower left side of the original magmatic water, and close to the metamorphic water area; throwing-dots at Hetaoping are located on the magma underwater side, compared with Luziyuan deposit, $\delta^{18}O_{H2O}$ composition is close, but δ DV-SMOW value drifts downwards. It may be because the test minerals in the Luziyuan mining area are skarn early-altered mineral rose-pyroxene, which is not mixed with the atmospheric precipitation, and the test minerals in the Hetaoping mining area are mineralization of quartz, so the value of δ DV-SMOW is decreased due to the atmospheric precipitation and the construction water mixture. The composition of the stable isotopes S, H and O shows that the ore-forming fluids of the lead-zinc deposits on the Baoshan block are mainly derived from magmatic water and mixed with some metamorphic fluids and precipitation.

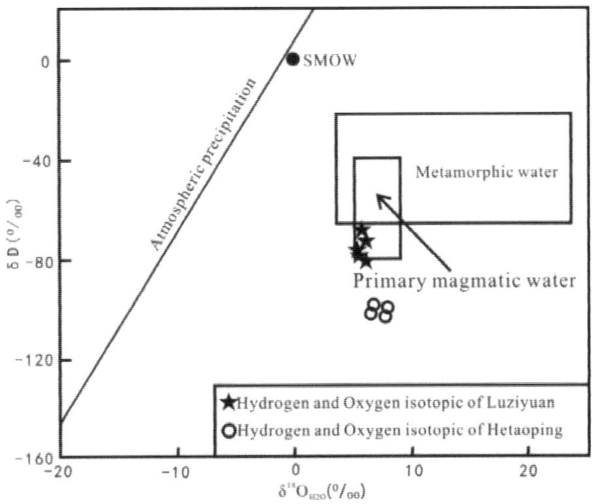

Figure 2. Hydrogen and Oxygen isotopic diagram of the Skarn Pb-Zn deposit.

3.3. Age of Rb-Sr mineralization

The skarn-type ore-bearing stratum of lead-zinc deposit in Baoshan block is Cambrian strata, probably formed before 500 Ma. The zircon U-Pb diagenesis age of the granite body exposed to the block is 126.7 ± 1.6 Ma, with the initial Sr isotopic Sr^{87} / Sr^{86} ratio to be 0.716810. The ore-forming age of Rb-Sr isochron of the metal sulfides in the Hetaoping deposit is 116.1 ± 3.9 Ma, with the initial Sr isotopic Sr87 / Sr86 ratio to be 0.711850. The ore-forming age of Rb-Sr isochron of the metal sulfides in the Luziyuan deposit is 141.9 ± 2.6 Ma, and the Sr isotopic Sr^{87}/Sr^{86} ratio is 0.714479. The above results indicate that the age of the skarn-type Pb-Zn deposits in the Baoshan block is close to that of the Yanshanqi Zhibenshan granite, which is very different from that of the Yanshanian granitoids. Therefore, it can be seen that the genesis of mineral deposit is related to the Yanshanian tectonic-magmatic rocks activities, belonging to sedimentary deposit. The initial Sr isotope ratio of the two deposits is similar to that of the Zhubenshan granite, suggesting that the ore-forming materials are mainly derived from deep buried magmatic rocks.

4. Conclusion

(1) The ore-bearing stratum of skarn-type lead-zinc ore deposits in Baoshan block belong to the upper Cambrian Hetaopinp and Shahechang groups. The ore-bearing rocks are mainly skarn and skarn marble with high manganese content. The ore minerals are sphalerite, galena, chalcopyrite, magnetite and pyrite, while gangue minerals are a set of manganese skarn construction consisting mainly of the garnet, diopside, rose pyroxene, epidote, chlorite, quartz and calcite.

(2) The ore bodies are stratified and appear to be layered along the interlayer faults, and the tectonic control is obvious. The mineralization shows the zoning characteristics of magnetite mineralization→ Pb-Zn-Cu mineralization →Pb-Zn mineralization from the deep to the shallow.

(3) Stable isotopes of S, H and O indicate that the ore-forming fluids are mainly derived from magmatic water, and some metamorphic water and meteoric water are mixed at the later stage.

(4) The initial Sr isotopic Sr^{87}/Sr^{86} ratio suggests that the ore-forming materials originate from deep concealed magmatic rocks. In conclusion, the Yanshanian tectonic-magmatic-fluid coupling mineralization is considered to be the main reason for the formation of the skarn-type Pb-Zn deposit in the Baoshan block based on the Rb-Sr mineralization age.

5. References

[1] Metcalfe I, 2011 *Gondwana Res.* **19(1)** 3-21.
[2] Deng J, Wang Q F, Li G J, Li C S and Wang C M 2014*Gondwana Res.* **26** 419-37.
[3] Wang X F, Metcalf I, Jian P, He l Q and Wang C S 2000 *J. Asian Earth Sci.* **18(6)** 675-90.
[4] Yu L, Li G J, Wang Q F and Liu X F 2014 *Acta Petrol. Sin.* **30(9)** 2709-24.

[5] Gu Y Q, Qan T H and Ye Z S 1998 *Acta Petrol. ET Miner.* **7(1)** 38-48.
[6] Tao Y, Hu R Z, Zhu F L, Ma Y S, Ye L and Cheng Z T 2010 *Acta Petrol. Sin.* **26(6)** 1760-72.
[7] Xue C D, Han R S, Yang H L, Yang Z M, Tian S H, Liu Y Q and Hao B W 2008 *Miner. Depos.* **27(2)** 243-52.
[8] Zhu F L, Tao Y, Hu R Z, Liao M Y, Wang Y X and Li Y B 2011 Bull. Mineral. Petrol. Geochem. **30(1)** 73-9.
[9] Ohmoto H and Rye R O 1979 *Isotopes of sulfur and carbon.* In Barnes H L, ed. Geochemistry of Hydrothermal Ore Deposits, New York: Wiley-Interscience, 509-67.

IOP Conf. Series: Materials Science and Engineering **269** (2017) 012064 doi:10.1088/1757-899X/269/1/012064

Adoption of Smart Structures for Prevention of Health Hazards in Buildings

Ayodeji Oke[1], Clinton Aigbavboa[1], Wiseman Ngema[1]

[1]Department of Construction Management and Quantity Surveying, University of Johannesburg, Doornfontein, 2028, South Africa

Abstract. The importance of building quality to the health and well-being of occupants and surrounding neighbors cannot be overemphasized. Smart structures were construed to proffer solution to various issues of sustainable development including social factors that is concerned with health and safety of people. Based on existing literature materials on building quality, smart structures and general aspect of sustainable developments, this study examined the benefits of smart structures in the prevention of various health issues in infrastructural buildings, which has been a concern for stakeholders in the architecture, engineering and construction industry. The criterion for indoor environmental quality was adopted and various health and bodily issues related to building quality were explained. The adoption of smart structure concept will help to manage physical, chemical, biological and psychological factors of building with a view to enhancing better quality of life of occupants.

1. Introduction

Chemicals of interest in the built environment [1] are volatile and semi-volatile organic compounds (VOC's), pesticides, and some chemicals produced during combustion such as carbon monoxide and nitrogen oxides. Health issues such as cancer, resulting from combustion occurring indoors have been a growing concern for researchers studying the phenomena of the sick building syndrome and indoor environmental quality. However, these studies have also included VOC'S (such as benzene, toluene, ethyl, benzene and xylene among others) and the health effects thereof, primarily because building occupants are continuously experiencing inexplicable discomfort in their homes. The reasons for this happenings need to be identified as being separated directly from the construction or physical attributes of the building [1-4].

A more chemical approach can be taken to understand illnesses such as cancer, asthma and the likes, as the chemical reactions occurring in and around the building could well be the contributing factor for their occurrence. However, these chemical reactions and the products thereof need to be taken into consideration whenever a survey of indoor environmental quality, which is an aspect of smart structure concept, is carried out. Leukaemia in children has also been attributed to the presence of plastic in the dust particles within the building that could have been obtained or transported in small fragments from the external environment, and the mixture attacks the children's weak immune system and causes chronic illness [1-3].

In the early 2000's, researchers embarked on studies comparing the relationship and connection between VOC's and children contracting leukemia as well as asthma contraction amongst building occupants. The study [5] showed that depending on the child's age, the likelihood to contract any chronic illness was dependent on the amount of VOC's the baby or toddler was exposed to. This contradicts a study performed in England [6] where it was found that the likelihood was not based on the exposure to VOC's excluding exposure to formaldehyde, but, rather the degree of dampness that the building had and the inhalation, or constant inhalation of the damp air in and around the building. There has been leading to higher illness amongst children in the UK as compared to teenagers and adults [1]. From the view of indoor environmental quality and based on existing literature materials, the purpose of this research is to discuss the implications of the adoption of Smart structure concept on health and safety of buildings occupants.

2. Health and bodily issues of building quality

This section discusses various health and safety risks associated with the quality of infrastructural buildings. They are explained based on their manifestations among the people living within or around a sick building.

2.1. Nasal manifestations

Nasal stuffiness: The sudden stuffiness of the nasal passage is a common occurrence in buildings with stuffy air but the sensation does not last as the stuffiness perishes soon after the person leaves the premise of the building and changes environments. However, for the duration of the stay in the building giving the sensation, the occupant will continue to experience the discomfort. Although, this stuffy sensation doesn't lead to itching, it is however linked to the temperature levels that prevail within the building [7, 8].

Nasal irritation and rhinorrhea: These nasal discomforts are not as common in buildings but are worth note as these are as a result of allergic conditions that the occupants have been experiencing, and are therefore not prevalent in the general population that enters a given space [8-10].

2.2. Ocular manifestations

The Ocular manifestation is related to the discomfort of the eyes. Sick building occupants experience an array of eye-burning sensations as a result of the breaking up tear film in the eyes but, fortunately these manifestations are prevalent on people who wear contact lenses rather than those who do not wear such [8].

2.3. Oropharyngeal manifestations

Although, examination of the throat does not reveal any inflammation, the most prevalent symptom of oropharyngeal infection is throat dryness. Studies also revealed that these can be resolved by the consumption of large quantities of water or any other liquid as the dampening of the throat will alleviate the pain experienced, and these manifestations generally have no serious aftermath [7, 9].

2.4. Cutaneous manifestations

Female occupants of sick buildings are privy to a certain phenomenon that causes dryness of the skin, which is generally contracted from the working environment over a period of time. These manifestations have been attributed to the heat prevalent in the building and how the building humidity affects air movements and subsequently transfers to the exposed skin of the building occupant, particularly the females [7-10].

2.5. Respiratory manifestations

Existing studies have shown that respiratory manifestations are not linked to asthma but, building related manifestations do cause the tightening of the chest and difficulties in breathing. However, like cutaneous manifestations, these can also be alleviated with a simple remedy, that is, stepping outdoors for a breath of fresh air [8, 9].

2.6. Constitutional diseases

Constitutional diseases include diseases such as eczema and sinusitis and these may be exacerbated in certain buildings. Sensations such as fatigue occur mostly during the second half of the working day, and cannot be alleviated by simply manipulating one's dietary intake at lunchtime. Previous studies have shown that constitutional diseases are often linked to high room temperatures [7, 8].

2.7. General manifestations

One of the most frequent experiences in buildings that create discomfort are headaches and excessive fatigue, which may be frequently experienced or on several days. They usually affect the frontal lobe of the brain in similar fashion to a migraine and are attributed mainly to the presence of artificial lighting and fluorescent lights in buildings. These can be alleviated by moving from the artificially lit room and or the wearing of sunglasses in order to prevent the sensation of these headaches which in effect do lead to fatigue [9, 10].

3. Benefits of smart structures

When considering the manner in which smart buildings can affect building occupants, it is essential to distinguish the factors that will be of significance to the occupants and how these factors can subsequently be grouped. Existing studies have classified the factors into four groups, which include physical factors, chemical factors, biological factors and psychological factors [8]. The focus of this study is on the physical factors since they have a more external effect on the building occupants of building and are subject to standards. These include aspects such as temperature, humidity, ventilation as well as other physical aspects of infrastructural buildings [5, 9, 11].

3.1. Temperature

In order for comfort to be ensured, the standard room temperature is to be set at a range between 20 and 26 and in order to compensate for the added heat on bodies due to clothing. The recommended temperature between the 6 degree range is 23 so as not to impair mental thinking ability and focus during the day as a result of added body heat [7].

High room temperatures lead to not only fatigue and reduced mental ability, but when overheated, the gaseous chemicals contained in materials begin to release and contaminate the air affecting the indoor air quality of the given building, and increases the likelihood of occupant dissatisfaction with the indoor environmental quality of the building and so too the perception of what constitutes the IEQ.

3.2. Relative humidity

Mold development due to rising temperatures causes discomfort through as a result of humidity level at varying times during day. As a result of this, many issues of discomfort especially in cold regions, the change in temperature reacts negatively with the building materials used to construct the house as these were used with the thought in mind that the conditions are a certain way and temperatures remain in a particular range. However, if these change, the building can react by means of cracking [8, 9, 11].

With the increase or decrease of the humidity, the experiences and reactions from both people and buildings are different, meaning that exposure to low humidity can result in fever and other ill responses by the body if it is used to more arid and humid temperatures, thus, if the humidity lowers or raises and remains that way for a while, the chemical reactions takes place, which may be injurious to the people and also affect the building negatively.

The reactions of the people and buildings will not surface until the predominating humidity level has adjusted and remained that way. Therefore, it can be deduced that while studying the concept of smart building, indoor environmental quality and or sick building syndrome, the researcher is obliged to be aware of the humidity situation of the area being surveyed. This will help to direct survey responses to either a lack of knowledge of humidity change and or the ignorance thereof, so the two will not be confused with the actual IEQ criteria being studied. Also, all responses from the former categories must be discarded as fruit of the poisonous tree.

3.3. Ventilation

Ventilation standards differ from areas with and without smokers but, a lack of ventilation is the main reason for the building occupants experiencing poor air quality as any and every contaminant that enters the building can create a stuffy atmosphere therein and the resulting discomfort [9, 11].

In non-smoker areas, a rate of ventilation of 8l/s/person would be sufficient to permit the transfer and removal of animal and human body odors from the room or building, therefore making for a pleasant atmosphere for all who are within it. Whereas, if the same rate of ventilation is applied in a room but the CO2 rate was 0.1%, a fifth of the people entering that given room would be displeased with the atmosphere meaning that the ventilation rate need be reduced as this will also reduce the amount of unwanted odors and air entering the room or building [7, 11]. In a smokers room, the contrary applies, as the building would be satisfying for any person who smokes or is used to smoking would find the place pleasant. If a person enters the building or room and finds it unpleasant, rather than reducing the ventilation rate, in this case it would be increased as a result of the presence of more unpleasant air [7]

Building ventilation contributes to the reduction of building contaminants' concentration caused by building materials and processes occurring within any given building [7, 8]. The best way to reduce building contaminants is through source control, and the understanding that ventilation alone is not prone enough to cause issues detrimental to building health such as draughts and odors [9]. Therefore, close attention must be observed when maintaining and using ventilation equipment so as to ensure a clean breeze, and not to impair the building beyond that which it is except is due to third-party contaminants other than ventilation problems.

3.4. Artificial light
The adoption of artificial light as compared to the conventional lighting that is used in homes, that is, fluorescent tube lighting, can result in the reduction of eye stress and discomfort. This is because the artificial light releases a lower light into the building thus, lower strain is imposed on the eyes and the prevalence of headaches as the conventional lighting creates a painful glare and strain which leads to unwanted health and challenges [8, 9].

3.5. Noise/Acoustic comfort
Excessive noise is attributed to fatigue and discomfort as the high noise levels create annoyance and frustration, thus affecting one's mental capabilities such as rational thinking, and can hamper one's quality of sleep [1, 9]. Although, dizziness and nausea are attributed to the prevalence of infrasound, it is not likely to affect the building occupants as these would be sounds that our ear cannot hear but, the noise created by a building consists of cracks and creeks and rumbling which creates the disturbances and unwanted noise due to outdoor contributions such as a loud neighborhood. A very good example is a railway track that produces loud vibrations [8].

People are sensitive to noise, and the building they reside has to provide them with a means of shelter from excessive noise in an attempt to create a peaceful indoor environment. Therefore, the amount of noise or the noise levels experienced within a house is fundamental aspect of the indoor environmental quality criteria that determines the satisfaction of occupants with a building.

3.6. Particles and fibers
Dust develops in buildings due to various reasons such as the degree of ventilation, the degree of hygiene or cleanliness within the home, the amount of traffic outdoors, the type of traffic (such as trucks), the proximity of vehicles from the building (such as how far the fence is from the roadway) as well as the degree of tobacco smoking in and around the building. These impact of these factors increases if there is a dirt road near the building as the dust particles from the outside will be transported into the building, creating chest tightness in time and wheezing as the dust particles stick to the chest [7, 9]. In dustier environment, the building occupant can inflict self-discomfort through contact with man-made fibers which can be contracted from the ceilings of buildings, this usually create eye discomfort and skin irritation through hand contact.

4. Conclusion
This study has been able to discuss various health hazards in buildings and the benefits of smart structure concept in addressing them. The study concluded that the adoption of smart structure concept will improve the indoor environment of infrastructural buildings, thereby enhancing the work performance of the building occupants. The research will also help construction professionals and other stakeholders in considering the provision of buildings that contributes to the effective

performance of building occupants in the work environment. This will help the architecture, engineering and construction sector of any economy to incorporate strategies that will result in improvement in the reduction of health and safety issues of infrastructural buildings.

5. References

[1] Mitchell C S, Zhang J, Sigsgaard T, Jantunen M, Lioy P J, Samson R and Karo M H 2007 *Current State of the Science: Health Effects and Indoor Environmental Quality Department of Environmental Health Sciences: Johns Hopkins Bloomberg School of Public Health Baltimore Maryland*: USA.

[2] Oie L, Hersoug L G and Madsen J O 1997 *Environ. Health. Perspect.* **105** 972–8.

[3] Fan Z, Lioy P, Weschler C, Fiedler N, Kipen H and Zhang J 2003 *Environ. Sci. Technol.* **37** 1811-21.

[4] Bornehag C G, Lundgren B, Weschler C J, Sigsgaard T, Hagerhed-Engman L and Sundell J 2005 *Environ. Health. Perspect.* 2005 113:1399–1404

[5] Weschler C J, 2004 *Indoor. Air.* **14(7)** 184–94.

[6] Venn A J, Cooper M, Antoniak M, Laughlin C, Britton J and Lewis S A 2003 *Thorax.* **58** 955–60.

[7] European Concerted Action 1989*Indoor Air Quality & its impact on man: Cost Project 613 Commission of the European Communities Directorate General for Science Research and Development Joint Research Centre*: Institute for the Environment.

[8] Jansz J 2011 *Theories and Knowledge About Sick Building Syndrome Department of Health & Safety Environmental Health*: Curtin University Perth WA 6845.

[9] Ibem E O 2012 *Nigeria Int. J. Qual. Reliabil. Manag.* **29(9)** 1000-18.

[10] National Heart 2016 *Lung and Blood Institute Hypersensitivity pneumonitis Available from: https://wwwnhlbinihgov/health/health-topics/topics/hp 2016*

[11] Smith A and Pitt M 2001 *Facilit.* **29(3/4)** 169–87.

IOP Conf. Series: Materials Science and Engineering **269** (2017) 012065 doi:10.1088/1757-899X/269/1/012065

The Topology Optimization Design Research for Aluminum Inner Panel of Automobile Engine Hood

Minhao Li[1, a*], Dongqing Hu[1], Xiangzheng Liu[1], Huanquan Yuan[1]

[1]GAC Automotive Engineering Institute, China-Mainland

Abstract. This article discusses the topology optimization methods for automobile engine hood design. The aluminum inner panel of engine hood and mucilage glue regions are set as design areas, and the static performances of engine hood included modal frequency, lateral stiffness, torsional stiffness and indentation stiffness are set as the optimization objectives. The topology optimization results about different objective functions are contrasted for analysis. And based on the reasonable topology optimization result, a suited automobile engine hood designs are raised to further study. Finally, an automobile engine hood that good at all of static performances is designed, and a favorable topology optimization method is put forward for discussion.

1. Introduction

The automobile engine hood is the important part for the car, and the static performances for automobile engine hood are crucial for the safety performance about the car [1]. Finite element analysis is an effective helper method for automobile engine hood design [2, 3]. The most important static performances that influence the design of the hood are: modal frequency, lateral stiffness, torsional stiffness and indentation stiffness [4]. Modal frequency is an important factor affecting the NVH performance; the performance of lateral stiffness and torsional stiffness can be evaluated as generated displacement when engine hood under certain external loads, and they are important factors to evaluate engine hood operation performance; indentation stiffness means the resistance capacity to deformation by unexpected forces, and it is related to pedestrian protection, suitable indentation stiffness can balance resistance capacity to deformation with pedestrian protection [5, 6].

With the development of Computer-Aided Manufacturing (CAM), topology optimization is widely used in modern manufacturing technology. Topology optimization is a conceptual design method, which aimed to find the optimal stiffener distribution of the structure, increasing the stiffness of the structure, without increasing the structure weight. And topology optimization can be used to find the best stiffener distribution of the engine hood inner panel [7, 8].

In recent years, energy saving has become the major research area in the field of automotive engineering. Weight reduction can reduce the energy consumption, because eliminate the influence of aerodynamic resistance, the energy required for moving a vehicle is mostly proportional to its weight [9, 10]. Using aluminum in engine hood could reduce vehicle weight, and it is of great influence to improve vehicle fuel economy under the premise of vehicle driving performance [11, 12].

In this article, the optimal stiffener distribution of different objective parameters is analyzed, and based on this, an automobile engine hood that good at all of static performances is designed. And the material of automobile engine hood inner panel is replaced by aluminum for reduce vehicle weight.

2. Optimization modal

The optimizing finite element modal of engine hood is shown in figure 1. The finite element modal is set up to topology optimization under help of Hypermesh. The outer and inner panels of engine hood are connected by mucilage glue. Hinge reinforcing plate and lock reinforcing plate are connected to inner panel by rigid element. The material of inner panel is aluminum, and the others are steel. In order to obtain a reasonable stiffener distribution of inner panel, outer panel, hinge reinforcing plate and lock reinforcing plate are keep unchanged, the voids between inner panel and outer panel are full with mucilage glue, inner panel and mucilage glue regions are set to design areas. Taking into account the manufacturing process requirement, the topology optimization considers are the symmetry restraint of structure.

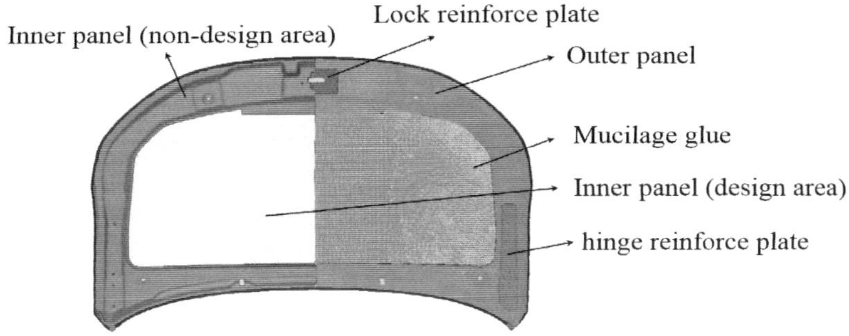

Figure 1. The optimization finite element modal of engine hood.

3. Topology optimization

Topography optimization is a conceptual design method, which aimed to find the optimal stiffener distribution of engine hood, increasing the stiffness of the structure, without increasing the structure weight. The optimization analyzing constraints and loads set regions are shown in figure 2. As shown in figure 2, when modal frequency, torsional stiffness and indentation stiffness are set to the optimization objectives, the freedom of lock catch area and hinge area are completely constrained. And when lateral stiffness is set to the optimization objective, the freedom of hinge area is completely constrained, the freedom of lock catch area is released. When torsional stiffness is set as the optimization objective, a load of 100N is constrained in Z direction at the yellow circle region of engine hood. When indentation stiffness is set as the optimization objective, 5 loads of 20N are set constrained in Z direction at the blue circle region of engine hood. And when lateral stiffness is set as the optimization objective, a load of 100N is constrained in Y direction at the lock catch area.

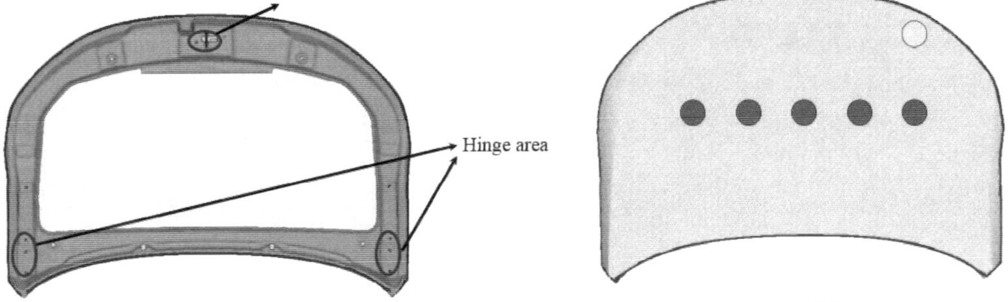

(a) (b)

Figure 2. The constraints and loads set regions.

When modal frequency is set as the optimization objective, the maximum volume fractions about inner panel and mucilage glue region are set as optimizing constrains, the maximum 1st torsional mode is taken as the optimization goal. When lateral stiffness, torsional stiffness and indentation stiffness are set to the optimization objectives, the maximum volume fractions about inner panel and mucilage glue region are set as optimizing constrains, the minimum compliance is taken as the optimization goal.

Just as shown in figure 3. (a. Element density distribution contours and b. Inner panel stiffener distribution as the value of element density sets upper than 0.5), when modal frequency is set to the optimization objective, the topology optimization results of inner panel is distributed as double herringbone. It means that this arrangement is beneficial to raise the frequency of 1st torsional mode.

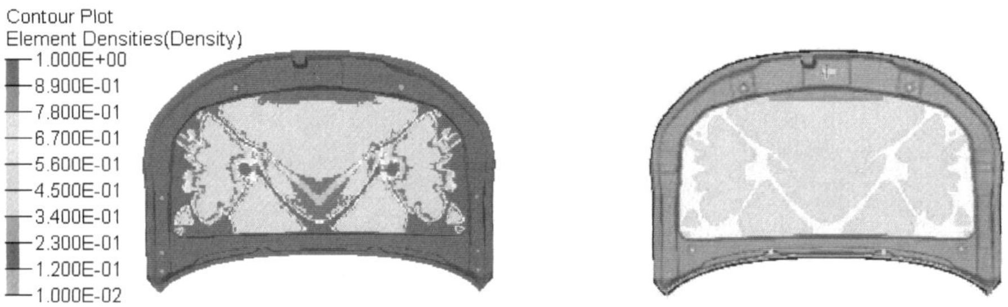

(a) (b)

Figure 3. The topology optimization result of inner panel as 1st torsional mode is taken as the optimization goal: (a) Element density distribution contours. (b) Inner panel stiffener distribution when the value of element density sets upper than 0.5.

As shown in figure 4. (a. Element density distribution contours and b. Inner panel stiffener distribution as element density sets upper than 0.5), when lateral stiffness is set to the optimization objective, the topology optimization results of inner panel are distributed as double forks. And this arrangement is similar with the results that shown in figure 3. This means that improving the torsional stiffness of engine hood is helpful to raise the frequency of 1st torsional mode.

(a) (b)

Figure 4. The topology optimization result of inner panel as lateral stiffness is taken as the optimization goal: (a) Element density distribution contours. (b) Inner panel stiffener distribution when the value of element density sets upper than 0.5.

As shown in figure 5. (a. Element density distribution contours and b. Inner panel stiffener distribution as element density sets upper than 0.5), when torsional stiffness is set as the optimization objective, the topology optimization results of inner panel is distributed as single herringbone. It shows that these two stiffeners are very important to the torsional stiffness of engine hood.

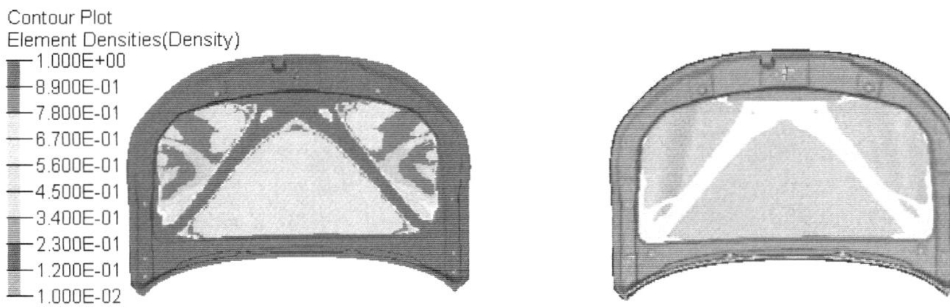

(a) (b)

Figure 5. The topology optimization result of inner panel as torsional stiffness is taken as the optimization goal: (a) Element density distribution contours. (b) Inner panel stiffener distribution when the value of element density sets upper than 0.5.

As shown in figure 6. (a. Element density distribution contours and b. Inner panel stiffener distribution as element density sets upper than 0.5), when indentation stiffness is set as the optimization objective, the topology optimization results of inner panel is distributed as a stiffener located in the middle of the panel.

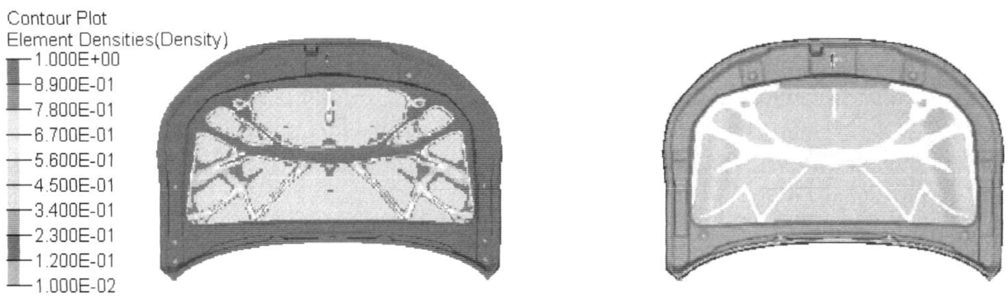

(a) (b)

Figure 6. The topology optimization result of inner panel as indentation stiffness is taken as the optimization goal: (a) Element density distribution contours. (b) Inner panel stiffener distribution when the value of element density sets upper than 0.5.

Muti-objective optimization analysis is proposed to obtain the most suitable optimal stiffener distribution of engine hood. Muti-objective function equation is enumerated based on compromise programming approach. The equation can be expressed as:

$$\min F(\rho)=\{w_f^2(\frac{f_{\max}-f(\rho)}{f_{\max}-f_{\min}})^2+w_l^2[\frac{U_l(\rho)-U_l^{\min}}{U_l^{\max}-U_l^{\min}}]^2+w_t^2[\frac{U_t(\rho)-U_t^{\min}}{U_t^{\max}-U_t^{\min}}]^2+w_i^2[\frac{U_i(\rho)-U_i^{\min}}{U_i^{\max}-U_i^{\min}}]^2\}^{\frac{1}{2}} \quad (1)$$

where, $F(\rho)$ is the muti-objective function; w_f is the weight of frequency objective function; w_l is the weight of lateral stiffness objective function; w_t is the weight of torsional stiffness objective function; w_i is the weight of indentation stiffness objective function; $f_{\max}$ is the optimal frequency; $f_{\min}$ is the original frequency; $f(\rho)$ is the frequency objective function; $U_l^{\max}$ is the strain energy of optimal lateral stiffness; $U_l^{\min}$ is the strain energy of original lateral stiffness; $U_l(\rho)$ is the strain energy of lateral stiffness objective function; $U_t^{\max}$ is the strain energy of optimal torsional stiffness; $U_t^{\min}$ is the strain energy of original torsional stiffness; $U_t(\rho)$ is the strain energy of torsional stiffness objective function; $U_i^{\max}$ is the strain energy of optimal indentation stiffness that limited by pedestrian protection; $U_i^{\min}$ is the strain energy of original indentation stiffness; $U_i(\rho)$ is the strain energy of indentation stiffness objective function.

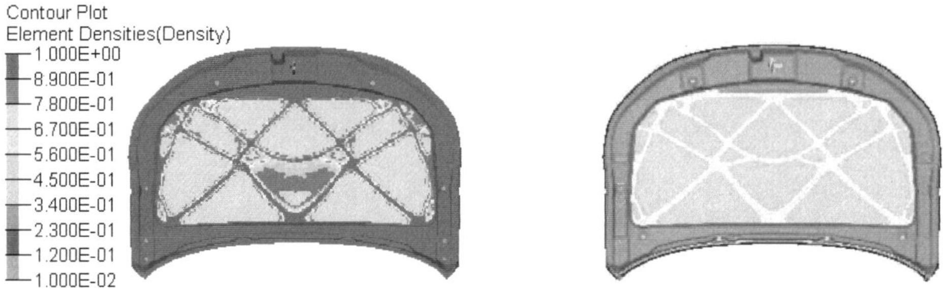

(a) (b)

Figure 7. The topology optimization result of inner panel as $F(\rho)$ is taken as the optimization goal: (a) Element density distribution contours. (b) Inner panel stiffener distribution when the value of element density sets upper than 0.5.

As shown in figure 7. (a. Element density distribution contours and b. Inner panel stiffener distribution as element density sets upper than 0.5), when $F(\rho)$ is set to the optimization objective, the topology optimization results of inner panel is distributed as double forks, and a stiffener across the middle of the panel.

4. Discussion

An automobile engine hood is designed based on the result of muti-topology optimization. More specifically, the result of muti-topology optimization can be simplified as 9 stiffeners, which can be marked as A, B, C, D, E, F, G, H, M, as shown in Figure 8. The stiffeners, which marked as A, B, C, D, are bigger than the other stiffeners. For the stiffeners, which marked as E, F, are smaller than the stiffeners that marked as G, H.

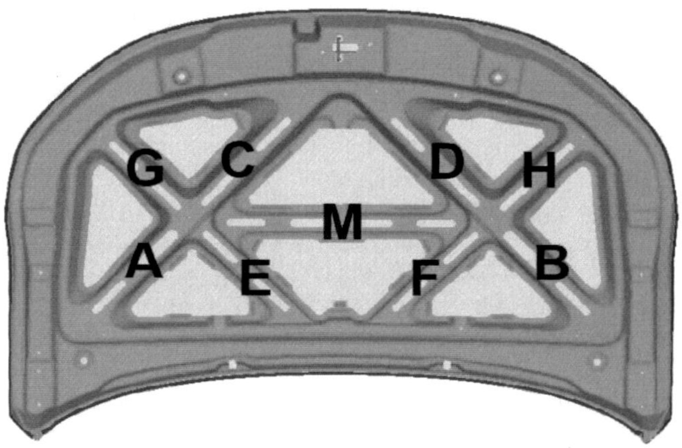

Figure 8. The designed automobile engine hood.

From the upward optimization analysis, it can be deduced that the stiffeners marked as A, B, E, F, G, H are helpful to raise the 1st torsional modal, the stiffeners marked as A, B, C, D, E, F, G, H are helpful to raise the lateral stiffness, the stiffeners marked as A, B, C, D are helpful to raise the torsional stiffness, and the stiffener marked as M is helpful to raise the indentation stiffness singly.

Table 1. Static performances of designed engine hood.

	Steel inner panel	Aluminum inner panel	Δ(Percentage difference)
1st torsional modal (HZ)	33.4	30.6	8.4%
Lateral stiffness (mm)	1.70	1.85	8.8%
Torsional stiffness (mm)	2.11	2.27	7.6%
Indentation stiffness (mm)	4.46	4.28	4.0%
Weight (kg)	22.3	17.8	20.2%

Analyzing static performances of designed engine hood, the results show that the designed engine hood is excellent for both stiffness performances. And just as shown in table 1, replacing the material of inner plate from steel to aluminum, and adding the thickness of inner panel from 0.7mm to 0.9mm, the frequency of 1st torsional modal only reduced 8.4%, the lateral stiffness only reduced 8.8%, the torsional stiffness only reduced 7.6%, and the indentation stiffness only reduced 7.6%, and all of that static stiffness performances are greater than the target values. It means that, replacing the material of inner plate form steel to aluminum can achieve engine hood weight reduction of 20.2%, and static stiffness performances keeping well.

5. Conclusions

This study explored topology optimization method for automotive engine hood. Stiffener distributions of different optimization objectives are examined in this work. In this article, it is first proposed that mucilage glue region should be considered as design region, when taking topology optimization analysis for automobile engine hood. An automobile engine hood, which is excellent for both stiffness performances, is designed based on the results of topology optimization. Replacing the material of inner plate from steel to aluminum, and comparing the difference of these two engine hoods, they have similar stiffness performances, and it shows that replacing the material of inner plate form steel to aluminum can achieve engine hood weight reduction of 20.2%. It is believed that this work has guiding significance for the future research of topology optimization method and the design of automotive engine hood.

6. Reference

[1] Ishitobi H 2010 Automobile hood: US, US7810877.
[2] Darwish S M, Elseufy and Ahmad A 2013 *Finite Element Analysis of an Automobile Engine Hood,* Iems 2013 International Conference on Industry, Engineering, and Management Systems.
[3] Darwish S M, Elseufy S M and Ahmad A 2014 *J. Manag. Eng.* **7(1)** 62.
[4] Shojaeefard M H, Najibi A and Ahmadabadi M R 2014 *Thin-Walled Struct.* **77(4)** 77-85.
[5] Zhou J, Wang F and Wan X 2015 *Mater. Today Proc.* **2(10)** 5029-36.
[6] Zhang J, Shen G Z, Du Y, et al. 2013 *Appl. Mech. Mater.* **281** 364-9.
[7] Krishnamoorthy R 2012 *Math. Geosci.* **44(6)** 651-72.
[8] Teng T L, Liang C C, Shih C J, et al. 2013 *Int. J. Vehic. Des.* **63(2/3)** 185-98.
[9] Mathaudhu S N, Luo A A, Neelameggham N R, et al. 2016 *A Lightweight Automobile Body Concept Featuring Ultra-Large, Thin-Wall Structural Magnesium Castings, Essential Readings in Magnesium Technology.* Springer International Publishing, 55-63.
[10] Li M H, Li Z R and Jiang L 2016 *Preliminary Study of Automobile Lightweight Technology,* International Conference on Mechanical Materials and Manufacturing Engineering.
[11] Kong C, Lee H and Park H 2016 *Polym. J.* **91(2)** 18-26.
[12] Li Y and Liu H 2010 *Material and Structural Optimization for Engine Hood Inner Panel of Car Body Aimed to Lightweight,* Wase International Conference on Information Engineering. IEEE Computer Society, 291-4.

Robust Optimization Design for Turbine Blade-Tip Radial Running Clearance using Hierarchically Response Surface Method

Chen Zhiying[1, a], **Zhou Ping**[1, b*]

[1]School of Energy and Power Engineering, Beihang University, Beijing 100191, China

Abstract. Considering the robust optimization computational precision and efficiency for complex mechanical assembly relationship like turbine blade-tip radial running clearance, a hierarchically response surface robust optimization algorithm is proposed. The distribute collaborative response surface method is used to generate assembly system level approximation model of overall parameters and blade-tip clearance, and then a set samples of design parameters and objective response mean and/or standard deviation is generated by using system approximation model and design of experiment method. Finally, a new response surface approximation model is constructed by using those samples, and this approximation model is used for robust optimization process. The analyses results demonstrate the proposed method can dramatic reduce the computational cost and ensure the computational precision. The presented research offers an effective way for the robust optimization design of turbine blade-tip radial running clearance.

1. Introduction

Turbine blade-tip clearance continues to be a concern in the design of gas turbines, because of a reasonable implemented active tip clearance control benefits to increase efficiency, reduce specific fuel consumption, and improve the performance and security of aeroengine [1-3]. The experiments and numerical simulation analysis of BTRRC have been studied [4-6]. In those researches, the effects of dynamic characteristics of mechanisms on turbine blade-tip clearance are detailed analyzed and experimented. Kypuros and Melcher [4] developed a simplified model from first principles that grossly captures the dynamic change in turbine tip clearance during take-off. Much of those work identified the influence of dynamic characteristics to BTRRC by experiments and deterministic analysis which predict initially the radial deformations of disk, blade and casing and then to estimate the BTRRC. To some extent, these efforts possess great blindness in BTRRC design due to scarcely consider the randomness of various impact factors [7]. Considering the influence of parameters uncertainty, Fei and Bai [8] presented distributed collaborative probabilistic design method for BTRRC. Compare with the deterministic analysis method, the probabilistic design method is more reasonable for design turbine blade-tip clearance. However, it is difficult for the probabilistic design to decrease the impacts of parameters variables to blade-tip clearance under operating conditions, so it is necessary to implement the BTRRC robust optimization design so that to decrease the degree of influence of parameters variation to BTRRC.

For complex mechanical assembly structure directly FE (Finite Element) method based robust optimization process has to be confronted with unaffordable efficiency resulting from too large computational load, so several surrogate based robust optimization methods such as Kriging method, polynomial response surface and artificial neural network have been studied [9-11]. However, the robust design of BTRRC is an inherently multi-component multi-discipline (MOMD) nonlinear dynamic design problem, which is involving multiple objects (turbine disk, blade and casing shroud), multiple disciplines (heat, centrifugal force, mechanical vibration, etc.). Therefore, it is difficult to directly construct a system approximation function model of parameters and blade-tip clearance. Bai and Fei [8, 12] presented a distributed collaborative response surface method (DCRSM) based on quadratic polynomial function for complex mechanical dynamic assembly reliability design, and the result reveals DCRSM has high efficiency and high precision in solving the complex structure assembly relationship probability analysis. The robust optimization objective is minimizing mean and standard deviation of objective response value [13], it means that robust optimization iteration process has to implement thousands of sampling so that to obtain mean and/or standard deviation of objective response. So traditional directly surrogate model based robust optimization process is very time-consuming.

In this paper a hierarchically response surface robust optimization algorithm is proposed for BTRRC robust optimization design. The theory idea and mathematic model of hierarchically response surface method is described in section 2. And then the robust optimization design of HPT BTRRC is to be accomplished. The efforts offer an effective way for the robust design of mechanical dynamic assembly and enrich mechanical robust optimization theory and method.

2. The mathematic model of hierarchically response surface method

In this section, the basic theory of hierarchically response surface robust optimization algorithm has been introduced briefly and the mathematical model of presented method is established.

2.1. Hierarchically response surface robust optimization algorithm.

The distributed collaborative response surface method is divided complex assemblage model into several sub-components model, and then the appropriate response surface method is selected to generate the approximation function model of each sub-component. Finally, the system response surface approximation model is constructed according to the relationship of each sub-model. The advantages of distributed collaborative response surface method are relative to the overall system “big” model, the number of parameters and the nonlinearity between input parameters and output response is greatly reduced of sub-model, so the accuracy can dramatic increase of each sub-structure response surface approximation model.

The key ideal of proposed method is using distributed collaborative response surface method to construct the System assemblage response surface model. A set of new samples of design parameters and robust objective response value is generated by using System response surface model and design of experiment (DOE) method, and then a new response surface model is constructed by using those samples. The basic flowchart of proposed method is shown in figure 1.

In figure 1, the $y(z_i, p_i)(i = 1,2,\cdots,n)$ is sub-model response surface approximation function, z_i and p_i are design parameters and random parameters vector, respectively. $\mu(z)$ and $\sigma(z)$ are approximation function of design variables and objective response mean and standard deviation.

The advantages of proposed method are summarized as follows:

(1) Compare with finite element method the distributed collaborative response surface method can dramatic decrease the computation cost to generate the samples of design parameters and robust optimization objective value (mean and standard deviation).

(2) The design parameters response surface model is applied to robust optimization process can convert indeterminacy optimization to c deterministic optimization that can save the optimization times.

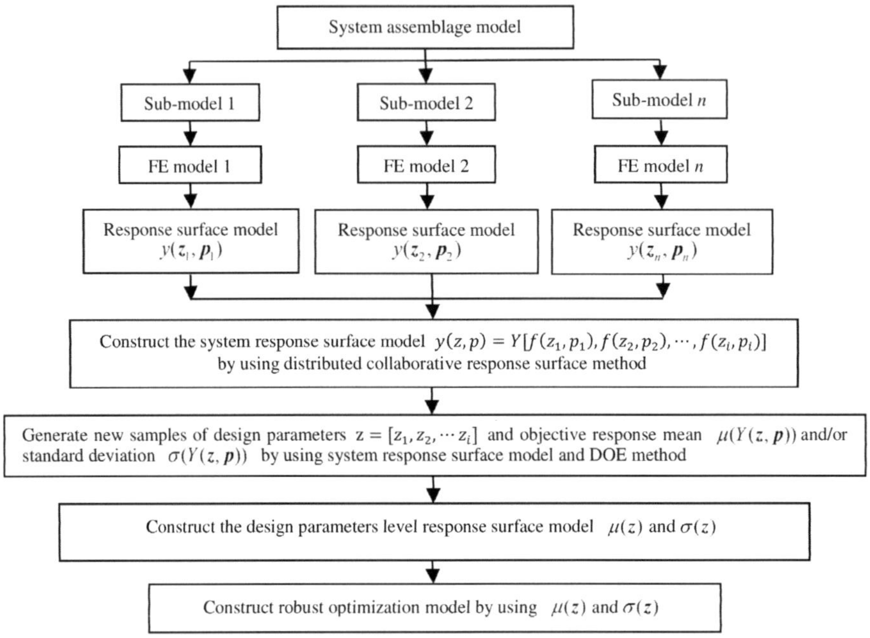

Figure 1. The basic flowchart of proposed method.

2.2. The mathematic model of presented method.

For many complex mechanical assembly structures like blade tip clearance, they are implicit and highly nonlinear function relations of input parameters and blade tip clearance. Compared with several flexible approximations model, the Kriging method is considered to be a reliable surrogate method, because that provides an accurate approximation model of a highly nonlinear function [14, 15]. In this work, the Kriging model is selected to construct the distributed collaborative response surface model.

Assuming that a mechanical assemblage comprises n ($n \in Z$) components and the each component constructs a sub-model via Kriging method [16].

$$\hat{y}_i(z_i,p_i) = f_i^T(z_i,p_i)\hat{\beta}_i + r_i^T(z_i,p_i)R_i^{-1}(Y_i - F_i\hat{\beta}_i) \quad (1)$$

where, $\hat{y}$ is output response, $z_i = [z_i^1, z_i^2, \cdots, z_i^j]\ j = 1,2,\cdots m$ and $\boldsymbol{p}_i = [p_i^1, p_i^2, \cdots, p_i^k]\ k = 1,2,\cdots q$ are design variables and random variables of each component, respectively.

Let a set of n known input samples $X=[x_1, x_2 \cdots x_n]$ with $x_i \in \Omega$ and output responses $Y=[y_1, y_2 \cdots y_n]$. The R is the correlation matrix of known samples, $r(x)$ is the vector of correlations between the sample $x_i=(i=1, \cdots, m)$ and sample x:

$$r(x)=[R(\theta, x_1, x), \cdots, R(\theta, x_m, x)]^T \quad (2)$$

The Gaussian function provides a relatively smooth and infinitely differentiable surface and defined with only one parameter θ, [17, 18]. So the Gaussian function is selected for this work and the mathematics as follows:

$$R(\theta, x_i, x) = e^{-\theta|xi-x|^2} \quad where\ \theta > 0 \quad (3)$$

The least-squares estimate of $\hat{\beta}$ is:

$$\hat{\beta} = (F^T R^{-1} F)^{-1} F^T R^{-1} Y \quad (4)$$

The vector F is generated by estimating $f(x)$ at the n known observations:

$$F = [f(x_1), f(x_2), \cdots, f(x_n)]^T \quad (5)$$

where, $f(x)$ is the known approximation function. In general $f(x)$ is assume a constant [19, 20], so we employ a constant term for $f(x)$ in this paper.

According to each component assembly relationship, the assemblage approximation mathematic model is constructed by using Kriging based distributed collaborative response surface model [8, 12]. The mathematic model as shown in equation (6).

$$\hat{y}(z,p) = f^T(z,p)\hat{\beta} + r^T(z,p)R^{-1}(Y - F\hat{\beta}) \tag{6}$$

where, $z = [z_1, z_2, \cdots z_i]$ and $\boldsymbol{P} = [p_1, p_2, \cdots p_i](i = 1,2,\cdots,n)$ are design variables vector and random variables vector of mechanical assemblage system.

The robust optimization objective is trying to choose properly levels of the design parameters to ensure structural performance insensitivity to change of the random parameters. It means that the robust optimization process just select a set of design parameters to minimize the mean value μ and standard deviation σ of objective response.

So the design parameters z design space is divided into k_i levels and the design of experiment method is used to generate a set of sample design point of design parameters. According to the number of parameters and the number of levels of each design parameter, full combination, uniform design or orthogonal array can be selected as the design of experiment method. And then, each experiment the design parameter vector $z_i(i = 1,\cdots,n)$ and random parameters vector $\boldsymbol{p}$ generate a set of sample points that as shown in table 1. The each experiment sample point substituting into equation (6) and be implemented thousands sampling by using Monte Carlo simulation. Finally, the objective response mean μ_i and standard deviation σ_i of each experiment are calculated by using mathematical statistics method.

Table 1. The sample point of design parameters and random parameters.

Run number	Design parameters	Random parameters	Mean	Standard deviation
1	z_1	$\boldsymbol{p}$	μ_1	σ_1
2	z_2	$\boldsymbol{p}$	μ_2	σ_2
⋮	⋮	⋮	⋮	⋮
n	z_n	$\boldsymbol{p}$	μ_n	σ_n

Once the samples of design parameters and objective response mean μ and standard deviation σ is generated. The approximation function model $\mu(z)$ and $\sigma(z)$ are constructed by select proper response surface approximation model and the robust optimization can be formulated as an optimization problem as follows:

$$\begin{array}{ll} \textit{Find} & z \\ \textit{Minimizing} & \alpha\mu(z)+(1-\alpha)\,\sigma(z) \\ \textit{Subject to} & z_L<z<z_U \end{array} \tag{7}$$

where, z is the design parameters vector, The α is weighting factor. z_L and z_U are the lower-limit and upper-limit of the control parameters. $\mu(z)$ and $\sigma(z)$ are the mean value and standard deviation of the objective response, respectively.

3. Robust optimization for BTRRC

3.1. The FE model and parameters study.

The robust optimization design of BTRRC involves three basic components – the casing shroud, turbine disk, and turbine blade [4]. The FE models of three components were established for their radial deformations analysis as shown in figure 2. The casing shroud is simplified as the bushing ring, and the axial cross section of bushing ring is only analyzed to build FE model shown in figure 2(a). The turbine blade model is shown in figure 2(b). The turbine disk structure is modeled shown in figure

2(c), disk mortises are assumed to be negligible and the loads and constraint conditions on disk are axisymmetric.

The input variables of blade and disk are consists of rotor speed ω, gas temperature t and corresponding material parameters are selected [8, 12]. These parameters are assumed to be mutually independent and obey a Gaussian normal distribution. These random variables mean and standard deviation are shown in table 2.

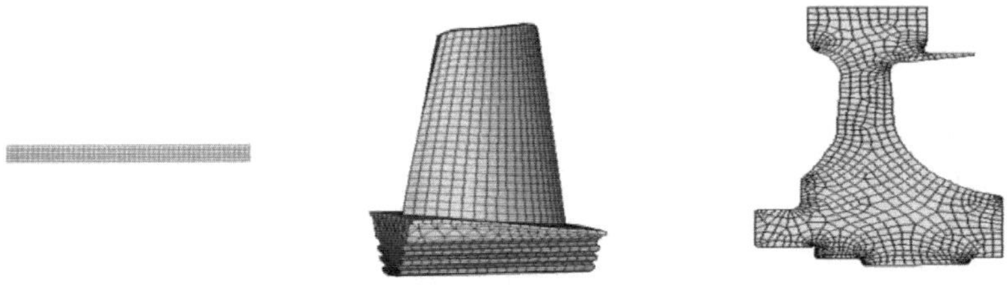

a b c

Figure 2. Finite element model of three components:
(a) Casing shroud, (b) Turbine blade, (c) Turbine disk.

Table 2. The variables mean and standard deviation of turbine blade and disk.

Turbine blade			Turbine disk		
variables	Mean	Standard deviation	variables	Mean	Standard deviation
ω/(rad/s)	1168	15	ω/(rad/s)	1168	15
t/(°C)	1050	12	t/(°C)	1050	12
E_b/(Gpa)	179	5	E_d/(Gpa)	184	6.4
K_b/(10^{-5}/°C)	1.549	0.0871	K_d/(10^{-5}/°C)	1.507	0.0871
λ_b/(w/(m·°C)	26.8	0.834	λ_b/(w/(m·°C)	20.8	0.834
α_b/(w/(m^2·°C)	11756	352.7	α_b/(w/(m^2·°C)	1500	45
ρ_b/(kg/m^3)	8500	125	ρ_b/(kg/m^3)	8200	123

In table 2, the E and ρ are Modulus of Elasticity and material density; α, K and λ are surface coefficients of heat transfer, Expansion Coefficients and Thermal Conductivity, respectively. The subscripts b and d are describe the turbine blade and turbine disk.

In this example the casing shroud is hoop-like structure, so the shroud radial deformation due to thermal stresses [4] that due to the temperatures differentials of outer and inner surfaces of the casing shroud. The selected random parameters of casing shroud are shown in table 3 [8, 12].

In table 3, the h is shroud thickness, t_1 is outside temperature of shroud, the α_c and α_c^* are inside and outside surface coefficients of heat transfer of casing shroud, respectively. The subscript c denoted the casing shroud.

Table 3. The variables mean and standard deviation of casing shroud.

variables	Mean	Standard deviation
$h/(mm)$	3	0.01
t/(°C)	1050	12
t_1/(°C)	320	6
K_c/(10^{-5}/°C)	1.8	0.39
α_c/(w/(m·°C)	6000	180
$\alpha_c^*/(w/(m^2\cdot°C)$	2600	78
λ_b/(w/(m·°C)	26.4	0.0834

3.2. Robust optimization model construct of BTRRC.

The mathematic of BTRRC is described:

$$\delta = \delta^{*} - Y \tag{8}$$

where, δ is the BTRRC, δ^{*} is initial assembly blade-tip clearance, respectively. In this example the initial assembly blade-tip clearance is 2.5(mm). Y is the radial deformations of blade-tip clearance, and that can be calculated by [4]:

$$Y = Y_1 + Y_2 - Y_3 \tag{9}$$

where, $Y_i (i = 1,2,3)$ are the radial deformations of disk, blade and casing, respectively, and those can be obtained using equation (1):

$$Y_i = \hat{y}_i(z_i, p_i) = f_i^T(z_i, p_i)\hat{\beta}_i + r_i^T(z_i, p_i)R_i^{-1}(Y_i - F_i\hat{\beta}_i) \qquad i = 1 \cdots 3 \tag{10}$$

According to proposed method the radial deformations Y of blade-tip clearance can be obtained by equation (6):

$$Y = \hat{\boldsymbol{y}}(z, \boldsymbol{p}) = [1,1,-1][\hat{y}_1(z_1, \boldsymbol{p}_1), \hat{y}_2(z_2, \boldsymbol{p}_2), \hat{y}_3(z_3, \boldsymbol{p}_3)]^T \tag{11}$$

The BTRRC is given by substituting equation (11) into equation (8):

$$\delta = \delta^{*} - \hat{y}(z,p) = [1,-1,-1,1][\delta^{*}, \hat{y}_1(z_1,p_1)\hat{y}_2(z_2,p_2)\hat{y}_3(z_3,p_3)]^T \tag{12}$$

According to the statistical characteristics of random variables in table 2 and table 3, the Latin Hypercube sampling (LHS) method is used to generate 150 input samples and corresponding output response is calculated using FE method. The Kriging approximation models are constructed by three components using 130 samples, and the others 20 samples are used to validate the Kriging model estimate accuracy. The result of Kriging model predict is shown in figure 3, and the predict value root mean square deviation (RMSE) of Kriging model can be seen in table 4.

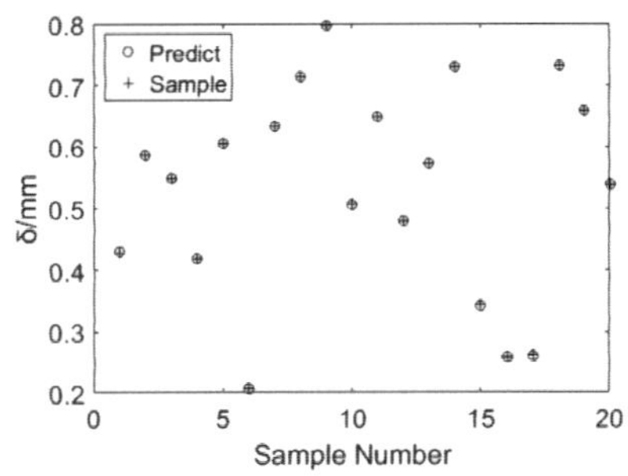

Figure 3. The validate samples and Kriging based DCRSM predict value of blade tip clearance.

Table 4. The RMSE of Kriging based DCRSM estimate value of blade tip clearance.

Method	Validate samples number	RSME
Kriging based DCRSM	20	6.5206×10-4

According to the characteristic of all parameters and the result of sensitivity analysis [8], the rotor speed ω, gas temperature t, shroud thickness h and outer temperature t_1 of shroud are selected as the design parameters, and those design parameters work point design interval is presented in table 5. The low-limit and upper-limit of table 5 are design parameters work point optimization interval.

Table 5. Design parameters optimization interval.

Design parameter	Low-limit	Upper limit
ω/(rad/s)	1100	1200
t/(°C)	1000	1100
h/(mm)	2.6	3.4
t_1/(°C)	300	350

The LHS method is selected to generate 120 design parameters samples in optimization space, and the equation (12) to implement 10^5 times sampling by using Monte Carlo method in each experiment run so that to calculate the corresponding mean value μ and standard deviation σ. Considering the parameters number and robust optimization iterative computing time, the polynomial response surface model is selected to fitting the approximation function model. The fitting result as shown in equation (13) and equation (14).

$$\begin{aligned}\mu=&3.6654+6.5478\times10^{-5}\omega-0.0029t-0.0105h-5.3401\times10^{-4}t_1\\&-2.1536\times10^{-7}\omega t-1.2911\times10^{-5}\omega h+5.6580\times10^{-7}\omega t_1+1.3244\times10^{-5}th\\&+9.9553\times10^{-8}tt_1-1.1673\times10^{-6}ht_1-5.7033\times10^{-7}\omega^2+7.3360\times10^{-7}t^2\\&+0.0022h^2+2.9914\times10^{-7}t_1^2\end{aligned} \tag{13}$$

$$\begin{aligned}\sigma=&0.1352-2.4763\times10^{-5}\omega-4.9894\times10^{-6}t-0.0149h-3.6864\times10^{-4}t_1\\&-1.3180\times10^{-8}\omega t-5.5909\times10^{-6}\omega h+2.3628\times10^{-7}\omega t_1+7.4843\times10^{-6}th\\&+5.6335\times10^{-8}tt_1+4.8232\times10^{-6}ht_1-2.1406\times10^{-8}\omega^2+2.8055\times10^{-8}t^2\\&-1.4935\times10^{-4}h^2+4.7324\times10^{-8}t_1^2\end{aligned} \tag{14}$$

The 10 samples are selected to validate estimate precision of polynomial response surface approximation model as shown in figure 4. The figure 4 (a) is mean μ predict value and the figure 4 (b) is standard deviation σ predict value. The RSME of estimated mean value μ and standard deviation σ are shown in table 6.

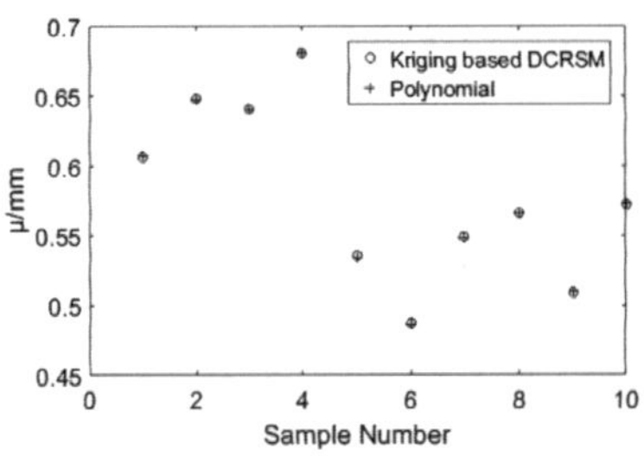

(a)

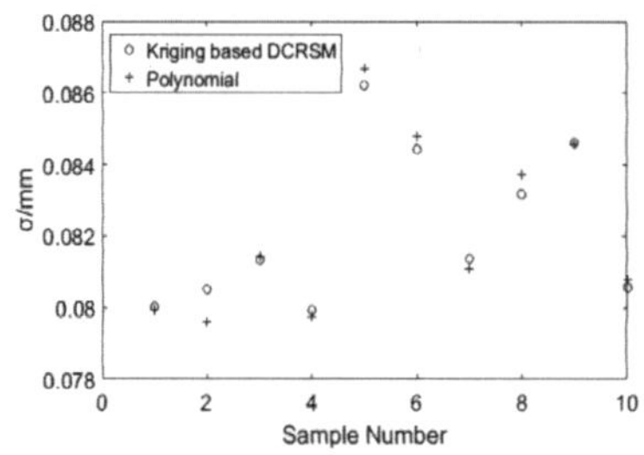

(b)

Figure 4. The Kriging based DCRSM and polynomial predict result: (a) Mean μ (b) Standard deviation σ.

Table 6. The RMSE of polynomial estimate value.

Objective	Validate samples number	RSME
μ/(mm)	10	7.4852×10^{-4}
σ/(mm)	10	4.0975×10^{-4}

The robust optimization of BTRRC can be formulated by substituting equation (13) and equation (14) into equation (7):

$$\begin{aligned}&\textit{Find} && z=[\omega,t,h,t_1]\\&\textit{Minimizing} && \alpha\mu(z)+(1-\alpha)\,\sigma(z)\\&\textit{Subject to} && z_L<z<z_U\end{aligned} \tag{15}$$

where, z is the design parameters vector that shown in table 5. The weighting factor α is 0.2. z_L and z_U

are the lower-limit and upper-limit of the design parameters that describes in Table 5. μ and α are the mean value and standard deviation of the objective response, respectively.

3.3. Result and discussion.
In order to discuss the effectiveness and precision of the presented method, the Kriging based DCRSM robust optimization algorithm is applied, and the numerical results are compared with the presented method. The Fruit Fly Optimization Algorithm is selected to simulation the robust optimization process in this article.

The optimization results of is shown in table 7. The "Error" means relative error of the proposed method estimated solution for the solution by directly using Kriging based DCRSM robust optimization method.

Table 7. Robust optimization result of Kriging based DCRSM and proposed method.

Methods	ω/(rad/s)	t/(°C)	h/(mm)	t_1/(°C)	μ/(mm)	σ/(mm)	Simulation times(s)
Kriging based DCRSM	1190.5	1043	3.1081	308	0.5211	0.0814	1785
Proposed method	1199.5	1035	3.105	305	0.5236	0.0813	2.89
Error	0.76%	0.77%	0.1%	0.97%	0.48%	0.12%	

As demonstrated by table 7, the result of optimized reveals that maximum relative error is 0.97%. This result illustrates the proposed method has higher calculate accuracy.

From table 7, the simulation time (2.89 s) of proposed method is far less than that (1785 s) of Kriging based DCRSM because of the polynomial can directly calculate the mean value μ and standard deviation σ not need implement 5000 times sampling to calculate the mean and standard deviation of objective response in each optimum iterative procedure. The results demonstrate that the proposed hierarchically response surface method is able to further improve computational efficiency for complex mechanical dynamic assembly relation.

4. Conclusion
This article presented a hierarchically response surface method based robust optimization algorithm for BTRRC robust optimization design. The Kriging based DCRSM is used to construct the system approximation model of overall parameters and blade tip clearance, and then a set of samples of design parameters and robust optimization objective (mean and /or standard deviation) is generated by using this system surrogate model and design of experiment method. Finally, the polynomial response surface is selected to construct the approximation model of design parameters and robust optimization objective (mean and /or standard deviation) by using those samples, and polynomial approximation model is used for robust optimization process.

By comparing the results with Kriging based DCRSM, the precision and calculation cost were discussed. Through the presentation of the numerical result, we can conclude that the presented method has strong advantages with high calculating efficiency and accuracy in complex mechanical assembly relation robust optimization. It means that the proposed method is suitable for industrial applications.

This paper provides a novel robust optimization idea and approach for the design of reasonable mechanical assembly relationship, as well as develops the theory of mechanical assembly relation robust optimization design.

Reference
[1] Nho Y C, Park J S, Yong J L, et al. 2012 *Int. J. Heat & Fluid Flow* **33(1)** 92-100.
[2] Rahman M H, Kim S I and Hassan I 2013 *Int. J. Comput. Meth. Eng. Sci. Mech.* **14(4)** 290-303.
[3] Nielsen A E, Moll C W and Staudacher S 2005 *J. Eng. Gas Turb. Pow.* **127(3)** 564-72.
[4] Kypuros J A and Melcher K J 2003 *A reduced model for prediction of thermal and rotational effects on turbine tip clearance,* NASA/TM. (2003–212226).

[5] Cao S Z, Duan F J and Zhang Y G 2006 *J. Phys.: Conf. Seri.* **48** 873-7.
[6] Jackson A J, Wheeler A P S and Ainsworth R W 2015 *Int. J. Heat Fluid Flow,* **56** 335-43.
[7] Hu D, Wang R and Tao Z 2011 *Aircraft Eng. Aerosp. Tech.* **83(4)** 199-207.
[8] Fei C W and Bai G C 2014 *Mech. Syst. Sign. Proc.* **49(1)** 196-208.
[9] Lee K H and Park G J 2005 *JSME Int. J.* **49(3)** 779-88.
[10] Dellino G, Kleijnen J P C and Meloni C 2009 *Inform. J. Comput.* **24(24)** 2012.
[11] Elsayed K and Lacor C 20104 *Appl. Math. Comput.* **236(4)** 325-44.
[12] Bai G and Fei C 2013 *Chinese J. Mech. Eng.* **26(6)** 1160-8.
[13] Beyer H G and Sendhoff B 2007 *Comput. Meth. Appl. Mech. Eng.* **196(33–34)** 3190-218.
[14] Sun G, Baek S, Baek S, et al. 2014 *Struct. Multidisc. Opt.* **49(6)** 897-913.
[15] Janusevskis J and Riche R L 2013 *J. Global Opt.* **55(2)** 313-36.
[16] Kleijnen J P C 2009 *European J. Operat. Res.* **192(3)** 707-16.
[17] Gano S E, Renaud J E, Martin J D, et al. 2006 *Struct. Multidisc. Opt.* **32(4)** 287-98.
[18] Martin J D and Simpson T W 2012 *AIAA J.* **43(4)** 853-63.
[19] Simpson T W, Poplinski J D, Koch P N, et al. 2001 *Eng. Comput.* **17(2)** 129-50.
[20] Li R and Sudjianto A. *Technometr.* **47(2)** 111-20.

Acknowledgments

This paper is supported by the National Natural Science Foundation of China (Grant no. 51275024).

A research on Performance Efficiency of Rubber Metal Support Structures

Oleg V. Mkrtychev[1], Artem A. Bunov[1*]

[1]Moscow State University of Civil Engineering, 26 Jaroslavskoje shosse, Moscow, 129337 Russian Federation

Abstract. The paper scrutinizes structural behavior of lead rubber bearings by a Chinese manufacturer subjected to a single-component seismic action. Several problems were solved using specialized software complexes, which conducted forth integration of motion equations through the explicit method or response spectrum method. Depending on the calculation method, the diagram of the bearing performance was assumed to be either an actual diagram approximated by an idealized non-linear diagram or an idealized linear diagram with a specific stiffness. The computational model was assumed to be a single-mass oscillator with a lumped mass. The effort undertaken facilitated the investigation of the patterns of horizontal displacement of the bearing top relative to bottom caused by earthquakes modeled as accelerograms with different spectral compositions. The behavior of the support structure was benchmarked against similar supports by another manufacturer. The paper presents the outcomes of the research effort and draws conclusions about the efficiency of using the bearings of this particular type and model.

1. Introduction

Buildings and structures intended for seismic areas often incorporate various kinds of seismic protection in their design. As of today, LRB support is one of the most wide spread systems enhancing seismic resilience of structures [1, 2]. Major manufacturers like Maurer Sohne, FIP Industriale etc. produce bearings of this type. They have a track record of supplying reliable and high-quality products, however, they often lose a certain market share to Chinese companies, whose produce is often of dubious reliability and effectiveness.

2. Problem statement

The present paper considers the efficiency of LRB types 600, 700 and 800 by a Chinese manufacturer and benchmarks them against their counterparts by FIP Industriale [3].

3. Basic calculation provisions

The research was based on the assumption that an isolated structure behaves as a completely rigid body. The computational models were thought of as single-mass linear and non-linear oscillators with lumped mass (figure 1). Such problem setting enables us to draw a general qualitative conclusion about the operational effectiveness of the support structures [4].

Figure 1. Computational model of the bearing.

To ensure accuracy of the results, the research was conducted in two separate software packages, LIRA 10.6 and Ansys (LS-Dyna) [5].

A verification of the materials and the types of finite elements was also undertaken as part of the research. The software-based numeric modeling, namely the work diagrams and natural periods of oscillation of LRBs, was compared with the outcomes of the in-situ testing. The degree of matching proved to be high.

The response of a building equipped with a seismic protection in the form of LRB support is known to be sensitive to the spectral composition of the seismic load. For this reason the research effort processed accelerograms with a varying spectral composition 1, 1.5, 2 and 3 Hz. Figures 2 and 3 present acceleration curves with the dominant frequencies of 1 and 3 Hz. The vertical load on the bearing was considered in accordance with its estimated bearing capability:

1) vertical load on LRB-D600 – 3393 kN;
2) vertical load on LRB-D700 – 4618 kN;
3) vertical load on LRB-D800 – 6032 kN.

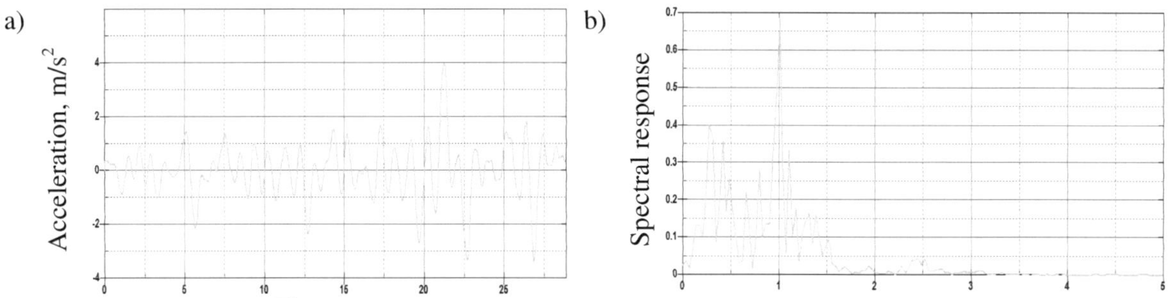

Figure 2 Accelerogram with the dominant frequency $f_1 = 1,0$ Hz (a) acceleration spectrum (b).

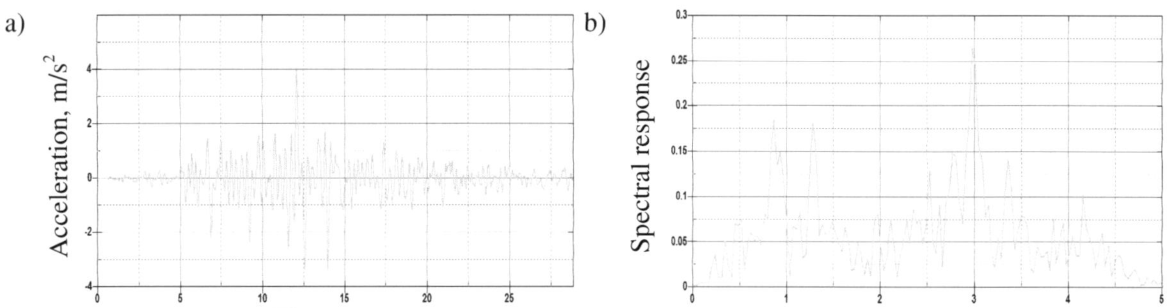

Figure 3. Accelerogram with the dominant frequency $f_4 = 3,0$ Hz (a) acceleration spectrum (b).

4. The results of calculations

The seismic impact calculation using LIRA 10.6 software made use of the spectrum response method. The research considered the LRB performance diagram to have an idealized linear character and to possess a specific stiffness.

IOP Conf. Series: Materials Science and Engineering **269** (2017) 012067 doi:10.1088/1757-899X/269/1/012067

The isofields and the displacement diagram of the support structure LRB-D600 are presented below (figure 4-5).

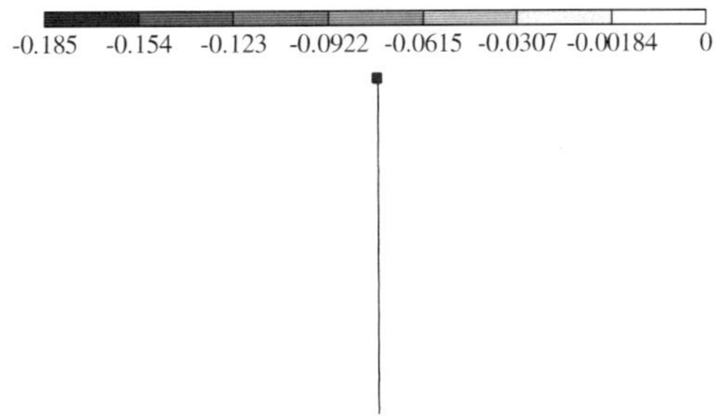

Figure 4. Isofields of LRB top displacements relative to the bottom along the X axis (m), for the accelerogram with the dominant frequency f_4=3,0 Hz.

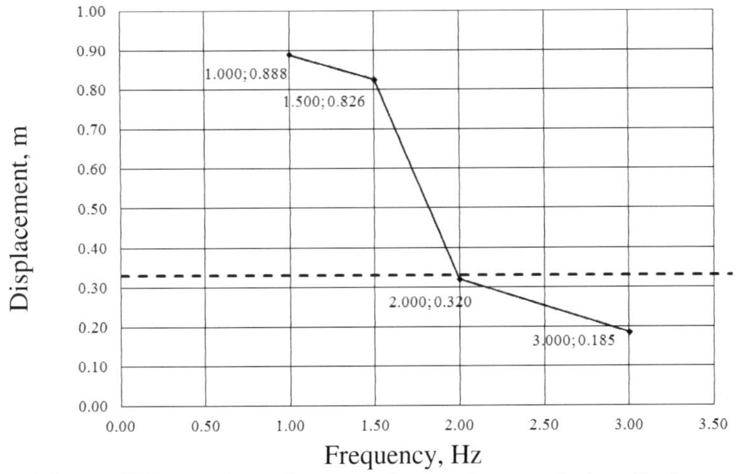

- Maximum allowable displacement of the top of the bearing 0,330 m

Figure 5. Dependency between the maximum relative displacement of the top of the isolator and the dominant frequency of the external effect.

The seismic impact calculation performed using the Ansys (LS-Dyna) software package utilized the central differences method, which in turn employs the explicit method of integrating motion equations. The LRB behavior diagram was set as an actual diagram approximated by an idealized non-liner diagram [6]. The computational model of the bearing in its general form is presented in figure 1.

The diagrams displaying the accelerations and displacements of the LRB-D600 are presented below (figure 6-8).

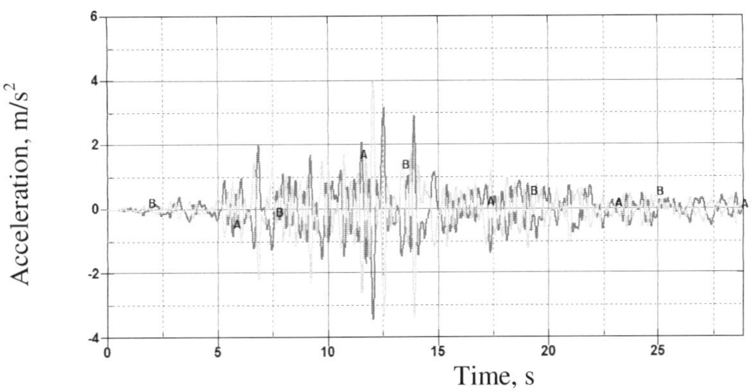

1) red curve – accelerations of the top of LRB;
2) green curve – the original accelerogram

Figure 6. Acceleration diagrams of LRB top and the original accelerogram with the dominant frequency f_4=3,0 Hz along the X axis (m/s^2).

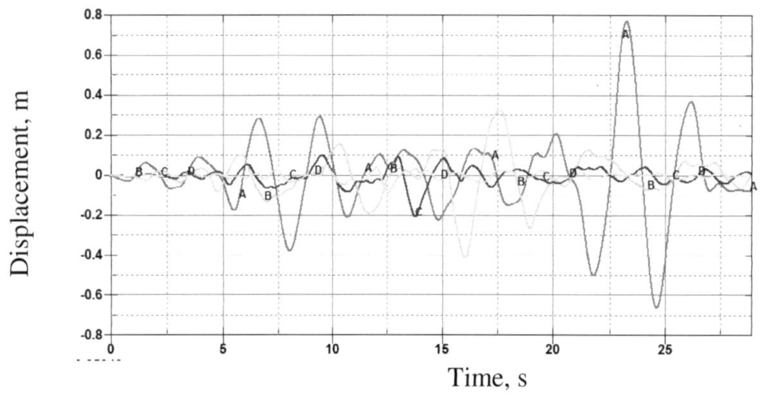

Dominant frequency of impact, Hz

1) Red curve – 1 Hz;
2) Green curve – 1,5 Hz;
3) Darkblue curve – 2 Hz;
4) Lightblue curve – 3 Hz

Figure 7. Curves of LRB top displacements relative to the bottom along the X axis (m) exposed to the dominant frequency f_4=3,0 Hz.

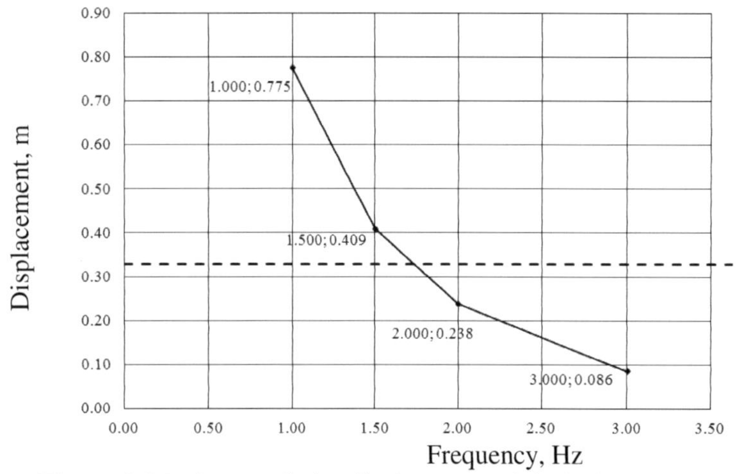

\- Maximum allowable displacement of LRB top 0,330 m

Figure 8. Maximum relative displacement of the isolator top and its dependence on the frequency of external effect.

Another research was dedicated to benchmarking of China-made LRB against LRB with a range of similar characteristics produced by a major well-known manufacturer FIP Industriale. The analogous bearings by FIP Industriale were selected by such characteristics as diameter, maximum horizontal displacements and rated bearing capacity.

The seismic impact modeling was performed using the software package Ansys (LS-Dyna). The LRB work diagram was set as the actual diagram approximated by the idealized non-linear diagram. The computational model of the bearing in its general form is presented in figure 1.

Table 1 presents the results of benchmarking the bearings.

Table 1. Benchmarking of LRB-D600, LRB-D700, LRB-D800 against FIP Industriale bearings.

Comparative characteristics	Dominant frequency of impact, [Hz]							
	1		1,5		2		3	
LRB model	LRB-D600	LRB-SN 600/180-150	LRB-D600	LRB-SN 600/180-150	LRB-D600	LRB-SN 600/180-150	LRB-D600	LRB-SN 600/180-150
LRB top accelerations, [m/s^2]	4,71	2,89	3,72	3,12	3,69	3,21	3,43	3,47
LRB top displacements, [m]	0,775	0,655	0,409	0,371	0,238	0,234	0,086	0,09

LRB model	LRB-D700	LRB-SN 700/203-170	LRB-D700	LRB-SN 700/203-170	LRB-D700	LRB-SN 700/203-170	LRB-D700	LRB-SN 700/203-170
LRB top displacements, [m/s^2]	4,76	2,93	3,86	3,16	3,87	3,21	3,40	3,46
LRB top displacements, [m]	0,733	0,688	0,440	0,384	0,234	0,240	0,089	0,093
LRB model	LRB-D800	LRB-S 800/200-175	LRB-D800	LRB-S 800/200-175	LRB-D800	LRB-S 800/200-175	LRB-D800	LRB-S 800/200-175
LRB top displacements, [m/s^2]	3,96	3,15	3,58	3,16	3,31	3,21	3,41	3,39
LRB top displacements, [m]	1,02	0,825	0,418	0,409	0,209	0,219	0,074	0,111

5. Summary

The research we have conducted gives rise to the following conclusions:

1. The values of the compared parameters vary depending on the calculation technique: the response spectrum method or non-linear dynamic methods. The response spectrum method overrates the actual values of displacement.
2. In case the spectrum composition of the earthquake accelerograms varies, the Chinese-made LRBs demonstrate a decrease in accelerations up to 21%.
3. The critical values of seismic action dominant frequencies have been obtained, which may cause LRB to fail as the displacement limit is reached, and the structure to collapse.
4. A benchmarking of LRB-D600, LRB-D700, LRB-D800 and FIP Industriale bearings has been conducted. It has been established that the difference intensifies as the dominant frequency reduces. The benchmarking by the principal parameters has shown close matching for exposures to dominant frequencies between 2 and 3 Hz. Benchmarking for the case of exposure to dominant frequencies below 2 Hz exhibits considerable dissimilarity.

6. References

[1] Murav'ev N P 2013 *Far East: Problems of development of architectural-construction complex.* **1** 212-8.

[2] Bubis A A, Yun A J, Petryashov S O and Petryashov N O 2012 *Earthq. Eng. Saf. Build. Seism. Constr.* **2** 31-5.

[3] Catalogue on lead rubber bearings series LRB, FIP Industriale S.P.A.

[4] Mkrtychev O V 2000 *Reliability of multi-element truss systems of engineering structures,* Dissertation for the degree of doctor of technical sciences. Moscow.

[5] LS-DYNA, Keyword user's manual, Volume I, II. Livermore Software Technology Corporation (LSTC).

[6] Mkrtychev O V, Dzhinchvelashvili G A and Bunov A A, 2014 *Proc. Eng.* **91** 48-53.

Acknowledgements

This study was performed with the support of the RF Ministry of Education and Science, grant №7.1524.2017/Project Part.

IOP Conf. Series: Materials Science and Engineering **269** (2017) 012068 doi:10.1088/1757-899X/269/1/012068

Parallel Calculations in LS-DYNA

Oleg Vartanovich Mkrtychev[1], Andrey Aleksandrovich Reshetov[1*]

[1]Moscow State University of Civil Engineering (National Research University), Moscow, Russia

Abstract. Nowadays, structural mechanics exhibits a trend towards numeric solutions being found for increasingly extensive and detailed tasks, which requires that capacities of computing systems be enhanced. Such enhancement can be achieved by different means. E.g., in case a computing system is represented by a workstation, its components can be replaced and/or extended (CPU, memory etc.). In essence, such modification eventually entails replacement of the entire workstation, i.e. replacement of certain components necessitates exchange of others (faster CPUs and memory devices require buses with higher throughput etc.). Special consideration must be given to the capabilities of modern video cards. They constitute powerful computing systems capable of running data processing in parallel. Interestingly, the tools originally designed to render high-performance graphics can be applied for solving problems not immediately related to graphics (CUDA, OpenCL, Shaders etc.). However, not all software suites utilize video cards' capacities. Another way to increase capacity of a computing system is to implement a cluster architecture: to add cluster nodes (workstations) and to increase the network communication speed between the nodes. The advantage of this approach is extensive growth due to which a quite powerful system can be obtained by combining not particularly powerful nodes. Moreover, separate nodes may possess different capacities. This paper considers the use of a clustered computing system for solving problems of structural mechanics with LS-DYNA software. To establish a range of dependencies a mere 2-node cluster has proven sufficient.

1. Introduction

Parallel computing is a way of organizing computerized calculations when a software program is executed as a set of interacting parts running in parallel (simultaneously). As of today, several modes of running parallel computing are available. E.g., a software program may represent a single multi-stream process executed within an operating system (OS). Such structure of computing provides that the simultaneously executed streams are interacting via shared memory (figure 1a). At the same time, actual parallelism is supposed to be achieved with multi-core CPUs. This way of structuring parallel computing is targeted at computing systems belonging to SMP architecture (Symmetric Multiprocessing). Another way of organizing parallel computing implies that a program is executed as a total of interacting processes. A single machine (as OS processes) or multiple networked machines may execute these processes. If the processes are executed by different machines (i.e. by a shared memory system), such configuration is called distributed computing. Processes interact via the so-called MPI (Message Passing Interface). The processes running on different machines exchange messages via network communication (figure 1b). This architecture of a computing system is referred to as MPP (Massive Parallel Processing). Good scalability of the system by extension makes for an advantage of MPP over SMP.

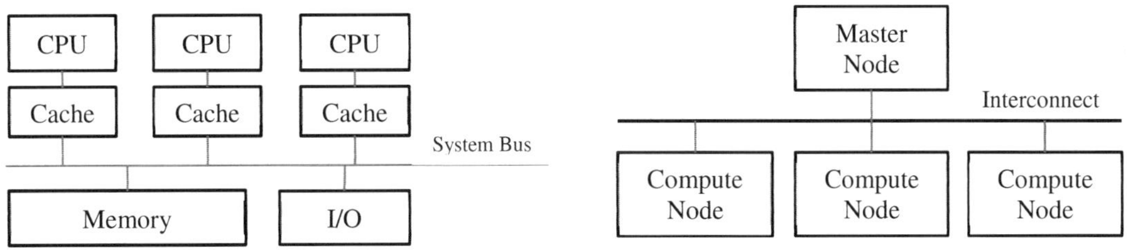

a) b)

Figure 1. a) SMP architecture, b) MPP architecture.

It should be noted that the executable program, or the way its code is structured, determines the correlation between the number of processors in the system and the capacity. The capacity can be assessed using Amdahl's law (figure 2):

$$S_p = \frac{1}{\alpha + \frac{1-\alpha}{p}} \tag{1}$$

where, α – the portion of the program code executed consecutively, p – number of computing units (cores, processors, nodes etc.)

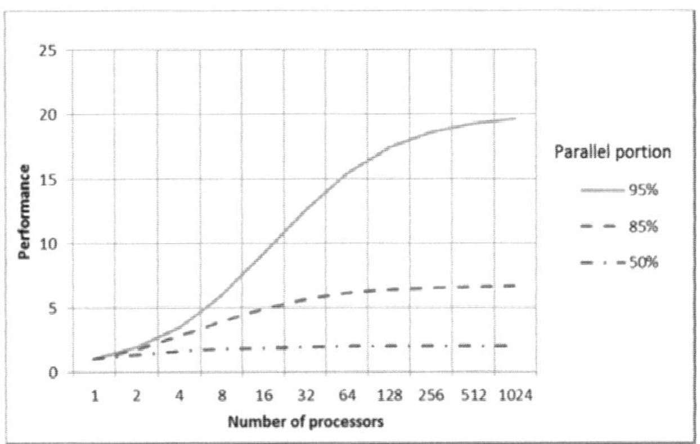

Figure 2. Amdahl’s law.

2. MPI applications in distributed computing

As mentioned previously, MPP architecture means that a software program is executed as a total of processes interacting via MPI interface. This in turn presumes that the software program in question was developed using MPI API (Application Programming Interface). MPI API provides software developers with a set of commands and features facilitating process synchronization, data transfer between processes via messages etc.

An MPI-program run on cluster suggests that the processes be distributed between the cores of the processors belonging to the cluster nodes. When synchronizing the data flows from different nodes the MPI interface transfers messages via network communication. Hence, for the system to operate effectively, a network with sufficient throughput is required. Besides, a key prerequisite for the efficiency of software is good balancing of load among the node processors, which is most evident in case of a cluster whose nodes have varying capacities.

Let us consider a generic configuration of MPI program’s launch and operation.

Firstly, an MPI interface must be installed on all nodes, and all the nodes must have an MPI interface service started (in the case of MS-MPI it is MS-MPI Launch Service). Secondly, all the nodes must feature identical credentials with identical passwords. The software can be started (for instance, with a name program1) from the master node with the command below:

IOP Conf. Series: Materials Science and Engineering **269** (2017) 012068 doi:10.1088/1757-899X/269/1/012068

<path to mpiexec>\mpiexec –machinefile hosts_list –wdir <path to workin folder> program1

where, hosts_list stands for the file with a list of nodes; the path to the working folder contains a network address of the storage accommodating the working folder. The file hosts_list contains data about the number of processes to be started on each of the nodes.

Program execution involving MPI is effected as follows:

1) The master node decomposes the task into subtasks and transfers them to the computing nodes based on the number of the processes, node capacity and other considerations (figure 3). The master node can at the same time act as a computing node and take over a subtask.

2) As the program is running, synchronization and file creation processes take place (e.g.creation of files with solutions) etc.

3) At the completion of the program all the created processes and the mpiexec program are terminated, the established connections are closed.

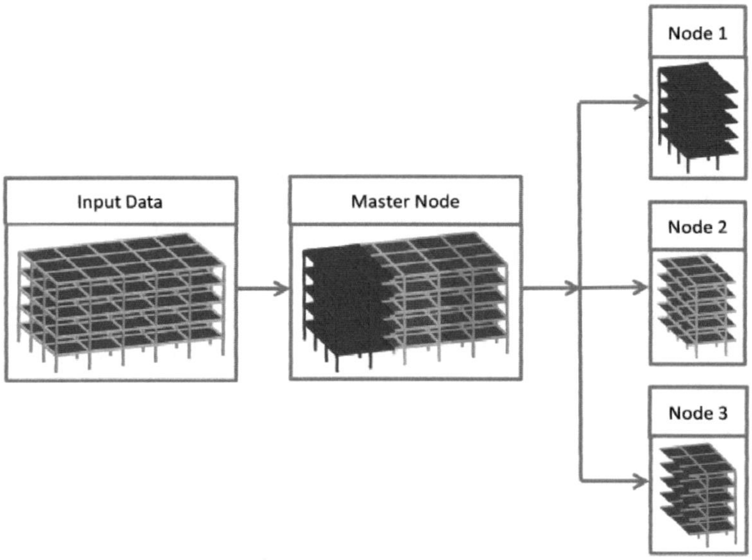

Figure 3. Task decomposition (LS-DYNA).

A more detailed account of parallel computing can be seen in [1-3].

3. Distributed computing in LS-DYNA

A cluster comprised of two nodes with varying capacities, HALL9000-5 and ANDREWR-X200LA, are going to be used for problem solving. Each node is equipped with a quad core processor (incl.hyperthreading) and a corresponding RAM: HALL9000-5 (Intel Core i5 2.9 GHz, 8 GB), ANDREWR-X200LA (Intel Core i3 1.7, GHz, 4 GB). The nodes are connected via a gigabit network. MPP single precision solver LS-DYNA R8.0.0 will be used. We will create one process per node for convenience of analyzing the dependencies. Figure 3 presents the problem to be solved. The structure consists of columns, rods and slabs. Columns and rods are modeled as beam finite elements, while slabs take the form of shells. The material is linear. External strains are set as inertia forces (own weight and earthquake accelerogram [4-6]). The explicit method will be used for solving. The solution will involve several patterns of task decomposition in order to determine the dependency between the utilization of processors, network traffic and the approach to decomposition.

The first approach to decomposition is as follows: we split the structure into two equal parts and assign them to the nodes for solution. The section will be effected along the rods and slabs (figure 4).

IOP Conf. Series: Materials Science and Engineering **269** (2017) 012068 doi:10.1088/1757-899X/269/1/012068

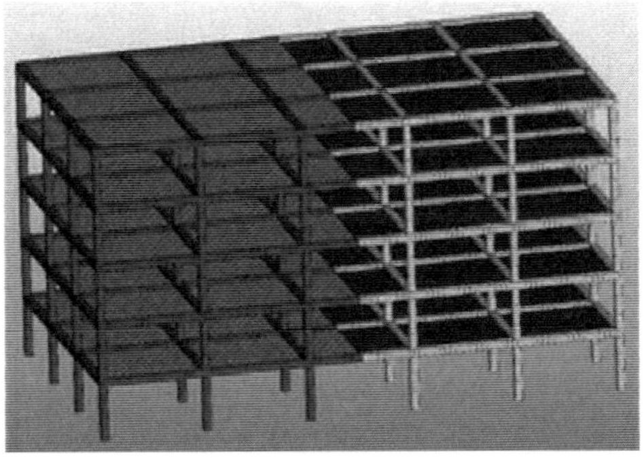

Figure 4. Task decomposition.

The outcome is presented in figure 5.

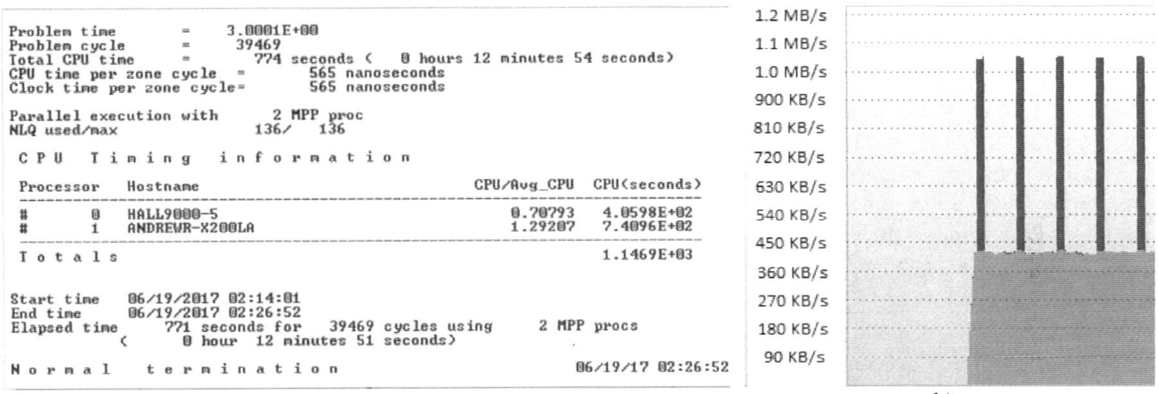

```
Problem time         =     3.0001E+00
Problem cycle        =      39469
Total CPU time       =        774 seconds (    0 hours 12 minutes 54 seconds)
CPU time per zone cycle  =          565 nanoseconds
Clock time per zone cycle=          565 nanoseconds

Parallel execution with       2 MPP proc
NLQ used/max                136/   136

 C P U   T i m i n g   i n f o r m a t i o n

 Processor   Hostname                                CPU/Avg_CPU  CPU(seconds)
 ----------------------------------------------------------------------------
 #       0   HALL9000-5                                  0.70793    4.0598E+02
 #       1   ANDREWR-X200LA                              1.29207    7.4096E+02
 ----------------------------------------------------------------------------
 T o t a l s                                                        1.1469E+03

Start time    06/19/2017 02:14:01
End time      06/19/2017 02:26:52
Elapsed time       771 seconds for   39469 cycles using      2 MPP procs
              (       0 hour  12 minutes 51 seconds)

N o r m a l   t e r m i n a t i o n                          06/19/17 02:26:52
```


a) b)

Figure 5. a) Computing report, b) Network diagram.

The outcome lets us conclude that the CPU utilization was balanced unevenly (figure 5a, CPU/Avg_CPU). The first node with higher capacity proved underutilized (0.70793) and the second overloaded (1.29207). The reason is that the task divided into two equal parts caused the first node to run calculations faster than the second one. At the same time, every iteration done by the first node for its portion of the task requires the other node to provide the boundary conditions obtained in the previous iteration. As the latter node solves its portion of the task slower, the former node idles, which slows down the performance of the entire clustered system.

Figure 6. Task decomposition.

Under the second mode of decomposition, we divide the structure into two uneven parts and assign the larger part to the more powerful node, while the less powerful one receives the smaller part (figure 6). The decomposition section was determined based on the weighting coefficients derived from the performance analysis of the cluster nodes.

The result is presented in figure 7.

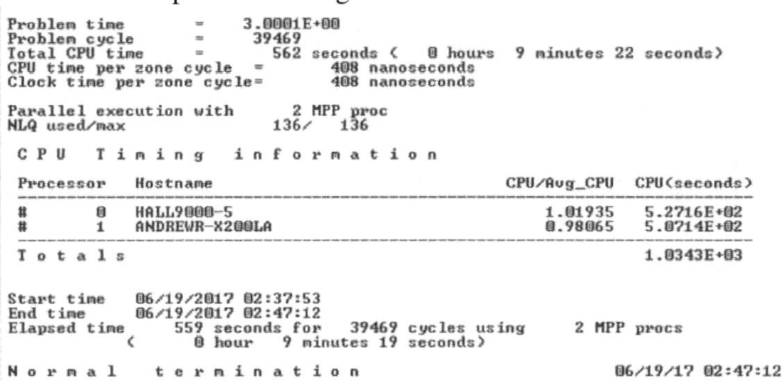

```
Problem time          =     3.0001E+00
Problem cycle         =      39469
Total CPU time        =        562 seconds (   0 hours   9 minutes 22 seconds)
CPU time per zone cycle   =          408 nanoseconds
Clock time per zone cycle=           408 nanoseconds

Parallel execution with         2 MPP proc
NLQ used/max                  136/   136

 C P U    T i m i n g    i n f o r m a t i o n

 Processor   Hostname                              CPU/Avg_CPU   CPU(seconds)
 ----------------------------------------------------------------------------
 #       0   HALL9000-5                              1.01935    5.2716E+02
 #       1   ANDREWR-X200LA                          0.98065    5.0714E+02
 ----------------------------------------------------------------------------
 T o t a l s                                                    1.0343E+03

Start time    06/19/2017 02:37:53
End time      06/19/2017 02:47:12
Elapsed time       559 seconds for    39469 cycles using       2 MPP procs
              (    0 hour    9 minutes 19 seconds)

N o r m a l     t e r m i n a t i o n                        06/19/17 02:47:12
```

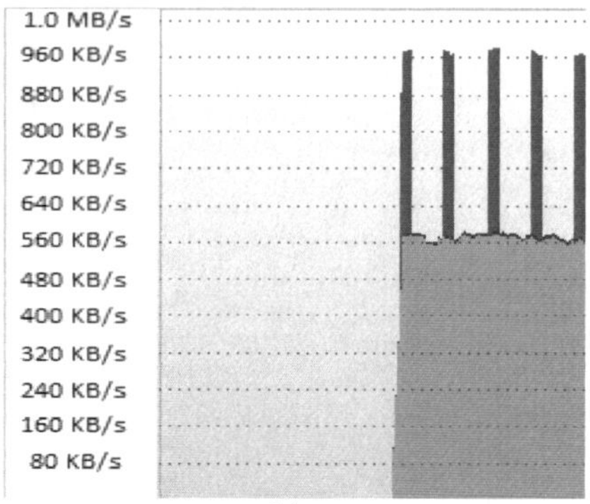

a) b)

Figure 7. a) Computing report, b) Network diagram.

The results give rise to the conclusion that the utilization was balanced in this case: HALL9000-5 (1.01935), ANDREWR-X200LA (0.98065).

Although the utilization of the CPUs proved balanced, this way of decomposition cannot be declared optimal. The fact is that the structure was sectioned along the beams and slabs. Once we draw the section along the columns, the number of boundary points between two subtasks will reduce and therefore the network diagram will shrink. Naturally, the network communication capability available for this task contains a large surplus, however, we will obtain a more robust task decomposition for research purposes. This task appeared to be incapable of being decomposed by the columns with CPU utilization remaining balanced (figure 8), since the resulting volumes of subtasks do not match the capacity of the nodes.

The optimal decomposition is achieved by combining the sections by columns, by beams and by slabs (figure 9). The outcome (figure 10) indicates that the CPU utilization is balanced and network traffic is less intense (figure 5b, 7b, 10b). This way of decomposition required the least computing time.

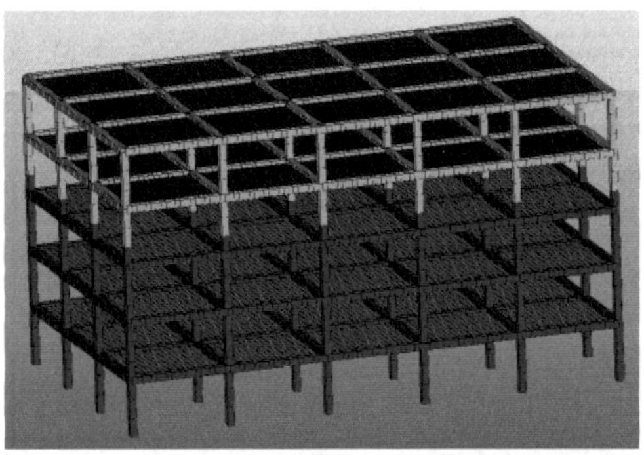

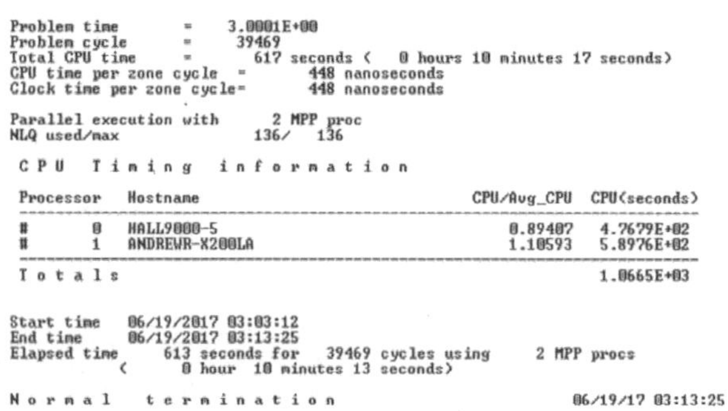

```
Problem time          =     3.0001E+00
Problem cycle         =      39469
Total CPU time        =        617 seconds (   0 hours  10 minutes 17 seconds)
CPU time per zone cycle   =          448 nanoseconds
Clock time per zone cycle=           448 nanoseconds

Parallel execution with         2 MPP proc
NLQ used/max                  136/   136

 C P U    T i m i n g    i n f o r m a t i o n

 Processor   Hostname                              CPU/Avg_CPU   CPU(seconds)
 ----------------------------------------------------------------------------
 #       0   HALL9000-5                              0.89407    4.7679E+02
 #       1   ANDREWR-X200LA                          1.10593    5.8976E+02
 ----------------------------------------------------------------------------
 T o t a l s                                                    1.0665E+03

Start time    06/19/2017 03:03:12
End time      06/19/2017 03:13:25
Elapsed time       613 seconds for    39469 cycles using       2 MPP procs
              (    0 hour   10 minutes 13 seconds)

N o r m a l     t e r m i n a t i o n                        06/19/17 03:13:25
```

a) b)

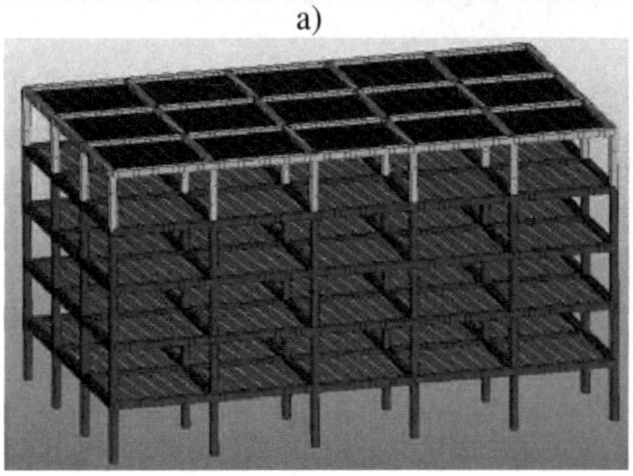

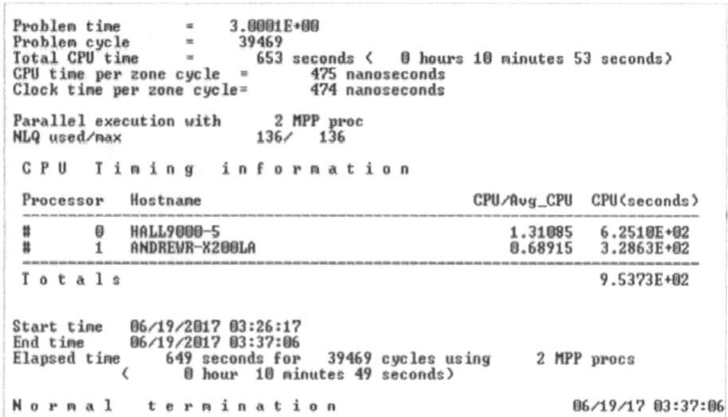

```
Problem time          =     3.0001E+00
Problem cycle         =      39469
Total CPU time        =        653 seconds (   0 hours  10 minutes 53 seconds)
CPU time per zone cycle   =          475 nanoseconds
Clock time per zone cycle=           474 nanoseconds

Parallel execution with         2 MPP proc
NLQ used/max                  136/   136

 C P U    T i m i n g    i n f o r m a t i o n

 Processor   Hostname                              CPU/Avg_CPU   CPU(seconds)
 ----------------------------------------------------------------------------
 #       0   HALL9000-5                              1.31085    6.2510E+02
 #       1   ANDREWR-X200LA                          0.68915    3.2863E+02
 ----------------------------------------------------------------------------
 T o t a l s                                                    9.5373E+02

Start time    06/19/2017 03:26:17
End time      06/19/2017 03:37:06
Elapsed time       649 seconds for    39469 cycles using       2 MPP procs
              (    0 hour   10 minutes 49 seconds)

N o r m a l     t e r m i n a t i o n                        06/19/17 03:37:06
```

c) d)

Figure 8. a, c) Task decomposition; b, d) Computing report.

Figure 9. Task decomposition.

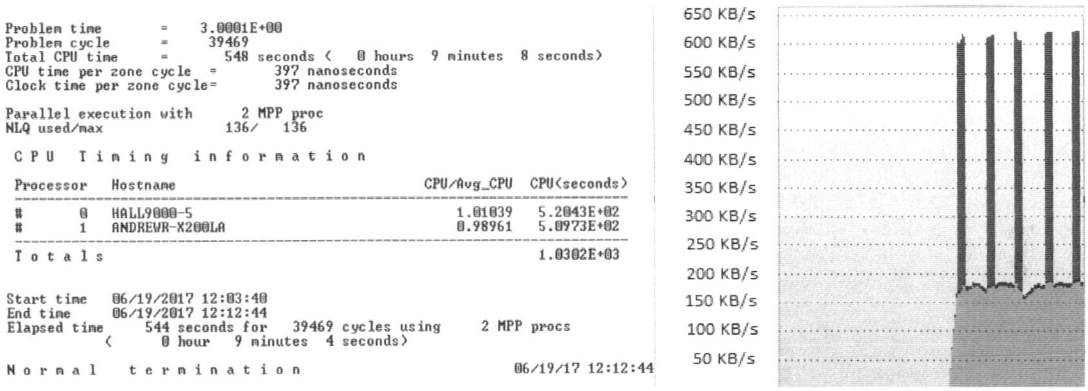

a) b)

Figure 10. a) Computing log, b) network diagram.

4. Conclusions

Therefore, efficient use of LS-DYNA software suite for distributed computing using clusters comprised of nodes with varying performance characteristics necessitates optimized methods of task decomposition. For decomposition to be optimized, node CPUs must be balanced more evenly and the number of messages exchanged between nodes must be reduced.

5. References

[1] Krysl P and Bittnar Z 2001 *Comput. Struct.* **79(3)** 345-60.
[2] París J, Colominas I, Navarrina F and Casteleiro M, 2013 *Comput. Struct.* **125** 62-73.
[3] Jin H Q, Jespersen D, Mehrotra P, Biswas R, Huang L and Chapman B 2011 *Parall. Comput.* **37(9)** 562-75.
[4] Mkrtychev O V and Myasnikova E S 2009 *Struct. Mech. Calculat. Struct.* **1** 38-40.
[5] Mkrtychev O V, 2000 *Reliability of multi-element rod systems of engineering structures*, The dissertation author's abstract on competition of a scientific degree of the doctor of technical sciences, Moscow.
[6] Mkrtychev O V, Dzhinchvelashvili G Q and Bunov A A 2014 *Proc. Eng.* **91** 48-53.

Acknowledgements

This study was performed with the support of the RF Ministry of Education and Science, grant №7.1524.2017/Project Part.

A New Model for Simulating Gas Metal Arc Welding based on Phase Field Model

Yongyue Jiang[1, 2*], Li Li[1] and Zhijiang Zhao[1]

[1]Beijing Time Technologies Co., Ltd., Kaituo Road 17, Beijing 100085, China
[2]School of Civil & Environmental Engineering, University of Science and Technology Beijing, Beijing 100083, China

Abstract. Lots of physical process, such as metal melting, multiphase fluids flow, heat and mass transfer and thermocapillary effect (Marangoni) and so on, will occur in gas metal arc welding (GMAW) which should be considered as a mixture system. In this paper, based on the previous work, we propose a new model to simulate GMAW including Navier-Stokes equation, the phase field model and energy equation. Unlike most previous work, we take the thermocapillary effect into the phase field model considering mixture energy which is different of volume of fluid method (VOF) widely used in GMAW before. We also consider gravity, electromagnetic force, surface tension, buoyancy effect and arc pressure in momentum equation. The spray transfer especially the projected transfer in GMAW is computed as numerical examples with a continuous finite element method and a modified midpoint scheme. Pulse current is set as welding current as the numerical example to show the numerical simulation of metal transfer which fits the theory of GMAW well. From the result compared with the data of high-speed photography and VOF model, the accuracy and stability of the model and scheme are easily validated and also the new model has the higher precieion.

1. Introduction

Gas metal arc welding (GMAW) is one kind of arc welding method which uses the arc between the moving welding wire and workpiece as the heat source and also shielding gas which is arc media to protect weld pool and high temperature melting metal in the welding zone. It is widely used in industry because of its high efficiency, excellent quality and low cost. In GMAW, the welding wire is the electrode and the tip of the wire would melt at high temperature coming from arc, Ohmic heating and electrode heating. As the welding wire keeps melting and other forces like electromagnetic forces acts on the droplet, the droplet will overcomes the resistance of surface tension and then break up and fall into weld pool. Metal transfer has been studied extensively both theoretically and experimentally as it plays an important role in GMAW. Mckelliget and Szekely [1] proposed a steady mathematical model to investigate velocity and temperature of electrode and weld pool in early time but not the dynamic process of metal transfer. Haidar and Lowke [2] investigated the influence of different currents on the size of droplets. Cheng, Wu and Lian [3] presented a dynamic model for metal transfer combined electromagnetic theory with volume of fluid (VOF) considering electromagnetic force, surface tension and arc plasma. A unified comprehensive model was developed by Hu and Tsai [4] to simulate the transport phenomena occurring during the gas metal arc welding. But still, they used VOF to simulate the interface of multi-phase fluids. Murphy [5] presented a self-consistent three dimensional model of GMAW which still a steady model without the consideration of metal transfer. Liu, Shan and Yuan [6] proposed a new formula for the calculation of effective radius of droplets that would be used in calculating drag force and plasma force. Also, other people do research about GMAW to analyze arc, welding current and fume formation [7-9].

Actually, the model for simulating the metal transfer used before contains NS equation, VOF (interface of multi-phase fluids) which just make volume fraction of fluid couple with velocity, energy equation and Maxwell equations. Thermocapillary effect plays an important role in the system where the surface tension becomes dominant [10-12]. One famous example of thermocapillary effects in the thermocapillary migration of drops as the drops set in a liquid possessing a temperature gradient will move towards the hot region due to the thermocapillary effects. As all we know that surface tension is an important resistance for stopping the droplet breaking up, thermocapillary effect should be considered in the numerical model due to the result in multi-phase fluids. In addition, the system of GMAW is a mixture composed of welding wire, melting metal and droplet. The mixture energy also should be considered in the model as it was introduced in phase field model. Unlike sharp interface models [13-15], two or more fluids in the phase field model are treated as one fluid and are mixed smoothly near the interfacial region, where the flow parameters should change rapidly. This method can make the process of modeling easier as the interface of multi-phase fluids is not boundary anymore and it can avoid dealing with the jump of boundary conditions [16-18]. In this paper, we propose a new model for simulating the metal transfer in GMAW which consider thermocapillary effect and mixture energy. In next sections, the fully discretized GMAW system with a continuous finite element method and a second order temporal scheme is presented. A Numerical example with pulse current is computed and the result would be discussed and compared with the result of VOF and high-speed photography experiments.

2. Mathematical model and weak forms

Let Ω be the welding region which is a bounded domain and Γ be the boundary of Ω. We present a new model for GMAW with a system composed of NS equations equation 1 and equation 2, phase field model equation 3 with thermocapillary effect and mixture energy, energy equation 4 and Maxwell equations 5 as follows:

$$\frac{\partial \rho}{\partial t}+\nabla\cdot(\rho\mathbf{v})=0 \tag{1}$$

$$\frac{\partial(\rho\mathbf{v})}{\partial t}+\nabla\cdot(\rho\mathbf{v}\otimes\mathbf{v})=-\nabla p+\nabla\cdot\left(\eta\left(\nabla\mathbf{v}+\nabla\mathbf{v}^{\mathrm{T}}\right)\right)-\nabla\cdot\left(\lambda_f\left(\nabla f\otimes\nabla f\right)\right)+\mu\nabla f-\rho\mathbf{G}+\mathbf{J}\times\mathbf{B} +\rho\mathbf{G}\beta_T\left(T-T_0\right) \tag{2}$$

$$\frac{\partial(\rho f)}{\partial t}+\nabla\cdot(\rho\mathbf{v}f)=\nabla\cdot(M\nabla\mu),\ \mu=\lambda_f\frac{f\left(f^2-1\right)}{\varepsilon}-\lambda_f\varepsilon\Delta f \tag{3}$$

$$\frac{\partial\left(\rho c_pT\right)}{\partial t}+\nabla\cdot\left(\rho c_p\mathbf{v}T\right)=\nabla\cdot\left(k\nabla T+M\mu\nabla\mu\right)-\rho H\frac{\partial f}{\partial t}+\frac{\mathbf{J}^2}{\sigma_e}+\frac{5k_b}{2e}\mathbf{J}\cdot\nabla T \tag{4}$$

$$\nabla\cdot\left(\sigma_e\nabla\Phi\right)=0,\ \mathbf{J}=-\sigma_e\nabla\Phi,\ \Delta\mathbf{A}=-\mu_0\mathbf{J},\ \mathbf{B}=\nabla\times\mathbf{A} \tag{5}$$

Here β_T, M, ε, ρ, η, k, c_p, k_b, e, σ_e are the thermal expansion coefficient, phenological mobility coefficient, thickness of interface, density, viscosity, thermal conductivity coefficient, specific heat, Stefan-Boltzman constant, electronic charge and electrical conductivity, respectively. μ is called chemical potential which represents the mixture energy as it has two parts contributing to separation and mixing, respectively. $\mathbf{v}$, $\mathbf{J}$ are the velocity and current density. f is the "phase" of the mixture ($f=1$: fluid and $f=-1$: metal) and T represents temperature. Φ, $\mathbf{B}$ are the electrical potential and self-induced electronmagnetic field. $\mathbf{A}, T_0$ are magnetic vector and initial temperature. λ_f is a temperature-dependent coefficient representing thermocapillary effects along the interface with $\lambda_f = \eta\varepsilon\sigma_0 - \eta\varepsilon\sigma_T(T-T_0)$. The boundary conditions and computing domain in [4] are used in this paper. The boundary conditions for f and μ are $\partial_{\mathbf{n}} f = 0, \partial_{\mathbf{n}}\mu = 0$. As the densities of fluid and metal are matched, it can be simplified in the form

$$\nabla\cdot\mathbf{v} = 0 \tag{6}$$

$$\rho\frac{\partial\mathbf{v}}{\partial t} + \rho(\mathbf{v}\cdot\nabla)\mathbf{v} = -\nabla p + \eta\Delta\mathbf{v} - \nabla\cdot\left(\lambda_f(\nabla f\otimes\nabla f)\right) + \mu\nabla f - \rho\mathbf{G} + \mathbf{J}\times\mathbf{B} + \rho\mathbf{G}\beta_T(T-T_0) \tag{7}$$

$$\frac{\partial f}{\partial t} + \mathbf{v}\cdot\nabla f = \nabla\cdot(M\nabla\mu),\ \mu = \lambda_f\frac{f(f^2-1)}{\varepsilon} - \lambda_f\varepsilon\Delta f \tag{8}$$

$$\rho c_p\frac{\partial T}{\partial t} + \rho c_p\mathbf{v}\cdot\nabla T = \nabla\cdot(k\nabla T + M\mu\nabla\mu) - \rho H\frac{\partial f}{\partial t} + \frac{\mathbf{J}^2}{\sigma_e} + \frac{5k_b}{2e}\mathbf{J}\cdot\nabla T \tag{9}$$

$$\nabla\cdot(\sigma_e\nabla\Phi) = 0, \mathbf{J} = -\sigma_e\nabla\Phi, \Delta\mathbf{A} = -\mu_0\mathbf{J}, \mathbf{B} = \nabla\times\mathbf{A} \tag{10}$$

We rewrite equation 8 as $\frac{\partial f}{\partial t} + \mathbf{v}\cdot\nabla f = M\Delta(\omega + cf)$ and $\omega + cf = \mu = \lambda_f\left(\frac{f(2f-1)(f-1)}{2\varepsilon^2} - \Delta f\right)$ where the positive constant c is introduced to enhance the stability [18, 20]. Also, we deal with equation 6 by the penalty formulation as the same as the idea in [18] $\nabla\cdot\mathbf{v} + \delta p = 0$ where δ is small penalty.

Denote $\mathbf{W}^{1,3}(\Omega) = \left(W^{1,3}(\Omega)\right)^3$, $\mathbf{L}^2(\Omega) = \left(L^2(\Omega)\right)^2$ and $L_0^2(\Omega) = \left\{p\in L^2(\Omega), \int_\Omega p\, d\mathbf{x} = 0\right\}$ as usual. Applying the idea in [18] to the present work, the weak form of the penalized GMAW system reads: Find $\mathbf{v}, \mathbf{J}, \mathbf{B}, \mathbf{A}\in\mathbf{W}^{1,3}(\Omega)$, $p\in L_0^2(\Omega)$, $f, \mu, \Phi, T\in W^{1,3}(\Omega)$ such that

$$\int_\Omega(\nabla\cdot\mathbf{v} + \delta p)q d\mathbf{X} = 0 \qquad \forall q\in L_0^2(\Omega) \tag{11}$$

$$\begin{aligned}\int_\Omega\Big(\rho\frac{\partial\mathbf{v}}{\partial t}\cdot\mathbf{u} + \rho(\mathbf{v}\cdot\nabla)\mathbf{v}\cdot\mathbf{u} - p(\nabla\cdot\mathbf{u}) + \eta\nabla\mathbf{v}:\nabla\mathbf{u} + \rho\mathbf{G}\cdot\mathbf{u} - (\mathbf{J}\times\mathbf{B})\cdot\mathbf{u} - \rho\mathbf{G}\beta_T(T-T_0)\cdot\mathbf{u} - (\omega+cf)\nabla f\cdot\mathbf{u}\\ + \lambda_f(\nabla f\otimes\nabla f):\nabla\mathbf{u}\Big)d\mathbf{X} + \int_\Gamma p(\mathbf{n}\cdot\mathbf{u})ds = 0 \qquad \forall\mathbf{u}\in\mathbf{W}^{1,3}\end{aligned} \tag{12}$$

$$\int_\Omega\left(\frac{\partial f}{\partial t}\gamma + (\mathbf{v}\cdot\nabla f)\gamma + M\nabla(\omega+cf)\cdot\nabla\gamma\right)d\mathbf{X} = 0 \qquad \forall\gamma\in W^{1,3} \tag{13}$$

$$\int_\Omega\left((\omega+cf)\chi - \lambda_f\frac{f(f+1)(f-1)}{\varepsilon}\chi - \lambda_f\varepsilon\nabla f\cdot\nabla\chi\right)d\mathbf{X} = 0 \qquad \forall\chi\in W^{1,3} \tag{14}$$

$$\begin{aligned}\int_\Omega\left(\rho c_p\left(\frac{\partial T}{\partial t} + \mathbf{v}\cdot\nabla T\right)\theta + \left(k\nabla T + M(\omega+cf)\nabla(\omega+cf)\right)\cdot\nabla\theta + \rho H\frac{\partial f}{\partial t}\theta - \frac{\mathbf{J}^2}{\sigma_e}\theta - \frac{5k_b}{2e}\mathbf{J}\cdot\nabla T\theta\right)d\mathbf{X} = 0\\ \forall\theta\in W^{1,3}\end{aligned} \tag{15}$$

$$\int_{\Omega}(\nabla\mathbf{A}:\nabla\boldsymbol{\alpha}+\mu_0\sigma_e\nabla\Phi\cdot\boldsymbol{\alpha})d\mathbf{X}-\int_{\Gamma}(\mathbf{n}\cdot\nabla\mathbf{A})\cdot\boldsymbol{\alpha}ds=0,\ \int_{\Omega}(\sigma_e\nabla\Phi\cdot\nabla\varphi)d\mathbf{X}-\int_{\Gamma}\sigma_e(\mathbf{n}\cdot\nabla\Phi)\varphi ds=0,$$

$$\int_{\Omega}(\mathbf{J}+\sigma_e\nabla\Phi)\cdot\mathbf{S}\,d\mathbf{X}=0,\ \int_{\Omega}(\mathbf{B}-\nabla\times\mathbf{A})\cdot\mathbf{L}\,d\mathbf{X}=0 \qquad \forall\boldsymbol{\alpha},\varphi,\mathbf{S},\mathbf{L}\in\mathbf{W}^{1,3} \quad (16)$$

3. Result and discussion

We use a modified mid-point scheme [18] to discretize the system and we compute a numerical example in this section. We choose stainless steel as the electrode whose diameter is 1.2mm. The shielding gas is pure argon and its arc length is 3mm. Pulse current (table 1) are set as the welding current. figure 1 gives the wave of pulse current and four points (0, 1.377, 4.116, 7.231ms in one period) are chosen to show the numerical result. Only the interface structure are presented in figure 2. From figure 2a to figure 2b, we can see that the droplet comes out and grows up due to the current increase quickly in peak time and also the necking effect. Compared with figure 2b and 2c, the necking effct become more clearly as the increase of gravity contributed by melting metal and the change of electromagnic force. After the project time, the droplet breaks up in base time (figure 2d) and becomes flat ellipse due to arc pressure gradient. The numerical result fits the theory of metal transfer well.

Table 1. The time and current of three stages in one period of pulse current.

	Time ([ms])	Current ([A])
Peak	2.4	327
Project	3.0	106
Base	9.5	24

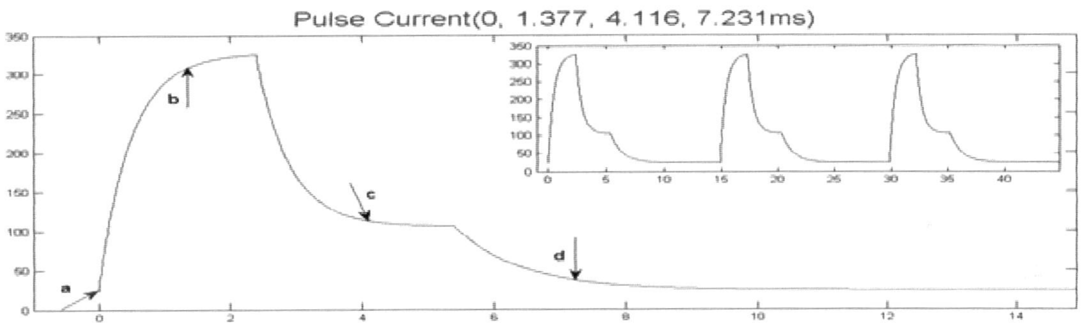

Figure 1. The wave of pulse current (Large: One period; Small: Three periods).

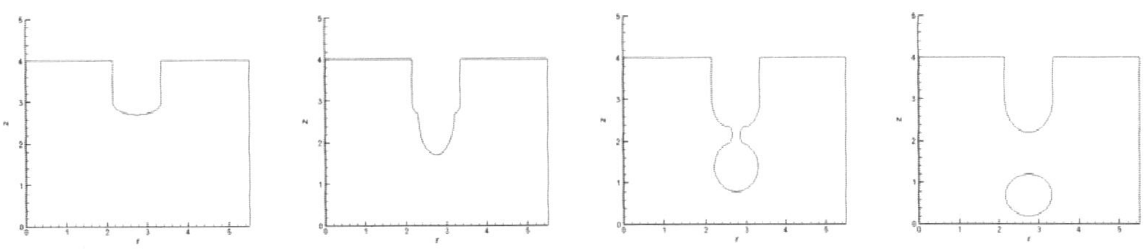

Figure 2. Interface structure of droplet and welding wire at t = 0, 1.377, 4.116, 7.231ms.

In order to show the validity of this new model, we also compare the numerical result of droplet size with the data of high-speed photography (figure 3) in the same conditions. We choose the horizontal and vertical diameter at the time just before the droplet touches the weld pool, respectively. The data of high-speed phototgraphy is the average value of 3000 periods. The compare is shown in table 2 and we also comput the same numerical example with VOF model without thermocapillary effect and mixture energy. The comparison shows that the result of new model agrees well with data of high-speed photography and the new model improve the presicion in predicting the droplet size compared with VOF.

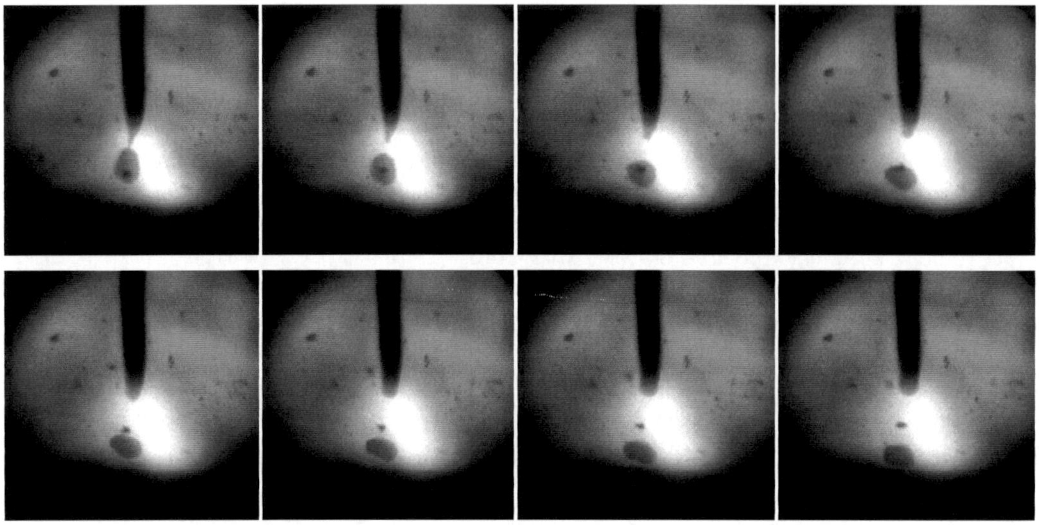

Figure 3. Image of metal transfer by high-speed photography.

Table 2. Compare among results of high-speed photography, new model and VOF.

	Horizontal Diameter	Vertical Diameter
High-speed photography	1.088 [mm]	0.962 [mm]
Numerical Simulations (New Model)	1.169 [mm]	1.053 [mm]
Relative Error (New Model)	7.444 %	9.459 %
Numerical Simulations (VOF)	1.287 [mm]	1.116 [mm]
Relative Error (VOF)	18.290 %	16.008 %

4. Conclusion

In this paper, a new model with the consideration of thermocapillary effect and mixture energy based on phase field model is proposed to analysis GMAW which can be more practical. The model composed of NS, energy equation, Maxwell's equation and phase field model is solved with a continuous finite element method and a modified midpoint scheme. The metal transfer in GMAW with pulse current is examined as its numerical result fits the theory of GMAW well. Compared with the numerical result of VOF and data of high-speed photography, the validity of this new model has been validated due to the relative error of droplet diameter below 10% and also it has the higher precision in predicting the droplet size than the previous work. This new model and method could provide guiding function in predicting how the droplet looks like which plays an important role in control of GMAW. In the future work, we will apply plasma force and more physical factor in this model and develop the practicability of this model.

5. References

[1] Mckelliet J and Szekely J 1986 *Metall. Mater. Trans. A* **17** 1139-48.
[2] Haidar J and Lowke J J 1998 *J. Phys. D: Appl. Phys.* **31** 1233-44.
[3] Chen M, Wu C and Lian R 2004 *Acta. Metall. Sin.* **40** 1227-32.
[4] Hu J and Tsai H 2007 *Int. J. Heat Mass. Trans.* **50** 833-46.
[5] Murphy A B 2011 *J. Phys. D: Appl. Phys.* **44** 252-60.
[6] Liu W, Shan Y, and Yuan Z 2016 *Hot Working Technol.* **45** 215-9.
[7] Lei Z L, Ni L C, Li B W and Zhang K Z 2017 *Opt. Laser Technol.* **88** 1-10.

[8] Tashiro S and Tanaka M 2011 *Trans. JWRI,* **40** 21-7.
[9] Haidar J 1998 *J. Phys. D: Appl. Phys.* **31** 1233-44.
[10] Borica R and Bestehorn M 2003 *Phys. Rev. E* **67** 28-51.
[11] Borcia R, Merkt D and Bestehorn M 2004 *Int. J. Bifurcat. Chaos.* **14** 4105-16.
[12] Verschueren M, van de Vosse F N and Meijer H E H 2001 *J. Fluid Mech.* **434** 153-66.
[13] Zhang C, Lin N, Tang Y and Zhao C 2014 *Comput. Fluids* **89** 214-31.
[14] Kapahi A, Sambasivan S and Udaykumar H S 2013 *J. Comput. Phys.* **241** 308-32.
[15] Friess M B and Kokh S 2014 *J. Comput. Phys.* **273** 485-519.
[16] Ding H, Spelt P D M and Shu C 2007 *J. Comput. Phys.* **226** 2078-95.
[17] Yue P, Zhou C and Feng J J 2010 *J. Fluid Mech.* **645** 279-94.
[18] Jiang Y, Lin P, Guo Z and Dong S 2015 *Commun. Comput. Phys.* **18** 180-202.
[19] Guo Z and Lin P 2015 *J. Fluid Mech.* **766** 226-71.
[20] Feng X, Feng J, Liu C and Shen J 2006 *J. Comput. Phys.* **218** 417-28.

Acknowledgements
Yongyue Jiang is partially supported by Beijing Postdoctoral Research Foundation.

Microstructural and Microanalytical Study on Concrete Exposed to the Sulfate Environment

Fang Qing[1], Li Beixing[2*], Yin Jiangang[3], Yuan Xiaolu[4]

[1]Hubei Electric Power Survey & Design Institute, Wuhan, 430040, China
[2]State Key Laboratory of Silicate Materials for Architecture, Wuhan University of Technology, Wuhan, 430070, China
[3]State Grid Hubei Electric Power Company, Wuhan, 430077, China
[4]College of Civil Engineering and Architecture, China Three Gorges University, Yichang 443002, China

Abstract. Microstructural properties have been examined to investigate the effect of mineral admixtures on the sulfate resistance of concrete. Concrete and cement paste specimens made with ordinary Portland cement (OPC) or ordinary Portland cement incorporating 20% fly ash (FA) or 30% ground blast furnace slag (GBFS), were made and exposed to 250 cycles of the cyclic sulfate environment. Microstructural and Microanalytical study was conducted by means of x-ray diffraction (XRD), scanning electron microscope (SEM), energy dispersive spectroscopy (EDS) and mercury intrusion porosimetry (MIP). Results indicate that the pore structure of concrete after sulfate exposure possesses the fractal feature. The OPC concrete presents more complex pore internal surface, higher porosity and less micro-pores than the concrete incorporating fly ash and GBFS. Portlandite in OPC concrete and OPC-FA concrete is mainly converted to gypsum; while for OPC-GBFS concrete, both gypsum and ettringite are formed. In the cyclic sulfate environment, repeated hydration and dehydration of sulfates produce the expansive stress in pores, aggravating the demolishment of concrete structure.

1. Introduction

Deterioration of concrete structure exposed to sulfate-bearing environments such as seawater, soils or groundwater containing high concentration of sulfates, is known as sulfate attack, which is commonly observed. Sulfate attack is a quite complicated process. Sulfates diffuse into concrete and react with cement matrix including the calcium hydroxide and the calcium aluminate hydrate, forming the expansive products such as the ettringite and gypsum. This induces the expansion and cracking of concrete. It is also reported that the sulfate attack causes the decomposition of hydrated products such as the decalcification of C-S-H, leading to the disintegration of cement matrix [1-6].

Incorporation of mineral admixtures in concrete decreases the C3A quantity in cementitious materials, which reduces the potential for the formation of harmful products such as ettringite. The pozzolanic reaction of mineral admixtures partially consumes calcium hydroxide, and produces the secondary C-S-H gel to densify the microstructure of concrete [7-12]. Therefore, the presence of mineral admixtures is known to affect the sulfate resistance of concrete. The physical and mechanical properties of concrete incorporating mineral admixtures under the sulfate exposure have been studied in prior work [13, 14]. In this study, microstructure analyses are performed to investigate the effect of mineral admixtures on the deterioration mechanism of concrete subjected to sulfate attack.

IOP Conf. Series: Materials Science and Engineering **269** (2017) 012070 doi:10.1088/1757-899X/269/1/012070

2. Experimental

2.1. Materials

Ordinary Portland cement (OPC) was used. Class I fly ash (FA) with the specific surface area of 620m^2/kg and ground blast furnace slag (GBFS) with the specific surface area of 450m^2/kg were used as supplementary cementitious materials. Chemical compositions of cement, FA and GBFS are given in table 1. BASF SP-8CR superplasticizer with the water reducing rate of 25% was used at different dosages in mixes to achieve the adequate workability in concretes and pastes. River sand with the fineness modulus of 2.9 and the coarse aggregate of crushed limestone with the size of 5mm~25mm were used in making concrete. The analytically pure sodium sulfate was used to prepare the sulfate solution.

Table 1. Chemical analysis for cement, fly ash and ground blast furnace slag (wt. %)

Oxide/phase	SiO_2	Al_2O_3	Fe_2O_3	CaO	MgO	SO_3	Na_2O	K_2O	C_3S	C_2S	C_3A	C_4AF
OPC	21.78	4.85	3.36	65.66	2.61	0.58	0.19	0.56	58.71	17.22	7.17	10.21
FA	48.91	28.05	8.85	2.65	1.03	0.86	0.78	2.08	—	—	—	—
GBFS	32.69	15.50	1.70	38.14	9.14	1.73	0.10	0.43	—	—	—	—

Table 2. Concrete proportions.

Samples	Materials (kg/m^3)						
	Cement	Water	Sand	Stone	FA	GBFS	Superplasticizer
L-1	460	154	724	1086	0	0	4.6
L-2	384	154	724	1086	96	0	4.8
L-3	336	154	724	1086	0	144	4.8

2.2. Specimens preparation.

Mixture proportions used in making concretes and pastes are listed in tables 2 and 3, respectively. Three different binders were used in this study: 100%OPC, and 80%OPC-20%FA, 70%OPC-30%GBFS. A 0.335 water/binder (w/b) ratio was used for pure OPC concrete, and the w/b for concrete mixes with mineral admixtures was 0.32. Concretes (100×100×100 mm) and pastes (20×20×20 mm) were made and cured until 28d in accordance with the standard curing method described in GB/T 50081-2002 (Standard test method for mechanical properties km ordinary concrete). Before exposure to the sulfate environment, mercury intrusion porosimetry (MIP) of L-1 concrete and x-ray diffraction (XRD) of OPC paste were measured.

Table 3. Paste proportions.

Binders	w/b	Superplasticizer (%)
100%OPC	0.30	0.25
80% OPC-20% FA	0.29	0.25
70% OPC-30% GBFS	0.29	0.25

2.3. Experimental methods.

To accelerate the process of experiment, the cyclic sulfate exposure was applied for both concrete and paste specimens. One cycle was composed of drying at 80℃ for 6h, cooling at room temperature for 1h, exposure to the 10% Na_2SO_4 solution at (20±3)℃ for 16h and airing at room temperature for 1h. The container of the Na_2SO_4 solution was sealed. The solution was mixed everyday with constant regulation. The sulfate solution was refreshed every 7 cycles.

After 250 cycles of exposure, the paste samples were examined for the chemical components through x-ray diffraction (XRD). The spalling concrete was analyzed by scanning electron microscope (SEM), energy dispersive spectroscopy (EDS) and mercury intrusion porosimetry (MIP).

3 Results and discussion

3.1. MIP analysis

Pore structure, which relates to the physical and mechanical properties of concrete including the strength and the durability, includes some vital parameters such as the porosity and the pore size distribution. Table 4, figure 1 and figure 3 show the porosity and the pore size distribution of concrete specimens. It can be seen that the porosity of L-1 concrete subjected to 250 cycles of sulfate attack is decreased by more than 30%, compared with that before exposure. However, there are less micro-pores (<50 nm) in L-1 after exposure than that before exposure. This is possibly because that the expansive products such as gypsum and ettringite produced by the sulfate attack fill in pores of concrete, reducing the porosity of concrete. But these expansive products are not densely packed, contributing to big pores (100~200 nm and >200 nm) in concrete. The porosity of concrete is decreased with the addition of mineral admixtures. Compared with L-1 whose pores are mainly concentrated in big sizes (50~100 nm, 100~200 nm and >200 nm), L-2 and L-3 achieve the decreased pore distribution in >50 nm. Both of them have a largest pore distribution in <20 nm, which do little harm to the durability of concrete. This indicates that the incorporation of mineral admixtures improves the sulfate resistance of concrete, which possesses the lower porosity and more micro-pores after 250 cycles of sulfate attack than the OPC concrete. It is probably because that the pozzolanic reaction of GBFS and fly ash reduces the contents of $Ca(OH)_2$ and calcium aluminate hydrate in the cement paste and promotes the hydration of Portland cement, which together with the filler effect of mineral admixtures, make the matrix denser to safeguard against the external ingressive sulfates.

Table 4. Pore structure parameters of concrete.

Samples	Porosity (by volume) %	Pore distribution (by volume) %				
		<20 nm	20~50 nm	50~100 nm	100~200 nm	>200 nm
L-1(0)	13.64	36.35	29.34	15.85	5.02	13.44
L-1	9.04	29.91	17.12	9.12	6.96	36.88
L-2	7.72	46.19	12.53	3.54	3.15	34.57
L-3	7.08	44.28	14.32	4.71	4.07	32.61

L-1(0): L-1 before sulfate exposure; L-1: L-1 after sulfate exposure.

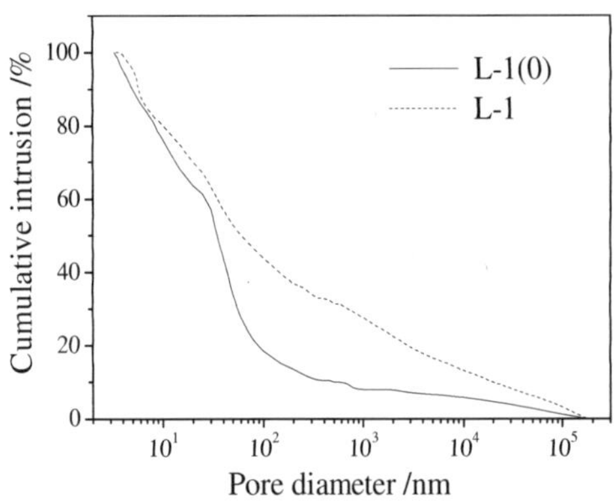

Figure 1. Pore size distribution of L-1 before and after exposure.

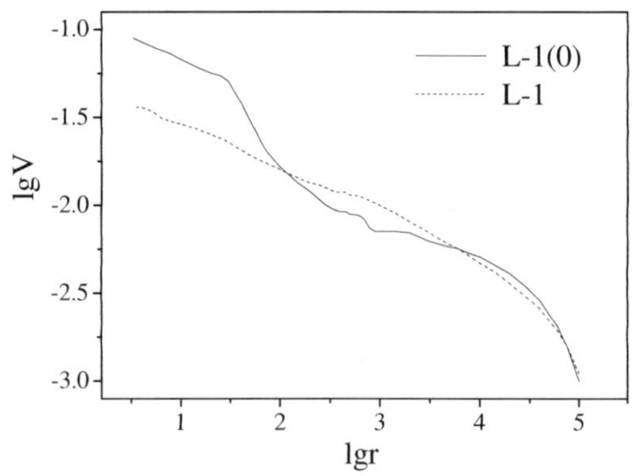

Figure 2. Fractal features of pore structures for L-1 before and after exposure.

The morphology of pores is also an important aspect for the pore structure, which can be used to analyze the complexity of the pore structure within concrete. Fractal dimension has been applied to quantitatively describe the morphology of pores, which can be obtained by the fractal structural model. In this paper, the Space-filling model is used to analyze the fractal features of the pore structure of concrete [15]:

$$\lg V = (3 - D)\lg r + \lg t \tag{1}$$

where V is the cumulative pore volume, r is the pore size, D is the fractal dimension, t is the constant.

According to the experimental results of MIP, $\lg V$ and $\lg r$ are calculated and applied to draw the fractal feature diagram of the pore structure, shown in figure 2 and figure 4. Regression analysis is made for the curves, and the fractal analysis results of the pore structure are shown in table 5. It can be seen that three regression equations have large R^2, which indicates that the curves conform to the equation (1). The fractal dimension D is a parameter to describe the complexity of the pore structure. The higher D represents the more complex internal surface of pores. From the fractal dimension D for L-1(0)~L-3 in Table 5, it is indicated that the OPC concrete before exposure possesses more complex pore internal surface than that after exposure. The OPC concrete after 250 cycles of exposure has more complex pore internal surface than the concrete incorporating fly ash and GBFS.

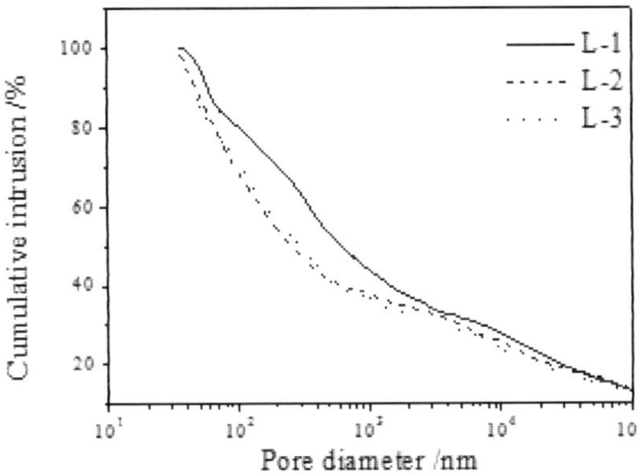

Figure 3. Pore size distribution of concrete after exposure.

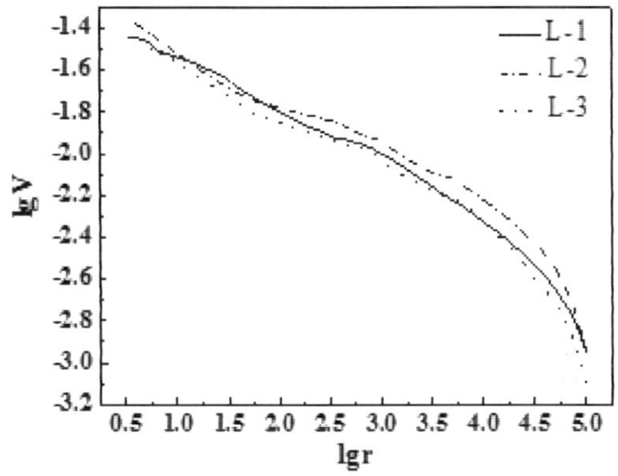

Figure 4. Fractal features of pore structures for concrete after exposure.

Table 5. Fractal analysis of pore structures.

Samples	Regression equation	R^2	D
L-1(0)	y = – 0.3930x – 0.8668	0.9575	3.3930
L-1	y = – 0.2941x – 1.2298	0.9686	3.2941
L-2	y = – 0.2702x – 1.2132	0.9512	3.2702
L-3	y = – 0.2614x – 1.2440	0.9522	3.2614

3.2. X-ray diffraction analysis

To study the corrosive products in the concrete caused by the sulfate attack, the X-ray diffraction analysis was conducted for the powder of paste samples, drawn from three types of mixes. The XRD diffractograms are presented in figures 5-8.

It can be seen that there were relatively low intensity peaks for C-S-H and portlandite in OPC sample subjected to 250 cycles of exposure (figure 6) compared with that obtained before exposure (figure 5). Some amounts of thenardite and gypsum were also detected in figure 6. Peaks for ettringite were intensified after exposure. This signifies that C-S-H and portlandite are consumed, and that gypsum, thenardite and ettringite are produced during cyclic sulfate exposure. There are relatively large intensity peaks for portlandite in OPC sample subjected to 250 cycles of exposure compared with those incorporating mineral admixtures, which is attributed to the pozzolanic reactivity of mineral admixtures. Large quantities of thenardite and gypsum, as well as limited ettringite, were detected in OPC paste. There is largest intensity peak of gypsum in OPC-FA sample, with large amounts of thenardite and little ettringite seen in figure 7; while large quantities of ettringite, thenardite and gypsum are found in the diffractogram of OPC-GBFS sample (figure 8).

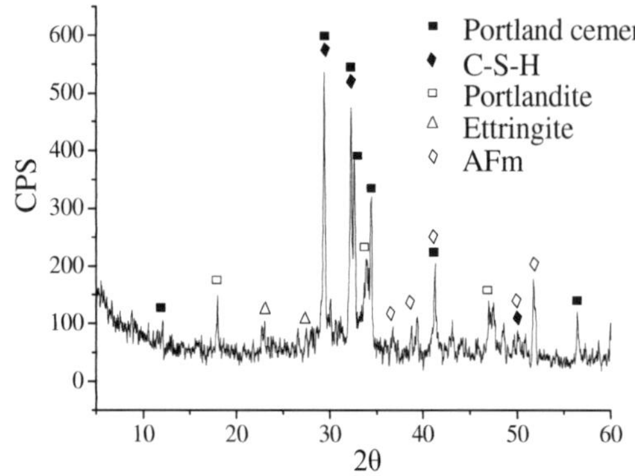

Figure 5. X-ray diffractogram of OPC paste sample before the sulfate exposure.

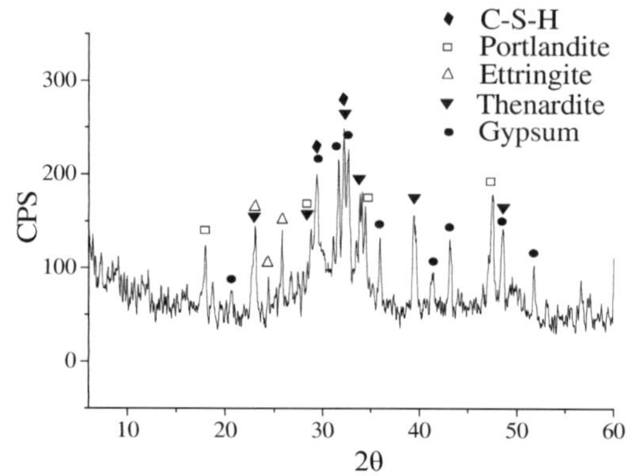

Figure 6. X-ray diffractogram of OPC paste sample subjected to 250 cycles of exposure.

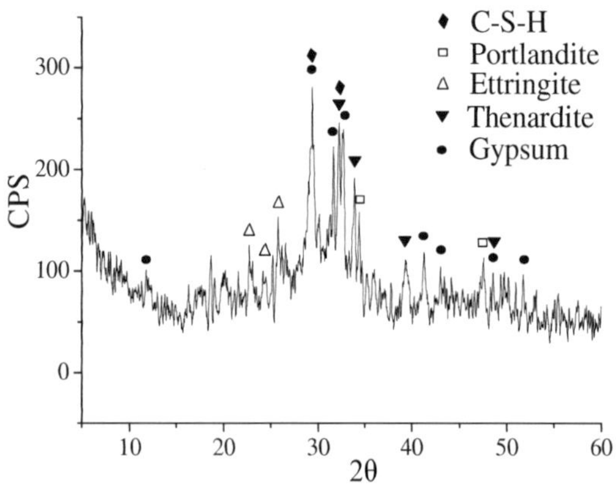

Figure 7. X-ray diffractogram of OPC-FA paste sample subjected to 250 cycles of exposure.

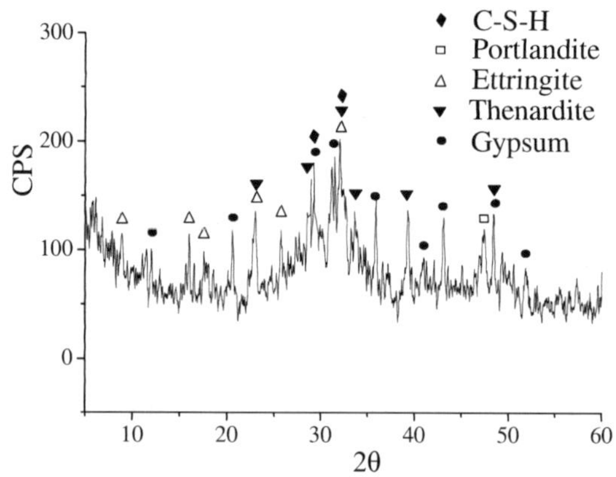

Figure 8. X-ray diffractogram of OPC-GBFS paste sample subjected to 250 cycles of exposure.

3.3. SEM and EDS analysis

To observe the microstructure of the concrete subjected to the sulfate environment, the SEM analysis was applied for the spalling concrete samples after 250 cycles of exposure. In addition, the EDS analysis was conducted to determine the existing compounds after reactions with external corrosive ions.

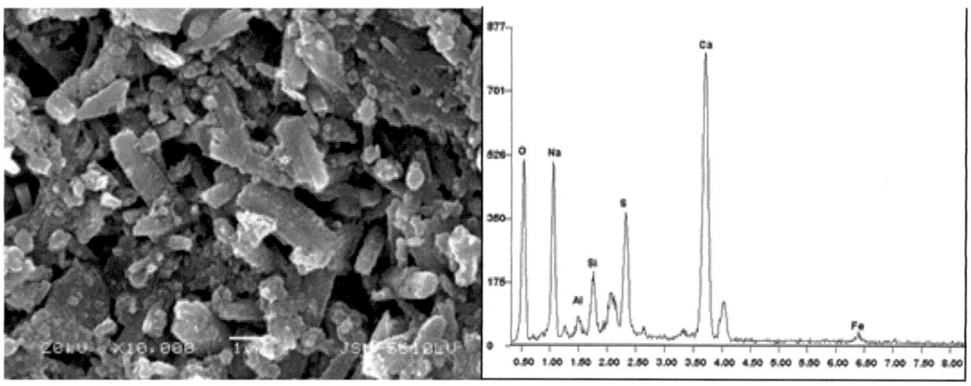

Figure 9. SEM and EDS of OPC concrete sample subjected to 250 cycles of exposure.

The SEM image of OPC concrete sample presents large amounts of white blocky crystals with the fine granular crystals on the surface, as well as small bar-like particles and fine cracks, shown in figure 9. The EDS analysis identifies the existence of sodium, aluminum, silicon, calcium and sulfur.

According to the XRD and EDS results, it is supposed that these white blocky crystals are gypsum and mirabilite, and the fine granular crystals are thenardite, and the bar-like particles are ettringite.

In the SEM image of OPC-FA concrete sample in figure 10, there are abundant white blocky crystals with the fine granular crystals on the surface, as well as the occurrence of fine cracks. The EDS analysis, together with the XRD analysis, confirms that the white blocky crystals are gypsum and mirabilite, and the fine granular crystals are thenardite. Low Al peak in the EDS image signifies the little amount of ettringite.

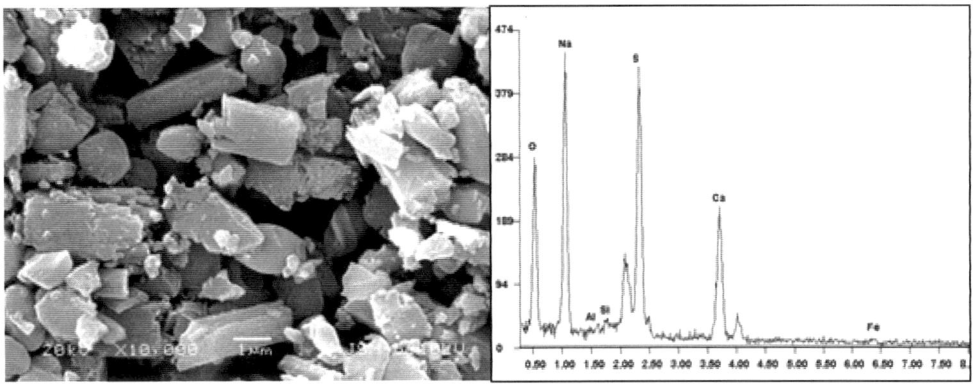

Figure 10. SEM and EDS of OPC-FA concrete sample subjected to 250 cycles of exposure.

Figure 11 shows the SEM and EDS images for the OPC-GBFS sample. There are large quantities of needle-like crystals, as well as some white crystals with the fine granular particles on the surface. According to the EDS and XRD analysis, these needle-like crystals are ettringite, and the white crystals are gypsum and mirabilite, and the fine granular particles are thenardite.

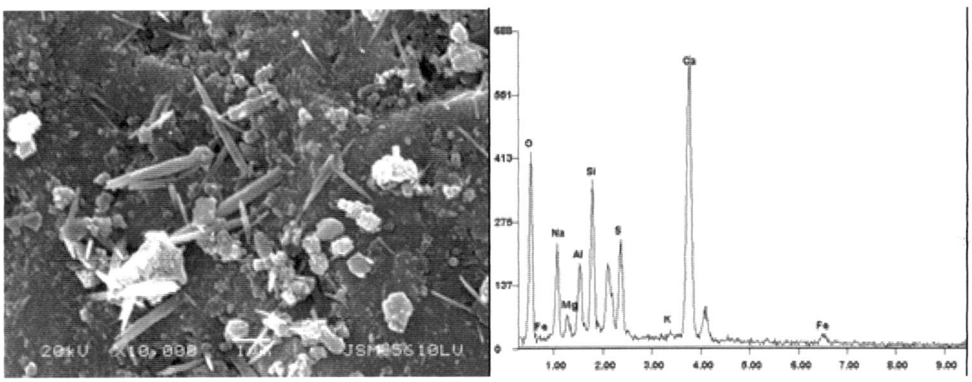

Figure 11. SEM and EDS of OPC-GBFS concrete sample subjected to 250 cycles of exposure.

Therefore, it can be concluded from the XRD, SEM, EDS analysis that sulfates react with the hardened cement paste in the concrete exposed to the sulfate environment. This produces various corrosive products, which is determined by the composition of the cementitious materials. The portlandite in the cement paste is mainly converted to gypsum by reaction with sulfates for OPC concrete and OPC-FA concrete; while for the OPC-GBFS concrete, a large quantity of portlandite reacts with sodium sulfate to form gypsum and ettringite. These expansive products accumulate to cause the occurrence of cracks in the structure, leading to the destruction of concrete.

It is also seen from the SEM images that mirabilite and thenardite are coexisted in various concrete subjected to 250 cycles of exposure. But only abundant thenardite is detected in X-ray diffraction analysis. This may be attributed to the dehydration of mirabilite when making paste samples. It was reported that mirabilite rapidly dehydrated at 20°C (RH<71%) to form thenardite [16]. Thus, it is indicated that repeated hydration and dehydration of sulfates occurred in concrete subjected to the cyclic sulfate exposure may produce additional expansive stress in pores, which aggravates the demolishment of concrete structure.

4. Conclusions

Incorporation of mineral admixtures improves the sulfate resistance of concrete, which possesses the lower porosity and more micro-pores after 250 cycles of sulfate attack than the OPC concrete. The pore structure of concrete after sulfate exposure possesses the fractal feature, and the OPC concrete after 250 cycles of exposure has more complex pore internal surface than the concrete incorporating fly ash and GBFS.

In the sulfate environment, the portlandite in the cement paste is mainly converted to gypsum by reaction with sulfates for OPC concrete and OPC-FA concrete; while for the OPC-GBFS concrete, a large quantity of portlandite reacts with sodium sulfate to form gypsum and ettringite.

Repeated hydration and dehydration of sulfates occurred in concrete subjected to the cyclic sulfate exposure, may produce additional expansive stress in pores, aggravating the demolishment of concrete structure.

5. References

[1] Mindess S, Young J F, Darwin D et al., 2004 *Concrete 2nd edition*, Chemical Industry Press, Beijing.
[2] Ganjiana E and Pouyab H S *2005 J. Cem. Concr. Res.* **35** 1332-43.
[3] Rafieizonooz M, Salim M R, Mirza J, et al, 2017 *J. Constr. Build. Mater.* **143** 234-46.
[4] Zhang J R, Sun M, Hou D S and Li Z J 2017 *J. Constr. Build. Mater.* **139** 365-73.
[5] Qi B, Gao J M, Chen F and Shen D 2017 *J. Constr. Build. Mater.* **138** 254-62.
[6] Tan Y S, Yu H F, Ma H Y, et al, 2017 *J. Constr. Build. Mater.* **141** 453-60.
[7] Sahmaran M, Erdem T K and Yaman I O 2007 *J. Constr. Build.* Mater. **21** 1771-8.
[8] Shafiq N, Nuruddin M F, Khan S U, Ayub T 2015 *J. Constr. Build.* Mater. **81** 313-23.
[9] Kalla P, Rana A, Chad Y B, et al, 2015 *J. Clean. Prod.* **87** 726-34.
[10] Yu R, Spiesz P and Brouwers H J H 2015 *J. Cem. Concr. Comp.* **55** 383-94.
[11] Yalçınkaya Ç and Yazıcı H 2017 *J. Constr. Build. Mater.* **144** 252-9.
[12] Mermerdaş K and Arbili M M 2015 *J. Constr. Build. Mater.* **94** 371-9.
[13] Yuan X L, Li B X, Cui G, Zhao S C and Zhou M K 2009 *J. Chinese Ceram. Soc.* **37** 1754-9.
[14] Yuan X L, Li B X, Cui G, Zhao S C and Zhou M K 2010 *J. Wuhan Univ. Tech. (Mater. Sci. Ed.).* **25** 1065-9.
[15] Jin S S, Zhang J X and Li S 2009 *J. Concr.* **9** 34-7.
[16] Rodriguez-Navarro C, Doehne E and Sebastian E 2000 *J. Cem. Concr. Res.* **30** 1527-34.

Acknowledgements

Authors gratefully acknowledge the supports from the New Technology Research Projects for 2016 infrastructure engineering of State Grid Corporation of China (No. 40).

Analysis of Surface and Subsurface Damage Morphology in Rotary Ultrasonic Machining of BK7 Glass

Wang Hong-xiang[1*], Wang Chu[1], Liu Jun-liang[1], Gao Shi[1], and Zhai Wen-Jie[1]

[1]School of Mechatronics Engineering, Harbin Institute of Technology, Harbin, Heilongjiang, China

Abstract. This paper investigates the formation process of surface/subsurface damage in the rotary ultrasonic machining of BK7 glass. The results show that during the milling using the end face of the tool, the cutting depth and the residual height between the abrasive grains constantly change with the high-frequency vibration, generating lots of cracks on both sides of the scratches. The high-frequency vibration accelerates the chips falling from the surface, so that the chips and thermal damage are reduced, causing the grinding surface quality better. A plastic deformation area is formed during the grinding, due to the non-uniform cutting force on the material surface, and the residual stress is produced in the deformation area, inducing the median/lateral cracks.

1. Introduction

During the process of rotary ultrasonic machining (RUM) of BK7 glass, the material usually produces plastic deformation and brittle fracture, and the median and lateral cracks are generated on the surface. The median cracks propagate downwardly in the loading direction to form subsurface damage (SSD), while the lateral cracks extend upwardly parallel to the surface of the workpiece, when it is extended to the surface, the material falls off in the form of chip debris. Markov AI proposed that the repeated impacts of diamond abrasive grains cause a large number of microscopic and macroscopic cracks to nucleate and extend, which are interwoven into a network, forming a weakened layer in the material [1]. Prabhakar D investigated the forward motion and rolling of the abrasive grains can cause the brittle fracture of the weakened layer [2]. Spur G et al. found that the real contact time was reduced by 50% due to the sinusoidal motion of the workpiece with ultrasonic assistance, resulting in an enormous reduction in frictional effects and normal forces [3]. Qu W et al. found that the lateral cracks would inhibit the formation of subsequent median cracks if the tangential displacement of abrasive grains is less than the extended length of the lateral crack, so the SSD depth is decreased [4]. Lv et al. did the scratching tests with a gradient load on the BK7 glass. The results showed that with a small scratching depth the scratches on the surface were discontinuous, and there were a few of cracks at the bottom of the scratches. As the scratching depth increasing, there were lateral cracks and median cracks on the both sides of the scratches, and the overlapped cracks caused the formation of the chips, which indicated the removal of the material in RUM is mainly caused by the crack propagation [5, 6].

As described above, in the RUM process, the abrasive grains have a great transient acceleration with the assistance of ultrasonic vibrations. Therefore, it is necessary to study the influence of high-frequency vibration on the surface and SSD characteristics in RUM of BK7 glass, in order to perfect the mechanism of RUM.

2. Formation process of BK7 glass surface by RUM

The RUM of BK7 glass is essentially a micro-cutting process on the surface of the workpiece using a large number of diamond abrasive grains. Abrasive grains press and scratch on the surface, ultimately

IOP Conf. Series: Materials Science and Engineering **269** (2017) 012071 doi:10.1088/1757-899X/269/1/012071

forming the machined surface. The RUM with different cutting depths is shown in figure 1, with increase of the cutting depth, not only the cutting takes place at the end face of the tool, the grains at the lateral of the tool are also involved in cutting. When the back engagement (a_p) at the end face of the tool is less than maximum protrusion height of grain (h_e), machining only occurs at the end face of the tool and the material is removed by milling (figure 1a); When a_p is greater than h_e, the grains at the lateral of the tool will cut the lateral surface of the workpiece, at the same time, the horizontal surface of workpiece is milled by grains at the end face of the tool (figure 1b).

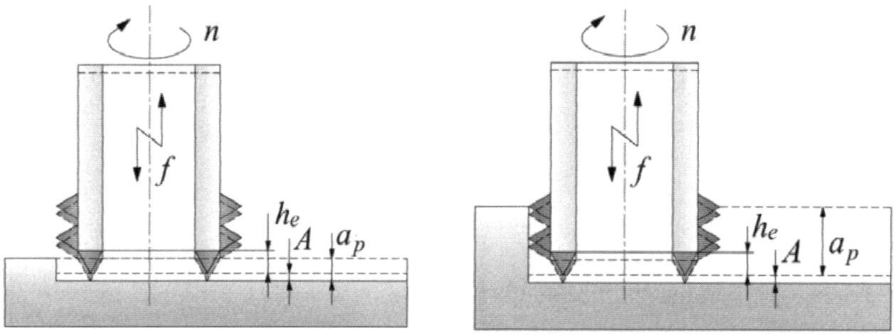

(a) $a_p < h_e$ (b) $a_p \geq h_e$

Figure 1. Illustrations of the RUM with different cutting depths.

It can be seen from figure 1, the RUM produces two machining surfaces, the milling surface perpendicular to the spindle axis and the grinding surface parallel to the spindle axis. Because the grains at the end and lateral faces have different trajectories, so it is necessary to do a separate investigation on the cutting effects of the grains on the end and lateral faces, and analyze the formation process of BK7 glass surface by RUM.

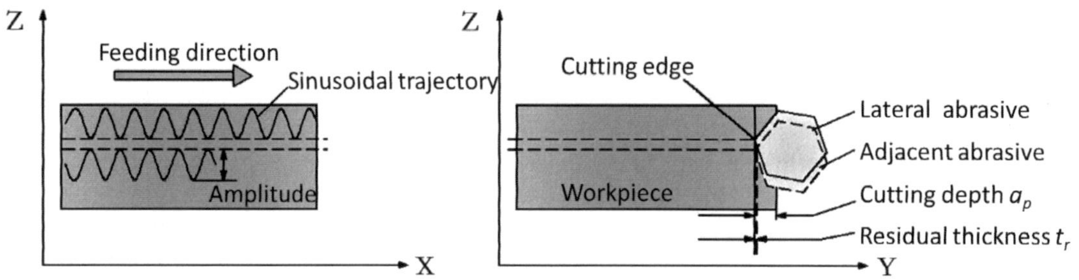

Figure 2. Illustrations of the end face milling.

The cutting process of grain at the end face is shown in figure 2, and it can be seen that the axial high-frequency vibration has an influence on the motion of grain, causing the cutting depth constantly changing. At the same time, the milling residual height (h_r) between the adjacent abrasive grains is also constantly changing, and resulting in a large number of microscopic cracks on the both sides of the scratches, and then they develop into the surface and subsurface damage.

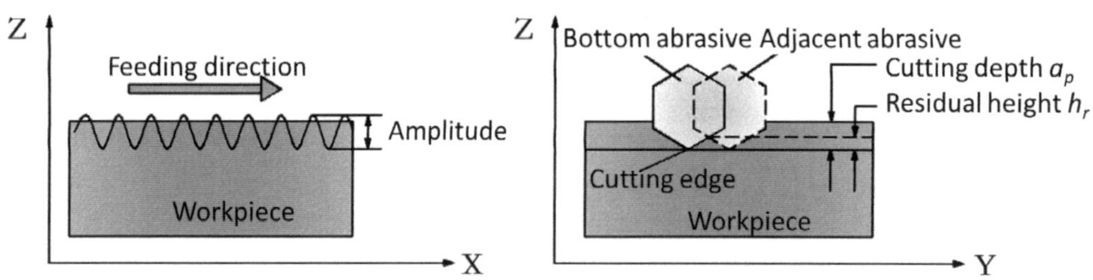

Figure 3. Illustrations of lateral face grinding.

The cutting process of grain at the lateral face is shown in figure 3. The axial high-frequency vibration causes the abrasive grains to move in a sinusoidal trajectory, and there are interferences between the adjacent abrasive grains causing the grinding residual thickness (t_r') to be less than the average grinding residual thickness (t_r), reducing the roughness of the lateral grinding surface.

3. Morphology of surface damage

In order to study the characteristics of surface damage in RUM, the milling experiments for BK7 glass were carried out. The experiments were performed on a RUM machine, and used hollow nickel-based electroplating diamond tool (outer diameter is 8mm, inner diameter is 6mm grain size is 91μm, grain concentration is 100%). The dimensions of the specimens are 30mm×15mm×6mm. Other experimental conditions are listed in table 1.

Table 1. Experimental conditions.

Parameter	Unit	Value
Spindle speed	[rpm]	6300
Feed rate	[mm/s]	20
Cutting depth	[μm]	200
Vibration frequency	[kHz]	20.689
Vibration amplitude	[μm]	5

In order to obtain the surface damage morphology, the specimens were observed by three-dimensional super-depth optical microscope (VHX-1000E). Figure 4 shows the morphologies of milling surface machined by end face and grinding surface machined by lateral face. Figure 4a shows the brittle fracture area is small on the milling surface (produced by end face), and there is a large amount of fine collapses and a small amount of pulverizable areas.

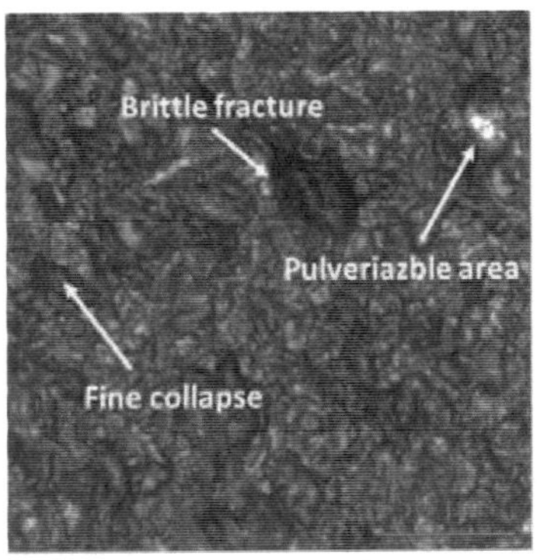

(a) Milling surface

(b) Grinding surface

Figure 4. Morphologies of the surface produced by end face and lateral face (by VHX-1000E).

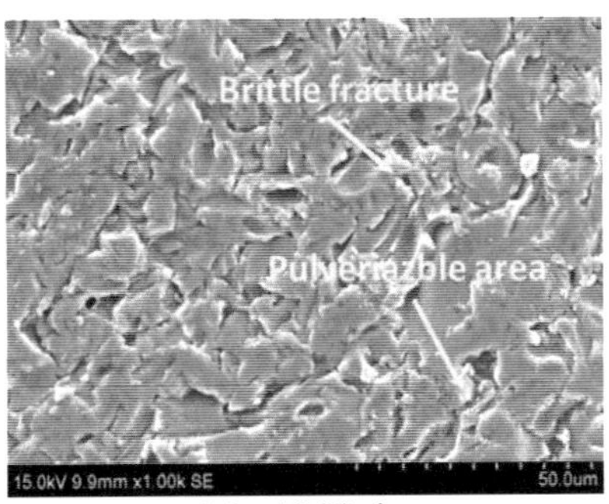

(a) x1000 times

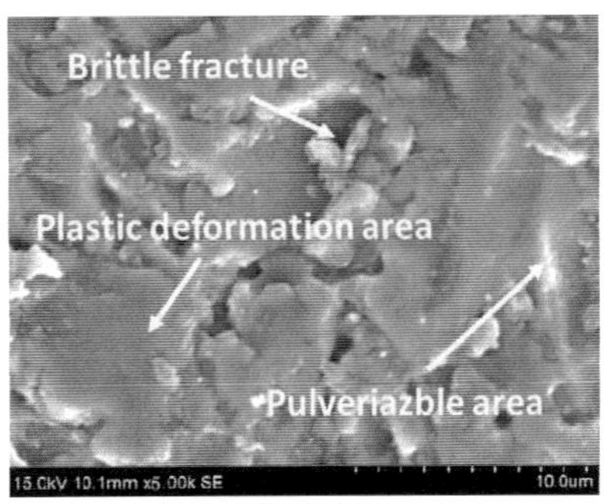

(b) x5000 times

Figure 5. Morphologies of the surface produced by end face milling (by SEM).

The morphology of surface damage produced by lateral face of the tool is shown in figure 4b, and the grinding surface has more small plate shaped pits rather than large plate shaped pits, causing fewer chips and better surface quality. This is due to the grains at the lateral face have high-frequency axial vibration, impacting the grinding surface and accelerating the separation of the chips and the workpiece. Thereby effectively reducing the thermal damage on the grinding surface. In addition, the periodic change of the contact area between the abrasive and the workpiece reduces the wear of the grain. In order to further reveal the effect of RUM on the surface damage, the microscopic surface morphologies produced by end face milling were observed by SEM (figure 5). As shown in figure 5(a), the pulverizable areas are stacked with small fragments of cluster distribution, and the pulveriazble areas are less than the fractured areas.

4. Morphology of subsurface damage

Prior to the formal machining tests, a bonded interface sectioning technique was selected to examine the morphology of subsurface damage [7]. In this method, the surfaces of two specimens (15mm×15mm×6mm) were first ground and polished until the surface roughness was 10nm or less in order to remove the subsurface damage caused in the preparation of the specimen. Subsequently, the two specimens were bonded together with an epoxy resin adhesive. To minimize edge chipping of the bonded surfaces, the clamping pressure was applied to ensure that a thin adhesive thickness was achieved. In order to investigate the influence of RUM, the bonded specimen was machined by RUM. After the machining they were separated by melting the adhesive and then cleaned with acetone in the ultrasonic cleaner. The SSD morphologies were observed by VHX-1000E (figure 6).

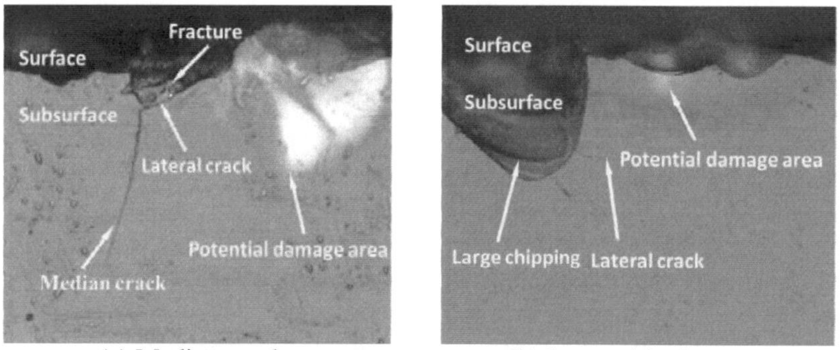

(a) Median crack (b) Large chipping

Figure 6. Morphologies of SSD caused by RUM.

Figure 6a shows the SSD caused by RUM mainly includes chipping layer and cracking layer. The chipping layer was composed of irregular surface defects mainly generated by chips falling; the cracking layer was composed of lateral cracks and median cracks. At the bottom of the shell-like defect there was a median crack extending downwardly perpendicular to the surface, at the lower right of the shell-like defect there was a lateral crack at the beginning of the extension, and there was a bright area above the lateral crack, which was a potential damage area. There were a large number of lateral cracks in the interior of the material under the potential damage area (shown in figure 6b). At the right side of the large chipping shown in figure 6b, there was a lateral crack extending downwardly under the potential damage area. This phenomenon is because in the RUM process a plastic deformation area was formed, and the cutting force was changing during the RUM process. Thus the residual stress was in the deformation area, which eventually induced median/lateral cracks.

5. Subsurface damage depth measurement

In this paper, we use an advanced finishing technique (magneto-rheological finishing or MRF), causing no SSD, to create a shallow groove on the BK7 glass surface to obtain the distribution of SSD. SSD depth is calculated by the deepest distance from the crack tip to the surface and the size of the groove. This method can measure the SSD depth by the measurement of the horizontal extension distance of SSD in millimeter instead of the measurement of the vertical depth of the SSD in micrometer in tens of microns, in other words, it can amplify the SSD depth in small scale, which is suitable for precise detection of SSD depth [8, 9].

The BK7 glass specimens (30mm×15mm×6mm) were polished by a RUM machine (DMG Ultrasonic70-5) with a hollow nickel-based electroplating diamond tool (outer diameter is 8mm, wall thickness is 1mm, grain size is 76μm, grain density is 100%), and other experimental conditions are shown in table 2.

Table 2. Experimental conditions.

Parameter	Unit	Value
Spindle speed	[rpm]	8300
Feed rate	[mm/s]	75
Cutting depth	[μm]	10
Vibration frequency	[kHz]	20.689
Vibration amplitude	[μm]	5

The procedures of the measurement method are briefly described here. (1) Using the MRF machine (shown in figure 7a), the specimen was polished a groove on the surface, and the middle of the groove must penetrate the entire SSD in order to ensure the deepest SSD can be observed; (2) etched in the 5% HF solution for 15min until the subsurface cracks exposed, and the etched specimens were cleaned by ultrasonic cleaner; (3) using the contourgraph to obtain the morphology of the groove, and observing the cracks by the optical microscope; (4) analyzed by image and morphology of the groove to determine the depth and length of SSD. The polished specimens are shown in figure 7b

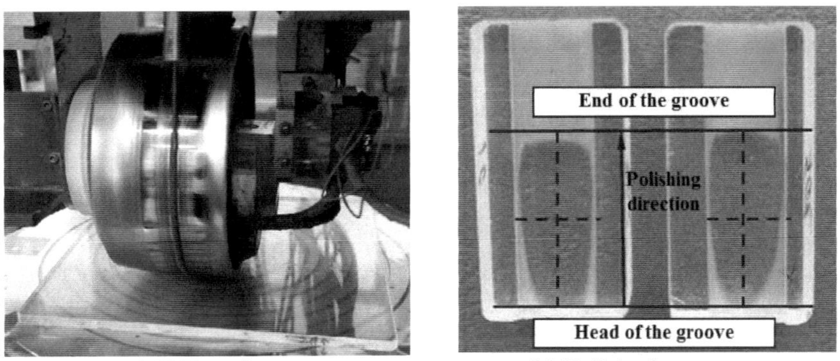

(a) MRF equipment (b) Polished groove

Figure 7. MRF equipment and the polished specimens.

Obtained by the contour graph and microscope, the cross-sectional profiles of the groove along the horizontal and vertical directions are shown in figure 8, and the morphologies and distributions of cracks are shown in figure 9.

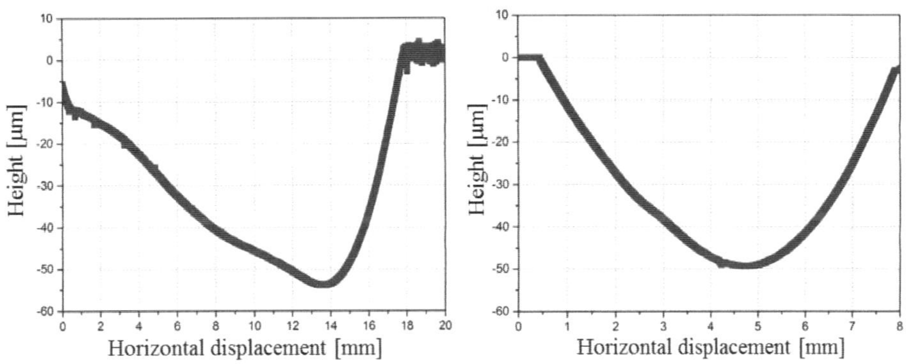

(a) Along the polishing direction (b) Vertical to the polishing direction

Figure 8. Cross-section profiles produced by MRF.

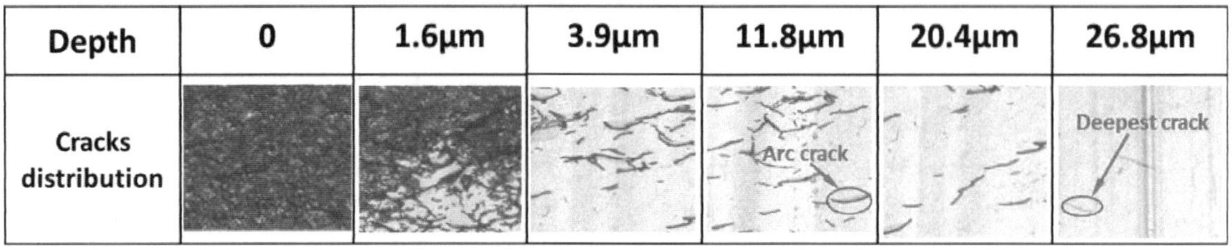

Figure 9. Cracks distributions at different depth.

It can be seen from figure 8 that the profiles of the head and the end of the groove were not the same, so the magnifications of SSD depth were different (75.47 times in head and 279.96 times in end). The polished groove were tested from head and end respectively, and the average value of the results was taken as the SSD depth. Finally, the maximum SSD depth of the BK7 glass specimen was 26.8 μm.

6. Conclusion

(1) The formation process of BK7 glass surface by RUM was investigated. It was found that the cutting depth and the residual height (h_r) between the adjacent abrasive grains changed with the high-frequency vibration, causing a large number of cracks on the both sides of the scratches, and the interferences between grains caused the roughness of the lateral grinding surface reduced

(2) There are a large amount of chips produced on the milling surface, the brittle fracture area was small, and there were a small amount of pulverizable areas. The grinding surface had a better surface quality, because the high-frequency vibration accelerated the chips falling and reduced the large chips and thermal damage at the same time.

(3) The morphology of SSD by RUM was analyzed, and there was a median cracks extending downwardly at the bottom of the fracture. At the right of the fracture there was a lateral crack which is parallel to the surface. The lateral and median cracks were induced by the residual stress, and the accumulation of these cracks formed SSD.

7. References

[1] Markov A I 1997 *J. Mach. Tool.* **48** 45-7.
[2] Prabhakar D 1992 *Machining Advanced Ceramic Materials Using Rotary Ultrasonic Machining Process*, D. University of Illinois at Urbana-Champaign, Illinois.
[3] Spur G, Holl S E 1996 *J. Mater. Proc. Tech.* **62** 287-93.
[4] Qu W, Wang K, Miller M H, et al. 2000 *J. Prec. Eng.* **24** 329-37.
[5] Lv D, Wang H, Tang Y, et al. 2013 *J. Prec. Eng.* **37** 839-48.
[6] Lv D, Huang Y, Wang H, et al. 2013 *J. Mater. Proc. Tech.* **213** 1548-57.
[7] Suratwala T, Wong L, Miller P, et al. 2006 *J. Noncryst. Solids.* **352** 5601-17.
[8] Menapace J A, .Davis P J, Steele W A, et al. 2005 *J. Proe. SPIE.* **5991** 39-49.
[9] Randi J A, Lambro Poulos J C, and Jacobs S D 2005 *J. Appl. Opt.* **44** 2241-9.

Acknowledgments

This research was financially supported by National Natural Science Foundation of China (Grant No.51475106, 51475119) and Science Challenge Project (Grant No.JCKY2016212A506-0503).

Investigation on Tensile Fatigue Characteristics of Meshed GUM Metal Plates for Bone Graft Applications

Koki Sekiguchi[1*], Jianmei He[1]

[1]Dept. of Mechanical Engineering, Kogakuin Univ. 1-24-2, Nishi-shinjuku, Shinjuku-ku, Tokyo, 163-8677 Japan

Abstract. GUM Metal has characteristics of lower elasticity rigidity, large elastic deformation, higher strength and biocompatibility etc. When it is used for implant applications, there is still problem like overloading on the natural-bone because of its high rigidity compared with the human bones. Therefore, the purpose of this study is to create more flexible meshed plates for implant applications from the viewpoints of elastic rigidity and volume density. Basic mesh shapes are designed, devised and applied for meshed GUM Metal plates using three dimensional (3D) CAD tools. Experimental evaluation on tensile fatigue characteristics of meshed GUM Metal plate specimens are carried out. Analytical approaches on stress evaluation are also executed through finite element method to obtain the S-N curve for fatigue characteristic evaluation.

1. Introduction

Recently in Japan, degenerative intervertebral discs have a sign of epidemic as one of diseases caused by aging and lifestyle habits. Degenerative intervertebral discs include lumber disc herniation, intervertebral disc herniation and cervical disc herniation. Table 1 shows the number of those patients having been increased year by year.

Table 1. Disc herniation the annual number of surgery (2014.4~2015.3).

Indivisual surgeries	Total case number	Average days in hospital
Surgical removal	7,152	11.5
Other surgeries	17,187	18.0
No surgeries	16,590	9.8
Total throughout the year	40,929	13.5

In this study, disc herniation as one type of disc defects is interested. Figure 1 shows some cases of the intervertebral disc protrudes beyond the normal intervertebral space to compress the nerve and cause pain. The treatment for disc herniation varies depending on different conditions. One of the currently practiced treatments is called spinal fusion surgery as also shown in figure 1. In this treatment, the upper and lower spinal cords of the defective disc are fixed using pure titanium or titanium alloy implant products. However, there is a problem that the load on the natural-bone of the human body is large and will cause overloading on healthy nature bones [1]. Therefore, in order to reduce the load on patient's natural-bone, creation of implant products matching the mechanism such as the elasticity and rigidity of natural bones are desired as much as possible.

In this study, mesh structure [2] applications are considered for GUM Metal plates. GUM Metal is one kind of titanium alloy having relatively low elastic modulus and large elastic deformation compared

IOP Conf. Series: Materials Science and Engineering **269** (2017) 012072 doi:10.1088/1757-899X/269/1/012072

with pure titanium and other titanium alloy. The purpose of this study is to design flexible meshed plate shapes and evaluate the tensile fatigue properties of meshed GUM Metal plate specimens for implant applications. Figure 2 shows the example of spinal fusion treatment using meshed GUM Metal plates.

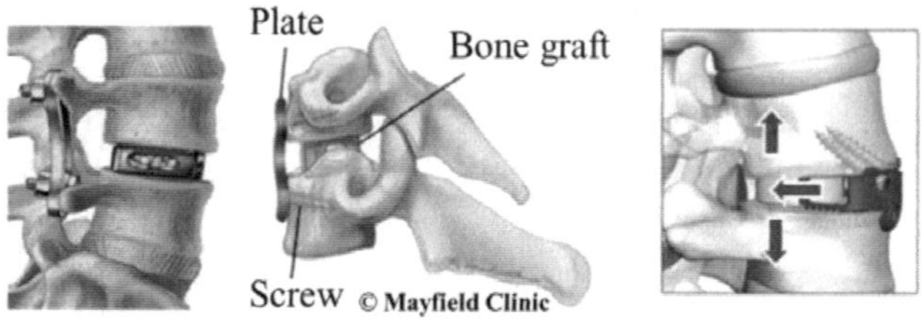

Figure 1. Fixation methods on spine for hernia of intervertebral discs using titanium implants.

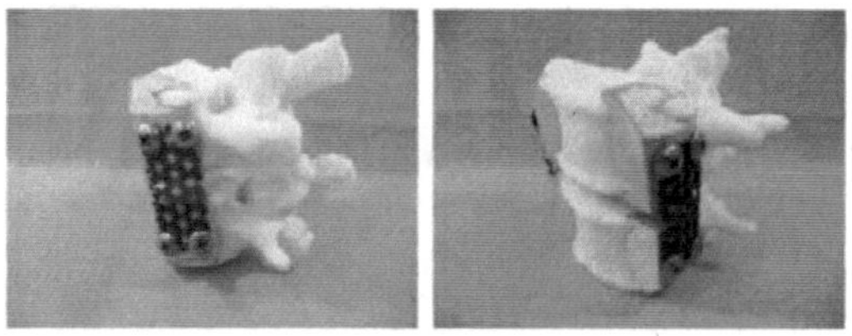

Figure 2. Sample application with meshed plates on spine for hernia of intervertebral discs.

2. Design of basic mesh shapes for meshed GUM metal plates

2.1. Mechanical properties of GUM metal

As shown in figure 3[3], GUM Metal is characterized by relatively low elasticity rigidity, high strength, large elastic deformability and high biocompatibility compared with other metals and metal alloys. Table 2 shows the mechanical properties of GUM Metal plates used in this study for experimental approach.

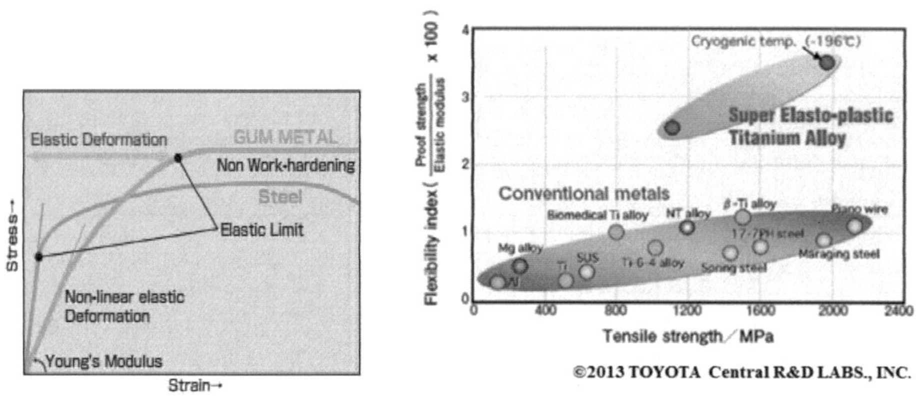

Figure 3. Comparison on tensile properties of GUM Metal with other metals and alloys.

Table 2. Mechanical property of GUM Metal.

Material property of GUM Metal	
Density [kg/m^3]	5600
Young's modulus [GPa]	91.0
Poisson's ratio	0.33

2.2. Designed basic mesh shapes and prototype model of meshed GUM metal plates.

Design concepts for basic mesh shapes are as follows considering of low cost, long life and higher flexibility etc.

(1) Single basic mesh shape construction for simplification of manufacturing and cost-down purpose

(2) Higher 3D flexibilities including expansion/contraction, bending and torsion for possibility of handily shape changes during surgery

(3) Easy-controllable mechanical properties like elastic modulus, bending stiffness etc. for approachability to natural-bone's mechanical properties

(4) Uniform mesh line width and non-angle smooth shape desired to minimize stress concentrations and lead for higher strengths and longer operating life

Basic mesh shapes based on the above design concepts are obtained using 3D CAD code Solidworks and sample ones for experimental evaluation are shown in figure 4. Two types of basic mesh shape are designed: 90° axisymmetric type based from regular tetragon and 60° axisymmetric type based from regular hexagon. For each type, three basic mesh shapes are designed having different sizes as design variables. Using these mesh basic shapes, prototype shape model of mesh GUM Metal plates are then created and shown in figure 5. These shape models are used and GUM Metal plate specimens are manufactured by laser-cutting processing. These meshed GUM Metal plate specimens are subjected to tensile fatigue experiment to evaluate their tensile fatigue characteristics and shape models are used for stress analysis.

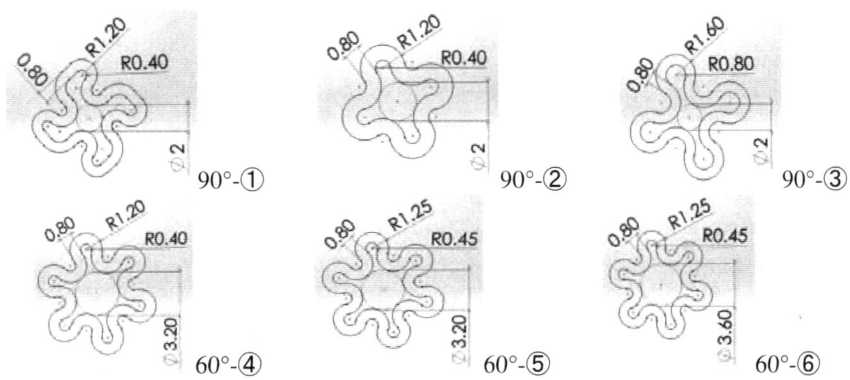

Figure 4. Basic mesh shapes for prototype meshed GUM Metal plates.

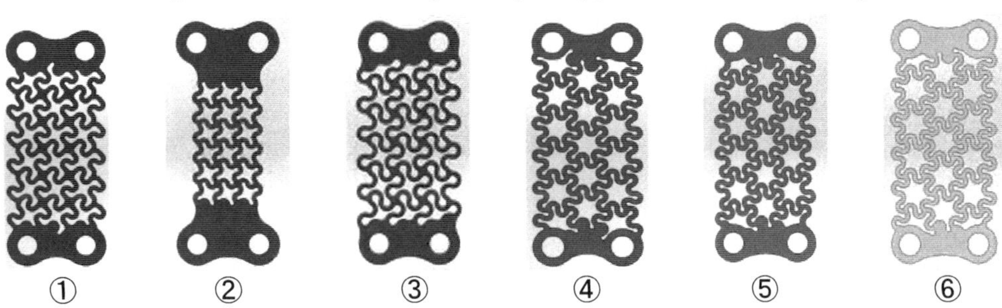

Figure 5. Meshed plate models for sample specimens (90° and 60° axisymmetric types).

3. Tensile fatigue characteristics of prototype meshed GUM metal plates

3.1. Tensile fatigue experiment of prototype meshed GUM metal plates.

Tensile fatigue tests were carried out on prototype meshed GUM Metal plate specimens. Total 36 specimens of 6 type of meshed GUM Metal plates are tested as shown in figure 6 with testing machine (Asahi Seisakusho FRS-20) and specimen installation. Tensile loads of 45N, 55N, 65N, 75N, 85N and 95N are applied for experimental investigation.

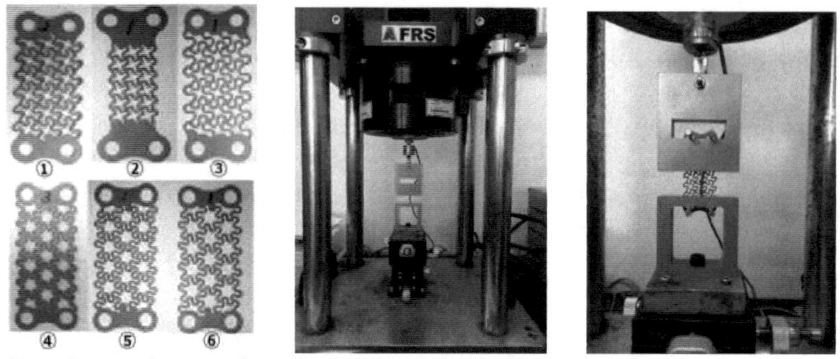

Figure 6. Sample specimens of meshed GUM Metal plates with fixtures and testing machine.

3.2. Stress evaluation of prototype meshed GUM Metal plates under tensile loading.

It is difficult to use the effective cross-sectional area of the prototype meshed plates for stress evaluation because of the complex shapes. Thus, ANSYS finite element analysis code were introduced and stress analyses were carried out in order to grasp the maximum Von Mises stress under the tensile loading of the tensile fatigue tests. Figure 7 shows an example model of the tensile fatigue test on meshed GUM Metal plate with the loading/constraint conditions and the stress contour plot of a 60° axisymmetric specimen. Table 3 shows the fixed finite element mesh for two type of meshed GUM Metal specimens. Material properties shown in table 2 are used for analytical approach. Stress results as shown in figure 7 confirmed that the locations where stress concentrations occurred didn't change under different tensile loading. Maximum Von Mises stress with respect to tensile loads is shown in figure 8.

Table 3. Analytical conditions for meshed GUM Metal plates.

	90° axisymmetric model	60° axisymmetric model
Element size [mm]	0.128	0.125
Number of mesh elements	220,836	248,500
Number of mesh nodes	169,235	194,454

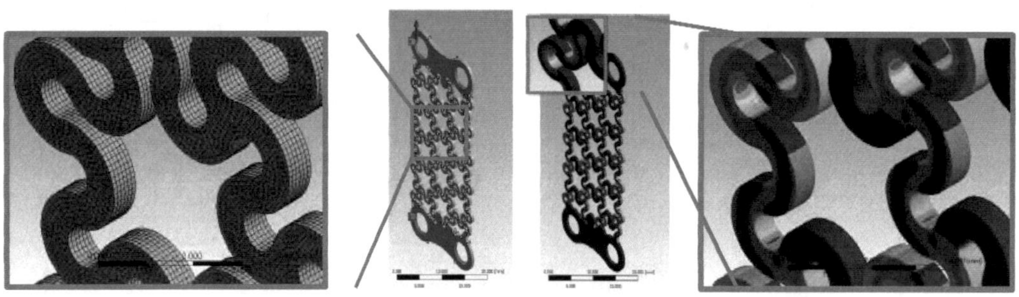

Figure 7. Stress analysis of prototype meshed GUM Metal plate (60° axisymmetric model).

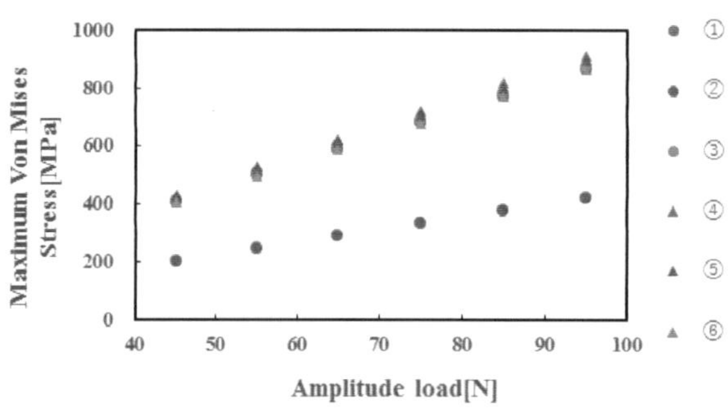

Figure 8. Maximum Von Mises stress results with respect to tensile loading.

3.3. Experimental and analytical results of tensile fatigue tests.
Figure 9 shows the fractured prototype meshed plates under different tensile loading after tensile fatigue tests. Comparing 90° with 60° axisymmetric prototype meshed plates, fractures are occurred at the inflectional locations with large curvature for both type of the specimens. These are coincident with the stress analysis results as shown in figure 7. Thus, it was possible to evaluate the tensile fatigue properties of meshed plates combined with experimental and analytical approaches introduced in this study.

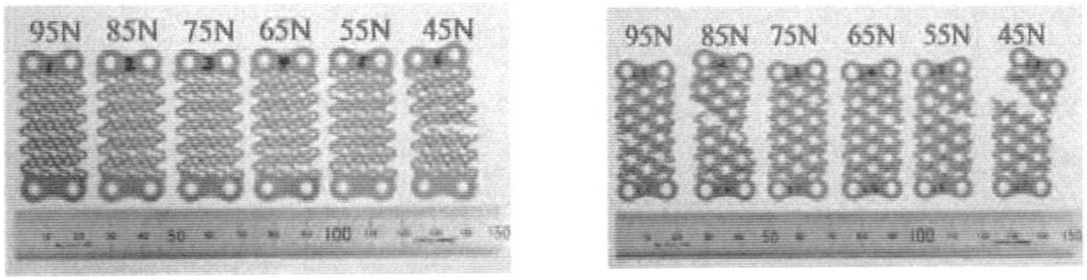

Figure 9. Typical meshed plate specimens after tensile fatigue tests (left: 90° specimen, right: 60° specimen).

S-N plots combined with analytical maximum Von Mises stress results of prototype meshed GUM Metal plates and tensile fatigue test results are then shown in figure 10. Horizontal axis represents the loading iteration number obtained from experiments and vertical axis shows the maximum Von Mises stresses under 45N, 55N, 65N, 75N, 85N and 95N tensile loads obtained from analysis.

From these results, one can see first that 90° axisymmetric specimen ② has most long life performance compared with other 5 models. Compared basic mesh shapes of prototype meshed GUM Metal plates as shown in figure 4, one can see that the curvature change in shape ② is the smallest one. Also from maximum Von Mises stress results with respect to tensile loading as shown in figure 8, 90° axisymmetric specimen ② shows the lowest stress value, which means that the curvature change in basic mesh shape will greatly affect the maximum Von Mises stress.

Additionally, if the maximum Von Mises stress under tensile loading shows near 200.0 MPa, it can endure approximately 10^6 times of tensile loading cycles. Therefore, the tensile fatigue limits of prototype meshed GUM Metal plates are under 200.0 MPa. Furthermore, in the results of the 60° axisymmetric meshed plate specimens, the difference in tensile fatigue endurance times due to the difference in basic mesh shape is small. However, it was found that changing the size (center radius) of the basic mesh shape does not significantly affect the tensile fatigue characteristics of meshed plates. Thus, it can be considered that reducing the curvature change of basic mesh shape will be most useful way to reach the tensile fatigue limit of meshed plates.

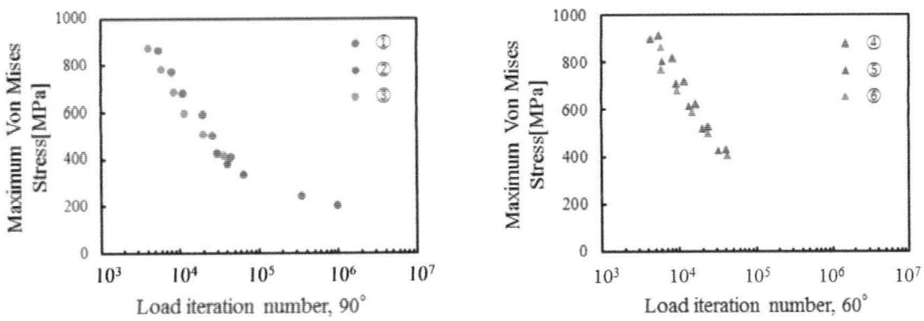

Figure 10. S-N curve of prototype 90° and 60° axisymmetric meshed plates.

4. Summary
Experimental and analytical evaluations on tensile fatigue characteristics of prototype meshed GUM Metal plates were executed. From the experimental and analytical results,

(1) Fractures were occurred at the inflectional locations with large curvatures for both type of the specimens, and coincident with the stress concentration locations obtained from analytical results.

(2) The change in curvatures of basic mesh shapes was found to greatly affect the maximum tensile stress results and then affect the tensile fatigue properties of the meshed plates.

(3) It was found that changing center radius of basic mesh shape would hardly affect the tensile fatigue characteristics of meshed plates.

Therefore, the curvature changes in the basic mesh shape are effective for improving the fatigue resistance characteristics of meshed plates under tensile loading.

5. References

[1] Mayfield Brain & Spine Q&A, http://www.mayfieldclinic.com/QA_ArtificialCervicalDisc.htm
[2] He J M, Suzuki S, Chung U I 2014 *Trans. JSME* **80(809)** 1-15.
[3] Introduction of GUM METAL http://www.nissey-sabae.co.jp/wp/wp-content/uploads/GUM METAL.pdf

IOP Conf. Series: Materials Science and Engineering **269** (2017) 012073 doi:10.1088/1757-899X/269/1/012073

Response Surface Methodology for Design of Porous Hollow Sphere Thermal Insulator

Nazanin Shohani[1, 2], Saeed Pourmahdian[2], Behzad Shirkavand Hadavand[1*]

[1]Department of Polymer Engineering, Amirkabir University of Technology, Tehran, Iran
[2]Department of Resin and Additives, Institute for Color Science and Technology, Tehran, Iran

Abstract. In this study, response surface method is used for synthesizing polystyrene (PS) as sacrificial templates and optimizing the particle size. Three factors of initiator, stabilizer concentration and also stirring rate were selected as variable factors. Then, three different concentration of tetraethyl orthosilicate (TEOS) added to reaction media and core-shell structure with PS core and silica shell was developed. Finally, core-shell structure was changed to hollow silica sphere for using as thermal insulator. We observed that increased initiator concentration caused to larger PS particles, increase the stirring rate caused the smaller PS and also with increased the stabilizer concentration obtained that particle size decrease then after 2.5% began to increase. Also the optimum amount of TEOS was found.

1. Introduction

Recently, the synthesis of inorganic spheres with hollow structures such as TiO_2, SiO_2, ZnO, calcium silicate [1-4] and etc. had attracted a lot of attention because of many extraordinary properties such as low density, low toxicity, good chemical, thermal stability and large surface area. Sacrificial templates such as polystyrene [5] were commonly used to prepare hollow spheres. Silica hollow spheres have a wide range of application for thermal insulation [6]. Polystyrene due to easily synthesis and uniform particles is attractive as a template. In many cases, spherical polymeric micro particles are synthesized by suspension polymerization as a template for core-shell structure [7].

For preparing silica shell sol-gel process was usually used. Sol-gel polymerization of TEOS, involves hydrolysis and condensation reaction. The pH of the reaction media, type of catalyst and proportion of solvent to precursor are important and effective parameters on preparation of silica particles [8]. In a research, mineral insulators in paint for wall coatings has been simulated for flow field and temperature distribution in a room [9, 10]. The results showed that the coating with mineral particles has leads to a 17% decrease in energy consumption.

In this research, response surface experimental design as an efficient way [11, 12] was used in suspension polymerization for synthesizing of polystyrene template and the effect of initiator concentration; stabilizer and stirring rate on polystyrene particle size were investigated. Then, the best sample was selected and core-shell structure with polystyrene core and silica shell with different shell thickness were synthesized by sol-gel process. Finally, silica hollow spheres were obtained by calcinations process and pouring it into THF solvent.

2. Experimental

2.1. Materials

Styrene as a monomer, poly(vinyl alcohol) (PVA, M_w~128000) used as a stabilizer, tetraethyl orthosilicate as a precursor, ethanol and HCl (37%) for sol-gel process and tetrahydrofuran (THF) were purchased from Merck company (Germany). Benzoyl peroxide as an initiator was purchased from Alfa Aeser Company (USA). Styrene was distillated before using and other chemicals used without any purification. Also deionized water (DI water) was used throughout all of the experiments.

2.2. Synthesis of PS template and preparation of hollow silica spheres

Polystyrene templates with various sizes were synthesized by suspension polymerization. Amount of initiator, stabilizer and stirring rate were investigated in 1-3 wt%, 1-4 wt% of monomer and 550-750 rpm, respectively. Table 1 shows experimental conditions for the synthesis of polystyrene. Suspension polymerization was carried out in 500 ml three-necked flasks equipped with a condenser styrene monomer (20gr) and benzoyl peroxide was poured in the reactor and nitrogen was purged for 20 minutes to remove dissolved oxygen. The vessel was heated to 90 °C and PVA was added to reactor after 10 minutes then the polymerization was continued for 8 hours under a stirring according to the condition of the table 1.

Table 1. Preparation conditions for polystyrene synthesis.

Sample	Initiator concentration (wt%)	Stabilizer concentration (wt%)	Stirring rate (rpm)
S-1	1	1	550
S-2	3	1	550
S-3	1	4	550
S-4	3	4	550
S-5	1	1	750
S-6	3	1	750
S-7	1	4	750
S-8	3	4	750
S-9	0.48	2.5	650
S-10	3.52	2.5	650
S-11	2	0.21	650
S-12	2	4.8	650
S-13	2	2.5	500
S-14	2	2.5	800
S-15	2	2.5	650
S-16	2	2.5	650

To achieve uniform shells, smooth and spherical cores silica spheres were prepared via a sol–gel process in a Stöber method [13]. At first, prepared PS templates were ultrasonically treated with 20 ml of ethanol for 15 minutes to obtain a well dispersed suspension. The suspension was transferred into a 250 ml flask and heated to 70 °C in a water bath and kept stirred for 90 minutes. Then, tetraethyl orthosilicate (TEOS) and ethanol solution was added drop wisely to the suspension in 180 minutes and stirring (250 rpm) was kept for 5 hours. The pH of reaction was adjusted to value about of 1.5 by adding HCl solution. The pH value of the reaction solution was kept at about 1.5 during the coating process by adding of the HCl. In table 2 the values of the synthesis conditions for the experiment were shown.

Table 2. Preparation conditions for SiO_2-coated PS templates.

Sample	PS (gr)	TEOS (mol)	Ethanol (ml)	pH
CS-1	2	0.024	20	1.5
CS-2	2	0.033	30	1.5
CS-3	2	0.043	45	1.5

Calcination is one of the methods for removing polystyrene cores from core-shell particles. Therefore, a drop of suspension in ethanol was placed on an aluminum sheet and dried. It was then placed in furnace and heated at a rate of 1 °C/min to 500 °C with air flow and kept at 500 °C for 3 hours. After this time, it was cooled to room temperature. For completely removal of polystyrene, hollow silica spheres were kept in THF solvent for 14 days to extract the residue polystyrene.

2.3. Characterization

The Malvern Master Sizer 2000 dynamic light scattering (DLS) was used to measure particle sizes in this study. Scanning electron microscopy (SEM) Leo 1455VP and field emission scanning electron microscopy (FE-SEM) Mire 3-XMU were used to investigate the morphologies of the polystyrene templates, the silica-coated latex template core-shell composites and the hollow silica spheres.

3. Results and discussion

3.1. Effect of different synthesize parameters on particle size

The particle size of polystyrene synthesized by suspension polymerization was obtained are between 100-320 µ. Effect of initiator concentration on particle size is shown in figure 1.a. It was found that by increasing the initiator concentration, larger polystyrene particle are obtained. This event can be attributed to balance the system and polymerization rate at the beginning of the reaction. When the initiator concentration increase in system, activated site and rate of polymerization were increased too.

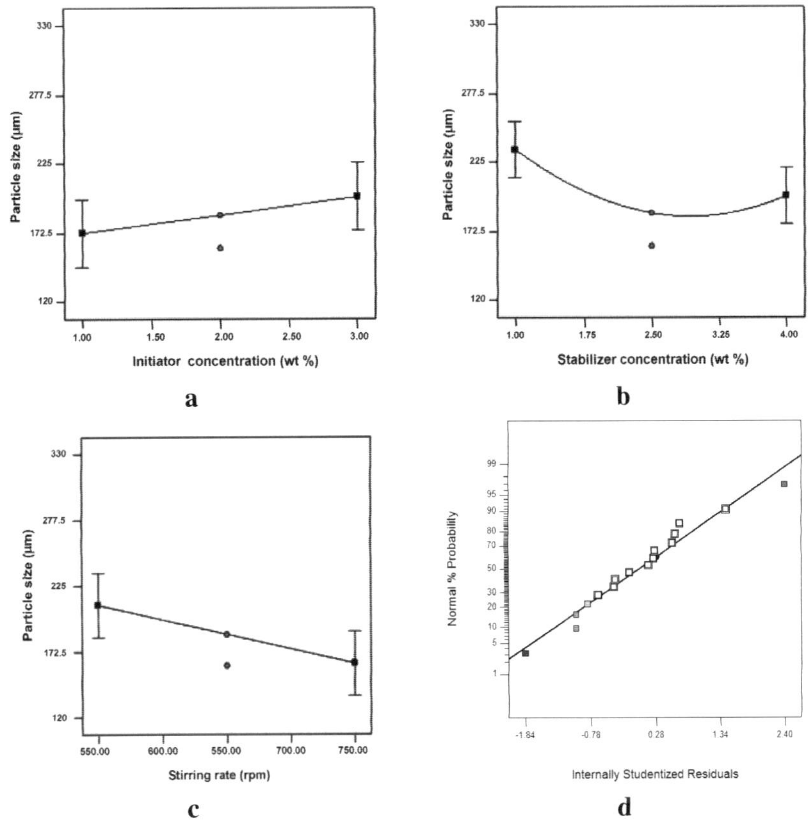

Figure 1. a) Initiator concentration (wt%)-particle size curve by constant value for stabilizer concentration (wt%) and stirring rate (rpm), b) Stabilizer concentration (wt%)-particle size curve by constant value for initiator concentration (wt%) and stirring rate (rpm), c). Stirring rate-particle size curve by constant value for initiator concentration (wt%) and stabilizer concentration (wt%) and d) Normal distribution curve.

The next parameter affecting is stabilizer concentration. The effect of this parameter on polystyrene particle size in range of investigating is shown in figure 1.b. Stabilizer creates a film around monomer droplets, so it causes droplets not to coagulate. Stabilizer leads to decreasing interfacial tension, then the possibility of colliding the droplets is decreased. As a result, smaller particles are obtained. However, it must be noticed that with increasing the stabilizer concentration to about 2.5 wt%, the particle sizes are increased. The reason of this phenomenon, stabilizers coagulate, deactivate in the media and the bridging effect is included.

As it can be seen in figure 1.c, in range of 550-750 rpm with enhancing the stirring rate, decreasing in the sizes of polystyrene particles can be observed. In suspension polymerization, each monomer droplet acts as a bulk reactor; therefore, final size of polymer was influenced by the size of initial monomer droplets. The monomer droplets sizes are decreased by increasing stirring rate because of the stress is created by stirrer.

According to sum of squares and P value, that amount of P factor is lower than other modeling, the linear model was selected as a selective model and all of curves and results were studied by this model. In the linear model, the amount of P value was 0.0134.

In this work, the normal distribution investigated to ensure the integrity of the chosen model. According to all of experimental results are around the line (figure 1.d), it can be conclude that suitable model was selected.

3.2. Morphology of particles

SEM images of polystyrene particles have spherical shape. The SEM images and DLS results were in the same manner about the particle size of this sample which was 100 μ and monodisperse in distribution. According to figure 2, When TEOS with amount of 5 (0.024 mol) was used, the shell created incompletely around the polystyrene template and after the calcinations process, the strength of shell was low and it was fractured. The amount of TEOS was increased to 7 gr (0.033 mol) in order to improve the strength of the shell. As can be seen in figure 3, with increasing amount of TEOS, more uniform and completely shell obtained. Also after calcination, broken hollow sphere was not observed. For 0.043 mol of TEOS, the thickness was not changed significantly and the silica particles were agglomerated.

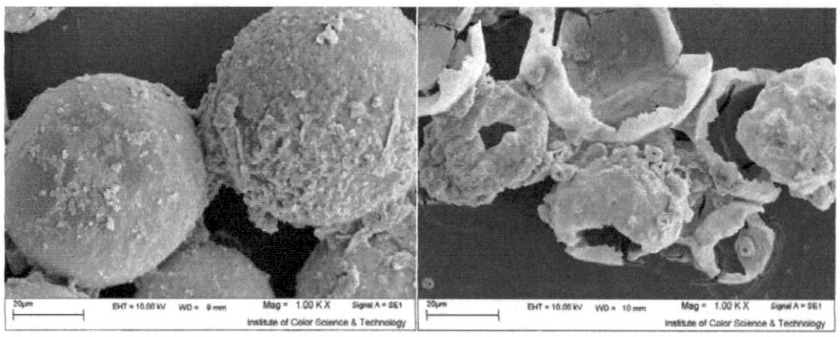

Figure 2. SEM image of CS-1 (left) and hollow sphere of CS-1 after calcination (right).

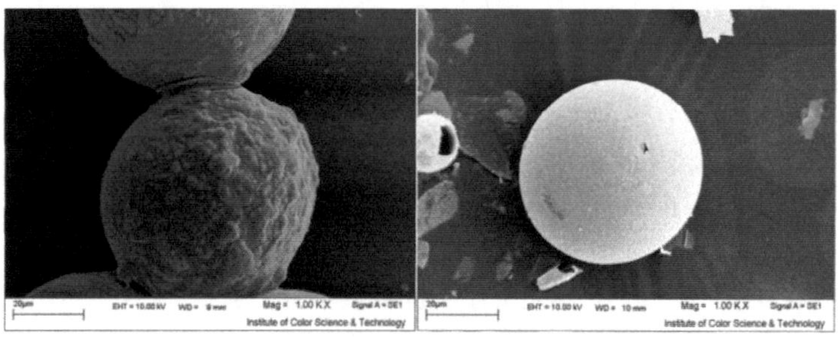

Figure 3. SEM image of CS-2 (left) and hollow sphere of CS-2 after calcination (right).

4. Conclusion

Different sizes of polystyrene particles were prepared by suspension polymerization. It was found out that particle size was increased by increasing the amount of initiator, decreasing the speed of stirrer and increasing stabilizer concentration up to certain level (2.5 wt%). For synthesis of PS/silica core-shell, three levels of TEOS for controlling of shell thickness and strength was investigated. We obtained that, when 0.033 mol TEOS was used, silica shell was created completely and after calcinations hollow silica spheres SEM image showed that hollow spheres had a no fracture and spherical shape of them was maintained.

5. References

[1] Zhao J, Yang Y, Cui C and Hu H 2016 *J. Alloys Compd.* **663** 211-6.
[2] Chun L, Heng-B.O Y, Ai-Li W and Zhan-A.O W 2012 *Trans. Nonferrous Met. Soc. China* **22** 1161-8.
[3] Zang Zh, Wen M, Chen W and Zeng Y 2015 *Mater. Des.* **84** 418-21.
[4] Toorisaka E, Nagamatsu S, Saruwatari Y, Hirata M and Hano T 2014 *Sci. Res.* **4** 114-21.
[5] Xu D, Ji X, Liu H, Wang M and Xu D Z 2012 *Mater. Lett.* **79** 61-4.
[6] Kim M K, Kim D W and Shin D W 2014 *Phys. Chem. Chem. Phys.* **17** 2416-20.
[7] Gorsd M N, Blanco M N and Pizio L R 2012 *Proc. Mater. Sci.* **1** 432-8.
[8] Chrusciel J and Slusarski L 2003 *J. Mater. Sci.* **21** 461-9.
[9] Azemati A A, Shirkavand Hadavand B, Hosseini H and Salemi Tajarrod A 2013 *Energy Build.* **56** 109-14.
[10] Khorasanizadeh H, Sheikhzadeh G A, Azemati A A and Shirkavand Hadavand B 2014 *Energy Build.* **78** 98-104.
[11] Shirkavand Hadavand B, Saeb M R, Najafi F and Malekian A R 2016 *J. Macromol. Sc. Part A,* **53** 741-9.
[12] Shirkavand Hadavand B, Najafi F, Saeb M R and Malekian A 2017 *High Perform. Polym.* **29** 651-62.
[13] Bai A, Song H and He G 2016 *Ceram. Int.* **42** 7583-92.

Surface Damage Characteristics of BK7 Glass in Ultrasonic Vibration Machining Based on Scratching Experiment

Gao Rui[1], Wang Hong-xiang[1*], Liu Jun-liang[1], Wang Chu[1] and Zhai Wen-jie[1]

[1]School of Mechatronics Engineering, Harbin Institute of Technology, Harbin, Heilongjing, 150001, China

Abstract. To further explore the material removal mechanism in ultrasonic vibration machining, a diamond Vickers indenter was used to carry out scratching experiment for BK7 glass specimen. The morphologies of scratches and removal mechanism of material were analysed under different conditions. The results showed that the damage mode of scratch was plastic deformation when the scratching depth was small enough, and no crack was observed. With increase of scratching depth, the intermittent and continuous scratches appeared in plastic removal area, and plastic flow phenomenon was obvious. With further increase of scratching depth, the median/radial cracks and lateral cracks were induced, and the material was removed by plastic flow and brittle-plastic mixed mode. When the indenter arrived at the brittle fracture removal area, cracks in scratched surface became denser, the lateral cracks extended from inside of material to workpiece surface, and the material was removed by brittle fracture.

1. Introduction

BK7 glass is a typical hard-brittle material and possesses some properties such as large brittleness and high hardness, which make it extensively applied in the aerospace, optical engineering and other fields. Brittle fracture frequently during processing causes many defects such as surface/subsurface damage, micro-cracks, and residual stress. The traditional method is difficult to achieve the desired results [1]. Under the high-frequency vibration of spindle, rotary ultrasonic machining (RUM) realizes the efficient processing of BK7 glass by nickel-based electroplated diamond tools [2]. However, it is difficult to accurately observe the surface damage caused by a single abrasive during the RUM process. The interaction between a single abrasive and workpiece surface is often simulated by scratching experiments [3-5]. Bulsara V H et al. photographed the deformation and fracture of scratches in brittle solids, and indicated that the compressive stress induced the nucleation of median/lateral cracks in loading process and the tensile stress induced the nucleation of lateral cracks in unloading process [6]. Qiu Z J et al. investigated the propagation of surface/subsurface crack by single and double scratching experiments of glass–ceramics, and found that the interaction between lateral cracks and radial cracks was the main mode of material removal [7]. Lv et al. observed the scratches on BK7 glass in the scratching tests with and without ultrasonic, and revealed that the ultrasonic vibration could significantly affect the nucleation and propagation of the crack [2, 8]. Cao et al. researched the material removal behaviours in ultrasonic-assisted scratching of SiC ceramics, and concluded that the tool's impact and cutting force on the machined surface were the main factors of removal [9].

In summary, the removal mechanisms of brittle materials have studied extensively based on the scratching experiment. However, relative movement of abrasive-material in rotary ultrasonic machining is the coupling of spindle rotary motion, linear feed motion, and axial high-frequency reciprocating motion. These alterations in movement mode have significant effect on the material removal mechanism, while no detailed investigations of these factors have been reported. To further explore the mechanism of BK7 glass in RUM, the workpiece surface was subjected to ultrasonic vibration scratching with various depths by the diamond Vickers indenter. Subsequently, the morphology of the scratched surface was observed with the optical microscope and AFM.

2. Experimental procedures

Schematic diagram of experimental device is illustrated in figure 1. The experiments were conducted on the RUM machine (DMG Ultrasonic 70-5-linear, Germany), which equipped the tool holder (HSK63) and the ultrasonic generator (USG2000), and the frequency range of ultrasonic vibration was 18~30kHz. The indenter with the vertex radius of 0.2μm and the opposite angle of 136° was mounted on the tool holder by the spring chuck. BK7 glass was adhered on a special steel plate which was securely fixed on a three-direction dynamometer (Kistler9256C2). In order to avoid the flatness error of specimen during installation, the three-point leveling system on the machine was used to adjust the workbench. To ensure the Vickers indenter could still axial high-frequency reciprocating vibration without spindle rotation, the ultrasonic vibration mode was adjusted to the manual mode.

The BK7 glass specimens with dimensions of 30mm×15mm×6mm were selected for this investigation. Prior to the scratching experiments, the specimens were successively polished with finer abrasive to facilitate the observation of optical microscopy. In all experiments, the scratching depth increased gradually along the scratching direction. The scratching length was 2000μm and scratching speeds were 10, 20, 40, 60mm/min, respectively. Finally, the BK7 glass specimen was observed with the three-dimensional optical microscope and AFM.

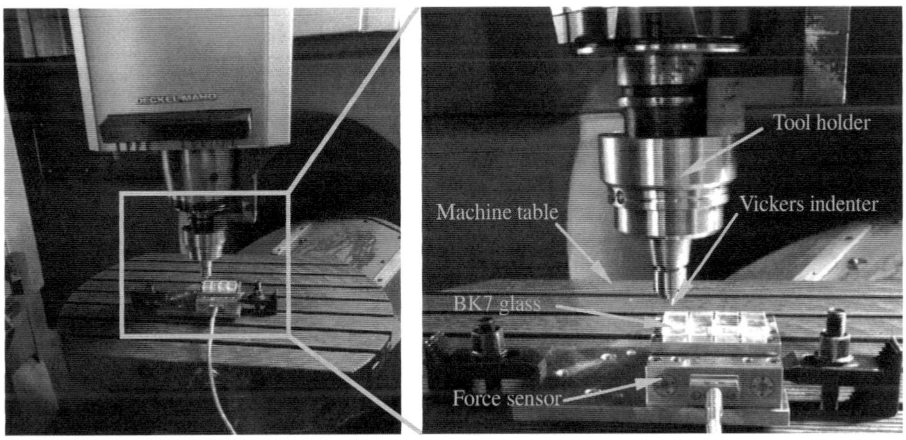

Figure 1. Experimental device for ultrasonic vibration scratching.

3. Different areas of ultrasonic vibration scratching

Different areas of ultrasonic vibration scratching with various scratching depth are shown in figure 2. Apparently, the periodic fluctuation of the abrasive trajectory causes the cutting process along the feed direction to be divided into two stages: 1) intermittent cutting stage, where abrasives intermittently cut into and cut off the workpiece, leaving a series of interrupted isolated scratches on the surface; 2) continuous cutting stage, where abrasives continuously contact the workpiece, forming continuous scratches on the surface.

In the process of scratching from the starting point S_{uvs} to the ending point E_{uvs}, the different morphologies of scratches are caused by different deformation and removal methods of the material, which are mainly composed of three ways: plastic deformation, plastic removal, and brittle fracture. In order to compare the different characteristics on surface damage morphology, the whole scratching process is divided into four areas: plastic deformation area A_{uvs}, plastic removal area B_{uvs}, brittle/plastic mixed removal area C_{uvs}, and brittle fracture area D_{uvs}.

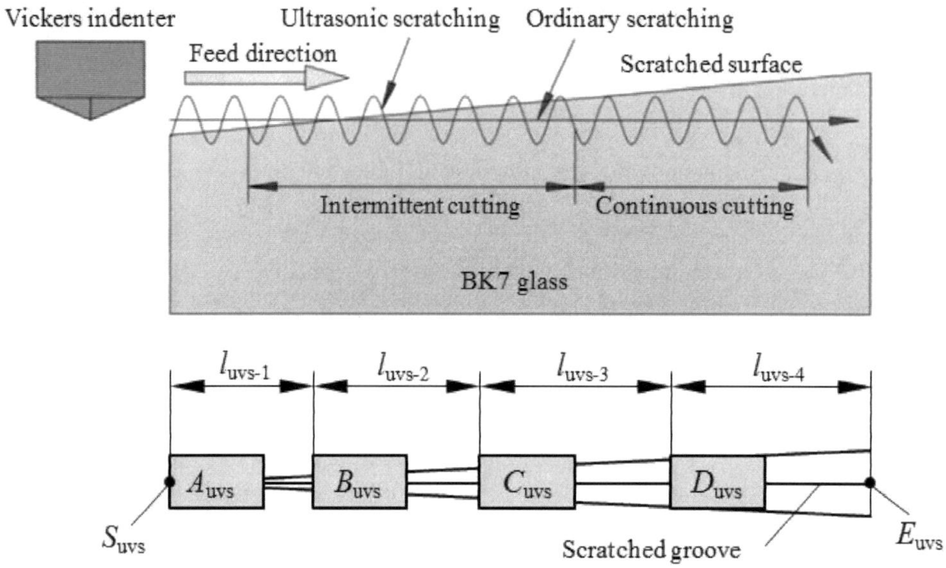

Figure 2. Schematic diagram of ultrasonic vibration scratching along the feed direction.

4. Experimental results and analysis

4.1. Surface damage of scratched surface

When the scratching speed v=10mm/min, the surface damage morphology with different scratching depth is shown in figure 3, and the four areas lengths were 0.1mm, 0.2mm, 0.3mm, and 1.2mm, respectively.

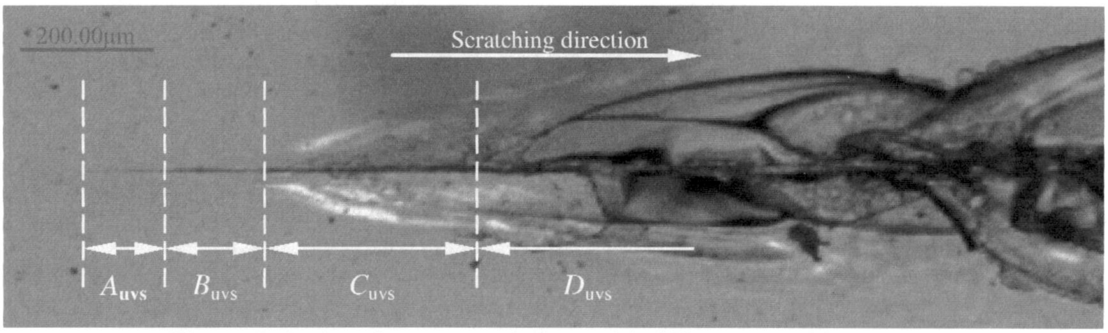

Figure 3. Surface damage morphology with different scratching depth.

The Vickers indenter intermittently cut into the specimen surface under the ultrasonic vibration, and a series of shallow pits with a certain space were impacted in the area A_{uvs}. Because the surface material of the pits was not subjected to sufficient compressive stress and shear stress, the damage of scratch was the pure plastic deformation, and no cracks were observed. With the increase of scratching depth, there were intermittent and continuous scratches in the area B_{uvs}. The plastic flow phenomenon was obvious on the right side of this area. The scratched surface was still relatively smooth, which indicated that the material in this area was mainly removed by plastic flow. With the further increase of scratching depth, there were large bright areas on both sides of the area C_{uvs}. The median/radial cracks extended perpendicularly to the scratched surface and the lateral cracks horizontally extended towards the specimen surface, which indicated that the material in this area was mainly removed by plastic flow and brittle/plastic mixed removal. When the Vickers indenter reached the area D_{uvs}, the cracks in the scratched surface area were denser. The lateral cracks on the local surface extended from the inside of workpiece to the surface under the large compressive stress and shear stress, resulting in the surface material falling from the workpiece. The material in this area was mainly removed by brittle fracture.

The AFM three-dimensional morphology in different material removal areas are shown in figure 4. In the plastic removal area (see figure 4a), the material on both sides of the scratch was extruded and deformed, resulting in plastic flow phenomenon. In the brittle/plastic mixed removal area (see figure 4b), the plastic flow of material, as well as the nucleation and propagation of cracks led to lateral cracks, ridge-like bulge and zigzag corrugation on both sides of the scratch. Simultaneously, the transition from plastic flow to brittle fracture occurred at the position C, and the scratching depth of this position was 2.23μm.

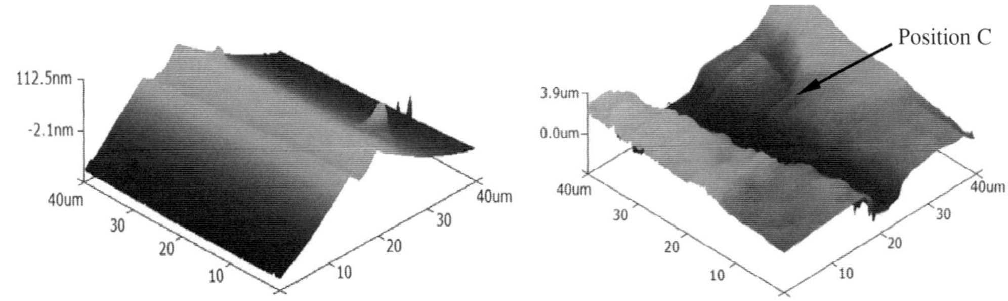

(a) Plastic removal area (b) Brittle/plastic mixed removal area

Figure 4. AFM three-dimensional morphology in different material removal area.

4.2. Change of scratch damage morphology

The three-dimensional morphologies of scratches were detected layer by layer along the scratching direction by AFM. The variations of the scratching depth with scratching length are demonstrated in figure 5.

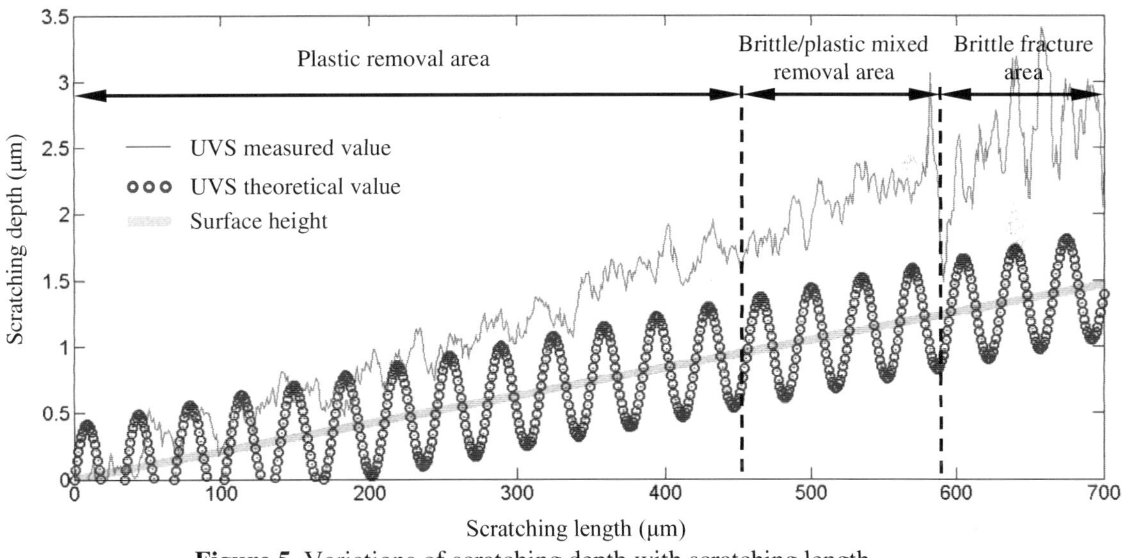

Figure 5. Variations of scratching depth with scratching length.

The repetitive positioning accuracy of machine tool was ±1μm. When the diamond Vickers indenter was mounted on the spindle and returned to the origin position again, the height error caused the theoretical scratching depth was less than the actual scratching depth. In the plastic removal area, the assistance of ultrasonic vibration changed the trajectory of indenter, which led to a slight periodic fluctuation of the scratching depth and the surface morphology. In the brittle/plastic mixed removal area, scratching depth began to fluctuate significantly at the scratching length of 500 ~ 600μm. With the further increase of scratching depth, the material produced brittle fracture and the scratching depth appeared irregular fluctuations. The increase of surface damage was induced by the propagation of radial/median cracks and the mutual overlap of many lateral cracks.

Under different scratching speeds, the cross-sectional area of the scratches increased with the scratching length (see figure 6a). The cross-sectional area first increased slowly, and then increased faster after reaching the inflection point. The width-to-depth ratio of the scratches decreased with the scratching length (see figure 6b). The width-to-depth ratio decreased at first, and then tended to be steady after reaching a certain length. When the scratching speed v=10mm/min, with the increase of scratching length, cross-sectional area and width-to-depth ratio of the scratches had a significant change in the position where the scratching length was about 580μm. This change was in good agreement with the results of optical microscopy and the variations of the scratching depth, which indicated that the inflection point was the critical point of the transition from plastic flow to brittle fracture.

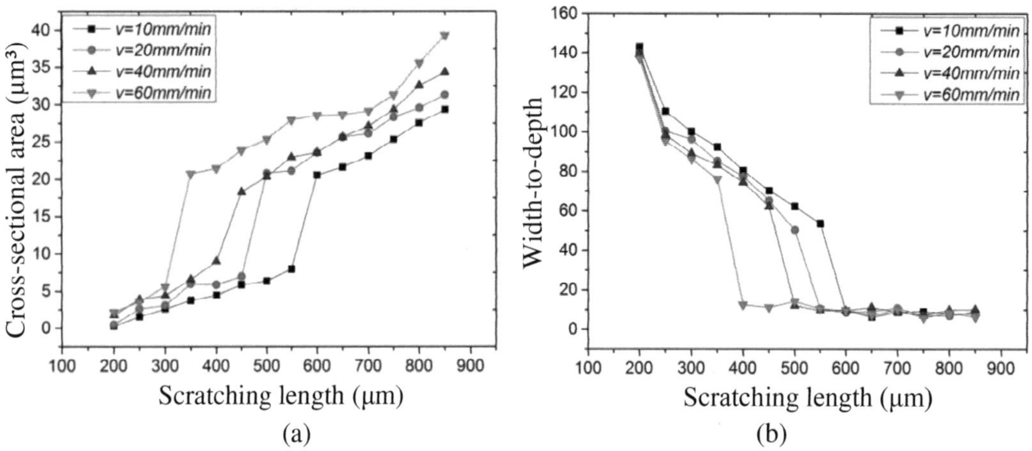

Figure 6. Variations of cross-sectional area and width-to-depth ratio with scratching length.

In addition, the scratching speed had a certain impact on the critical point of the transition from plastic flow to brittle fracture. The variations of the critical scratching depth with the scratching speed are shown in figure 7. With the increase of scratching speed, the critical scratching depth of the transition from plastic flow to brittle fracture was gradually reduced.

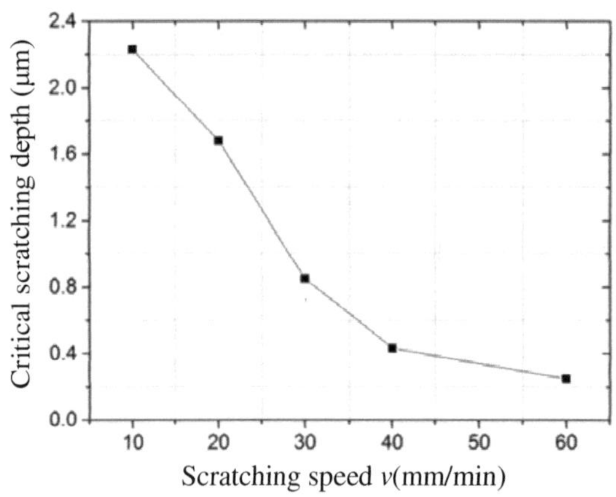

Figure 7. Variations of critical scratching depth with scratching speed.

5. Conclusions

The present work analyzed the morphologies of scratches under different conditions and material removal mechanism of BK7 glass in scratching experiments with ultrasonic vibration. Following conclusions can be drawn from this investigation:

1. The deformation and removal methods of the material divide the morphologies of scratches into plastic deformation area, plastic removal area, brittle/plastic mixed removal area, and brittle fracture area.

2. With the increase of scratching depth, the removal of materials was changed from pure plastic deformation to brittle/plastic mixed mode, and finally became the brittle fracture under the compressive stress and shear stress.

3. The critical point of the transition from plastic flow to brittle fracture was obtained by analyzing the variations of cross-sectional area and width-to-depth ratio, and the critical scratching depth was gradually reduced with the scratching speed.

6. References

[1] Lee P H and Lee S W 2011 *Int. J. Mach. Tool. Manuf.* **51** 201-9.
[2] Lv D X, Tang Y J, Wang H X, et al. 2013 *Appl. Mech. Mater.* **427** 187-90.
[3] Gu W B and Yao Z Q 2011 *J. Mech. Sci. Tech.* **25** 1167-74.
[4] Gu W B, Yao Z Q and Liang X G 2011 *J. Wear.* **270** 241-6.
[5] Petit F, Ott C and Cambier F 2009 *J. European Ceram. Soc.* **29** 1299-307.
[6] Bulsara V H and Chandrasekar S 1997 *Proc. SPIE.* **3060** 76-88.
[7] Qiu Z J, Liu C C, Wang H R, et al. 2016 *J. Mech. Behave. Biomed. Mater.* **64** 75-85.
[8] Lv D X, Wang H X, Tang Y J, et al. 2013 *J. Prec. Eng.* **37** 839-48.
[9] Cao J G, Wu Y B, Lu D, et al. 2014 *Int. J. Mach. Tool. Manuf.* ***79*** 49-61.

Acknowledgments

This research was financially supported by National Natural Science Foundation of China (Grant No.51475106, No.51475119) and Science Challenge Project (Grant No.JCKY2016212A506-0503).

Evaluation on Compressive Characteristics of Medical Stents Applied by Mesh Structures

Kazuki Hirayama[1*], Jianmei He[2]

[1]Kogakuin Univ. Dept. of Mechanical Engineering, Nishishinjyuku 1-24-2, Shinjyuku-ku, Tokyo, 163-8677 Japan

Abstract. There are concerns about strength reduction and fatigue fracture due to stress concentration in currently used medical stents. To address these problems, meshed stents applied by mesh structures were interested for achieving long life and high strength perfromance of medical stents. The purpose of this study is to design basic mesh shapes to obatin three dimensional (3D) meshed stent models for mechanical property evaluation. The influence of introduced design variables on compressive characteristics of meshed stent models are evaluated through finite element analysis using ANSYS Workbench code. From the analytical results, the compressive stiffness are changed periodically with compressive directions, average results need to be introduced as the mean value of compressive stiffness of meshed stents. Secondly, compressive flexibility of meshed stents can be improved by increasing the angle proportional to the arm length of the mesh basic shape. By increasing the number of basic mesh shapes arranged in stent's circumferential direction, compressive rigidity of meshed stent tends to be increased. Finaly reducing the mesh line width is found effective to improve compressive flexibility of meshed stents.

1. Introduction

In recent year, Lumens in living body, for example blood vessels, bile ducts and tracheas, may narrow and cause various diseases. For mainly treatment of blood vessels, the stent placing surgery has been using widely. The lumens have very complex structures. In addition, they have severe movements such as bending, twisting and stretching. Furthermore, The Rejection may occur when foreign matter enters. From these reasons, it is desirable for the medical stent to have flexible structure, higher biocompatibility and follow ability to blood vessel wall.

On the other hands, previous medical stents have concerns that repetitive strains accompanying pulsation cause strength reduction and fatigue failure as shown in figure 1 [1-5]. There is the MISAGO meshed stents shown in figure 2 (Terumo corp.) [6], which has relatively low stress and high durability, the problems of strength reduction and fatigue failures are still existed [7-9]. More flexible and higher stretchy medical stents which can cope with harsh movements are desired. For these problems, it is considered to solve them by applying mesh structures [10] to the surface of nickel titanium cylinder to obtain meshed stents with higher flexibility and strength.

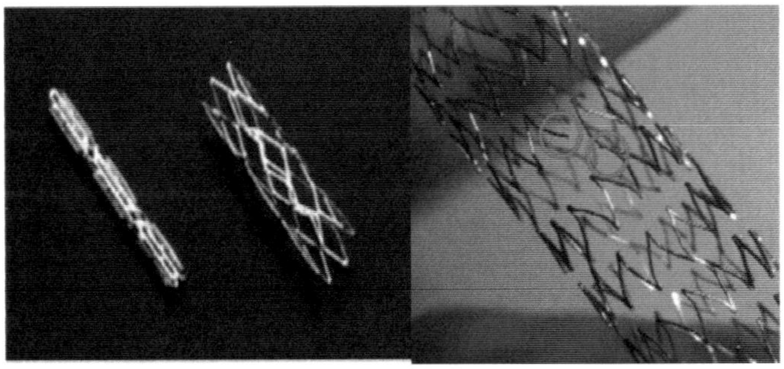

Figure 1. Fatigue failure of previous stents.

Figure 2. MISAGO stent.

1.1. Designing basic mesh shapes for 3D meshed stent models.
Basic mesh shape based on regular hexagon, which has higher in-plane isotropy and lower out-of-plane elasticity, is designed for meshed stent application [10]. Design variables like the angle affecting to arm length (hereinafter called "θ"), the number of basic mesh shape in the stent circumferential direction (hereinafter called "n") and mesh line width (hereinafter called "w") of the basic mesh shape and meshed stent as shown in figure 3 and figure 4 are introduced for analytical investigation.

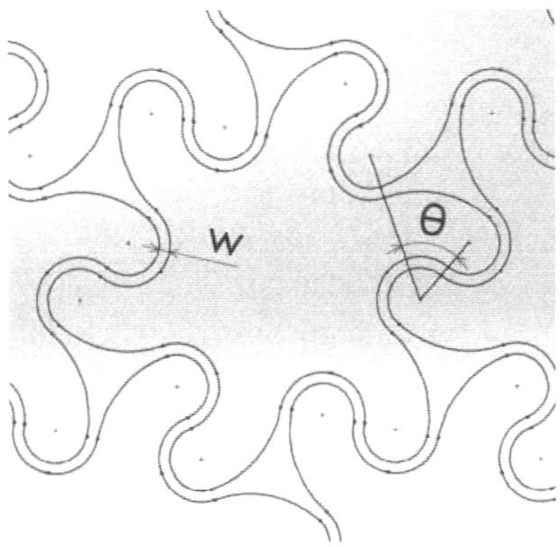

Figure 3. Design variables: angle affecting to arm length θ and mesh line width w.

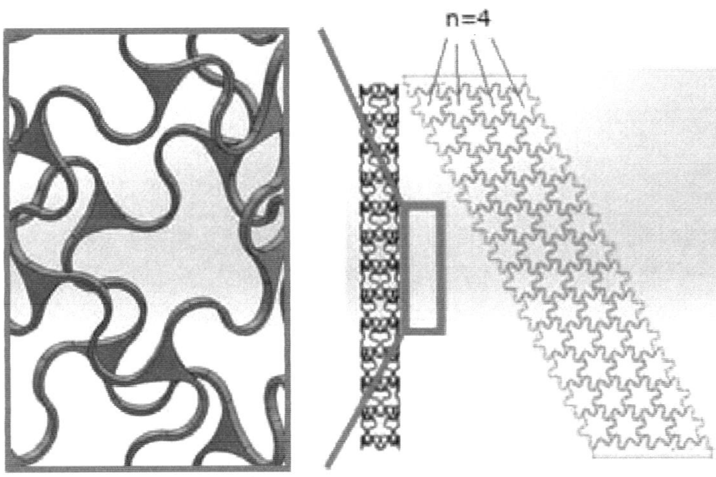

Figure 4. 3D meshed stent model with basic mesh shape sketch (n=4).

The diameter of the mesh stent cylinder is 6.0 mm and the strut thickness is 0.18 mm for sample model analysis. These size are same as the medical stent used in femoral artery. The basic mesh shapes and 3D meshed stent models are designed and created using 3D CAD software Solidworks.

2. Evaluation of compression characteristics of the mesh stent

Analytical evaluation on compressive stiffness of 3D meshed stent models are carried out by using ANSYS Workbench software. Meshed stent model with compressive jigs under compressive loading are introduce and shown in figure 5. Compressive stiffness of mesh stents can be obtained from the maximum displacement with respect to the applied compressive load obtained by analysis.

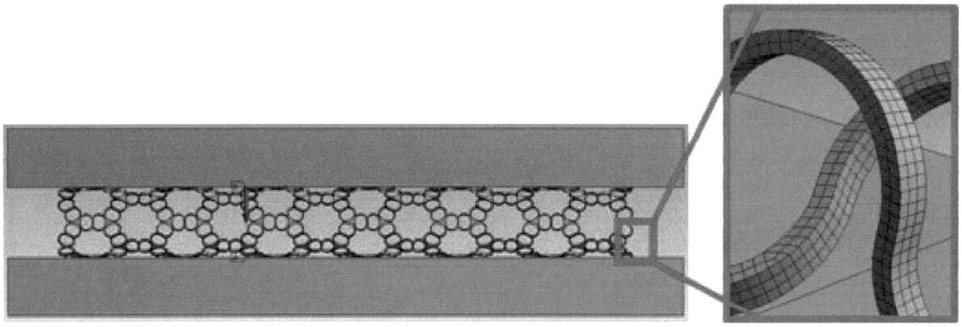

Figure 5. FEA model evaluating compressive characteristics of meshed stents.

Material properties of nickel titanium and finite element mesh settings of FEA model of meshed stents are shown in table 1 and sample finite element mesh image is also shown in figure 5 for 3D analytical evaluation.

Table 1. Material properties and element division settings.

Material properties	Longitudinal elastic modulus [GPa]	75
	Poisson's ratio	0.3
Finite element mesh	Size of elements [mm]	0.06 ~ 0.09
	Number of element in thickness direction	2 or 3

2.1. Periodicity due to the basic mesh shapes based from regular hexagon.

In this section, periodicity of compressive property of meshed stent models using regular hexagon based mesh shapes is to be evaluated with mesh line width w=0.1 mm, different basic mesh shape number n and angle affecting to arm length θ. Analyses are carried out with changing the compressive directions by rotating the meshed stents round the cylindrical axis from 0° to an angle obtained by dividing 360° by basic mesh shape number n. Obtained results are shown in figure 6. From these results, it is found that the periodicity of compressive stiffness of meshed stents is confirmed by changing the compressive directions from 0° to the angle. The periodicity of compressive stiffness can be considered from the application of regular hexagon based mesh shapes.

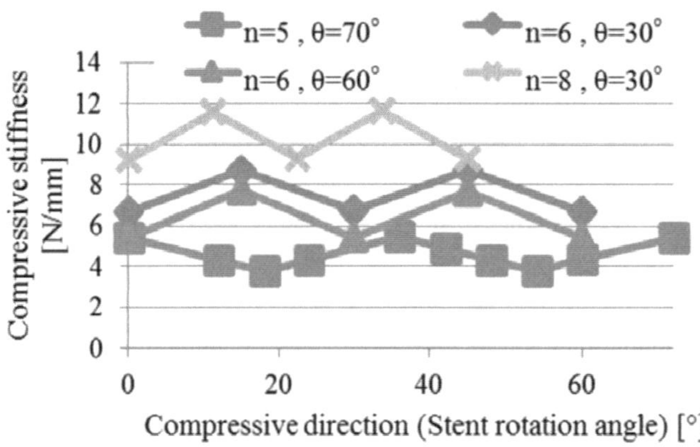

Figure 6. Periodicity of compressive stiffness by stent rotation angle.

2.2. Effects of the angle affecting arm length to compressive stiffness.
In this section, effects of the angle affecting arm length to the compressive property of meshed stent models using regular hexagon based mesh shapes are to be evaluated with basic mesh shape number n setting at 5, 6 and 8, fixed mesh line width w of 0.1 mm. Analyses are carried out with changing the angle affecting to arm length θ between 0°and 90°. Obtained results are shown in figure 7. From these results, it is found that the more the angle affecting to arm length, the less the compressive stiffness of meshed stents.

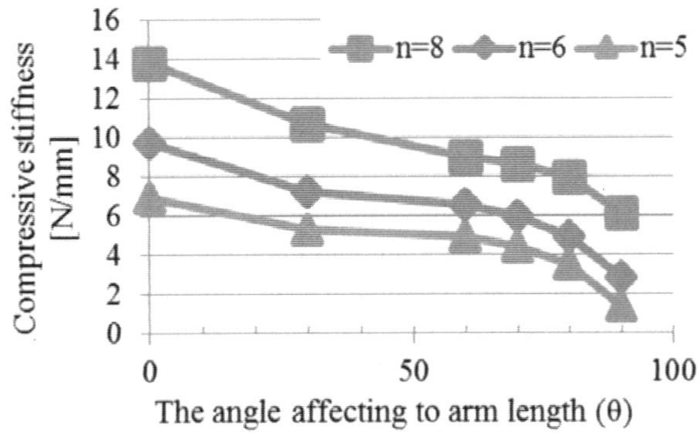

Figure 7. Compressive stiffness by angle proportional to arm length θ.

2.3. Effects of basic mesh shape number in the circumferential direction.
In this section, the effects of the basic mesh shape number in the circumferential direction on the compressive property of meshed stent models using regular hexagon based mesh shapes are to be evaluated, with the angle affecting arm length θ setting at 70° and 30°, mesh line width w=0.1 mm. Analyses are carried out with changing the basic mesh shape number n between 3 and 8. Obtained results are shown in figure 8. From these results, it is found that the more the basic mesh shape number, the more the compressive stiffness of meshed stents.

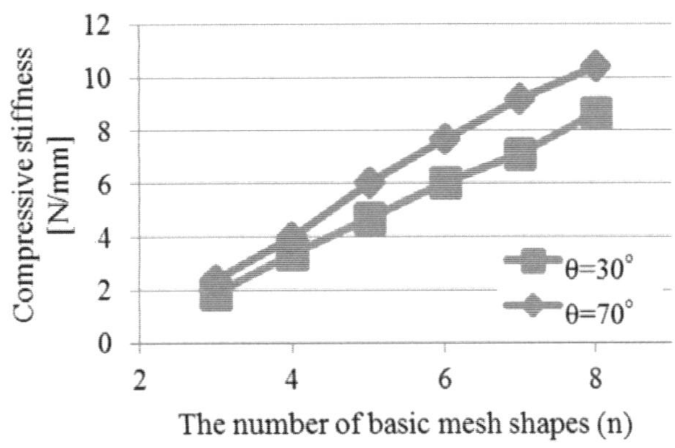

Figure 8. Compressive stiffness by number of basic mesh shapes n.

2.4. Effects of the mesh line width w on compressive stiffness of meshed stents.
In this section, the effects of the mesh line width w on the compressive property of meshed stent models using regular hexagon based mesh shapes are to be evaluated, with the angle affecting arm length θ fixed at 30° and basic mesh shape number n setting at 3, 4 and 5. Analyses are carried out with changing the mesh line width w between 0.05mm and 0.5mm. The obtained results are shown in figure 9. From these results, it is found that the narrower the mesh line width, the less the compressive stiffness of meshed stent models.

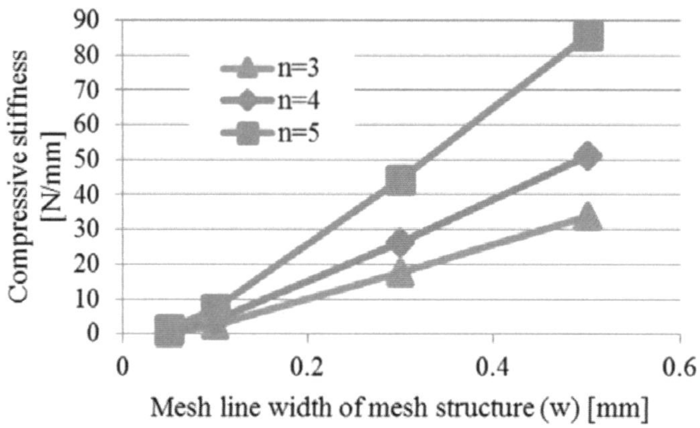

Figure 9. Compressive stiffness by mesh line width *w*.

3. Summary

Meshed stent models using regular hexagon based mesh shapes are designed and created for medical devise applications. The influences of different design variables on compression characteristics of meshed stent models are analytically evaluated by using finite elements method. From the obtained results,

(1) The compressive stiffness are changed periodically with compressive directions, average results need to be introduced as the mean value of compressive stiffness of meshed stents.

(2) Compressive flexibility of meshed stents can be improved by increasing the angle proportional to the arm length of the mesh basic shape.

(3) By increasing the number of basic mesh shapes arranged in stent's circumferential direction, compressive rigidity of meshed stent tends to be increased.

(4) Finaly reducing the mesh line width is found effective to improve compressive flexibility of meshed stents.

4. References

[1] RJaff M, Dake M, Pompa J, Ansel G and Yoder T 2007 *Catheter. Cardiovasc. Interv.* **70** 460-2.

[2] Scheinert D, Scheinert S, Sax J, Piorkowski C, Bräunlich S, Ulrich M, et al. 2005 *J. Am. Coll. Cardiol.* **45** 312-5.

[3] Schlager O, Dick P, Sabeti S, Amighi J, Mlekusch W, Minar E, et al. 2005 *J. Endovasc. Ther.* **12** 676-84.

[4] Zhao H Q, Nikanorov A, Virmani R, Jones R, Pacheco E and Schwartz L B 2009 *Cardiovasc. Intervent. Radiol.* **32** 720-6.

[5] Yazdani S K and Berry J L 2009 *J. Vasc. Interv. Radiol.* **20** 101-6.

[6] Terumo Home Page. <http://www.terumo.co.jp/medical/equipment/me240.html> (accessed at 25th July. 2017)

[7] Müller-Hülsbeck S, Schäfer P J, Charalambous N, Yagi H, Heller M and Jahnke T 2010 *J. Endovasc. Ther.* **17** 767-76.

[8] Schulte K L, Müller-Hülsbeck S, Cao P, Becquemin J P, Langhoff R, Charalambous N, et al. 2010 *Euro Interv.* **5** 687-91.

[9] Schulte K L, Kralj I, Gissler H M, Bagnaschino L A, Buschmann I, Pernès J M, et al. 2012 *J. Endovasc. Ther.* **19** 774-84.

[10] Jianmei H E, Suzuki S and Chung U I 2014 *Trans. JSME* **80(809)** 1-15.

IOP Conf. Series: Materials Science and Engineering **269** (2017) 012076 doi:10.1088/1757-899X/269/1/012076

Conceptual Design Study on Bolts for Self-Loosing Preventable Threaded Fasteners

Atsushi Noma[1*], **Jianmei He**[1]

[1]Dept. of Mechanical Engineering, Kogakuin Univ. 1-24-2, Nishi-shinjuku, Shinjuku-ku, Tokyo, 163-8677 Japan

Abstract. Threaded fasteners using bolts is widely applied in industrial field as well as various fields. However, threaded fasteners using bolts have loosing problems and cause many accidents. In this study, the purpose is to obtain self-loosing preventable threaded fasteners by applying spring characteristic effects on bolt structures. Helical-cutting applied bolt structures is introduced through three dimensional (3D) CAD modeling tools. Analytical approaches for evaluations on the spring characteristic effects helical-cutting applied bolt structures and self-loosing preventable performance of threaded fasteners were performed using finite element method and results are reported. Comparing slackness test results with analytical results and more details on evaluating mechanical properties will be executed in future study.

1. Introduction

Threaded fasteners using bolts are inexpensive and easy to be attached and detached. Therefore, they are used not only in industrial field but also in various fields such as precision instruments and medical field. There are various kinds of fastening methods using bolts, such as a method of using bolts and nuts together or a method of using tapping screws, and a designer can select assuming the environment in which different products are used.

However, loosening problems of threaded fasteners are unavoidable through the bolt tightening method, and many accidents are actually occured and reported. For example, the railroad derailment accident occurred in the UK in 2002 is one of the most famous accident. This accident occurred because the screw used for the railroad switch was loosed and fatigue failure due to the vibration occurred. In order to prevent such kind of accidents, bolts or screws having locking functions are designed and developed.

In this study, spring characteristic effect as shown in figure 1 are interested. The purpose is to realize self-loosing preventable threaded fasteners by applying spring characteristics to bolts. Spring characteristic effects can be applied to bolt structures by introducing helical-cutting process to ready-made bolt to obtain self-loosing preventable theaded fasteners. The spring characteristic effects are swelling effect under rotating conterclockwise (loosing direction) shown in figure 1 [a] and shrinking effect under rotating clockwise (tightening direction) shown in figure1 [b].

[a] Swelling effect in loosing direction [b] Contraction effect in tightening direction

Figure 1. Conceptual diagram of spring characteristic effects.

2. Design of bolt structures for self-loosing preventable threaded fasteners

Structural design of bolts for self-loosing preventable threaded fasteners was investigated using 3D CAD software SolidWorks. Helical cutting with cross-sectional shape as cutting shape and cross thread turning in reverse direction to thread rotation direction of screw is introduced to the general hexagon bolt shown in figure 2(a) and (b). As the result, bolt structure provided by plurality of springs cross each other are designed with spring characteristic effects surposed to be imparted.

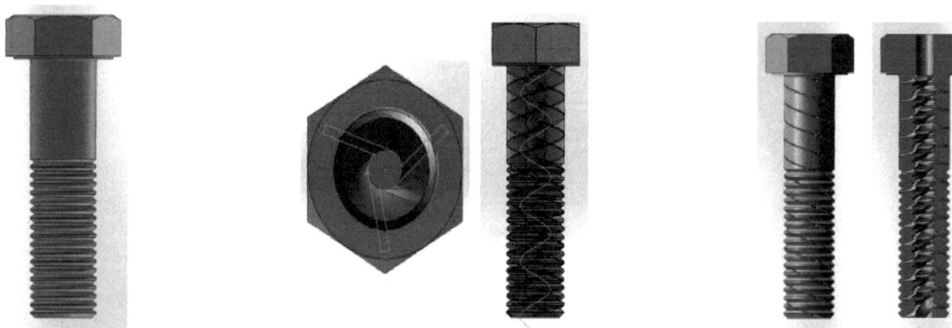

(a) Convetional bolt (b) Cutting shape & cross thread turning (c) Helical-cutting applied bolt

Figure 2. Image of helical-cutting applied bolt for self-loosing preventable threaded fasteners.

3. Analytical evaluation on spring characteristic effects of helical-cutting applied bolts

Analytical evaluation on spring characteristic effects of helical-cutting applied bolt structures with material properties of structural steel are carried out. Young's modulus of 200.0 GPa and Poisson's ratio of 0.3 of structural steel are applied to finite element analysis model of designed bolt structures through ANSYS software.

As loading and constraint conditions, the end of designed bolt was fixed and a torque of 5.4 Nm corresponding to 110% of 4.9 Nm fastening torque of general M6 bolt was applied. Finite element mesh of analytical model of helical-cutting applied bolt is shown in table 1. Figure 3 shows image of finite element mesh of bolt model for self-loosing preventable threaded fastener.

Table 1. Finite element mesh of analytical bolt model.

Minimum Surface Size	0.002mm
Maximum Surface Size	0.35mm
Maximum Size	0.40mm
Number of Element	219695
Number of Node	138410

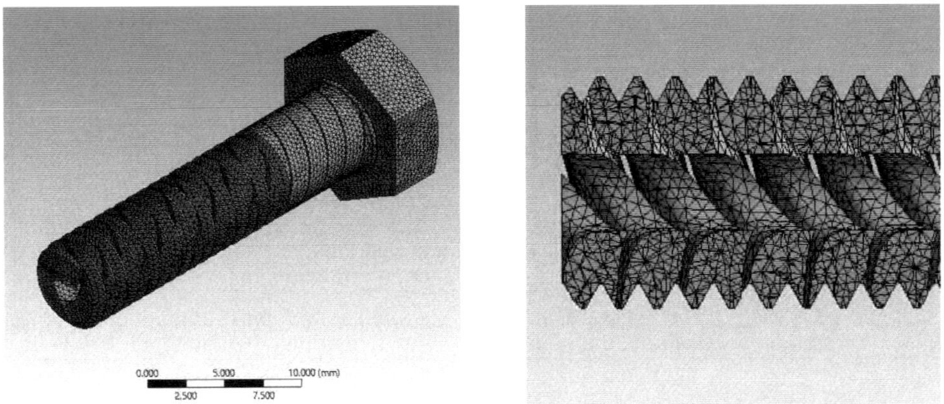

Figure 3. Image of finite element mesh of bolt for self loosing-preventable threaded fastener.

Figure 4 shows the analytical result on spring characteristic effect of helical-cutting applied bolt structures with self-locking performance for threaded fastener application. From this result, spring characteristic effect is confirmed by the obtained swelling effect under rotating loads conterclockwise.

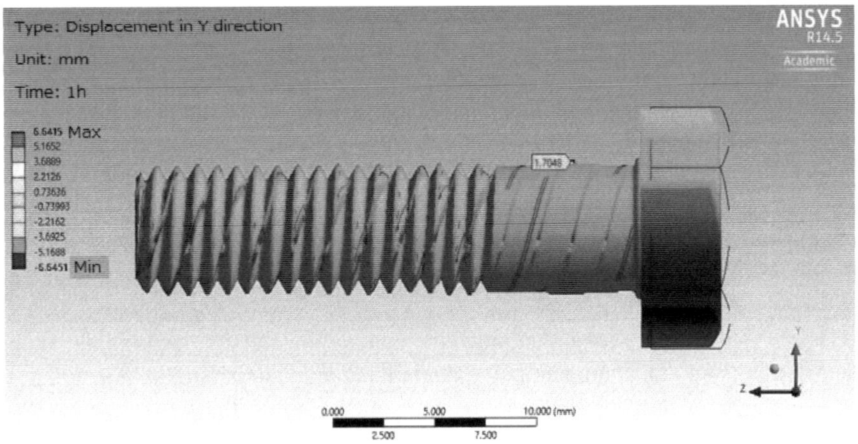

Figure 4. Analytical result on spring characteristic effect of helical-cutting applied bolt.

4. Analytical evaluation on self-loosing preventable performance of threaded fasteners

Self-loosing preventable performance of threaded fastener using helical-cutting applied bolts are analytically examined by comparison with general hexagon bolt. Junker vibration test under ISO 16130 standards are modeled including bolt, nut, fixed plate and vibration plate as shown in figure 5. Material properties of structural steel are also used for bolt structure in this model.

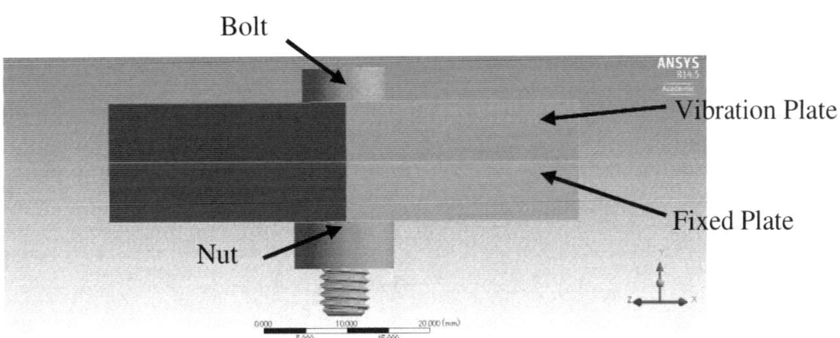

Figure 5. Analytical model of threaded fastener using different bolt structures.

General M6 hexagon bolts with or without helical-cutting processing are introduced for self-loosing preventable threaded fastener evaluation. The friction generated on the contact surface between the stationary plate and the diaphragm and nut was considered negligible and the friction coefficient was set to 0.17 by using experimental conditions of Yamamoto et al [1]. Contact of each surface was defined as contact element based on the penalty method [2, 3]. Table 2 shows the constraint conditions applied for analytical model of threaded fastener.

Table 2. Restraint conditions on self-loosening preventable threaded fastener model.

Test Type	Junker Vibration Test: ISO 16130
Number of Vibrations	1000
Direction of Vibration	x Direction : ±0.3mm
Fixed	Side of Fixed Plate
	Bolt End : x and z Directions
Fastening Axial Force	4330N

Junker vibration test is a test in which a diaphragm and a fixing plate are fastened by using of bolt and vibration of the diaphragm is applied in the direction perpendicular to bolt axis. In this analytical model, fixations in x and z directions of the fixed plate and the end of the bolt were fixed, the fastening axial force was given in the first step, and then the diaphragm was oscillated with the amplitude of 0.3 mm in the x direction along perpendicular direction to bolt axis.

To analyzing the Junker vibration test, very long time calculation is needed because of the analysis type and model sizes. Analysis time can be shortened by increasing the element sizes of the diaphragm and fixing plate, which are considered having little influence on the analytical result. Table 3 shows the number of elements and nodes for self-loosing preventable threaded fastener model.

Table 3. Number of elements and nodes on threaded fastener model.

Threaded fasteners using general Bolt	
Number of Elements	16800
Number of Nodes	36751
Self-loosing preventable threaded fasteners	
Number of Elements	15762
Number of Nodes	36526

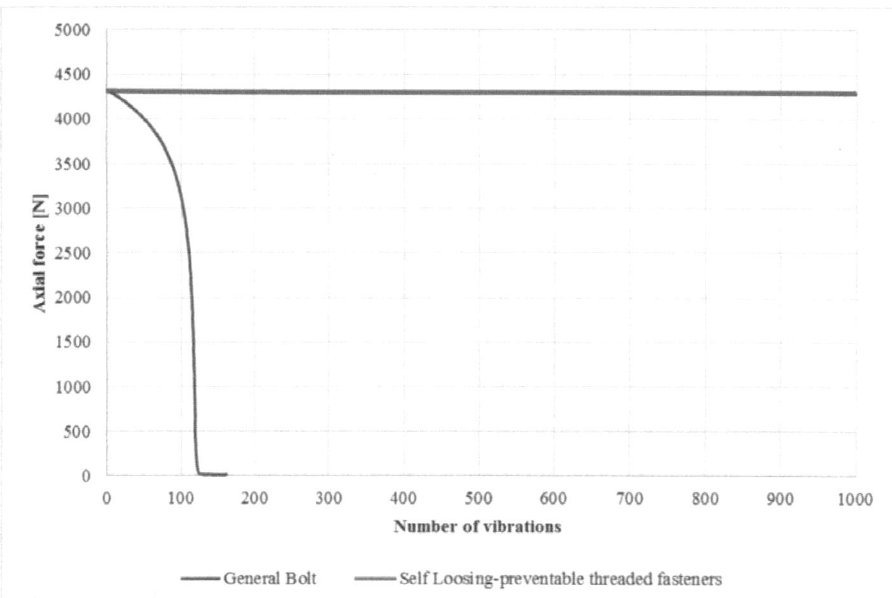

Figure 6. Analytical results of Junker vibration test on threaded fasteners.

Figure 6 shows the change in fastening axial force for each number of oscillations obtained. Horizontal axis shows the number of vibration and vertical axis represents the fastening axial forces. Comparison between general and helical-cutting applied bolts are executed and self-loosing preventable performance is confirmed.

As shown in figure 6, unlike ordinary hexagon head bolts, for threaded fastener using helical-cutting applied bolt with self-locking effect, it is possible to maintain the initial fastening axial force even when the number of vibrations reaches 1000.

Figure 7 shows the hysteresis loop of translational load with translational displacement of 5 cycles, respectively. From the hysteresis loop of self-loosing preventable threaded fasteners shown in figure 7, almost gentle gradient portion in which slip occurs on the entire threaded surface can't be confirmed. In addition, it can be seen that the flat part where the slip occurs over the entire surface of the bearing surface is minute.

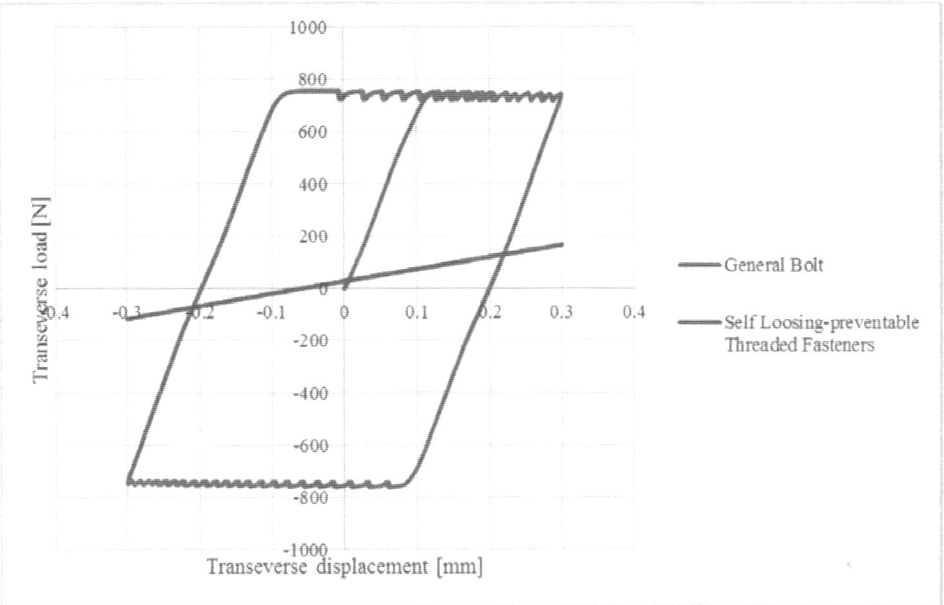

Figure 7. Relationship between translational force and displacement of analytical results.

5. Summary

Self-loosing preventable threaded fasteners using helical-cutting applied bolt structures are designed in this study. Spring characteristic effects of helical-cutting applied bolts and self-loosing preventable performance of threaded fasteners are examined through finite element analysis. From analytical results,

(1) Spring characteristic effect of helical-cutting applied bolt is confirmed by the analytical results of swelling effect under rotating loads conterclockwise.

(2) Self-loosing preventable performance of threaded fasteners using helical-cutting applied bolts were performed using finite element method and confirmed. Comparing slackness test results with analytical results and more details on evaluating mechanical properties will be executed in future study.

6. References

[1] Yamamoto A and Kasei S 1977 *J. Japan Soc. Prec. Eng.* **43(4)** 470-5.

[2] Izumi S, Yokoyama T, Iwasaki A and Sakai S 2005 *Trans. Japan Soc. Mech. Eng. Series(A),* **71(702)** 204-12.

[3] Kamiya S, Izumi S, Sakai S and Yamada Y 2012 *Trans. Japan Soc. Mech. Eng. Series(A),* **78(795)** 99-107.

Design Investigation on Applicable Mesh Structures for Medical Stent Applications

Shoji Asano[1*], Jianmei He[1]

[1]Dept. of Mechanical Engineering, Kogakuin Univ. 1-24-2, Nishi-shinjuku, Shinjuku-ku, Tokyo, 163-8677 Japan

Abstract. In recent years, utilization of medical stents is one of effective treatments for stenosis and occlusion occurring in a living body's lumen indispensable for maintenance of human life such as superficial femoral artery (SFA) occlusion. However, there are concerns about the occurrence of fatigue fractures caused by stress concentrations, neointimal hyperplasia and the like due to the shape structure and the manufacturing method in the conventional stents, and a stent having high strength and high flexibility is required. Therefore, in this research, applicable mesh structures for medical stents based on the design concepts of high strength, high flexibility are interested to solve various problem of conventional stent. According to the shape and dimensions of SFA occlusion therapy stent and indwelling delivery catheter, shape design of the meshed stent are performed using 3-dimensional CAD software Solid Works first. Then analytical examination on storage characteristics and compression characteristics of such mesh structure applied stent models were carried out through finite element analysis software ANSYS Workbench. Meshed stent models with higher strength and higher flexibility with integral molding are investigated analytically. It was found that the storage characteristics and compression characteristics of meshed stent modles are highly dependent on the basic mesh shapes with same surface void ratio. Trade-off relationship between flexibility and storage characteristics is found exited, it is required to provide appropriate curvatures during basic mesh shape design.

1. Introduction

In recent years, stenosis and obstruction have occurred in body lumen such as blood vessels, bile ducts, trachea and other like indispensable for maintenance of human life as in the superficial femoral artery (SFA) occlusion, which is caused by disease. As an effective treatment for such stenosis and occlusion, indwelling of medical stents is performed. On the other hand, the inside of the living body lumen where the stents are placed is a corrosive environment with complicated structure. The difficult environment may cause a rejection reaction when foreign matter intrudes. Therefore, it is required that the stent placed in the living body lumen has high flexibility, excellent biocompatibility, high strength and durability etc.

However, ready-made conventional stents have concerns that the design and manufacturing method of the shape structure causes stress concentration, and then fatigue fractures due to pulsation, or causes neointimal hyperplasia due to strong radial forces generated on the inner wall of the stents [1-9]. Therefore, there is the need for medical stents having high strength and high flexibility capable of following severe movements such as bending, torsion, expansion and contraction in the femoral artery.

Therefore, the purpose of this research is to design applicable mesh structures [10] for medical stent application to solve the problems of conventional stents. Meshed stent models with higher strength and higher flexibility with integral molding are design and investigated analytically. The storage characteristics and compression characteristics of meshed stent models are examined through finite element analysis.

2. Design of meshed stent model with mesh structures

2.1. Evaluation of surface void contents of meshed stent models.
Based on the design concepts for mesh structures [10] of high strength and high flexibility, the meshed stent models are obtained as shown in figure 1 according to the shape and dimensions of the SFA occlusion treatment stents shown in table 1. For the five types of mesh basic shapes as shown in figure 1, the basic mesh shape number arranged in the stent circumferential direction (hereinafter referred to as "N") is changed from 5 to 8, and the mesh line width is changed from 0.1 mm to 0.2 mm. Total of 40 kinds of meshed stent models are designed for all combinations between mesh type, basic mesh shape number and mesh line width. These basic mesh shapes are composed only with a tangent circular arc, and the S-shape curve bending is provided to achieve high strength and high flexibility.

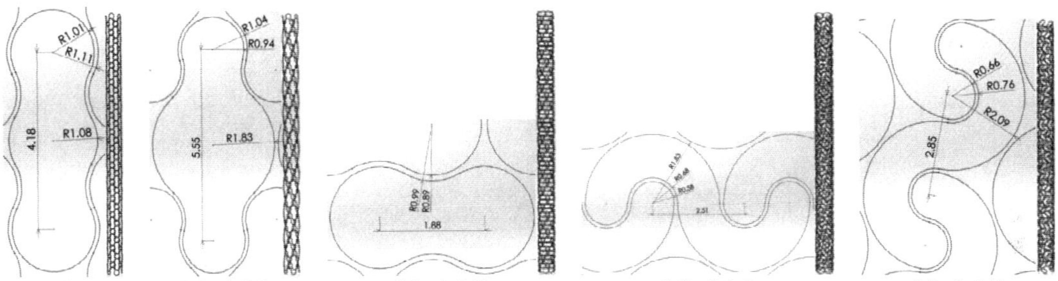

Figure 1. Basic mesh shapes and three-dimensional meshed stent models.

Table 1. Dimensions of desired stents.

Material	Nickel-titanium alloy
Stent Length [mm]	~100.0
Mesh Line Width [mm]	0.1, 0.2
Strut Thickness [mm]	0.18
Pre-storage Diameter [mm]	6.0
Post-storage Diameter [mm]	~2.0
Surface Void Content [%]	around 80

For the designed 40 meshed stent models, surface void contents were evaluated. Surface void content is defined as the surface area removed before and after mesh structure's application. It is desirable to have around 80% from medical point of view. Surface void content can be calculated by equation (1), and meshed stent models with surface void content around 80% are shown in Table 2 for further analytical investigations.

$$a = 100 - \frac{b}{c} \times 100 \tag{1}$$

where a: Surface void content of meshed stent models [%]
b : Cylindrical surface area after application of mesh structure [mm^2]
c: Cylindrical surface area before application of mesh structure [mm^2]

3. Evaluation of Storage Characteristics of Meshed Stent Models

The storage characteristics were analyzed and evaluated using finite element analysis software ANSYS Workbench 14.5 for meshed stent models having surface void content of around 80% shown in table 2. Material properties of nickel titanium alloy shown in table 3 were applied and constraint conditions were applied as 2.0 mm storage diameter of meshed stent model. Figure 2 shows typical finite element mesh of meshed stent model for analysis. Figure 3 shows examples of the storage characteristic evaluation results of designed meshed stent models.

Table 2. Surface void content results of meshed stent models.

Mesh shape - Mesh line width	N Number (Surface void content [%])
Model 1-0.2mm	N5(82.65), N6(80.21), N7(77.77)
Model 2-0.2mm	N7(81.58), N8(79.84)
Model 3-0.1mm	N8(82.55)
Model 3-0.2mm	N5(79.23)
Model 4-0.1mm	N5(82.25), N6(81.01), N7(78.81)
Model 5-0.1mm	N6(81.59), N7(80.51), N8(78.77)

Table 3. Material properties of nickel-titanium, number of nodes and number of elements.

Density [g/cm^3]	6.45
Poisson's ratio	0.3
Modulus of longitudinal elasticity [GPa]	75
Number of nodes	130260~176620
Number of elements	79827~103362

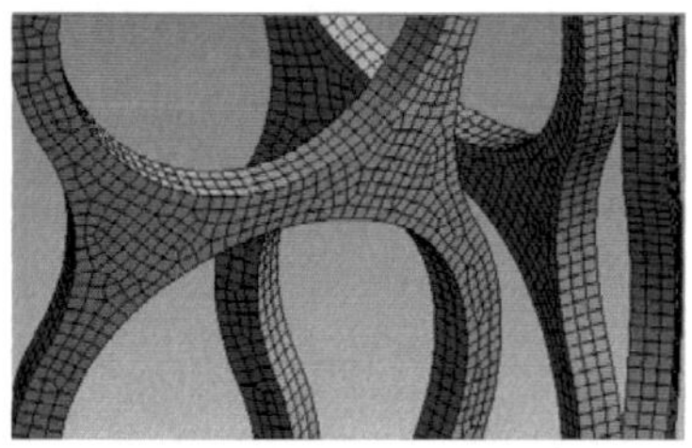

Figure 2. Finite element mesh for meshed stent model 1-N5.

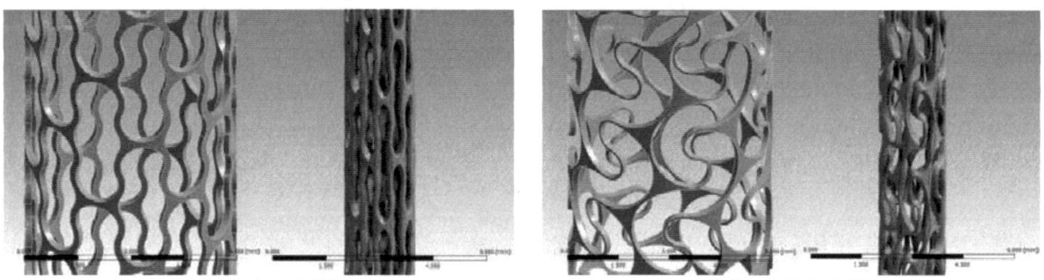

Model 1-N7 Model 5-N7

Figure 3. Sample analytical results of storage meshed stent models.

From these analytical results,

(1) Stents having basic mesh shapes with deep S-shaped curve (Model 5) can be expected to greatly reduced stress concentration during storage, but the mesh lines tend to come into contact during storage and tend to be inferior.

(2) Stents composed of basic mesh shapes with a shallow S-shape curve (Model 1, 2) can be expected to have higher storability.

(3) The storage characteristics are not necessarily dependent on surface void contents. In the case of same surface void contents of meshed stent models, the storage characteristics are highly dependent on basic mesh shapes.

4. Evaluation of compression characteristics of meshed stent models

Similarly to the storage characteristic analysis, compression characteristics were analyzed and evaluated using finite element analysis software ANSYS Workbench 14.5 for meshed stent models having surface void contents of around 80% shown in table 2. Meshed stent models are sandwiched between flat jigs and applied by compressive load of 10 N in the direction perpendicular to stent cylinder axis. Table 4 shows the number of nodes and elements of meshed stent model for analysis. Figure 4 shows the image of typical finite element mesh of meshed stent model including compressive fixtures.

Table 4. Number of nodes and elements in analysis of cylindrical axis vertical direction.

Number of nodes	186090~251505
Number of elements	94578~125192

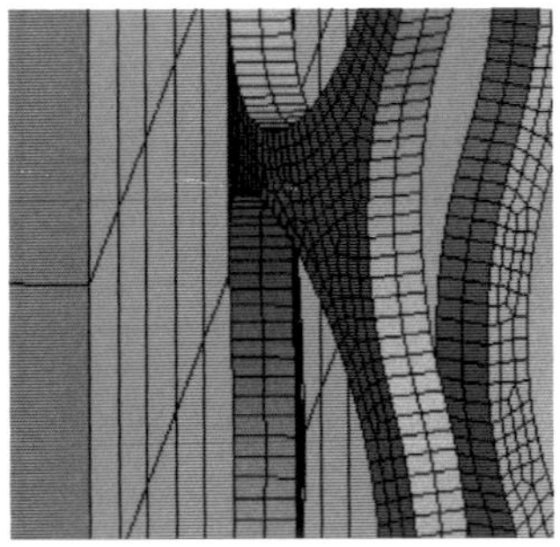

Figure 4. Finite element mesh of meshed stent model with compressive fixtures (Model 1-N5).

Figure 5 shows sample un-deformed and deformed results of meshed stent model under compressive loads. Compressive stiffness k is then calculated by fitting the deformation amount δ with compressive load P applied to the meshed stent models to equation (2). Figure 6 shows the relationship between compressive stiffness and surface void content of different meshed stent models.

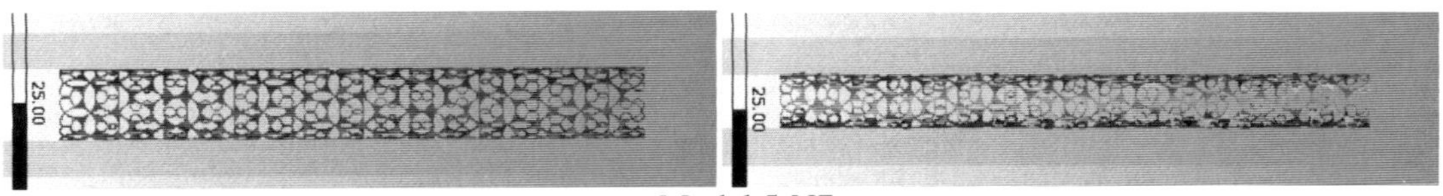

Model 5-N7

Figure 5. Un-deformed and deformed results of meshed stent model with compressive fixtures.

$$k = \frac{P}{\delta} \tag{2}$$

where k: Compressive stiffness [N/mm]
P: Compressive load [N]
δ: Displacement [mm]

(1) Stents having longitudinally elongated basic mesh shapes tend to have lower compressive stiffness than elongated basic mesh shapes in lateral direction.

(2) Similarly to the storage characteristic evaluations, the compression characteristics of meshed stent models are not necessarily dependent on their surface void contents. With same surface void contents, the compressive stiffness of meshed stent models is highly dependent on basic mesh shapes.

(3) Compressive stiffness of a ready-made conventional stent with a diameter of 10.0 mm is to be about 0.3 N / mm, and the compressive stiffness of the meshed stent models designed in this research tend to be considerably higher overall. Improvement on the meshed stent models need to be executed.

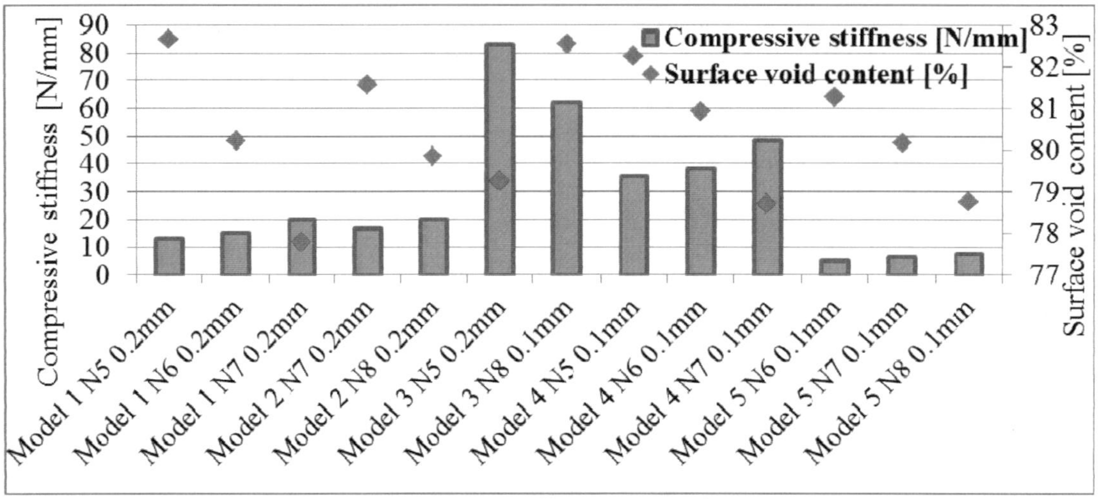

Figure 6. Relevance between compressive stiffness and surface void content of meshed stent models from these analytical results,

5. Summary

Meshed stent models with higher strength and higher flexibility with integral molding are investigated analytically. It was found that the storage characteristics and compression characteristics of meshed stent modles are highly dependent on the basic mesh shapes with same surface void content. Trade-off relationship between flexibility and storage characteristics is found exited, it is required to provide appropriate curvatures during basic mesh shape design.

6. References

[1] Jaff M, Dake M, Pompa J, Ansel G and Yoder T 2007 *Catheter. Cardiovasc. Interv.* **70** 460-2.
[2] Kralj I, Boos I and Müller-Bühl U 2013 *Vasa* **42** 340-9.
[3] Müller-Hülsbeck S, Schäfer P J, Charalambous N, Yagi H, Heller M and Jahnke T 2010 *J. Endovasc. Ther.* **17** 767-76.
[4] Scheinert D, Scheinert S, Sax J, Piorkowski C, Bräunlich S, Ulrich M, et al. 2005 J. Am. Coll. Cardiol. **45** 312-5.
[5] Schlager O, Dick P, Sabeti S, Amighi J, Mlekusch W, Minar E, et al. 2005 *J. Endovasc. Ther.* **12** 676-84.
[6] Schulte K L, Kralj I, Gissler H M, Bagnaschino L A, Buschmann I, Pernès J M, et al. 2012 *J. Endovasc. Ther.* **19** 774-84.
[7] Schulte K L, Müller-Hülsbeck S, Cao P, Becquemin J P, Langhoff R, Charalambous N, et al. 2010 Euro Interv. **5** 687-91.
[8] Yazdani S K and Berry J L 2009 *J. Vasc. Interv. Radiol.* **20** 101-6.
[9] Zhao H Q, Nikanorov A, Virmani R, Jones R, Pacheco E and Schwartz L B 2009 Cardiovasc. Intervent. Radiol. **32** 720-6.
[10] He J, Suzuki S and Chung U I 2014 *Trans. JSME* **80(809)** DOI:10.1299/taransjsme.2014smm0003.

Evaluation on Bending Properties of Biomaterial GUM Metal Meshed Plates for Bone Graft Applications

Hiromichi Suzuki[1*], Jianmei He[1]

[1]Dept. of Mechanical Engineering, Kogakuin Univ. 1-24-2, Nishi-shinjuku, Shinjuku-ku, Tokyo, 163-8677 Japan

Abstract. There are three bone graft methods for bone defects caused by diseases such as cancer and accident injuries: Autogenous bone grafts, Allografts and Artificial bone grafts. In this study, meshed GUM Metal plates with lower elasticity, high strength and high biocompatibility are introduced to solve the over stiffness & weight problems of ready-used metal implants. Basic mesh shapes are designed and applied to GUM Metal plates using 3D CAD modeling tools. Bending properties of prototype meshed GUM Metal plates are evaluated experimentally and analytically. Meshed plate specimens with 180°, 120° and 60° axis-symmetrical types were fabricated for 3-point bending tests. The pseudo bending elastic moduli of meshed plate specimens obtained from 3-point bending test are ranged from 4.22 GPa to 16.07 GPa, within the elasticity range of natural cortical bones from 2.0 GPa to 30.0 GPa. Analytical approach method is validated by comparison with experimental and analytical results for evaluation on bending property of meshed plates.

1. Introduction

People who have lost a part of their bones due to accidents, injuries or illnesses are subjected to bone graft surgery. This study focuses on metallic implants used for artificial bone grafts. Figure 1 shows an X-ray photograph of fixation by a metal plate in actual bone graft implantation. Conventionally used implants of titanium alloys have the problems that the burdens on the patient are increased due to their excessive rigidity and overweight than human bones. Therefore, in this study, GUM Metal plates with low elastic modulus, high strength and high biocompatibility was interested to improve these problems. The purpose of this research is to evaluate the out-of-plane bending characteristic of meshed GUM Metal plates with higher bending flexibility near to the natural bones.

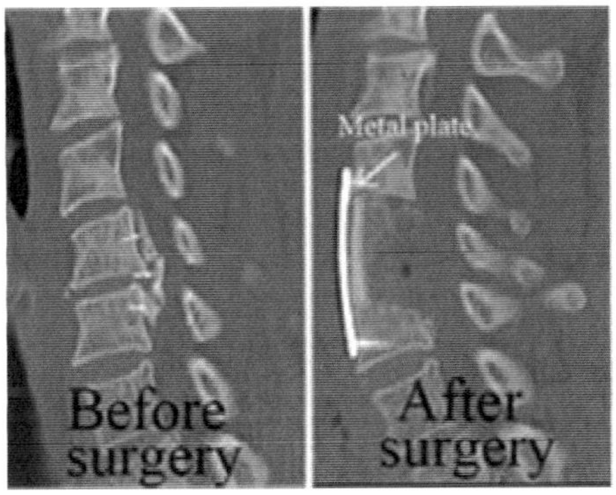

Figure 1. X-ray photograph of artificial bone graft.

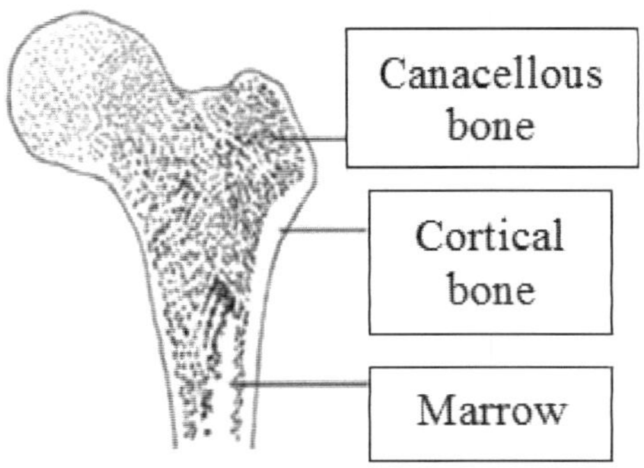

Figure 2. Cortical bone in the femur.

Basic mesh shapes are designed using 3D CAD software Solidworks. And meshed GUM Metal plate specimens are fabricated by laser-cutting processing for experimental evaluations. The mechanical properties of meshed plate implants should be close to the mechanical properties of human's cortical bones to minimize damages on patients. Natural cortical bones as shown in figure 2 have volume density range from 0.5 to 1.1 g/cm^3 and elastic modulus range from 2.0 to 30.0 GPa. On the other hand, analytical approaches on bending stiffness of meshed GUM Metal plate specimen are also carried out through finite element method and comparison with experimental results are used to validate analytical method proposed in this study.

2. Design of basic mesh shape and preparing of meshed plate specimens

2.1. Characteristics of GUM Metal plate used for meshed plate specimens

It is necessary to use materials with high biocompatibility for medical implants. Human bodies do not recognize pure titanium and titanium alloy as foreign matter, and new generated bones will catch around the titanium implants. GUM Metal is one kind of titanium alloy made by melting and recrystallizing powders of pure titanium, niobium, tantalum and zirconium. GUM Metal has mechanical performances like low elastic modulus, high strength and high biocompatibility/bioaffinity. Tensile test based on JIS Z 2241 standared was performed on GUM Metal plate used for prototype meshed plates. From experimental results, tensile elastic modulus and Poisson's ratio of GUM Metal were calculated and shown in table 1. These materical properties can be used for analytical approach on prototype meshed GUM Metal plates.

Table 1. Material properties of GUM Metal plate obtained by tensile test.

Young's modulus [GPa]	91.2
Poisson's ratio	0.33

2.2. Design of basic mesh shapes for prototype meshed plates

Basic mesh shapes are designed following design concepts shown below using 3D cad modeling tools.

① One unified basic mesh shape to reduce designing and manufacturing costs.

② Reduce stress concentrations to reach high strength nag long operating life, by eliminating corners in the basic mesh shape and approximating the mesh line width uniformly.

③ Easy control of in-plane stretchability and out-of-plane bending characteristics by providing S-shaped curves with changed curvatures in the basic mesh shape.

Figure 3 shows the basic mesh shapes designed for prototype meshed plate applications. From left to right, 180° axisymmetric shape, 120° axisymmetric shape and 60° axisymmetric shape are obtained for experimental and analytical evaluations.

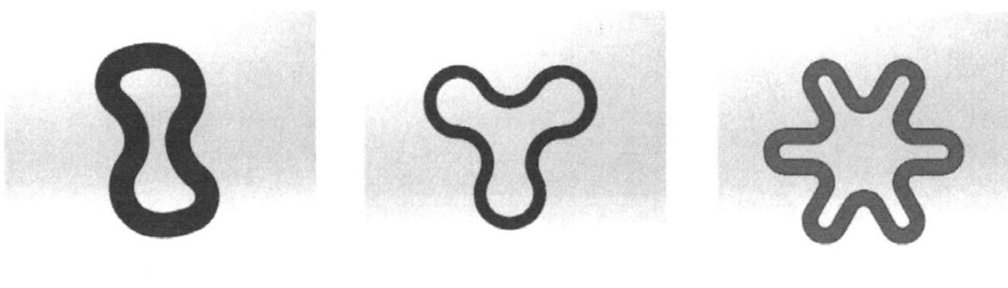

180° axisymmetric shape 120° axisymmetric shape 60° axisymmetric shape

Figure 3. Basic mesh shapes designed for experimental and analytical evaluations.

2.3. Design of prototype meshed plate specimens

Four types of prototype meshed plate specimens were designed using the basic mesh shapes shown in the previous section. Figure 4 shows the respective sizes and shapes of prototype meshed plate models. All of them have same mesh line width of 0.8 mm. 180° prototype model is obtained by rotating 60 ° of 180° axisymmetric basic mesh shape and having most in-plane isotropy performance. Difference between 120° and 120° (revers) prototype models is reversed of the central array of 120° axisymmetric basic mesh shape.

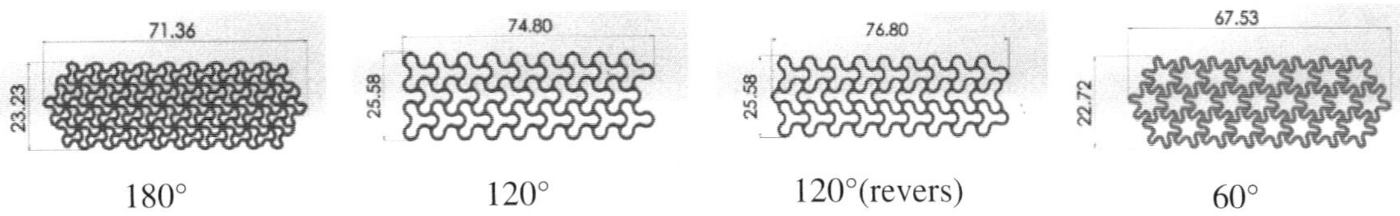

Figure 4. Prototype meshed plate models for experimental and analytical evaluations.

2.4. Evaluation on volume density of prototype meshed plates

Prototype meshed plate models become light-weight structures because of the applied basic mesh shapes. Then the volume densities of meshed GUM Metal plate models can be calculated from equation (1) and the results are shown in figure 5 comparing with no-meshed GUM Metal plate.

$$\text{Volume density} = \frac{g}{V}\left(\begin{array}{l} g\text{: Mass after mesh processing} \\ V\text{: Volume before mesh processing} \end{array}\right) \quad (1)$$

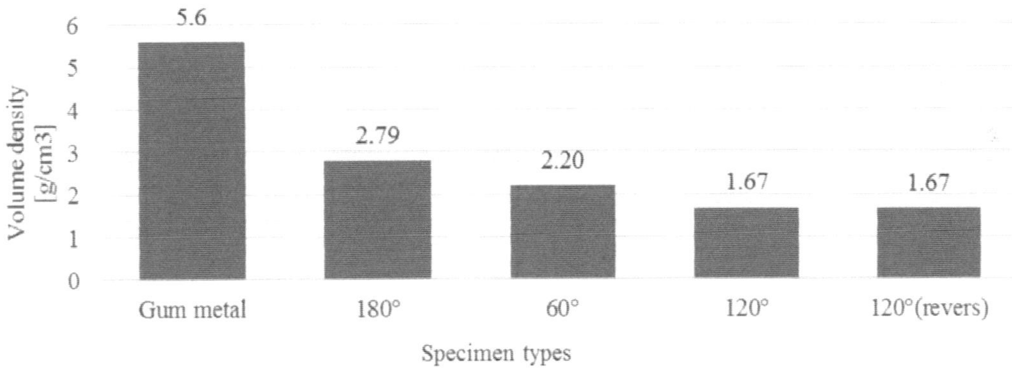

Figure 5. Meshed plate shape models for sample specimen evaluations.

From figure 5, one can see that prototype GUM Metal Plate models are significantly reduced in weight by mesh processing. It is confirmed that the weight characteristics of meshed GUM Metal plate models are approached close to the volume density of human cortical bones range from 0.5 to 1.1 g/cm^3 through applying the basic mesh shapes.

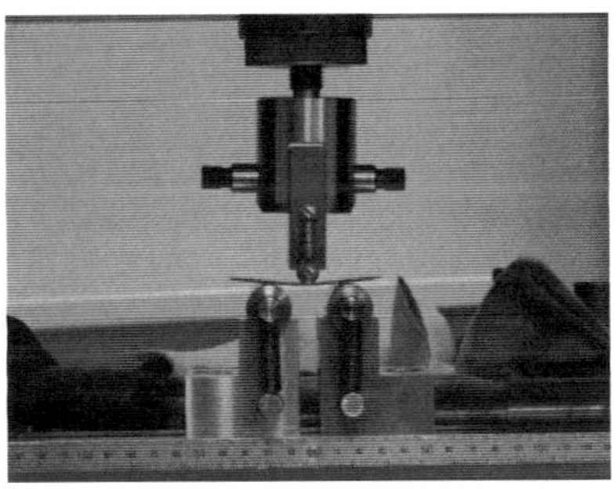

Figure 6. Three-point bending test.

Figure 7. Meshed plate specimens for 3-point bending test.

3. Bending property evaluation of prototype meshed GUM Metal plates

Three-point bending tests for three kinds of prototype meshed GUM Metal plates were carried out based on JIS Z 2248 (Metal material bending test method) standard. Figure 6 shows the photo image of three-point bending test and figure 7 shows the prototype meshed plate specimens.

Based on the loading-deflection result of three-point bending test, bending pseudoelastic modulus E of sample meshed plate specimens can be calculated from equation (2).

$$E = \frac{Pl^3}{48I\delta} \tag{2}$$

where P represents bending load, δ represents bending deflection, l represents the distance between surporting points of three-point bending test and P / δ = tan θ can be obtained from loading-deflection diagram. The angle θ shows the inclination of the load-deflection diagram in the linear elastic range of the three-point bending test result.

4. Analytical evaluation on bending properties of meshed plate specimens

Analytical approach on evaluation of bending characteristics for meshed plate specimens were executed through finite element method. Figure 8 shows analytical model of three point bending test on meshed plate specimen with surport fixture and loading jig. Bending load is set at 1.0 N perpendicular to the meshed plate. Further, the contact conditions between surport fixture and meshed plate specimen were defined as "having friction". The values of friction coefficient and vertical stiffness were taken into account. On the other hand, contact condition between loading jig and meshed plate specimen was set as "solid bond". Table 2 shows detailed friction coefficients and vertical stiffness coefficient for different prototype meshed GUM Metal plates. Material model of GUM Metal plates are based on the tensile test results as shown in table 1.

Figure 9 and table 3 show details about finite element meshes of different prototype meshed plates. The bending pseudoelastic modulus of prototype meshed plates were calculated using the analytical results of deflection under 1.0 N bending load and equation (2) shown in previous section.

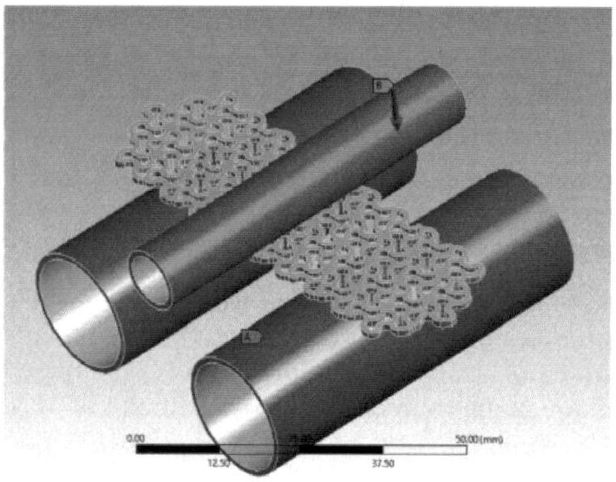

Figure 8. Analytical model of 3-point bending test.

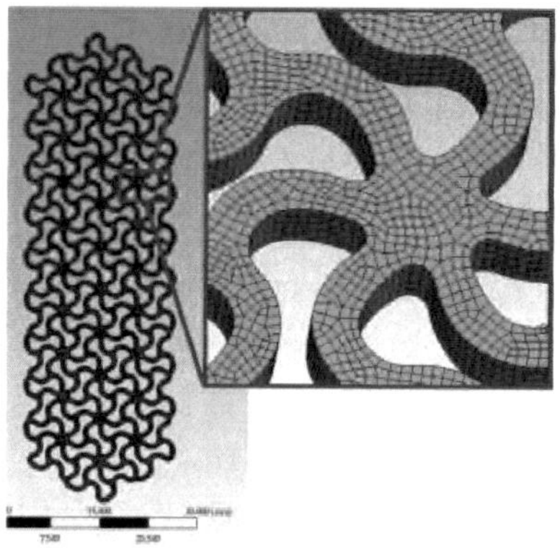

Figrue 9. Meshed specimen.

Table 2. Contact condition of model analysis.

Types of analysis models	Friction coefficient	Vertical stiffness coefficient
180°	0.08	0.00015
120°	0.07	0.00004
120°(revers)	0.09	0.00007
60°	0.009	0.00001

Table 3. Details of mesh split.

Types of analysis models	Node quantities	Element quantities	Sweep split quantities	Element size [mm]
180°	171792	117688	5	0.2
120°	237966	170987	5	0.15
120°(revers)	239478	171597	5	0.15
60°	136824	92296	5	0.2

Figure 10 shows the analytical bending pseudoelastic modulus calculated with experimental results for compareson. From the results shown in figure 10, one can see that the 180 ° axisymmetric meshed plate specimen shows the highest bending pseudoelastic modulus compared with other axisymmetric meshed plate specimens. The range of bending pseudoelastic modulus obtained from these four type meshed plate specimens are from 4.22 GPa to 16.07 GPa, which is closed to the elastic modulus of natural cortical bones range from 2.0 to 30.0 GPa. In addition, diviations between analytical and experimental results on 3-point bending test of prototype mehsed GUM Metal plates are very small. That validate the proposed analytical method here and more detail parameteric examinations can be carried out on bending property evaluation through analatical approaches.

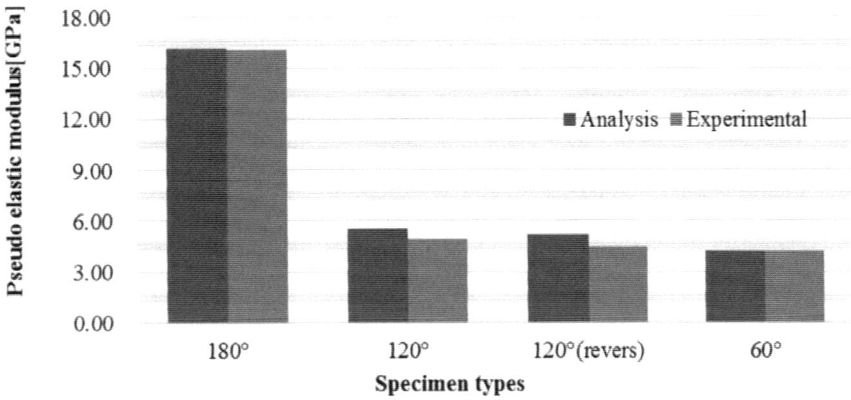

Figure 10. Experimental and analytical bending pseudo elastic modulus of meshed plate specimens.

5. Summary

Based on three type of axisymmetric basic mesh shape designed in this study, prototype meshed GUM Metal plates are fabricated for experimental evaluation. From the bending pseudoelastic modulus results of meshed plate specimens obtained experimentally and analytically, the following conclusions are reached.

(1) It is possible to process a lightweight meshed GUM Metal plate for bone graft application. More than 50% of volume density reduction can be obtained by applying the basic mesh shapes on GUM Metal plates and close to the value of natural cortical bones.

(2) The range of bending pseudoelastic modulus obtained from the four type meshed plate specimens introduced are from 4.22 GPa to 16.07 GPa with higher flexibility, which is closed to elastic modulus of natural cortical bones range from 2.0 to 30.0 GPa.

(3) Diviations between analytical and experimental results on 3-point bending test of prototype mehsed GUM Metal plates are very small. That validate the proposed analytical method here and more detail parameteric examinations will be carried out on bending property evaluation through analatical approaches in future study.

6. References

[1] Ohara H 2017 *Study on bending property of titanium alloy meshed plate for implant,* Graduate School of Engineering.
[2] Kokubo K, Goto Y, Mori Y and Tateno T *Mechanical Engineering Basic Course Material Mechanics* Maruzen Co., Ltd.
[3] Introduction of GUMMETAL <http://www.nissey-sabae.co.jp/wp/wp-content/uploads/gummetal.pdf>, (Reference Date July 10, 2017)
[4] Moriyama Neurosurgery
[5] <https://www.moriyama-neurosurgical-clinic.com/>,(Reference Date July 27, 2017)
[6] Astellas Pharma
[7] <https://www.astellas.com/jp/>,(Reference Date July 27, 2017)

IOP Conf. Series: Materials Science and Engineering **269** (2017) 012079 doi:10.1088/1757-899X/269/1/012079

Experimental Evaluation on Fatigue Strength of Adhesive-Bonded Sandwich Panels with CFRP Faces and Honeycomb Cores

Fumihiko Sato[1*], **Jianmei He**[1]

[1]Dept. of Mechanical Engineering, Kogakuin Univ. 1-24-2, Nishi-shinjuku, Shinjuku-ku, Tokyo, 163-8677 Japan

Abstract. There is no existed standard testing method suitable to evaluate the fatigue strength of thick adhesive bonded structures like honeycomb sandwich panels. Fatigue strengths of adhesive-bonded sandwich panels constructed with aluminum honeycomb cores and CFRP (Carbon Fiber Reinforced Plastics) laminate faces were interested in this study. Fatigue experiments on sample cantilever beams of such honeycomb sandwich panels were carried out to validate the experimental method by improving the configurations of specimens, testing fixtures and instruments etc. Based on the improved testing method, influences of design parameters, such as specimen length and height, foil thickness and cell size of honeycomb cores, on the fatigue strength of adhesive bonded sandwich panels will be investigated in future experimental study.

1. Introduction

Honeycomb sandwich panels constructed with thin surfaces and honeycomb cores have characteristics like light-weight, higher stiffness and shock absorbency etc. Therefor they are usually developed for large-scale structure applications in various industrial fields, especially light-weight purpose desired fields like aerospace and architecture as shown in figure 1.

On the other hand, recognition on the durability of adhesive-bonded structure is very important for such honeycomb sandwich panels. Adhesive peeling between panel surfaces and honeycomb cores made from different materials such as light-weight FRP laminate surfaces and aluminum honeycomb cores need to be avoided to maintain the function of such sandwich panels [1]. Especially, the de-bonding problem caused by repetitive loading will cause the problem of strength degradation of honeycomb sandwich panel structures and lead to fractures. There is no existed standard testing method suitable to evaluate the fatigue strength of thick adhesive-bonded structures like honeycomb sandwich panels.

In this study, fatigue strengths of adhesive-bonded sandwich panels constructed with aluminum honeycomb cores and CFRP (Carbon Fiber Reinforced Plastics) laminate faces were interested [2]. Fatigue tests on sample cantilever beams of honeycomb sandwich panels were carried out. Improvements of configurations of specimens, testing fixtures and instruments etc. are executed to validate the experimental method.

Figure 1. Honeycom sandwich panel and its applications [3, 4].

2. Fatigue experiment of honeycomb sandwich panels

2.1. Specimen configuration

Sample honeycomb sandwich panels with 1.0mm thick CFRP laminate, and 1/8 inch cell size, 0.001 inch foil thickness and 30 mm height aluminum honeycomb core are introduced for fatigue experimental investigations. Two-liquid-component type, room-temperature cured adhesive consisting Epoxy and Polyamidoamine was introduced for bonding CFRP laminate faces and aluminum honeycomb cores, to minimize the residual stress caused by the linear thermal expansion during curing. The specimen configurations with panel shape and sizes are shown in figure 2 and table1. Figure 3 shows the detail of honeycomb specification.

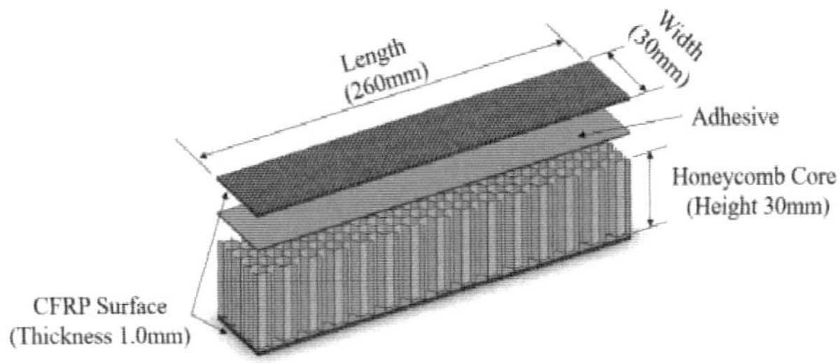

Figure 2. Honeycomb sandwich panel structure.

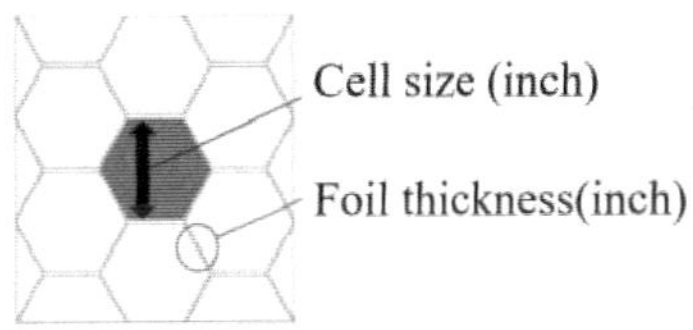

Figure 3. Honeycomb specification.

Table 1. Specification of composite panels.

Honeycomb Core	Material	A5052
	Cell size [inch]	1/8
	Foil thickness [inch]	0.001
	Height (t_c) [mm]	30
Panel Surface	Material	CFRP
	Thickness (t_f) [mm]	1.0
	Bending modulus [GPa]	42.3
Specimens size Width (b)×Length×Height [mm]		30×260×32

2.2. Experimental configuration

Cantilever bending fatigue tests on sample honeycomb sandwich panel specimens were introduced for experimental evaluations. Figure 4 shows the testing machine introduced for this study. This testing machine configuration are basically along the ASTM D 671-63 T standard (Standard Test Method for Flexural Fatigue of Plastics by Constant-Amplitude-of-Force). The left end of specimen is fixed and a repeated bending deflection is applied on the right end as shown in figure 5. Two strain gauges are installed on both sides of specimen surfaces for measurements. The deflections of cantilever bending tests are set from 2.0 mm to 6.0 mm with a number of loading cycle of 10^6 for sample examinations. Test conditions are described in table 2.

Figure 4. Repeated-Flexural-Tester.

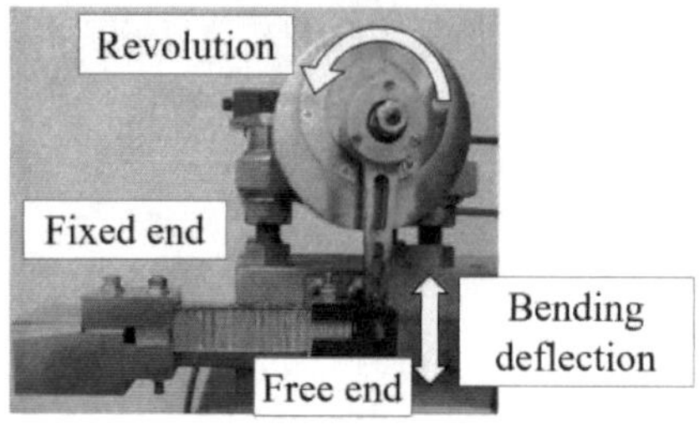

Figure 5. Photo image of bending fatigue test.

Table 2. Conditions of bending fatigue test of composite panels.

Rotating speed [rpm]	1700
Sampling period [Hz]	500
Number of data point [-]	18000000
Number of cycle [-]	10^6
Bending deflection [mm]	2.0~6.0

2.3. Improved design of fixture for cantilever beam's fatigue test

Based on the ASTM D 671-63 T standard, the testing configuration need to be improved because of the honeycomb sandwich panel's thickness. Figure 6 shows the improved fixture for cantilever bending fatigue test. It has enough bending stiffness compared with the specimens and durability for the load occurred by bending deformation. By fixing the upper lid to not only the specimen but also the base of the fixture with two bolts, stability of fixing the specimen can be obtained. For the bending fatigue test, it is desirable that the center axis in the height direction of specimen and bending deflection are coincident. By using a height adjusting sheet material according to the height of panel specimen, it is possible to correct the deviation from the center of bending deflection with the specimen center line. Figure 6 shows an image diagram of the improved fixture in 3D CAD model with panel specimen fixed.

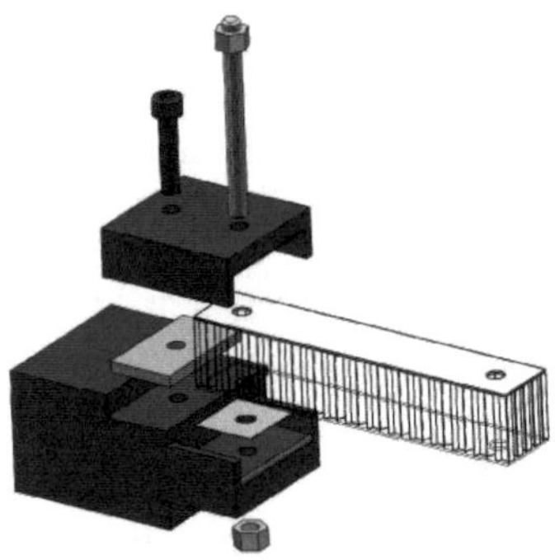

Figure 6. Designed fixed-end fixture with composite panel specimen (Exploded diagram).

2.4. Evaluation on modulus of rigidity of honeycomb sandwich panel

Three-point bending test for honeycomb sandwich panels were carried out. Figure 7 and table 3 shows the three-point bending test and test conditions. Based on the result of three-point bending test, the modulus of rigidity G of honeycomb sandwich panel was calculated at 152.0 MPa using equation (1) and (2) [5, 6]. These results are needed for theoretical strain calculation used for comparison with experimental results in next section.

Table 3. Test conditions of three-point bending test.

Specimen size [mm]		160×30×32
CFRP	Thickness (t_f) [mm]	1.0
	Bending modulus (E) [GPa]	43.4
Span (L) [mm]		100
Diameter of indenter [mm]		5
Feed speed [mm/min]		1.0

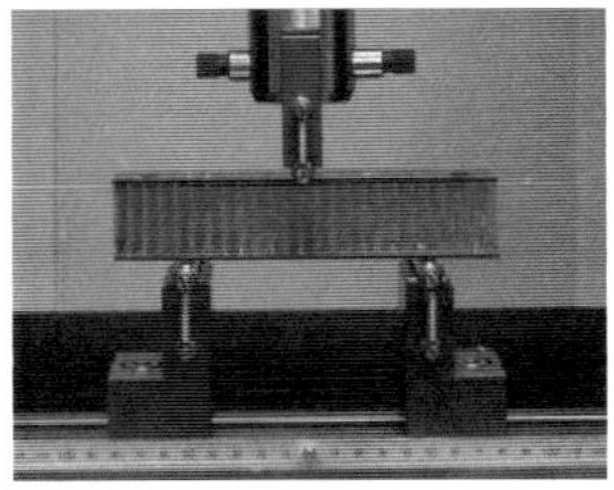

Figure 7. Three-point bending test.

$$G = \frac{L}{\frac{4}{P/\delta} - \frac{L^3}{12EI}} \times \frac{t_c}{b(t_f+t_c)^2} \quad (1)$$

$$I = \frac{bt_f}{6}\left(3t_c^2 + 6t_f t_c + 4t_f^2\right) \quad (2)$$

2.5. Experimental results

Measured strain results from the cantilever bending fatigue test of the honeycomb sandwich panels are analyzed using DIAdem software. Figure 8 shows firstly the typical close-up strain data obtained from data manipulation. Horizontal axis represents the number of loading cycle and vertical axis represents the strains obtained from the strain gauges. The accurate strain value with respect to the frequency can be obtained from the rain-flow counting analysis and sample result is shown in figure 9.

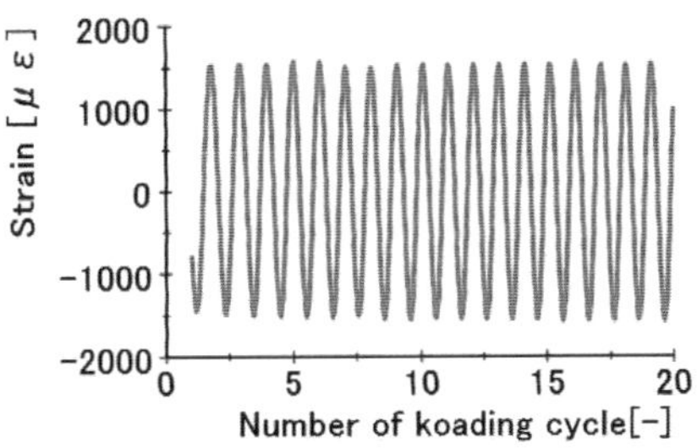

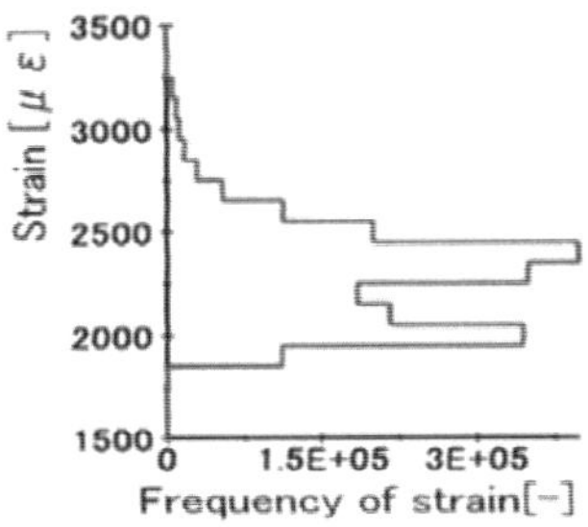

Figure 8. Example of analyzed strain data. **Figure 9.** Result of rain flow counting.

The typical strain data measured from upper strain gauges (CH1) under 2.0mm, 3.0 mm, 4.0 mm and 6.0 mm bending deflections are shown in figure 10 with logarithmic representation of loading cycle number at horizontal axis. At the 6mm deflection case, the test was discontinued because fracture of core at the fixed-end was confirmed as shown in figure 11.

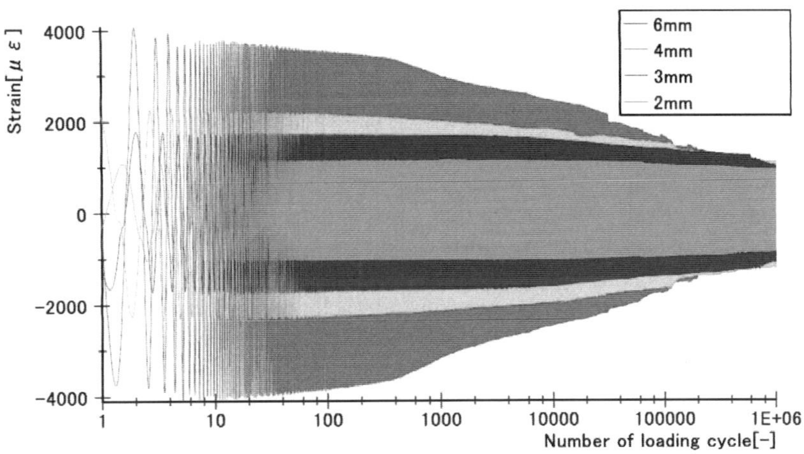

Figure 10. Compilation of upper surface strain data.

Figure 11. Fracture of honeycomb core.

Further, the theoretical strain value of the panel surface corresponding to different bending deformations can be calculated from the equation (5) and the results are shown in Table 4 compared with experimental results. Equation (5) can be derived from equation (3) and (4), which are the basic formulas for strength of materials as function of bending deflection δ and modulus of rigidity G [5, 6].

$$\delta = \frac{PL^3}{3EI} + \frac{PL}{U} \tag{3}$$

$$\mathrm{U} = \frac{(t_f+t_c)^2}{t_c} Gb \tag{4}$$

$$\varepsilon = \frac{3U\delta y}{UL^2+3EI} \tag{5}$$

Table 4. Theoretical and experimental value of bending fatigue test.

	Theoretical value		Experimental value			
Bending deflection [mm]	Bending moment [N·mm]	Strain [με]		Maximum strain (Theoretical deviation) [με]	Most frequent strain [με]	Minimum strain [με]
3.0	0.73×10^5	1918.5	CH1	1762.9 (-8.1%)	1222.7	975.5
4.0	0.97×10^5	2558.0	CH1	2144.4 (-16.2%)	1144.0	1072.2
6.0	1.46×10^5	3837.0	CH1	3746.7 (-2.4%)	1251.6	1048.3

Looking at the experimental strain data obtained from bending fatigue tests shown in figure 10, it can be confirmed the attenuation of strain due to bending deflections. From table 4, one can see that strain values immediately after test starts were close to the results calculated from theoretical formula, but they decreased sharply thereafter and gradually stabilized at about 1200με.

From sectional observation on tested specimens, no noticeable changes were observed at 2.0mm, 3.0 mm and 4.0 mm bending deflection cases. At 6.0 mm, honeycomb core near the fix end was found broken instead of fractures at adhesive bonds. From figure 10, it can be seen that the start point of attenuation differs with bending deflections. At relatively small bending deflection cases such as 2.0 mm, attenuation of strain value can hardly be confirmed until the number of load cycles exceeds 10^6. However, as the deflection increases, the number of load cycles at attenuation beginning decreases. This can be considered to be caused by fatigue phenomenon due to the bending moment generated when bending the specimen. A reaction force of bending deflection acts on the fixed-end of the specimen and causes compression effects on honeycomb core. With repeated bending deflections, deformation and breakage of inner honeycomb core are considered to be occurred mainly in low cycle fatigue region up to 10^3 times.

Utilization of the designed fixture for bending fatigue tests of thick honeycomb sandwich panels confirmed that the maximum strain value increased proportionally with the bending deflection increasing.

3. Summary

Testing method for the fatigue property evaluation on adhesively bonded honeycomb sandwich panels is investigated through experimental approach. Fatigue bending tests on sample aluminum/CFRP

IOP Conf. Series: Materials Science and Engineering **269** (2017) 012079 doi:10.1088/1757-899X/269/1/012079

honeycomb sandwich panels was carried out. From experimental results the following conclusions are obtained.

(1) By comparison, the deviation between experimental maximum strain and calculated static strain was mainly less than 10%, and the proportional relationship between bending deflection and measured maximum strain could validate the testing method proposed in this study. However, there are cases with deviation of more than 10%, which means improvement of installing strain gauges need to be done.

(2) Inner deformation or breakage of honeycomb core is considered to be occurred due to repeated compressive loads caused by large bending deflections. The destruction phenomenon can be confirmed from the decreased loading cyclic numbers with increased bending deflections.

(3) Based on the proposed fatigue bending test method, honeycomb cores at the fixed-end are deformed or broken before debonding of panel structure occurred. Therefor, thick adhesive structures can not be loaded stably. It is necessary to generate same bending strains with smaller bending moments by changing shape of panel specimens and reducing their bending rigidities.

4. Reference

[1] Sentanzairyougizyutukyoukai ed.,: Hanikamukouzouzairyou No Ouyou 2002 (Application of Honeycomb Structure Materials-). CMC.

[2] Mitsui T and He J 2014 *Experimental Investigation on Fatigue Performance of Light-weight Honeycomb Sandwich Panels*, The 5th TSME International Conference on Mechanical Engineering, AMM **019** 12.

[3] JAXA Homepage URL:http://www.jaxa.jp/projects/sat/index_j.html (28.07.2017)

[4] ANA Homepage URL: https://www.ana.co.jp/ (28.07.2017)

[5] Nihonzairyougakkai ed.,: Sentanzairyousiri-zu Hakai To Zairyou – (Tip Materials Series Destruction and Materials-). Shoukabou.

[6] Kunio K, Yoshiki G, Takao M, Masayoshi T eds,.: zairyourikigaku (Strength of Materials-). Maruzen Co. Ltd

Compression Characteristics of Honeycomb Sandwich Panels to Improve Their Impact Resistances

Kentaro Takagi[1*], Jianmei He[1]

[1]Dept. of Mechanical Engineering, Kogakuin Univ. 1-24-2, Nishi-shinjuku, Shinjuku-ku, Tokyo, 163-8677 Japan

Abstract. Honeycomb sandwich panels are kind of composite panels constructed by bonding core material of honeycomb structures with thin plate surfaces together. Since honeycomb sandwich panels (Briefly called composite panels) have the light-weight and high strength performances, they are widely utilized in many fields, especially light-weight required aerospace field. However, data on impact resistant properties of such composite panels are not widely publicized, and the testing method is not clearly specified. Therefore, in this study, the local buckling tendency of composite panels was experimentally evaluated with respect to various design variables, such as the foil thickness, cell size and specimen dimension of honeycomb structures. As conclusion, under local compressive loading for honeycomb sandwich panels, honeycomb cores are buckled before reaching yielded points and cause the danger of panles being degraded from their original levels.

1. Introduction

Honeycomb sandwich panels are kind of composite panels constructed by adhering thin surface plates such as CFRP (Carbon Fiber Reinforced Plastic) laminates or metal sheets with honeycomb cores made of aluminum or aramid materials. Figure 1 shows the image of the honeycomb core structure with different design variables. Honeycomb sandwich panels have light-weight and high strength performances. So they are widely used in many fields such as the transportation equipment field, especially for light-weight mainly required aerospace field. On the other hand, the merit of the honeycomb sandwich panels is effective only in the limited environment. For example, it is strong only when uniform loads are applied on the surfaces of the panels, while honeycomb cores show vulnerability with local concentrated loads applied. Under concentrated loads, it is possible that the strength of whole panel should be reduced due to local buckling failure inside the honeycomb core. However, no data on local buckling properties of honeycomb structures are widely publicized and testing methods are not clearly stipulated.

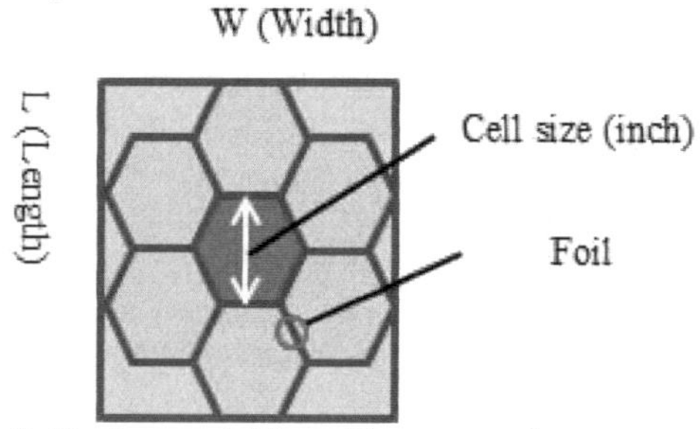

Figure 1. Design variables of honeycomb structure model.

IOP Conf. Series: Materials Science and Engineering **269** (2017) 012080 doi:10.1088/1757-899X/269/1/012080

Therefore, the purpose of this study is to grasp the local buckling tendency of honeycomb structures affected by design variables like foil thickness, cell size and specimen size (length, width, height dimension) of honeycomb structures. Compression tests on honeycomb sandwich panels having different design variables are executed to examine buckling performance under local loading.

2. Compression tests of composite panels with different design variables

Sample composite panel specimens with different design variables were fabricated by bonding aluminum honeycomb cores and 1.0 mm thick CFRP laminate faces. As shown in table 1, 12 kinds of panel specimens were introduced having honeycomb cores having three different cell sizes and four kinds of specimen sizes. 10 specimens were prepared for each kind of panel specimens indicated by the numbers as shown in table 1. The foil thickness of specimens is 0.003 inch for specimen number ⑤ and ⑦, and 0.001 inch for the other specimens. In the compression tests, a circular indenter with diameter of 100.0 mm was utilized to the tensile/compression tester. The test speed of compression loading was set at 1.0 mm/min for composite panel specimens. Figure 2 shows the photo images of the compression test and composite panel specimen.

Table 1. Dimensions of honeycomb cores for composite panels.

Cell size [inch]		1/8	3/16	1/4
Dimensions [mm] (Length × width × height)	30×30×30	①	⑤	⑨
	30×30×20	②	⑥	⑩
	20×20×30	③	⑦	⑪
	20×20×20	④	⑧	⑫

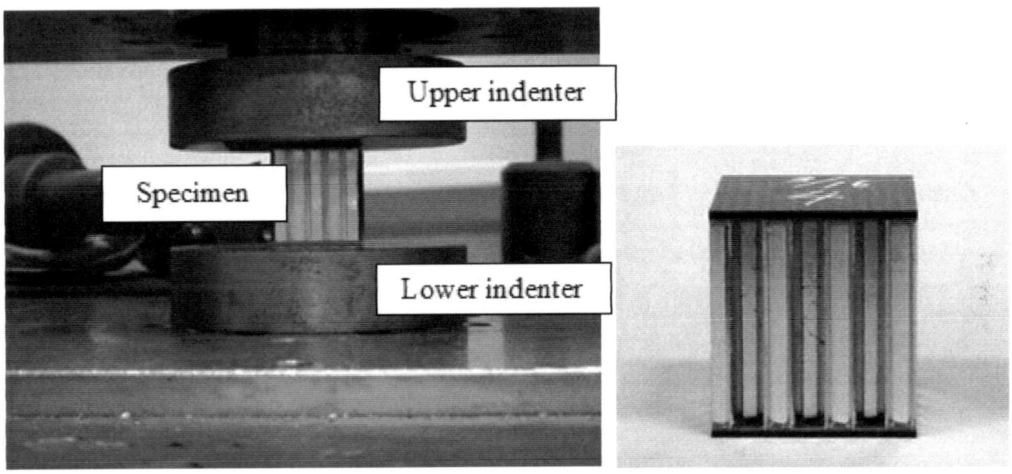

Figure 2. Photo image of compression test and composite panel specimen.

3. Compression characteristics of composite panels

3.1. Compression test results of composite panels

Typical compressive load-displacement diagrams obtained from the compression tests of composite panel specimens are shown in figure 3 to figure 5. These results represent three patterns of compression characteristics for different composite panel specimen groups. Figure 3 shows the type results of specimen number ⑧, ⑨, ⑩, ⑪ and ⑫. Figure 4 shows the type results of specimen number ③ and ⑥. Figure 5 shows the type results of specimen number ①, ②, ⑤ and ⑦. Specimen number ④ shows different compression characteristic with all trends of these three patterns appeared.

It can be seen from figure 3 first that the honeycomb structures buckled starting at the unstable yield loads. Figure 4 and figure 5 show the honeycomb structures buckling after stable yield points, and also

IOP Conf. Series: Materials Science and Engineering **269** (2017) 012080 doi:10.1088/1757-899X/269/1/012080

different behaviors can be observed from these two type results. Regarding the overall results, the buckling behaviors due to compressive load-displacement curves tends to be more stable with the cell size of honeycomb structure is smaller. When considering the local compression effects of honeycomb sandwich panels, compression strength of honeycomb structure will become lower than usual when buckling started at yielding loads as shown in figure 3. So, it is important to pay attention to the cell size of honeycomb structures as applied for composite panels under local compression loading.

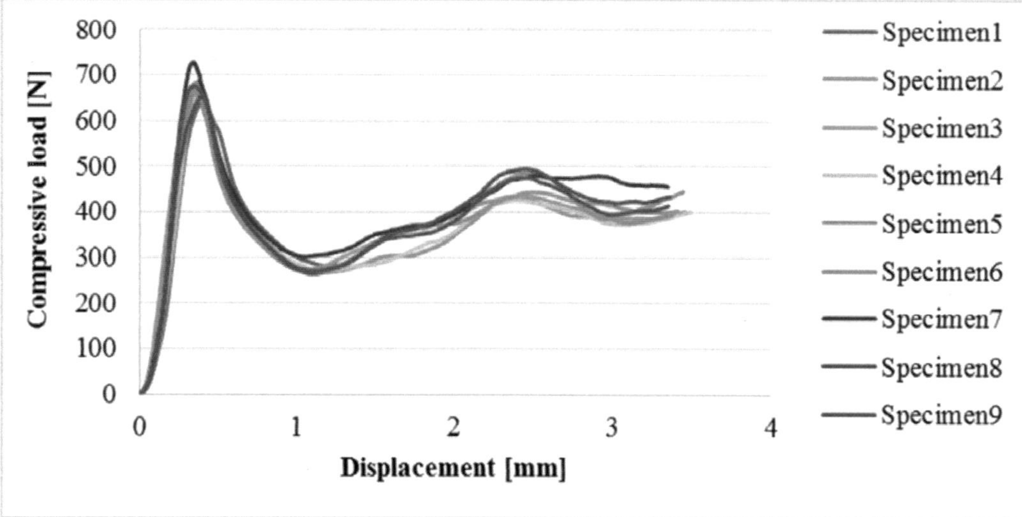

Figure 3. Typical compresion test results (specimen number ⑩).

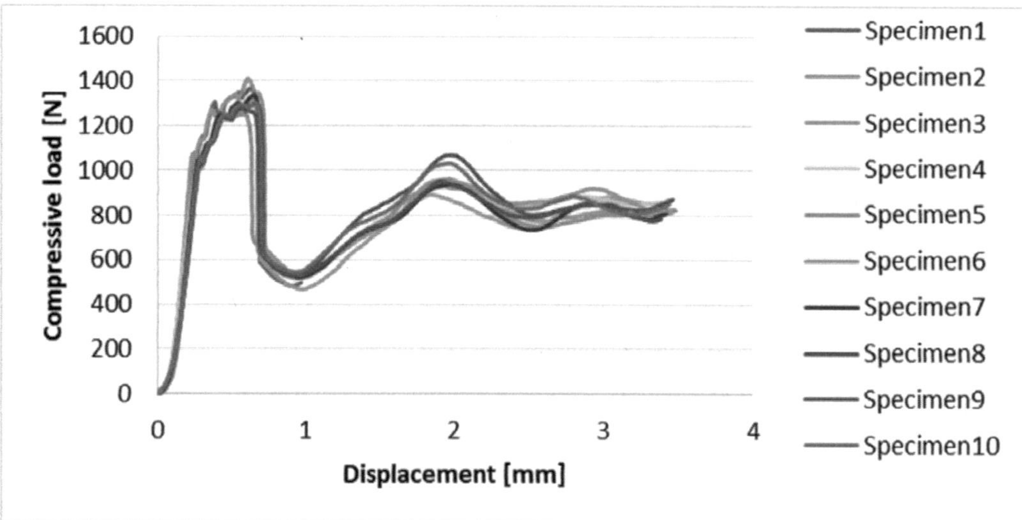

Figure 4. Typical compresion test results (specimen number ⑥).

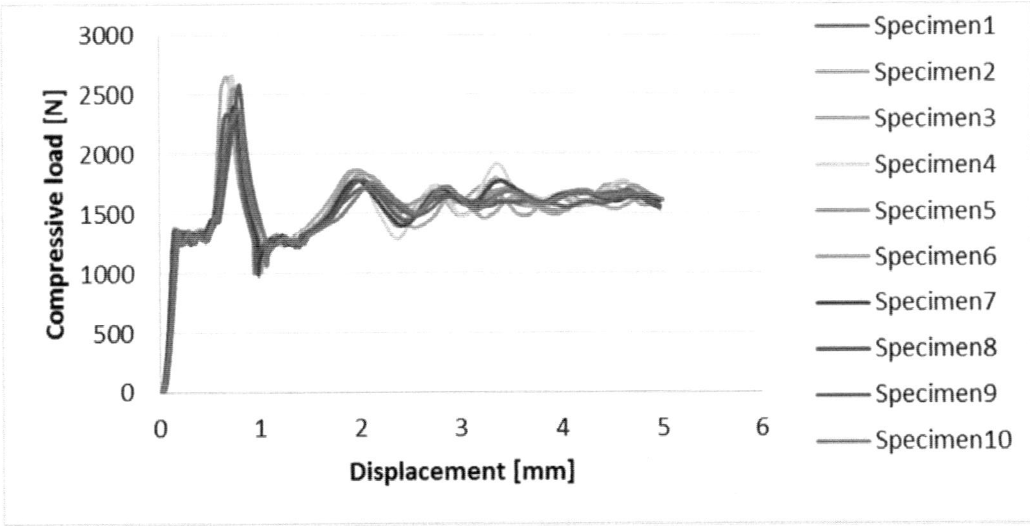

Figure 5. Typical compresion test results (specimen number ②).

3.2. Compression characteristics of honeycomb structures

Average yield loads derived from the compressive load-displacement diagrams are shown in figure 6. In addition, yield stresses calculated from the average yield load results shown in figure 6 divided by the exact cross-sectional area of honeycomb cores are shown in figure 7.

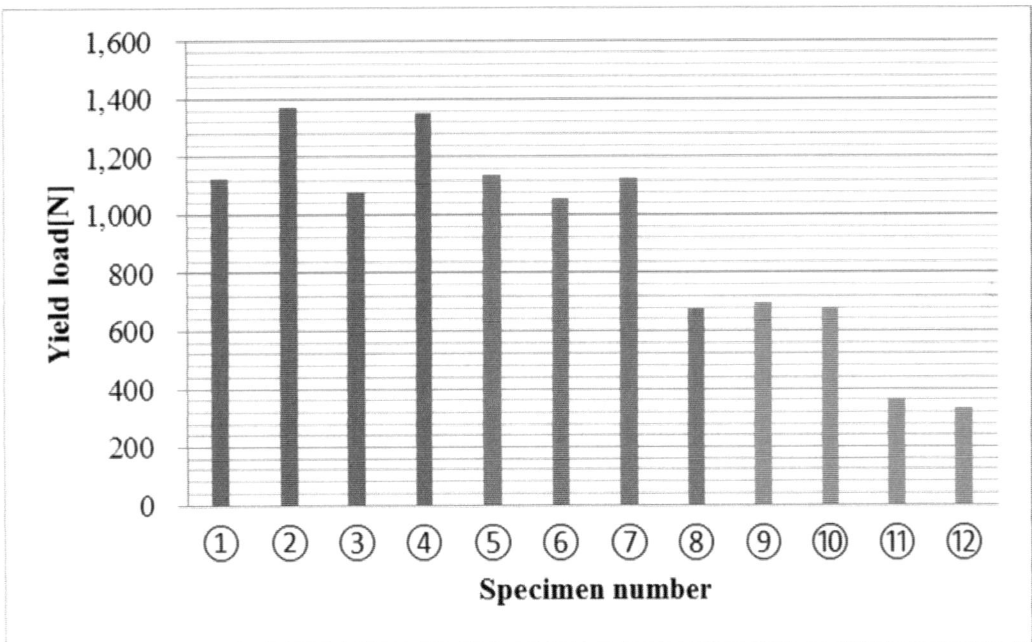

Figure 6. Average yield loads of different composite panel specimens.

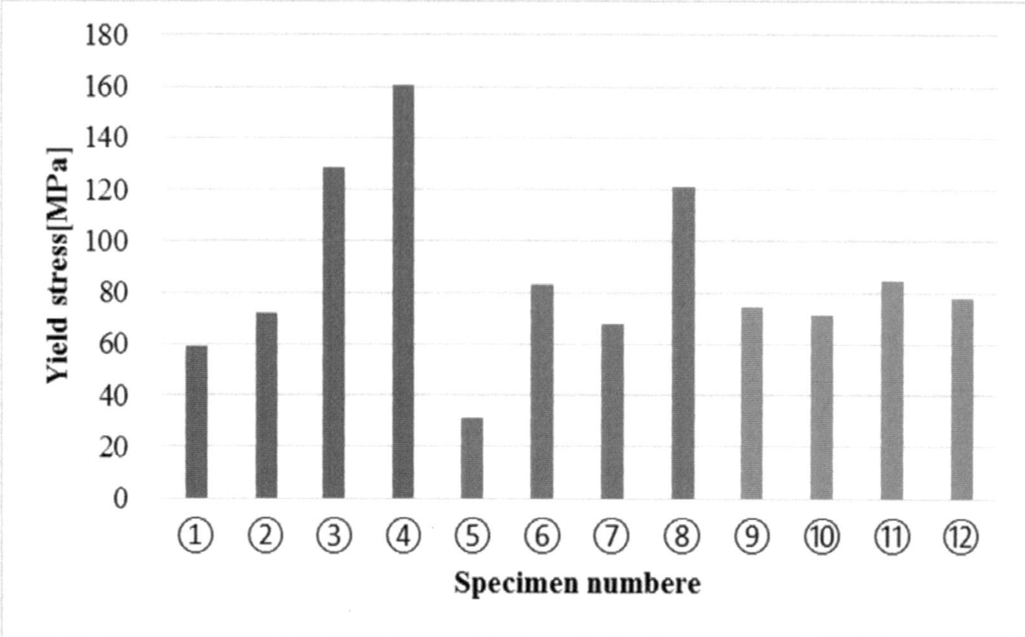

Figure 7. Average yield stresses of different composite panel specimens.

From figure 6, it can be seen that the yield loads tends to increase as the height of honeycomb core is decreased, regardless of the compression area of composite panel specimens of ①, ②, ③ and ④ with 1/8 inch cell size. However, from the yield stress results which take into account of the cross-sectional area of the honeycomb core, it can be seen that the stress level becomes larger with smaller compressive area and lower height of specimen. From the results of specimen number ⑤ and ⑦, one can see that the yield stresses become smaller with respect to the yield load results. This is just because of the 0.003 inch foil thickness which is thicker than the others.

Comparing the test results of specimen number ⑨, ⑩, ⑪ and ⑫ with honeycomb cell size of 1/4 inch, one can see that the yield loads become proportional to the panel specimen area and not affected by specimen height. However, the yield stress results of specimen number ⑨, ⑩, ⑪ and ⑫ show different behaviours. This is considered that buckling of 1/4 inch panel specimens occurred at yielding point, which is different from other two type composite panel specimens.

4. Summary

Local buckling tendency of composite panels was experimentally evaluated with respect to various design variables, such as the foil thickness, cell size and specimen dimension of honeycomb structures. From experimental results, the following conclusions are obtained.

(1) Buckling behavior of the honeycomb structure due to compressive load stabilized with smaller cell size. Composite panels with large cell size honeycomb cores will be caused unstable buckling behavior under compression loading no matter with the local impact sizes.

(2) When the cell size of honeycomb core is large and unstable buckling occurred at yield point, yield loads become proportional to the panel specimen area and hardly affected by specimen height. However, the yield stress results of such composite panels show smaller differences due to different design variables.

(3) For composite panels under local compression loads or impacts, honeycomb structures having different design variables will be selected due to local size of compression loads or impacts.

5. References

[1] Shimizu A 1969 *The material dynamics*, KYORITSU SHUPPAN Co., Ltd., Tokyo.
[2] Sato T 2002 *Application of honeycomb structured material* CMC Publishing Co., Ltd., Tokyo.

IOP Conf. Series: Materials Science and Engineering **269** (2017) 012081 doi:10.1088/1757-899X/269/1/012081

Structure and Mechanical Properties of the AlSi10Mg Alloy Samples Manufactured by Selective Laser Melting

Xiaodan Li[1], Jiaqiang Ni[1], Qingfeng Zhu[2], Hang Su[1], Jianzhong Cui[2], Yifei Zhang[3], Jianzhong Li[3*]

[1]Shenyang aircraft industry (Group) Co., Ltd., Shenyang, 110819, China
[2]Key Laboratory of Electromagnetic Processing of Materials, Ministry of Education Northeastern University, Shenyang, 110819, China
[3]School of Metallurgy, Northeastern University, Shenyang, 110819, China

Abstract. The AlSi10Mg alloy samples with the size of 14×14×91mm were produced by the selective laser melting (SLM) method in different building direction. The structures and the properties at -70°C of the sample in different direction were investigated. The results show that the structure in different building direction shows different morphology. The fish scale structures distribute on the side along the building direction, and the oval structures distribute on the side vertical to the building direction. Some pores in with the maximum size of 100 μm exist of the structure. And there is no major influence for the build orientation on the tensile properties. The tensile strength and the elongation of the sample in the building direction are 340 Mpa and 11.2 % respectively. And the tensile strength and the elongation of the sample vertical to building direction are 350 Mpa and 13.4 % respectively

1. Introduction

As one kind of additive manufacturing method, selective laser melting (SLM) is now widely used for various materials [1-11]. Recently, the SLM power-bed technology has been widely concerned by aerospace manufacturers for it can realize the "rapid manufacturing" of complicated metallic components [1, 10, 11]. There have been an increasing number of reports on SLM of Al-alloys recently, because of the demand from the aircraft industry for lightweight structures with complex geometries [9, 11]. Among the various aluminum alloys, the AlSi10Mg seem better suited to the SLM technology for the good mechanical properties and process performance. And various reports have been published to study the structure and properties of SLM-preparation AlSi10Mg [8, 11-13].

This paper focuses on the mechanical properties of the selective laser melting AlSi10Mg alloy at low temperature. The influence of the building direction (vertical and horizontally built samples) on the tensile properties at -70°C was investigated. In addition, the fracture morphology and the structure near the facture also were investigated.

2. Experimental procedures

The chemical composition of the AlSI10Mg power is shown in table 1. Figure 1 shows a SEM (scanning electron microscope) micrograph of the power. It is shown that the particles of the power are not all spherical. Many irregular morphology particles with different size distribute in the power. The size range of the particles was 15-70 μm, and the average size of the particles is about 40 μm.

Table 1. Chemical composition of the investigated AlSI10Mg alloy (wt. %).

Al	Si	Fe	Cu	Mn	Mg	Zn	Cr	Ni	Ti
89.6	9.80	0.158	0.005	0.004	0.363	0.0015	0.007	0.005	0.004

The specimens with the size of 14×14×91 mm (as shown in figure 2) were preparation by a SLM (laser powder-bed) system. And the argon was employed to protect the whole building processing. In order to get the specimens in different building direction, the specimens were built along Z and Y direction respectively, as shown in figure 2. This means that the long boundary of the sample is perpendicular and parallel to the building direction respectively. The specimen was cut in the transverse direction (X-Z plane in figure 2). And then, the specimen was polished and etched for metallurgical analysis.

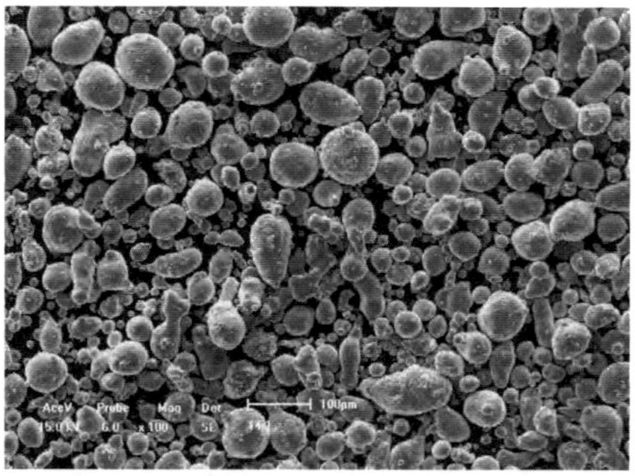

Figure 1. Morphology of the AlSi10Mg power under SEM.

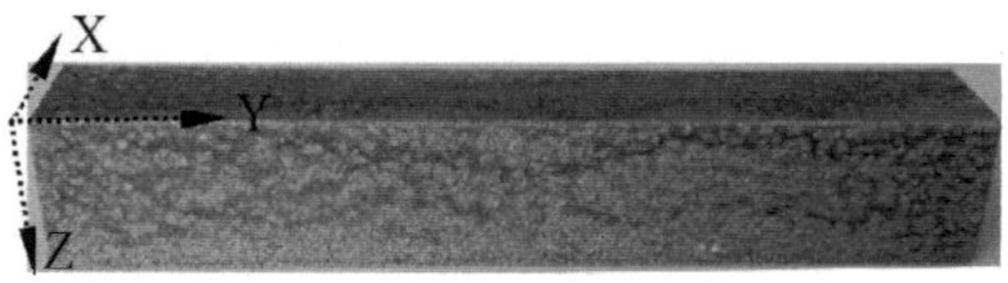

Figure 2. The specimen fabricated by the SLM.

3. Mechanical testing

The specimens in different building direction were machined to standard tensile samples. The tensile test geometry is shown in figure 3. Tensile tests were conducted on a Sans type tensile testing machine with the tensile speed of the 1 mm/min at -70°C. And all the mechanical test results are the average of 3 samples. The fracture surface was analyzed by a Shimadzu SSX-550 scanning electron microscope. And the microstructure near the fracture was investigated by the optical microscope.

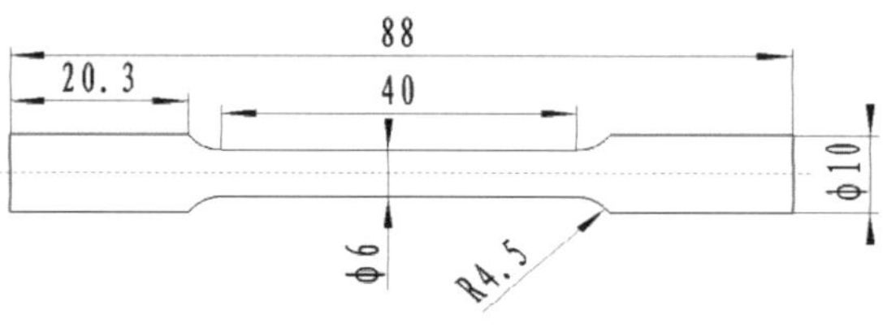

Figure 3. Tensile test samples and sample geometry.

4. Results and discussion

Figure 4 shows the three dimensional macrostructure of the specimen. B is surface formed after the last laser scanning. It is shown that the traces of laser scanning parallel distribute along the scanning direction on surface B. C is the original side surface built during laser scanning layer by layer. This surface is pretty rough, and many of cavities and unmelted AlSi10Mg particles exist on surface C. Surface A is the transverse section (which cut along the building direction) of the specimen. The inner structure of the specimen is revealed in this section. It is shown that the "fish scale" structure distribute on this section.

Figure 5 shows the microstructure of the specimen in different transverse section. It is shown that the microstructure in different transverse section is difference. The oval structures along different direction distribute on the section parallel to the scanning surface, and many black pores with the size less than 50 μm exist in this section as shown in figure 5 (a). While, the fish scale structure distribute on the section vertical to the scanning surface, and many black pores (with the size about 100 μm) exist in this section as shown in figure 5 (b). Oval structures along different direction show the top view of the melt pool and suggest the scanning direction of each laser scanning layer is difference. The arc edge of the fish scale is actually consisted by the bottom of the melt pool of each laser scanning. The porosities and pores are easily formed on the edge of the fish scale during remelting and solidification process.

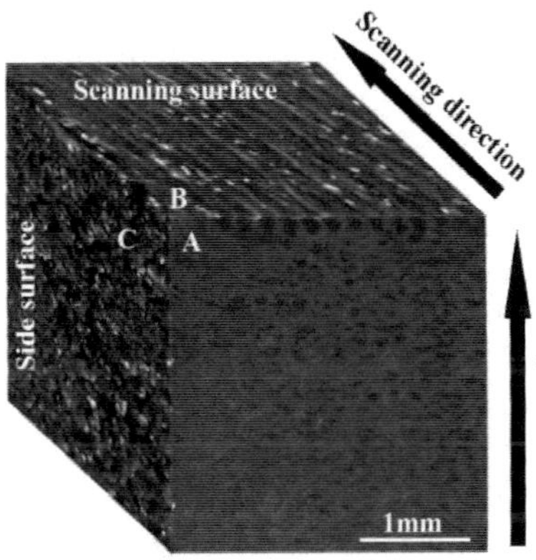

Figure 4. Optical micrographs from three orthogonal planes showing the macrostructures of the specimen.

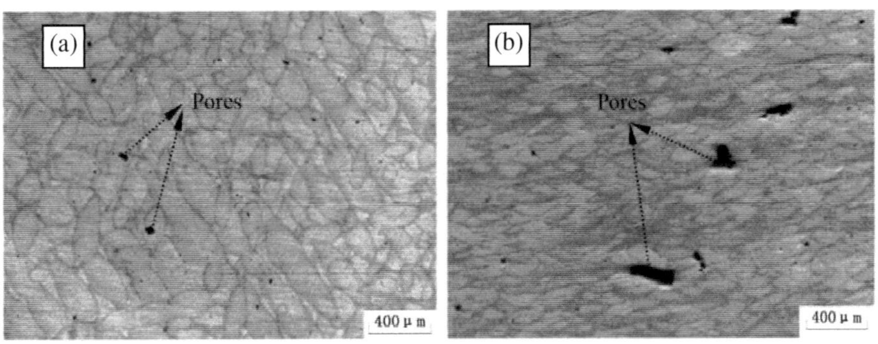

(a) parallel to surface B in figure 4, (b) the transverse section parallel to surface A in figure 4.

Figure 5. Microstructure in different t transverse section of the specimen.

The tensile strength and the elongation parallel to the building direction of the sample are 350 MPa and 11 %, respectively. The tensile strength and elongation vertical to the building direction of the sample are 340 MPa and 13 %, respectively. This indicates that the building direction has little influence on the properties of the SML specimens at low temperature.

Figure 6 shows the fracture morphology of the specimen. It is shown that the fracture morphology corresponding to different direction are quite difference. The fracture surface of the specimen pulled along the building direction displays a step like morphology, as shown in figure 6 (a). The average step size is about 100 μm, which corresponds to the size of the laser tracks. At high magnifications, the fracture surface appears to contain a mix of small dimples and smooth areas, as shown in figure 6 (b) and (c). These dimples in Al matrix are related to the pulling out the brittle Si particles. The smooth areas may be related with the pores in the specimen. The fracture surface of the specimen pulled vertical the building direction is rather irregular, as shown in figure 6 (d), and many cavities distribute on the surface of this fracture. At high magnifications, the fracture surface appears to contain a mix of small dimples and deep cavities, as shown in figure 6 (e) and (f). Moreover, there are no un-melted powders particles are observed on the surface.

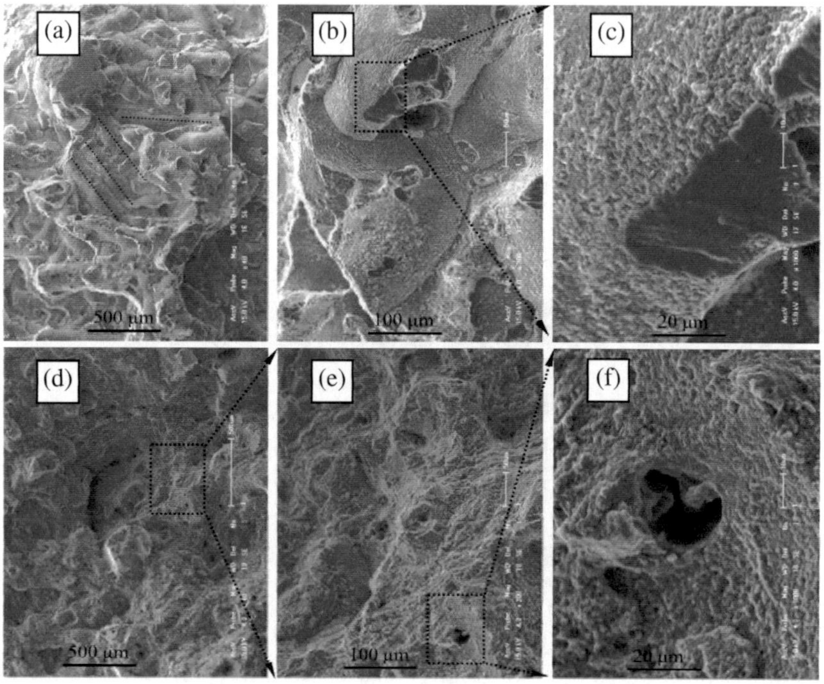

Figure 6. Fracture morphology of the specimen pulled along different direction (a) the typical fracture morphology of the specimen pulled parallel to the building direction, (b) partial enlargement of the (a), (c) partial enlargement of the (b), (d) the typical fracture morphology of the specimen pulled vertical to the building direction, (e) partial enlargement of the (d), (f) partial enlargement of the (e).

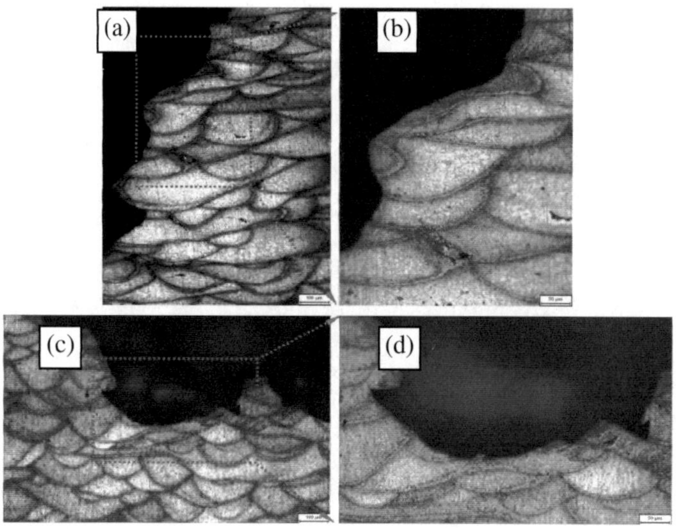

Figure 7. Microstructure near the fracture of the tensile samples. (a) microstructure of the specimen pulled vertical to the building direction, (b) partial enlargement of (a) (c) microstructure of the specimen pulled parallel to the building direction, (d) partial enlargement of (c).

Figure 7 (a) and (c) show the microstructure near the fracture of the tensile samples pulled vertical and parallel the building direction, respectively. It is can be seen that the cracks grow along the boundaries of the "fish scale" structure in two pulling direction. This suggests that the boundaries of the "fish sale" structure act as preferential sites for failure. These boundaries of the "fish sale" structure actually are the bonding zones formed during laser scanning layer by layer. In fact, many of curved surfaces consisted by the bonding zones exist in the SLM specimens and parallel to the scanning direction. When pulling along the building direction, the sample fractures along the curved surface and forms a step like morphology as shown in figure 6 (a). When the pulling direction is vertical to the building direction, the bonding zones fracture vertical to the curved surface and form rather irregular fracture morphology, as shown in figure 6.

5. Conclusion

(1) The structure in different building direction shows different morphology. The fish scale structures distribute on the side along the building direction, and the oval structures distribute on the side vertical to the building direction. Some pores in with the maximum size of 100 μm exist of the structure.

(2) There is no major influence for the build orientation on the tensile properties. The tensile strength and the elongation of the sample in the building direction are 340 Mpa and 11.2 % respectively. And the tensile strength and the elongation of the sample vertical to building direction are 350 Mpa and 13.4 % respectively.

6. Reference

[1] Amato K N, Gaytan S M, Murr L E, et al. 2012 *Acta Mater.* **60** 2229-39.
[2] Kruth J P, Froyen L, Van Vaerenbergh J, et al. 2004 *J. Mater. Process. Tech.* **149** 616-22.
[3] Thijs L, Verhaeghe F, Craeghs T, et al. 2010 *Acta Mater.* **58** 3303-12.
[4] Rombouts M, Kruth J P, Froyen L, et al. 2006 *CIRP Ann-Manuf. Techn.* **55** 187-92.
[5] Abe F, Osakada K, Shiomi M, et al. 2001 *J. Mater. Proc. Tech.* **111** 210-3.
[6] Vandenbroucke B and Kruth J P 2007 *Rapid. Prototyp. J.* **13(4)** 196-203.
[7] Louvis E, Fox P and Sutcliffe C J 2011 *J. Mater. Proc. Tech.* **211** 275-84.
[8] Brandl E, Heckenberger U, Holzinger V, et al. 2012 *Mater. Des.* **34** 159-69.
[9] Dadbakhsh S and Hao L 2012 *J. Alloy. Comp.* **541** 328-34.
[10] Gu D, Wang Z, Shen Y, et al. 2009 *Appl. Surf. Sci.* **255** 9230-40.
[11] Read N, Wang W, Essa K, et al. 2015 *Mater. Des.* **65** 417-24.
[12] Thijs L, Kempen K, Kruth J P, et al. 2013 *Acta Mater.* **61** 1809-19.
[13] Kempen K, Thijs L, Van Humbeeck J, et al. 2012 *Phys. Proc.* **339** 439-46.

Acknowledgement

The Project Supported by National Natural Science Foundation of China (No. 51204053), Central University Basic R & D Operating Expenses (No. N130409005).

Dynamic Pressure Distribution due to Horizontal Acceleration in Spherical LNG Tank with Cylindrical Central Part

Dae-Eun Ko[1, a*], **Sang-Hoon Shin**[2, b]

[1]Dept. of Naval Architecture & Ocean Engineering, Dong-Eui University, Busan, Korea
[2]Dept. of Aero Mechanical Engineering, Kyungwoon University, Gyeongbuk, Korea

[a]deko@deu.ac.kr, [b]shshin@ikw.ac.kr

Abstract. Spherical LNG tanks having many advantages such as structural safety are used as a cargo containment system of LNG carriers. However, it is practically difficult to fabricate perfectly spherical tanks of different sizes in the yard. The most effective method of manufacturing LNG tanks of various capacities is to insert a cylindrical part at the center of existing spherical tanks. While a simplified high-precision analysis method for the initial design of the spherical tanks has been developed for both static and dynamic loads, in the case of spherical tanks with a cylindrical central part, the analysis method available only considers static loads. The purpose of the present study is to derive the dynamic pressure distribution due to horizontal acceleration, which is essential for developing an analysis method that considers dynamic loads as well.

1. Introduction

Because of the many advantages of spherical LNG tanks, especially when it comes to structural safety, these tanks are used as the cargo containment system in LNG carriers, even though their material cost is higher than that of membrane-type LNG tanks. To satisfy the demands of ship owners, the ship-builder must be able to design and construct spherical LNG tanks of various sizes. However, it is practically difficult to fabricate perfectly spherical tanks of different sizes in the yard because the number of fabrication shops required for making spherical LNG tanks is proportional to the radius of the tanks to be constructed. Thus, an efficient method of increasing the cargo capacity is to extend conventional spherical LNG tanks by inserting a cylindrical shell structure within them.

There have been several studies on spherical tanks, including studies on their initial design using finite element analysis [1], initial design using the simplified analysis method [2], fatigue strength and crack propagation behaviour [3, 4], and sloshing characteristics [5]. In a study on spherical tanks with a cylindrical central part, Shin and Ko developed and verified a simplified analysis method that considers static loads such as the overpressure, tank weight, and liquid cargo weight [6].

Because a spherical tank is a rotating body not only on the vertical axis but also on the horizontal axis, the pressure distribution due to horizontal acceleration is also axially symmetric. In contrast, a spherical tank with a cylindrical central part is a rotating body on the vertical axis but not on the horizontal axis. Further, no guidelines or classification rules exist for such tanks. Hence, in the present study, we determined the mathematical expressions for the pressure distribution caused by horizontal acceleration, as these are essential for the design and structural analysis of LNG carriers with a spherical cargo tank with a cylindrical central part.

2. Pressure distribution due to horizontal acceleration

The equations for the pressure distribution are determined for the case where the tank is subjected to horizontal acceleration and gravitational acceleration simultaneously. The pressures corresponding to horizontal acceleration alone were then obtained by subtracting gravitational acceleration from the calculated pressures.

Figure 1 show the direction of the resultant acceleration when the horizontal acceleration and the gravitational acceleration act simultaneously. Here, a_H is the dimensionless horizontal acceleration obtained by dividing the horizontal acceleration by the gravitational acceleration, g, and a_R is the dimensionless resultant acceleration obtained by dividing the resultant acceleration by g. The dimensionless resultant acceleration acts in the axial direction, which is β° off the vertical axis, as can be seen from the figure. This yields equations (1)-(3).

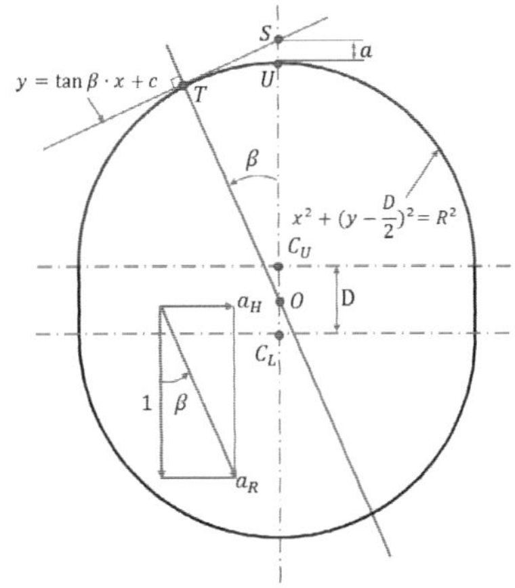

Figure 1. Resultant acceleration.

$$\cos\beta = \frac{1}{a_R} \tag{1}$$

$$\sin\beta = \frac{a_H}{a_R} \tag{2}$$

$$a_R = \sqrt{{a_H}^2 + 1} \tag{3}$$

The slope of the tangent at point T is tan β, and the value of c (the length of $\overline{SO}$) is $R\sec\beta + D/2$. Thus, the value of a (the length of $\overline{SU}$) can be expressed as shown in equation (4).

$$a = R(\sec\beta - 1) = R\left(\sqrt{{a_H}^2 + 1} - 1\right) \tag{4}$$

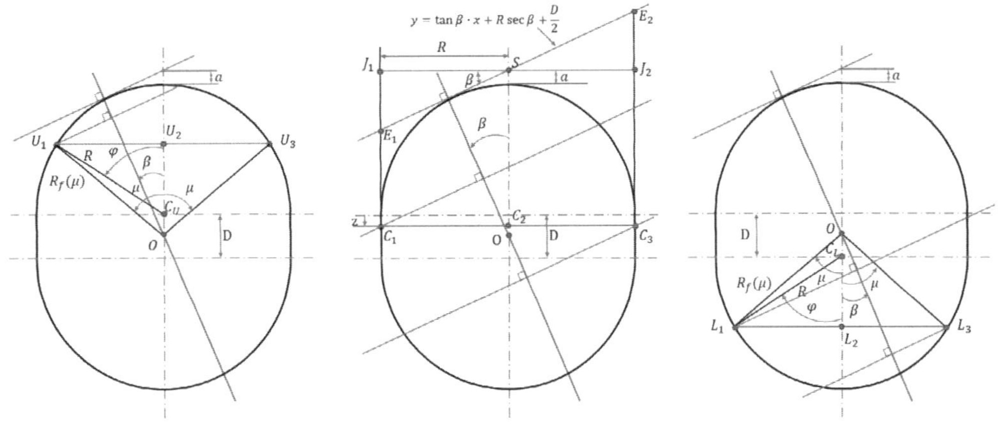

(a) Upper hemisphere (b) Cylindrical part (c) Lower hemisphere

Figure 2. Water head in each part.

2.1. Pressure distribution in upper hemisphere

In order to calculate the water head in the upper hemisphere, it is convenient to introduce the distance, R_f, from the central point of the cross-section to a point on the tank boundary, as shown in figure 2(a). R_f is a function of μ based on point O.

The water head at locationsU_1, U_2, and U_3 can be obtained from equations (5)-(7), respectively.

$$h_{U,\theta=0} = \left(R + \frac{D}{2} + a\right)\cos\beta - R_f(\mu)\cos(\mu - \beta) \quad (5)$$

$$h_{U,\theta=\frac{\pi}{2}} = \left(R + \frac{D}{2} + a\right)\cos\beta - R_f(\mu)\cos\mu\cos\beta \quad (6)$$

$$h_{U,\theta=\pi} = \left(R + \frac{D}{2} + a\right)\cos\beta - R_f(\mu)\cos(\mu + \beta) \quad (7)$$

By substituting equations (1) and (2) in equations (5)-(7), we get equations (8)-(10).

$$h_{U,\theta=0} = \frac{1}{a_R}(R + a - R\cos\varphi - a_H R\sin\varphi) \quad (8)$$

$$h_{U,\theta=\frac{\pi}{2}} = \frac{1}{a_R}(R + a - R\cos\varphi) \quad (9)$$

$$h_{U,\theta=\pi} = \frac{1}{a_R}(R + a - R\cos\varphi + a_H R\sin\varphi) \quad (10)$$

Referring to equations (8)-(10), the water head in the upper hemisphere can be expressed through a single equation, which is given by equation (11), for the range $0 \le \theta \le \pi$.

$$h_U = \frac{1}{a_R}(R + a - R\cos\varphi - a_H R\sin\varphi\cos\theta) \quad (11)$$

Since the resultant acceleration is dimensionless, a pressure due to the resultant acceleration can be calculated by multiplying $\rho g a_R$ and the water head at the location in question. Therefore, the pressure due to the resultant acceleration in the upper hemisphere can be expressed as follows:

$$P_U = \rho g(R + a - R\cos\varphi - a_H R\sin\varphi\cos\theta) \quad (12)$$

By subtracting gravitational acceleration $\rho g R(1 - \cos\varphi)$ from equation (12), the pressure corresponding to horizontal acceleration alone is as follows:

$$P_{HU} = \rho g(a - a_H R\sin\varphi\cos\theta) \quad (13)$$

2.2. Pressure distribution in cylindrical part

Figure 2(b) shows the points and line segments that are used for determining the water head in the cylindrical central part. The length of $\overline{J_1C_1}$, $\overline{SC_2}$ and $\overline{J_2C_3}$ is $R + a + D$, while the length of $\overline{J_1E_1}$ and $\overline{J_2E_2}$ is $R\tan\beta$. That is, the length of $\overline{E_1C_1}$ is $R + a + D - R\tan\beta$, and the length of $\overline{E_2C_3}$ is $R + a + D + R\tan\beta$.

The water head at locationsC_1, C_2, and C_3 can be obtained from equations (14)-(16), respectively.

$$h_{C,\theta=0} = (R + a + z - R\tan\beta)\cos\beta \quad (14)$$

$$h_{C,\theta=\frac{\pi}{2}} = (R + a + z)\cos\beta \quad (15)$$

$$h_{C,\theta=\pi} = (R + a + z + R\tan\beta)\cos\beta \quad (16)$$

By substituting equations (1) and (2) in equations (14)-(16), we get equations (17)-(19), respectively.

$$h_{C,\theta=0} = \frac{1}{a_R}(R + a + z - a_H R) \quad (17)$$

$$h_{C,\theta=\frac{\pi}{2}} = \frac{1}{a_R}(R + a + z) \quad (18)$$

$$h_{C,\theta=\pi} = \frac{1}{a_R}(R + a + z + a_H R) \quad (19)$$

With reference to equations (17)-(19), the water head in the cylindrical part can be expressed using a single equation, which is given by equation (20), for the range $0 \le \theta \le \pi$.

$$h_C = \frac{1}{a_R}(R + a + z - a_H R \cos\theta) \tag{20}$$

If equation (20) is multiplied by $\rho g a_R$, it yields the pressure due to the resultant acceleration in the cylindrical part:

$$P_C = \rho g(R + a + z - a_H R \cos\theta) \tag{21}$$

By subtracting gravitational acceleration $\rho g(R+z)$ from equation (21), the pressure corresponding to horizontal acceleration alone is as follows:

$$P_{HC} = \rho g(a - a_H R \cos\theta) \tag{22}$$

2.3. Pressure distribution in lower hemisphere

Next, let us determine the water head at locationsL_1, L_2, and L_3 in figure 2(c).

The procedure for determining the water head in the lower hemisphere is nearly the same as that in the case of the upper hemisphere. On the basis of a comparison of figures 2(a) and 2(c), $R_f(\mu)\cos(\mu-\beta)$ is subtracted from the expression for the water head at a location where θ is zero, whereas $R_f(\mu)\cos(\mu+\beta)$ is added in the case of the lower hemisphere. Furthermore, $R_f(\mu)\cos\mu\cos\beta$ is subtracted from the expression for the water head in the upper hemisphere at a location where θ is $\pi/2$ whereas the value is added in the case of the lower hemisphere. Finally, $R_f(\mu)\cos(\mu+\beta)$ is subtracted from the expression for the water head at a location where θ is π in the upper hemisphere whereas $R_f(\mu)\cos(\mu-\beta)$ is added in the case of the lower hemisphere. Therefore, water head at locations L_1, L_2, and L_3 can be obtained from equations (23)-(25), respectively.

$$h_{L,\theta=0} = \left(R + \frac{D}{2} + a\right)\cos\beta + R_f(\mu)\cos(\mu+\beta) \tag{23}$$

$$h_{L,\theta=\frac{\pi}{2}} = \left(R + \frac{D}{2} + a\right)\cos\beta + R_f(\mu)\cos\mu\cos\beta \tag{24}$$

$$h_{L,\theta=\pi} = \left(R + \frac{D}{2} + a\right)\cos\beta + R_f(\mu)\cos(\mu-\beta) \tag{25}$$

If equations (23)-(25) are arranged using the same method as in Section A, they can be expressed as shown in equations (26)-(28), respectively.

$$h_{L,\theta=0} = \frac{1}{a_R}(R + a + D + R\cos\varphi - a_H R \sin\varphi) \tag{26}$$

$$h_{L,\theta=\frac{\pi}{2}} = \frac{1}{a_R}(R + a + D + R\cos\varphi) \tag{27}$$

$$h_{L,\theta=\pi} = \frac{1}{a_R}(R + a + D + R\cos\varphi + a_H R \sin\varphi) \tag{28}$$

With reference to equations (26)-(28), the water head in the lower hemisphere can be expressed through a single equation, which is given by equation (29), for the range $0 \le \theta \le \pi$.

$$h_L = \frac{1}{a_R}(R + a + D + R\cos\varphi - a_H R \sin\varphi\cos\theta) \tag{29}$$

Multiplying equation. (29) with $\rho g a_R$ yields the pressure due to the resultant acceleration in the lower hemisphere:

$$P_L = \rho g(R + a + D + R\cos\varphi - a_H R \sin\varphi\cos\theta) \tag{30}$$

By subtracting gravitational acceleration $\rho g(R + R\cos\varphi + D)$ from equation (30), the pressure corresponding to horizontal acceleration alone is as follows:

$$P_{HL} = \rho g(a - a_H R \sin\varphi\cos\theta) \tag{31}$$

3. Conclusions

In this study, the equation for the distribution of the pressure resulting from horizontal acceleration was derived for spherical tanks with a cylindrical central part for use as a cargo containment system of LNG carriers, as shipbuilders in the world have no prior experience with the design and construction of such tanks.

For spherical LNG tanks with a cylindrical central part, the pressure distributions in the upper and lower hemispheres are similar to those in a perfect spherical tank if the magnitude of the horizontal acceleration is the same, and the maximum pressure occur at the cylindrical central part of the tank.

The equation for the pressure distribution in the cylindrical central part was derived clearly. This equation should be useful to designers for structural analysis of spherical LNG tanks with cylindrical extension.

For the future work, by using the equations derived in this study, a simplified analysis method suitable for dynamic load conditions will be developed and combined with a previously developed simplified analysis method for static loads [6] to develop a comprehensive simplified analysis method that allows for precise initial estimates and is more efficient than time-consuming finite element analyses.

4. References

[1] Kim J W 2000 *J. Korean Weld. Soc.* **18(4)** 28-37.
[2] Shin S H 2015 *J. Soc. Nav. Archit. Korea* **52(2)** 125-34.
[3] Nho I S, Nam Y Y and Lee H S 1993 *J. Soc. Nav. Archit. Korea* **30(2)** 132-40.
[4] Nho I S, Nam Y Y and Lee H S 1993 *J. Soc. Nav. Archit. Korea* **30(4)** 74-82.
[5] Noh B J 2005 *Sp. Iss. Soc. Nav. Archit. Korea* June pp 22-30.
[6] Shin S H and Ko D E 2016 *Int. J. Nav. Archit. Ocean Eng.* **8(6)** 530-6.

IOP Conf. Series: Materials Science and Engineering **269** (2017) 012083 doi:10.1088/1757-899X/269/1/012083

A Study on the Application of New Lining Material for Ship Propulsion Shafting System

Dae-Eun Ko[1, a*], Sang-Hoon Shin[2, b] and Yong-Jin Cho[1, c]

[1]Dept. of Naval Architecture & Ocean Engineering, Dong-Eui University, Busan, Korea
[2]Dept. of Aero Mechanical Engineering, Kyungwoon University, Gyeongbuk, Korea
[a]deko@deu.ac.kr, [b]shshin@ikw.ac.kr, [c]cyjdeu@deu.ac.kr

Abstract. Exothermic accident at the rear part of the after stern tube journal bearing is mainly caused by an excessive local compressive pressure at the rear part as the propeller weight deflects the propulsion shaft. By using a material with lower stiffness than white metal, which is used for the lining of stern tube bearings, it would be possible to decrease the local compressive pressure by increasing the contact area with the shaft, eliminating the cause of exothermic accident. We analyzed the shafting system in order to investigate the compressive pressure distribution characteristics and allowable pressure when a new, less rigid material is applied. The analysis results confirmed that a low-stiffness product can significantly decrease the local compressive pressure compared to an existing product when applied to an actual ship. Furthermore, the pressure distribution changes were small even when the loading condition was changed, allowing for a robust design.

1. Introduction

Journal bearing supports the smooth rotation of a cylindrical rotating shaft and receives a load perpendicular to the rotating shaft. It has a long cylindrical shape to support large loads, and uneven loading to the bearing can lead to an exothermic accident. Stern tube is attached where the propeller shaft penetrates the hull, preventing seawater intrusion into the ship as well as supporting the propeller shaft. Generally, stern tube bearings are placed at the front and rear of the stern tube, transmitting torque to the propeller and supporting the shaft [1]. The bearing at the front of the stern tube is called a forward stern tube bearing, and the one at the rear is called an after stern tube bearing. Both are long cylindrical journal bearings.

The most frequent damages in ship machinery occur in propeller shaft bearing and main engine bearing, and stern tube bearings are damaged the most among them [2]. Most exothermic accidents occur at the rear part of the long after stern tube bearing. Their main cause is an excessive local compressive pressure at the rear part as the propeller weight deflects the propulsion shaft.

Recently, after stern tube bearings are installed with a consideration of rake, which slightly elevates the front part to distribute some of the rear load to the front. Load distribution by rake has limitations, in that the evenness of contact between the shaft and lining material for stern tube bearing (white metal) is not sufficient. After stern tube bearings whose lining materials have much lower stiffness than white metals are emerging as a solution, but most shipyards are still using stern tube bearings made of white metal.

This study aims to investigate the compressive pressure distribution characteristics and determine the allowable pressure for the new material through shafting system analysis before applying bearings made of materials with a low stiffness. The results could also become the performance test requirement criteria for guaranteeing product performance from the manufacturer. The Hertzian contact condition assuming half-elliptical pressure distribution along the contact width was adopted in this study to calculate local compressive pressure [3]. We chose an actual LNG carrier for the analysis, and the analysis data of this ship were used for the shafting system bearing offset and hull deflection. We introduced the gap element used in [4] to simulate bearing rake and calculate reactions.

2. Design criteria for shafting system alignment

Shafting system alignment design is a design process to set the bearing reaction to a desired level by adjusting the spacing, length, and offset of bearings supporting the shafting system [5]. The only design criterion in the past was a mean pressure of 8bar on the stern tube bearing in a shafting system alignment design, and there has been active research to add a local pressure design criterion to the mean pressure criterion in order to prevent exothermic accidents [6]. A number of shipyards are using local pressure under 100bar as the design criterion.

The Hertzian contact condition assuming half-elliptical pressure distribution along contact width was adopted in this study to calculate the local pressure, as used in [6]. The reaction per unit length (Q) can be expressed as a relationship between contact width (b) and material properties as shown in (1), where E^* and R are defined in (2) and (3), respectively.

$$Q = \frac{\pi b^2 E^*}{2R} \tag{1}$$

$$E^* = \left(\frac{1-\nu_1{}^2}{E_1} + \frac{1-\nu_2{}^2}{E_2}\right) \tag{2}$$

$$R = \frac{D_1 D_2}{D_1 - D_2} \tag{3}$$

In (2), ν_1 and ν_2 are the Poisson's ratios of the bearing material and shaft material, respectively, and E_1 and E_2 are their respective Young's modulus. In (3), D_1 and D_2 are bearing inner diameter and shaft diameter, respectively.

Because of the half-elliptical contact pressure distribution, the maximum pressure is expressed as in (4), and substituting (1) here yields (5).

$$P_{max} = \frac{bE^*}{R} \tag{4}$$

$$P_{max} = \sqrt{\frac{2}{\pi}}\sqrt{\frac{QE^*}{R}} \tag{5}$$

The reactions computed in the shafting system analysis are substituted in (5) to get the maximum local compressive pressure, and the value must not exceed the tolerance of the material.

This study refers to bearing material properties in order to calculate the local pressure and determine the allowable pressure to be used in the new product, and the properties of relevant materials are listed in table 1.

Table 1. Material properties.

Material	Young's modulus [N/mm^2]	Poisson's ratio
Shaft	206,900	0.3
White metal	53,000	0.3
New lining material	490	0.45

3. Shafting system analysis

As heating problems in the shafting system occur during operation of the engine, propeller weight in an operating engine, thermal effect and hull deflection were taken into consideration. We chose the representative ship loading conditions i.e. "ballast" condition, "full load" condition, and "full load with wave" condition. The model for shafting system analysis is shown in figure 1.

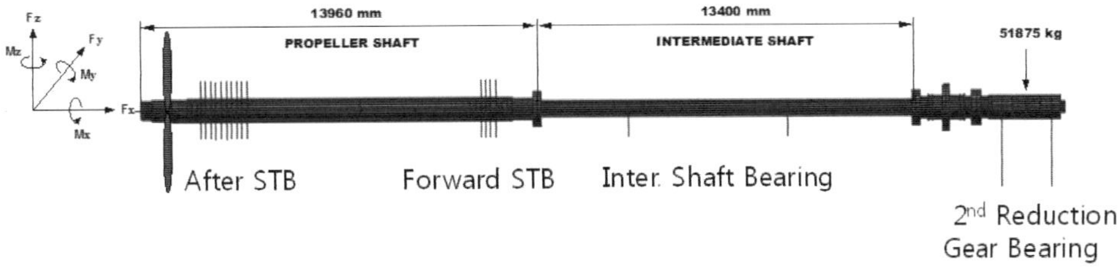

Figure 1. FE model for shaft analysis.

3.1. Hull deflections

There has been much research on hull deflection and shafting system analysis, using finite element analysis results to calculate hull deflection [6], constructing a hull deflection database and applying it [7], and using approximation curves for the hull deflection [8] for example. Most recent shafting system analyses consider the effect of hull deflection. In this study, the finite element analysis was used to analyze the shafting system considering hull deflection.

3.2. Bearing contact model and offset

The gap element was applied to calculate the compressive pressure in the after stern tube bearing. As shown in figure 2, 10 elements were placed in the left and right width directions and up and down directions of the ship, adding up to 40.

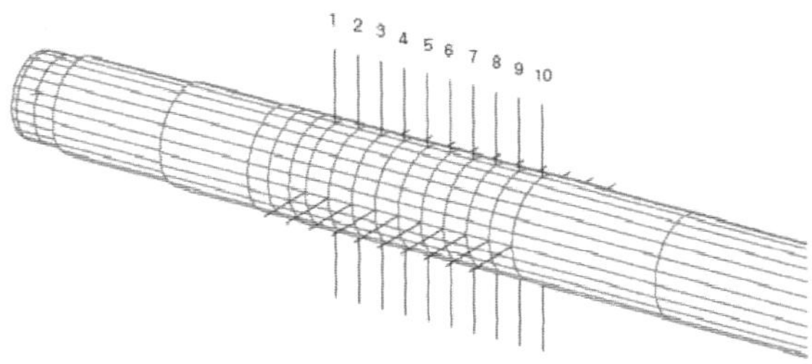

Figure 2. Gap element for simulation of after stern tube bearing.

The offsets in shafting system analysis of the actual ship were used with no modification, and the details for operating engine condition are outlined in table 2. Table 3 lists the rake values in the actual ship. There is no rake in the forward stern tube bearing, and the rake applies only for the after stern tube bearing.

Table 2. Bearing offsets.

Aft. STB [mm]	Fwd. STB [mm]	Inter. Shaft Bearing [mm]		2nd Reduction Gear Bearing [mm]	
		Aft.	Fwd.	Aft.	Fwd.
0.0	0.0	-0.35	-2.70	-5.40	-5.40

Table 3. Rake of stern tube bearing.

Bearing	Aft. Offset [mm]	Fwd. offset [mm]	Rake [10^{-4} rad]
Aft. STB	-0.12	0.13	1.42
Fwd. STB	0.0	0.0	0.0

4. Analysis result

The total reactions on forward and after stern tube bearings determined through a shafting system analysis for the new product with lower stiffness are listed in table 4. The results of the new product are nearly identical to the existing actual ship results. Calculated mean pressure for new product are 6.0bar and 1.4bar for after and forward stern tube bearings, respectively.

Table 4. Bearing reaction at stern tube bearing.

Bearing	Bearing reaction force [ton] New product / Existing ship		
	"ballast"	"full load"	"full load with wave"
Aft. STB	77.0 / 77.4	76.9 / 77.3	77.4 / 78.2
Fwd. STB	5.8 / 5.2	5.8 / 5.2	3.9 / 3.8

The Hertzian contact condition assuming half-elliptical pressure distribution along the contact width was applied to calculate the local pressure, and table 5 shows the results.

Calculation results for the new product are largely different from the actual ship shafting system analysis results for the existing product. The biggest difference is the decrease of maximum local pressure to 1/10 of the actual ship in all loading conditions, possibly due to a significant increase in the contact width in each position. It also displays lower sensitivity to the loading conditions. This can be an advantage of robust design, since it means a very small change in pressure distributions even when the loading condition changes.

Table 5. Maximum pressure distribution of stern tube bearing.

Station	Maximum local pressure [bar] New product / Existing ship		
	"ballast"	"full load"	"full load with wave"
1	6.8 / 79.9	6.8 / 80.9	6.8 / 76.3
2	6.7 / 66.2	6.7 / 67.4	6.8 / 70.6
3	6.6 / 52.3	6.6 / 54.0	6.7 / 55.0
4	6.5 / 38.3	6.5 / 41.2	6.6 / 38.5
5	6.5 / 47.2	6.5 / 30.5	6.5 / 43.4
6	6.5 / 44.3	6.5 / 46.3	6.5 / 35.8
7	6.5 / 47.6	6.5 / 49.2	6.5 / 52.2
8	6.5 / 56.2	6.5 / 57.3	6.5 / 57.1
9	6.5 / 68.4	6.5 / 69.2	6.5 / 66.5
10	6.6 / 82.8	6.6 / 83.8	6.6 / 87.9

The pressure distribution on the entire after stern tube bearing under "full load with wave" condition is plotted in figure 3, and it shows that the shaft and bearing make contact evenly overall.

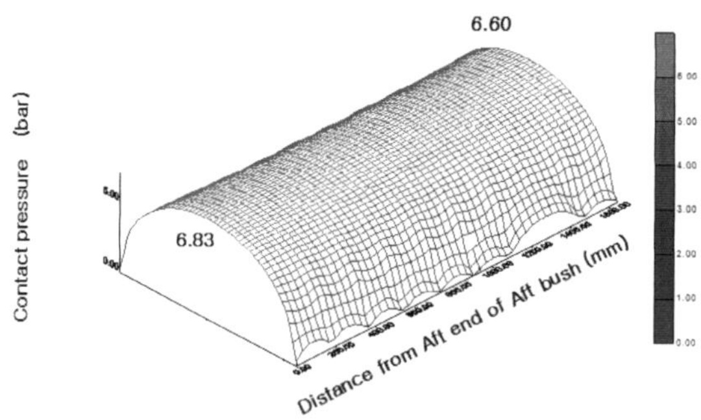

Figure 3. Pressure distribution under "full load with wave" condition.

The maximum local pressure for an actual ship was observed under "full load with wave" condition with a value of 87.9bar, 14% lower than the allowable pressure. To apply the new product, manufacturers should guarantee with a reliable experimental data the durability for local compressive pressure of over 7.8bar, up 14% from 6.8bar. If the new product cannot withstand a local compressive pressure of over 7.8bar, it is desirable to use an existing product with proven material properties.

5. Conclusion

A shafting system analysis and local compressive pressure calculation were performed in this study for a journal bearing lining product with a low stiffness which enables a larger contact area between the shaft and the stern tube bearing, and the following results were obtained.

Bearing products made of materials with a low stiffness significantly lower the local compressive pressure by increasing the contact width with the shaft.

Products with a low stiffness are less sensitive to changes in loading condition, and provide the advantage of robust design because they show small changes in pressure distribution even when the loading condition changes.

When manufacturers guarantee a proper level of allowable pressure, the application of products with a low stiffness will be possible.

In order to apply the new product, it is mandatory to perform shafting system analysis and calculation of local compressive pressure and present the product performance specifications that the manufacturers need to guarantee.

6. References

[1] Cho k H, Lee J H and Kim Y G 2010 J. Korean Soc. Marine Eng. **34(5)** 653-60.
[2] Vartdal B J, Gjestland T and Arvidsen T I 2009 *Lateral Propeller Forces and their Effects on Shaft Bearings,* First International Symposium on Marine Propulsors, Trondheim, Norway, 475-481.
[3] Johnson K L 1989 *Contact Mechanics,* Cambridge University Press.
[4] Shin S H, Sung Y J, Park J Y and Han B W 2016 *J. Soc. Nav. Archit. Korea* **53(4)** 282-9.
[5] Choung J M, Choe I H and Shin S H 2005 *J. Soc. Nav. Archit. Korea* **42(3)** 259-67.
[6] Shin S H and Choe I H 2004 *J. Soc. Nav. Archit. Korea* **41(3)** 35-40.
[7] Choung J M and Choe I H 2006 *J. Soc. Nav. Archit. Korea* **43(4)** 512-20.
[8] Sun J S, Lee Y J and Kim U K 2009 *J. Korean Soc. Marine Eng.* **33(1)** 28-36.

IOP Conf. Series: Materials Science and Engineering **269** (2017) 012084 doi:10.1088/1757-899X/269/1/012084

A Study on the Optimum Design of Corrugated Bulkhead for Product Carrier

Dae-Eun Ko[1, a*], **Sang-Hoon Shin**[2, b]

[1]Dept. of Naval Architecture & Ocean Engineering, Dong-Eui University, Busan, Korea
[2]Dept. of Aero Mechanical Engineering, Kyungwoon University, Gyeongbuk, Korea
[a]deko@deu.ac.kr, [b]shshin@ikw.ac.kr

Abstract. For a product carrier, in general, longitudinal bulkhead as well as transverse bulkhead is corrugated type and the intersection part of bulkheads is utilized for a pipe trunk. Since lower and upper stools are to be connected with all of longitudinal and transverse bulkheads, they have a uniform height respectively. The purpose of this study is the development of design technique for the minimization of total weight of longitudinal and transverse bulkheads at the initial design stage. In this study, optimum design of corrugated bulkhead for product carrier was conducted by applying evolution strategy method (ESM) as an optimization method and beam elements based on generalized slope deflection method (GSDM) in structural analysis. Furthermore, the installations of pipe trunks and the interference of stool diaphragms and brackets were considered to conform to the actual ship design. The optimization results are compared with the existing 70K product carrier designed in accordance with the DNV classification requirements and showed about 3% reduction in weight.

1. Introduction

Corrugated bulkheads are used often in ships due to such advantages as easy maintenance, convenience in cargo handling, and flexibility towards expansion and contraction by thermal load. Many studies have been conducted to minimize the weight of such corrugated bulkheads and its importance is being magnified these days because of high material price.

For studies on the optimum design of corrugated bulkhead, Yum used the Hooke and Jeeves method for minimum weight design of corrugated bulkhead [1]. He focused on the optimization method rather than workability, which was sensitive to initial values, and he could obtain useful results only when the solution could be predicted to some degree. Shin and Nam conducted the optimum design of the watertight bulkhead using the evolution strategy method (ESM) which is insensitive to initial values and excellent for searching the global minimum [2]. Shin and Nam used the generalized slope deflection method (GSDM) developed by Jang and Na to conduct the optimum design of deep tank bulkhead which requires repeated structural analysis in optimization process [3, 4]. The targets of optimum design were, until now, limited to the corrugated bulkheads of bulk carriers or oil tankers whose longitudinal bulkheads are not the type of corrugated.

In this study, optimum design of corrugated bulkhead for product carrier was conducted. For the optimization method, ESM which is a global optimization method was applied. Product carrier has the corrugated type bulkheads in both longitudinal and transverse direction. The beam element model based on GSDM was applied to the structural analysis of longitudinal bulkheads (L.BHDs) and transverse bulkheads (T.BHDs) in order to conduct the minimum weight design of every corrugated bulkhead in a product carrier. Furthermore, the installations of pipe trunks and the interference of stool diaphragms and brackets were considered to conform to the actual ship design.

2. Evolution strategy method

As optimization methods such as gradient method and direct search method are weak at global minimum search, the application of stochastic search method evolved from the direct search method is increasing. But the stochastic search method have disadvantage in searching time, researches have been conducted to overcome this problem. Lee and Kim presented a hybrid search method that combines the Hooke and Jeeves method which is a direct search method with the genetic algorithm which is a stochastic search method to reduce searching time [5]. Na developed a multi-objective function optimization method based on the Pareto optimum points which combines only the advantages of the direct search method and the stochastic search method [6]. The optimization method used in this study is the ESM which is one of the stochastic search methods. The problem of searching time can be solved by using beam elements based on GSDM in structural analysis. The results were obtained within just a few minutes.

ESM was first introduced by Rechenberg and later developed and systematized by Schwefel [7, 8]. It is similar to the genetic algorithm based on the principles of the survival of the fittest and natural selection. It is largely divided into plus strategy which is of the form $(\mu + \lambda)$ and the comma strategy which is of the form (μ, λ). Difference between these two strategies is the participation of the parent entity in the selection step. In the plus strategy, both the parent and child entities participate in the selection, whereas in the comma strategy, only the child entity participates in the selection. In general, the plus strategy is used often, in which μ parent entities form λ child entities, and out of all of $(\mu + \lambda)$ entities, new μ parent entities with good compatibility are formed. The child entities are formed through the process of recombination and mutation.

3. Objective function

The weight of the corrugated part of the longitudinal and transverse bulkheads was set as the objective function. That is the weight of the corrugated bulkheads per one cargo hold excluding the stool structures.

In case of the deep tank bulkhead of bulk carrier, a pipe trunk is installed at the corrugated bulkhead that meets with the outer shell of the ship, but in case of product carrier, it is installed at the intersection between the longitudinal and transverse bulkheads. So the installations of pipe trunks were considered to conform to the actual ship design.

4. Design variables

The longitudinal bulkhead of the product carrier is generally installed at the ship centerline, and the heights of the upper and lower stools are identical for both longitudinal and transverse directions because they are installed in both directions.

As shown in figure 1, six design variables were set: corrugation depth of T.BHD $(X1)$, corrugation flange width of T.BHD $(X7)$, projection length of corrugation web of T.BHD $(X8)$, corrugation depth of L.BHD $(XX1)$, corrugation flange width of L.BHD $(XX7)$, and projection length of corrugation web of L.BHD $(XX8)$. The thickness of each plate according to these dimensions was determined by the classification rules. For the dimensions of upper and lower stools, input values by designer were used. Discrete values were used for all variables to allow application to design.

5. Constraints

The following constraints were applied. The objective functions that violate the constraints were given penalties to intentionally exclude from competition.

1) Limitation of the slope angle of the inclined plate: This angle must be maintained at 60 degrees for deep tank bulkheads of bulk carrier for which the ABS classification regulations require the calculation of transverse strength if it is smaller than 60 degrees.

2) Limitation of the corrugation depth: For the design of watertight bulkheads, 2.5 times the corrugation depth must be smaller than the bottom width of the lower stool as a condition of the IACS

UR regulation. Furthermore, the corrugation depth was set not to be greater than the top width of the upper stool.

3) Limitation of production: The thickness of corrugation plates must be equal to or smaller than 25 mm which is the bending limit.

4) Limitation by structural analysis: The compressive stress of the corrugation center must be smaller than the critical buckling stress, and the stress of the bottommost part of the corrugation must satisfy the permissible stress of each classification rule.

5) Limitation in consideration of workability: Brackets are installed at the bottom stool along the corrugation web (inclined plate) in product carrier. In this case, the brackets must not interfere with the diaphragm, and they must be separated by at least the minimum distance to enable manufacturing.

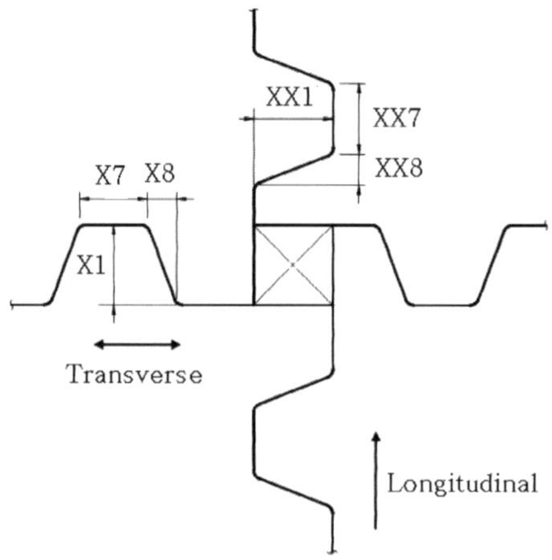

Figure 1. Design variables for optimum design.

Many transverse bulkheads have a lower stool that is perpendicular to the slope surface as shown in figure 2, and if a pipe trunk is additionally installed, many more cases must be considered compared to longitudinal bulkheads. This is because if the lower stool is perpendicular to the slope surface, the coordinate where the bracket meets with longitudinal differs between odd and even number of corrugation pitches, and if a pipe trunk is added, the starting points of left and right corrugations around the centerline become different. These cases were all considered in this study.

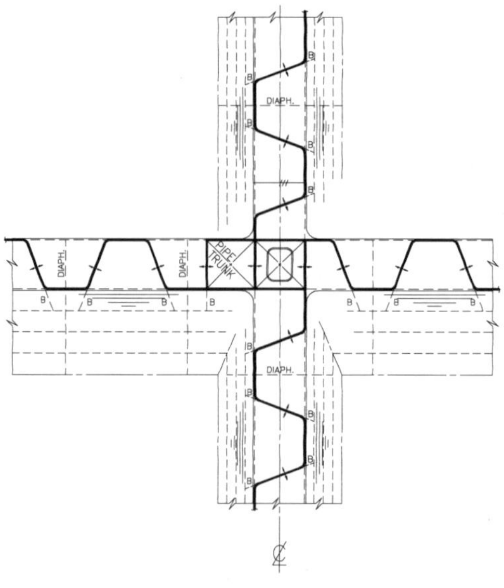

Figure 2. Design variables for optimum design.

Figure 3 shows an example of optimization results in case a pipe trunk is additionally installed, containing the design variables when the total weight of longitudinal and transverse bulkheads become the minimum, the optimum positions of bulkheads, the stress and permissible stress values of interested parts.

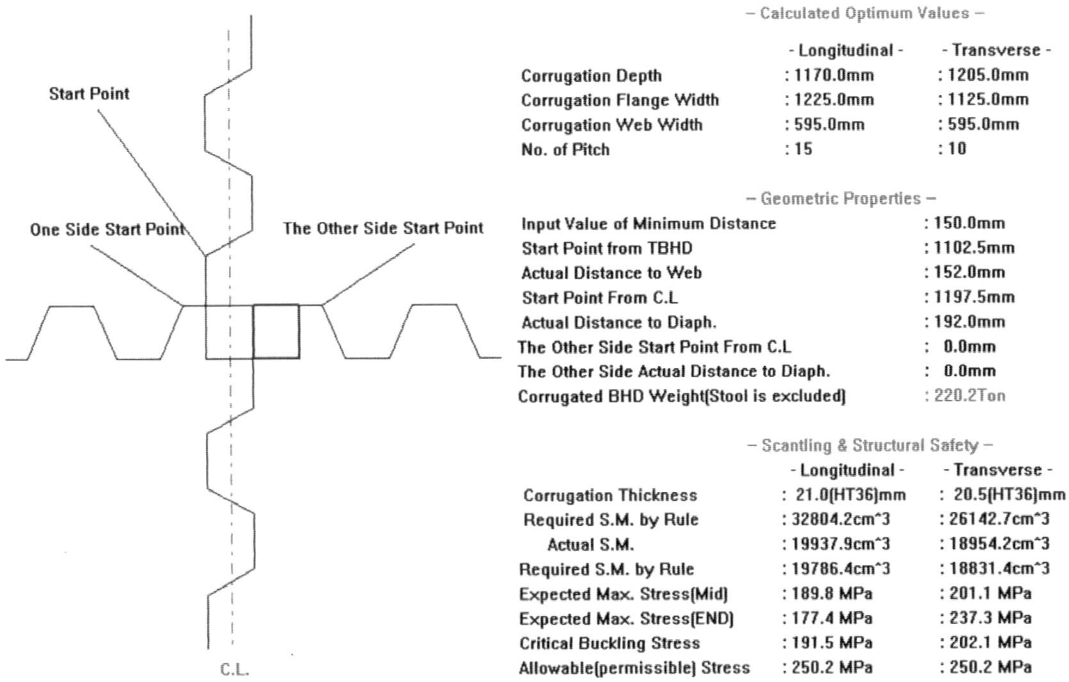

Figure 3. Example of optimization results in case a pipe trunk is additionally installed.

6. Results and discussion

The optimization results and original designed dimensions of the existing ship (70K product carrier) are compared and summarized in table 1. The optimization result showed about 3% reduction in weight compared to the existing ship designed in accordance with the DNV classification requirements. This effect is meaningful because the workability was also taken into consideration.

Table 1. Comparison of optimum results.

Design Variables	Dimension [mm]	
	Optimized Results	Original Design (DNV class)
X1	1,120	1,150
X7	1,115	1,150
X8	455	430
XX1	1,130	1,150
XX7	1,215	1,240
XX8	500	480
Weight [ton]	162.5	167.2
Ration [%]	97.2	100

As mentioned above, weight in table 1 is for the corrugated bulkheads per one cargo hold excluding the stool structures. The weight reduction effect per ship of 70K product carrier is about 26 ton. Comparison with the existing ship designed in accordance with the requirements of other classifications need to be studied.

7. Conclusions

In this study, optimum design of corrugated bulkhead for product carrier was conducted. And the following conclusions were obtained:

The problem of search time for global optimum point was solved by applying ESM which is excellent for global optimum point search and beam elements based on GSDM in repeated structural analysis. The results were obtained within just a few minutes.

The proposed optimization technique in this study improved the practicality by making even the design variables that give minimum weight to be eliminated if they violate workability.

The weight reduction of 26 ton per ship of 70K product carrier was obtained. It is also expected that proposed technique will considerably contribute to the saving of design man-hours.

8. References

[1] Yum J S 1990 *Minimum Weight Design of Corrugated Bulkheads Using the Generalized Slope Deflection Method* M S Thesis Seoul National University.
[2] Shin S H and Nam S K 2003 *J. Soc. Nav. Archit. Korea* **40(4)** 53-8.
[3] Shin S H and Nam S K 2003 *J. Soc. Nav. Archit. Korea* **40(6)** 12-9.
[4] Jang C D and Na S S 1992 *J. Soc. Nav. Archit. Korea* **29(4)** 202-13.
[5] Lee D K and Kim S Y 1996 *J. Ship Prod.* **12(3)** 141-52.
[6] Na S S 2005 *J. Soc. Nav. Archit. Korea* **42(2)** 175-82.
[7] Rechenberg I 1973 Evolution Strategy: Optimization of Technical Systems Based on the Principles of Biological Evolution Frommann-Holzboog Verlag Stuttgart.
[8] Schwefel H P 1981 *Numerical Optimization of Computer Models,* Wiley Chichester, 389.

A Study on the Application of New Shaft System for VLCC

Dae-Eun Ko[1, a*], Sang-Hoon Shin[2, b]

[1]Dept. of Naval Architecture & Ocean Engineering, Dong-Eui University, Busan, Korea
[2]Dept. of Aero Mechanical Engineering, Kyungwoon University, Gyeongbuk, Korea
[a]deko@deu.ac.kr, [b]shshin@ikw.ac.kr

Abstract. Some of the most frequent damages in the ship-mounted machinery occur in the propeller shaft system, and the most frequent damage is due to the exothermic accident occurs in the after stern tube bearing (STB). Recently, engine power and shaft diameter of very large crude-oil carriers (VLCCs) tend to increase whereas the distance between forward and after STB becomes shorter. In this study, the applicability of new shaft system for VLCCs adopting single STB system was investigated. The study results show that single STB system is quite possible by applying proper shaft arrangement and partial slope at the white metal. Also, a considerable economic effect is expected by simplifying related shaft system.

1. Introduction

The stern tube is mounted at the place where propeller shaft is penetrated through the hull. It plays a role in supporting propeller axis as well as preventing sea water from entering to the inside of the ship. Generally, two stern tube bearings (STBs) are installed. A bearing installed at the forward of the stern tube is called forward STB, and a bearing installed at the rear side of the stern tube is called after STB.

Some of the most frequent damages in the ship-mounted machinery occur in the shaft system, and the most frequent damage occurs in the after STB [1]. Exothermic accidents at the after STB whose length is long are caused by excessive local compressive pressure at the rear part as propeller load squeezes the propeller shaft.

In recent design of very large crude-oil carriers (VLCCs), the engine horsepower and shaft diameter tend to increase, whereas the shaft length becomes shorter and shaft diameter tends to increase as the location of engine is moved to a stern. Furthermore, as the distance between forward and after STBs becomes relatively shorter, shaft system flexibility decrease. This can aggravate the state of exothermic problem according to hull deflection.

In this study, the applicability of the shaft system where only after STB was installed while removing the forward STB was investigated as a measure to solve above mentioned problem. If the forward STB can be removed, additional benefit such as simplification of building the complex stern tube which takes much time can be achieved.

2. Shaft system arrangement

The proposed shaft system arrangement for 318K VLCC after removing forward STB is shown in figure 1. An intermediate shaft bearing is moved to the stern side compared to that in the existing ship to reduce a load burden on the after STB what is increased due to the removal of the forward STB.

In the past, the forward and after STBs were installed to have a rake to be set to "0", but recently, a front part of the after bearing is raised and a rear part is lower. The reason for this is to prevent excessive local pressure occurred at the rear part of the bearing via squeezing the propeller shaft by propeller load.

Length of the after bearing in the proposed design is 2000 mm, which is 200 mm longer than that of the existing ship. By doing this, a load burden on the after bearing can be alleviated and lakes are arranged similarly between them.

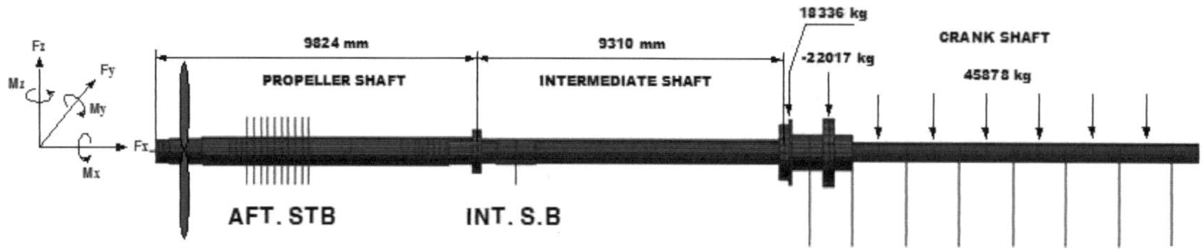

Figure 1. Proposed shaft system arrangement for VLCC.

3. Shaft system analysis

The exothermic problem in the shaft is generated during engine running state. Thus, propeller load, heat effect, and hull deflection at the engine running state are considered. For ship loading conditions, besides the typical loading conditions i.e. normal ballast condition and full load condition, full load with 6m wave condition is considered additionally.

3.1. Hull deflection

There have been a number of studies on hull deflection and shaft analysis as follows: a study on hull deflection using the finite element analysis results [2], a study on construction and application of hull deflection database [3], and a study on approximate curve of hull deflection [4]. Recently, most studies on shaft analysis consider hull deflection. In this study, shaft system analysis was conducted considering hull deflection using the finite element analysis (FEA).

The relative hull deflection calculated using FEA is shown in figure 2. This relative deflection is obtained by fixing the end location of the rear part of the after STB and the location of girder in the main engine to "0".

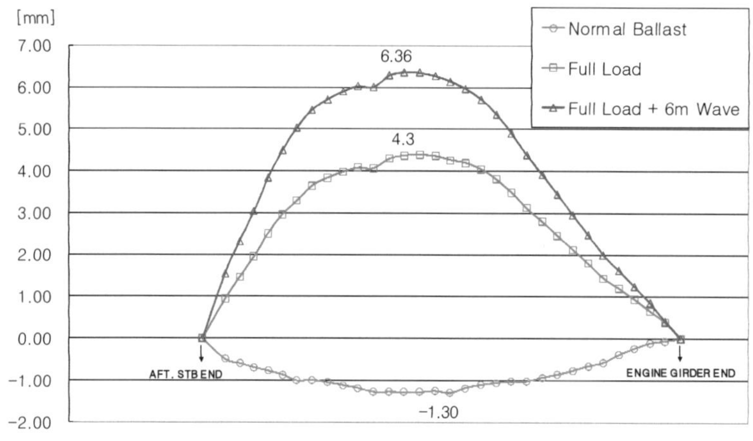

Figure 2. Relative hull deflection.

3.2. Bearing contact model

Gap element (contact element of NASTRAN) was applied to calculate the compressive pressure in the after STB. As shown in figure 3, total numbers of 40 elements, 10 each in the vertical and horizontal directions of the ship, were arranged. The location of the element in the stern side was set to No. 1 followed by assigning a number sequentially to distinguish the elements.

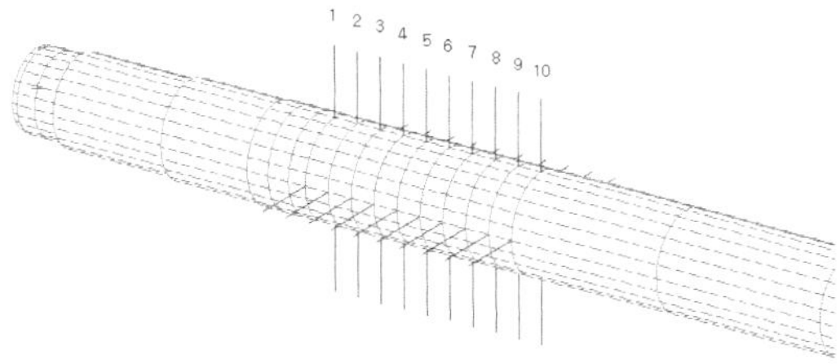

Figure 3. Gap element for contact simulation.

4. Calculation of pressure

The Hertzian contact condition assuming half-elliptical pressure distribution along contact width was adopted in this study to calculate the local pressure, as used [2]. The reaction per unit length (Q) can be expressed as a relationship between contact width (b) and material properties as shown in (1), where E^* and R are defined in (2) and (3), respectively. In (2), ν_1 and ν_2 are the Poisson's ratios of the bearing material and shaft material, respectively, and E_1 and E_2 are their respective Young's modulus. In (3), D_1 is bearing inner diameter and D_2 is shaft diameter.

$$Q = \frac{\pi b^2 E^*}{2R} \quad (1)$$

$$E^* = \left(\frac{1-\nu_1^2}{E_1} + \frac{1-\nu_2^2}{E_2}\right) \quad (2)$$

$$R = \frac{D_1 D_2}{D_1 - D_2} \quad (3)$$

The maximum pressure is expressed as in (4).

$$P_{max} = \frac{bE^*}{R} = \sqrt{\frac{2}{\pi}}\sqrt{\frac{QE^*}{R}} \quad (4)$$

The reaction forces computed in the shaft system analysis are substituted in (4) to get the maximum local compressive pressure. If the obtained value does not exceed 100 bar, which is the maximum allowable value of the material, a shaft system that installs only the after STB can be applicable. Table 1 summarizes the compressive pressure distribution of the after STB at each loading condition according to the shaft system analysis.

Table 1. Pressure distribution of after STB.

Station	Maximum pressure [bar]		
	Ballast	Full load	Full load with 6m wave
1	119.6	116.8	112.9
2	94.3	91.1	80.2
3	69.6	51.5	55.9
4	45.6	0	0
5	0	0	0
6	0	0	0
7	0	0	0
8	12.2	27.4	23.5
9	28.7	64.0	71.9
10	64.0	84.9	98.7

The calculation results showed that all pressures at the rear part in the after STB exceeded 100 bar at all loading conditions, and the pressure at full load with 6m wave condition was also close to the maximum allowable value at the front part of the bearing. The shaft analysis was conducted with some possible offsets and only the result under the full load with 6m wave condition did not satisfy the goal. Although there was some relieving effect of load burden on the after STB by moving the intermediate shaft bearing into the stern side, it did not obtain the effect that lengthened the length of the after STB. This was because the center of the after STB and the shaft did not contact with each other. To solve this problem, a measure that builds a partial slope was applied [2, 5]. This method aims to reduce the local pressure by distributing contact evenly with the shaft by means of a slope in the white metal.

The rear part of the bearing was assumed to have a partial slope and the shaft analysis was conducted. Currently, shipbuilding yards can process up to 0.01 mm precision, and 0.15mm of partial slope height was assumed.

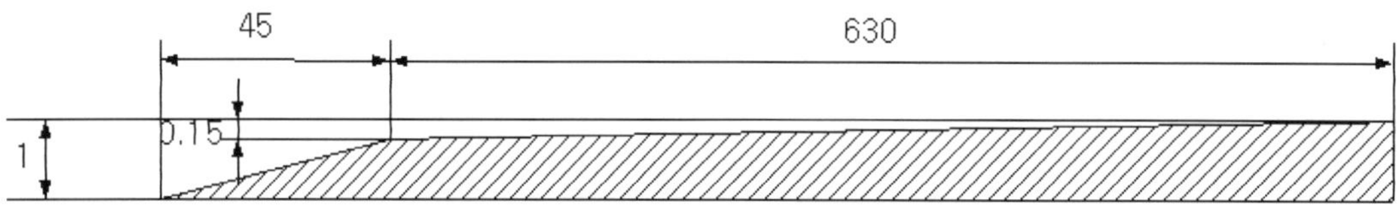

Figure 4. Partial slope of white metal in after STB.

Table 2 summarizes the compressive pressure distribution of the after STB at each loading condition when 0.15 mm partial slop is applied. The calculation results showed that the maximum local pressure of the after STB did not exceed 100 bar at all loading conditions. The maximum pressure (94.5 bar) was generated at the ballast condition, and large pressure was generated at No. 4 location overall. This was due to the high height of the partial slope so that load burden was significantly exerted on the center part of the bearing.

Table 2. Pressure distribution of after STB with partial slope height of 0.15 mm.

Station	Maximum pressure [bar]		
	Ballast	Full load	Full load with 6m wave
1	68.6	63.2	56.1
2	68.7	64.8	48.7
3	77.9	63.0	67.3
4	94.5	83.2	80.5
5	57.2	55.0	57.9
6	27.8	26.9	14.1
7	0	0	18.7
8	0	24.5	20.0
9	0	52.4	61.8
10	35.4	68.9	85.4

The analysis results showed that even if the forward STB was removed, a problem of shaft heating can be solved by making the contact surface between the after STB and the shaft distribute evenly, and a good flexible shaft system can be constructed. Furthermore, lengthening the length of the after STB by 11% compared to that of existing ship, and moving the intermediate shaft bearing into the stern side to lessen the loading burden on the after STB contributed also to the possibility of removal of the forward STB.

5. Conclusions

This study aimed to ensure shaft flexibility against hull deflection with regard to the proposed shaft system for VLCC. The applicability of the shaft system by removing the forward STB and installing only the after STB was investigated, and the following conclusions were made.

The movement of the intermediate shaft bearing into the stern side contributed to the effect of relieving the load burden on the after STB, which was increased due to the removal of the forward STB.

Only lengthening the length of the after STB did not achieve the reduction effect of local compressive pressure but the additional construction of partial slope achieved even distribution of compressive pressure over the entire bearing length.

The several additional measures proposed in this study verified that the forward STB was able to be removed from the shaft system of existing VLCC and a significantly visible effect can be expected due to the simplification of the system.

6. References

[1] Vartdal B J, Gjestland T and Arvidsen T I 2009 *Lateral Propeller Forces and their Effects on Shaft Bearings First* International Symposium on Marine Propulsors, Trondheim, Norway, 475-81.
[2] Shin S H and Choe I H 2004 *J. Soc. Nav. Archit. Korea* **41(3)** 35-40.
[3] Choung J M and Choe I H 2006 *J. Soc. Nav. Archit. Korea* **43(4)** 512-20.
[4] Sun J S, Lee Y J and Kim U K 2009 *J. Korean Soc. Marine Eng.* **33(1)** 28-36.
[5] Choung J M and Choe I H 2007 *J. Part. Slope Bearing Bush J. Soc. Nav. Archit. Korea* **44(6)** 666-74.

IOP Conf. Series: Materials Science and Engineering **269** (2017) 012086 doi:10.1088/1757-899X/269/1/012086

Development of Static Balance Measurement and Correction Compound Platform for Single Blade of Controllable Pitch Propeller

Zhang Chao[1, a], Su Shijie[1, b*], Yang Yilin[1, c], Wang Guofu[1, d] and Wang Chao[1, e]

[1]School of Mechanical Engineering, Jiangsu University of Science and Technology, Zhenjiang, China

Abstract. Aiming at the static balance of the controllable pitch propeller (CPP), a high efficiency static balance method based on the double-layer structure of the measuring table and gantry robot is adopted to realize the integration of torque measurement and corrected polish for controllable pitch propeller blade. The control system was developed by Microsoft Visual Studio 2015, and a composite platform prototype was developed. Through this prototype, conduct an experiment on the complete process of torque measurement and corrected polish based on a 300kg class controllable pitch propeller blade. The results show that the composite platform can correct the static balance of blade with a correct, efficient and labor-saving operation, and can replace the traditional method on static balance of the blade.

1. Introduction

Controllable pitch propeller (CPP) consists of multiple blades and hubs. Thanks to its advantage on achieving rapid thrust, large-scale changes by rotation of the blades around its own axis to change their spacing distribution [1, 2], CPP gradually become the mainstream on market. As the key factor in the life and vibration of the ship, static unbalanced moment of the pitch is required to meet the requirements of ISO484 and GJB3546. Therefore, it is necessary to carry out static balance measurement on CPP. Meanwhile, in order to ensure the Interchangeability between the blades, a single blade must be subjected to a static balance measurement. The static balance of single blade is the basis of the static balance of whole CPP, and its test accuracy will directly affect the accuracy of the static balance measurement on CPP [2, 3].

At present, the commonly used method of measuring the static balance of the pitch is mainly: the fulcrum method, the torque method and the center of gravity method. Chubb L B [4] uses a fulcrum method to design a blade static balance measuring device, its structure is simple, but the measurement method depends on the technical level and experience of the operator, and has the disadvantages of low efficiency and high labor intensity. Chen Bin [2] using the torque method to design a blade static balance device. The device solves the problem of low degree of automation and complicated operation process, but there are problems that the equipment has poor versatility, high manufacturing precision and high cost. Yan Xincun [5] have developed a method for the measurement of torque, which is mainly used in the overall balance measurement of aircraft, the feasibility can be demonstrated. The method, as the main measurement method, is gradually replacing the fulcrum method and torque method based on its better equipment versatility and operability. At home and abroad, some of the equipment based on this method have been put into application testing, but their structure is relatively simple, and there are problems such as vulnerable sensors and complicated calculation need to be solved. As the measuring device and grinding device independent of each other, after the completion of the torque measurement, operators need disassemble the blade across and then polished it. Due to the special working condition of the blade, it should only be removed part of the material to adjust its static balance. In practice, the selection of grinding volume is generally conservative. Therefore, it is necessary to disassembly positioning and measurement grinding repeatedly. So now process of static balance on CPP still exists many issues such as difficult to locate, repeatedly moving positioning, low grinding efficiency, manpower loss to be solved.

In this paper, a composite platform based on static balance measurement and correction is proposed, which uses the method of double weighing and the algorithm of attitude compensation in space to realize the accurate measurement on unbalanced torque of the blade. On the one hand, combined with the flexible grinding method of the robot, the platform is able to achieve efficient grinding; On the other hand, by the integrated design on measurement and correction structure, the platform can reduce the number of blade handling and the overall processing time.

2. Study on the process and principle of static balance for single blade

The process of static balance of CPP includes two parts: torque measurement and modified grinding. The phase of torque measurement is mainly to measure the deviation of the unbalanced torque (M_z and M_Y) of a single blade in a set of blades. The phase of modified grinding is to correct compound platform for the single blade so that the unbalanced torques of the whole blades are in the same range. At present, the domestic and foreign shipyards generally use the device of center of gravity measuring and artificial correction for the static balance. The main problems are as follows: (1) the requirements of blade positioning are very high and the installation is difficult; (2) the position of weighing fulcrum is ambiguous and the repetition of precision is very low; (3) there is lateral force at the fulcrum, affecting the accuracy of measurement; (4) artificial correction led to high labor intensity.

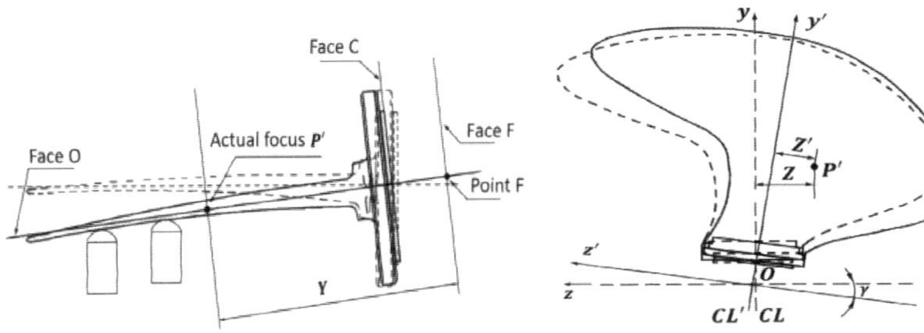

(a) Paddle posture compensation (b) Actual focus

Figure 1. Schematic diagram of the measurement of space posture.

As shown in figure 1, for the problems about positioning and installation of blade, this composite platform uses the displacement sensor and magnetic railing ruler to scan and calculate the edge of blade. Specific steps are as follows:

(1) Measure the coordinates of the different edges of the blade flange and the edge of the positioning hole, establish the coordinates of the flange surface, and calculate the center coordinate of the flange surface and the center coordinates of the positioning hole.

(2) Perpendicular to flange plane and through the center of flange and positioning hole, create a plane in which the center of gravity of the blade located by deflecting a fixed angle.

(3) Through the theoretical focus measured, create a line which perpendicular to the YZ surface, the intersection (p') of the line and the center of gravity of the blade, that is, the actual coordinates.

Through above methods, combined with software of the algorithm developed, you can directly calculate the space posture of blade with computer, to achieve compensation to the space posture of blade. In this way, the measurement error caused by installation and positioning can be eliminated, meanwhile the requirement and difficulty of blade placement are reduced.

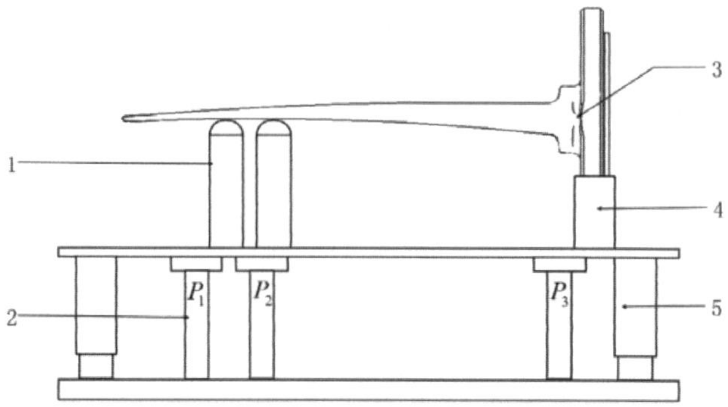

1. Support column 2. Weighing sensor 3. Vane 4. V-type clamp body 5. Guide components

Figure 2. Schematic diagram of the secondary weighing structure.

In this paper, through the design of t the double - layer platform structure (figure 2). The lateral force is transformed into the internal force of the platform by the secondary weighing method, and the lateral component force is eliminated. And the introduction of the guide component of the weighing point to limit the point to ensure that the secondary weighing process in the support point positioning accuracy and repeatability. This solves the problem of the fissure slip and the ambiguity of the traditional measurement. At the same time, this paper designed a real-time compensation of the flexible grinding mechanism, the grinding machine at the bottom of the linear guide and double-layer measurement of the lower channel steel frame connected to achieve the measurement of grinding automation and integration, reducing the number of blade disassembly and labor Strength, shortening the processing cycle relative to the traditional single Blade static balance device greatly improved efficiency and accuracy.

3. Design of composite platform

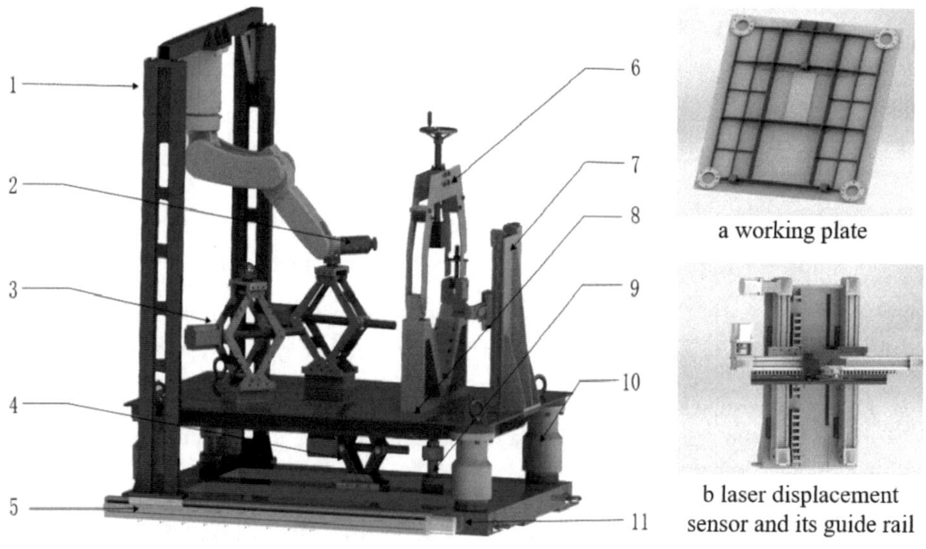

1. Gantry manipulator 2. Flexible air float spindle 3. Upper lifting mechanism 4. Lower lifting mechanism 5. Linear slide 6. V-type clamp body 7. Laser displacement sensor and its rail frame 8. Work board 9. Force sensor 10. Guide assembly 11. Slot frame

Figure 3. Adjusting and modifying the composite platform for the static balance of the single blade.

In this paper, a composite platform for measuring and correcting static balance of CPP blade is shown in figure 3. The device consists of a double-layer measuring platform, the platform for the measurement of the need to load the first no-load and a gantry manipulator. After no-load weighing and load weighing, the quality (P) of the blade and unbalanced moment (M_z and M_Y) can be got by the quadratic weighing algorithm, which means the static balancing of a CPP blade can get rid of iterative disassembly and assembly and just need plan the grinding process by host computer, then control the gantry manipulator to grind the blade to reach the standard.

3.1. The design of double-layer measuring platform

The double-layer measuring platform mainly includes work board, channel frames, guiding components, upper elevating mechanisms, lower elevating mechanism and V type clamp. When carrying out grinding, the lower elevating mechanism lifts and struts the work board to avoid excessive impact on the force sensors when hoisting and grinding, ensuring the precision and durability of the sensors.

The work board is the main bearing part of the double-layer measuring platform, whose strength and stiffness directly affect the measuring accuracy and stability of the composite platform. In this paper, ANSYS Workbench is used to optimize the configuration of stiffeners, and obtain the structure of work board which is shown in figure 3 (a). When the work board is at a total load of 317.4kg, its maximum strain is 0.024mm.

3.2. The design of laser displacement sensor and its rail frame

As is shown in figure 3 (b), laser displacement sensor and its rail frame are mainly composed of synchronous belt modules, motors, laser sensor, grating rulers. The laser sensor can move in two directions of space X and Y through two sets of synchronous belt modules and motors to measure and position the points on the flange of the blade.

Since the laser sensor positioning accuracy directly affects the accuracy of the attitude compensation and the precision of calculated moment, the requirement of accuracy can't be met only by synchronous belt modules, consequently the magnetic grating rulers and position sensors are installed on the rail. The points can be measured when the sensors read magnetic grating rulers, whose positioning accuracy can reach 0.002mm, so as to ensure the accuracy of the final measurement of the blade's gravity center is less than 0.01mm.

3.3. The design of grinding device

The grinding device is mainly composed of four parts, the portal frame, the manipulator, the air-spindle and the grinding head. The portal frame is arranged at the channel frames through the linear slide rails to move straight, providing more working space for the manipulator where different grinding requirements of variable blade specifications can be met.

An air-spindle is installed on the top of the manipulator, whose maximum buoyancy of the air bearing spindle is 60N, and the maximum float is 14mm. A displacement sensor is also installed on the spindle for data collection. When the grinding weight is out of the scope of upper and lower limit protection value, the PC controls manipulator to move in opposite directions to compensate the grinding weight, avoiding the grinding weight out of tolerance. At the same time, the spindle itself has a floating value, and the grinding weight is compensated passively under the sudden change of the blade surface, so as to ensure the smoothness of the blade after grinding.

4. Implementation of compound platform

The cost and versatility of the 300kg class are selected as the main processing objects, and the prototype of the composite platform is measured and modified according to figure 4 (a). The prototype can be used for the static range of 102.5kg-440kg single blade static balance measurement weighing and data recording, and can be based on the data after the completion of the paddle grinding path planning to achieve the gantry robot on the paddle automatic grinding.

At present, the commonly used control software development methods include Labview or configuration software and other graphical programming environment for the development and use of advanced programming language for the development of two [6]. In this paper, Visual Basic language is used to realize the development of man-machine interface of composite platform control system in Visual Studio 2015 software. The preparation of PLC program is carried out using ISPsoft.

As shown in figure 4 (b), the host computer includes load cell calibration and data acquisition, laser displacement sensor data acquisition and blade attitude adjustment and other functional interface. The left half of the interface is used to record the no-load, load measurement of the weighing data and the edge of the coordinates, after the measurement can be obtained directly through the interface paddle center of gravity coordinates and unbalance torque values; the right part of the path planning and Surface grinding function interface, grinding area and grinding mode can be selected to achieve a simple and efficient static balance measurement and correction.

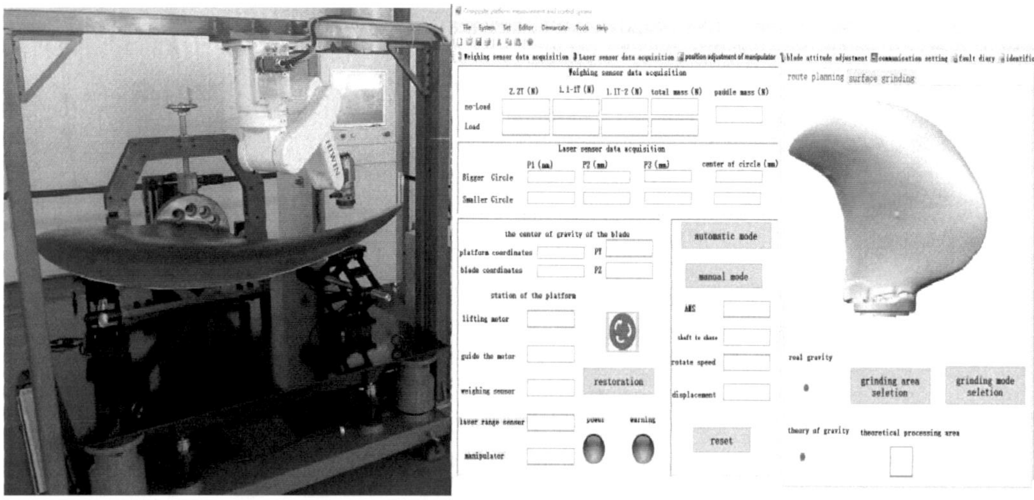

(a) Prototype in kind (b) software interface

Figure 4. Composite platform load weighing and its software interface.

5. Experiment

The static balance measurement and correction test were carried out by using the developed static balance measurement and correction composite platform. The test object was a nickel leaf and aluminum leaf with a diameter of 1.4m and a weight of about 320kg.

The specific test procedure is as follows:

(1) According to the blade quality, size and shape, the upper lifting mechanism of the location of coarse adjustment, and lock the V-clamp body; then the platform initialization and reset the laser displacement sensor, down the work board no load measurement, after completion Rise platform;

(2) Open the V-clamp body, the blade lifting and posture adjustment, and lock the V-clamp body; again down the work board for load measurement, after the completion of the platform, repeat the measurement of three sets of data;

(3) Fine-tune the attitude of the blade, repeat (2) the process three times, and get three different paddle posture under the nine sets of data;

(4) The host computer automatically calculates the parameters, according to the blade quality and unbalanced torque value grinding path planning, and automatic blade grinding.

(5) Decrease the working plate, load the load again, if the blade quality and unbalanced torque values meet the requirements, then stop; otherwise, repeat (2), (4) until the process reaches the requirements.

Table 1 for the first load of the treated load was repeated for nine times the weight of the blade, the unbalanced torque before and after the compensation. Among them, the paddle quality (P) before space attitude compensation and the paddle quality (P') after space attitude compensation are: kg; the unbalanced moment (M_z and M_Y) before space attitude compensation, and the unbalanced moment (M'_z and M'_Y) after space attitude compensation are: kg·mm.

Table 1. Test the measurement results.

Group No	Serial number	P	M_Z	M_Y	P'	M'_z	M'_Y
	1	316.90	37973	231210	316.90	37481	232660
I	2	316.90	37981	231353	316.90	37524	232673
	3	316.90	37896	231048	316.90	37528	232583
	4	317.00	36932	232538	317.00	37518	232744
II	5	316.90	36847	232733	316.90	37412	232664
	6	316.90	36839	232869	316.90	37539	232663
	7	316.90	38568	231616	316.90	37444	232571
III	8	317.00	38734	231515	317.00	37438	232657
	9	317.00	38649	231510	317.00	37443	232727
Average Value		316.92	37824	231821	316.93	37481	232660

The unbalanced moments (M_z , M_Y , M'_z and M'_Y) before and after the spatial posture compensation are selected, and the unbalanced moments and the data are respectively made with the mean value of the posture compensation, and the Y-direction and Z-direction unbalanced moments are plotted before and after the spatial posture compensation. Posture compensation after the paddle quality (P') and spatial posture compensation (M'_z and M'_Y) after the unbalanced torque, the data, respectively, with its average value is poor, according to the deviation obtained by plotting the posture compensation after the paddle quality, unbalanced torque deviation curve, as shown in figure 5 .

It can be seen from figure 5 (a) that the deviation of the Z-direction unbalance torque is less than ± 1000kg · mm and the Y-direction unbalanced torque deviation is about ±900kg · mm when the measured data of the platform is not compensated by spatial attitude. Unbalanced torque measurement results have a great impact, after the spatial attitude compensation, Z-direction unbalanced torque deviation range is less than ± 100kg · mm, Y-direction unbalanced torque deviation range of about ± 70kg · mm system has a high measurement accuracy, The basic elimination of the attitude of the blade on the unbalanced torque measurement results.

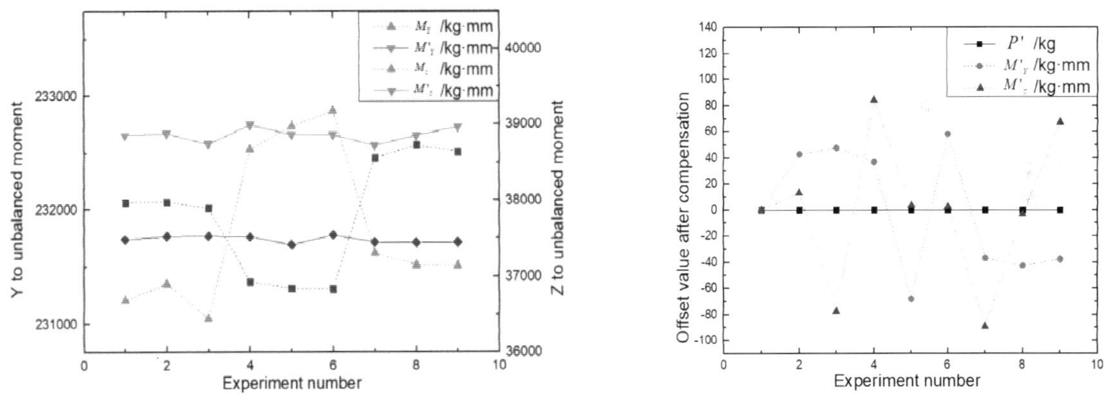

(a) Y-direction and Z-direction unbalance torque deviation (b) results after compensation

Figure 5. Deviation analysis chart of paddle quality and unbalance torque deviation.

It can be seen from figure 5 (a) and (b) that the range of the Z-direction unbalance torque deviation is less than ± 100 kg·mm and the Y-direction unbalance torque deviation range is ±60 kg·mm, indicating that the system has a high measurement accuracy and good repeatability.

6. Summary

1) The design of a double-layer measuring table structure, through the laser displacement sensor and magnetic grid scale to achieve the blade space attitude measurement and compensation algorithm, combined with the secondary weighing algorithm can be measured method of the center of gravity pad

to solve the existing Method of the existence of the paddle positioning difficult, the measurement results are not accurate problems;

2) The grinding machine of the gantry robot is developed, and the grinding force is compensated by installing the displacement sensor on the grinding spindle, and the real-time passive compensation of the grinding amount is realized by combining the floating amount of the flexible air-floating spindle itself. Combined with the double-layer to achieve the integration of measurement and grinding, simplifying the static balance process of the blade to solve the problem of low efficiency, large human consumption;

3) The experimental application shows that the measurement principle, mechanical structure and control software of the composite platform are reasonable. The results show that the composite platform can correctly perform the static balance operation of the paddle blade, which can replace the traditional method the traditional static balance device, the composite platform in efficiency, cost, operational and so have a greater advantage.

7. References

[1] Zhang Y Z 1991 *Large Cast. Forg.* **1** 83-7.
[2] Chen B 2016 *Chinese Foreign Ship Tech.* **4** 21-3.
[3] Jiang Z F 2004 *Ship Sci. Tech.* **26(4)** 11-4.
[4] Chubb L B 1948 *Apparatus for testing static balance of propeller blades* U.S. Patent 2,436,096.
[5] Yan X C et al. 2014 *Geod. Geodyn.* (Geodesy and Geodynamics) **5(4)** 68-72.
[6] Su S J, Gao L L and Wang X Y 2011 *J. Weld.* **32(11)** 96-9.

Research on the Cross Section Precision of High-strength Steel Tube with Rectangular Section in Rotary Draw Bending

Hongliang Yang[1], Hao Zhao[1] and Zhongwen Xing[1*]

[1]School of Mechanical and Electrical Engineering, Harbin Institute of Technology, Harbin, Heilongjiang Province, 150001, P. R. China

Abstract. For the demand of energy conservation and security improvement, high-strength steel (HSS) is increasingly being used to produce safety related automotive components. However, cross-section distortion occurs easily in bending of HSS tube with rectangular section (RS), affecting the forming precision. HSS BR1500HS tube by rotary draw bending is taken as the study object and a description method of cross-section distortion is proposed in this paper. The influence on cross-section precision of geometric parameters including cross-section position, thickness of tube, bend radius etc. are studied by experiment. Besides, simulation of the rotary draw bending of HSS tube with rectangular section by ABAQUS are carried out and compared to the experiment. The results by simulation agree well with the experiment and show that the cross-section is approximately trapezoidal after distortion; the maximum of distortion exists at 45~ 60° of the bending direction; and the absolute and relative distortion values increase with the decreasing of tube thickness or bending radius. Therefore, the results can provide a reference for the design of geometric parameters of HSS tube with rectangular section in rotary draw bending.

1. Introduction

In recent years, for the demand of energy conservation and security improvement, HSS is increasingly being used to produce safety related automotive components. HSS tubes with rectangular section (RS) process many advantages such as light weight, high strength, material saving, energy efficiency and environmental protection. Therefore, they are the key components in the frame structures, with a wide range of applications in fields of construction, automotive, marine and aerospace. The tensile strength of HSS BR1500HS can reach up to 1500 MPa after strengthening by heat treatment, which can improve the structural strength and reduce structural weight greatly.

However, cross-section distortion occurs easily in bending of HSS tube with rectangular section, affecting the forming precision. Some studies on sectional distortion of RS tubes in bending process have been carried out by domestic and foreign scientists. Corona established a theoretical model of sectional distortion and the model was verified by experiment [1]. Paulsen and Welo have studied the influence law of cross-sectional shape on the cross-sectional distortion of RS tubes in bending process [2]. Mandrel was used by Utsumi for reducing the cross-sectional distortion in rotary draw bending of RS tubes [3]. Liu established a three-dimensional FE model to study the influence of process parameters on distortion of thin-walled RS tubes in bending process [4]. Peng et al. have compared and analyzed the sectional distortion of overall bending and multi-point bending for the RS tubes, and studied the influence of friction coefficient, wall thickness, bend radius on the cross-section distortion in multi-point bending [5]. The previous research objects are mostly ordinary steel or aluminum alloy, however, less studies focus on the bending of high-strength steel RS tubes. Compared with the ordinary steel, HSS has some characteristics such as poorer plasticity, larger forming resistance and bigger springback, etc. In this paper, the combination of pulling and rotary drawing bending of high-strength

IOP Conf. Series: Materials Science and Engineering **269** (2017) 012087 doi:10.1088/1757-899X/269/1/012087

steel RS tubes is carried out for studying the influence law of geometric parameters on the cross-sectional distortion.

2. Experimental device and FE model

According to the principle of rotary drawing bending, the bending molds consist of rotary draw bending die, gap clamp block, clamp block, boosting block, mandrel and other components. The geometric model, the corresponding numerical model and the experimental device of the molds are respectively shown in figure 1 A, B and C.

Figure 1. Rotary drawing bending molds:
A) geometry model, B) FE model and C) experimental device..

40Cr was taken as the material of molds while high-strength boron steel BR1500HS of which the stress-strain curve before strengthened (as-delivered) is shown in figure 2 was used for the RS tubes. In FE simulation, the molds were treated as discrete bodies and meshed by four nodes curved shell elements S4R; the RS tube was defined as deformable body and meshed by eight nodes linear hexahedral element C3D8R.

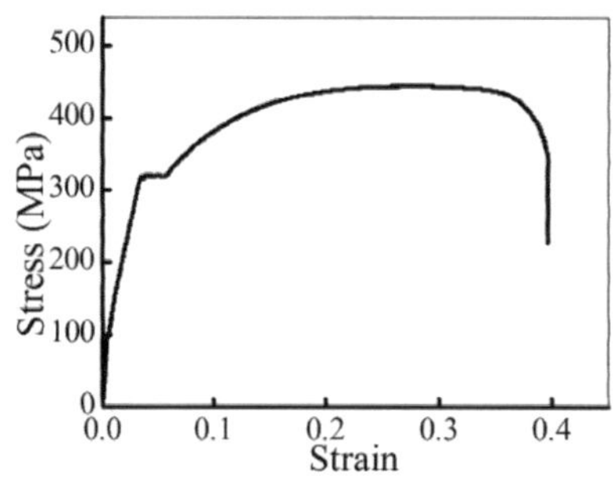

Figure 2. The stress-strain relationship of BR1500HS as-delivered.

The process parameters for bending were as follows, bending angle (α) was 90°, bending speed (ω) was 1.576 rad/s, mandrel gap was 0.3 mm, friction coefficient was 0.1, the elasticity modulus was 2.11×105 MPa and Poisson's ratio was 0.284.

3. The defining method of cross-section distortion

The outer wall of the RS tube is under tensile stress in the tangential direction, compressive stress in the radial direction and tensile stress in the width direction during the bending process. However, the inner wall is under compressive stress in the three directions. Under the combined action of the tangential stresses on both inner and outer wall, distortion occurs in the cross-section. As a consequence, the normal size will reduce, the size of inner wall in the width direction will increase. Besides, the size of outer wall will reduce because of the sunken appears in the central portion.

The position of cross-section on the tube after bending is shown in figure 3, the position of the bent portion tangent to the straight edge is defined as 0 cross section, and the other sections are 15° from one to the next. The size distribution is shown in figure 4, in which the original width and height of the tube is B_0 and H_0; after bending, the minimum width of outer wall, intermediate width, maximum width of inner wall, side height and central height are B_1, B_2, B_3, H_1 and H_2, respectively.

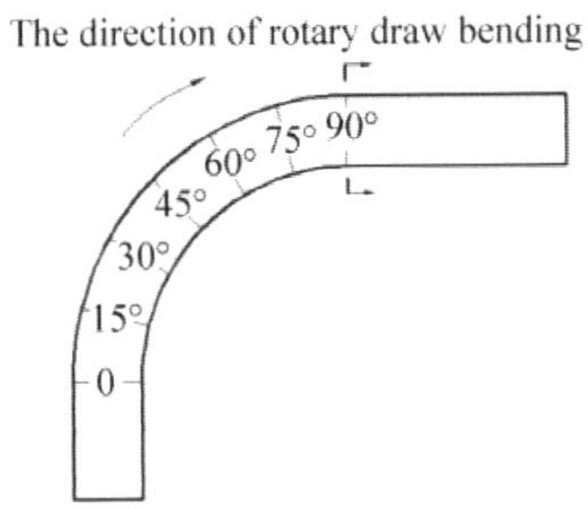

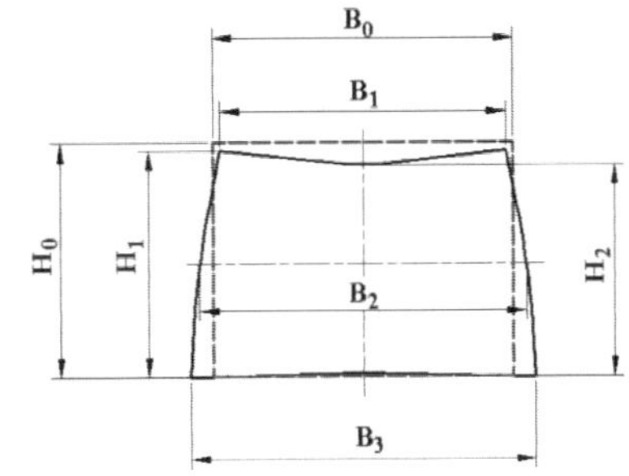

Figure 3. Distribution of measure position. **Figure 4.** The definition of cross-section size.

The amount of absolute distortion of height and width after bending can be expressed as:

$$\Delta H_i = H_0 - H_i \quad (i = 1, 2). \tag{1}$$

$$\Delta B_j = B_0 - B_j \quad (j = 1, 2, 3). \tag{2}$$

The maximum amount of relative distortion of height and width after bending can be expressed as:

$$\Delta H = H_1 - H_2 \quad \text{and} \quad \Delta B = B_3 - B_1. \tag{3}$$

The amount of absolute distortion represents the sectional deformation of RS tube after bending, in which ΔB_2 is close to 0 and thus not concerned. The amount of relative distortion represents the size difference produced by distortion in the same direction after bending and offers better reflection for the degree of sectional distortion.

4. Analysis of experimental and simulated results

4.1. Analysis of experimental results

First is the influence of thickness on the absolute distortion. Figure 5 shows different cross-sectional shapes of 0~ 90° with the aspect ratio B/H= 40/30, bending radius R= 90 mm and wall thickness t= 2 mm. The cross-sectional shapes can match up with the shape that defined in figure 4, and they are approximately trapezoidal. The outer wall has inward deflection, the inner wall remains substantially straight, and the side wall is not obviously convex. The values of sectional distortion at all positions change with the thickness of wall, as shown in figure 6. Obviously, the maximum cross-sectional distortion occurs at the position between 45~ 60°. The distortion values along the directions of height and width increase with the decrease of the wall thickness, and the value ΔH_1 along the height direction is greater than ΔH_2. In the width direction, the size of outer wall reduces while the inner wall increases. Besides, the position of section has little influence on the size of outer wall.

Figure 5. Cross-sectional shapes of the RS tube: A) 0, B) 15°, C) 30°, D) 45°, E) 60°, F) 75° and G) 90°.

Then is the influence of the bending radius on the absolute distortion. Figure 7 shows the RS tubes after bending with the aspect ratio B/H= 40/30, wall thickness t= 2 mm and the bending radius R= 75, 90, 105 and 120 mm. The influence of bending radius on the cross-sectional distortion is shown in figure 8.

The maximum cross-sectional distortion occurs at the position between 45~ 60° along the bending direction. Besides, the distortion value decreases with the increasing of bending radius.

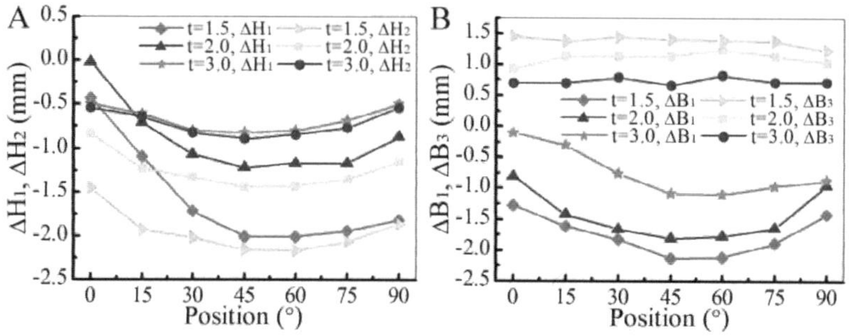

Figure 6. Influence of the thickness on the absolute distortion: A) height variation and B) width variation.

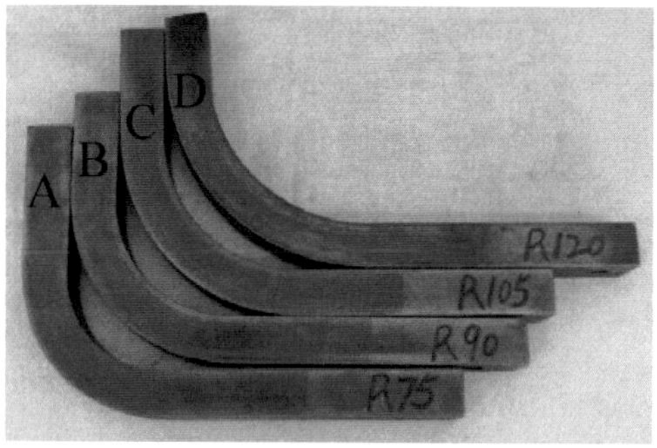

Figure 7. Test samples with different bending radius: A) 75°, B) 90°, C) 105°, D) 120°.

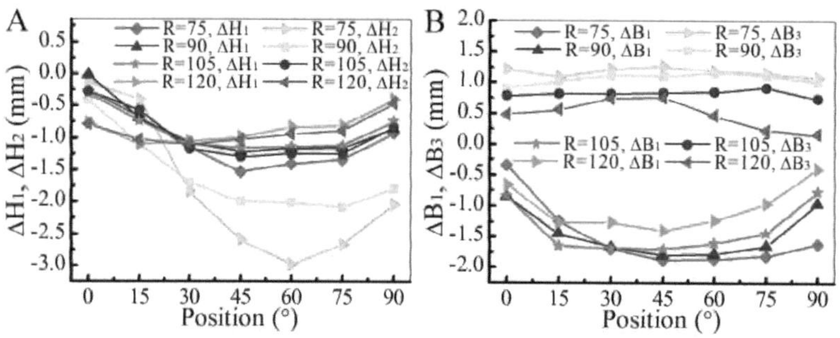

Figure 8. Influence of the bending radius on the absolute distortion: A) height variation and B) width variation.

Last is the influence of thickness on the relative distortion. Figure 9 (A) shows the values of relative distortion at each section with aspect ratio B/H= 40/30, bending radius R= 90 mm and wall thickness t= 1.5, 2, 3 mm; and figure 9 (B) shows the valus with aspect ratio B/H= 40/30, wall thickness t= 2 mm, bending radius R= 90, 105, 120 mm. Similarly, the maximum value of the relative distortion appears mostly at the position between 45~ 60°. The values of relative distortion along the directions of width and height decrease mainly with the increasing of thickness or bending radius. However, the distortion value in the height direction changes a little when the wall thickness varies from 1.5 to 2 mm, and meanwhile the values with the wall thickness 2 mm are larger. This phenomenon appears due to the side wall is too thin, and it is difficult for the edges to maintain a straight state in distorting process.

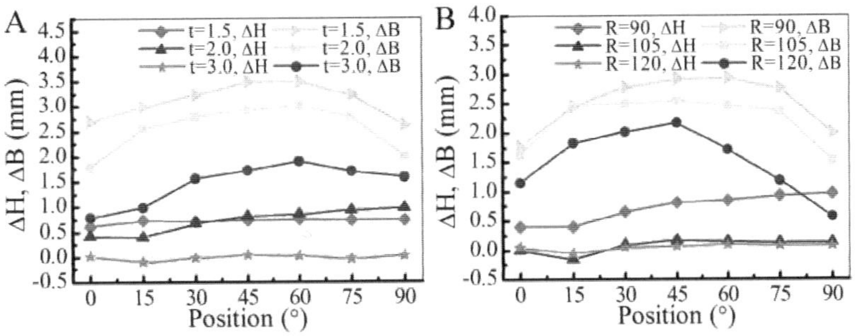

Figure 9. The values of relative distortion change under the influence of:
A) thickness and B) bending radius.

Besides, erratically fluctuation appears in the data at the position between 0~ 30°, as shown in figure 5, 7 and 8. This phenomenon appears because the wrinkling trend appears at that position, which would affect the sectional size of distortion.

4.2. Comparison between simulated and experimental results

Figure 10 shows the simulated cross-sectional shapes at different sections according to figure 3 with the aspect ratio B/H= 40/30, t= 2 mm, R= 90 mm. Obviously, the cross-sectional shapes have the same variation trend with the experiments as shown in figure 5.

Figure 10. Cross-sectional shapes of the simulated RS tube:
A) 0, B) 15°, C) 30°, D) 45°, E) 60°, F) 75° and G) 90°.

The comparison of each sectional size between experimental and simulated values is shown in figure 11. The absolute distortion amount of height and width after bending by experiments (ΔH_1, ΔH_2, ΔB_1, ΔB_3) are compared with the corresponding simulated values $\Delta H_1'$, $\Delta H_2'$, $\Delta B_1'$, $\Delta B_3'$, with the consistent trend and the average error of 13.2%. Therefore, the FE model can predict and analyse the cross-sectional distortion of RS tubes in the bending process with good accuracy and reliability.

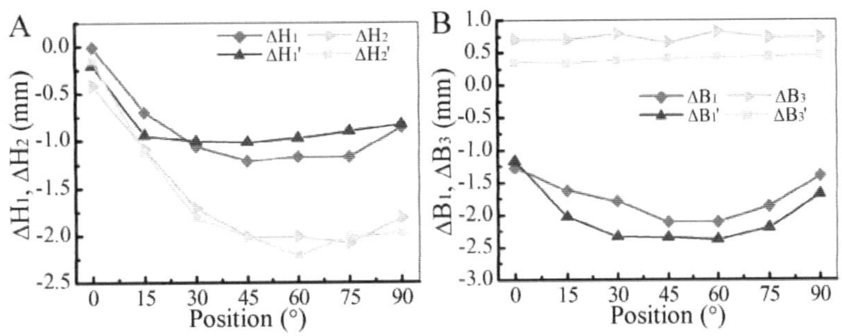

Figure 11. Comparison between the experimental and simulated data of:
A) height variation and B) width variation.

5. Conclusion

After bending, the cross-sectional distortion of high strength steel RS tube is approximately trapezoidal and different with the drum cross-section distortion of the ordinary steel or aluminium alloy RS tube.

The maximum values of absolute and relative cross-sectional distortion appear at the position between 45~ 60° along the bending direction. The inner wall appears wrinkling tendency at the position between 0~ 30°, which would affect the shape and size of the section distortion.

Increasing the wall thickness can enhance the ability of anti-sectional distortion, and the value of distortion decreases with the increasing of thickness or bending radius.

The FE model can predict and analyse the cross-sectional distortion of RS tubes in the bending process with good accuracy and reliability. Besides, the result of this study provides a reference for the geometric parameters selection for the bending process of high-strength steel RS tubes.

6. References

[1] Corona E 2004 *Int. J. Mech. Sci. Oxford.* **46** 433-48.
[2] Paulsen F and Welo T 2001 *Int. J. Mech. Sci. Oxford.* **43** 131-52.
[3] Utsumi N and Sakaki S 2002 *J. Mater. Process. Technol.* **123** 264-9.
[4] Liu Y L, Yang H, Cui S S and Zhao G Y 2007 *J. Plast. Eng.* **14** 72-5.
[5] Peng H L, Li M Z, Fu W Z and Lan Y W 2012 *Forg. Stamp. Tech.* **37** 87-90.

Study on Finite Element Method of Stress Field in Aluminum Alloy High-Speed Milling Process

Wei Zhang[1*] Shunming Li[2], Qijun Wu[1], Zenghui An[2]

[1]China Ship Development and Design Center Wuhan, China
[2]College of Energy and Power Engineering Nanjing University of Aeronautics and Astronautics Nanjing, China

Abstract. Three-dimensional numerical simulation model has been built by means of Advantage FEM. Perform simulation the stress fields of 7050-T7451 aluminum alloy in high speed milling process at the speed range of 628 m/min~5946 m/min. The dynamic change and speed's influence of stress fields and residual stress in machined layer is systematically analyzed. Some conclusions were drawn. With the cutting process development, the stress field converts to the stress state that crushing stress occupies a leading position. The magnitudes of crushing stress in all directions reduce with milling processes as the effect of Thermal-Mechanical-Coupled weakens; With the cutting speed increasing the magnitudes of crushing stress in all directions fluctuate near -950Mpa first, and then increase at the speed of 3000m/min; The residual pulling stress beneath the surface 0.03mm has been found and the magnitude increases with the cutting speed. A good agreement was obtained between predictions and experiments.

1. Introduction

High speed cutting generally involves a large amount of plastic deformation and is coupled high temperature, high strain rate and large strain processes [1, 2]. It always comes with non-uniform Thermal-Mechanical-Coupled stress field on machined layer [3, 4]. The large plastic deformation process and sever friction condition in the chip-tool-workpiece contact zone has important influence on mechanism of deformation and microstructure, work-hardening and residual stress in machined layer [5, 6].

High speed milling is a process which is typical periodicity interrupted cutting [7]. Machined surface is arisen from the process of plastic deformation as the result of cutting force and the changing temperature process which is heating and cooling. The machined surface of high speed milling process relative to conventional cutting appears the non-uniform serious stress field introduced from the frictional behavior between cutting tool flank and workpiece [8]. The theory that the non-uniform serious stress field form on the machined surface is based on the deformation theory and tribology of high speed milling process [9, 10]. The study of non-uniform Thermal-Mechanical-Coupled stress field on machined layer play an important role in the research on chip formation mechanism, the failure mechanism of cutting tool and improving the surface quality of workpiece [11].

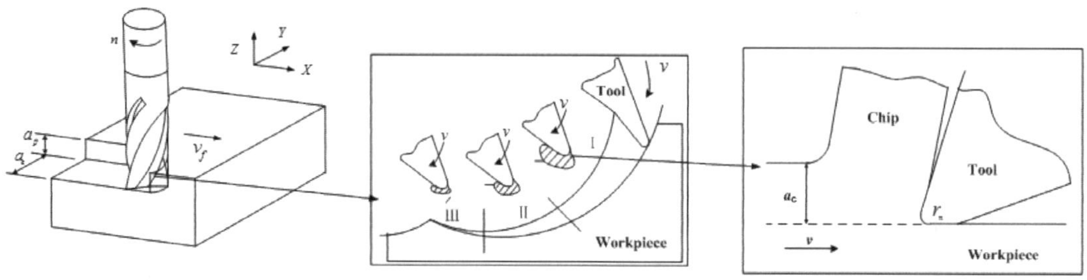

(a) (b) (c)

Figure 1. Three-dimensional model of high speed milling process.

As figure 1(b) shows, when viewed from a narrow perspective, single-point tool similar to single edge oblique cutting. There are three phases of deformation in high speed milling process: shear, extrusion and slippage. This cutting process could be reduced to the orthogonal cutting model (figure 1(c)). But the simplified model can cause inaccurate results. Three-dimensional numerical simulation has been introduced in this work. The cutting speed within 628~5946m/min is of main interest. The study can provide data basis for the prediction of the flow stress, the research on the failure mechanism of cutting tool and so on.

2. Finite element aspects and material modeling

2.1. Material modeling

050-T7451 is Al-Zn-Mg-Cu-Zr alloy with high strength, high resistance to stress-corrosion cracking and good fracture toughness. The physical mechanical properties and composition of 7050-T7451 are provided in table 1, 2 [12].

Table 1. Physical mechanical properties of 7050-T7451 aluminum alloy.

Thermal Conductivity W/(m·K)	Heat Capacity J/(kg·K)	Density Kg/m^3	Alpha 10^{-5}/K	Poisson's Ratio	Young's Modulus 10^{10}Pa
180	860	2830	2.54	0.33	7.17

Table 2. Chemical composition of 7050-T7451 aluminum alloy.

w(Si)	w(Fe)	w(Cu)	w(Mn)	w(Mg)	w(Cr)	w(Zn)	w(Ti)	w(B)	w(Al)
0.057	0.119	2.2	0.016	2.21	0.012	6.28	0.0332	0.0006	Bal.

For the description of the deformation localization during high speed load the model of Johnson-Cook [13] (equation 1) turned out to be appropriate because of its viscoplastic flowing regulation in the form of an exponential function.

$$\sigma = \left[A + B(\varepsilon)^n\right]\left[1 + C\ln\left(\frac{\dot{\varepsilon}}{\dot{\varepsilon}_0}\right)\right]\left[1 - \left(\frac{T - T_0}{T_m - T_0}\right)^m\right] \tag{1}$$

where coefficient A (MPa) is the yield strength; B (MPa) the hardening modulus; C the strain rate sensitivity coefficient; n the hardening coefficient; m is the thermal softening coefficient. $\dot{\varepsilon}_0$ the reference plastic strain rate (s^{-1}); σ is the flow stress; ε the plastic strain; $\dot{\varepsilon}$ the strain rate (s^{-1}); T (°C) the workpiece temperature; T_m (635°C) the melting temperature of the workpiece material; T_0 (25 °C) is the room temperature. Table 3 reports the material constants obtained by Fu Xiuli's [14] researches.

Table 3. Johnson Cook material parameters.

Variable	$\dot{\varepsilon}_0$ (s^{-1})	A(MPa)	B(MPa)	C	n	m
value	1	463.4	319.5	0.027	0.32	0.99

IOP Conf. Series: Materials Science and Engineering **269** (2017) 012088 doi:10.1088/1757-899X/269/1/012088

2.1.1. Cutting parameters. The cutting parameters have been selected based on the high speed experiment. Detailed parameters go as follows (table 4)

Table 4. Cutting condition.

Cutting Speed	628~5946 m/min
Other Parameters	f_z=0.15mm/z, a_p=3mm, a_e=12mm
tool	YG6X γ_0=25°, α_0=6°, d=30mm, r_n=0.02mm

3. Near-surface stress field of high speed milling process

3.1. Dynamic analysis of stress field

Figure 2 illustrates the stress distribution in all directions when the cutting speed is 5026m/min. As you can see from this figure, the non-uniform serious stress field is found mainly in primary shear zone near the tool tip. The stress field is very complex during the initial phase of the cutting process, which shows the phenomenon of interlacing pulling stress and crushing stress. With the cutting process development, the stress field converts to the stress state that crushing stress occupies a leading position. This stress field introduced from high-speed milling could help to improve the surface quality.

The variation on stress field of machined layer in the whole high speed process is shown in figure 3(a). Obviously, the magnitude of Stress-XX is slightly bigger than Stress-YY. The magnitudes of crushing stress in all directions reduce with milling processes. A trend can be derived from figure 3(b), which is that the temperature and cutting force all reduce with milling processes. This is because the practical cutting thickness attenuates in the whole high speed process as the result of the trochoid theory of down milling. This trend leads to the effect of Thermal-Mechanical-Coupled weaken. The stress field of layer is the net result.

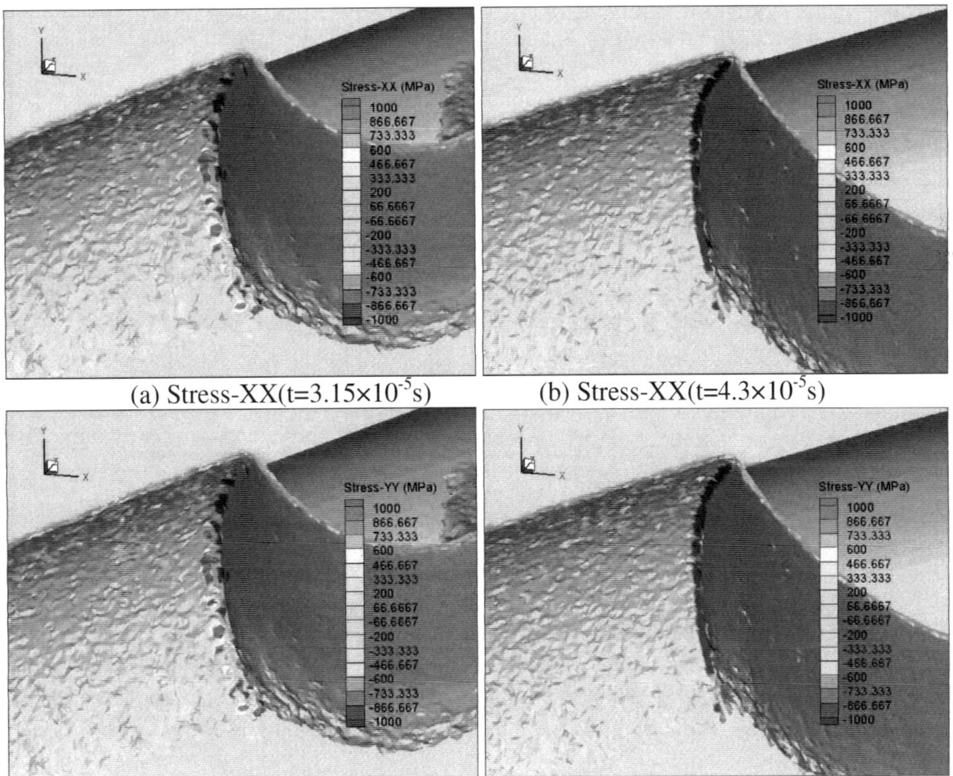

(a) Stress-XX(t=3.15×10^{-5}s) (b) Stress-XX(t=4.3×10^{-5}s)

(c) Stress-YY(t=3.15×10^{-5}s) (d) Stress-YY(t=4.3×10^{-5}s)

Figure 2. Stress distribution in all directions (v=5026m/min).

IOP Conf. Series: Materials Science and Engineering **269** (2017) 012088 doi:10.1088/1757-899X/269/1/012088

3.1.1. The influent of cutting speed for the stress field. As is shown in figure 4, with the cutting speed increasing the magnitudes of crushing stress in all directions fluctuate near -950Mpa first, and then increase at the speed of 3000m/min. The magnitude of Stress-XX is slightly bigger than Stress-YY.

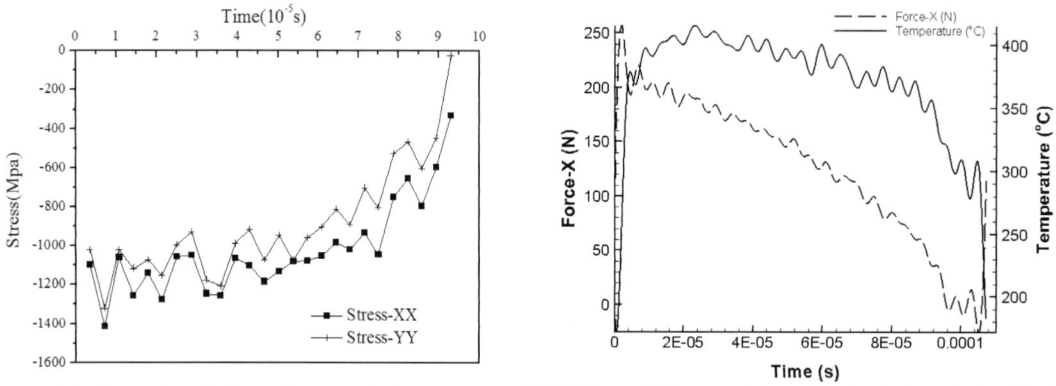

(a) Stress in all directions with time (b) Effect of Thermal-Mechanical-Coupled with time

Figure 3. Dynamic distribution of the physical parameters.

Figure 5 illustrates the change of residual stress in machined layer with the cutting speed. The layer 0.03~0.2mm beneath the surface introduces the residual crushing stress and the magnitude is increase with the cutting speed. The residual pulling stress 0.03mm beneath the surface has been found and the magnitude also increases with the cutting speed. This may because the temperature increases with the cutting speed. The residual pulling stress caused by thermal stress occupies a leading position. But the cutting forces reduce with cutting speed within certain limits. The plastic deformation, therefore, reduces. So the magnitude of the residual pulling stress 0.03mm beneath the surface increases with the cutting speed.

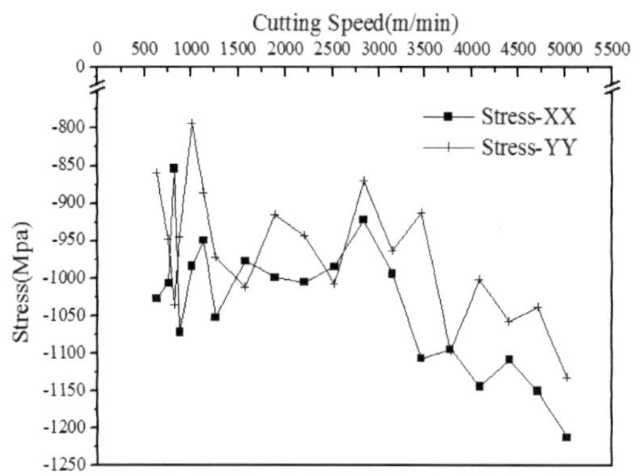

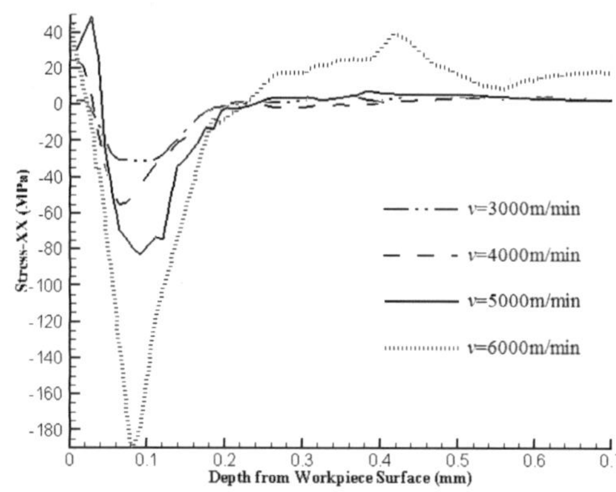

Figure 4. Stress with cutting speed. **Figure 5.** Residual stress with depth from surface.

3.1.2. Experimental verification. Comparing the cutting force of Numerical Simulation with cutting tests' results, the modeling of the system is reasonable. Figure 6 is the cutting force experiment at a speed of 1500m/min. Figure 7 is the curve of the vibratory cutting force getting through the experiments and finite element simulation of cutting force.

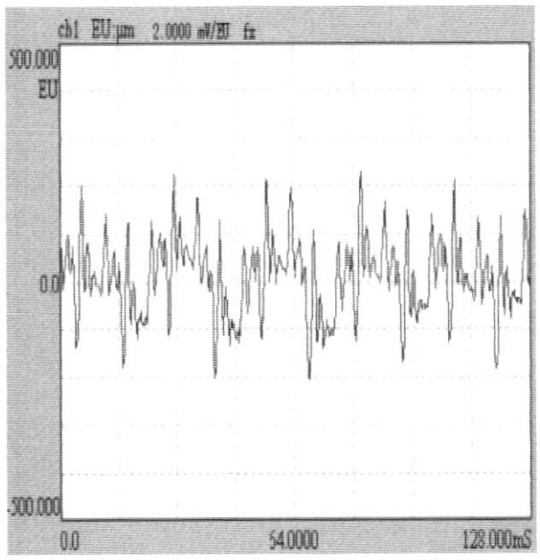

Figure 6. Cutting force of experiment.

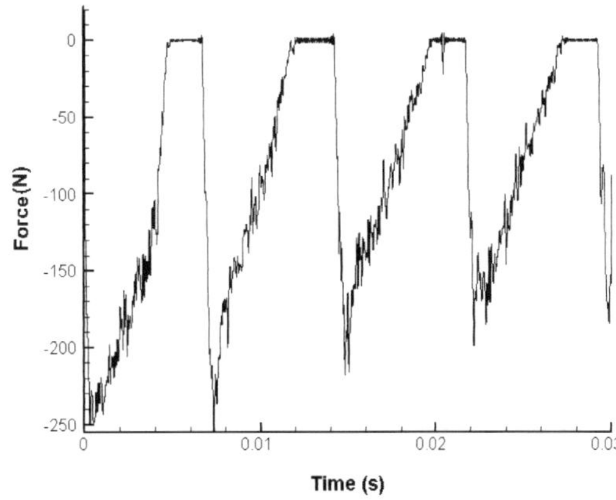

Figure 7. Cutting force of FEM.

During milling process, the continuous cutting of cutting tool is the repeat of this cutting period. The cutting force curve form the test in figure 7 is a periodic change of the milling tools' in and out. And because of the vibration of the tool and machine tool, the cutting force is existent instead zero in theoretical. Simulation of cutting process is a continuous cutting process. Comparing with the experiment, we can get the conclude that the trends change over time in cutting force curve by experiment and simulation are in good agreement.

4. Conclusions

This work builds three-dimensional numerical simulation model by means of FEA software to analysis the stress field of the machined layer. Conclusions are as following:

(1) The stress field is very complex during the initial phase of the cutting process, which shows the phenomenon of interlacing pulling stress and crushing stress. With the cutting process development, the stress field converts to the stress state that crushing stress occupies a leading position;

(2) The magnitudes of crushing stress in all directions reduce with milling processes as the effect of Thermal-Mechanical-Coupled weakens;

(3) With the cutting speed increasing the magnitudes of crushing stress in all directions fluctuate near -950Mpa first, and then increase at the speed of 3000m/min;

(4) The residual pulling stress 0.03mm~0.2mm beneath the surface has been found and the magnitude increases with the cutting speed.

5. References

[1] Sun Y Z, Liu H T and Lu Z S 2011 J. Mech. Eng. **47(1)** 187-93.
[2] Yang K, Liang Y C, Zheng K N, Bai Q S and Chen W Q 2011 Int. J. Adv. Manuf. Tech. **52** 905–12.
[3] Yang G, Saldana C, Compton W D, et al. 2011 *Acta Mater.* **59** 4538-47.
[4] Yang G, Efe M, Moscoso W, et al. 2012 *Scrip. Mater.* **66** 235-8.
[5] Che-Haron C H and Jawaid A. 2005 J. Mater. Proc. Tech. **166(2)** 188-92.
[6] Rotella G, Dillon O W, Umbrello D, et al. 2013 J. Manuf. Proc. **15** 87-95.
[7] Yang G, Saldanal C, Mann J B, et al. 2011 Adv. Mater. Res. **223** 325-31.
[8] Zhang H J, Wen W D and Cui H T 2011 Trans. Nanjing Univ. Aeronaut. Astronaut. **28(1)** 81-6.
[9] Zain A M, Haron H and Sharif S 2010 Exp. Syst. Appl. **37** 4650-9.
[10] Danisa I, Wojtowicz N, Monies F et al. 2013 *Proc. Eng.* **63** 36-44.
[11] Chan C Y, Lee W B and Wang H 2013 Int. J. Mach. Tool. Manuf. **73** 62-70.
[12] Fu X L, Ai X, Liu Z Q and Wan Y 2007 China Mech. Eng. **18**(2) 220.
[13] Johnson G R and Cook W H, 1985 Eng. Fract. Mech. **21**) 31-48.
[14] Fu X L, Ai X, Wan Y and Zhang S 2007 Trans. Nanjing Univ. Aeronaut. Astronaut. **24(2)** 139-44.

Acknowledgments

The research was supported by National Natural Science Foundation of China (51675262) and also supported by the Advance research field fund project of China (6140210020116H K02001) and the Project of National Key Research and Development Plan of China "New energy-saving environmental protection agricultural engine development" (2016YFD0700800).

IOP Conf. Series: Materials Science and Engineering **269** (2017) 012089 doi:10.1088/1757-899X/269/1/012089

A Study on the Saving Method of Plate Jigs in Hull Block Butt Welding

Dae-Eun Ko[1, a*]

[1]Dept. Naval Architecture & Ocean Engineering, Dong-Eui University, Busan, Korea
[a]deko@deu.ac.kr

Abstract. A large amount of plate jigs is used for alignment of welding line and control of welding deformations in hull block assembly stage. Besides material cost, the huge working man-hours required for working process of plate jigs is one of the obstacles in productivity growth of shipyard. In this study, analysis method was proposed to simulate the welding deformations of block butt joint with plate jigs setting. Using the proposed analysis method, an example simulation was performed for actual panel block joint to investigate the saving method of plate jigs. Results show that it is possible to achieve two objectives of quality accuracy of the hull block and saving the plate jig usage at the same time by deploying the plate jigs at the right places. And the proposed analysis method can be used in establishing guidelines for the proper use of plate jigs in block assembly stage.

1. Introduction

For the welding work of hull block joint, plate jigs are used to adjust the misalignment and to prevent welding deformation. At each shipyard, according to its scale, some hundreds thousands of plate jigs are used in a year, and the working time required for installation, removal, and finishing work is also huge. According to a survey, an around of 38,000 plate jigs are used for a 50K tanker [1]. Therefore, for the productivity improvement in the assembly and erection stages of hull blocks, the effort to reduce the usage of plate jigs as well as accuracy management through the control of welding deformation is very important.

Regarding the butt joint welding, there have been many analytical and experimental researches since the past. Recently, thanks to the development of the computer, there have been researches to precisely evaluate the deformation and residual stress of the welding joint using finite element method (FEM). However, most of the researches have been limited in a range of simple test specimen level [2, 3]. Simulation of welding deformation by FEM has a problem of generating a pretty large difference from the experimental results at the deformation point of view, in spite of a large calculation time for considering non-linearity of materials [4]. It is known that accuracy of the solution is largely relies on the adequate considerations of the melting zone of high temperature [5].

This study was performed with objectives of securing the quality accuracy of hull block and saving plan of plate jig usage. As research contents, the restraint effect of the plate jig on the deformation during butt welding was investigated quantitatively through a series of welding experiment. Further, a calculation method was established for the welding deformation of butt joint where plate jigs were attached by using a thermal elasto-plastic method based on FEM. The results were compared and verified with the experimental results. Lastly, an example simulation was performed for actual panel block joint to investigate the effect of deformation control according to the number of plate jigs installed. Results show that it is possible to achieve two objectives of quality accuracy of the hull block and saving the plate jig usage at the same time by deploying the plate jigs at the right places.

2. Experiment

Welding experiment was performed for the three specimens as shown in figure 1 to quantitatively compare the welding deformations by the butt welding according to the number of plate jigs installed. The size of base plate is 940 x 1100 x 15 (mm) and the size of plate jig is 200 x 100 x 8 (mm). The other conditions except the number of plate jigs are same in all the experiment cases. Flux cored arc welding was used and the details of welding condition are summarized in table 1. Welding experiment was performed in the order of butt welding, plate jig removal, and deformation measurement. It was confirmed that both the results of monitoring of welding conditions and appearance examination of welding beads were good.

Figure 1. Specimens for welding experiment.

Table 1. Butt welding conditions.

Item	Value	Unit
Current	201 – 270	A
Voltage	27 – 31	V
Speed	15	cm / min
V-groove angle	35	Degree
Root gap	8	mm

Deflections due to welding were measured for total of 504 points marked in advance with a gap of 50mm (for the welding line, a gap of 20mm) for each specimen by using a contact type 3D measurement system after removing the plate jigs. Figure 2 shows the relative deflections obtained by fixing the end values of measure line to "0". Deflections measured in transverse direction along center line are plotted in figure 2(a) and deflections measured in welding direction along edge line are plotted in figure 2(b).

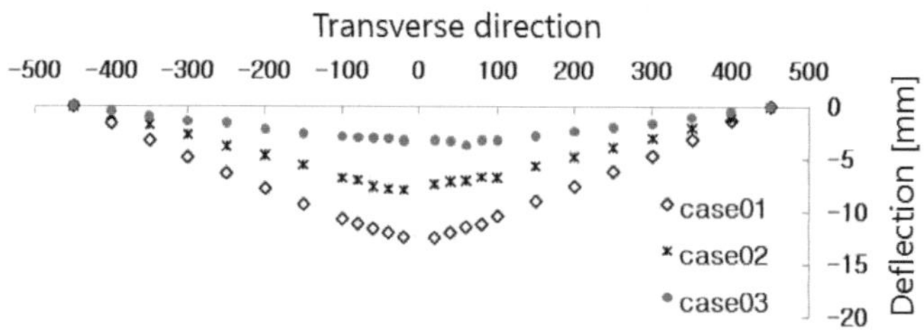

(a)

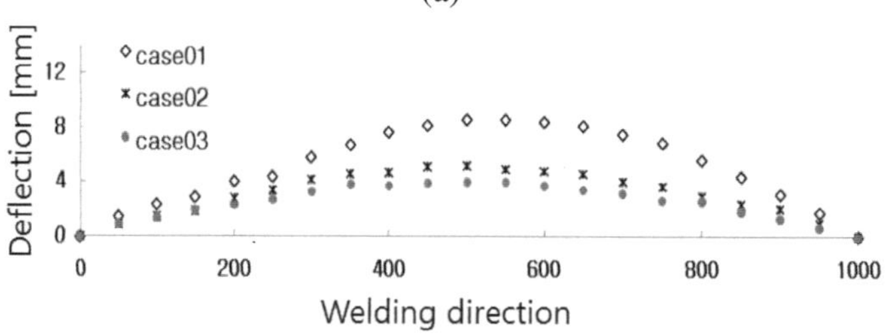

(b)

Figure 2. Deflections results of welding experiment.

The welding experiment as above yielded the data for the welding deformation according to the number of plate jigs installed. These results would be utilized for the comparison and verification to establish the welding deformation analysis method.

3. Analysis method for welding deformation

In this study, a thermal elasto-plastic analysis method based on FEM was used. As mentioned above, it is known that simulation of welding deformation by FEM has a problem of generating a pretty large difference from the experimental results at the deformation point of view, in spite of a large calculation time for considering non-linearity of materials. To improve the accuracy of deformation results by analysis, analysis parameters were decided based on the experiment results. The major items for analysis are listed as below;

(1) MSC MARC was used as an analysis code. Shell element having nine integral points was used to consider the temperature gradient in thickness direction.

(2) A half-model from the weld line was created as shown in figure 3. Symmetric boundary condition was applied along welding line and fix conditions were applied at the nodes of tag welding area to prevent rigid body motion.

(3) Temperature-dependent material properties were applied by referring to the research of Shin et al. [2]. Welding conditions for analysis were set as current of 248 (A), voltage of 26 (V), and welding speed of 15 (cm/min) by referring to the values in table 1.

(4) The moving heat source was used to simulate the 3D effect of welding deformation. The modeling technique where the welding beads of high temperature are gradually generated by elapse of time was implemented.

(5) Analysis was repeatedly carried out by changing the heat flux radius to find out heat input condition matching to deformation results of the experiment. At this time, experiment result of case 1 was used as the reference. Both angular deformation and longitudinal deformation showed the best similarities when the heat flux radius was 135mm.

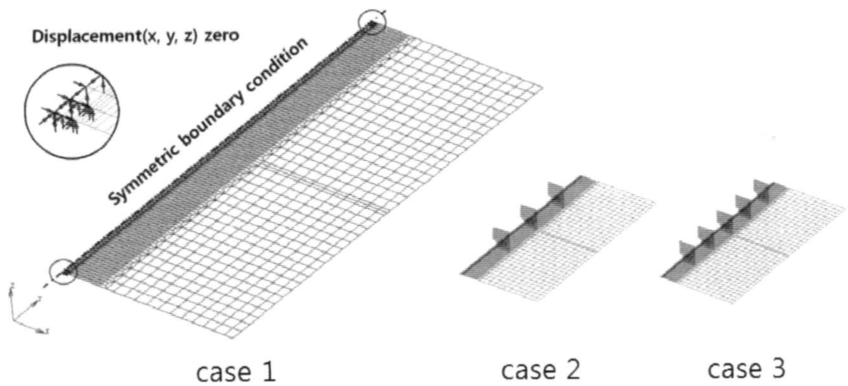

Figure 3. Finite element models and boundary conditions.

Similarities between experiments and analysis results are summarized in table 2. Match tolerance of 2mm was applied for the evaluation of similarities. Angular deformations were well matched in both experimental and analysis results for all the cases. The deflections along welding direction were also pretty similar in both experimental and analysis results except the case 3. For the difference of deflections along welding direction in case 3, it might be due to the analysis condition of heat flux radius determined from the case 1 where horizontal restraint does not exist. Fortunately, case 3 where the installation distance of plate jigs is 180mm is not the case of our interest since the installation distance of plate jigs at the site is normally more than 300mm.

Table 2. Similarity between experimental and analysis results [%].

Item	Case 1	Case 2	Case 3
Angular deformation	100	100	100
Deflection along welding direction	98	95	56

4. Welding deformation analysis of hull block joint

Using the proposed analysis method for welding deformation, an example simulation was performed for actual panel block joint to investigate the saving method of plate jigs. Figure 4 shows the panel block composed of 4 sub blocks B01~B04 and finite element model for welding deformation analysis. Analyses are conducted for 3 cases according to the number of plate jigs used as shown in table 3. Case 1 is selected to decide the deformation value for the lower limit, case 2 is the present method of construction in shipyard, and case 3 is an attempt to reduce the number of plate jigs.

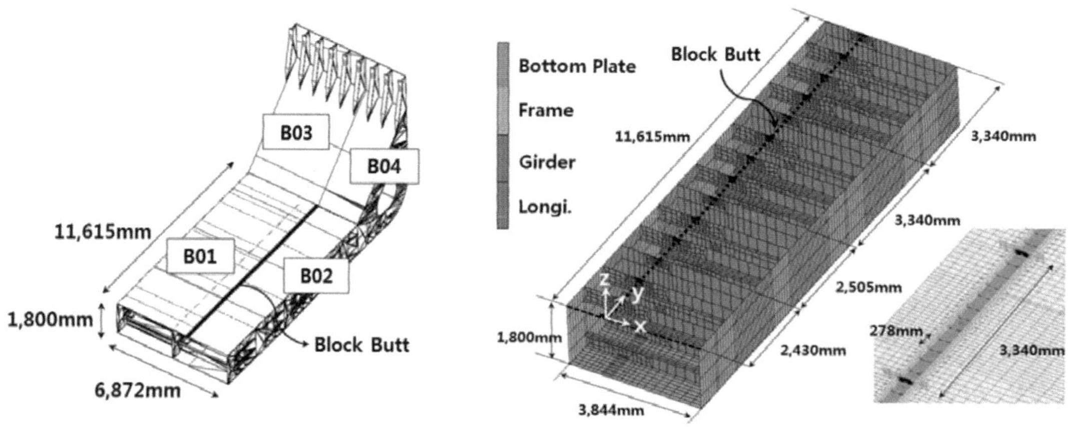

Figure 4. Panel block and Finite element analysis model.

Table 3. Analysis cases for panel block joint.

	Case 1	Case 2	Case 3
No. of plate jigs installed	0	39	20

Deflections along y-direction are plotted for all analysis cases as shown in figure 5. Result of case 3 is quite similar to that of case 2, but the plate jigs are need to be added at the location of $6,605<y<7,440$ and $9,945<y<10,780$ since the deflections exceed the values of case 1. Finally, it can be concluded that 22 plate jigs are sufficient for the butt joint welding of panel block and 17 plate jigs can be saved compared to the present method of construction in shipyard.

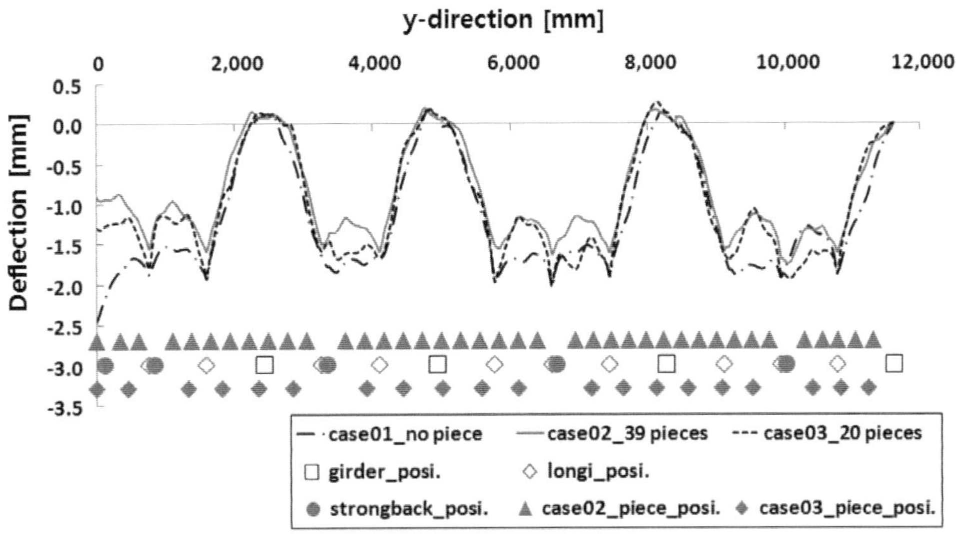

Figure 5. Analysis results for panel block joint.

5. Conclusions

This study was performed with the objectives of securing quality accuracy of block and of saving method of plate jigs used during butt welding of hull blocks. The major research results are as follows;

(1) By carrying out a series of welding experiment, quantitative data were obtained to be utilized for the comparison and verification in establishing the welding deformation analysis method.

(2) A deformation analysis method for the butt welding with plate jig setting was established using the thermal elasto-plastic analysis method based on FEM.

(3) An example simulation was performed for actual panel block joint and the results show that it is possible to achieve two objectives of quality accuracy of the hull block and saving the plate jig usage at the same time by deploying the plate jigs at the right places.

(4) The proposed analysis method is expected to be used as an analysis tool for the preparation of plate jig deployment drawing and its installation guide.

6. References

[1] Noh H W, Kim K J, Leem R S and Kim H K 2008 *Sp. Iss. Soc. Nav. Archit. Korea* 136-43.
[2] Shin D H, Shin S B and Lee J S 2007 *J. Soc. Nav. Archit. Korea* **44(6)** 598-604.
[3] Shin S B, Lee D J and Park D H 2010 *J. Korean Weld. Join. Soc.* **28(4)** 380-6.
[4] Ko D E, Jang C D, Seo S I and Lee H W 1999 *J. Soc. Nav. Archit. Korea* **34(4)** 116-27.
[5] Koichi M 1980 *Analysis of Welded Structures* Pergamon Press.

IOP Conf. Series: Materials Science and Engineering **269** (2017) 012090 doi:10.1088/1757-899X/269/1/012090

Blocking Mechanism Study of Self-Compacting Concrete Based on Discrete Element Method

Xuan Zhang[1*], Zhida Li[1], Zhihua Zhang[1]

[1]School of Transportation, Wuhan University of Technology, Wuhan 430063, China

Abstract. In order to study the influence factors of blocking mechanism of Self-Compaction Concrete (SCC), Roussel's granular blocking model was verified and extended by establishing the discrete element model of SCC. The influence of different parameters on the filling capacity and blocking mechanism of SCC were also investigated. The results showed that: it was feasible to simulate the blocking mechanism of SCC by using Discrete Element Method (DEM). The passing ability of pebble aggregate was superior to the gravel aggregate and the passing ability of hexahedron particles was bigger than tetrahedron particles, while the tetrahedron particle simulation results were closer to the actual situation. The flow of SCC as another significant factor affected the passing ability that with the flow increased, the passing ability increased. The correction coefficient λ of the steel arrangement (channel section shape) and flow rate γ in the block model were introduced that the value of λ was 0.90-0.95 and the maximum casting rate was 7.8 L/min.

1. Introduction

In recent years, with rapid development and extensive application of SCC, more and more construction projects are beginning to use SCC. In some special conditions, such as coarse aggregate size large or reinforced configuration too dense, it may still happen blocking phenomenon and not fill with lead to structure within the dense area [1]. Many scholars have conducted the slump or slump flow to study the relationship between the yield stress of concrete, measured slump or slump flow is determined by material yield stress, and has nothing to do with the plastic viscosity [2]. Khayat [3] adopted slump flow test, V-funnel test and filling capacity test (parallel steel, spacing is 34mm) to study fiber reinforced SCC and showed that V-funnel test and filling capacity test had no special relationship. Xie [4] used Multi-scale method to study the filling capacity of SCC, it was concluded that application scope of these two models and studied affect of the stone size, mortar flow to filling capacity.

Compared with numerical simulation and laboratory test, it can save a lot of manpower and material resources and financial resources, which can effectively avoid error caused by the test conditions, personnel, operating and experimental environmental factors. A large number of researchers have started to adopt the method of numerical calculation was carried out on the concrete simulation studies [5]. Numerical methods mainly include Discrete Element Method (DEM) [6], the Lattice Boltzmann Method (LBM) [7], Computational Fluid Dynamics (CFD) [8], Smoothed Particle Hydrodynamics (SPH) [9] and so on, among them, the Discrete Element Method can be used for simulating the macro flow state of concrete, but also in the process of flow within the aggregate particles collide with each other between the friction phenomenon. Puri [10] applied a combination unit of mortar wrap the aggregate particles to simulate concrete, and made a 2D simplified simulation of the relationship between the injection pressure and the rebound quantity of the shotcrete. SCC was also simplified as coarse aggregate and mortar units, the slump, V-funnel and the U-box experiment of

SCC were simulated [11]. A new DEM model for the simulation of SCC flow was presented that it was described as an assembly of composite particles made of spherical hard grains representing coarse aggregates surrounded by concentric spherical layers representing mortar, rheological simulations showed that the rheological behavior of simulated concretes can be approximated with the Bingham model [12]. Mechtcherine [13] used DEM define the rheological model of SCC, then the main focus was directed at establishing an algorithm to derive the model parameters related to yield stress according to the Bingham model, the results of the numerical analysis agreed well with the results predicted by analytical solutions for all parameter combinations. Huang [14] conducted a quick contact searching algorithm of HACell and the DEM constitutive model of self-compacting mortar (SCM) and SCC, the rheological parameters of SCM element were used in the simulation of flow ability and viscosity of SCM.

In this paper, SCC is simplified as the coarse aggregate and the mortar two-phase medium, a 3D DEM model for SCC is established and validated and the rheological model adopts the Bingham model, then Roussel's granular blocking model is verified and extended by establishing the discrete element model of SCC and the influence of different parameters on the filling capacity and blocking mechanism of SCC are also investigated.

2. Discrete element model of SCC

The discrete element method is a method to simulate rigid ball movement and interaction based on discrete element theory. A fundamental assumption of this method is that the material consists of separate discrete particles and the particles are treated as rigid bodies. The contacts between neighboring particles occur only at one point at a given time, the calculations alternate between the application of Newton's Second Law with respect to the motion of particles and the force–displacement law at the contacts. Thus the displacements and rotations of the particles are calculated according to the following governing equations (1) and (2).

$$F_i = m \cdot (\ddot{x}_i - g_i) \tag{1}$$

$$M_i = I \cdot \dot{\omega}_i \tag{2}$$

where F_i is the resultant force, m is the total mass of the particle, $\ddot{x}_i$ is the translational acceleration and g_i is the acceleration of gravity. M_i is the resultant moment acting on the particle, the moment of inertia I and the angular acceleration of the particle $\dot{\omega}_i$.

The contact force vector F_i which represents the action of ball-ball contact, and the action of the ball on the wall for ball-wall contact can be resolved into normal and shear components with respect to the contact plane according to equation (3).

$$F_i = F_i^n + F_i^s \tag{3}$$

where F_i^n and F_i^s denote the normal and shear component vectors, respectively. The contact stiffnesses relate the contact forces and relative displacements in the normal and shear directions via equations (4) and (5), which are repeated here.

$$F_i^n = K^n U^n n_i \tag{4}$$

$$\Delta F_i^s = -k^s \Delta U_i^s \tag{5}$$

where K^n and k^s denote the normal and shear stiffness, U^n and ΔU_i^s denote the normal and shear relative displacement at the contact, n_i denotes the normal vector of contact plane.

Concrete is a composite material composed of a variety of materials. In the discrete element simulation, it is obviously not realistic to use a variety of particles to simulate the various materials of concrete. This article summarizes the research results of scholars, the SCC is simplified as two phases of coarse aggregate and mortar, respectively used two kinds of particles with different materials properties. The mortar is treated as a kind of discrete element which used parallel bond model, and coarse aggregate is treated as another kind of discrete element which used linear contact model, the discrete element model of SCC is established by adding different contact relations.

3. Rheological model of SCC

Concrete can be considered as a fluid, provided that a certain degree of flow can be achieved and that the mix is quasi-homogeneous with regularly dispersed aggregate particles. There are many models describing the rheological properties of concrete, Bingham model Herschel-Bulkley model are best suited to the real situation [15, 16]. Bingham model consists of Newton fluid model and St. Venant solid model, rheological equation is given by equation (6).

$$\begin{cases} \tau = \tau_0 + \mu_p \dot{\gamma} & |\tau| > \tau_0 \\ \dot{\gamma} = 0 & |\tau| < \tau_0 \end{cases} \tag{6}$$

where τ_0 is the yield stress, μ_p is the plastic viscosity and $\dot{\gamma}$ is the shear rate. Both of these two models contain two key parameters: τ_0 and μ_p. τ_0 is the maximum internal force of material internal resistance to plastic deformation, it is caused by adhesion and friction force between each phase, and determines the deformation of the material, has nothing to do with the external factors. μ_p is a characteristic that prevents its shear deformation from inside the material. It represents the resistance of the flow of the material which the flow layers to prevent its flow and determines the velocity of the material flow.

4. Study on passing ability and blocking mechanism of SCC

The French scholar Roussel proposed the granular blocking model of SCC, it is verified and extended by establishing the discrete element model of SCC and the influence of different parameters on the filling capacity and blocking mechanism of SCC are also investigated. The blocking parameter P applicable to concrete was obtained to predict the probability that the concrete would be blocked by obstruction.

$$P = (2.6\Omega / D_{max}^3)(0.1\phi)^{0.85\delta^2/0.81(\beta D_{max})^2} \tag{7}$$

where D_{max} is the diameter of the coarsest particles, ϕ is the volume fraction of the aggregate particles, Ω is the total amount of concrete, obstacle spacing δ and coarse aggregate shape β.

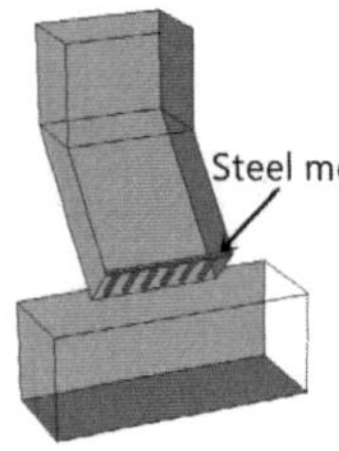

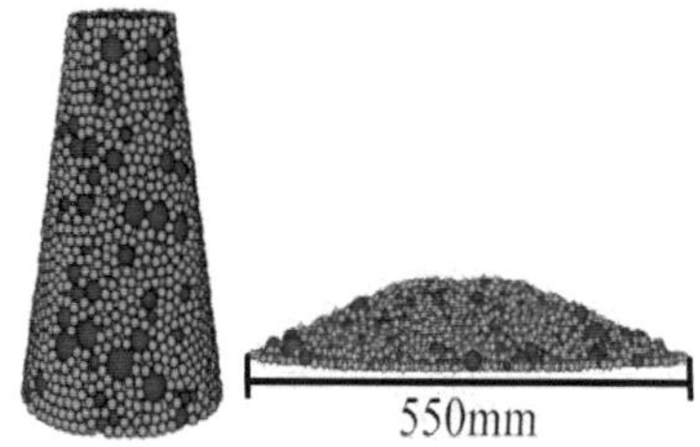

Figure 1. Schematic of model size. **Figure 2.** Final shape of slump flow 500×500mm.

4.1. Building of model

Based on literature [4] to establish a model size as shown in figure 1. In order to study the different parameters on the properties of SCC through the influence of test selection flow (690×690mm, 550×550mm), coarse aggregate size (5-10mm, 5-20mm), shape of coarse aggregate (gravel, pebble), steel spacing (20-40mm) and casting rate (1.5-4L/min) as the main parameters. Simulation results compare with the literature [4] test results analysis the blocking mechanism of SCC. The mixture ratio and parameter information of SCC used in simulation test are shown in table 1.

Table 1. Mixture ratio of SCC.

Group number	C (Kg)	W (Kg)	S (Kg)	G (Kg)	Type	Distribution (mm)	Slumpflow (mm)
No. 1	491	205	859	694	gravel	5-20	690×690
No. 2	491	205	859	694	gravel	5-20	550×550
No. 3	491	205	859	694	gravel	5-10	690×690
No. 4	491	205	859	694	pebble	5-10	690×690

4.2. Calibration of microscopic parameters

Particles microscopic parameters will directly affect macroscopic rheological properties of SCC. However the microscopic parameters between particles cannot be directly obtained through macroscopic mechanical tests such as laboratory tests, and can only be obtained through numerical tests. According to the slump flow of table 1, slump simulation test is carried out on microscopic parameters (figure 2) preliminary Calibration, the simulation tests debug repeatedly and reference literature [13], checking to calculate different slump flow microscopic parameters of SCC, microscopic parameters of slump flow 500× 500mm are shown in table 2.

The simulation is carried out according to distribution curve of the literature [4]. The distribution curve is shown in figure 3, the slump simulation used the same distribution curve. Due to the influence of discrete element calculation efficiency, it is impossible to calculate relatively small particles. Therefore, it is possible to treat the part that is less than 4.75mm as an aggregate (4.75mm) to process the mortar of SCC. The coarse aggregate and mortar materials are simulated with different radius spherical particle, and the coarse aggregate particles generated by the middle limit of distribution are 553, and the mortar particles are 5247.

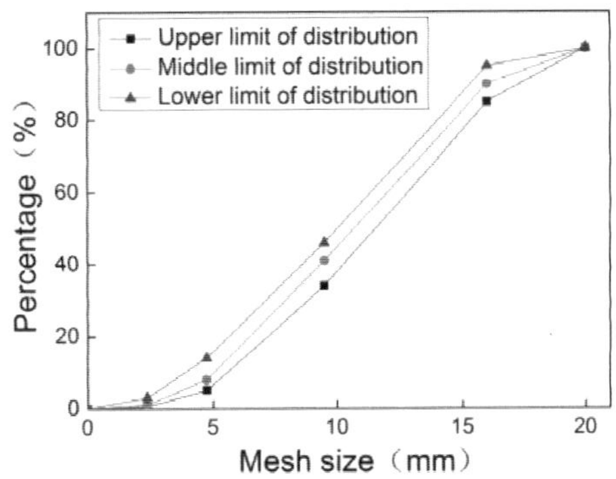

Figure 3. The distribution curve of mixed aggregate.

Table 2. Parameters of parallel bond model (550× 550mm).

Parameters	Aggregate-Aggregate	Aggregate-Mortar	Mortar-Mortar
Normal strength ($N \cdot m^{-2}$)	——	4×10^4	4×10^3
Shear strength ($N \cdot m^{-2}$)	——	2×10^4	3×10^3
Normal stiffness ($N \cdot m^{-1}$)	——	1×10^4	2×10^3
Shear stiffness ($N \cdot m^{-1}$)	——	5×10^3	1.5×10^3
Bond radius	——	0.35	0.25
Local damping	0.70	0.70	0.70

5. Discrete element numerical simulation and analysis

5.1. Effect of steel spacing

Effect of changing the steel spacing on SCC shows in table 3.With the decrease of steel spacing, the passing ability of SCC decreases gradually to 30% after the steel spacing is less than 30 mm, it has a serious blocking phenomenon that could not satisfy the filling requirement. Blocking probability P also shows the same trend. It can be seen from the comparison between physical test and simulation test that it is feasible to use discrete element to simulate the passing ability of SCC. Figure 4 shows the different situation diagram of the model when the steel spacing is 20mm, initial situation (figure 4(a)) and final situation (figure 4(b)) of model, the local illustration (figure 4(c)) near the steel are shown. The main reason for the blocking is the accumulation of coarse aggregate in the vicinity of the steel, with the passing of mortar particles, coarse aggregate particles gradually accumulate near the steel and size of coarse aggregate is greater than that of the steel spacing, and the blocking of natural formation is consistent with the actual experiment.

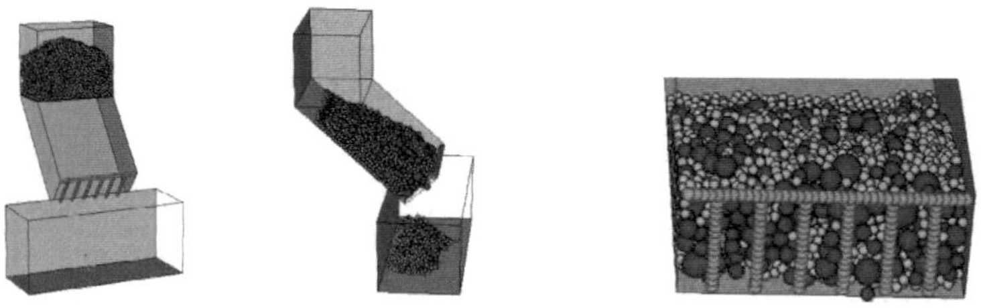

(a) Initial situation (b) Final situation (c) Local illustration of final situation

Figure 4. Different situation diagram of steel spacing 20mm.

Table 3. Test results of changing steel spacing.

Test number	SCC number	Steel spacing (mm)	Steel arrangement	Passing ability test/simulation (%)	Parameter P test/simulation
1-1	No. 1	40	parallel	100/100	0.0006/0.0042
1-2	No. 1	35	parallel	100/100	0.02/0.06
1-3	No. 1	30	parallel	76.6/81.5	0.4/0.8
1-4	No. 1	25	parallel	25.1/23.7	4.9/3.8
1-5	No. 1	20	parallel	8.2/10.5	39.6/42.4

5.2. Effect of steel arrangement

Figure 5 shows different arrangement forms of steel spacing 30mm, three groups of steel arrangement (2-1 & 2-2, 2-3 & 2-4, 2-5 & 2-6) were compared with physical test and simulation test (table 4). For the SCC, if parallel steel changes to grid steel, the passing ability of SCC is significantly reduced, the passing ability of grid steel significantly lower than the parallel steel. In the verification test, Roussel [1] used parallel steel and did not discuss the situation of grid steel. To the application of comprehensive Roussel particles blocking model, steel arrangement (channel section shape) correction coefficient λ is introduced into parameter P, to reflect the steel arrangement way of particles blocking probability, the influence of steel spacing into $\lambda\delta$ at this time, the particles blocking model formula is:

$$P = (2.6\Omega / D_{\max}^{3})(0.1\phi)^{0.85(\lambda\delta)^2/0.81(\beta D_{\max})^2} \tag{8}$$

According to the physical test result, the value λ was given in 0.87 to 0.97 [4], considering the concrete test generally have larger randomness and discreteness, in order to accurately determine the value of λ and develop the advantage of discrete element numerical simulation. After several other groups of numerical simulation test, calibration value λ is 0.90 to 0.95.

(a) (b)

Figure 5. Different forms of steel arrangement of (a) parallel; (b) grid.

Table 4. Test results of changing steel arrangement forms.

Test number	SCC number	Steel spacing (mm)	Steel arrangement	Passing ability test/simulation (%)
2-1	No. 1	30	parallel	76.6/81.5
2-2	No. 1	30	grid	43.3/38.4
2-3	No. 2	30	parallel	36.9/32.4
2-4	No. 2	30	grid	24.2/19.3
2-5	No. 2	35	parallel	78.8/80.8
2-6	No. 2	35	grid	72.4/64.7

5.3. Effect of coarse aggregate shape

In order to reflect the effect of the shape of the aggregate, the spherical particles are used to simulate the impact of pebble aggregate, and the spherical particles are replaced by the same volume of tetrahedron and hexahedron to simulate gravel aggregate (figure 6). The comparison tests (4-1, 4-2 & 4-3) and (4-4, 4-5 & 4-6) show that the SCC is better passing ability to use the pebble aggregate than the gravel aggregate, then the shape of aggregate also influences the probability of passing and blocking of SCC (table 5). The different shapes of the gravel aggregate have different passing ability, and passing ability of the hexahedral particles are more than tetrahedral particles. The reason is to generate particles when a relatively large gap between spherical particles with mortar particles, the gap is mostly filled with mortar particles after replacement of polyhedron particles, then the contact of polyhedron particles in the process of movement is more closely with the mortar particles than the spherical particles, that leads to increased flow. But the tetrahedron particle simulation results are closer to the actual situation, it is better able to simulate gravel aggregate.

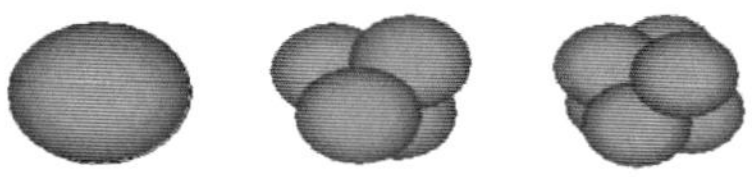

(a) (b) (c)

Figure 6. Different shape of coarse aggregate of (a) pebble; (b) gravel (1); (c) gravel (2).

Table 5. Test results of changing coarse aggregate shape.

Test number	SCC number	Steel spacing(mm)	Aggregate size(mm)	Steel arrangement	Shape aggregate	Passing ability test/simulation (%)
4-1	No. 3	25	5-10	parallel	Gravel (1)	36.1/42.9
4-2	No. 3	25	5-10	parallel	Gravel (2)	36.1/46.3
4-3	No. 4	25	5-10	parallel	Pebble	49.4/57.6
4-4	No. 3	20	5-10	parallel	Gravel (1)	13.1/17.2
4-5	No. 3	20	5-10	parallel	Gravel (2)	13.1/20.1
4-6	No. 4	20	5-10	parallel	pebble	15.9/22.3

5.4. Effect of SCC flow

Three groups of flow contrast tests (5-1 & 5-2, 5-3 & 5-4, 5-5 & 5-6) show the flow of SCC can also affect the passing ability, the simulation results with the test results also reflect the same rule (table 6). It is pointed out that the probability of blocking is independent of rheological properties when it is stable, the blocking model Roussel proposed only for the case of particle blocking, and does not consider the effect of rheological properties. The decrease passing ability of SCC is completely the result of the decrease of its own flow, which has nothing to do with particle blocking probability. Therefore, the flow of SCC will also affect by test results, the effect is not achieved by changing the probability of particle blocking, but by changing the relationship between the yield stress and shear stress.

Table 6. Test results of changing SCC flow.

Test number	SCC number	Steel spacing (mm)	Steel arrangement	Slump flow(mm)	Passing ability test/simulation (%)
5-1	No. 1	30	parallel	690×690	76.6/81.5
5-2	No. 2	30	parallel	550×550	36.9/32.4
5-3	No. 1	30	grid	690×690	43.3/38.4
5-4	No. 2	30	grid	550×550	24.2/19.3
5-5	No. 1	35	grid	690×690	100/83.6
5-6	No. 2	35	grid	550×550	72.4/64.7

5.5. Effect of casting rate

The casting speed of concrete cannot be directly set in the discrete element, it can be simulated by giving the wall a certain speed [17]. The real physical time T of Simulation test is calculated by $\text{timestep} \times \text{step}$, after running a certain number of steps ($\text{timestep}=1.6\times10^{-4}\text{s}$), the volume of concrete particles is calculated and it is equal to that volume of different casting rate.

Table 7. Test results of changing casting rate.

Test number	SCC number	Steel spacing (mm)	Steel arrangement	Casting rate (L/min)	Passing ability test/simulation (%)
6-1	No. 1	30	parallel	1.5	76.6/81.5
6-2	No. 1	30	parallel	3.0	66.0/58.7
6-3	No. 1	35	grid	1.5	100/83.6
6-4	No. 1	35	grid	4	46.0/53.4

Table 7 shows the influence of casting rate on passing ability of SCC, Compared with test 6-1 and 6-2, it can be seen that when the casting rate is 1.5L/min, the passing ability of SCC is higher than the casting rate of 3.0L/min. Compared with test 6-3 and 6-4, SCC can smoothly pass through the 35mm grid steel when the casting rate is 1.5L/min, when casting speed increased to 4 L/min, passing ability of SCC drops significantly. The increase of casting rate leads to the increase of flow rate, in fact, the rate of casting can influence the probability of particle blocking by changing the actual flow rate of SCC. According to the Bingham model, the plastic viscosity of SCC determines the speed of its flow, so the actual flow rate of SCC can also be affected by its plastic viscosity. Equation (7) does not take into account the flow rate of particles blocking probability, the influence of flow rate γ shall be increased in the equation (7), the flow rate actually changes that local particle concentration of in front of the obstacles, so the parameter ϕ is changed to $\gamma\phi$, the particles blocking model formula is:

$$P=(2.6\Omega / D_{\max}^{3})(0.1\gamma\phi)^{0.85\delta^{2}/0.81(\beta D_{\max})^{2}} \tag{9}$$

When flow rate is less than a certain critical value which the impact can be ignored, shows that the SCC has enough time to adjust its position when it meets obstacles, and will not lead to local aggregate content increased dramatically; when the flow rate is greater than this critical value, the change of casting rate will affect the passing ability of SCC. Therefore, parameter γ has an upper limit, when the flow rate is greater than a certain critical value, even if the casting rate continues to increase, the local particle content of in front of the obstacle will not change, and the probability of particle blocking will also not change. The flow rate is determined by the casting rate and the plastic viscosity, and it is controlled by changing the casting rate when the plastic viscosity is constant. The lower limit of the casting speed of Xie [4] is 1.5L/min, and the maximum casting rate is capped at 7.8 L/min by using the conversion of physical time in the previous simulation test.

6. Summary

(1) The way of steel arrangement will have an effect on the passing ability and particle blocking probability of the SCC, steel arrangement (channel section shape) correction coefficient λ was introduced into parameter P and the value λ of the simulated test was 0.90-0.95.

(2) The passing ability of pebble aggregate was superior to the gravel aggregate, passing ability of hexahedron particle was bigger than tetrahedron particle, but the tetrahedron particle simulation results were closer to the actual situation.

(3) The flow of SCC can also affect the passing ability, the effect was not achieved by changing the probability of particle blocking, but by changing the relationship between the yield stress and shear stress.

(4) The moving wall was used to simulate the casting speed of concrete, the flow rate parameter γ was introduced and the maximum casting rate was capped at 7.8 L/min by using the conversion of physical time in the previous simulation test when the plastic viscosity was constant.

7. References

[1] Roussel N and Nguyen T L H 2009 *Cem. Concr. Res.* **39(3)** 227-32
[2] Kasemchaisiri R and Tangtermsirikul S 2008 *Mag. Concr. Res.* **60(2)** 93-108
[3] Khayat K H 2004 *Aci. Mater. J.* **101(2)** 168-76
[4] Xue Y T 2014 Experimental study on the filling capacity of SCC and the microscopic properties of the interfaces in RFC. A thesis of Tsinghua University, Beijing, 54-66 (In Chinese)
[5] Roussel N, Gram A and Cremonesi M 2016 *Cem. Concr. Res.* **79** 265-71
[6] Mechtcherine V, Gram A and Krenzer K 2014 *Mater. Struct.* **47(4)** 615-30
[7] Chen S G and Zhang C H 2014 *Sci. China Phys. Mech. Astron.* **57(3)** 532-40
[8] Vasilic K, Schmidt W and Kühne H C 2016 *Cem. Concr. Res.* **88** 1-6
[9] Dhaheer M S A and Kulasegaram S 2016 *Cem. Concr. Res.* **89** 27-34
[10] Puri U C nad Uomoto T 2002 *J. Mater. Civil Eng.* **14(2)** 137-44
[11] Noor M A 1999 *Numer. Simul. Fresh Concr. IIS J.* **51(4)** 25-8
[12] Remond S and Pizette P 2014 *Cem. Concr. Res.* **58(15)** 169-78
[13] Mechtcherine V and Shyshko S 2015 *Cem. Concr. Comp.* **55** 81-90
[14] Huang M S 2010 The Application of DEM on Filling Performance Simulation of SCC in RFC. A thesis of Tsinghua University, Beijing, 70-80 (In Chinese)
[15] Chidiac S E and Mahmoodzadeh F 2009 *Cem. Concr. Comp.* **31(8)** 535-44
[16] Larrard F D, Ferraris C F and Sedran T 1998 *Mater. Struct.* **31(7)** 494-8
[17] Nguyen T L H and Roussel N 2006 *Cem. Concr. Res.* **36(10)** 1789-96

IOP Conf. Series: Materials Science and Engineering **269** (2017) 012091 doi:10.1088/1757-899X/269/1/012091

A Study on the Heat Flow Characteristics of IRSS

Yong-Jin Cho[1, a], Dae-Eun Ko[1, b*]

[1]Dept. Naval Architecture & Ocean Engineering, Dong-Eui University, Busan, Korea
[a]cyjdeu@deu.ac.kr, [b]deko@deu.ac.kr

Abstract. The infrared signatures emitted from the hot waste gas generated by the combustion engine and generator of a naval ship and from the metal surface around the funnel are the targets of the enemy threatening weapon system, thereby reducing the survivability of the ship. Such infrared signatures are reduced by installing an infrared signature suppression system (IRSS) in the naval ship. An IRSS consists of three parts: an eductor that creates a turbulent flow in the waste gas, a mixing tube that mixes the waste gas with the ambient air, and a diffuser that forms an air film using the pressure difference between the waste gas and the outside air. This study analyzed the test model of the IRSS developed by an advanced company and, based on this, conducted heat flow analyses as a basic study to improve the performance of the IRSS. The results were compared and analyzed considering various turbulence models. As a result, the temperatures and velocities of the waste gas at the eductor inlet and the diffuser outlet as well as the temperature of the diffuser metal surface were obtained. It was confirmed that these results were in good agreement with the measurement results of the model test.

1. Introduction

Due to the changes in the global security environment and the dramatic advances in maritime weapon systems, the modern battlefield environment in the ocean exhibits various complexities. To overcome these complexities, technologies related to special performances for survivability have been improved [1]. Such technologies include radar cross section (RCS), which determines the size, direction, and distance of a naval ship by detecting the energy of the electromagnetic waves that were radiated from the radar and reflected by the naval ship, underwater radiated noise (URN), which detects mechanical noise emitted from a ship to the surrounding fluid, and infrared (IR), which detects the heat energy emitted from the surface of a naval ship.

As for the studies related to the IR stealth technology, a study has been conducted on the calculation of the IR signatures from a naval ship and the sensitivity analysis of the marine environmental variables [2], and studies on the measurement and verification of the IR from the actual naval ships have been conducted to secure the reliability of analytical results [3]. In addition, the establishment of the reference marine environmental conditions for signature analysis [4] and the effect of the waste gas on the infrared signatures of a naval ship [5] were also studied. There have been also many studies on the equipment development for reducing IR signatures from a naval ship, and studies on the development of analysis software that can calculate IR signatures more accurately have been consistently conducted [6].

The IR signature reduction systems for naval ships include the hull seawater cooling system, which reduces the contrast radiant intensity (CRI) against the surroundings by spraying seawater onto the hull of a naval ship, and the IRSS, which reduces the hot waste gas by optimizing the shape of the waste pipe. An IRSS consists of three parts: an eductor that creates a turbulent flow in the waste gas, a mixing tube that mixes the waste gas with the ambient air, and a diffuser that forms an air film using the pressure difference between the waste gas and the outside air.

This study analyzed the test model of the IRSS developed by an advanced company and, based on this, conducted heat flow analyses as a basic study to improve the performance of the IRSS. For the heat flow analysis, a commercial analysis program was used and the results were compared and analyzed considering various turbulence models.

2. Model test

Manufacturers produce a miniature model with the same geometry as the actual system and conduct a performance test to verify the performance of a designed IRSS. In this instance, the main performance evaluation items include the back pressure applied to the engine, the temperature of the waste gas, and the metal surface temperature of the waste pipe. The temperature, static pressure, and total pressure of the waste gas that is flowed in the equipment are measured using Pitot tubes at the eductor inlet and the diffuser outlet to verify the test conditions and results. Figure 1 shows the schematic diagram of the IRSS covered in this study.

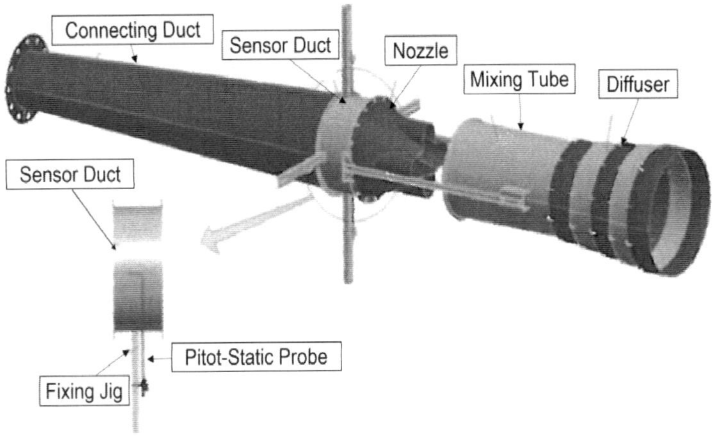

Figure 1. Schematic of IRSS.

The Pitot tubes are placed in the transverse direction of the eductor inlet and the diffuser outlet, and measures the temperature, static pressure, and total pressure at the same time. The measurement positions of the Pitot tubes are 21 points at the eductor inlet and the diffuser outlet, respectively. The areas of the eductor inlet and the diffuser outlet are divided into five equal areas and the measurement was performed in the up, down, left, and right directions. The metal surface temperature of the waste tube is measured through the temperature measuring instruments that use thermocouples, which were welded on both ends of each diffuser ring at 5mm intervals. The IRSS has five diffuser rings and the number of the temperature measurement points is 20 in total.

Once the system for test evaluation is constructed, the IRSS model is connected to the thermal wind tunnel tester and the heat flow conditions are set. The operation is maintained for a certain period of time until the measurement values are stabilized, and then the stabilized measurement values are recorded. table 1 shows the test conditions of the miniature model [7].

Table 1. Model test conditions.

Mass flow [kg/s]	Gas Temperature [°C]	Gas Density [kg/m^3]
1.55	525	0.4424

3. Heat flow analysis

In this study, STAR-CCM+ 11.02, software for flow analysis, was used. For more precise analysis, a 3D analysis model was implemented using the 3D shape modification function and the automatic mesh creation function provided by the software.

The continuity equation and the Reynolds averaged Navier-Stokes (RANS) equation used as the governing equations of the flow analysis are as shown in equations (1) and (2), respectively.

$$\frac{\partial \overline{u_i}}{\partial \overline{x_i}} = 0 \tag{1}$$

$$\rho \overline{u_i} \frac{\partial \overline{u_i}}{\partial x_j} = \rho \overline{f_i} + \frac{\partial}{\partial x_j}\left[-\overline{p}\delta_{ij} + \mu\left(\frac{\partial \overline{u_i}}{\partial x_j} + \frac{\partial \overline{u_j}}{\partial x_i}\right) - \rho \overline{u_i' u_j'}\right] \tag{2}$$

In the equations, $\overline{u_i}$ is the velocity in each direction in the Cartesian coordinate system, ρ the density, $\overline{f_i}$ the mean volume force, $\overline{p}$ the isotropic pressure, δ_{ij} the deformation of the fluid particle, and μ the friction coefficient [8]. The last term in equation 2 is the Reynolds stress, which refers to the shear stress generated by turbulence. In this study, K-Epsilon, shear stress transport (SST) K-Omega, Reynolds stress turbulent (RST) models were applied as turbulence models to obtain the Reynolds stress, the results of each model were compared and analysed.

The domain for the heat flow analysis was set large enough to prevent the energy from being accumulated in the domain and affecting the analysis results. In other words, the length of the domain was set to 100 times the diameter of the diffuser tip and the width was 50 times the diameter of the diffuser tip. Structured grids were used in the gas region of the domain, and approximately six million grids were created around the IRSS by increasing the grid density. In the solid region of the domain, about 150,000 polyhedral grids were created using polyhedrons with 12-16 faces to reduce the analysis time. If structured grids are used in the solid region corresponding to the thin plate in the analysis model, too many grids can be created, which significantly increases the analysis time. Figure 2 shows the magnified part of the created grids.

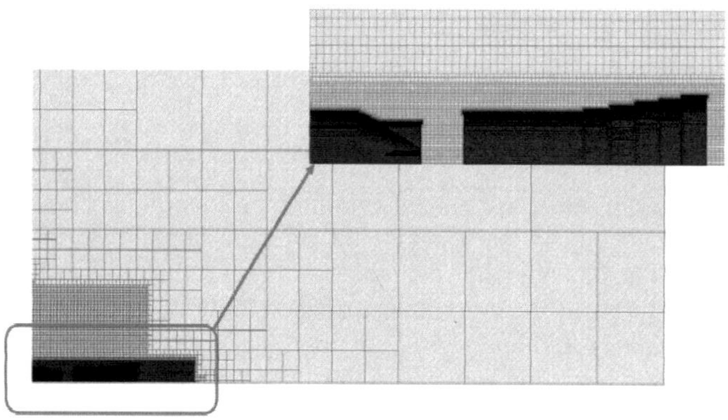

Figure 2. Grids of analysis domain.

4. Analysis results

4.1. Waste gas temperature

The numerical analysis results of the temperature and velocity of the waste gas at the eductor inlet were compared with the test results and were shown in figure 3. In this study, the temperature and velocity error rates of the turbulence model at the eductor entrance compared to the test measurement values were minimized by creating polynomial functions from the test measurement values and setting them as the initial conditions of the numerical analysis. The temperature and velocity at the eductor inlet were highest in the center of the eductor. This is because the flow created by the thermal wind tunnel tester gradually changed into a fully developed flow, resulting in the highest flow velocity in the center and the temperature increase due to convection.

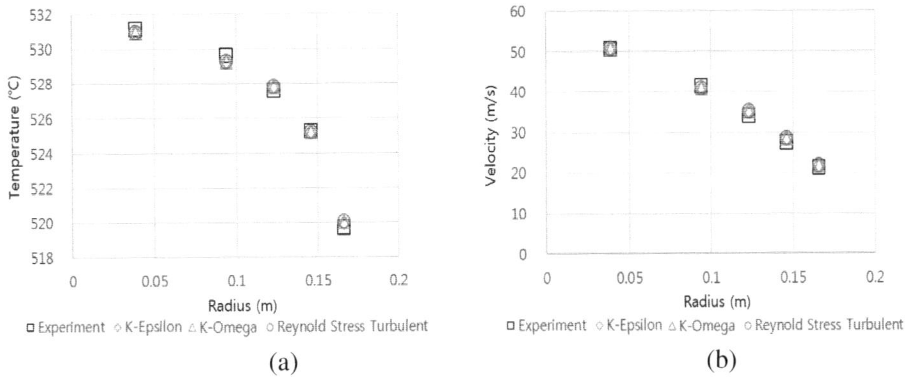

Figure 3. Temperature and velocity of the waste gas at the eductor inlet.

The numerical analysis results of the temperature and velocity of the waste gas at the diffuser outlet were compared with the test results and were shown in figure 4. The temperature and velocity in the radial direction were similar to those at the eductor inlet, but the error rates of the turbulence model compared to the test measurement values were relatively high. In this study, the results of the K-Epsilon model, in which the maximum error rates of the temperature and velocity were 8% and 21%, exhibited best agreement with the model test results.

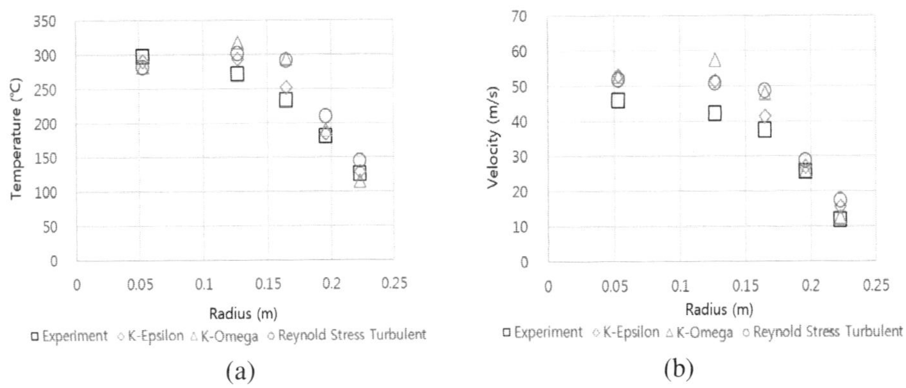

Figure 4. Temperature and velocity of the waste gas at the diffuser outlet.

4.2. Metal surface temperature of the diffuser

The spray pattern of the hot waste gas varies depending on the shape of the eductor outlet. The eductor outlet in this study had the lobe shape and the temperatures of the mixing tube and the diffuser metal surface caused by the waste gas sprayed from the eductor are shown in figure 5(a). When the hot waste gas passes through the diffuser, the low-temperature outside air is introduced inside through the diffuser ring gap due to the pressure difference. In this case, the temperature of the lower portion of the diffuser ring is relatively decreased compared to the upper portion. Figure 5(b) shows the velocity vectors of the outside air introduced through the diffuser ring gap. Unlike the analysis results, the temperature difference between the upper portion and lower portion of each ring was not clear in the test results that used thermocouples.

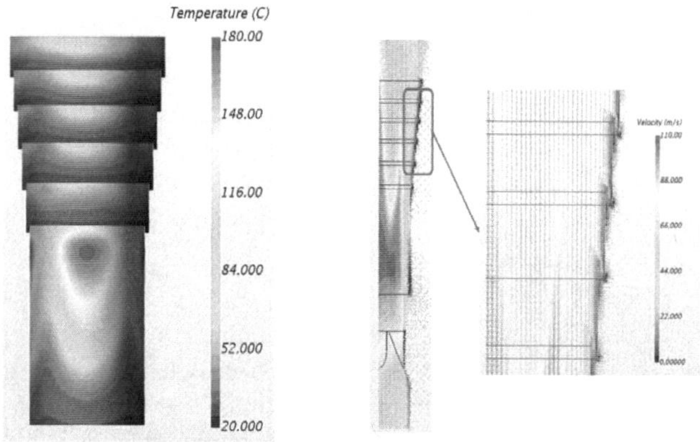

(a) (b)

Figure 5. Temperature of the metal surface and vector of air flow.

5. Conclusion

This study is a basic study to investigate the heat flow characteristics and improve the performance of the IRSS for naval ships. Below are the results of the heat flow analysis.

1) The test model of the IRSS developed by an advanced company and installed in naval ships were analyzed and the heat flow analysis conditions were identified.

2) Various turbulence models were applied to the heat flow analysis and the results of each analysis were compared and analyzed. In this study, the results of the K-Epsilon model, among the turbulence models, exhibited best agreement with the model test results.

3) Through the heat flow analysis, it was possible to identify the phenomena including the temperature distribution in the diffuser ring caused by the outside air introduction effect, which could not be observed by the model test results alone.

6. References

[1] Cho Y J 1995 *A Survey on Stealth Technology for a Surface Warship* Study report.

[2] Cho Y J 2005 *An Analytical Study on the Sensitivity of Ship's IR Signature According to the Meteorological Environment in Korean Waters* D.S Thesis, Chung-nam National University, 2005.

[3] Cho Y J 2009 *Study on Calibration and Verification Method of Naval ship Infra-red Measurement Results* Study Report.

[4] Han S I 2013 *A study on the sampling methods of ocean meteorological for ship IR signature* M.S Thesis, Dong-Eui University.

[5] Han K I 2013 *Study on the effect of exhaust plume gas on IR image characteristics of a naval ship* M.S Thesis, Chung-Ang University.

[6] Thompson J, Vaitekunas D and Birk A M 1998 *Infrared Signature of Modern Naval Ship* American Society of Naval Engineers 21th, Century Combatant Technology Symposium.

[7] Sung Shin ENSTEC 2011 *Procedure of Performance Test Evaluation.*

[8] CD-Adapco, Star-CCM+ User Guide

A Study on the Characteristics of Design Variables for IRSS Diffuser

Yong-Jin Cho[1, a], Dae-Eun Ko[1, b*]

[1]Dept. Naval Architecture & Ocean Engineering, Dong-Eui University, Busan, Korea
[a]cyjdeu@deu.ac.kr, [b]deko@deu.ac.kr

Abstract. In modern naval ships, infrared signature suppression systems (IRSS) are installed to decrease the temperature of waste gas generated in propulsion engine and the metallic surface temperature of heated exhaust pipes. Generally, IRSS is composed of eductor, mixing tube, and diffuser. Diffuser serves to reduce the temperature by creating an air film using the pressure difference between internal gas and external air. In this study, design variables were selected by analyzing the diffuser and the characteristics of design variables that affect the performance of diffuser were examined using Taguchi experiment method. For the diffuser performance analysis, a heat flow analysis technique established in previous research was used. The IRSS performance evaluation was carried out based on the average area value of the metal surface temperature and the temperature of the exhaust gas at the outlet of the diffuser, which are variables directly related to the intensity of infrared signature in naval ships. It was verified that the exhaust gas temperature is greatly affected by changes in the diameter of the diffuser outlet, and the metal surface temperature of diffuser is greatly affected by changes in the number of diffuser rings.

1. Introduction

Advanced naval powers are establishing plans for the development and building of new naval ships through in-depth analysis of operational requirements for future naval battles and development trends of military science and technology [1]. In Korea, special performance technologies started to be applied from the 1990s during naval ship design for survivability improvement. In the 2000s, research on infrared stealth technologies started to be conducted through the acquisition of advanced foreign technologies [2].

Studies on infrared signature suppression systems (IRSS) include a research on the optimization of nozzle shape for infrared signature reduction in the field of aviation [3], a research on the identification of critical design factor of DRES-Ball type IRSS [4], a research on the classification of infrared signature sources in naval ships and its reduction [5], and a research on the development of analysis software that can provide more precise calculations of infrared signature of naval ships [6]. Almost researches are limited to the subjects of calculation and evaluation of infrared signature because the design of IRSS is a security related subject.

The types of IRSS developed in advanced companies are Cheese Grater type, Eductor/Diffuser type, Eductor/Bliss type, and DRES-Ball type. Normally, Korean naval ships have equipped the Eductor/Diffuser type IRSS composed of eductor, mixing tube, and diffuser. Among the components of Eductor/Diffuser type IRSS, diffuser is the most possible to be detected by threatening weapon systems of enemy.

In this study, the characteristics of design variables for diffuser were investigated. Design variables were set by analyzing the shape of standard model and examined through smaller-the-better characteristics and Taguchi experiment method [7]. In the performance analysis, a heat flow analysis technique established in previous research [8] was used.

2. Variables and analysis cases for diffuser design

2.1. Design variables

Exhaust gas of high temperature generated in the propulsion engine of naval ships produces turbulent flow as it passes through IRSS eductor and mixes with ambient air at an extent which depends on the size of the mixing tube. The purpose of this research is to identify the characteristics of design variables that affect the performance of diffuser without any changes in eductor and mixing tube.

Design variables were derived by analyzing the shape of standard model of diffuser. Selected design variables of the diffuser are: 1) number of rings (NoR), which is directly connected to the production cost and machinability of the product, 2) tip size (TS) of the ring that forcibly forms an air film on the metal surface of diffuser, 3) diameter (D) of diffuser outlet, which serve to control the gap created by diameter differences between the rings that are related to the internal and external pressure and intake flow, and 4) difference of ring height (DRH), which affects the internal and external pressure difference depending on the correlation between the Bernoulli's equation and the flow equation.

The shape and dimensions of the standard IRSS model considered in this research are shown in figure 1. These dimensions are from the test model reduced by a factor of 2 from the actual IRSS. Regarding the level of the diffuse design variables, the number of rings was set to 3-5 considering production costs and machinability of the product. The tip size was set to 1-2 times the size of standard model to force the formation of air film. The diameter of diffuser outlet was set to 440-500mm for air suction flow control. The height of rings was set so that it decreases gradually as rings get closer to the diffuser outlet. Here, the difference of ring height was set so that the values of pressure difference with external air would be as uniform as possible.

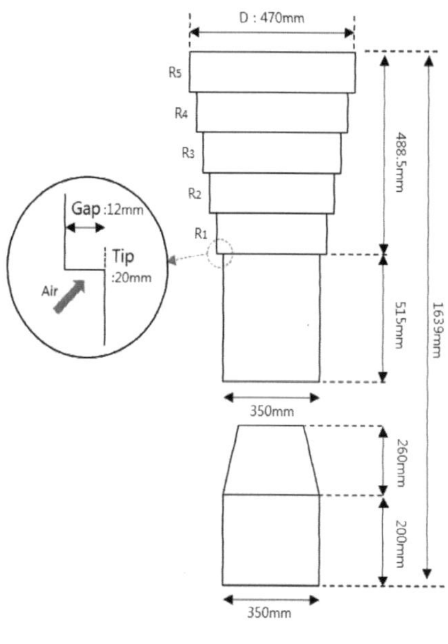

Figure 1. Standard IRSS model.

2.2. Analysis cases

Total of 81 analysis cases are produced from 4 design variables and 3 levels for each design variable. Since a lot of time is required to handle these all cases, Taguchi experiment method was used to reduce analysis cases without impacts on characteristics of design variables [7].

Using an orthogonal array table $L_9(3^4)$ of the design variables and their levels, the final 9 analysis cases are derived excluding unnecessary analysis cases as shown in table 1.

Table 1. Analysis cases.

Case	NoR [EA]	TS [m]	D [mm]	DRH [mm]
1	3	20	440	15.00
2	3	30	470	20.00
3	3	40	500	25.00
4	4	20	470	15.00
5	4	30	500	18.25
6	4	40	440	11.25
7	5	20	500	15.00
8	5	30	440	9.00
9	5	40	470	12.00

3. Heat flow analysis

In this research, a heat flow analysis technique established through previous research [8] was used, and initial input conditions reflected in the analysis are shown in table 2.

Table 2. Initial condition of heat flow analysis.

Mass flow [kg/s]	Gas Temperature [°C]	Gas Density [kg/m^3]
1.55	525	0.4424

Figure 2 shows the results of cases 2, 4, and 9, which have different amounts of diffuser rings but same diffuser outlet diameter to verify temperature distribution trends in the metal surface of diffuser.

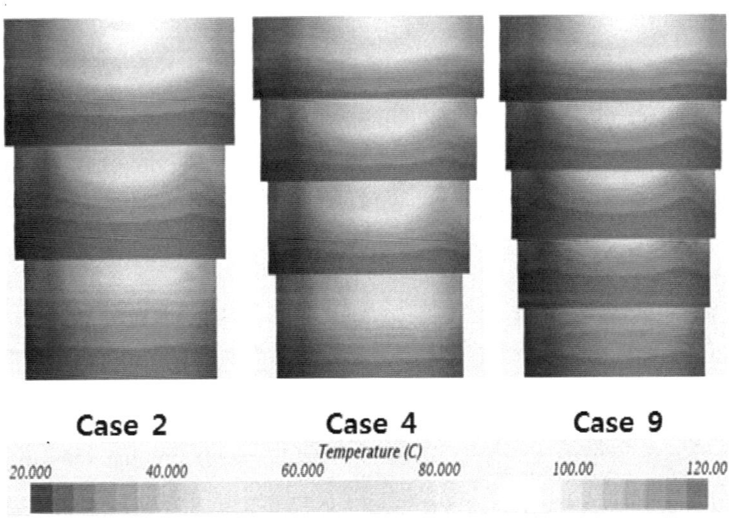

Figure 2. Temperature distribution of metal surface according to the number of rings.

The temperature of the exhaust gas at the diffuser outlet and the metal surface temperature are summarized for 9 analysis cases as in table 3.

IOP Conf. Series: Materials Science and Engineering **269** (2017) 012092 doi:10.1088/1757-899X/269/1/012092

Table 3. Temperature of exhaust gas and metal surface.

Case	Gas Temp. [℃]			Metal Surface Temp. [℃]		
	Min.	Max.	Mean	Min.	Max.	Mean
1	104	444	236	28	126	83
2	89	431	226	28	100	74
3	66	451	219	28	75	68
4	82	417	226	28	88	50
5	72	433	218	28	80	39
6	104	445	237	28	115	52
7	65	451	219	28	77	44
8	102	434	238	28	106	58
9	66	449	226	28	93	42

4. Results analysis

To identify the impact of changes in the sensitivity and level of the diffuser design variables, analysis results were analyzed using the concept of signal to noise (S/N) proposed by Taguchi. In this research, smaller-the-better characteristics shown in equation 1 were used.

$$S/N = -10\,Log\left(\frac{1}{n}\Sigma({y_i}^2)\right) \quad (1)$$

Here, n is the number of times the analysis was performed, and y_i is the analysis result.

From the heat flow analysis results, the mean values for exhaust gas temperature and metal surface temperature were converted into S/N ratio using equation (1), and cases in which the factor level of design variables was the same were averaged and indicated in table 4 and 5, respectively. Through table 4 and 5, it can be seen that the design variable that affects the most on the exhaust gas temperature is the diffuser outlet diameter, and the design variable that affects the most on the metal surface temperature is the number of rings.

Table 4. S/N of exhaust gas temperature.

S/N	NoR	TS	D	DRH
Level 1	-47.1	-47.1	-47.5	-47.1
Level 2	-47.1	-47.1	-47.1	-47.1
Level 3	-47.1	-47.1	-46.8	-47.1
Sum of Squares	0.00	0.00	0.74	0.00
Contribution Ratio [%]	0.17	0.04	99.50	0.30

Table 5. S/N of metal surface temperature.

S/N	NoR	TS	D	DRH
Level 1	-37.5	-35.1	-36.1	-34.3
Level 2	-33.5	-34.9	-34.7	-34.9
Level 3	-33.6	-34.6	-33.8	-35.4
Sum of Squares	31.4	0.5	7.5	1.7
Contribution Ratio [%]	76.4	1.2	18.3	4.1

5. Conclusion

In this research, the main design variables that affect the performance of IRSS diffuser were drawn and its characteristics examined for infrared signature reduction in naval ships. The contents and results of this research are as follows.

1) Design variables were derived for the diffuser, which is the most possible to be detected by threatening weapon systems of enemy, by analyzing the shape of standard model. Selected design variables are number of rings, tip size, diameter of diffuser outlet, and difference of ring height.

2) Analysis cases to examine the design variables were optimized using Taguchi experiment method, and heat flow analyses were performed for the optimized 9 analysis cases.

3) Analysis results were analyzed using the smaller-the-better characteristics of the signal to noise concept. Through this, it was verified that the exhaust gas temperature is greatly affected by changes in the diameter of diffuser outlet, and the metal surface temperature of diffuser is greatly affected by changes in the number of rings of the diffuser.

4) Since this research results are for specific values of exhaust gas flow rate and initial temperature, additional research is required for different conditions. Also, research on the design of mixing tube and eductor connected to the turbulent flow of exhaust gas is required.

6. References

[1] The Society of Naval Architects of Korea, Naval Ship, Textbooks, 2012.

[2] Cho Y J 2013 J. Soc. Nav. Archit. Korea **50(3)** 182-9.

[3] Oh J G 2013 *A Study on Optimization of Nozzle Shape for IR Signature Reduction* M.S Thesis, Gyeong-sang National University.

[4] Lee S I 2006 *A Study for identification of critical design factor of DRES-Ball and for efficient cooling of exhaust gas by water spray in terms of numerical analysis* M.S Thesis, Seoul National University.

[5] Thompson J, Vaitekunas D and Birk A M 1998 *Infrared Signature of Modern Naval Ship,* American Society of Naval Engineers 21th, Century Combatant Technology Symposium.

[6] Vaitekunas D A 1996 *SHIPIR/NTCS: A Naval Ship Infrared Signature Countermeasure and Threat Engagement Simulator* Proceedings for Society of Photo-Optical Instrumentation Engineers (SPIE) Infrared Technology XXII.

[7] Korea Advanced Institute of Science and Technology, Experiment and Analysis: Taguchi method and Utilization of Orthogonal Tables, 2005.

[8] Yoon S T, Jung H S, Cho Y J and Ko D E 2017 *Numerical Simulation of Infra-Red Signature Suppression on the Naval Ship* The Korea Institute of Military Science and Technology.

IOP Conf. Series: Materials Science and Engineering **269** (2017) 012093 doi:10.1088/1757-899X/269/1/012093

Tensile behaviors of three-dimensionally free-formable titanium mesh plates for bone graft applications

Jianmei He[1]

[1]Dept. of Mechanical Engineering, Kogakuin Univ. 1-24-2, Nishi-shinjuku, Shinjuku-ku, Tokyo, 163-8677 Japan

Abstract. Present metal artificial bones for bone grafts have the problems like too heavy and excessive elastic modulus compared with natural bones. In this study, three-dimensionally (3D) free-formable titanium mesh plates for bone graft applications was introduced to improve these problems. Fundamental mesh shapes and patterns were designed under different base shapes and design parameters through three dimensional CAD tools from higher flexibility and strength points of view. Based on the designed mesh shape and patterns, sample specimens of titanium mesh plates with different base shapes and design variables were manufactured through laser processing. Tensile properties of the sample titanium mesh plates like volume density, tensile elastic modulus were experimentally and analytically evaluated. Experimental results showed that such titanium mesh plates had much higher flexibility and their mechanical properties could be controlled to close to the natural bones. More details on the mechanical properties of titanium mesh plates including compression, bending, torsion and durability will be carried out in future study.

1. Introduction

Tailor-made tricalcium phosphate bone implants fabricated using a 3D ink-jet printer and used in clinical trial have mechanical properties such as very poor stiffness & fracture strength and have been applying only for the non-load implant cases (Tessier et al. 2005 [1], Eppley et al. 2005 [2], Igawa et al. 2010 [3]). Then titanium plate implants can be used as the reinforcements of such Tailor-made tricalcium phosphate bone implants for many implant cases. On the other hand, the present titanium plates for implant applications have the problems like too heavy, mismatch-elasticity and excessive-strength compared with natural bones (Satoh et al. 2011 [4], Wakui et al. 2012 [5], Seki et al. 1994 [6]). In this study, improved titanium mesh plates with higher 3-dimensional flexibility (Abiko et al. 2011 [7], J. He et al. 2014 [8]) were interested to solve such kind of problems.

Fundamental mesh shapes and patterns were designed under different base shapes (triangle, quadrangle and hexagon) and design parameters through three dimensional CAD tools from higher flexibility and strength points of view. Based on the designed mesh shapes with different base shapes (triangle, quadrangle and hexagon) and mesh line widths, sample specimens of titanium mesh plates were manufactured through the laser beam machining for experimental and analytical studies. Tensile experiments on sample titanium mesh plates were carried out to evaluate the effects of design parameters on the mechanical properties like tensile elasticity and volume density etc. On the other hand, analytical approaches on mechanical properties of titanium mesh plates were also carried out using ANSYS finite element analysis code. Comparisons between experimental and analytical results are used to validate the analytical approach method.

2. Fundamental mesh shape designs for 3D free-formable titanium mesh plates

To solve the too heavy, mismatch-elasticity and excessive-strength problems compared with natural bones in the present metal plate implants, mechanical properties of improved plate implants need to be closed to the natural bones. Then fundamental mesh shapes are considered under the following structural design conceptions:

(1) Single fundamental mesh shape construction for simplification of manufacturing processing and cost-down purpose

(2) Higher three-dimensional flexibilities including expansion/contraction, bending and torsion for possibility of handily shape changes during surgery

(3) Easy-controllable mechanical properties like elastic modulus, bending stiffness etc. for approachability to natural-bone's mechanical properties

(4) Uniform mesh line width and non-angle smooth shape desired to avoid the stress concentrations and lead for higher strengths and longer operating life

Figure 1 shows three basic mesh shapes with basic five design parameters introduced in this study based on the above mentioned design conceptions.

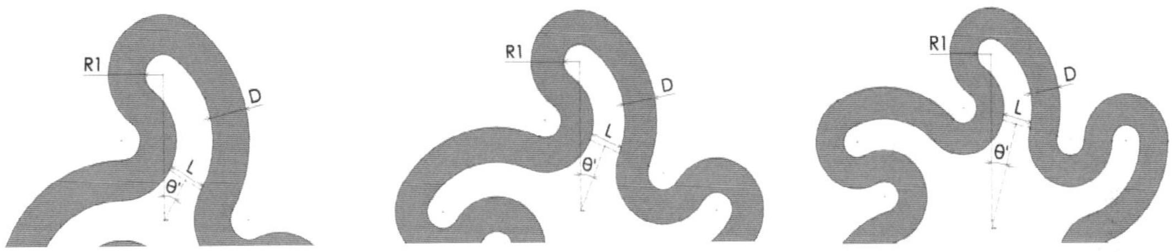

Triangle Quadrangle Hexagon

Figure 1. Mesh shapes with five basic design parameters: n: triangle=3, quadrangle=4 and hexagon=6; D: mesh line width; R1: mesh space radius; L: minimum space length; θ: bending angle.

It is possible to create different varieties of mesh shapes from these five basic design parameters as shown in figure 2a. Other dimensions shown in figure 2b can be deduced from the following equations with five basic design parameters.

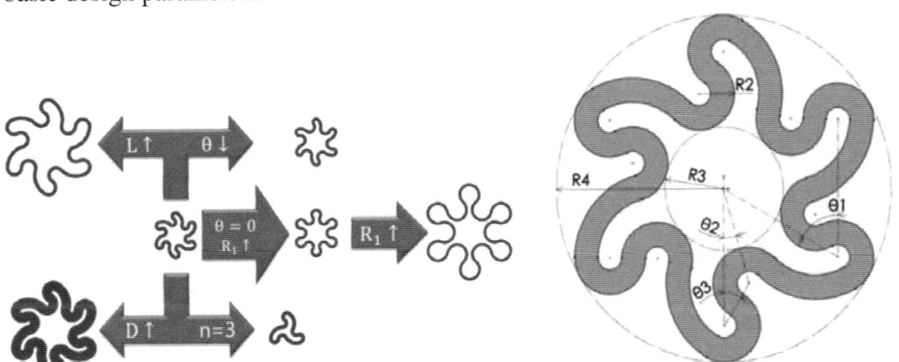

(a) (b)

Figure 2. Different varieties and dimensions of mesh shapes.

$$R_2 = R_1 + D \quad (1)$$

$$\theta_1 = \frac{90\cdot(n-2)}{n} \quad (2)$$

$$\theta_2 = \frac{180}{n} - \theta' \quad (3)$$

$$R_3 = \left(R_2 + \frac{L}{2}\right)\sec\theta_1 - R_2 \quad (4)$$

$$R_3 = \left(R_2 + \frac{L}{2}\right)\sec\theta_1 - R_2 \tag{5}$$

$$R_4 = (R_1 + R_2)\cos\theta_3 + (R_2 + R_3)\cos\theta_2 + R_2 \tag{6}$$

Above mentioned basic mesh shapes are designed from regular triangle, quadrangle and hexagon shapes, they have 120°, 90° and 60° axial symmetry with respect to the plate plane. Then three dimensional meshed plate models using the designed mesh shapes with two different mesh line widths (0.4 mm and 0.8 mm) are obtained and the CAD models for sample titanium mesh plates are shown in figure 3 to figure 5. From these models, one can see that the meshed plates have uniformed mesh line widths and smooth shapes just along the requirements of design conceptions.

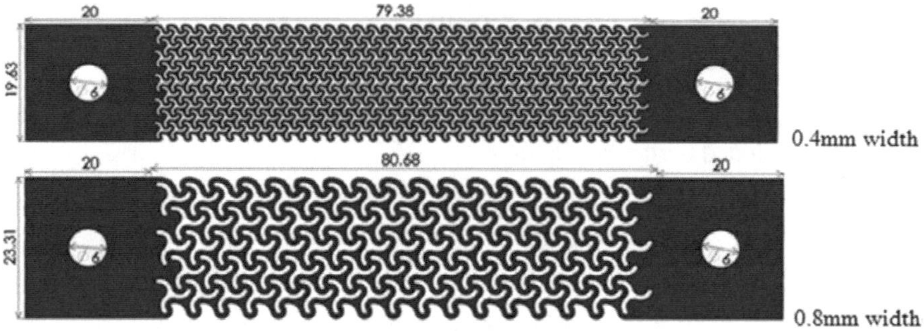

Figure 3. Meshed plate model from triangle shape.

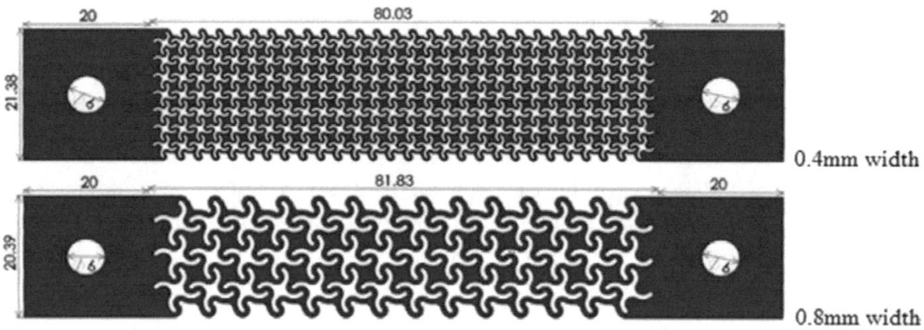

Figure 4. Meshed plate model from quadrangle shape.

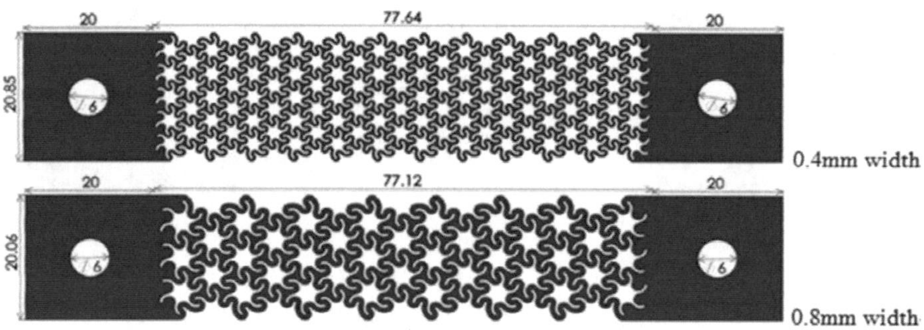

Figure 5. Meshed plate model from hexagon shape.

3. Experimental evaluations on mechanical properties of sample titanium mesh plates

Based on the mesh shapes shown from figure 3 to figure 5, sample titanium mesh plate specimens with 0.6 mm plate thickness are fabricated for experimental evaluations by laser cutting processing. Six kinds of titanium mesh plate specimens are shown in figure 6. Volume densities of these sample titanium mesh plates are then obtained by dividing the specimen's total weight with the corresponding total unmeshed plate volume and taking the area ratio of the meshed part to total specimens and the results are shown in figure 7. The comparison results of the volume densities shown in figure 10 indicate that the volume densities of sample titanium mesh plates are reduced to about 30% of same titanium plate and can be controlled to close to the natural bone's densities ranged from 0.5 g/cm^3 to 1.1 g/cm^3.

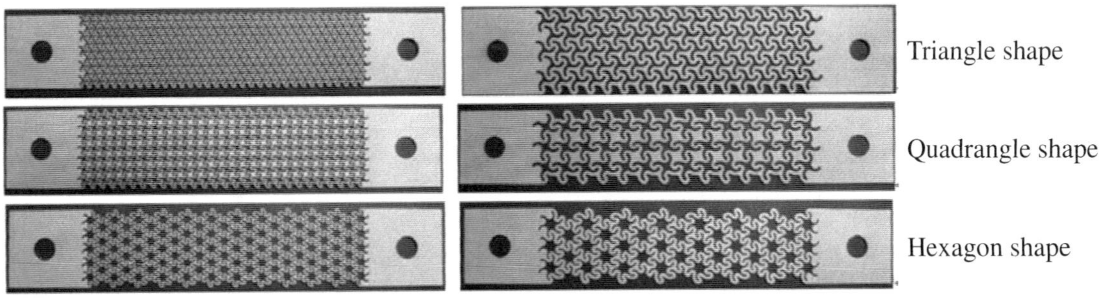

Figure 6. Tensile specimens of titanium mesh plates (0.4mm width and 0.8mm width).

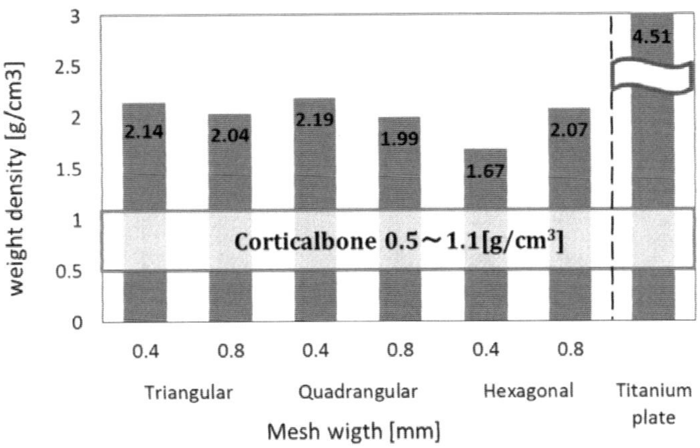

Figure 7. Volume densities of titanium mesh plates.

Tensile properties of sample titanium mesh plates such as elastic stiffness and elastic modulus are evaluated through tensile tests. Figure 8(a) shows typical measured load-displacement diagram of sample titanium mesh plates obtained from triangle shape specimen's experiment and figure 8(b) shows the approached tensile elastic modulus of sample titanium mesh plates obtained by using the unmeshed cross-sectional areas.

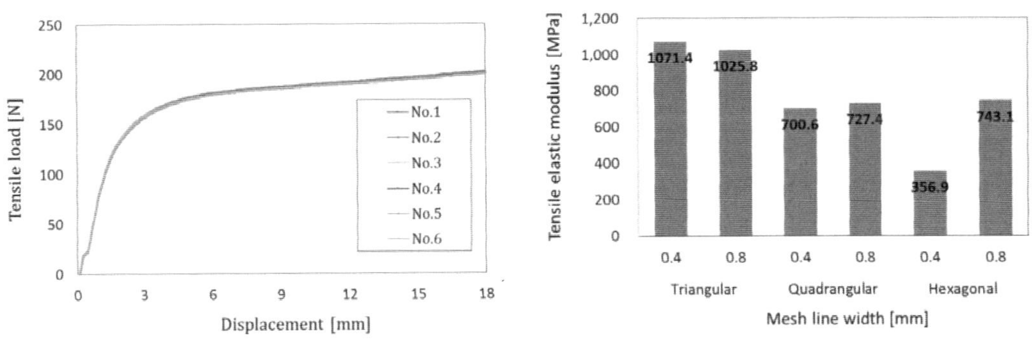

(a) Load-displacement diagram (b) Approached elastic modulus

Figure 8. Typical tensile experimental results (triangle 0.4 mm width specimen).

4. Analytical approaches on tensile properties of sample titanium mesh plates

Analytical approaches on tensile experiments of the sample titanium mesh plates are carried out here using Solidworks 3D CAD software and ANSYS finite element analysis code. 3D meshed plate models with the same shapes and sizes of sample titanium mesh plate specimens on experimental evaluations are used and material properties of titanium plate shown in table 1 are used for analytical inputs. Figure 9 shows the loading and boundary conditions of the analysis model.

Table 1. Material property input for analysis.

Material	Elastic modulus (GPa)	Poison's ratio	Density (kg/mm^3)
Titanium	105.0	0.3	4.514

Figure 9. 3D analytical model on titanium mesh plate tensile tests.

Comparisons with experimental and analytical results on tensile properties of sample titanium mesh plates are shown in figure 10. From these results, more than 10% deviations between experimental and analytical tensile elastic modulus of titanium mesh plates are existed and these indicated that improvements of analytical method are needed in further study.

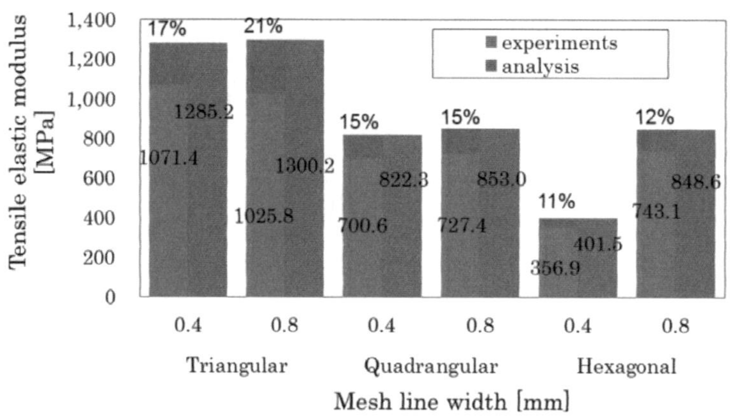

Figure 10. Comparisons on tensile elastic modulus of titanium mesh plates.

5. Summary

Experimental and analytical assessments of tensile properties on sample titanium mesh plates with higher 3-dimensional flexibility were executed and the following conclusions are obtained.

(1) Mesh plates with volume densities closed to natural bone's densities can be ontained by changing the fundamental design parameters of basic mesh shapes.

(2) Experimental tensile elasticity results of sample titanium mesh plates show the higher flexibility compared with unmeshed titanium plates and natural bones. It shows the possibility to controll the mechanical properties of such titanium mesh plates to close to natural cortical bones.

(3) Analytical approach methods for tensile property evaluations of titanium mesh plates are introduced. Comparisons between experimental and analytical results indicate that improvement of analytical approach method need to be done in future study.

6. References

[1] Tessier P, Kawamoto H, Matthews D, Posnick J, Raulo Y, Tulasne J F and Wolfe S A 2005 *Plast. Reconstruct. Surg.* **116** 92S–94S.

[2] Eppley B L, Pietrzak W S and Blanton M W 2005 *J. Craniofac. Surg.* **16(6)** 981–9.

[3] Igawa K and Chung U I et al. 2010. *Chin augmentation with inkjet-printed custom-made tricalcium phosphate implant,* Selected papers of the Congress of the European Association of Cranio-Maxillo-Facial Surgery, MEDIMOND/M914 **C0611** 135-9.
[4] Satoh K, Murata J, Tomioka M, Seino Y, Yokota M, Mizuki H and Miura H 2011 *Dent. J. Iwate Med. Univ.* **36(1)** 46-52.
[5] Wakui D, Nagashima G, Ueda T, Takada T, Tanaka Y and Hashimoto T 2012 *Japanese J. Neurosurg.* **21(2)** 138-42.
[6] Seki Y, Bessho K, Sugatani T, Kageyama T, Inui M and Tagawa T 1994 *Japanese J. Oral Maxillofacial Surg.* **40** 892-6.
[7] Abiko R, Zama K, Sinoda T, Suzuki S and He J 2011 *A Primary Study on the Titanium mesh plate Structures for Reinforcement of Tricalcium Phosphate Bone Implants,* International Conference on Advanced Technology in Experimental Mechanics 2011 (ATEM'11, 2011), Kobe, Japan.
[8] He J M, Suzuki S and Chung U I 2014 *Trans. JSME (in Japanese)* **80(809)** 1-15.

Experiment Research on Power Generation Performance of Double Piezoelectric Vibration Energy Harvester

Zhijuan Zhanga[1, *], Chao Nib[1] and Xiangyu Zhangc[1]

[1]School of Electrical and Electronic Engineering, North China Electric Power University1, Baoding 071003, P.R. China

Abstract. In order to collect the vibration energy in the environment, the cantilever double piezoelectric vibration energy harvester is designed to collect the mechanical vibration energy of the motor. The output voltage of the upper and lower piezoelectric ceramic wafer of the bimorph piezoelectric vibrator and the output voltage of the cantilever suspension length at different rotational speeds of the motor are investigated. At last the experimental results are verified.

1. Introduction

Mechanical vibration are ubiquitous in the environment, the mechanical vibration energy harvesting environment together into a microelectronic device for low-power electrical energy, take the place of the traditional chemical batteries supply for wireless sensor network node, portable electronic devices MEMS and other low power consumption electronic products [1]. At present, there are many types of vibration energy harvesting technology: piezoelectric, electrostatic, composite, electromagnetic and others [2]. The piezoelectric vibration energy harvester is to convert the mechanical energy of vibration into the electric energy that can be used for low energy consumption electronic products [3].

The piezoelectric vibration energy harvester is usually composed of cantilever piezoelectric bimorph or piezoelectric unimorphs structure, in order to improve the conversion efficiency, the end of the cantilever beam is usually a fixed amount of mass to reduce the energy harvester natural frequency to increase vibration inertial force of the cantilever beam [4]. The vibration energy in the environment is unstable, so the piezoelectric power generator can only transform the vibration energy into smaller electrical energy, the power generated by the mechanical vibration is often difficult to meet the demand of low power consumption for a long time [5]. In order to effectively improve the energy conversion efficiency of the high voltage power generation device, it is necessary to predict, analyze and optimize the parameters of the double piezoelectric vibration energy harvester and bimorph piezoelectric parameters of the cantilever beam [6]. At present, the piezoelectric vibration energy harvester of cantilever beam has been widely studied at home and abroad. In 2015, Hongjun Liu and Dingfang Chen et al. [7, 8] established the electromechanical coupling model of piezoelectric vibration energy harvester, obtained the influence of different geometric dimensions on structural response characteristics and output performance. In 2017, Binqiang Yang, Wentan Xu et al. [9] used Hamilto principle and Raleigh-Ritz method to establish the electromechanical coupling model of the piezoelectric energy harvester, numerically analyzed the influence of the parameters such as the mass ratio, the stiffness ratio and the damping ratio of the energy collector on the vibration characteristics and output characteristics of the system. Cantilever beam piezoelectric vibration energy harvester and bimorph piezoelectric structure parameters are all explored through the establishment of an ideal model.

2.Theoretical analysis

Figure 1 shows the sketch of model of double piezoelectric energy harvester with cantilever beam. Due to the same properties of the two piezoelectric ceramic materials on the bimorph piezoelectric, the upper and lower surfaces of the piezoelectric substrate are respectively pasted on the upper surface of the substrate, the upper and the lower piezoelectric ceramics are symmetrical on the X axis, therefore, the piezoelectric cantilever beam is forced to vibrate together with the vibration source when the mechanical vibration is excited, the piezoelectric cantilever beam will be deformed, and the change in strain and stress in the piezoelectric layer will be the same. For the piezoelectric effect of the bimorph, the open circuit voltage and the electrical energy of the upper and lower piezoelectric ceramic plates are equal, but the open circuit voltage is opposite.

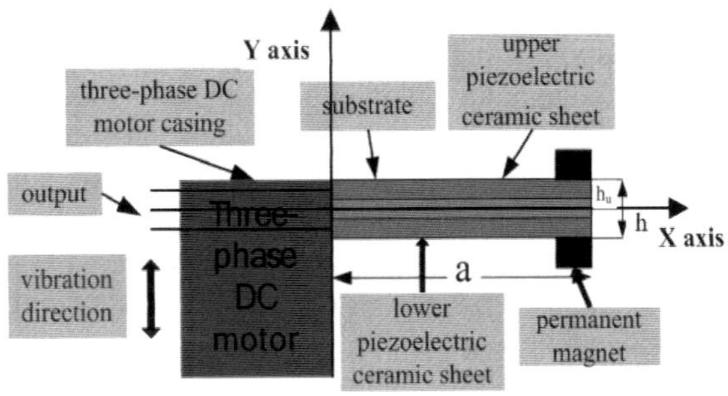

Figure 1. Sketch of model of double piezoelectric vibration energy harvester with cantilever beam.

According to the literature [10], the relation between the charge quantity and the exciting displacement of a single bimorph piezoelectric under the action of external force is deduced:

$$Q = \frac{-3bABg_{31}E_P h^2}{16aAh_u(E_P g_{31}^2 + \beta_{33}^T) - 6haB^2 E_P g_{31}^2}\delta t \tag{1}$$

The capacitance of bimorph piezoelectric vibrator is:

$$C = \frac{8Aab}{8Ah_u(E_p g_{31}^2 + \beta_{33}^T) - 3hB^2 E_P g_{31}^2} \tag{2}$$

where $A = 1-\alpha^3+\alpha^3\beta$, $B = 1-\alpha^2$, $\beta = E_m / E_p$ δt - the excitation displacement, α - the ratio of the thickness of the metal substrate to the total thickness of the cantilever beam, β - the young's modulus ratio, E_m - the young's modulus of the metal substrate, E_p - the young's modulus of piezoelectric ceramics, g_{31} - the piezoelectric constant.

The power on the load is:

$$P = I^2 \times R = \frac{V^2}{\frac{(R-r)^2}{R} + 4r} \tag{3}$$

3. Experiment

3.1. Experimental principle

The vibration energy structure principle of the double cantilever piezoelectric vibration energy harvester is shown in figure 1, double cantilever piezoelectric vibration energy harvester is composed of a bracket, a bimorph piezoelectric plate and two permanent magnets, the double piezoelectric element operates in the d31 mode. The double piezoelectric ceramic crystals of bimorph piezoelectric bonds adhere to both sides of the elastic substrate to form an electric energy output anode, the surface of the upper and lower piezoelectric ceramic crystal plates respectively constitutes the positive electrode of the upper and lower piezoelectric plates. This paper used the positive piezoelectric effect of piezoelectric ceramics. The cantilever bimorph is fixed on the shell of the three-phase DC motor at one end, the permanent magnet on the free end fixed a certain quality, the piezoelectric cantilever is forced to vibrate in conjunction with the vibration source when the mechanical vibration of the motor generating excitation, the permanent magnet will be affected by the inertial force and occur the piezoelectric cantilever deformation, which will cause the changes of piezoelectric layer strain and stress , there will have a change of potential difference output from a positive and negative electrode on bimorph because of the positive piezoelectric effect between the bimorph.

3.2. Experimental device structures
The double piezoelectric vibration energy harvester is a bimorph piezoelectric cantilever with two piezoelectric ceramic layers. One end of the piezoelectric energy harvester is fixed on the shell of the three-phase synchronous DC motor, and the other end is fixed with a permanent magnet strip of 2g, the motor will produce mechanical vibration when it is running, and the piezoelectric vibration energy harvester will change the mechanical vibration energy of the motor into electrical energy. As shown in figure 2, the AC motor test bench controls the speed (N) of the three-phase synchronous DC motor, the speedometer shows the motor speed N, the oscilloscope measures the waveform and size of the output voltage of the piezoelectric vibration energy harvester. Table 1 is the material parameters of bimorph piezoelectric.

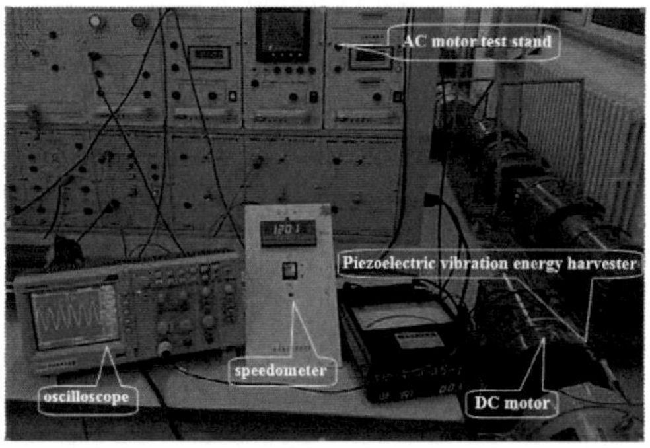

Figure 2. Experimental device diagram.

Table 1. Material parameters of bimorph piezoelectric.

parameters	value
Resonance impedance/oHm	*<70*
Static capacitance/nF	*380--430*
Piezoelectric ceramic materials	*PZT P8-79Y*
Substrate material	*Kover#4J29*
Resonance impedance	*<90 oHm*
Qm	*400*
eT33	*1300*
Kt33	*4500*

4. Experimental results

4.1. Output voltage of upper and lower piezoelectric ceramic wafers
Figure 3 is the output voltage of the upper and lower piezoelectric wafers at different speeds of the motor. It can be seen from the curve that the output voltage of the upper and lower piezoelectric wafers has a maximum peak value with the increase of the motor speed. And both reach the maximum at the motor speed of 1200r/min, the maximum output voltage of the upper piezoelectric wafer can reach 756mv, the maximum output voltage of the lower piezoelectric wafer can reach 728mv. The speed of motors are the

same when the upper piezoelectric wafer output voltage is close to lower and reaches the maximum open circuit voltage.

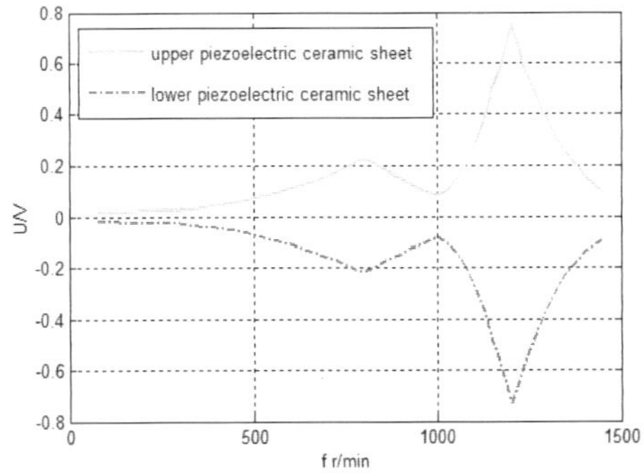

Figure 3. Output voltage of upper and lower piezoelectric bimorph at different speeds.

4.2. Suspension length of cantilever beam piezoelectric bimorph influence on output voltage

Figure 4 is the output voltage diagram of a cantilever beam bimorph with different suspension lengths at different speeds of the electromotor. It can be seen from the figure that the motor speed increases with the cantilever suspension length increasing to reach the maximum output voltage, but the output voltage of the peak size are different, there is a maximum output voltage, the relationship between the maximum output voltage are V4.5 > V4 > V5 > V3.5. In the figure, the maximum output voltage is 801Mv when the cantilever piezoelectric plate is suspended by 4.5cm. The maximum output voltage will decrease when the length of the cantilever beam is increasing or decreasing, however, the speeds of the motor are different when reaching the maximum speed, the speed of the motor increases with the increase of the length, which means, the vibration frequency will increase when the maximum is reached.

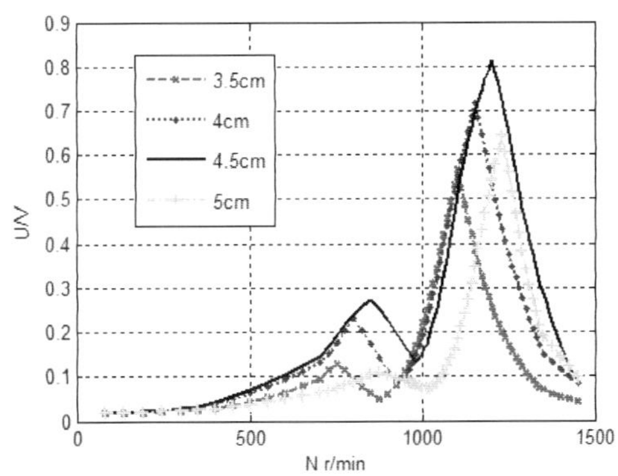

Figure 4. Output voltage of cantilever piezoelectric bimorph under different suspension length when the motor is at different rotational speeds.

5. Conclusions

This paper set up an experimental platform for mechanical vibration energy harvester, the mechanical vibration energy of the motor is converted into electric energy by means of a cantilever beam piezoelectric generator. The output voltage of the upper and lower piezoelectric ceramic wafer of the bimorph piezoelectric vibrator, the output voltage of the cantilever suspension length at different rotational speeds of the motor are investigated. The experimental results show that the upper and lower output voltage of the bimorph piezoelectric are equal, both are sinusoidal AC voltage; the frequency of the mechanical vibration increases with the increase speed of the motor, and there is an optimum suspension length for the bimorph cantilever beam, at which the output voltage is the maximum. Therefore, there is a best paste position on the vibration source of the bimorph piezoelectric cantilever beam to make the double piezoelectric vibration energy harvester get an optimal voltage output characteristic, which provide a reference for the study of mechanical vibration motor piezoelectric energy harvesting.

6. References

[1] Toprak A and Tigli O 2014 *Appl. Phys. Rev.* **1(3)** 031104-1-14.
[2] Liu X J and Chen R W 2012 *J. Vibr. Shock.* **31(16)** 2-4.
[3] Du X Z, Zeng X W and Feng B 2014 *Micro nano Electr. Tech.* **51(3)** 2-3.
[4] Wang G Q, Liu C and Zhang W 2015 *Chinese J. Sens. Actuat.* **28(6)** 2-3.
[5] Hofmann H F, Ottman G K and Lesietre G A 2002 *IEEE Trans. Power Electr.* **18(2)** 696-703.
[6] Kan J W, Tang K H and Wang S Y 2008 *Opt. Prec. Eng.* **16(1)** 71-4.
[7] Xu Y Y, Gong J J and Song Z L 2012 *Mech. Eng. Autom.* **1** 80-1.
[8] Liu H J, Chen D F and Tao M L 2015 *J. Mech. Des. Res.* **6** 28-30.
[9] Yang B Q, Xu W T and Lu G L 2017 *China Mech. Eng.* **28(2)** 127-34.
[10] Hu S J, Li X Q and Li L 2016 Mach. Manuf. **53(620)** 18-21.

Acknowledgment
This work was supported by the natural science foundation of Hebei Province (Grant No. F201450204)